Fundamentals of Drug Delivery

Fundamentals of Drug Delivery

Edited by

Heather A. E. Benson
Curtin University of Technology, School of Pharmacy, Perth, Australia

Michael S. Roberts
University of Queensland, School of Medicine, Brisbane, Australia

Adrian C. Williams
University of Reading, School of Chemistry, Food and Pharmacy, Reading, UK

Xiaowen Liang
University of Queensland, Faculty of Medicine, Brisbane St Lucia, Australia

WILEY

This edition first published 2022

Registered Office
John Wiley & Sons, Inc., 111 River Street, Hoboken, NJ 07030, USA

Editorial Office
111 River Street, Hoboken, NJ 07030, USA

For details of our global editorial offices, customer services, and more information about Wiley products visit us at www.wiley.com.

Wiley also publishes its books in a variety of electronic formats and by print-on-demand. Some content that appears in standard print versions of this book may not be available in other formats.

Library of Congress Cataloging-in-Publication Data

Names: Benson, Heather A. E., editor. | Roberts, Michael S., 1949- editor. | Williams, Adrian C., 1963- editor. | Liang, Xiaowen, editor.
Title: Fundamentals of drug delivery / edited by Heather A. E. Benson, Michael S. Roberts, Adrian C. Williams, Xiaowen Liang.
Description: Hoboken, NJ : Wiley, 2022. | Includes bibliographical references and index.
Identifiers: LCCN 2021033395 (print) | LCCN 2021033396 (ebook) | ISBN 9781119769606 (cloth) | ISBN 9781119769651 (adobe pdf) | ISBN 9781119769675 (epub)
Subjects: MESH: Drug Delivery Systems | Drug Administration Routes
Classification: LCC RS199.5 (print) | LCC RS199.5 (ebook) | NLM QV 785 | DDC 615/.6–dc23
LC record available at https://lccn.loc.gov/2021033395
LC ebook record available at https://lccn.loc.gov/2021033396

Cover Design: Wiley
Cover Image: © shutterstock\Yurchanka Siarhei

C9781119769606_210921

MIX
Paper from responsible sources
FSC www.fsc.org FSC® C013604

Contents

Preface

Effective, controlled drug delivery has the potential to greatly impact the therapeutic outcome, clinical benefit, and safety of drugs in a wide range of diseases and health conditions. There are a large number of potentially useful drugs with limited effectiveness and/or safety concerns due to poor drug delivery. This may occur because a physiologically relevant concentration is not delivered to the target site, does not remain in contact for a sufficient period, or causes adverse effects because of the resulting high blood concentrations associated with indiscriminate release. Controlled drug delivery systems are designed to carry the drug and release it at the target site in a timely manner, facilitating its absorption and optimizing its physiological action. An effective controlled drug delivery system improves efficacy and safety by controlling the rate, time, and place of drug release within the body, thereby minimizing dose requirements and the potential to interact with non-target body sights that can contribute to undesirable side effects. Drug delivery has evolved from relatively simple systems to modern technologies designed to personalize medicines that have biologically precise drug release in response to real-time monitoring of body parameters.

Controlled drug delivery system development is rapidly evolving with an ever-increasing focus on advanced technologies that bring together a wide range of skilled professions including pharmaceutical scientists, chemical, mechanical, and electrical engineers, chemists, physicists, and clinicians. It is an exciting field that has helped to advance clinical outcomes in almost every health condition, ranging from negating the need for cold-chain storage thus allowing medicines to be transported to the most remote parts of the world, to precision targeting of drugs in cancer treatment. This book is designed to provide an insight into the fundamentals of drug delivery and the important processes in the development of controlled drug delivery systems.

The book is divided into three parts.

Part 1 (Chapters 1–8) introduces the concept of drug delivery and provides a perspective into the challenges, opportunities, and fundamental processes involved in the development of controlled drug delivery systems. It includes a historical perspective and a peek into the future of drug delivery. There is a focus on the drug development process, including the selection of pharmaceutical candidates and evaluation of their physicochemical characteristics with emphasis on the relevance to dosage form design. The role and application of mathematical modeling and the influence of drug transporters in pharmacokinetics and drug disposition complete this section.

Part 2 (Chapter 9–13) is focused on particular challenges in controlled drug delivery and advanced delivery technologies. This includes delivery systems for biologicals, an increasing drug category that presents enormous therapeutic opportunities and equally enormous delivery challenges. The application and recent advances in cell-mediated drug delivery are discussed, and there is a series of chapters on nanotechnology that include fundamentals, applications for targeted delivery, and discussion of the toxicological and safety issues.

Jinsong Hao
Department of Pharmaceutical Sciences
College of Pharmacy & Health Sciences
Campbell University
Buies Creek, NC, USA

Ann Mari Holsæter
Drug Transport and Delivery Research Group
Department of Pharmacy, Faculty of Health Sciences
University of Tromsø The Arctic University of Norway
Tromsø, Norway

May Wenche Jøraholmen
Drug Transport and Delivery Research Group
Department of Pharmacy, Faculty of Health Sciences
University of Tromsø The Arctic University of Norway
Tromsø, Norway

Panayiotis Kattou
Department of Chemical and Process Engineering
University of Surrey
Guildford, UK

Vitaliy V. Khutoryanskiy
Reading School of Pharmacy
University of Reading
Reading, UK

Li Li
Australian Institute for Bioengineering and Nanotechnology (AIBN)
The University of Queensland
Brisbane, QLD, Australia

Lingyi Li
College of Engineering
China Agricultural University
Beijing, P.R. China

S. Kevin Li
Division of Pharmaceutical Sciences
James L Winkle College of Pharmacy
University of Cincinnati
Cincinnati, OH, USA

Guoping Lian
Unilever Research Colworth
Bedford, UK

and

Department of Chemical and Process Engineering
University of Surrey
Guildford, UK

Xiaowen Liang
Therapeutics Research Group
The University of Queensland Diamantina Institute
The University of Queensland
Translational Research Institute
Brisbane, QLD, Australia

Julien Meissonnier
Science and Technology
Catalent Pharma Solutions
Beinheim, France

Javier O. Morales
Center of New Drugs for Hypertension (CENDHY)
Santiago, 8380494, Chile

and

Advanced Center for Chronic Diseases (ACCDiS)
Santiago, 8380494, Chile

and

Department of Pharmaceutical Science and Technology
School of Chemical and Pharmaceutical Sciences
University of Chile
Santiago, 8380494, Chile

Peter W.J. Morrison
Reading School of Pharmacy
University of Reading
Reading, UK

Preface

Effective, controlled drug delivery has the potential to greatly impact the therapeutic outcome, clinical benefit, and safety of drugs in a wide range of diseases and health conditions. There are a large number of potentially useful drugs with limited effectiveness and/or safety concerns due to poor drug delivery. This may occur because a physiologically relevant concentration is not delivered to the target site, does not remain in contact for a sufficient period, or causes adverse effects because of the resulting high blood concentrations associated with indiscriminate release. Controlled drug delivery systems are designed to carry the drug and release it at the target site in a timely manner, facilitating its absorption and optimizing its physiological action. An effective controlled drug delivery system improves efficacy and safety by controlling the rate, time, and place of drug release within the body, thereby minimizing dose requirements and the potential to interact with non-target body sights that can contribute to undesirable side effects. Drug delivery has evolved from relatively simple systems to modern technologies designed to personalize medicines that have biologically precise drug release in response to real-time monitoring of body parameters.

Controlled drug delivery system development is rapidly evolving with an ever-increasing focus on advanced technologies that bring together a wide range of skilled professions including pharmaceutical scientists, chemical, mechanical, and electrical engineers, chemists, physicists, and clinicians. It is an exciting field that has helped to advance clinical outcomes in almost every health condition, ranging from negating the need for cold-chain storage thus allowing medicines to be transported to the most remote parts of the world, to precision targeting of drugs in cancer treatment. This book is designed to provide an insight into the fundamentals of drug delivery and the important processes in the development of controlled drug delivery systems.

The book is divided into three parts.

Part 1 (Chapters 1–8) introduces the concept of drug delivery and provides a perspective into the challenges, opportunities, and fundamental processes involved in the development of controlled drug delivery systems. It includes a historical perspective and a peek into the future of drug delivery. There is a focus on the drug development process, including the selection of pharmaceutical candidates and evaluation of their physicochemical characteristics with emphasis on the relevance to dosage form design. The role and application of mathematical modeling and the influence of drug transporters in pharmacokinetics and drug disposition complete this section.

Part 2 (Chapter 9–13) is focused on particular challenges in controlled drug delivery and advanced delivery technologies. This includes delivery systems for biologicals, an increasing drug category that presents enormous therapeutic opportunities and equally enormous delivery challenges. The application and recent advances in cell-mediated drug delivery are discussed, and there is a series of chapters on nanotechnology that include fundamentals, applications for targeted delivery, and discussion of the toxicological and safety issues.

Part 3 (Chapters 14–20) provides a "top to bottom" critique of the common administration routes for controlled drug delivery. Each chapter begins with a short introduction and then a more detailed discussion of the physiology pertinent to each administration route, focusing on the barriers to drug delivery. Controlled drug delivery systems that have been evaluated for each route are then discussed before some conclusions summarizing the state-of-the-art and potential future developments. Each chapter includes comprehensive references at the time of writing for those wishing to read the primary literature. *Controlled* drug delivery systems imply that control over dosing resides in the formulation with control over drug release and predictable drug delivery. However, it is apparent that, given the complexities of the biological barriers present for the administration routes, in several cases control over drug delivery arises predominantly from the biological barrier. Although strategies have been developed to reduce these barriers – for example the use of penetration enhancers – it is contentious whether these approaches truly allow *controlled* drug delivery. However, in seeking to provide a comprehensive critique of the current literature, such partially controlled systems have been considered.

We express our thanks to the authors who have contributed to this book. In each case, the chapters are authored by well-respected researchers in the field who have generously provided their knowledge and experience, and continue to contribute to advancing research in their fields. We are also grateful to Jonathan Rose and his team at Wiley who have brought the concepts and chapters to fruition, and shown remarkable patience in dealing with editors who agree with Albert Einstein that "time is an illusion".

Australia
June 2021

Heather A. E. Benson
Michael S. Roberts
Adrian C. Williams
Xiaowen (Tina) Liang

List of Contributors

Heather A.E. Benson
Curtin Medical School
Curtin Health Innovation Research Institute
Curtin University
Perth, WA, Australia

Annette L. Bunge
Chemical and Biological Engineering Department
Colorado School of Mines
Golden, CO, USA

Nazareth E. Ceschan
Departamento de Biología
Bioquímica y Farmacia, UNS
Bahía Blanca, Argentina

and

Planta Piloto de Ingeniería Química (PLAPIQUI)
Consejo Nacional de Investigaciones Científicas y Técnicas (CONICET) – Universidad Nacional del Sur (UNS)
Bahía Blanca, Argentina

Tao Chen
Department of Chemical and Process Engineering
University of Surrey
Guildford, UK

Barbara R. Conway
Department of Pharmacy
University of Huddersfield
Huddersfield, HD1 3DH, UK

Gøril Eide Flaten
Drug Transport and Delivery Research Group
Department of Pharmacy, Faculty of Health Sciences
University of Tromsø The Arctic University of Norway
Tromsø, Norway

Muhammad U. Ghori
Department of Pharmacy
University of Huddersfield
Huddersfield, HD1 3DH, UK

Shivakumar H.N
Department of Pharmaceutics
K.L.E College of Pharmacy
Bengaluru
India

Lujia Han
College of Engineering
China Agricultural University
Beijing, P.R. China

Jinsong Hao
Department of Pharmaceutical Sciences
College of Pharmacy & Health Sciences
Campbell University
Buies Creek, NC, USA

Ann Mari Holsæter
Drug Transport and Delivery Research Group
Department of Pharmacy, Faculty of Health Sciences
University of Tromsø The Arctic University of Norway
Tromsø, Norway

May Wenche Jøraholmen
Drug Transport and Delivery Research Group
Department of Pharmacy, Faculty of Health Sciences
University of Tromsø The Arctic University of Norway
Tromsø, Norway

Panayiotis Kattou
Department of Chemical and Process Engineering
University of Surrey
Guildford, UK

Vitaliy V. Khutoryanskiy
Reading School of Pharmacy
University of Reading
Reading, UK

Li Li
Australian Institute for Bioengineering and Nanotechnology (AIBN)
The University of Queensland
Brisbane, QLD, Australia

Lingyi Li
College of Engineering
China Agricultural University
Beijing, P.R. China

S. Kevin Li
Division of Pharmaceutical Sciences
James L Winkle College of Pharmacy
University of Cincinnati
Cincinnati, OH, USA

Guoping Lian
Unilever Research Colworth
Bedford, UK

and

Department of Chemical and Process Engineering
University of Surrey
Guildford, UK

Xiaowen Liang
Therapeutics Research Group
The University of Queensland Diamantina Institute
The University of Queensland
Translational Research Institute
Brisbane, QLD, Australia

Julien Meissonnier
Science and Technology
Catalent Pharma Solutions
Beinheim, France

Javier O. Morales
Center of New Drugs for Hypertension (CENDHY)
Santiago, 8380494, Chile

and

Advanced Center for Chronic Diseases (ACCDiS)
Santiago, 8380494, Chile

and

Department of Pharmaceutical Science and Technology
School of Chemical and Pharmaceutical Sciences
University of Chile
Santiago, 8380494, Chile

Peter W.J. Morrison
Reading School of Pharmacy
University of Reading
Reading, UK

S. Narasimha Murthy
Pharmaceutics and Drug Delivery
The University of Mississippi
113 Faser Hall, MS, USA

and

Institute for Drug Delivery and Biomedical Research
Bengaluru, India

and

Department of Pharmaceutics
K.L.E College of Pharmacy
Bengaluru, India

and

Institute for Drug Delivery and Biomedical Research
Bengaluru, India

José das Neves
i3S – Institute for Research and Innovation in Health
University of Porto, Portugal

Zhi Qi
Australian Institute for Bioengineering and Nanotechnology (AIBN)
The University of Queensland
Brisbane, QLD, Australia

María V. Ramírez-Rigo
Departamento de Biología
Bioquímica y Farmacia, UNS
San Juan 670, 8000 Bahía Blanca, Argentina

and

Planta Piloto de Ingeniería Química (PLAPIQUI)
Consejo Nacional de Investigaciones Científicas y Técnicas (CONICET) - Universidad Nacional del Sur (UNS)
Bahía Blanca, Argentina

Michael S. Roberts
Therapeutics Research Centre
Basil Hetzel Institute for Translational Medical Research
The Queen Elizabeth Hospital
Adelaide, SA, Australia

and

School of Pharmacy and Medical Sciences
University of South Australia
Adelaide, SA, Australia

and

Diamantina Institute
The University of Queensland, Translational Research Institute
Brisbane, QLD, Australia

and

Therapeutics Research Group
The University of Queensland Diamantina Institute
The University of Queensland, Translational Research Institute
Brisbane, QLD, Australia

Qi Ruan
Therapeutics Research Group
The University of Queensland Diamantina Institute, The University of Queensland Translational Research Institute
Brisbane, QLD, Australia

Bruno Sarmento
INEB – Institute for Biomedical Engineering
University of Porto, Portugal

Ronak Savla
Science and Technology
Catalent Pharma Solutions
Somerset, NJ, USA

Monika Schäfer-Korting
Institute of Pharmacy (Pharmacology and Toxicology)
Freie Universität Berlin
Berlin, Germany

Yan Shu
Department of Pharmaceutical Sciences
School of Pharmacy, University of Maryland at Baltimore
Baltimore, MD, USA

Nataša Škalko-Basnet
Drug Transport and Delivery Research Group
Department of Pharmacy, Faculty of Health Sciences
University of Tromsø The Arctic University of Norway
Tromsø, Norway

Hugh D. C. Smyth
College of Pharmacy
The University of Texas at Austin
2409 West University Avenue
Austin, TX, United States

Sarasija Suresh
Institute for Drug Delivery and Biomedical Research
Bengaluru
India

Selenia Ternullo
Drug Transport and Delivery Research Group
Department of Pharmacy, Faculty of Health Sciences
University of Tromsø The Arctic University of Norway
Tromsø, Norway

Lionel Trottet
DMPK (Drug Metabolism and PharmacoKinetics)
Galapagos, Romainville, France

Sitaram Velaga
Kenox Pharmaceuticals Inc
Monmouth Junction
NJ, 08852, USA

Parameswara R. Vuddanda
Research Centre for Topical Drug Delivery and Toxicology
University of Hertfordshire
Hertfordshire, AL10 9AB, UK

Haolu Wang
Therapeutics Research Group
The University of Queensland Diamantina Institute, The University of Queensland
Translational Research Institute
Brisbane, QLD, Australia

Adrian Williams
University of Reading
Reading, UK

Klintean Wunnapuk
Toxicology Division
Department of Forensic Medicine, Faculty of Medicine
Chiang Mai University
Chiang Mai, Thailand

Haotian Yang
Therapeutics Research Group
The University of Queensland Diamantina Institute, The University of Queensland
Translational Research Institute
Brisbane, QLD, Australia

Hong Yang
Department of Pharmaceutical Sciences
School of Pharmacy
University of Maryland at Baltimore
Baltimore, MD, USA

Senpei Yang
College of Engineering
China Agricultural University
Beijing, P.R. China

Christian Zoschke
Institute of Pharmacy (Pharmacology and Toxicology)
Freie Universität Berlin
Berlin, Germany

Part I

Product Design, the Essence of Effective Therapeutics

1

Challenges and Innovations of Controlled Drug Delivery

Heather A.E. Benson[1] *and Michael S. Roberts*[2,3,4]

[1] *Curtin Medical School, Curtin Health Innovation Research Institute, Curtin University, Perth, WA, Australia*
[2] *Diamantina Institute, The University of Queensland, Translational Research Institute, Brisbane, QLD, Australia*
[3] *School of Pharmacy and Medical Sciences, University of South Australia, Adelaide, SA, Australia*
[4] *Therapeutics Research Centre, Basil Hetzel Institute for Translational Medical Research, The Queen Elizabeth Hospital, Adelaide, SA, Australia*

1.1 Background

A key health care advance over the last century has been the development of drug delivery systems that not only enable effective, safe, and reproducible delivery for optimal therapeutic benefit, but are also chemically and physically stable, aesthetically acceptable, convenient to use, cost-effective and optimized for the most appropriate mode of administration in their target population. There is a wide range of drug delivery systems available for a range of routes of administration, with the aim of achieving local or systemic effects, as shown in Figure 1.1. The most common route of administration is the oral route due to its convenience, cost, and patient acceptance. Injectables include the more common parenteral routes which are intradermal, subcutaneous (SC), intramuscular (IM), and intravenous delivery (Figure 1.1). In addition to the oral route, the most common nonparenteral routes of delivery are topical/transdermal, nasal, pulmonary, ocular, rectal, and vaginal (Figure 1.1). Drug absorption processes for nonparenteral administration generally involve passive and/or active transport across an epithelial barrier and carriage away into the systemic circulation by either local blood flow and/or via the lymphatic circulation. Topical delivery is used for a local effect, such as the treatment of pain, inflammation, or infection of the oropharynx (including nasal, buccal, and sublingual), eye, skin, lung, rectum, and vagina or for systemic delivery.

In this chapter, we introduce the routes of administration of therapeutic and diagnostic compounds to the body, the considerations in designing drug delivery systems, and some examples of the innovations that have contributed new controlled delivery products. The aim is to provide an overview of the topic, with considerable detail provided in the subsequent chapters in this and other volumes in this series.

1.2 Parenteral Dosage Forms

The parenteral route enables precise dosing for drugs that have a narrow therapeutic index, complete dosing for those with poor oral bioavailability to maintain optimal therapeutic concentrations, ready access in patients with swallowing difficulties, or an unconscious patient, and

Fundamentals of Drug Delivery, First Edition.
Edited by Heather A. E. Benson, Michael S. Roberts, Adrian C. Williams, and Xiaowen Liang.

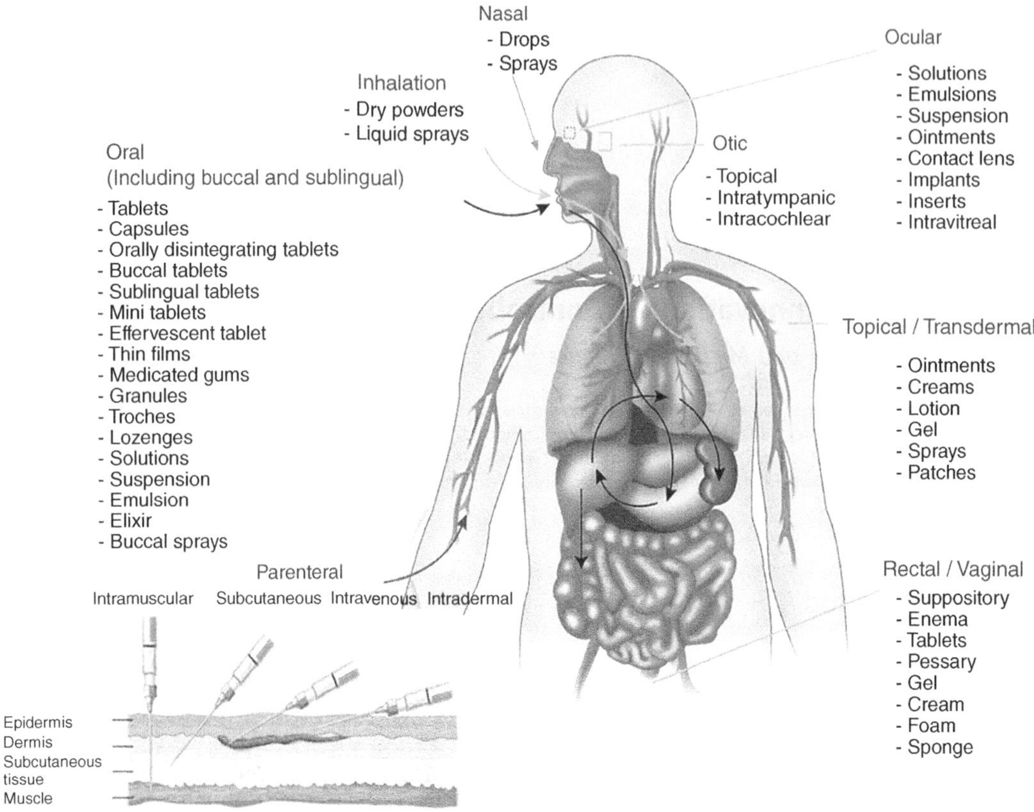

Figure 1.1 Routes of drug administration and associated dosage forms. Source: Khan and Roberts [1]. © 2018 Elsevier.

immediate blood levels of drugs in an emergency. It is used to deliver intravenous fluid replacement and parenteral nutrition fluids and is the preferred delivery route for emerging novel therapeutic molecules, such as proteins, peptides, and biologics. However, parenteral administration requires specialized personnel, and there is a risk of tissue damage, infection, and immune reactions. There is a requirement for sterility in processing, storage, and administration making it more expensive than other conventional drug products. The most common routes used for systemic delivery of a wide range of molecules including biopharmaceuticals, are SC, intravenous (IV), and IM [2] as shown in Figure 1.1. While many parenteral products are solutions, controlled release from formulations such as microparticles or implants composed of biodegradable polymers is also available.

1.2.1 Intravenous Route (IV)

The IV route is used to administer a drug directly into a vein, either as a bolus or by infusion, thus allowing administration of relatively large volumes of fluid. It avoids first pass metabolism and is the preferred choice for rapid pharmacological action, especially in emergencies and anesthesia. However, administration requires medical/nursing supervision, and there are associated risks of infection via the injection site, coagulation of infusion lines, and the possibility of red blood cell hemolysis, pain, and phlebitis. Effective sterile manufacture and appropriate storage is required to prevent contamination with microbes, pyrogens, and inert particles.

1.2.2 Intramuscular Route (IM)

IM injection enables systemic dosing, high bioavailability, and a rapid onset of action of drugs such as antibiotics, steroids, and narcotic analgesics. It provides a low-cost alternative to IV dosing as demonstrated by Milkovich and Piazza, who reported 10 times lower cost associated with IM compared to IV antibiotics and the opportunity for self-administration and earlier hospital discharge [3]. The IM route avoids many of the administration-related risks of IV administration and can provide a safe, rapid drug administration option in the management of anaphylaxis, where IM dosing of epinephrine was associated with significantly less cardiovascular adverse events compared with IV bolus epinephrine (1.3% vs. 10%) [4] and can be administered rapidly by a teacher, relative, or member of the public. IM injection is associated with pain from needle insertion and care is required to avoid piercing arteries or causing nerve damage.

1.2.3 Subcutaneous Route (SC)

The SC route is cost-effective, amenable to self-injection, minimally invasive, and associated with fewer side effects than the IV route [5]. While adverse reactions to SC dosing are uncommon, edema, pain, inflammation, infection, abscesses, and injection-site reactions may be more common in the elderly [6]. The SC route offers the opportunity for placement of implants that can provide controlled drug release over prolonged periods. Portable devices for SC injection such as insulin pumps, auto injectors, and pen injectors allow optimal dose delivery and self-administration in the management of a range of conditions. Examples include insulin, epinephrine, interferon β-1a (multiple sclerosis), and Enbrel® (auto injectors containing the biologic etanercept for use in arthritis) (https://www.enbrel.com/rheumatoid-arthritis/about-enbrel-for-ra, in).

1.2.4 Other Parenteral Routes

Less common parenteral routes for drug delivery include epidural, intrathecal, intracerebroventricular (ICV), intra-arterial, intra-articular, intracavernous, intralesional, intraosseous, and intravesical injection and infusion. Epidural (outside the dura/spinal fluid sac) and intrathecal (inside sac) routes are used primarily for anesthesia/analgesia. ICV involves injection into the cerebrospinal fluid in cerebral ventricles to bypass the blood–brain barrier. It can be used for drug administration in neurodegenerative disorders such as spinal muscular atrophy or chemotherapeutic agents in gliomas. It is also used as a research tool. Intra-articular injections (into a joint) of anti-inflammatory drugs such as steroids are generally used for the treatment of inflammatory joint conditions such as arthritis, tendinitis, gout, and Carpal Tunnel Syndrome. Vasodilators and vasoconstrictors are administered by intracavernous injection (into the penis) to treat erectile dysfunction. Intralesional steroid therapy is most commonly used in the treatment of hypertrophic and keloid scars, acne cysts, and alopecia areata. Intraosseous influsion (IO) involves injecting directly into the marrow of a bone as an alternative to provide drugs or fluid when IV injection is not available, such as trauma patients with compromised IV access. A needle is inserted through the hard bone cortex into the soft bone marrow to provide immediate access to the vascular system [7]. Intravesicular injection involves administration into the bladder using a catheter, used primarily to instil immunotherapeutic and chemotherapeutic agents in the treatment of bladder cancer.

1.3 Oral Route and Delivery Systems

More than 75% of all doses are given as tablets and capsules due to their ease and low cost of manufacture, stability, convenience in carrying and storing, ease of use, effective taste-masking strategies, and less dosing errors [8]. However, frequent dosing of three to four times a day may be required for drugs with short or immediate elimination half-lives, potentially creating a high pill burden and adherence issues particularly in the older population who may be taking a number of medications to manage multiple conditions. There is clearly enormous potential for controlled drug delivery via the oral route, hence, we have devoted a volume in this series to this topic. Here we provide an overview of the formulation challenges and opportunities offered by the oral route.

While most oral formulations are swallowed intact, as shown in Figure 1.1, immediate release and soluble tablets offer quick therapeutic effect and are particularly suitable for people with swallowing problems such as children and the elderly. Formulation options include tablets that are allowed to effervesce in water, and oral disintegrating tablets, medicated gums, and oral films that can be sucked or chewed before swallowing. Sublingual (placed under the tongue) and buccal (placed between the gum and cheek) tablets, sprays, and patches deliver the drug through the mucosal surfaces of the mouth, providing the added advantage of avoiding first-pass metabolism.

Oral modified release dosage forms include delayed, extended, enteric coated, and other delivery systems that can improve the stability, efficacy, safety, and convenience for patients. They can offer reduced dosing frequency, with improved adherence, clinical outcomes, and a reduction in the need for further clinical intervention such as hospital readmissions [9]. Controlled release technologies include the use of biodegradable polymers, release rate-controlling membranes, microencapsulation, and bioadhesives to enhance drug contact time [10]. Multiparticulate controlled release formulations offer more flexibility for pediatric or elderly patients who have trouble swallowing as they can be filled into sachets or compressed into minitablets [11].

1.4 Nasal Drug Delivery

Intranasal delivery systems are commonly used for treating local conditions (Figure 1.1) such as nasal allergy, congestion, and nasal infections, and are increasingly being used for delivery of opioids, respiratory medicines, vaccines, and other medicines for systemic effect. The thin and highly vascularized nasal mucosal layer enables a fast onset of drug action and avoidance of gastrointestinal tract and hepatic first-pass metabolism [12], and the nasal-associated lymphoid tissue (NALT) offers opportunities for vaccine delivery. The most common delivery systems are nasal sprays and drops administered to the anterior site of the nasal cavity, with the former showing twofold to threefold greater drug bioavailability [13]. The nasal cavity is designed to protect the lungs by filtering inhaled air via its hair follicles and mucous membrane, then transporting those particles down through the esophagus to the stomach by a mucociliary clearance process, thus exposing them to the degradative processes of the gastro-intestinal tract. This clearance process represents a challenge to nasal bioavailability as drug residence time in the nasooral cavity is decreased, resulting in reduced transport into the blood stream. Adhesive polymers, enzyme inhibitors, or the use of novel drug delivery formulations such as liposomes, nanoparticles, microspheres, microemulsions, and penetration enhancers can enhance drug retention and/or penetration across the nasal mucosa [14]. The most favorable candidates for nasal delivery, as shown in Table 1.1, are small lipophilic molecules such as propranolol (M_W = 295.8, $\log P_{o/w}$ = 2.7), fentanyl (M_W = 336.5, $\log P_{o/w}$ = 4.1), and estradiol (M_W = 296.4, $\log P_{o/w}$ = 4.3) [15].

Table 1.1 Properties and indications of nasally delivered drugs.

Drug	M_W	Log P	Indication
Butorphanol	327.5	3.7	Migraine
Estradiol	296.4	4.3	HRT
Fentanyl	336.5	4.1	Cancer pain
Naloxone	327.4	0.6	Opioid overdose
Nicotine	162.2	1.2	NRT
Sumatriptan	295.4	0.8	Migraine
Zolmitriptan	287.3	1.6	Migraine

HRT, hormone replacement therapy; NRT, nicotine replacement therapy; M_W, molecular weight.
Source: Fortuna et al. [15]. © 2014 Elsevier.

1.5 Pulmonary Drug Delivery

Pulmonary delivery takes advantage of the favorable properties of the lungs for drug delivery: large surface area of the alveoli, rich supply of blood vessels, thin (0.1–2 μm) alveolar membranes, limited enzymatic degradation, and avoidance of first-pass hepatic metabolism [16]. Consequently, relatively high bioavailability of small and large molecules, such as human growth hormone, interferon α, immunoglobulin G (IgG), and insulin (M_W approximately 22, 19, 150, and 5.8 kDa, respectively), can be achieved. The most commonly used inhaler devices are pressurized metered-dose inhalers (pMDIs: delivering liquid droplets) with or without a spacer (large volume container into which the aerosol is actuated allowing regular breaths for inhalation), dry powder inhalers (DPIs) and nebulizers (aerosolized liquid droplets) [17, 18]. Each device has advantages and disadvantages, and the utility of each type can vary, with some patient groups more commonly experiencing coordination issues or lung disease that can affect bioavailability from particular devices. Metered dose inhalers (pMDIs) deliver 1–3 μm particles to the respiratory surfaces of the lung with a typical efficiency of 10–20% of total expelled dose in young healthy adults, and even less in children, the elderly and in the presence of reduced lung function [19]. pMDIs with spacer are less portable but do increase bioavailability by removing the need to coordinate actuation with inhalation. DPIs have an ease of use advantage but still require minimum inspiratory flow and the presence of the dry powder may be irritating to lung tissue [20]. Nebulizers are particularly useful for children and the elderly as they provide a constant aerosol flow within a facemask. New technologies such as the Hailie™ (formerly known as the Smartinhaler™: Adherium, San Mateo, CA) allow remote monitoring of inhaler medication use and dosing reminders to improve adherence and treatment outcome.

1.6 Transdermal Drug Delivery

Drugs are applied to the skin for superficial (e.g. cosmetic, sunscreen, moisturization, and repelling insects), local (e.g. skin infection, inflammation, and disease), appendageal (e.g. acne and hair loss), deep tissue (e.g. pain and inflammation) or systemic (e.g. smoking prevention, hormone replacement, and contraceptive patches) effects (Figure 1.2) [21]. While the skin offers a large, readily accessible site for drug administration, it also provides a formidable barrier to permeation. The

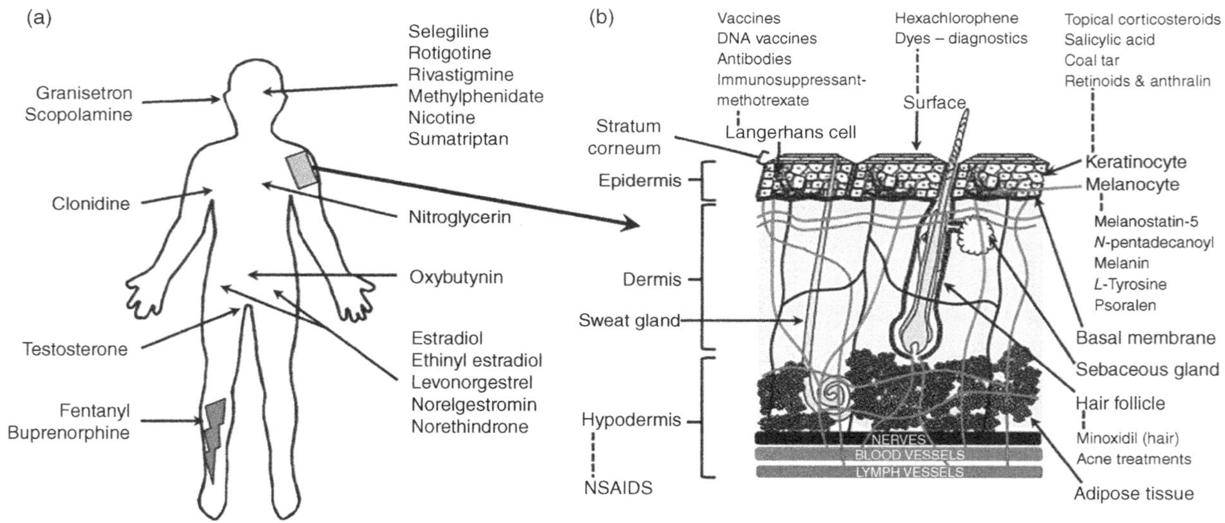

Figure 1.2 (a) Approved and under development of transdermal drugs and their recommended site of application. (b) Close-up of human skin and various sites and targets within the skin. Source: Benson et al. [21]. © 2019 Eureka Science.

epidermis is a protective layer to prevent ingress of environmental contaminants while also preventing water loss from the body. It is in a constant state of renewal with formation of a new cell layer of keratinocytes at the stratum basale, loss of their nucleus, and other organelles to form desiccated, proteinaceous corneocytes on their journey toward desquamation, which occurs from the skin surface at the same rate as formation in normal skin (Figure 1.2). The outermost layer, the stratum corneum (SC), consists of a brick wall-like structure of corneocytes in a matrix of intercellular lipids, with desmosomes acting as molecular rivets between the corneocytes. It is widely accepted that permeation across the SC is primarily via the lipid domains, even for polar compounds; therefore, solute diffusion models and permeation enhancement techniques are primarily focused on this region.

The range of products applied for local effects in the skin and underlying tissues include solutions, powders, ointments, pastes, gels, foams, creams, and vesicles-based formulations. Transdermal dosage forms are primarily patches, although other technologies such as the "patchless patch," a metered-dose spray that forms a reservoir within the skin, developed by Acrux (Melbourne, Australia) offer an alternative. The ideal drug candidate for skin delivery should have high potency with a daily dose requirement of a few milligrams, small size ($M_W < 500$ Da), be moderately lipophilic (log P of 1–3), and uncharged [22]. Clearly, this imposes a severe limitation on the number of drugs that can be developed as a transdermal product. There has been a major focus on development of skin permeation enhancement approaches that range from formulation-based (including inclusion of chemical, solvents, and nano-systems), energy-based (including iontophoresis, sonophoresis, magnetophoresis, and stratum corneum ablation), and minimally invasive technologies (including jet propulsion systems and microneedles). Transdermal patches are now widely used for systemic therapy of pain, angina, smoking cessation, cognitive disorders, and hormone replacement therapy (HRT). Table 1.2 summarizes the drug properties and indications of some of the key products now in use [22, 23]. As the transdermal route avoids first-pass metabolism-reduced dosages can provide the same therapeutic effect, thus reducing side-effects and increasing safety. Transdermal products also increase patient acceptability, adherence, and satisfaction as was demonstrated in Alzheimer's patients where those using patches were reported to comply better with treatment than patients using oral dosage forms over a six-month study period [24]. The most common side-effects found with the application of transdermal patches are mild to moderate erythema or itch at the patch site.

1.7 Ocular Drug Delivery

Topical administration of eye drops and ointments are the mainstay of drug delivery to the anterior segment of the eye; however, bioavailability is limited due to wash out by tear fluid [25]. Some devices have been developed to increase residence time such as the Lacrisert® ophthalmic insert (Bausch + Lomb) (http://www.bausch.com/ecp/our-products/rx-pharmaceuticals/rx-pharmaceuticals/lacrisert) which is a sterile, translucent, rod-shaped, water-soluble, preservative-free, slow-release lubricant composed of hydroxypropyl cellulose that is placed into the inferior cul-de-sac of the eye. One insert per day provides continuous lubrication. The Ocusert implant (Sandoz) (https://www.drugs.com/cons/ocusert-pilo.html) is also placed in the cul-de-sac of the eye and releases pilocarpine over seven days to treat glaucoma. The latter device is not biodegradable and must be removed and replaced weekly (Figure 1.3).

Drug delivery to the posterior segment of the eye for the treatment of retinal and choroid disease is challenging due to multiple barriers [26]. Conjunctival systemic absorption and tear turnover processes effectively limit topical drug delivery and cannot be prevented; therefore,

Table 1.2 Physicochemical, pharmacokinetic properties and indications of currently marketed transdermal drugs.

Drug	Dose/day (mg)	M_W (Da)	MP (°C)	Log *P*	S_{aq} (mg/ml)	Clearance (*cl*) l/h	$t_{1/2}$ (h)	Oral *F* (%)	Plasma concentration	Estimated J_{skin} required (μg/h)	*In vivo* J_{skin} (μg/cm²/h)	Product name	Indication
Buprenorphine	0.12–1.68	468	209	3.8	0.047, 0.008 (32 °C)	77 i.v. 55 i.m.	22–36	51 sI 28 bc	0.1–0.4	7.7	0.8	Butrans	Moderate to severe pain
Clonidine	0.1–0.3	230	130	2.7	0.17, 13.58	15	6–20	95	0.2–2.0	3–30	1.2	Catapres-TTS	Hypertension
Ethinyl oestradiol	—	296	141–146	4.3	0.039, 0.0092 (25 °C)	70	7.7 p.o, 17 t.d.	55	0.025–0.075	1.75–5.25	0.07	Ortho Evra	Female contraception
Fentanyl	0.288–2.400	337	83–84	3.9	0.15, 0.2 (30 °C), 0.2 (25 °C)	27–75	3–12	50	1–3	27–225	2.4	Duragesic	Chronic pain
Granisetron	3.1	312	152–154	2.6	0.017	33–76	4–6	60	3.9 (t.d. Mean *C* max)	129–296	2.5	Sancuso	Chemotherapy-induced emesis
Glyceryl trinitrate	2.4–15.0	227	13, liquid	1	0.66, 1.3 (30 °C)	216–3270	0.03–0.05	< 1	0.1–5	4.32–1308	20,30	Nitro-Dur	Angina
Estradiol and Levonorgesterel	0.050, 0.0007–0.015	272, 312	173–179 235–237	4.2 3.8	0.003 (30 °C) 0.0015 (25 °C) 0.017	600–800 5.7	Approx. 1p.o. 28t.d 19.3p.o.	3–5	0.03–0.05 0.1–0.2	24–48 1	0.63, 0.09, 0.12, 0.18, 0.14, 0.23, 0.42, 0.17 0.2 0.03	Climara Pro	Female contraception
Methylphenidate	26.0 – 80.0	233	74–75, liquid	2.1	1.8	12(d); 21(1) (children 30 kg)	1.5–5	5–20	5–25	60–315	88	Daytrana	ADHD
Nicotine	7.0–21.0	162	−79, liquid	1.1	62, 1085 (30 °C)	77	2	20–45	5–30	385–2310	29, 69, 31, 40	Nicoderm	Smoking cessation
Norelgestromin	0.2	327	110–130	3.67	0.0043	—	28	—	0.6–1.2	—	0.31	Ortho Evra	Female contraception
Norethindrone acetate	0.125–0.250	341	161–162	3.2	0.0065	20.6	6–8	60	0.5–0.8	10.3–16.5	0.65	Combipatch	Female HRT
Oxybutynin	3.9	358	56–58	4.3		10–64	2	6	0.5–5	5–192	4.2	Oxytrol	Enuresis
Oestradiol	0.025–0.100	272	173–179	4.2	0.003 (30 °C) 0.0015 (25 °C)	600–800	0.05	3–5	0.04–0.06	24–48	0.2, 0.17, 0.14, 0.12, 0.42, 0.18, 0.63, 0.23, 0.09	VivelleDot	Female HRT
Rivastigmine	4.6–9.5	250	Oil at 25 °C	2.3	25	108	1.5	36	2.5–20	270–2160	39	Exelon	Alzheimer's disease
Rotigotine	1.0–3.0	316	75–77	4.7	0.017	600	5–7	—	0.4–2	240–1200	8.3	Neupro	Parkinson's disease and restless leg syndrome
Scopolamine	0.3	303	55, liquid	0.8	1.8, 75 (30 °C)	65–121	1–5	4–27	0.04	3.25–6.05	5.6	Transderm scop	Travel sickness
Seligiline	6.0–12.0	187	Liquid at 25 °C	2.7	0.73	84	10	4,10	2	168	12.5	Emsam	Depression
Testosterone	0.3–5.0	288	155	3.6	0.02	41	0.17–1.7	<1, 7	3–10.5	123–430.5	13.9	Androderm	Hypogonadism

Notes: i.v., intravenous; i.m., intramuscular; p.o., per oral; t.d., transdermal; s.I., sublingual; bc, buccal.
Source: Adapted from Refs. [22, 23].

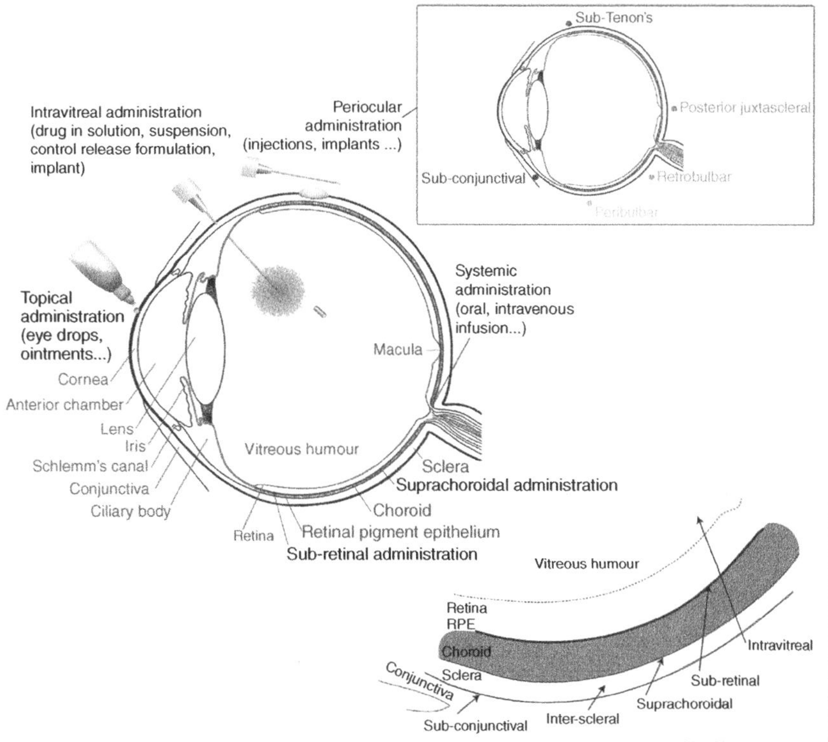

Figure 1.3 Routes for administration for retinal drug delivery. Source: Del Amo et al. [26]. © 2017 Elsevier.

alternative delivery systems have been developed to improve bioavailability. Intravitreal injection of anti-vascular endothelial growth factor (VEGF) proteins (bevacizumab, ranibizumab, and aflibercept) is the treatment of choice for management of wet age-related macular degeneration (AMD). However, compliance is poor, and therefore, ideal injection frequency is rarely maintained [27]. There are two main factors in improving drug delivery to the posterior segment of the eye: extend retention time of the delivery system to increase drug availability; overcome the barriers to increase drug penetration and target drug to a specific site within the eye. Controlled release approaches have included biodegradable polymeric particles and implants, nonbiodegradable implants, *in situ* gelling formulations, and encapsulated cell technology (ECT: entrapment of genetically engineered cells within a semipermeable matrix to immunologically isolated them from the host's body) [28].

1.8 Drug Delivery System Development Process

The development of a new drug delivery system to control or target delivery to a therapeutic site involves many steps from basic research through to clinical application as summarized in Figure 1.4 [29]. The early steps involve characterization of the drug which may be a small molecule, peptide/protein or gene-derived, and choice of delivery system design and materials. The delivery system is then evaluated for drug release and delivery *in vitro* and *in vivo* prior to clinical studies in volunteers and patient groups. The processes involved in developing a new controlled release dosage form from concept to consumer are addressed in detail in later chapters of this Encyclopedia.

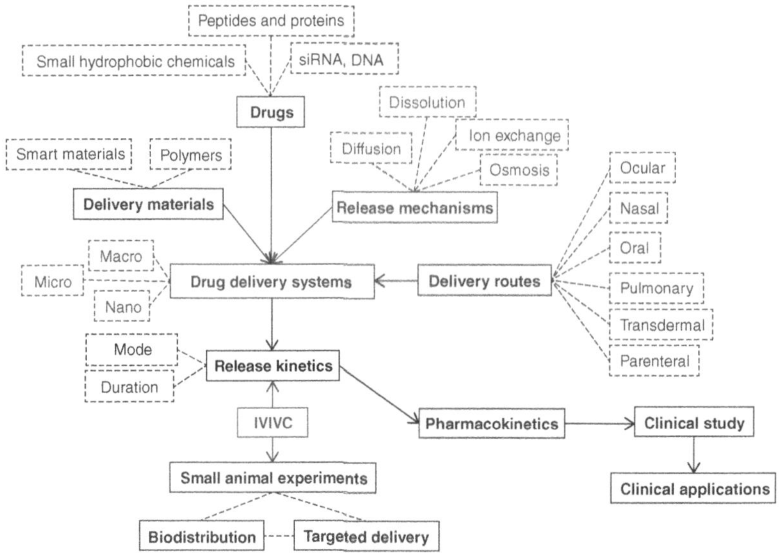

Figure 1.4 Overview of drug delivery system development from basic research to clinical applications. The main components are shown in bold-face and solid-box, and subsections of each component are shown in a dashed box (IVIVC: *in vitro–in vivo* correlation). Source: Park [29]. © 2014 Elsevier.

1.9 Conclusion

There is clearly enormous scope for controlled drug delivery to optimize therapeutic outcomes in a wide range of clinical conditions for human and veterinary application. Great improvements have been achieved but the myriad of different research directions being pursued globally suggests that there will be exponential growth in these delivery systems coming to market in the future. The aim of this series is to provide an insight into the state of the art of controlled drug delivery, from the underlying principles of formulation development (Volume 1), to in depth examination of controlled drug delivery for the oral (Volume 2) and other routes of administration (Volume 3), and a focus on the application of pharmacokinetic and pharmacodynamic aspects in the development of controlled drug delivery technologies (Volume 4).

References

1 Khan, M.S. and Roberts, M.S. (2018). Challenges and innovations of drug delivery in older age. *Adv. Drug Delivery Rev.* 135: 3–38.
2 Flynn, G.L. and Roberts, M.S. (2015). *Physical and Biophysical Foundation of Pharmacy Practice*. MI, USA: Michigan Publishing.
3 Milkovich, G. and Piazza, C.J. (1991). Considerations in comparing intravenous and intramuscular antibiotics. *Chemotherapy* 37 (Suppl 2): 1–13.
4 Campbell, R.L., Bellolio, M.F., Knutson, B.D. et al. (2015). Epinephrine in anaphylaxis: higher risk of cardiovascular complications and overdose after administration of intravenous bolus epinephrine compared with intramuscular epinephrine. *J. Allergy Clin. Immunol. Pract.* 3: 76–80.

5 Dychter, S.S., Gold, D.A., and Haller, M.F. (2012). Subcutaneous drug delivery: a route to increased safety, patient satisfaction, and reduced costs. *J. Infusion Nurs.* 35: 154–160.

6 Fonzo-Christe, C., Vukasovic, C., Wasilewski-Rasca, A.F., and Bonnabry, P. (2005). Subcutaneous administration of drugs in the elderly: survey of practice and systematic literature review. *Palliative Med.* 19: 208–219.

7 Luck, R.P., Haines, C., and Mull, C.C. (2010). Intraosseous access. *J. Emergency Med.* 39: 468–475.

8 Wilkinson, G.R. (1997). The effects of diet, aging and disease-states on presystemic elimination and oral drug bioavailability in humans. *Adv. Drug Delivery Rev.* 27: 129–159.

9 Toh, M.R., Teo, V., Kwan, Y.H. et al. (2014). Association between number of doses per day, number of medications and patient's non-compliance, and frequency of readmissions in a multi-ethnic Asian population. *Preventive Med. Rep.* 1: 43–47.

10 Wilson, C.G. and Crowley, P.J. (2011). *Controlled Release in Oral Drug Delivery*. Springer.

11 Rajabi-Siahboomi, A.R. (2017). *Multiparticulate Drug Delivery: Formulation, Processing and Manufacturing*. New York: Springer.

12 Bitter, C., Suter-Zimmermann, K., and Surber, C. (2011). Nasal drug delivery in humans. In: *Topical Applications and the Mucosa* (eds. C. Surber, P. Elsner and M. Farage), 20–35. Basel: Karger.

13 Harris, A.S., Nilsson, I.M., G. Wagner, Z., and Alkner, U. (1986). Intranasal administration of peptides: nasal deposition, biological response, and absorption of desmopressin. *J. Pharm. Sci.* 75: 1085–1088.

14 Illum, L. (2012). Nasal drug delivery - recent developments and future prospects. *J. Controlled Release* 161: 254–263.

15 Fortuna, A., Alves, G., Serralheiro, A. et al. (2014). Intranasal delivery of systemic-acting drugs: small-molecules and biomacromolecules. *Eur. J. Pharm. Biopharm.* 88: 8–27.

16 Labiris, N.R. and Dolovich, M.B. (2003). Pulmonary drug delivery. Part I: Physiological factors affecting therapeutic effectiveness of aerosolized medications. *Br. J. Clin. Pharmacol.* 56: 588–599.

17 Labiris, N.R. and Dolovich, M.B. (2003). Pulmonary drug delivery. Part II: The role of inhalant delivery devices and drug formulations in therapeutic effectiveness of aerosolized medications. *Br. J. Clin. Pharmacol.* 56: 600–612.

18 Stein, S.W. and Thiel, C.G. (2017). The history of therapeutic aerosols: a chronological review. *J. Aerosol Med. Pulmonary Drug Delivery* 30: 20–41.

19 Pavia, D., Thomson, M.L., Clarke, S.W., and Shannon, H.S. (1977). Effect of lung function and mode of inhalation on penetration of aerosol into the human lung. *Thorax* 32: 194–197.

20 Wallin, M., Tagami, T., Chen, L. et al. (2018). Pulmonary drug delivery to older people. *Adv. Drug Delivery Rev.* 135: 50–61.

21 Benson, H.A.E., Grice, J., Mohammed, Y. et al. (2019). Topical and transdermal drug delivery: from simple potions to smart technologies. *Curr. Drug Delivery* 16 (5): 444–460.

22 Wiedersberg, S. and Guy, R.H. (2014). Transdermal drug delivery: 30+ years of war and still fighting! *J. Controlled Release* 190: 150–156.

23 Pastore, M.N., Kalia, Y.N., Horstmann, M., and Roberts, M.S. (2015). Transdermal patches: history, development and pharmacology. *Br. J. Pharmacol.* 172: 2179–2209.

24 Molinuevo, J.L. and Arranz, F.J. (2012). Impact of transdermal drug delivery on treatment adherence in patients with Alzheimer's disease. *Expert Rev. Neurother.* 12: 31–37.

25 Ozdemir, M. and Temizdemir, H. (2010). Age- and gender-related tear function changes in normal population. *Eye* 24: 79–83.

26 Del Amo, E.M., Rimpela, A.K., Heikkinen, E. et al. (2017). Pharmacokinetic aspects of retinal drug delivery. *Prog. Retinal Eye Res.* 57: 134–185.

27 Holz, F.G., Tadayoni, R., Beatty, S. et al. (2015). Multi-country real-life experience of anti-vascular endothelial growth factor therapy for wet age-related macular degeneration. *Br. J. Ophthalmol.* 99: 220–226.

28 Kim, Y.C., Chiang, B., Wu, X., and Prausnitz, M.R. (2014). Ocular delivery of macromolecules. *J. Controlled Release* 190: 172–181.

29 Park, K. (2014). Controlled drug delivery systems: past forward and future back. *J. Controlled Release* 190: 3–8.

2

Challenges in Design of Drug Delivery Systems

S. Narasimha Murthy[1], Shivakumar H.N[2], and Sarasija Suresh[3]

[1] *Pharmaceutics and Drug Delivery, The University of Mississippi, 113 Faser Hall, MS, USA*
[2] *Department of Pharmaceutics, K.L.E College of Pharmacy, Bengaluru, India*
[3] *Institute for Drug Delivery and Biomedical Research, Bengaluru, India*

A rational drug therapy for most chronic disorders demands prolonged treatment regimens during which maintenance of ***optimal plasma concentration*** is a prime necessity in order to achieve a satisfactory therapeutic effect. Considering this need, continuous efforts have been made during drug discovery to identify new chemical entities (NCEs) that would elicit reproducible and prolonged effective plasma levels. However, a good number of therapeutic agents and NCEs discovered are inherently short-acting requiring multiple doses in a day to maintain a constant blood level, when administered as conventional or immediate release (IR) dosage forms. Such dosage forms need frequent dosing in order to maintain the plasma concentration of the drug in the therapeutic range. However, frequent dosing poses compliance issues in chronic diseases where most often the treatment would be life-long. Considering these limitations of conventional dosage forms, a number of ***extended release drug delivery systems (EDDS)*** have come into existence. As per the Food and Drug Administration (FDA), extended release products are those that release the drug over an extended or prolonged period of time and therefore allow reduction in the dosing frequency [1]. These products are known to maintain a constant plasma concentration throughout a 24-hour period thereby minimizing the blood levels fluctuations that are commonly encountered during repeated administration of IR dosage forms. ERDDS are further classified into sustained release and controlled release drug delivery systems based on the pattern of drug release.

Sustained release drug delivery systems (SDDS) refer to any dosage forms that release the medications over an extended time period [2]. The drug release from these systems is often concentration-dependent and therefore in a slow first-order fashion. A ***first-order*** release usually refers to the release kinetics that is usually dictated by the amount of drug in the dosage form. These systems, synonymously termed as ***prolonged release*** or ***time-release*** by different manufacturers may not be able to precisely maintain a constant plasma concentration though they are able to prolong the drug release. Sustained release is the terminology that is generally confined to oral dosage forms. Representative examples of sustained release products include Isoptin SR (verapamil; Abott), Trental (Pentoxifyline; Aventis), Cardizem® SR (diltiazem HCl; Aventis Pharmaceuticals), and Ritalin SR (Methylphenidate, Novartis), and Dilatrate SR (ISDN, Schwartz).

Controlled release drug delivery systems (CDDS) are those formulations that are designed to maintain predictably constant plasma concentrations that would be independent of the biological environment of application site [1]. These systems actually maintain a steady plasma concentration

Fundamentals of Drug Delivery, First Edition.
Edited by Heather A. E. Benson, Michael S. Roberts, Adrian C. Williams, and Xiaowen Liang.

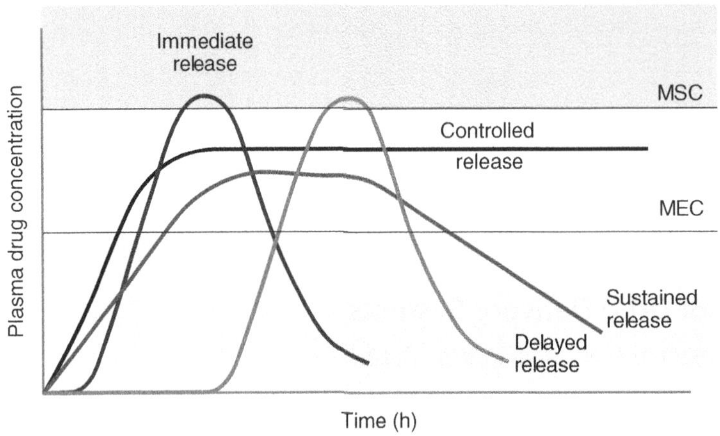

Figure 2.1 Plasma concentration profiles of conventional and extended release dosage forms.

over an extended time period, rather than merely extending the drug release from the dosage forms. To accomplish this, the drug release from CDD systems must invariably follow a zero-order fashion. A ***zero-order*** release refers to the release kinetics that is usually not affected by any physiological variable or the amount of drug in the delivery system. In such case, the *in vitro* release can be precisely used to predict the rate of release *in vivo* as the two rates typically would be the same. Oral osmotic (OROS) delivery systems are CDDS containing therapeutic agents that release the drug by zeroorder kinetics. Contrary to SDDS, CDDS systems are designed to be administered by various routes such as peroral, transdermal, ocular, intrauterine. Representative examples of controlled release systems include OROS® osmotic therapeutic systems of nifedipine (Procardia XL®, Alza Corporation, Pfizer), Norplant® subdermal implants of levonorgestrel (Wyeth), Dilacor® XR (diltiazem hydrochloride, Rhone Poulenc Rorer), Coreg CR™ (carvedilol phosphate; Glaxo Smith Kline) and Ditropan XL® (oxybutynin chloride; Ortho-McNeil), Transderm-Nitro (Nitroglycerine, Alza Corporation), and Vaginal Ring® insert (medroxy progesterone; Upjohn) [3].

The controlled drug delivery systems (CRDDS) developed between 1950s and 1980s were referred as the ***first-generation*** systems that include the oral and the transdermal systems. The ***first-generation*** controlled drug delivery systems have successfully made into the market. The success of these systems was ascribed to the excellent correlation observed between the ***in vitro*** drug release kinetics and the ***in vivo*** bioavailability [4]. The systems developed between 1980s and 2010s are termed the second-generation systems that may include the injectable depot formulations of proteins, pulmonary delivery of insulin, tumor-targeted drug delivery using nanoparticles, etc. However, many of the ***second-generation*** controlled drug delivery systems are still in the investigational stage, as they have not been translated into clinical applications [5]. The CDDS evolved subsequently are reported to fall under the ***third-generation*** systems that would include the feedback-regulated systems.

The typical plasma concentration profiles of a conventional dosage form and different modified release dosage forms are portrayed in Figure 2.1.

The plasma concentration attained is usually dependent on the type of dosage form, dosing frequency, drug clearance rates, and the route of administration. A typical plasma drug concentration vs. time profile is depicted in Figure 2.1. The profile for an IR dosage form is characterized by a peak and trough following each dosing. The plasma drug concentrations remain tolerable as long as it is within the ***therapeutic window*** that corresponds to the region between ***maximum safe concentration (MSC)*** and a ***minimum effective concentration (MEC)***. The drug concentrations exceeding the MSC represent the toxic levels, while the concentrations below the MEC indicate the subtherapeutic concentrations. Untoward side effects are evident when plasma drug concentrations exceed the MSC, while no therapeutic effect would be observed when the plasma concentrations are below the MEC.

The plasma concentration profile of IR dosage form containing short-acting drugs are usually characterized by several peaks and troughs in 24-hour time period owing to high frequency of administration and therefore sometimes prone to be either ineffective or tend to cause drug-level-associated adverse effects. For such formulations, the dosing interval (DI) of drug is a function of the ***half-life*** ($t_{0.5,el}$) and the ***therapeutic index*** (MSC/MEC) as indicated by Eq. (2.1) [6]. Equation (2.1) thus indicates that drugs with short-life invariably need frequent administration/dosing compared to their long-acting counterparts:

$$\mathrm{DI} = 1.44 \times t_{0.5,el} \times \ln \frac{\mathrm{MSC}}{\mathrm{MEC}} \tag{2.1}$$

The major challenge involved in the design of controlled release drug delivery systems of short-acting drugs is to equate the rate of ***drug delivery*** to the ***rate of elimination*** of the drug from the body so as to maintain a steady plasma concentration within the therapeutic window. As the rate of elimination is governed by rate of ***drug disposition***, the rate of drug delivery can be easily modulated by the ***system design*** [7]. The rate of drug delivery that is expressed as the ratio of dose (D) to the dosing interval (T) is the product of plasma clearance (CL) and steady plasma concentration ($C_{p,ss}$) as shown in Eq. (2.2):

$$\text{Rate of delivery} = \frac{D}{T} = \text{Rate of elimination} = \mathrm{CL} \times C_{p,ss} \tag{2.2}$$

$$\text{Rate of delivery} = \frac{D}{T} = V_d \times K_{el} \times C_{p,ss} \tag{2.3}$$

The rate of drug delivery can be ultimately expressed as the product of three biopharmaceutical factors including volume of distribution (V_d), first-order elimination rate constant (K_{el}) and steady plasma concentration (C_{pss}) as per Eq. (2.3). Modulating the drug delivery would improve the pharmacokinetics and enhance the therapeutic value and thus the clinical utility of a several short-acting therapeutic agents.

A ***controlled release dose*** typically is composed of two parts namely a ***priming dose*** or a ***booster dose*** and a ***maintenance dose***. The ***priming dose*** is the fraction of the controlled release dose that is released immediately following administration of the dosage form. The priming dose is responsible to elicit a therapeutic response by increasing the plasma drug concentration so as to exceed the MEC. The maintenance dose is the portion of the dose that is released subsequently to maintain the plasma concentration in a steady state within the therapeutic window for prolonged period of time. Maintenance dose can be easily computed from the product of the ***rate of drug delivery*** (D/T) and the ***dosing interval*** (T). The relative proportion of the priming and the maintenance doses is usually determined by the intended duration of action and the biological half-life of the therapeutic agent [8].

The goal of an oral controlled drug delivery system is to achieve a zero-order kinetics that ensures a constant release of a therapeutic agent throughout the entire day. These systems are expected to overcome the fluctuating peak and trough concentrations associated with the IR dosage forms. Reduction of side effect profile of therapeutic agents and improvement in the patient compliance is the need of the day in clinical practice, as many chronic diseases demand lifelong therapy [9, 10].

Advantages of controlled drug delivery systems

- By virtue of the ability to release the drug in a constant manner for prolonged time periods, CRDDS maintain a steady plasma concentration in therapeutic range that is free from oscillations.
- Avoidance of undesirable peak or troughs in the plasma concentrations minimizes the toxic and subtherapeutic concentrations that reduces the systemic adverse effects.
- Having the ability to prolong the drug release, these systems are expected to reduce the frequency of administration of several short-acting drugs.
- The infrequent administration is likely to decrease the daily dose and hence lead to better utilization of a number of therapeutic agents.
- The decreased dose tends to reduce side effects associated with many therapeutic agents and substantially cut down the treatment cost in management of chronic disorders.
- A CRDDS would improve the medication adherence and convenience thereby enhancing the patient compliance that could eventually benefit the treatment outcomes.
- In addition, transdermal CRDDS are known to have several benefits such as
  - Provide controlled delivery of drugs with narrow therapeutic indices thereby the toxic side effects
  - Avoid the first pass metabolism and thereby the high presystemic clearance of drugs
  - Prevent the gastrointestinal side effects and possible gastrointestinal degradation of certain drugs
  - Substantially reduce the dosing frequency and therefore enable better drug utilization
  - Medication can be delivered over prolonged time periods because the delivery is not limited by the time taken by the dosage form to transit the gastrointestinal tract.
  - Enhance the therapeutic value of potent drugs that undergo extensive first-pass metabolism
  - Permit termination of treatment by removal from the skin if needed
  - Transdermal drug delivery systems are likely to substantially improve the patient compliance

Limitations of controlled drug delivery systems

- Ineffective in case the dosage form fails to release the drug.
- Therapeutic failure leading to adverse effects due to dose dumping owing to instant release of the drug (because of failure of the drug delivery system or mechanical damage or chewing/mastication) can pose risk of toxicity.
- Inter-subject variation and variation in the drug profile with food intake is common with CRDDS
- Oral CRDDS may not suitable for the drugs that undergo extensive first-pass metabolism.
- CRDDS may not be appropriate for the drugs that possess extremely short or long half-life.
- CDDS pose dose inflexibility during prescribing
- Increased product cost owing to technologies and specialty chemicals used
- In addition, transdermal CRDDS may have the below limitations:
  - Drugs having a dose exceeding 10 mg may not be suitable candidates
  - The therapeutic agents having a molecular weight of more than 400 Da may not be well suited for transdermal delivery
  - Drugs having the log of partition coefficient ($\log P$) value of below 2 to or above 4 are unsuitable.
  - Therapeutic agents having high melting point may not be suitable candidates.
  - Drugs that are likely to irritate or sensitize the skin are unsuitable

2.1 Drug Properties to be Considered in Design of Controlled Release Products

The ***physicochemical***, ***biopharmaceutical***, and ***pharmacokinetic*** properties of therapeutic agents need to be thoroughly evaluated in order to assess the feasibility to develop a controlled release product. The feasibility studies are undertaken upfront to evaluate the physicochemical, biopharmaceutical, and pharmacokinetic properties of drug candidate in order to determine its suitability for controlled release formulation development [11]. The choice of the controlled drug delivery technology depends on the physicochemical properties of the drug, consistency of absorption throughout the gastrointestinal (GI) tract, as well as pharmacokinetics [12].

2.2 Physicochemical Factors that Need to be Considered in Design of CRDDS

Physicochemical properties are the properties of the drug molecules that are typically determined by the *in vitro* experiments. The physicochemical characterization involves ***solid*** and ***solution state characterization*** that can include but not limited to ***solubility***, ***permeability***, and ***stability profiling***. These properties would in turn determine the ***biopharmaceutical attributes*** of the drugs and therefore the ***in vivo*** performance. Though the critical physicochemical parameters have been identified and the empirical ranges for each have been elucidated for development of extended release products, product development continues to be a challenging task. However, progresses in technological capabilities and better process understanding have assisted the formulating scientist to meet these challenges to large extent.

2.2.1 Dose Size

Generally, the biological half-life of the drug determines its frequency of administration and thereby the size of the dose. A dose of about ***0.5–1.0 g*** has been considered as the maximum dose size for a conventional dosage form. The same dose size was considered to be ideal for a sustained release product [2]. However, a maximum dose of ***250–300 mg*** was later proposed as a soft limit for controlled release products of late [11]. In case the drug requires a larger dose, it needs to be administered in multiple doses. Nevertheless, small doses are known to pose content uniformity issues during processing. In contrast to oral CRDDS, the ideal dose for other systems such as transdermal controlled drug delivery systems would be less than 10 mg. Similarly, the dose for a controlled release ocular insert would be few milligrams.

The physicochemical properties of the drug that need to be considered during feasibility assessment with the available options are portrayed in Table 2.1.

2.2.2 Molecular Weight/Size

The ***rule-of-5*** is considered as fundamental rule for design of orally absorbed compounds. The rule is based on ***computational evaluation*** of compounds which enables to identify important physicochemical factors that can influence oral drug absorption [13]. As per the rule, the intestinal absorption is likely to be good if:

- The molecular weight of the compound is less than 500 Da
- The $\log P$ value of the compound is less than 5

Table 2.1 Physicochemical properties influencing product development.

	Physicochemical properties	Proposed ranges	Remarks
1	Dose size	<1 mg	Prone to drug content uniformity issue
		10–250 mg	Number of technology option available for controlled release development
		>300 mg	May need more than one unit to accommodate the drug load
2	Molecular weight	<500 Da	Controlled release development feasible
		>500 Da	Controlled release development difficult
3	Dose to solubility ratio	<1 ml	Number of technology option available for controlled release development
		1–100 ml	Controlled release development feasible
		100–1000 ml	Controlled release development difficult
		>10 000 ml	Controlled release development impossible
4	Log of partition coefficient	<1	Controlled release development difficult
		1–3	Controlled release development feasible
		>5	Controlled release development impossible
5	Physicochemical stability	Compound demonstrating compatibility issues with common excipients	Controlled release development difficult
		Compound prone to be unstable in solid or solution state	Controlled release development difficult

- The number of H-bond donors in the compound is less than 5 (sum of OH and NH groups)
- The number of H-bond acceptors in the compound is less than 10 (sum of N and O atoms)

Molecular weight plays a key role in absorption of the compounds from transdermal, ocular, buccal, or sublingual dosage forms. It has to be noted that all drugs developed into transdermal products have molecular weights below 350 Da [14]. For instance, the molecular weight of nicotine, nitroglycerine, clonidine, estradiol, testosterone, scopalamine, and fentanyl, which are available as transdermal products have a molecular weight of 162, 227, 230, 272, 288, 303, and 337 Da respectively. Nitroglycerine is a direct acting vasodilator used in the treatment of ***ischemic heart disease*** and ***congestive cardiac failure***. Considering the high plasma clearance and extensive first-pass metabolism, nitroglycerine is designed as a transcutanous patch. Likewise, pilocarpine-loaded controlled release ocular inserts are used to reduce the intraocular pressure in the management of glaucoma. Pilocarpine (244.72 Da) is found to permeate with ease through the cornea into aqueous humor present in the anterior chamber of the eye.

Molecular weight also plays a critical role in release of the drug through a rate controlling polymeric membrane or a matrix of a drug delivery system. The ability of the therapeutic agent to diffuse through polymers is termed diffusivity that is said to be a function of the molecular weight as per Eq. (2.4) [15].

$$\log D = S_V \cdot \log M_V + K_V = S_M \cdot \log M_W + K_M \tag{2.4}$$

where M_V and M_W are the molar volume and the molecular weight of the drug, S_V, S_M, K_V, and K_M represent the constants. The value of diffusivity depends on the shape and size of the drug as well as the cavities. Generally, the diffusivities are known to range from 10^{-6} to 10^{-9} cm^2/s for drugs having a molecular weight that range between 150 and 400 Da. However, high molecular weight drugs (>500 Da) are likely to display a very slow release from controlled release drug delivery systems as the diffusivity for these compounds appear to be too small ($<10^{-12}$ cm^2/s). Likewise, owing to their high molecular weight, therapeutic macromolecules fail to permeate biological membranes.

2.2.3 Aqueous Solubility

Oral-controlled release formulations are expected to release the drug over the entire length of the GI tract. The release rate from these formulations would be directly dependent on the aqueous solubility. Generally, a drug has to be in ***soluble*** and ***unionized*** state to get well-absorbed from the GI tract. The least aqueous solubility that a drug has to possess in order to be formulated as a controlled release product is identified to be 0.1 mg/ml [2]. However, determination of solubility of 200 peroral products indicated that about 40% of these were ***poorly soluble*** as they were found to possess a solubility of less than aforesaid limit [5]. Such ***poorly soluble*** compounds (e.g. salicyamide, griseofulvin, digoxin, and warfarin) display ***dissolution-limited absorption*** and are said to be poor candidates for designing CRDDS. Moreover, aqueous solubility of drug is likely to influence the choice of the mechanism that can be employed to develop a controlled release system. For instance, ***diffusional systems*** are deemed to be a poor choice for slightly soluble compounds, as high drug concentrations that may be required to act as driving force for diffusion may not be attained. However, formulation approaches such as particle size reduction, nanosizing, salt forms, cosolvents, solid dispersions, cyclodextrin complexation, pH adjustment, micellar solubilization, and salting in have been used of late to address the solubility issues [5, 16].

On the other hand, drugs with very high solubility are equally difficult to formulate as controlled release dosage forms considering the difficulty associated in controlling the dissolution rate. Designing a matrix system with high drug load (>80%) is considered quite challenging, though newer techniques like hot melt extrusion have offered solutions to increase the drug load beyond 90% [17]. Development of a nondisintegrating matrix tablet or using buffering agents to modulate the microenvironment pH would be the other effective strategies employed to overcome the limitations owing to high drug solubility.

The drugs that display good and pH independent solubility are considered as ideal candidates for oral controlled release drug delivery systems. However, the vast majority of the therapeutic agents are either weak acids or weak bases that exhibit ***pH-dependent solubility***. Depending on the ionizable groups, these therapeutic agents are known to exhibit a ***pH-dependent solubility profile*** thereby posing considerable challenges in formulation of controlled release products. The design of CRDDS with such drugs is quite challenging as the solubility is found to vary with the physiological pH prevalent in the gastrointestinal tract (GIT). In such cases, the ***aqueous solubility*** of these drugs would be a function of the ***acid dissociation constant*** (***pK_a***) and the ***physiological pH***. In this context, the dissociation constant and the aqueous solubility of the drugs deserve prime consideration in design of controlled release products. Hence, the ***pH solubility profiling*** of the drug over entire physiological range (pH 1–7.5) is an integral part of preformulation studies prior to designing oral-controlled release product.

Generally, the unionized form of these drugs is lipophilic and therefore tends to exhibit high membrane permeability, while the ionized form though water soluble, displays poor permeability. Hence, these drugs though remain ***unionized*** at the absorption site, display poor aqueous solubility [18]. For instance, a weak acidic drug may exhibit poor solubility in acidic environment of the stomach, whereas a basic drug is likely to exhibit solubility issues in basic environment of the intestine. The variation in the solubilities could be attributed to differences in the concentration of the ionized and the unionized species in physiological pH range (pH 1–7.5). By and large, at the passive absorption site in the GI tract, most drugs would be unionized and therefore least soluble. The fraction of the ionized and unionized species for an acidic and basic drug at the given pH is represented by Eqs. (2.5, 2.6), respectively.

$$\mathrm{pH} - pK_a = \log_{10} \frac{\text{ionized}}{\text{unionized}} \tag{2.5}$$

$$\mathrm{pH} - pK_a = \log_{10} \frac{\text{unionized}}{\text{ionized}} \tag{2.6}$$

Similarly, the total solubility of the acidic drug and basic drug is represented by Eqs. (2.7, 2.8), respectively [15].

$$S_t = S_u \left(1 + \frac{K_a}{[H]}\right) = S_u(1 + 10^{pH-pK_a}) \tag{2.7}$$

$$S_t = S_u \left(1 + \frac{[H]}{K_a}\right) = S_u(1 + 10^{pK_a-pH}) \tag{2.8}$$

where $S_{t,}$ represents the total solubility of the acid or the base, S_u is the solubility of the unionized form, K_a is the acid dissociation constant, and pH represents the pH of the medium. For instance, the weakly acidic drug such as aspirin ($pK_a = 3.5$) is invariably absorbed from the stomach as a sizable fraction (>99%) is found to exist as unionized form. However, by virtue of its poor solubility, aspirin is likely to precipitate in the acidic pH of stomach and therefore need to be formulated as ***enteric-coated*** tablets.

On the contrary, basic drug like verapamil hydrochloride ($pK_a = 8.6$) displays a solubility of more than 80 mg/ml at pH below 3.0 which drops below 1 mg/ml at pH above 7.32 [19]. The problem associated with the poorly water-soluble free base is that it tends to dissolve to a greater extent in the acidic environment of the stomach, while it is likely to precipitate in the small intestine owing to higher pH. Likewise, dipyridamole, a weak base ($pK_a = 6.05$) exhibits a pH-dependent solubility [20]. The dissolution of the compound was found to double when the media was changed from pH 6.8 to pH 5.0. As a consequence of protonation at pH 5.0, the solubility and intrinsic dissolution was found to increase. Since the drug dissolves rapidly in stomach but incompletely in intestine, designing a gastroretentive system or modulation of the microenvironment pH has proven to be logical approach to enhance the drug absorption than developing a conventional controlled release formulation. The pH-dependent solubility of weak bases is a major issue that needs to be dealt with in the design of controlled release formulations. In designing CRDDS, a good aqueous solubility at pH values above 6.8 is mandatory for the drug, especially considering the long residence time in the intestine and limited amount of fluid available in the colon.

Based on exhaustive data gathered from extensive literature review, ***biopharmaceutical classification system (BCS)*** was proposed to categorize the therapeutic entities based on certain fundamental properties [21]. As per the scheme, oral absorption of a drug is determined by two key factors: the *in vitro* dissolution and the membrane permeability [22]. Good aqueous solubility and membrane permeability would ensure good systemic exposure in the absence of any presystemic metabolism. The system classified about 130 drugs into four different classes based on certain physicochemical properties such as aqueous solubility and permeability that determine the oral absorption. The objective of the system was to enable the accurate prediction of *in vivo*

Table 2.2 BCS classification of drugs.

		Solubility	
		High	**Low**
Permeability	High	**Class I – Highly suitable for Sustained Release/Controlled Release** Examples: Propranolol; Verapamil; Theophylline; Diltiazem, Metformin hydrochloride; Metoprolol, etc.	**Class II – Suitable for Sustained Release/Controlled Release** Examples: Danazol; Itraconazole; Mefanamic acid; Nifedipine; Atovaquone; Griseofulvin; Glibenclamide; Carbamazepine; Nifedipine; Felodipine, etc.
	Low	**Class III – Problematic for Sustained Release/Controlled Release** Examples: Atenolol, Metformin, Neomycin B; Captopril; Enalprilat; Alendronate; Atenolol; Cimetadine; Ranitidine, etc.	**Class IV – Unsuitable for Sustained Release/Controlled Release** Examples: Chlorthiazide; Furosemide; Tobramycin; Cefuroxime; Cyclosporin, etc.

Source: Based on Corrigan [23].

pharmacokinetics of drug product based on the *in vitro* data of ***aqueous solubility*** and ***intestinal permeability***.

The four classes of drugs as per the classification were the following:

- *Class 1*: drugs with ***high solubility*** and ***high permeability***
- *Class 2*: drugs with ***low solubility*** and ***high permeability***
- *Class 3*: drugs with ***high solubility*** and ***low permeability***
- *Class 4*: drugs with ***low solubility*** and ***low permeability***

A drug substance is considered as ***highly soluble*** if the highest marketed dose is found to be soluble in 250 ml of aqueous medium over a pH range of 1–7.5 at 37 °C. A drug substance is said to be ***highly permeable*** when the extent of absorption in humans is found to be greater or equal to 90% of the administered dose based on mass balance or in comparison to an intravenous reference dose. In other words, drugs with effective ***permeability coefficient*** above 4×10^{-4} cm/s are classed as highly permeable, while those below 1×10^{-4} cm/s are considered to have low permeability [23]. Table 2.2 provides a compilation of representative drugs classified as per BCS classification.

In formulation development, BCS is considered as a valuable tool to identify the rate-determining step in oral absorption, determine the possible oral absorption issues, and evaluate developmental challenges and the associated risks [22]. The ***US FDA***-accepted BCS as a science-based approach to allow waivers of *in vivo* bioavailability and bioequivalence testing for IR products of drugs belonging to class 1, when such products exhibit rapid dissolution [24]. Though the BCS classification was initially proposed for IR products, some considerations and modifications of the system was later suggested for sustained release products as well [23, 25].

For compounds belonging to class I, there is no rate-limiting step for oral absorption. Therefore, these drugs are regarded to be the ***highly suitable*** candidates for designing sustained release products as they do not exhibit any solubility nor permeability issues. Vast majority of the drugs in this class do not pose any difficulty in design of sustained release products. A good ***in vitro-in vivo correlation (IVIVC)*** can be expected from these drugs in case the dissolution rate is slower than the gastric emptying rate. Thus, ***waivers*** of the *in vivo* studies in certain cases of controlled release products are applicable for BCS class 1 drugs [26].

The absorption of class 2 drug is found to be rate-limited by dissolution as they are known to demonstrate solubility issues. Slow and incomplete dissolution often represents the commonly encountered challenge in developing any oral product. However, these drugs are considered "suitable" for sustained release only if the rate of dissolution is modulated by suitable techniques. Taking into account the low solubility of class 2 drugs, efforts are directed toward increasing the solubility and subsequently the dissolution rate. Some of the techniques that can be proposed to modulate the aqueous solubility and hence the dissolution rate includes crystal modification, particle size reduction, pH modification, and amorphization [27].

A good IVIVC can be expected primarily from extended release formulations of class 2 drugs if the ***in vitro dissolution rate*** is similar to the ***in vivo dissolution rate***. This is because deconvolution of plasma profiles to "*in vivo* dissolution" rather than "absorption" can result in more predictive IVIVC models that take into account both critical physiological variables and formulation attributes influencing drug absorption characteristics. However, obtaining *in vivo* dissolution rate from plasma concentration requires separation of multiple processes involved in drug absorption such as GI transit, permeation, and first-pass effects that determine *in vivo* systemic input rate from *in vivo* dissolution rate. Physiological-based pharmacokinetics (PBPK) deconvolution models, on the other hand, can disentangle these processes and estimate *in vivo* dissolution rather than absorption allowing establishment of more robust and transparent IVIVCs. Software are available to handle considerable amount of data and data manipulation to process for establishing PB-IVIVC. Examples of validated software include Simcyp Advanced Dissolution Absorption and Metabolism (ADAM) model (Simcyp Simulator, Certara) and Advanced Compartmental Absorption and Transit (ACAT) model in GastroPlus (v8.0; Simulations Plus, Lancaster) [28]. These software employ sophisticated gastrointestinal models to try and simulate oral absorption in the gastrointestinal tract considering that as a compound passes along the GI tract lumen and is exposed to the apical membrane of the enterocyte layer, it will cross this barrier at a rate that is influenced by physiological (size, surface area), physicochemical (pH, dissolution rate) and biochemical (metabolism enzymes, transporters) processes that vary along the GI tract. From data obtained from the literature, *in vitro* studies and *in vivo* studies, the PB-IVIVC software improve the predictive performance of the IVIVC model and result in a simpler linear IVIVC in comparison to the conventional IVIVC approaches [29].

Drugs belonging to ***class 3*** are known to possess high solubility but are likely to exhibit low and ***site-specific permeability*** in the GI tract. It is generally observed that compounds of very high polarity are likely to be poorly permeable through biological membranes. Despite high solubility, the bioavailability of drugs belonging to this class is limited by the membrane permeability in the GI tract. The probable reason could be the absorption of these drugs in the large intestine is limited or in some cases completely absent. Therefore, drugs belonging to class 3 of BCS are generally considered to be ***problematic*** for design of sustained release products. Generally, it is noted that poor permeability is more challenging to resolve with formulation and processing technologies. Even though, some absorption enhancers help to overcome the permeability issues, no efficacious and safe dosage options for clinical use have been proposed for this class of drugs.

Drugs of class 4 are known to display solubility as well as permeability issues in the GI tract and are therefore regarded as ***unsuitable*** drugs for development of sustained release formulations. The only strategy to improve the bioavailability of class 4 compounds is to revisit the lead optimization stage of drug discovery and modify the structure of drug appropriately so as to obtain the ideal physicochemical properties [30]. Limited or no ***IVIVC*** is expected from drugs belonging to this class 3 and class 4 of BCS.

The dose to solubility ratio is important index used to assess the suitability of drug candidate for design as controlled release products [11]. The index represents the highest dose divided by the lowest solubility of the drug in pH range of 1–7.5. The implication of the index is outlined in Table 2.1. The development of controlled release formulation happens to be quite challenging in case the dose to solubility ratio exceeds 100 ml.

2.2.4 Lipid Solubility and Partition Coefficient

Partition coefficient is a measure of lipid solubility of a compound that can play an important role in the product development. Partition coefficient is an important physicochemical parameter that influences the ***drug ADME*** and therefore determines the selection of drug candidates for the design of controlled release products. Partition coefficient (P) is defined as the fraction of the drug in n-octanol to that in aqueous phase as per Eq. (2.9).

$$P = \frac{C_o}{C_w} \tag{2.9}$$

where C_o represents the concentration of the drug in octanol, while C_w stands for the drug concentration in water.

On oral administration, the drug has to pass through a number of biological barriers to reach the site of action and elicit the therapeutic efficacy. As most of the biological membranes are lipoidal in nature, it would be necessary for the drugs to possess an ***optimum partition coefficient*** to penetrate these barriers. Generally, lipoidal compounds passively diffuse through the biological membranes by transcellular pathway, whereas ***polar molecules*** owing to the ***low partition coefficient*** have problems in passively diffusing the biological membranes. However, for ionizable drugs, the unionized form is more lipophilic and thereby more permeable across most biological membranes compared to their ionic counterparts. A simple empirical equation (Eq. (2.10)) relates the permeability coefficient with the partition coefficient and the molecular weight of the permeant [31].

$$Log\, K_p = -2.72 + 0.71 \ \log\, P - 0.0061\, M_W \tag{2.10}$$

where K_p represents the permeability coefficient, P is the octanol-water partition coefficient, and M_W is the molecular weight of the permeant.

Generally, the logarithm of the partition coefficient ($\log P$) of the drug indicates its absorptive potential. In case the drug is ionizable, the partition coefficient would be pH-dependent. In such cases, $\log P$ is normally determined and expressed at physiological pH of 7.4.

A ***log*** of ***partition coefficient (log P)*** value of 1–3 is considered ideal for a therapeutic entity in order to design a sustained drug product [32]. The same values of partition coefficient would be applicable for development of transdermal products. The $\log P$ values of scopolamine, nitroglycerine, estradiol, fentanyl, and testosterone, and that are available as transdermal products are found to be 1.24, 2.05, 2.49, 2.93, and 3.31, respectively [14]. Normally, the drugs listed in classes 1 and 2 of the BCS are known to possess the optimum partition coefficient and therefore less likely to exhibit any intestinal permeability issues. At the other extreme, compounds with partition coefficient exceeding 5 are found to be too lipophilic to dissolve in any of the biological fluids are therefore more likely to pose bioavailability issues.

The partitioning characteristic of the drug also determines the choice of the polymer that can be used in the design of sustained release product. In case the drug has a structure similar to that of the polymer used in the design the sustained release product, the drug tends to partition in favor of the polymer and therefore may fail to get easily and completely released during the gastrointestinal transit. Hence, partition coefficient of the drug is likely to limit the choice of the polymer and other excipients that can be used by the formulator in product development.

2.2.5 Physicochemical Stability

The physical and chemical stability of a therapeutic agent is critical in ***formulation design***, ***process development***, ***scale-up***, and ***routine manufacturing***. As a part of preformulation, ***forced degradation*** studies need to be performed to assess the drug stability under extreme conditions of ***pH***, ***heat***, ***oxidation***, ***hydrolysis***, and ***light***. The data can be computed to assess the condition of maximum stability which could prove vital for ***formulation design*** and ***product development***. For instance, ***solution state stability*** is one of the key issues that need to be addressed during product development of drug such as captopril [33]. The drug is found to be most stable below pH 3.5, while above pH 4.0, the drug undergoes oxygen facilitated ***first-order free radical oxidation*** at its thiol position, yielding captopril disulfide [33]. Considering the issues associated with the solution state stability, the drug is usually formulated as tablets and care is taken to store the drug in a dry state in airtight containers during processing in order to maintain the drug stability.

The ***drug excipients compatibility*** studies would help the formulator to select the right excipient for the drug during the product development. In the ***process development***, if the drug is moisture sensitive, the wet granulation process is assumed to be critical. For example, in the processing of aspirin tablets, ***granulation*** being a crucial step is usually performed using an ***alcoholic solution*** of the binder thereby eliminating the direct exposure of the drug to water in order to minimize hydrolysis. Similarly, adequate protection has to be ensured for photosensitive drugs exemplified by dihydropyridines such as nisoldipine; nifedipine; and so forth, during development, manufacturing, and subsequent storage. Likewise, drugs that are thermolabile are poor candidates for techniques such as hot melt granulation or extrusion. The functional groups that are prone to hydrolysis would be lactones group (pilocarpine), lactams (penicillin), and amides (proteins or peptides). Groups that are susceptible for oxidation include phenols (steroids), catechols (dopamine), and thioethers (phenothiozines such as chlorpromazine).

2.3 Biopharmaceutical Properties that Deserve Consideration in Design of Controlled Release Products

Feasibility assessment needs to be conducted to evaluate the ***biopharmaceutical*** and ***pharmacokinetic*** properties of drug candidate in an effort to determine its suitability for controlled release formulation development. The design of any controlled release product calls for a ***comprehensive review*** of the drug disposition process in the body. Hence, a thorough evaluation of the ***absorption***, ***distribution***, ***metabolism***, ***and elimination (ADME)*** characteristics of the therapeutic agent following multiple dosing is mandatory prior to the product development. Key attributes of a drug including the ***pharmacological safety***, ***preclinical pharmacokinetics***, and ***biopharmaceutical*** properties have to be considered before nominating a drug from development into a modified release product. Precise ranges for some of the ***critical pharmacokinetic*** or ***biopharmaceutical parameter*** have been proposed in literature for the development of extended release products beyond which the product design is quite challenging [8, 34]. Despite these constraints, better understanding of the ***drug pharmacokinetics*** and ***drug disposition*** process has enabled the formulator to breach through the proposed ranges to certain extent.

2.3.1 Biological Half-life

The duration of action of a drug that governs the ***dosing interval*** is quantitatively measured in terms of ***biological half-life*** of the drug. The half-life is a key biological factor that needs to be considered in design of any modified release drug delivery system. The goal of any controlled drug delivery system is to maintain constant therapeutic ***plasma concentrations*** over prolonged period of time. In order to ensure a steady therapeutic blood levels, the ***rate*** of ***drug delivery*** must be equal to the ***rate*** of ***elimination*** of the drug from the body. Since the ***first-order elimination rate constant*** is inversely related to the ***biological half-life*** ($t_{0.5,el}$) as depicted in Eq. (2.11). Each drug is said to have a characteristic half-life which is determined by the ***absorption***, ***elimination***, ***metabolism***, and ***distribution*** pattern of the drug in the body.

$$\text{Rate of delivery} = \frac{D}{T} = \frac{0.693 \times C_{P,ss} \times V_d}{t_{0.5,el}} \tag{2.11}$$

Drugs with short half-life need to be frequently dosed in order to minimize the fluctuations in the plasma concentrations, if they are to be administered as conventional dosage forms. In such cases, controlled release products are most desirable in view to reduce the ***dosing frequency*** and enhance the ***patient compliance***. Generally, drugs having a ***short half-life*** of two to four hours are considered to be the ideal candidates for design into oral sustained release products [8]. The extended release systems allow these drugs to be administered ***once*** or ***twice*** a day with comparable therapeutic efficacies and minimum side effects compared to the conventional dosage forms. The ***therapeutic value*** of short-acting drugs such as procainamide ($t_{0.5}$ = 3 hours) is enhanced by development of extended release forms. Procainamide is antiarrhythmic agents that get rapidly eliminated from the systemic circulation, resulting in a number of peaks and troughs in the plasma concentration profiles. These fluctuations are likely to be ***minimized*** and the ***plasma concentration profiles smoothened*** by administration as oral controlled release formulations that allow the slow release of the actives over extended time periods.

On the other hand, drugs having a half-life of less than two hours are considered as poor candidates for controlled release. Ultra-short-acting drugs having biological half-lives of less than two hours are deemed to be unsuitable to design any modified release products as they would invariably require ***high doses*** or need ***high frequencies*** of administration to sustain the efficacy. However, the exception being drugs like nitroglycerine ($t_{0.5}$ = 0.04 hours) that are designed as extended release formulations. ***Extended release nitroglycerine*** owing to its long duration of action is indicated for prophylactic treatment of angina pectoris.

Therapeutic agents with half-life greater than eight hours are inherently long-acting, and there is no need to design a sustained release system. In these cases, sustained ***release*** forms of long acting drugs are less likely to offer any additional therapeutic benefit over the conventional counterparts. For instance, drugs such as phenytoin ($t_{0.5}$ = 22 hours), diazepam ($t_{0.5}$ = 54 hours) are usually long-acting and therefore are considered as poor candidates for design of sustained release systems. Most of these drugs can be conveniently dosed once a day as conventional dosage forms without any need to develop sustained release systems. The biopharmaceutical properties of the drug that need to be considered during feasibility assessment are shown in Table 2.3. Half-life is a key biopharmaceutical consideration in the design of transdermal drug delivery systems as well. For instance, the half-life values of scopolamine, clonidine, fentanyl, and nicotine that have been developed as transdermal products are found to be 3, ~6, 3, and 2 hours, respectively [14]. However, unlike oral controlled release drug delivery systems, the duration of retention of the transdermal drug delivery system on the skin would extend from few to several days.

2.3.2 Absorption

A thorough knowledge of anatomy and physiology of the gastrointestinal tract would be essential in development of CRDDS. The drug has to first be released from the system prior to dissolution in

Table 2.3 Biopharmaceutical properties influencing product development.

	Physicochemical properties	Proposed ranges	Remarks
1	Half-life	0–2 h	Half-life too short for development of controlled release
		1–4 h	Best suited for controlled release development
		>8 h	Controlled release development may not be essential
2	Absorption (regional/colonic permeability)	Poor absorption ($P_{app\,Caco\text{-}2} < 10^{-6}$ cm/s, $K_a < 0.01\ \text{min}^{-1}$)	Controlled release formulation for prolonged delivery may not be feasible. Less likely to be bio equivalent to a IR
		Moderate absorption ($P_{app\,Caco\text{-}2} < 10^{-6} - 10^{-5}$ cm/s)	Controlled release formulation for prolonged delivery may be challenging. Less likely to be bio equivalent to a IR
		Good absorption ($P_{app\,Caco\text{-}2} < 10^{-6}$ cm/s, $K_a > 0.01\ \text{min}^{-1}$)	Controlled release formulation for prolonged delivery feasible. Most likely to be bioequivalent to a IR
3	Metabolism and efflux	High metabolism	Relative bioavailability of controlled release product is low
		Compounds prone to P-gp efflux	Performance of the controlled release formulation difficult to predict
4	Margin of safety	Therapeutic index less than 10	Controlled release formulation design difficult
		Narrow therapeutic concentration range	Controlled release formulation design difficult
		Drugs with broad therapeutic range	Controlled release formulation design feasible

the GI fluid before absorption through the GI tract. Therefore, the absorption characteristic of the drugs ***over the entire length of the GI tract*** is known to determine the feasibility of a sustained release product. Generally, the drugs that exhibit good oral bioavailability are considered to be suitable candidates for oral controlled release drug delivery systems. In addition, the drugs that are absorbed uniformly throughout the entire length of the gastrointestinal tract are considered to be ideal candidates.

A simple approach to quantify the intestinal drug absorption was the concept of ***Maximum absorbable dose (MAD)*** [35, 36]. The MAD of a drug through the intestine can be calculated by Eq. (2.12).

$$\text{MAD} = S \times K_a \times \text{SIWV} \times \text{SITT} \tag{2.12}$$

where S is the drug solubility at pH 6.5, K_a is the absorption rate constant, SIWV is the small intestinal water volume, and small intestinal transit time (SITT).

The absorption rate constant is usually obtained from rat intestinal perfusion methods, whereas the SIWV and SITT are assumed to be 250 ml and 4.5 hours, respectively. Eventually, the MAD would be a function of solubility of the drug and intestinal permeability. In the design of any controlled release products, it is always necessary that the ***rate of release*** is much slower than the ***rate of absorption***. Development into an oral controlled release product is considered to be feasible if the permeability is more than 1.5×10^{-5} cm/s [12]. Similarly, a minimum absorption rate constant

K_a value of ***0.17–0.23 h^{-1}*** is assumed to be necessary for ***80–95%*** of the oral dose to get absorbed, considering the GI transit time for the dosage form to be between ***8*** and ***12 hours*** and the half-life of the drug to be ***3–4 hours*** [2]. These calculations are true, only if the drug is ***uniformly*** absorbed ***throughout the entire length*** of the GI tract. Oral controlled release formulations of such drugs can be conveniently designed for once a day administration, though it is assumed that the release profiles of most oral controlled release products is found to span for the only ***eight hours***. Owing to the heterogeneous nature of the GI tract, the permeability of a drug is found to vary from the small intestine to colon. ***Site-specific differences*** in the permeability and GI transit times have resulted in high intra- and intersubject variation. ***Computational methods*** as well as ***in vitro*** and ***in vivo animal models*** have been utilized to predict the intestinal permeability. Determining and predicting the drug absorption as a function of position of the drug in the GI tract is considered to be vital in design of controlled release formulation. Such information helps the formulator to propose ***frequency of dosing*** of the controlled release preparations [11].

Drugs such as metformin and atenolol are absorbed at specific sites in the intestine by active transport. These drugs are thus known to have a ***narrow absorption window*** located in the upper part of the small intestine resulting in incomplete absorption from the gastrointestinal tract. Therefore, such drugs may require ***more than*** one dosing in a day unless the release of the drug is controlled at the absorption site by special design [17]. However, these drugs are not good candidates for conventional sustained release dosage forms, despite having a short half-life. Instead, gastroretentive dosage forms that remain in the stomach for longer period of time and release the drug slowly for prolonged duration are suitable for delivery of drugs with narrow absorption window [37]. Prolonged residence of the drug delivery system in the stomach increases the absorption and bioavailability of such drugs. Generally, drugs having a narrow absorption window in the upper GI tract seem to be good candidates for gastroretentive technologies. For instance, ***Medopar®*** is a gastroretentive hydrodynamically balanced system containing ***levodopa***. Development of a gastroretentive system for levodopa is justified as the drug is predominantly absorbed from the absorption window situated in the duodenum and jejunum [1]. Some drugs are known to utilize a ***carrier mediated active transport mechanism*** to get absorbed from ***specific site*** in the GI tract. For instance, angiotesin converting enzyme (ACE) inhibitors are known to exploit the ***peptide transporters*** in the small intestine to get absorbed from the GI tract. In such cases, increasing the residence time of the drug delivery system in the stomach or small intestine would be advantageous. Likewise, a gastroretentive system would be beneficial for the drugs that display pH-dependent solubility or stability. For example, ***Valrelease®*** is hydrodynamically balanced system of diazepam that was developed considering its poor solubility at higher pH values [1]. However, the design of any ***site-specific controlled drug delivery systems*** warrants a thorough knowledge of the ***regional permeability*** and ***absorption pattern*** of drugs in different segments of the GI tract.

On the other hand, drugs that such as aspirin that may have a tendency to irritate the stomach lining or those that are unstable in the acidic environment of the stomach need to be administered as ***enteric coated formulations***. Aspirin is usually designed as enteric-coated tablets for chronic administration considering its tendency to cause gastric lesions especially during long-term use [38].

Owing to the long colonic transit time that may vary from 10 to 20 hours, ***regional permeability*** of a drug in the colon is considered to be vital in determining the feasibility of a modified release product [11]. An apparent permeability as predicted by the Caco-2 cell lines exceeding ***1.5 × 10^{-5} cm/s*** suggests the feasibility to develop a controlled release product. Such a product is likely to be ***bioequivalent*** compared to the IR product. An apparent permeability value ranging between ***10^{-5}*** and ***10^{-6} cm/s*** indicates the feasibility of development controlled release product though it may not be ***bioequivalent*** to the IR preparation. However, an apparent permeability value below 10^{-6} cm/s rules out the feasibility of development of a long acting controlled release product.

In the design of specialized drug delivery systems like chronotherapeutic drug delivery systems, the absorption of therapeutic agents is ***deliberately delayed*** in order to deliver the drugs in a predictable manner as per the body's chronobiological needs in certain diseases [39]. The biopharmaceutical feasibility of such modified release formulations is largely determined by the ***colonic permeability*** of the drug. However, a number of factors such as ***solubility*** or ***permeability limited absorption***, constraints due to ***insufficient water*** in colon, or ***bacterially triggered degradation*** are known to interfere with the colonic absorption of drugs that eventually lead to ***poor bioavailability***. Despite these constraints that limit colonic absorption, drugs such as verapamil hydrochloride and propranolol hydrochloride have been successfully developed as chronotherapeutic formulations.

By and large, drugs that display high permeability along the entire length of the gastrointestinal tract and do not undergo first pass metabolism are likely to be good candidates for oral controlled release systems (Malaterre et al. 2009). On the contrary, the drugs that display good skin permeability but are prone to first-pass metabolism are considered to be good candidates for transdermal controlled release systems.

2.3.3 Metabolism

The metabolism of a drug in the body can influence its bioavailability and thereby the ***in vivo performance*** of a controlled release product. The metabolic pattern of a drug most often can dictate the route of administration in many cases. A good number of drugs are usually degraded by ***CYP mediated oxidative metabolism*** that takes place in the gut wall or the liver. The bioavailability of orally administered drug may be compromised by the metabolizing enzymes ***cytochrome P450 (CYP)*** especially the ***isozyme CYP3A4***. The levels of the intestinal metabolizing enzymes rise from duodenum to jejunum while they decline at the ileum and colon [37]. However, most of the enzyme-mediated metabolic processes are saturable as the enzymatic degradation often follow Michaelis–Menten kinetics. For example, aloprenolol is found to be get extensively metabolized in the wall of the intestine when administered as ***sustained release formulation*** when compared to a ***conventional preparation*** [40]. The extensive drug metabolism when administered as sustained release formulation can be attributed to the ***failure to saturate*** the intestinal enzymes as the drug is presented at a slow rate to the enzymatic breakdown. On the contrary, the better bioavailability of the IR formulations can be attributed to the ability of these forms to ***saturate*** the enzymes, when the entire dose of the drug was presented to the enzymes at once. Hence, drugs that are metabolized in the gut are considered to be poor candidates for designing oral controlled release formulations. Generally, if a drug exhibits ***variable blood levels*** owing to ***gut wall metabolism*** or due to the ***first pass effect*** in the liver, it is considered to be a poor candidate for developing sustained release products [41]. However, ***CYP3A4 expression*** is found to decrease in the colon thereby increasing the absorption of the substrates from the large intestine. Therefore, modified products like colon-specific drug delivery systems are likely to be more beneficial than drugs that undergo presystemic and first-pass metabolism.

An ***active drug*** can get transformed to an ***inactivate form*** or an ***inactive drug*** can get converted into an ***active metabolite*** during metabolism. Therefore, the enzyme-mediated metabolic process is desirable in a few cases to activate the parent compound in the body. For instance, propranolol is metabolized in the liver to the biologically active metabolite ***4-hydroxypropranolol***. However, owing to its short half-life (three to six hours), the drug is formulated as a sustained-release product. Similarly, procainamide gets metabolized to its active metabolite ***N-acetylprocainamide*** that is

Table 2.4 Biopharmaceutical drug disposition classification system (BDDCS).

		Solubility	
		High	**Low**
Metabolism	High	**Class I – Highly suitable for Sustained Release/Controlled Release** Examples: Propranolol; Verapamil; Theophylline; Diltiazem, Abacavir, Metoprolol, etc.	**Class II – Suitable for Sustained Release/Controlled Release** Examples: Danazol; Itraconazole; Naproxen; Griseofulvin; Carbamazepine; etc.
	Low	**Class III – Problematic for Sustained Release/Controlled Release** Examples: Atenolol, Metformin, Captopril; Atenolol; Cimetadine; Ranitidine, etc.	**Class IV – Unsuitable for Sustained Release/Controlled Release** Examples: Chlorthiazide; Furosemide; Ciprofloxacin, etc.

Source: Based on Wu and Bennet [42].

found to be equipotent with the parent molecule as an antiarrhythmic agent. However, if any therapeutic agent on chronic usage tends to either ***increase*** or ***inhibit*** the synthesis of the ***metabolizing enzymes***, it may not be feasible to design a sustained release product, considering the difficulty encountered in maintaining a steady plasma concentration [8].

Considering the elimination pathway differences of BCS class 1 and class 2 vs. class 3 and class 4 and the difficulty encountered to quantify the percent of drug absorbed in humans at a level of 90%, a revised classification scheme termed ***Biopharmaceutical Drug Disposition Classification System (BDDCS)*** was proposed [42]. The system classified the drugs into four classes based on the aqueous solubility and extent of metabolism so that it would be easier and less ambiguous. Class 1 and 2 compounds are eliminated primarily by ***metabolism***, while class 3 and 4 compounds are ***primarily eliminated*** unchanged into the urine and bile. The more permeable lipophilic compounds are ***good substrates*** for ***cytochome P450 system*** and therefore more prone to ***undergo extensive metabolism***. A drug is said to undergo ***extensive metabolism*** in case ~70% of the oral dose is metabolized, while the term ***poor metabolism*** is used when ~50% drug is excreted unchanged. The drugs categorized as per the BDDCS is portrayed in Table 2.4.

Despite being well absorbed from the GI tract, the oral bioavailability of several therapeutic agents such as nitroglycerine, diclofenac, and salbutamol are considerably hampered, owing to the ***extensive metabolism***. Such drugs that tend to undergo ***extensive first pass metabolism*** in the liver are considered to be ***poor candidates*** for oral-controlled drug delivery unless their metabolites are found to be pharmacologically active [43]. Therefore, efforts are made to identify alternate routes of administration or devise newer drug delivery systems in view to circumvent the ***bioelimination*** process. For instance, nitroglycerine is found to be an excellent candidate for rate-controlled transdermal drug delivery systems. Transdermal nitroglycerine is found to provide effective prophylaxis against coronary artery spasm and atherosclerotic coronary heart disease. The transdermal system is able to overcome the undesirable pharmacokinetics of nitroglycerine such as short elimination half-life owing to rapid plasma clearance by evading the first-pass metabolism [6]. Similarly, transdermal clonidine indicated once a week ensures long-lasting antihypertensive effect despite the short half-life of the drug (three to five hours). Though several such drug delivery systems have been proposed for such molecules, only few ventures have culminated in successful products.

2.3.4 Presystemic Clearance

The presystemic clearance of the drugs can determine the feasibility to design a modified release product. The poor bioavailability of orally administered drug may be due to ***hydrolysis in the stomach***, ***enzymatic degradation*** in the ***GI fluids***, ***intestinal efflux*** or by the ***microbial breakdown*** in the colon. A drug has to evade a number of transporters and mechanisms in the GI tract before getting absorbed into general circulation. The absorption of a good number of lipophilic drugs is more often limited by the ***intestinal efflux mechanism*** [44]. The ***efflux transporter P-glycoprotein (Pgp)*** is present on the mucosal surface of the intestinal enterocytes. The efflux transporter is considered to be responsible for the low and variable bioavailability of antineoplastic agents and antiretrovirals. However, it is noted that the Pgp expression is known to increase from ***proximal*** to the ***distal region*** of the ***small intestine*** resulting in a gradual decrease in the absorption of the drugs during the transit through the small intestine. Normally, the drugs that are Pgp substrates are considered to be a poor choice for a ***controlled release products***. Instead, a drug delivery system that is likely to release most of the drug in the upper part of the GI tract is considered to be more appropriate in such cases.

Colon is considered to have a less hostile environment in terms of density and diversity of the ***metabolizing enzymes***. Despite the mild conditions prevalent, a number of therapeutic agents are metabolized by the microorganisms in the large intestine. The microbial breakdown, often resulting in ***variable systemic bioavailability***, is known to occur under anaerobic conditions by the action of reductases. For instance, drugs such as levodopa, metoprolol, and digoxin [45] are metabolized by the gut microbial flora prior to absorption. In such cases, it may not be feasible to develop oral ***sustained release*** products or a ***chronotherapeutic*** formulation for drugs metabolized by the ***microbial flora***. Instead, a gastroretentive drug delivery system would be a viable option for such drugs, as confining the drug delivery system in the upper part of GI tract is likely to obviate the ***microbial break down*** and improve the bioavailability. Similarly, designing a gastroretentive system would be justified for captopril as the drug is known to exhibit a ***pH-dependent stability***. Increasing the ***residence time*** of the product in the stomach or intestine is reported to improve its oral bioavailability [44].

2.3.5 Margin of Safety

The ***therapeutic index*** of the drug gives a rough estimate of the safety of the drug. The therapeutic index is the ratio of the median toxic dose to the median effective dose. It is observed that ***larger*** the therapeutic index, the ***safer*** would be the drug. A drug is considered to be relatively safer if its therapeutic index exceeds a value of ***10***. Since the therapeutic index is more a relative, rather than an absolute value, the margin of safety of the drug is more clearly defined based on the therapeutic index as well as the safe and effective plasma concentration range of the drug [8]. The examples of drugs with narrow therapeutic ranges include digoxin (0.02–2 ng/ml), lidocaine (1.5–4 ng/ml), quinidine (2–5 ng/ml), procainamide (4–8 ng/ml), theophylline (10–16 μg/ml), phenytoin (10–20 ng/ml), etc. Drugs with ***broad therapeutic indices*** are considered to be ideal for development into a controlled release systems. On the other hand, it would be a big challenge to design sustained-release products for drugs with ***low therapeutic index*** or ***narrow therapeutic indices***. ***Accidental dose dumping*** of such drugs owing to failure of the drug delivery system would result in high-circulating blood levels with possible ***fatal consequences***. The therapeutic index and the margin of safety of the drug would determine the design of CRDDS. For instance, a simple matrix system such as a drug in adhesive transdermal patch would be sufficient for drugs that display a large therapeutic range despite variation in the permeability of the skin. On the other hand, for drugs with a narrow therapeutic index, a more precise control of the plasma concentration is desirable. In such cases, a rate-controlling membrane is generally introduced between the drug reservoir and the adhesive layer to effectively control on the drug release [46].

2.3.6 Adverse Effects

A drug may have a tendency to induce ***adverse effects*** which may be either ***local*** or ***systemic***. The ***local adverse effect*** in the GI tract usually results from the direct contact of the drug with the gastrointestinal mucosa due to rapid release of an irritant drug, while the systemic adverse effects are noticed when the plasma concentration exceeds the maximum safe concentration. Controlled release products have the potential to reduce the incidence of local side effects as it is generally assumed that controlling the release rate would minimize the gastrointestinal irritation [47]. For instance, the sustained release sodium chloride reduced the incidence of nausea and vomiting commonly associated with conventional formulations in sodium therapy. Similarly, sustained-release products by virtue of their ability to control the plasma concentrations minimize the systemic adverse effects. Controlled release levodopa was found to lower the incidence of drug-induced dyskinesis so that the patients were able to better tolerate higher drug doses [8]. ***Controlled release*** products are known to maintain ***therapeutic concentrations*** of nifedipine for prolonged time periods and thereby reduce the side effects associated with the IR forms. The drug is rapidly absorbed and metabolized from conventional formulations requiring multiple doses to be administered every day. Therefore, the conventional formulations have been often associated with higher incidence of ***vasodilatory*** adverse effects. Rapid absorption of nifedipine following the ***instant drug release*** from the ***IR form*** is usually characterized by ***high-peak*** and ***low-trough plasma concentrations*** of nifedipine. The fluctuating plasma levels of nifedipine were found to produce greater ***hemodynamic effect*** that result in ***reflex tachycardia*** or little ***flushing without reduction*** in the blood pressure. However, nifedipine-GITS (gastrointestinal therapeutic system), a controlled release formulation was more effective in lowering the ***systolic*** and ***diastolic blood pressure***. The reduction in the peak-to-trough ratio of plasma concentration was considered as the probable reason for the reduced vasodilatory side effects such as flushing, dizziness, and reflux tachycardia of nifedipine-GITS [10]. Likewise, by their ability to tightly control the plasma concentration, controlled release transdermal drug delivery systems eliminate the systemic side effects associated with most of the drugs [6]. For instance, ***transdermal scopolamine*** effectively controls nausea while reducing the undesirable side effects such as dry mouth, drowsiness, and tachycardia. Similarly, ***transdermal clonidine*** that is indicated once a week ensures long-lasting antihypertensive effect that is devoid of anticholinergic and sedative effects [48].

2.3.7 Therapeutic Need

The drugs that are meant for management of certain acute disorders are considered to be poor candidates for extended release dosage forms. For instance, a drug like β lactamase antibiotic is considered to be poor choice developed for controlled release drug delivery systems as prolonged elevated plasma levels may lead to emergence of resistant strains of bacteria [17]. Likewise, drugs that are meant for immediate effect, for instance, to relieve ***acute pain*** of angina may not be a suitable candidate for controlled release DDS.

However, the design of CRDDS for a drug is well justified in case it is used in management of chronic disease or disorder. Generally, controlled release drug delivery systems are developed for management of chronic disorders such as rheumatoid arthritis, diabetes mellitus, cardiovascular disorders, nocturnal asthma, CNS disorders. A controlled drug delivery system would be advantageous in maintaining therapeutic concentrations of drugs like diclofenac throughout the night to relieve the morning stiffness in management of rheumatoid arthritis. Glucotrol XL is a controlled release system of glipizide used to provide round-the-clock control of blood glucose level in management of diabetes. For example, Progestasert® is a controlled release intrauterine device containing progesterone [49]. The device releases progesterone at a rate of 65 μg/d for one year to achieve contraception. Similarly, Norplast is a controlled release subcutaneous implant that is fabricated to release 30 μg of progesterone per day for nearly seven years for the same purpose [1].

Sustained release products continue to play a ***vital role*** in ***management*** of cardiovascular disease such as hypertension [50]. Though such products have proved to be valuable in ***prophylactic treatment*** of ***diseases*** like angina pectoris, they fail to have any value in relieving the ***acute angina pain***. For example, the ***extended release*** product of isosorbide mononitrate is indicated for ***prophylactic treatment*** of angina pectoris. Once a day administration of the product is said to result in plasma levels that are high enough to offer ***adequate protection*** against angina during the day while low enough later on to avoid the development of tolerance. However, the onset of action was ***not rapid enough*** for ***treatment*** of ***acute angina episodes***. Hence, ***sublingual*** administration has been useful to provide high circulating levels of nitrates to relieve ***acute pain*** of angina, whereas the ***controlled release product*** could be used as ***prophylaxis*** [51]. However, the ***prophylactic treatment*** of ***angina*** is known to suffer from the limitations as it is likely to obscure the ***warning symptoms*** of pain, thereby leading to ***over-exertion***, and eventually ***fatal outcomes*** [8]. Likewise, Catapres® is a transdermal therapeutic system of clonidine that seems to be quite valuable in the management of chronic hypertension. Similarly, a transdermal controlled drug delivery system is justifiable in case the patient is forgetful and lack awareness as in case of ***Alzheimer's disease*** [12]. Likewise, Ocusert® is an ophthalmic insert designed by Alza corporation, CA, USA [49] to release 20 or 40 μg of the pilocarpine per hour for a period of seven days. The system has proved to be an excellent choice for regional delivery of the drug to the eye in the management of glaucoma.

2.3.8 Role of Circadian Rhythm

Several diseases are known to be influenced by the body's circadian rhythm. The circadian variation in the intensities of certain fatal diseases has prompted the design of specialized drug delivery systems called chronotherapeutic drug delivery systems. For example, the incidence of several cardiac events including ***angina***, ***myocardial infarction***, ***stroke***, and ***sudden cardiac death*** are more frequent in the morning and driven by the body's circadian rhythm [52]. The staging of these events is related to the time-dependent peaks in the ***blood pressure*** that in turn is influenced by the ***hormonal levels***, ***heart rate***, ***peripheral vascular resistance***, ***myocardial oxygen demand***, and ***blood coagulation***. The blood pressure exhibits ***circadian dependency*** and normally ***peaks*** in early mornings when individuals wake up from sleep. The early morning surge in the blood pressure corresponds to increased secretions of ***catecholamines*** and ***plasma renin activity*** that in turn would augment the ***vascular tone*** and ***total peripheral resistance*** [53].

Most of the sustained release products result in ***safe*** and ***effective lowering*** of the ***blood pressure***. However, the static mode of drug release from these products is not tailor-made to meet the ***daily physiological need*** of the body in the early morning hours. In contrast, a ***chronotherapeutic drug delivery system*** delivers the drug in a predictable manner as per the ***body's chronobiological need*** and thereby confers ***adequate protection*** against most cardiovascular risks. Moreover, these drug delivery systems that complement the ***body's daily physiological needs*** are known to enhance the ***therapeutic value*** of several ***short-acting cardiovascular drugs*** [54].

A typical chronotherapeutic drug delivery system is the ability to maintain ***higher plasma*** level of the drug during the ***critical morning hours*** and lower level during night. This dynamic mode of drug delivery is likely to maximize the ***control of BP*** in the mornings and minimize the risk of ***excessive drop*** of ***BP*** during night [55]. Maintaining adequate BP during sleep is considered critical in elderly patients and those with ***peripheral vascular disease***. Bedtime administration of ***conventional dosage forms*** of antihypertensives is known to heighten the risk of ***nocturnal hypotension*** and subsequent ***ischemic stroke*** and ischemia induced neuropathy of optic nerve [56].

Covera HS is a controlled onset extended release system containing verapamil hydrochloride for management of hypertension [55]. Similarly, Verelan-PM containing verapamil although designed for bedtime administration, ensures peak drug concentration in the morning when typically the blood pressure is likely to surge. Another example of chronotherapeutic system to treat hypertension is InnoPran XL™ composed of propranolol hydrochloride which provides peak drug concentration at 10 a.m., although it is administered at bedtime. Likewise, Cardizem XL is a chronotherapeutic extended release system of diltiazem designed to lower the blood pressure between 6 a.m. and 12 noon when dosed at 6 p.m. [57].

2.4 Conclusion

Controlled drug systems have been successful in transforming sizable number of the short-acting drugs used to treat chronic disorders to once-a-day or once-a-week formulations while maintaining constant plasma concentrations. Controlled release technologies have been instrumental in increasing the therapeutic value of a number of short-acting drugs by rendering them more effective. As these formulations maintain fairly consistent plasma concentration during the treatment period, they have the potential to improve the clinical outcome. By virtue of the fewer side effects, these systems have proved to be more patient-compliant and therefore display enhanced adherence to therapy. Cutting edge CR technologies such as printing technology, hot melt extrusion, injection molding, and dry coating have raised hopes of formulation scientist to develop CR formulations of challenging molecules that display solubility or permeability or metabolic issues. It is most likely that in near future, many of the existing conventional dosage forms would be gradually replaced by the controlled release drug delivery systems. Although, transdermal drug delivery technologies have offered viable solutions for delivery of potent, lipophilic, short-acting, extensively metabolizable therapeutic agents with low therapeutic index, a pipeline of drugs still await product development. In addition, efforts are on to identify alternate routes of administration and develop specialized drug delivery systems to circumvent the ***elimination*** process in the body. Development of controlled release drug delivery systems for delivery of macromolecules and biopharmaceuticals is likely to become a reality with the evolution of enabling technologies and smart materials.

References

1 Perrie, Y. and Rades, T. (2010). *Pharmaceutics-Drug Delivery and Targeting*, p92. London, UK: Pharmaceutical Press.

2 Jantzen, G.M. and Robinson, J.R. (2002). Sustained and controlled release drug delivery systems. In: *Modern Pharmaceutics* (eds. G. Banker and C.T. Rhodes), 501–528. New York, NY: Marcel Dekker Inc.

3 Jain, K.K. (2011). Drug delivery to cardiotherapeutic system. *Appl. Biotechnol. Cardiovasc. Ther.*: 29–56.

4 Park, K. (2014). Controlled drug delivery: past forward and future. *J. Controlled Release* 190: 3–8.

5 Yun, Y.H., Lee, B.K., and Park, K. (2015). Controlled drug delivery: historical perspectives for the next generation. *J. Controlled Release* 219: 2–7.

6 Breimer, D.D. (1984). Rationale for rate controlled drug delivery of cardiovascular drugs by the transdermal route. *Am. Heart J.* 108: 196–200. (in innovative DDS in CV medicine).

7 Uchizono, J.A. (2006). Application of pharmacokinetics and pharmacodynamics in the design of controlled delivery systems. In: *Design of Controlled Drug Delivery Systems* (eds. X. Li and B.R. Jasti), 1–40. New York, NY: MacGraw Hill.

8 Vincent, H.K., Li, K., and Robinson, J.R. (2005). Influence of drug properties and routes of drug administration on the design of sustained and controlled release systems. In: *Controlled Drug Delivery Fundamentals and Applications* (eds. J.R. Robinson and V.H. Lee), 3–94. New York, NY: Marcel Dekker Inc.

9 Haynes, R.B. (1981). The effect of therapeutic regimen on patient compliance and the possible influence of controlled-release dosage forms. In: *Controlled Release Pharmaceuticals* (ed. J. Urquhart), 121. Washington, DC: American Pharmaceutical Association.

10 Katz, B., Rosenberg, A., and Fisherman, W.H. (1995). Controlled release drug delivery system for cardiovascular medicine. *Am. Heart J.* 129 (2): 359–368. (CRDDS).

11 Thombre, A.G. (2005). Assessment of the feasibility of oral controlled release in an exploratory development settings. *Drug Discovery Today* 10 (17): 1159–1166.

12 Crowley, P.J. and Martini, L.G. (2004). Formulation design: new drugs from old. *Drug Discovery Today Ther. Strategies* 1 (4): 537–542.

13 van de Waterbeemd, H. (2000). Intestinal permeability: prediction from theory. In: *Oral Drug Absorption. Prediction and Assessment* (ed. B. Dressman Jennifer), 31–50. New York, NY: Dressman Marcel Dekker Inc.

14 Delgado Charro, M.B. and Guy, R. (2001). Transdermal Drug Delivery. In: *Drug Delivery and Targeting for Pharmacist and Pharmaceutical Students* (eds. A.M. Hillery, A.W. Lloyd and J. Swarbrick), 189–214. New York, NY: Taylor and Francis.

15 Ding, X.B., Alani, A.W.G., and Robinson, J.R. (2016). Extended release and targeted release drug delivery systems. In: *Remington: the Science and Practice of Pharmacy*, Chapter: 47, 21e (ed. R. Hendrickson), 939–964. Lippincott Williams & Wilkins.

16 Voelker, M. and Hammer, M. (2012). Dissolution and pharmacokinetics of a novel micronized aspirin formulation. *Immunopharmacology* 20: 225–231.

17 Wen, H. and Park, K. (2010). Introduction and overview of oral controlled release formulation design. In: *Oral Controlled Release Formulations Design and Drug Delivery Theory to Practice* (eds. H. Wen and K. Park), 1–20. New Jersey, NJ: Wiley.

18 Attwood, D. and Florence, A.T. (2008). Drug absorption. In: *FASTtrack: Physical Pharmacy* (eds. A.T. Florence and D. Attwood), 97–124. London, UK: Pharmaceutical Press.

19 Vogelpoel, H., Welink, J., Amidon, G.L. et al. (2004). Biowaiver monographs for immediate release solid oral dosage forms based on biopharmaceutical classification system (BCS) literature data: verapamil hydrochloride, propranolol hydrochloride, and atenolol. *J. Pharm. Sci.* 93: 1945–1956.

20 McNamara, D.P. and Amidon, G.L. (1988). Reaction plane approach for estimating the effects of buffers on the dissolution of acidic drugs. *J. Pharm. Sci.* 77: 511–517.

21 Amidon, G.L., Lennernas, H., Shah, V.P., and Crison, J.R. (1995). A theoretical basis for biopharmaceutics drug classification: the correlation of in vitro drug product dissolution and the in vivo bioavailability. *Pharm. Res.* 12: 413–420.

22 Zhou, D. and Qiu, Y. (2009). Understanding biopharmaceutics properties for pharmaceutical product development and manufacturing I-Oral absorption and the biopharmaceutics classification system. *J. Validation Technol.* 15 (4): 62–72.

23 Corrigan, O.I. (1997). The biopharmaceutic drug classification and drug administered in extended release (ER) formulations. *Adv. Exp. Med. Biol.* 423: 111–128.

24 Food and Drug Administration (2000). *Guidance for Industry: Waivers of in vivo Bioavailability and Bioequivalence for Immediate Release Solid Oral Dosage forms Based on Biopharmaceuticl Classification System*. Rockville, MD: Food and Drug Administration. Retrieved from www.fda.gov/cder/guidance/index.htm.

25 Wilding, I.R. (1999). Evolution of the biopharmaceutical classification system (BCS) to oral modified release (MR) formulations; what do we need to consider? *Eur. J. Pharm. Sci.* 8: 157–159.

26 Shah, R.B. and Khan, M.A. (2010). The science and regulatory perspectives of emerging controlled release dosage forms. In: *Oral Controlled Release Formulation Design and Drug Delivery* (eds. H. Wen and K. Park), 337–350. New Jersey, NJ: Wiley.

27 Kawabata, Y., Wada, K., Nakatani, M. et al. (2011). Formulation design of poorly soluble drugs based on biopharmaceutical classification system: basic approach and practical applications. *Int. J. Pham.* 420: 1–10.

28 Mario AG, Douglas FS, Use of IVIVC in the Development of Oral Extended-Release Formulations: A Personal Perspective, Dissolution Technologies, 2015, 35-42.

29 François, B., Kathryn, B., Nathalie, P., and Bernard, W. (2012). Physiologically based pharmacokinetic (PBPK) modeling tools: how to fit with our needs? *Biopharm. Drug Dispos.* 33: 55–71.

30 Fahr, A. and Liu, X. (2007). Drug delivery strategies for poorly water soluble drugs. *Expert Opin. Drug Delivery* 4 (4): 403–416.

31 Potts, R.O. and Guy, R. (1992). Predicting skin permeability. *Pharm. Res.* 9: 663–669.

32 Akala, E.O. (2005). Oral controlled release solid dosage forms. In: *Theory and Practice of Contemporary Pharmaceutics* (eds. T.K. Ghosh and B.R. Jasti), 333–366. LLC, USA: CRC Press.

33 Kadin, H. (1982). Captopril. In: *Analytical Profiles of Drug Substances*, vol. 11 (ed. K. Florey), 79–137. London: Academic Press.

34 Heinig, R., Ahr, G., Hayauchi, Y., and Kuhlmann, J. (1997). Pharmacokinetics of the controlled-release nisoldipine coat-core tablet formulation. *Int. J. Clin. Pharmacol. Ther.* 35 (8): 341–351. In Drug design Kinam.

35 Curatolo, W. (1998). Physical and chemical properties of oral drug candidates in the discovery and explanatory settings Pharm. *Sci .Technol. Today* 1: 387–392.

36 Johnson, K.C. and Swindell, A.C. (1996). Guidance in the setting of drug particle size specifications to minimize variability in absorption. *Pharm. Res.* 13 (12): 1795–1798.

37 Ali, R., Rajabi-Siahboomi, R.-A.R., Rane, M.S., and Felton, L.A. (2013). Oral modified release drug delivery system. In: *Remington: Essentials of Pharmaceutics* (ed. L. Felton), 623–632. United Kingdom: Pharmaceutical Press. Chapter 32.

38 Hoon, J.R. (1974). Bleeding gastritis induced by long-term release aspirin. *JAMA* 229: 841–842.

39 Lin, S.-Y. and Kawashima, Y. (2012). Current status and approaches to developing press-coated chronodelivery drug systems. *J. Controlled Release* 57 (3): 331–353.

40 Johansson, R., Regardh, C.G., and Sjogren, S. (1971). *Acta Pharm. Suec.* 8: 59–70.

41 Reidenberg, M.M., Drayer, D., Demarco, A.L., and Bello, C.T. (1973). Hydrazaline elimination in man. *Clin. Pharmacol. Ther.* 14: 970–977.

42 Wu, C.-Y. and Bennet, L.Z. (2005). Predicting drug disposition via application of BCS: transport/absorption/elimination and development of a biopharmaceutics drug disposition classification system. *Pharm. Res.* 22: 11–23.

43 Ritschel, W.A. (1989). Biopharmaceutical and pharmacokinetic aspects in the design of controlled release peroral drug delivery systems. *Drug Dev. Ind. Pharm.* 15: 1073–1103.

44 Davis, S.S. (2005). Formulation strategies for absorption windows. *Drug Delivery Today* 10 (4): 249–257.

45 Saha, J.R., Butler, V.P. Jr., Neu, H.C., and Lindenbaum, J. (1983). Digoxin inactivating bacteria: identification in human gut flora. *Science* 220 (4594): 325–327.

46 Siegal, R.A. and Rathbone, M.J. (2012). Overview of controlled release mechanisms. In: *Fundamentals and Application of Controlled Drug Delivery* (eds. S.R.A. Siepmann and M.J. Rathbone), 19–43. Springer.

47 Florence, A.T. and Jani, P. (1994). Novel oral drug formulations: their potential in modulating adverse effects. *Drug Safety* 10 (3): 233–266. (NoDDS).

48 Prisant, M.L., Bottini, B., and DiPiro, J.T. (1992). Novel drug-delivery systems for hypertension. *Am. J. Med.* 93 (2): S4–S55.

49 Hoffman, A.S. (2008). The origin and evolution of "controlled" drug delivery systems. *J. Controlled Release* 132: 153–163.

50 Schnepp, A.R. and Nykamp, D. (2010). Pharmacotherapy using cardiovascular agents: focus on top 200 drugs. *J. Pharm. Technol.* 26: 213–219.

51 Awan, N.A., Amsterdam, E.A., Vera, Z. et al. (1976). Reduction of ischemic injury by sublingual nitroglycerine in patients with acute myocardial infarction. *Circulation* 54 (5): 761–765. (In Kinam Book).

52 Smolensky, M.H. and Peppas, N.A. (2007). Chronobiology drug delivery and chronotherapeutics. *Adv. Drug Deliv. Rev.* 59: 828–851.

53 Tofler, G.H., Brezinski, D., Schfer, A.I. et al. (1987). Concurrent morning increase in platelet aggregability and the risk of myocardial infarction and sudden cardiac deaths. *N. Eng. J. Med.* 316: 1514–1518.

54 Ohdo, S. (2010). Chronotherapeutic strategy: rhythm monitoring, manipulation and disruption. *Adv. Drug Delivery Rev.* 62: 859–875.

55 Smith, H.G.D. (2001). Pharmacology of cardiovascular chronotherapeutic agents. *Am. J. Hypertens.* 14: 296S–301S.

56 Kario, K., Matsuo, T., Kobayashi, J. et al. (1996). Nocturnal fall of blood pressure and silent cerebrovascular damage in elderly hypertensive patients. *Hypertension* 27: 130–135.

57 Prisant, M.L. and Elliott, W.J. (2003). Drug delivery systems for treatment of systemic hypertension. *Clin. Pharmacokinet.* 42 (11): 931–940.

3

Drug Delivery of the Future (?)

Adrian Williams

University of Reading, Reading, UK

Prediction is very difficult, especially if it's about the future!

Niels Bohr, Nobel Laureate in Physics

3.1 Introduction

The history of making medicines can be traced back over 5000 years through the ancient civilisations of Mesopotamia, Babylonia, Egypt, China, Greece, and Rome and others. Over the centuries, formularies were defined, dosages became normalized into, for example, pills, poultices, or salves, and dosage intervals were understood requiring patients to take a course of their medicines. While dosage form design was recognized as an important element of medicines manufacture, it was only in the 1950s that the first controlled release formulation was introduced. The Spansule® technology developed by Smith Kline and French coated micropellets with different thicknesses of a water-soluble wax, poly(ethylene oxide) which provided prolonged release of a drug for 12 hours. This allowed twice daily dosing, at patient-convenient times as compared to the historical three to four times daily dosing for most existing oral dosage forms. The birth of controlled (or sustained/timed/extended) release drug delivery technologies led to a range of innovative products with delivery extended up to seven days from, for example, transdermal patches. Eloquently described by Park [1, 2], numerous clinical products were developed through this first generation of controlled drug delivery systems between 1950 and 1980.

A second wave of innovations (second-generation controlled drug delivery systems) was seen between 1980 and 2010, [1, 2] largely focused on zero-order drug release systems and longer-term sustained release, for example, from depot injections such as Zoladex® which was introduced in 1989 to deliver goserelin acetate for one month and three months. Despite significant advances in excipient (largely polymers) development to support controlled, sustained, and bioresponsive drug delivery, fewer commercially viable clinical products were developed during this second generation than through the first generation. Simplistically, the first generation technologies succeeded through modifying or mitigating issues arising from the physiochemical properties of the drug, whereas second-generation systems tackled issues arising from the biological barriers to drug delivery. The later decade of the second wave of dosage forms focused largely on development

Fundamentals of Drug Delivery, First Edition.
Edited by Heather A. E. Benson, Michael S. Roberts, Adrian C. Williams, and Xiaowen Liang.

of nanotechnology-based drug delivery systems. This leads onto the third generation, stretching forward to 2040, predicted to generate modulated drug delivery systems, with smart and responsive properties that precisely target the drug active within the body.

Given the rapid progress in dosage form design over the last 60 years, it is somewhat foolish to predict outcomes over the next 20 years or so. However, recent scientific and technological achievements perhaps hint as to the direction of travel in controlled drug delivery systems.

3.2 Therapeutic Indicators

Genomics, and genomic medicine, is now regarded as mainstream and perfectly illustrates the pace of change in both technology and clinical practice. The Human Genome Project – to sequence the three billion base pairs in our 23 pairs of chromosomes – took 13 years to complete, cost approximately $3 billion and drew on expertise from universities and research centers in the United States, the United Kingdom, France, Germany, Japan, and China. Reported in 2003, the last 15 years have seen remarkable advances such that today, the whole human genome can be sequenced in a few days for under $1500. Home-based "spit tests" are now commercially available to provide limited genetic information (for example, ancestry or some genetic predispositions to diseases) for less than $100.

As a consequence, genome-association studies are growing, from those within companies to national initiatives such as the UK 100 000 genome project. Nearly 2000 disease-associated genes have been described, and there are a similar number of genetic tests. Accompanying the diagnostic capabilities from genomics, nearly 400 biotechnology-based products are currently in clinical trials. However, genomic impacts on clinical practice are already apparent. These range from improvements in the use of existing drugs – for example, more precise dosing of warfarin – through to pharmacogenomic testing to identify, for example, whether a breast cancer patient will respond to the drug trastuzumab (Herceptin, a monoclonal antibody targeting the HER2 receptor; the HER2 gene is amplified in 20–30% of early stage breast cancers).

To date, genome-wide association studies (GWAS) have perhaps had greatest impacts in cancer treatment. For example, the BRAF gene, also known as the proto-oncogene B-Raf, codes for the B-Raf enzyme (a protein kinase) that is involved in directing cell growth. Over 30 mutations of the BRAF gene have been associated with human cancers including liver and kidney cancers, and BRAF mutations are found in over 80% of melanomas. Vemurafenib, the first drug developed by fragment-based drug discovery, is effective in melanoma patients whose cancer has a V600E BRAF mutation; V600E refers to a mutation at amino acid number 600 on the B-Raf protein, where valine is replaced by glutamic acid. About 60% of melanomas have this mutation.

Beyond cancers, GWAS have identified mutations associated with many common conditions ranging from cardiovascular disease, age-related macular degeneration, diabetes, rheumatoid arthritis, bipolar disorder, and psoriasis. However, as is apparent from the above, while the genetic code mutations and associations are useful in identifying the causes of disease and so can be used to select appropriate therapies, the genetic code is translated into proteins.

Transcriptomics essentially explores RNA expression, most commonly centered on the levels of mRNA. Other than mutations as described above, the genome is typically fixed for a particular cell. In contrast, the transcriptome is dynamic, reflecting the genes that gives rise to not only considerable diversity but also potentially damaging isoforms implicated in diseased states [3, 4]. Relatively minor changes to mRNA expression can have significant impacts on protein levels within a cell, but, to date, transcriptomics has a minor role in the diagnosis or treatment of diseases.

Proteomics, and the proteome, describes the study of the proteins found within a system or a body. Clearly, drawing on genomic data and that from transcription, proteomics examines not only the levels of proteins expressed but also their structure, activity, and posttranslational modifications. As a discipline, proteomics has benefitted greatly from advances in instrumentation and now typically relies on advanced mass spectrometry coupled with protein fractionation and subsequent bioinformatics. The benefits of a proteomic approach to understanding disease can be illustrated by a study into atopic dermatitis [5]. Filaggrin (FLG) is a major constituent of the skin epidermis. Loss-of-function mutations in *FLG gene* strongly predispose to the development of atopic dermatitis and secondary atopic conditions, including asthma and allergic rhinitis [6]. More than 40 different population-specific FLG mutations have been identified, but the precise molecular changes occurring directly as a result of FLG loss are relatively unknown. In the study, 17 proteins were significantly differentially regulated after FLG knockdown in a human epidermal skin equivalent, confirmed in biopsies from patients with atopic dermatitis. The authors [5] noted "Lack of correlation between the protein and mRNA expression profiles of cofilin-1 (CFL1) and peptidylpropyl isomerase A (PPIA) emphasize the value of proteomic studies" and explained that "A proteomics-based approach was used because protein changes largely represent the functional end point of cell signaling, and recent studies have identified an important disconnect between the transcriptome and proteome, suggesting that significant post-transcriptional regulation occurs."

Metabolomics moves beyond proteins to define the entire complement of small molecular weight molecules (metabolites) within a biological matrix of interest. Analysis of the metabolome focuses on identifying small molecules that may be present or at elevated levels in diseased states, primarily seeking to extract and analyze these for diagnostic purposes. An open access database of the human metabolome is available (http://www.hmdb.ca/) and contains over 110 000 metabolite entries including both water-soluble and lipid soluble metabolites as well as metabolites regarded as either abundant (>1 μM) or relatively rare (<1 nM).

The ***microbiome*** essentially considers the microorganisms found in and on the human body; the term is used for both the genetic material and/or the microorganisms themselves and is also used interchangeably with the term "microbiota." It is often quoted that there are ten times more microorganism on or in the human body than there are human cells, but this has recently been shown to be a overestimate [7]; for a "standard" man (70 kg), there are an estimated 3.0×10^{13} human cells and 3.8×10^{13} bacteria – i.e. we are approximately 56% prokaryotic. While the mass of the bacteria is ~200 g, there are over 1000 different bacterial species providing over 3.3 million genes, dwarfing the ~20 000 human genes. Though controversial, lateral gene transfer between prokaryotic and eukaryotic cells has been reported [8].

Clearly, the microbiota and imbalances can lead to numerous diseases and infections, but the microbiome has extensive metabolic activities and in the gut is essential to aid digestion. Microbial communities in the gut tend to stabilize around three years of age, but other communities are more dynamic. For example, the skin microbiota restructures during puberty (with sebum increases) and leads to expansion of "lipophilic" microorganisms such as *Propionibacterium* spp. and fungal *Malassezia* spp. However, some age-related changes in the skin microbiota have been associated with various conditions; cases of *Staphylococcus*-associated atopic dermatitis decline in the majority of children before puberty, whereas *Malassezia*-associated tinea versicolor is more common in adults than children. Further, the microbiota plays important roles in educating the innate and adaptive arms of the immune system and there is a long-established immune interface with gut microbiota. Indeed, "gut-brain" axis studies have shown that the gut microbiota influences neurological processes such as serotonin signaling.

The "omics" advances briefly described above, and our growing understanding of the roles of genes, their translation, metabolism, and the microbiome in health and disease are already impacting on clinical medicine. We have rapidly progressed molecular diagnostics with improved sensitivity for earlier diagnosis and for early interventions. Stratified medicine – identifying subgroups of patients with distinct mechanisms of disease, or particular responses to treatments – is routine in many areas of cancer therapeutics. For example, screening and prophylaxis based on genetic associations is common with inherited mutations to the *BRCA1* or *BRCA2* genes, which provide substantially elevated risks of breast and ovarian cancer, and warrant extensive and regular surveillance, chemoprevention or risk-reducing surgery. Personalized medicines are expanding – again based on genomics as well as metabolomics and – drawing on the successes from cancer therapeutics – will expand to other clinical areas.

However, these technical advances require new specialisms in the use and extraction ("mining") of information form large data sets – at the individual genome level or in terms of genomewide association studies. The ability to integrate knowledge from the differing 'omic areas remains problematic but is essential if we are to fully exploit the advances in these areas to benefit patient care (Figure 3.1).

The consequences of the 'omic advances are seen in the new chemical entities being developed as therapeutic agents. Section 3.3 considers the nature of these "future drugs" which then raises issues regarding their delivery and the ambition to target only the desired site of action with no "off target" effects – often termed precision medicine – as described in Section 3.4.

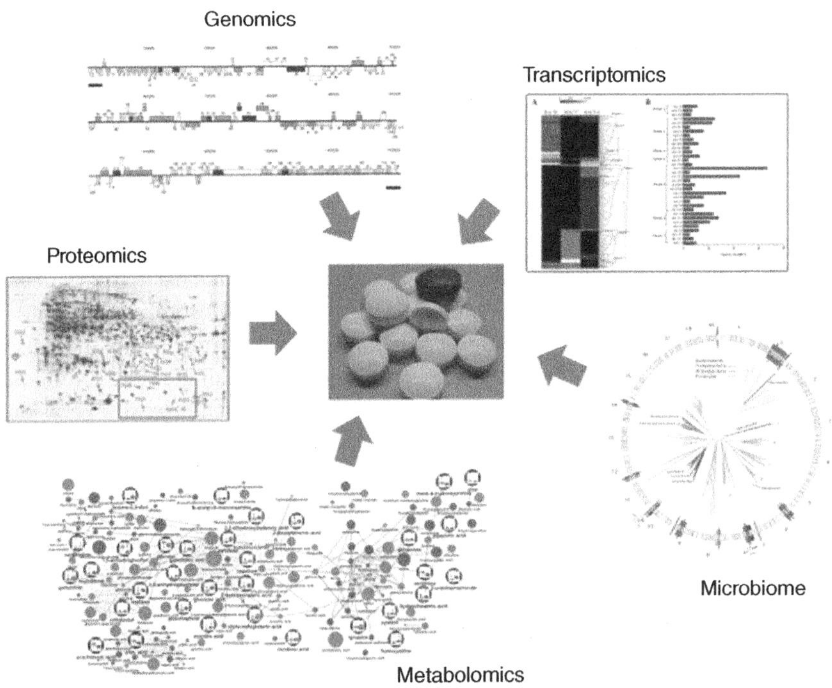

Figure 3.1 Assimilating knowledge from various 'omic datasets will bring tangible benefits in diagnosis and personalized treatment, but drawing together the information from the different disciplines to extract greatest benefits remains problematic. Source: (b) blackboard1965/Adobe Stock.

3.3 Drugs of the Future

Other than a few examples such as insulin, for many decades, drug discovery and development focused on relatively small organic molecules. Produced through chemical synthesis, these were (typically) well-defined single entities with high chemical purity that could be well- and fully characterized by analytical methods providing clear physiochemical properties and standards could be established.

In contrast, biologic drugs are typically produced in cell lines, are of high molecular weight with complex physiochemical properties. They tend to be less stable than the conventional small organic drugs, and are sensitive to heat, shear, and aggregation. They can be heterogeneous mixtures and are difficult to characterize and to standardize. Pharmaceutical biotechnological drugs are a relatively recent advance; Genentech were early pioneers in the late 1970s, and in 1985, the first recombinant human insulin (from recombinant human DNA) was registered by Genentech and Eli Lily. A year later, the first therapeutic monoclonal antibody (mAb) was approved; Muromonab-CD3 (Orthoclone OKT3) is an immunosuppressant to reduce organ rejection marketed by Janssen-Cilag. The first humanized mAb (daclizumab) was approved in 1994 followed by the first fully human mAb (Humira) in 2002. Gene therapy was first approved in 2012 (Glybera) though was not a commercial success. The pace of development of these biological drugs has increased radically. There are currently ~80 monoclonal antibodies approved by the EMA and FDA for conditions ranging from cancers to psoriasis to multiple sclerosis and age-related macular degeneration. Interestingly, several first-generation mAbs are now off-patent; biosimilars began to emerge in 2006 and in 2013 the first antibody biosimilar was approved (inflectra based on infliximab). The regulation of biosimilars has necessitated a different approach to simply matching the physical and chemical properties of the innovator, the drug assays, and manufacturing processes.

As the technologies described above have developed rapidly, drug targets have evolved and been refined with consequent effects on our therapeutic agents. This is illustrated in Table 3.1, showing changes to the top selling drugs over the last 10 years, from 2007 to 2012 and 2017 (Table 3.1). In 2007, the top selling drugs were small organic molecules such as statins (Lipitor topped the list with sales >$6 billion), proton pump inhibitors (Nexium, sales > $4 billion), and antipsychotic medicines. By 2012, four of the top ten selling drugs were biological. Last year, by sales revenue, Humira (a mAb approved to treat psoriasis and rheumatoid arthritis) was the top selling drug generating >$18 billion, despite the FDA approval of a Humira biosimilar. Eight of the top 10 selling drugs in 2017 were "biologicals"; six were monoclonal antibodies, predominantly for immunological (e.g. transplant rejection or arthritis) and oncology use, alongside a vaccine and long-acting insulin [9]. The two "conventional drugs" were Revlimid (lenalidomide), a derivative of thalidomide to treat myeloma and Lyrica (pregabalin) indicated for epilepsy, neuropathic pain, fibromyalgia, and generalized anxiety disorders.

It is readily apparent that the trend for increasing numbers of biological macromolecular drugs will continue. The rise in therapeutic antibodies has been remarkable. As described above, production of biological drugs is far more complex than for conventional organic drugs but in relative terms, protein-based therapeutics are the most accessible biologic drugs to develop since they are relatively stable. However, there are significant benefits to accrue from oligonucleotide therapies – from small interfering RNA (siRNA) to aptamers to gene therapy. Issues remain to be resolved; oligonucleotides are less stable in the blood than antibodies and generally have to cross cell and nuclear membranes to be effective.

Table 3.1 The top 10 drugs (by global sales) in 2007, 2012, and 2017, illustrating the rise in the value of biological drugs.

	2007	2012	2017 [9]
Rank by sales	**Product name (active ingredient)**	**Product name (active ingredient)**	**Product name (active ingredient)**
1	Lipitor (Atorvastatin)	Nexium (Esomeprazole magnesium)	**Humira (Adalimumab)**
2	Nexium (Esomeprazole magnesium)	Abilify (Aripiprazole)	Revlimid (Lenalidomide)
3	Advair Diskus (Fluticasone and salmeterol)	Crestor (Rosuvastatin)	**MabThera/Rituxan (Rituximab)**
4	Prevacid (Lansoprazole)	Advair Diskus (Fluticasone and salmeterol)	**Herceptin (Trastuzumab)**
5	Plavix (Clopidogrel)	Cymbalta (Duloxetine)	**Avastin (Bevacizumab)**
6	Singulair (Montelukast)	**Humira (Adalimumab)**	**Remicade (Infliximab)**
7	Seroquel (Quetiapine)	***Enbrel (Etanercept)***	***Prevnar 13 (Pneumococcal 13-valent Conjugate Vaccine)***
8	Effexor XR (Venlafaxine)	**Remicade (Infliximab)**	***Enbrel (Etanercept)***
9	Lexapro (Escitalopram)	Copaxone (Glatiramer acetate)	***Lantus (Insulin glargine)***
10	Actos (Pioglitazone)	***Neulasta (Pegfilgrastim)***	Lyrica (Pregabalin)

Bold = monoclonal antibodies; bold & italics = other "biological" drugs; plain font = standard "small" organic drugs.

Despite these difficulties, the first RNA-interference therapy ONPATTRO™ (patisiran) has recently been approved in the United States and the European Union for hereditary ATTR amyloidosis. Previously, only supported through palliative care, this condition results from mutations affecting the transthyretin protein, which is produced in the liver and normally carries Vitamin A. The mutant form forms amyloid fibrils which appear as toxic protein deposits in the heart and nervous system. The RNA interference (RNAi) effectively blocks production of the transthyretin protein which is otherwise inaccessible to conventional drugs.

Alongside RNA interfering therapeutics, gene therapy has also had a difficult journey to clinical efficacy. There have been numerous reports of serious adverse effects from direct gene therapy, but the number of clinical studies is increasing. Again, supported by technological advances such as Crispr gene editing, gene therapies are now coming to market. In 2017, the FDA approved Luxturna, the first *in vivo* gene therapy for the treatment of blindness due to Leber's congenital amaurosis. Luxterna is the first adeno-associated virus (AAV) vector gene therapy approved in United States. An alternative approach (CAR-T treatment) was also approved by the FDA in 2017. Kymriah is a treatment for B-cell acute lymphoblastic leukemia in those under 25 years. T-cells (an immune cell) are removed from the patient. A virus vector is then used to insert a gene coding for a protein called a chimeric antigen receptor (CAR) into these cells. The cells are then reinfused into the patient where the added protein helps these modified T-cells to bind to, and kill, leukemia cells. In trials, the approach resulted in an 83% remission rate over three months in people who hadn't responded to other treatment options. Gene therapy is being trailed for numerous cancers and other chronic conditions such as Hemophilia A.

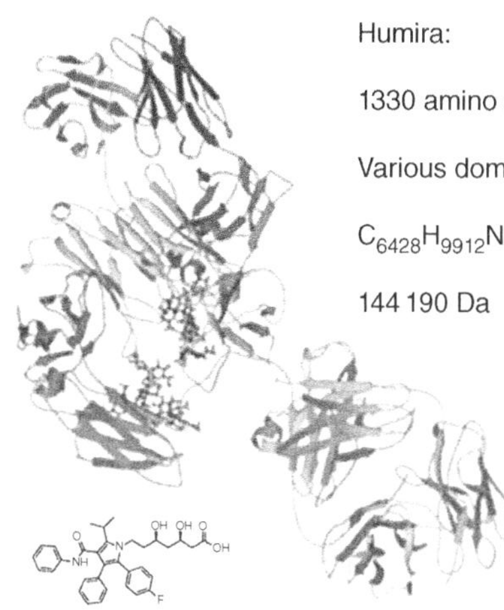

Figure 3.2 The changing nature of drugs. Lipitor (atorvastatin) was the top-selling drug of 2007, replaced by Humira (adalimumab) in 2017. A comparison of these molecules illustrates the new challenges faced in delivering them to their site of action.

In summary, our emerging drugs are increasingly biological – large molecular weight materials without the desirable attributes that are well-recognized for conventional therapeutics. We have been well-served by Lipinski's rule of five [10] when evaluating the likely suitability of a conventional chemical compound for oral dosing in humans. In essence the rule of thumb says that such molecules may be viable if there is only one (or no) violation of the rules:

- molecular weight less than 500 Da;
- log $P_{octanol/water}$ lower than 5;
- fewer than 5 hydrogen bond donors;
- fewer than 10 hydrogen bond acceptors.

Clearly, the rules were not developed for biological drugs, which cannot be orally dosed, but the emerging therapeutics offer very different physiochemical properties to conventional medicines. The potency of biologic therapeutics, and the need to deliver to specific (and often difficult to access) sites is highly challenging. But to realize the promise of these new agents, we need to dose precisely and in a controlled manner if the antibodies, RNA, and genes are to develop into medicines with an equivalent ease of administration to our current tablets, capsules, and creams (Figure 3.2).

3.4 Delivering the Drugs of the Future

Following the trends described in the Introduction, the next-generation drug delivery systems will seek to target our therapeutic agents and control/sustain delivery for extended periods, perhaps up to a year. This is a challenging ambition for our small molecule and stable therapeutic agents, and clearly more so for the biologics with relatively short half-lives and that are sensitive to biological environments.

Presently, most biological drugs, such as monoclonal antibodies, are administered by clinical staff through infusion with advantages over dosing for conditions such as cancer. However, there are alternatives such as with the dimeric fusion protein etanercept (934 amino acids, molecular weight 51 235 Da), indicated for moderate-to-severe rheumatoid arthritis or plaque psoriasis. The drug is available in vials, prefilled syringes or, following training, can be self-administered by some patients using an auto-injector system. Likewise, a prefilled syringe with certolizumab (a humanized pegylated Fab fragment) or adalimumab (Humira) from a pen or syringe can be self-administered.

At present, approximately 50% of all new drugs cannot be administered orally. Injections, even with advances such as click pens/auto injectors, require careful training and use by patients and are not suitable for all biological drugs. These challenges have led to some innovative drug delivery devices, though they are yet to reach the market.

Naturally, drug delivery systems will vary according to the desired site of action. For example, monoclonal antibodies (adalimumab – Humira) have been developed to target the tumor necrosis factor, implicated in psoriasis. In addition, other mAbs have been developed and received approval such as ustekinumab and guselkumab targeting interleukins such as IL-17 and IL-23 involved in the inflammatory cascade, and there is a strong pipeline of further mAbs. In treating psoriasis, it is desirable to treat locally since global suppression of immune responses can clearly lead to other problems such as infections. However, these large molecules are unable to traverse the skins stratum corneum barrier. Microneedles, typically a multiarray of short protrusion ~100–150 μm long are able to penetrate the stratum corneum but without stimulating the pain receptors deeper in the skin strata. They can be formed hollow, or coated with a drug, or can swell and deliver drug from an overlying patch. As they effectively generate micropores within the skin barrier, they can permit local dosing of, for example, vaccines and biological drugs. Indeed, currently (July 2018) a clinical trial is underway entitled "A Randomized, Double Blind, Placebo-controlled, Double-dummy Study to Assess Microneedle Delivery in Comparison to Subcutaneous Injection of Adalimumab in Healthy Volunteers," sponsored by the Centre for Human Drug Research, Netherlands. Interestingly, the microneedle approach to overcoming absorption barriers has been extended to consider oral dosing, using a "microneedle pill" to improve the bioavailability of macromolecules (insulin in this case) [11].

For less-accessible target sites, directed (also termed guided or focal) drug delivery systems have been explored. For example, magnetic particles can be preferentially guided to a target site or alternatively radiowaves or ultrasound have been employed as guidance modalities. The drug carrier systems are typically at nanoscale (solid lipid nanoparticles, liposomes, etc.) and can be used to focus therapeutic agents to the kidneys, lungs, gut, or even across the blood–brain barrier. An advantage of this approach is that the same guiding technology can often be used diagnostically, offering the opportunity for "theranostics" [12].

However, while strategies are emerging to deliver these large therapeutic agents, with some element of targeting or "precision," they do not remove the need for multiple and frequent dosing. Accompanying technological advances in other areas, such as electronics and their miniaturization, it seems clear that programmed and programmable delivery systems will develop to meet the therapeutic needs of the new drugs.

A microchip-based implant for prolonged dosing has been described and evaluated [13]. The device contains drug within micro reservoirs that are hermetically sealed. Delivery from each reservoir is triggered through a wireless signal delivered by a patient, and advanced systems are programmable to allow dosage adjustment by a doctor. A first-in-human clinical trial with the device evaluated the pharmacokinetic profiles of a human parathyroid hormone fragment aiming to treat osteoporosis. For this, current dosing is by daily injections and bone mineral density increases depend on pulsatile fragment delivery and so cannot be easily attained by sustained release from an implant. The microchip device was implanted into eight osteoporotic postmenopausal women for four months and was programmed to release doses from the device once daily for up to 20 days. Using the wireless capability, a remote computer-based programmer established a wireless communication link with the implant to program the dosing schedule and to receive implant status during the study to ensure that the device was operating as expected. The study reported that dosing from the microchip device produced similar pharmacokinetics to multiple injections but had the benefit of lower variability. Importantly, there were no toxic or adverse events due to the device or drug, and patients stated that the implant did not affect quality of life. Other microchip-based drug delivery devices have been described for areas where dosing is problematic, such as for brain cancer chemotherapeutics.

3.5 A View to the Longer Term?

The overhead section attempts to illustrate some of the emerging trends of our future therapeutic agents. From a greater understanding of pathophysiology, including genetic associations through proteomics to the microbiome, new and highly specific drug targets have – and will continue – to emerge. These targets require biological drugs that are inherently difficult to produce, formulate, and administer; they need to be delivered precisely to avoid, for example, the knockdown of broader immune response and for extended periods of time. This clearly requires new delivery systems and will likely draw on technological solutions such as implantable microchip arrays.

However, considering the rapid emergence of the new class of biological agents that now dominate the top-selling drugs list, perhaps therapeutics will progress in alternative directions.

Epidermolysis bullosa describes a series of conditions that causes blistering of the skin and mucous membranes. In essence, the disease is associated with mutations in the gene laminin β3 (LAMB3), which encodes an epidermal anchoring protein; genetic mutations prevent normal epidermal resistance or anchoring, making the skin fragile. Though there are varying degrees of severity and many variants, mechanical stress and minor trauma provoke epidermal fragmentation or detachment from the dermis, causing skin blistering and ulcers. This in turn leads to chronic, painful and untreatable wounds, and ultimately leads to skin cancers, infection, and sometimes death. There is currently no cure.

A remarkable study was reported late 2017 by Hirsch et al. [14] (Figure 3.3). A seven-year-old child with an extremely severe form of epidermolysis bullosa had lost 80% of his skin. A 4 cm^2 biopsy was taken from an unaffected part of his skin and the cells then genetically corrected using a retroviral vector carrying fully functioning LAMB3. The corrected cell population was grown to obtain 0.85 m^2 of transgenic epidermal grafts. Through three separate operations, 80% of the patient's skin was replaced with the grafts. After 21 months, the child seemed to have made a full recovery with no further blistering; the skin was resistant to stress and healed normally.

Single or multiple gene mutations are associated with many skin and other conditions. FLG gene mutations are associated with atopic dermatitis. In psoriasis, mutations cause the CARD14 protein to activate a family of nuclear proteins that control inflammatory gene expression; with a gain-of-function mutation, CARD14 is altered so that it continuously activates a molecular pathway that drives inflammation. Thus, while new monoclonal antibodies have been developed to inhibit inflammatory cascades (tumor necrosis factor [TNF], interleukins, etc.), the preceding study raises the real prospect of not treating – with the need for long-term therapy and adverse effects from suppressing global immune responses – but replacing. And, of course, replacement is not new – hips, knees, or heart valve replacements are widespread and currently use synthetic or, in some cases

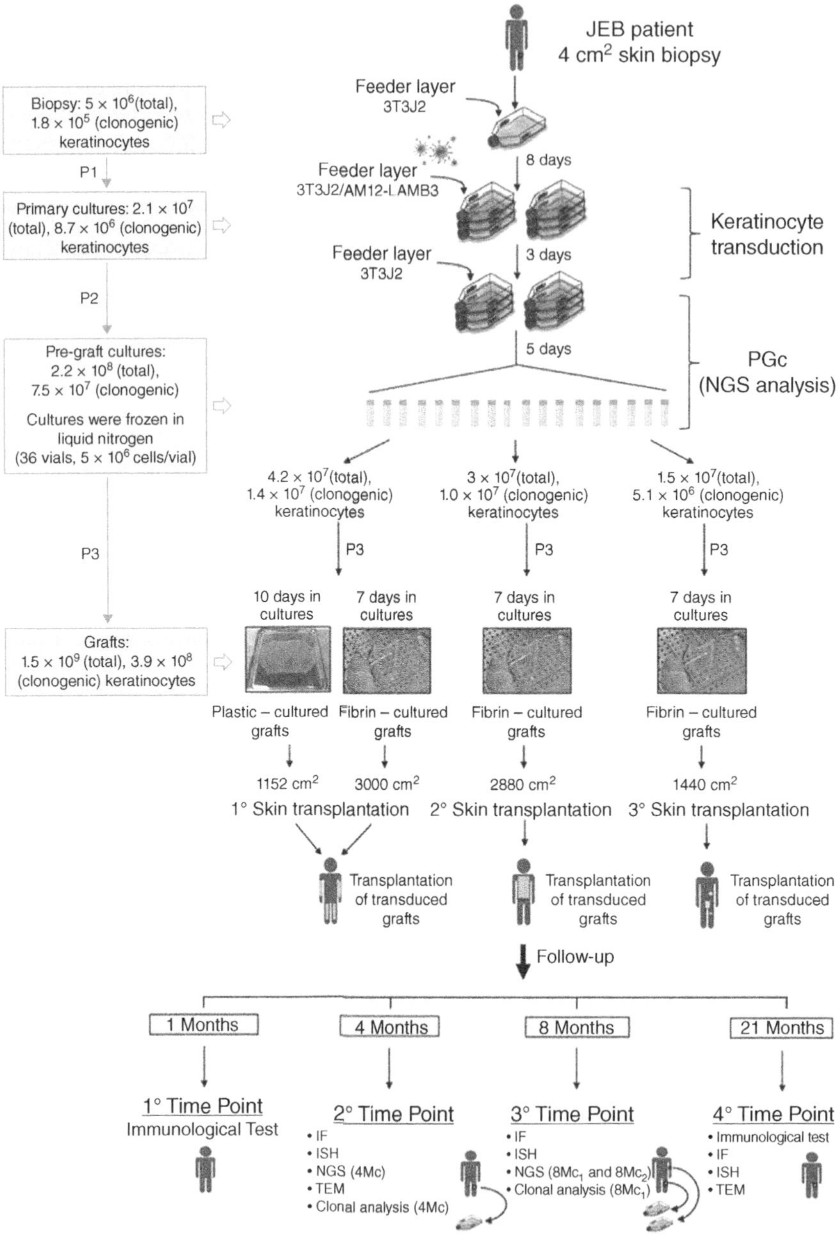

Figure 3.3 Schematic representation of the approach taken by Hirsch et al. to regenerate the entire human epidermis from transgenic stem cells. Source: Figure taken from Hirsch et al. [14]. © 2017 Springer Nature.

xenograft, replacements. The prospect overhead shows extension of this well-established principle to using and correcting an individual's own cells with benefits for reducing rejection.

A further important finding from Hirsch's study was that the in vitro growth of epidermal cells produced three types of colony; holoclones, which are proliferative and contain stem cells, differentiated paraclone colonies and meroclones, which are in an intermediate state of differentiation. Sheets of transgenic epidermis derived from these three colonies were transplanted. However, after four months, most paraclone and meroclone colonies from the initial transplant had been lost, and by eight months, almost the entire skin was derived from the initial holoclones. Thus, the skin graft was maintained by (relatively few) stem cells.

So, perhaps in the (not too distant) future, taking a biopsy, performing in vitro gene correction, cell expansion, and transplantation will itself become "old-fashioned." Stem cells have already been used to engineer various tissues *in vitro* – replacement of trachea or urethra to replace those damaged by cancers, cornea, and cartilage. And work is progressing to grow organs. For example, extracted stem cells have been used to grow skin to treat burn victims, attempts are ongoing to generate pancreas (for diabetes), hearts (postarrest), and even brains; currently stem cells are being grown into "mini-brains" – tens of thousands of cells containing all endogenous genes and so expressing "natural" proteins, focused on studying neurodegenerative conditions. And, again, technology is developing to help exploit the scientific advances with 3D cell printing and the use of biocompatible scaffolds currently supporting tissue growth from stem cells.

Stem cells have also been used to grow tissues *in vivo*; vascular precursor stem cells extracted from humans were implanted onto the surface of mice brains. Within two weeks, the cells formed a network of blood-perfused vessels which functioned as well as "natural" vessels. A recent *in utero* stem-cell transplant was reported for a fetus with the genetic blood disorder thalassemia, taking blood stem cells from the bone marrow of the mother.

Perhaps the next step is then to exploit stem cells *in situ*. Nanobots (nanoscale robots) were the stuff of science fiction; so too were laser guns or needleless injections (administered by Dr. Leonard "Bones" McCoy, chief medical officer of the *USS Enterprise* in Star Trek in the 1960s television series), both of which are now science facts. The 2016 Nobel Prize in Chemistry was awarded "for the design and synthesis of molecular machines" and nanocar races have been held. Drug containing nanobots have been designed to target cancer cells, nanobots have been constructed from DNA and so, perhaps, it is not such a leap of imagination that a DNA carrying nanobot with a surface recognition moiety could target stem cells, release DNA that would be taken into the stem cell to then differentiate into healthy cells? (Figure 3.4).

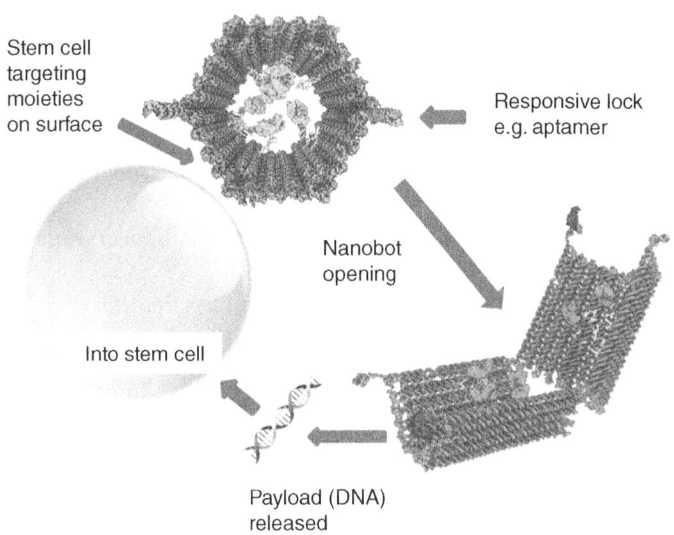

Figure 3.4 Could nanobots target stem cells to deliver DNA and then trigger differentiation into healthy cell populations to treat, *in situ*, conditions such as diabetes?

3.6 Conclusion

It is relatively easy to make predictions, or projections, from current trends. Our advancing knowledge of genetics, proteomics, transcriptomics, metabolomics, and from understanding the microbiome are generating new drug targets. The rise of biological drugs is linked to these advances, and it is clear that this trend will continue, perhaps progressing from monoclonal antibodies to oligonucleotide therapeutics. The new targets are specific, often difficult to reach, and require an element of personalized dosing (stratified medicine). Thus, new delivery systems that are increasingly technological will be required.

Is there a role for the "old-fashioned" small organic drug molecules? Clearly yes, and new chemically synthesized drugs continue to come to market. Conventional drugs will remain widely used for many indications; for example, nonsteroidal anti-inflammatory drugs for back pain are unlikely to be superseded by an RNAi therapeutic. The investment in, and cost of, biological drugs again prohibits application to all conditions, and so conventional drug delivery and technologies will be required to deliver medicines over extended time periods, with greater control and precision.

But therapeutics is not limited to our traditional pharmaceutical industry and the sector is joining alternative therapeutic ventures. From a chequered past, gene therapy now offers remarkable opportunities to cure (rather than manage) chronic conditions and will rapidly progress, from conditions where there are currently no cures to conditions that are poorly served by drug treatment and then beyond. Similarly, stem cell therapies have progressed from early clinical failures and are beginning to offer prospects for both cures and for organ replacement. Such a future vision both excites and scares – opportunities to "fix" consequences of our genetics, avoiding the need for organ donors by "growing your own" but also offering the prospect of "designer organs," perhaps bigger, better, and faster than nature intended. Maybe Albert Einstein was right (of course!), "I never think of the future, it comes soon enough."

References

1 Park, K. (2013). Facing the truth about nanotechnology in drug delivery. *ACS Nano* 7: 7442–7447.
2 Park, K. (2014). Controlled drug delivery systems: past forward and future back. *J. Controlled Release* 190: 3–8.
3 Faustino, N.A. and Cooper, T.A. (2003). Pre-mRNA splicing and human disease. *Genes Dev.* 17: 419–437.
4 Wang, E.T., Sandberg, R., Luo, S. et al. (2008). Alternative isoform regulation in human tissue transcriptomes. *Nature* 456: 470–476.
5 Elias, M.S., Long, H.A., Newman, C.F. et al. (2017). Proteomic analysis of filaggrin deficiency identifies molecular signatures characteristic of atopic eczema. *J. Allergy Clin. Immunol.* 140: 1299–1309.
6 Palmer, C.N.A., Irvine, A.D., Terron-Kwiatkowski, A. et al. (2006). Common loss-of-function variants of the epidermal barrier protein filaggrin are a major predisposing factor for atopic dermatitis. *Nat. Genet.* 38: 441–446.
7 Sender, R., Fuchs, S., and Milo, R. (2016). Revised estimates for the number of human and bacteria cells in the body. *PLoS Biol.* 14: e1002533.
8 Sieber, K.B., Bromely, R.E., and Dunning Hotopp, J.C.". (2017). Lateral gene transfer between prokaryotes and eukaryotes. *Exp. Cell. Res.* 358: 421–426.

9 Urquhart, L. (2018). Top drugs and companies by sales in 2017. *Nat. Rev. Drug Discovery* 17: 232.
10 Lipinski, C.A., Lombardo, F., Dominy, B.W., and Feeney, P.J. (1997). Experimental and computational approaches to estimate solubility and permeability in drug discovery and development settings. *Adv. Drug Delivery Rev.* 23: 3–25.
11 Traverso, G., Schoelhammer, C.M., Schroeder, A. et al. (2015). Microneedles for drug delivery via the gastrointestinal tract. *J. Pharm. Sci.* 104: 362–367.
12 Thanou, M. (ed.) (2018). *Theranostics and Image Guided Drug Delivery*. London, UK: Royal Society of Chemistry.
13 Farra, R., Sheppard, N.F., McCabe, L. et al. (2012). First-in-human testing of a wirelessly controlled drug delivery microchip. *Sci. Transl. Med.* 4 (122): 122ra21.
14 Hirsch, T., Rothoeft, T., Teig, N. et al. (2017). Regeneration of the entire human epidermis using transgenic stem cells. *Nature* 551: 327–332.

4

The Pharmaceutical Drug Development Process: Selecting a Suitable Drug Candidate

Lionel Trottet

DMPK (Drug Metabolism and PharmacoKinetics), Galapagos, Romainville, France

In the pharmaceutical industry, getting to the stage of drug candidate selection is a relatively early part of the full process of discovering a new medicine. It is generally reached after 3–4 years, while the full process leading up to new product launch will last about 10–12 years and can be divided in five stages [1], as shown in Figure 4.1. During the first part, the target is studied to assess the value of its mechanism of action compared to existing targets for a given indication. If there is reasonably strong confidence that such a target is of added value, the project will move to a drug discovery stage (hit and lead optimization), where many molecules will be synthesized and tested in order to find the drug candidate. After that, the selected molecule will progress into a preclinical development stage where various efforts especially in safety assessment, pharmaceutics development and chemical development will occur. Then the clinical phases will start and eventually, if successful, a New Drug Application (NDA) will be filed to regulatory bodies. Once approved, the new medicine will be launched.

Reaching the candidate selection milestone is a critical step that maybe underestimated when looking at the full drug discovery process, where one may tend to focus on the late clinical phases results and not to the decisions made earlier in the process. This is, however, the critical one as once the choice of the candidate molecule has been made, its quality will not improve in later phases. Indeed, the role of such late phases is only to demonstrate the clinical benefits vs. its risks of the selected molecule to obtain regulatory approval. If the quality of the candidate molecule is poor, this will not be changed during the clinical phases and the molecule is unlikely to make it to the market.

4.1 The Oral Drug Candidate: How to Get There and Questions to Answer

For an oral drug development project, once the target has been identified and the project is moving into the drug discovery phase, the know-how and knowledge in the industry are well set to guide the project team toward the candidate selection milestones. This drug discovery phase will not be easy and some projects will not be able to get a candidate, but, at least, the path and screening cascade to get there are well known.

Fundamentals of Drug Delivery, First Edition.
Edited by Heather A. E. Benson, Michael S. Roberts, Adrian C. Williams, and Xiaowen Liang.

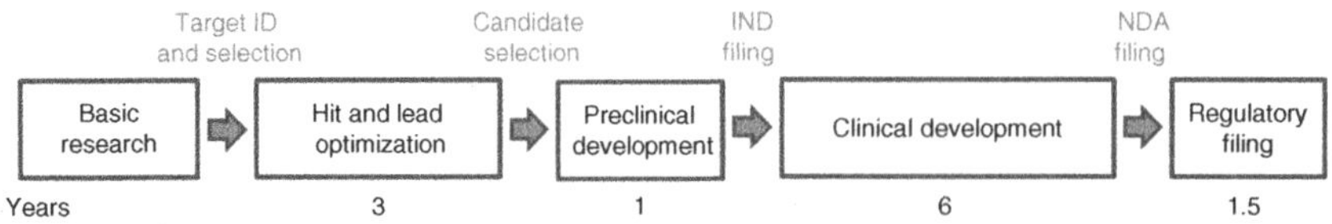

Figure 4.1 Drug discovery process from target selection through to filing of a compound and the approximate timescale for these processes (IND, Investigational New Drug; NDA, New Drug Application). Source: Modified from Hughes et al. [1].

Right at the beginning of the drug discovery phase, a screening of a large bank of compounds will be performed in order to find hit molecules which are able to interact with the identified target. The hit molecules obtained will not be suitable candidate medicine as their potency will be too low, their selectivity too poor and they will often have undesirable pharmacokinetic properties such as poor bioavailability and high clearance. In order to optimize such properties a Hit/Lead Optimization phase will start. This optimization phase is an iterative phase – as described in Figure 4.2 – where new molecules will be synthesized and optimized on selectivity, potency, and pharmacokinetics properties in order to obtain, ideally, a selective molecule with once daily administration with a projected dose up to 100 mg.

The iterative phase (Figure 4.2) will vary slightly from project to project and from company to company but the same principles will be applied. It will consist of assessing the *in vitro* potency of the new synthesized molecules, then to assess their selectivity, physicochemical, and *in vitro*

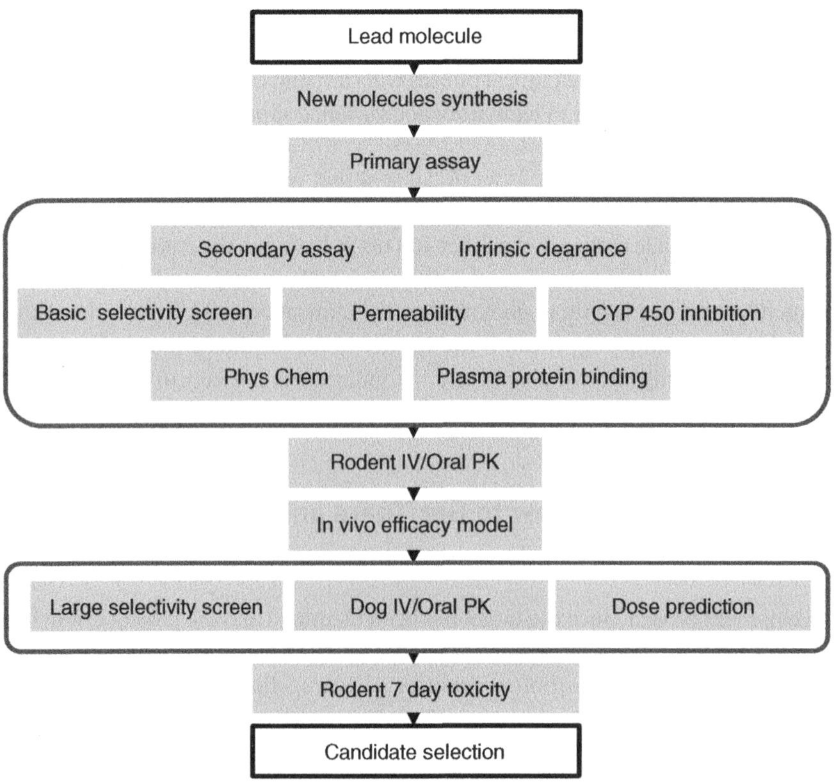

Figure 4.2 Example of a screening cascade during the lead optimization phase for an oral drug.

Drug Metabolism and PharmacoKinetics (DMPK) properties. The compounds presenting good or improved properties compared to previously synthesized compounds will be progressed into *in vivo* pharmacokinetic studies and then into *in vivo* pharmacology studies. For compounds starting to present near-drug candidate properties, dose prediction in man will be made and short-term toxicological studies will complete the candidate selection package.

The drug candidate will need to meet many properties, of which three are particularly important:

1. Strong potency against the desired target
2. High selectivity against other targets
3. Good pharmacokinetic properties (such as good bioavailability and low clearance)

In addition to the intrinsic properties of the compound itself, the project team will need to address the following four questions:

1. Target engagement and desired pharmacology have been observed in preclinical animal models? [2]
2. The target plasma (free) concentration to reach desired pharmacology is known? (obtained from target engagement studies in animal models) [2]
3. The target plasma concentration in human is expected to be reachable? (obtained by predicting the pharmacokinetic properties of the drug candidate in human)
4. The therapeutic window (or therapeutic index) is known and considered large enough? (obtained by comparing the target plasma concentration and the lowest concentration giving no adverse effect in preclinical species such as rat and dog)

These four questions are particularly important as they set the goals for the first clinical phase studies, and they assess the risks associated with the candidate compound. If such questions are poorly answered, the candidate molecule will move into clinical development with large unknown risks, and its chance to be successfully developed as a medicine will be low. For that reason, over the years, with the large cost associated with the clinical phases [3], the industry has increased its requirements on the quality of the drug candidate. In particular, more and more focus has been given to the definition of the desired target concentration to reach [2] as this is required to assess the therapeutic window, the dose in man and the goals of the Phase 1 study.

Volume 2 of this series is devoted to oral drug delivery. Here we follow the pharmaceutical drug development process for an alternate route of administration, using the dermal route as an example.

4.2 Challenges for Selecting a Topical Drug Candidate

With dermal drug development, the knowledge and know-how is poorly spread with regards to the properties that the candidate molecule should have. Of the three key properties required for the candidate (stated above), the potency and selectivity are understood, but the concept of good pharmacokinetic properties for a dermal drug is much more elusive for many. Should the molecule have high concentration in the skin tissues and not in the receiving fluid when performing *in vitro* studies? Should the protein binding in skin, in plasma be high or low?

Of the four questions to answer about the candidate (stated above), this gets even worse as only one question is answered: target engagement and desired pharmacology have been demonstrated in an animal model such as a mouse model. The other three questions are not answered. They are not, because measuring free concentration in skin after topical application is still relatively mysterious or poorly done, and the three other questions depend on the measurement of this target tissue concentration.

For an oral drug, this is much simpler as plasma concentration is used as a surrogate concentration of the target tissue concentration, as it is recognized to be in equilibrium with all tissues after systemic administration. As well, the free plasma concentration is the pharmacologically active concentration [4], it will be the concentration to look for in animal models and then subsequently in man. Plasma being easily sampled and methods to define plasma protein binding being well established, the assessment of plasma free concentration is assessed without issues.

This is not the case with dermal drugs, as the link in-between free plasma concentration and free skin concentration is not well established. In addition, because there is the possibility to generate skin tissue concentration *in vitro* using diffusion cells, this tends to be the way one would use such data to get some pharmacokinetics endpoints. There are, however, two potential issues with that approach. The first one is that this is the total concentration in skin that is measured while one would like to get the free concentration. To circumvent that, one can measure skin tissue binding, such as measurement of dermis binding [5]. The second issue, however, is more problematic and it is linked with the contamination of the skin tissue samples recovered during an *in vitro* skin permeation study due to the unremoved dose applied.

There are indeed three reasons why one might want to move away from measuring skin concentration after topical application. The first reason is that the impact of a small percentage of the dose unremoved before skin tissue collection can be substantial, as the dose applied is very large vs. the small quantity of compound that can be retrieved in the lower skin tissues.

In the example given in Table 4.1 only 0.12% of the dose is crossing the skin. The washing procedure would therefore need to be very effective to get confidence in the dose recovered within the skin tissues.

The second reason is that it is theoretically not possible to validate that the washing procedure gives a near 100% recovery. There are two ways to perform the validation of the dose removal. Option#1: To perform the dose removal check just after dose application. The problem here is that the topical will still be in a semisolid stage undried so more easily removed than if like a dried-film. Whatever the percentage dose applied removed calculated then, it will overestimate the percentage dose applied removed compared to if the semisolid had dried. Option#2: To perform the dose removal check once the formulation has dried on the skin surface. The problem here is that part of the dose will have penetrated into the stratum corneum so one will not be able to recover 100% of the applied dose, so calculating the real percentage of dose applied removed will not be possible.

Finally, from one formulation to another, for example comparing an aqueous solution and a semisolid formulation with strong ability to stick to the skin surface, the same washing procedure might give a near 100% of dose applied removed in the case of the solution, but a much worse

Table 4.1 Example of percentage of dose crossing the skin.

Drug concentration in formulation (% w:w; i.e. g/100 ml)	1%
Formulation applied (mg/cm^2)	10
Dose applied (μg/cm^2)	100
Average percutaneous flux over 24 h (ng/cm^2/h)	5
Permeated compound in 24 h/cm^2 (μg)	0.12
Dose permeated vs. dose applied (%)	0.12%

one for the sticky formulation. This last reason is the likely cause of observations that many in the skin permeation field have made when comparing different formulations of the same compound, where one observes a lack of proportionality in between viable epidermis or dermis concentration and the percutaneous flux. Such lack of proportionality is not expected, as skin retention (below the stratum corneum tissues) caused by the formulation is not theoretically explainable [6].

4.3 Percutaneous Flux as a Surrogate Measurement of Skin Tissue Concentration

Interestingly, the last guidance drafted by the US Food and Drug Administration (FDA) about topical bioequivalence [7], does not refer to skin tissue concentration measurement but to the measurement of the percutaneous flux. This may suggest that the difficulties in measuring direct skin concentration are recognized by regulatory bodies too. The measurement of the percutaneous flux is indeed not prone to the quality of the removal of the dose applied, so such data can be considered as valid. However, this measurement is not a concentration, and this is the likely reason why the pharmaceutical industry has been lukewarm to consider such a measurement as a pharmacokinetic endpoint. This may, however, evolve as there are equations linking flux and skin tissue concentration. Equations linking flux and skin tissue concentration have been established since the 1970s and 1980s[8, 9]. One difficulty with past mathematical models is that some of the input data were not practically measurable. More recently, the same equation, obtained by different approaches, all using Fick's law of diffusion, is now being described in the literature which uses more easily measurable parameters [10–12].

$$C_{free} \text{ viable skin tissue (at depth } x) = \frac{\text{Flux}}{D} * h * \left(1 - \frac{x}{h}\right) \tag{4.1}$$

where

Flux = percutaneous flux ($ng/cm^2/h$)
D = diffusion coefficient in the tissue (cm^2/h)
C = free drug concentration (ng/cm^3)
x = viable membrane depth (cm)
h = total thickness of the viable membrane (cm)

Note that this concentration corresponds to a free concentration (the pharmacologically active concentration) as only the diffusing molecules are taken into account. Bound molecules to protein would travel much less readily because of a much smaller diffusion coefficient and would contribute only marginally to the percutaneous flux measured. Obtaining the free skin concentration is ideal as this would answer, as discussed earlier, the very question missing at candidate selection stage for a dermal drug candidate.

One should however note that using the percutaneous flux as a surrogate of free skin concentration makes the assumption that neither binding, nor clearance from the viable tissue and dermis are rate limiting and that there is no systemic recirculation.

One can go further by integrating the first equation at half thickness:

$$C_{free} \text{ (average)} = \frac{\text{Flux}}{D} * \frac{h}{2} \tag{4.2}$$

If one focuses its effort in the upper part of the dermis, h could be set to 400 μm. Several authors [12–14] have looked at the determination of the diffusion coefficient in the dermis, and they found that for small molecules $D \sim 10^{-6}$ cm^2/s.

Table 4.2 Link in between flux and free concentration in the different skin compartments for various skin diseases.

	Free concentration (ng/ml) = Flux (ng/cm^2/h) * X (h/cm), where X = ...				
	Stratum corneum	Bottom of epidermis	Upper dermis	Bottom of dermis	Sub-cutaneous
Atopic dermatitis			1.8		
Psoriasis		0.73	0.18		
Acne (face)			3	0.6	
Fungal infection (athlete's foot)	666				
Dermal anesthesia			1.8	0.37	
Muscle pain					0.028

This would lead to Eq. (4.3) for the free concentration in upper dermis assuming no active capillary clearance occurs:

$$C_{free\ in\ upper\ dermis\ in\ vitro}\ [\mathrm{ng/ml}] = \mathrm{Flux}\ [\mathrm{ng/cm^2/h}] * 5.5\ [\mathrm{cm/h}] \quad (4.3)$$

Trottet [12] suggested a step further by adjusting Eq. (4.3) with various hypothesis and literature data to account for skin target site locations, skin disease barrier conditions, and capillary clearance.

$$C_{free\ skin\ target\ site\ in\ vivo}\ [\mathrm{ng/ml}] = Flux\ [\mathrm{ng/cm^2/h}] * X\ [\mathrm{cm/h}] \quad (4.4)$$

where X is described in Table 4.2, such a table would need to be used with caution and more as a guiding tool as many hypotheses were made.

To simplify it further and to apply it broadly for drug acting in the upper part of the dermis, one could set $X = 1$. A simple link would therefore be established between the percutaneous flux and free skin tissue concentration:

$$C_{free\ upper\ dermis\ in\ vivo}\ [\mathrm{ng/ml}] = \mathrm{Flux}\ [\mathrm{ng/cm^2/h}] * 1\ [\mathrm{cm/h}] \quad (4.5)$$

4.4 Learnings from Past Topical Drug Development of Factors Affecting Efficacy

The history of the discovery of dermal drugs can provide quite useful information about the key elements driving efficacy and success rate for a dermal drug.

A particular one is the success rate of first in class topical drugs [15].

From Table 4.3, it appears that for some drug classes, the development was relatively easy (such as the antivirals – aciclovir and penciclovir both oral compounds successfully developed later as topicals) and the antifungals (etonam (1st azole): effective – ketoconazole (oral ⇒ topical): effective – terbinafine (1st allyl): effective), while for some other classes, they followed a difficult drug development path (the retinoids (vitamin A: ineffective in psoriasis), the vitamin D3 derivatives (Calcitriol: systemic exposure issue), the immunosuppressors (cyclosporin: ineffective – tacrolimus developed later: effective in atopic dermatitis (A.D.) but not or poorly in psoriasis), the corticosteroids (cortisone: ineffective – hydrocortisone: effective in A.D. but poorly effective in psoriasis)).

Table 4.3 Success rate of oral drug developed topically and first in their class topical drugs.

Drug class	Success	Mixed results (positive but limited results or issues)	Failure
Antifungals [16–19]	Etonam Ketoconazole Terbinafine		
Antivirals [20–25]		Aciclovir Penciclovir	
Retinoids [26]			Vitamin A (psoriasis)
Vitamin D3 derivatives [27, 28]		Calcitriol (effective but systemic side effect)	
Immuno-modulators [29–40]	Tacrolimus (A.D., facial psoriasis)		Cyclosporin A – Tacrolimus (psoriasis)
Corticosteroids [41–44]		Hydrocortisone (A.D.)	Cortisone

A.D., atopic dermatitis.
Source: Trottet and Maibach [15]. © 2017, Springer Nature.

Looking into this finding, as shown in Table 4.4, with the focus on the location of the target site and integrity of the skin barrier, it appears that the ease of developing a topical drug for some skin diseases depends on two things:

1. If target site is in the stratum corneum (=top of the skin) (e.g. antifungals) ⇒ easy development, but if target site is deeper, development is more difficult.
2. If skin is severely damaged (e.g. antiviral) ⇒ easy development, but if skin is not damaged development is more difficult.

In Figure 4.3, one can schematically explain qualitatively how the drug concentration after dermal application can evolve in various skin tissues and depending on the skin barrier integrity.

Stratum corneum being at the surface of skin, drug concentration is higher than in the deeper tissues, hence a higher rate of success for drugs targeting this tissue. Similarly, with damaged skin, local concentrations are higher for the same target tissue, hence a higher success rate for drugs targeting a skin disease where the barrier is impaired.

This can be an important element to bear in mind for cosmeceutical approaches when active ingredients tend to be of low potency. In that case, reaching effective concentrations in skin tissues lower than the stratum corneum can prove very challenging.

Drug potency is indeed another key element that appears to drive topical efficacy as reported [15] and exemplified for some existing topical drugs from three dermal drug classes in Tables 4.5–4.7.

It can be seen from these three examples that the drug potency can have a substantial impact on the topical efficacy of a topical drug candidate.

Finally, another element seems to be playing a role: the percutaneous flux. It is difficult to drive a conclusion about percutaneous flux when one tries to compare one drug to another, as the drug potency may also play a role, as discussed previously. However, one can compare various formulations for the same drug. In Figure 4.4, the percutaneous flux of three clobetasol propionate formulations is compared showing that Temovate Ointment delivers more drug than Temovate Cream, which delivers more drug than Temovate Cream E.

Table 4.4 Barrier condition of skin diseases and success rate of topical drug development.

Drug class	Disease	Easiness of topical drug development[a)]	Target site	Skin barrier
Antifungals	Fungal infection	+	Stratum corneum	Not damaged
Antivirals	Coldsore	+	Bottom epidermis	Severely damaged
Corticosteroids	1. A.D. 2. Psoriasis	–[b)] – –[b)]	Dermis Epidermis/Dermis	Partly damaged[c)] Not damaged[c)]
Retinoids	1. Psoriasis 2. Acne	– –	Epidermis/Dermis Sebocyte	Not damaged Not damaged
Vitamin D3 derivatives	Psoriasis	–	Epidermis/Dermis	Not damaged
ImmunoModulators	1. A.D. 2. Psoriasis	– – –	Dermis Dermis	Partly damaged Not damaged

a) + for easy, – for difficult, – – for very difficult topical drug development.
b) Mild or moderately potent corticoids have poor clinical scores in psoriasis while they do better in A.D.
c) There is good evidence that A.D. skin barrier property is damaged [45–53]; however, the evidence that the psoriasis barrier is damaged is less clear. When the barrier property of psoriatic plaque compared to normal skin was studied with hydrocortisone in human *in vivo* [54], it was concluded that the barrier property of psoriasis plaque skin was comparable to normal skin.

Source: Trottet and Maibach [15]. © 2017, Springer Nature.

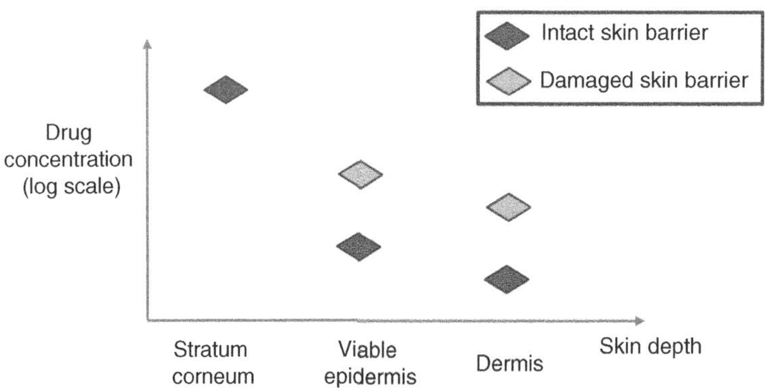

Figure 4.3 Schematic skin concentration in intact and damaged skin. Source: Modified from Trottet and Maibach [15]. © 2017, Springer Nature.

In the clinic, the corticosteroids blanching assay gives the same ranking [73, 74], which suggests that percutaneous flux is likely to play a role in the efficacy of a topically applied drug.

To conclude on the learnings from past dermal drug development, four elements seem to be strong drivers of topical efficacy:

#1 – Target location for the skin disease
#2 – Barrier properties of the skin disease
#3 – Potency of the test compound
#4 – Percutaneous flux of the test compound

Table 4.5 *In vitro* potency of some anesthetics, corticosteroids, and immunosuppressors.

Target class	Drug	Potency (nM)[a]
Anesthetics	Benzocaine [55]	910 000
	Lidocaine [55, 56]	155 000
	Prilocaine [55, 56]	125 000
	Tetracaine (amethocaine) [55, 56]	3500
Corticosteroids	Hydrocortisone [57–62]	14
	Bethametasone Valerate [57–62]	0.2
Immunosuppressors	Cyclosporin A [63–66]	11
	Tacrolimus[65, 67, 68]	0.17

a) The reported potency data in Table 4.5 are, for most of them, the average (geometric mean) of at least two data sets.
Source: Trottet and Maibach [15].

Table 4.6 Clinical efficacy comparison of some anesthetics, corticosteroids, and immunossupressors.

Target class/drugs	Clinical comparison
Anesthetics (Benzocaine, lidocaine, prilocaine, tetracaine)	In a pinprick model [69], Ametop (tetracaine) is compared with EMLA® (lidocaine + prilocaine) showing superiority of Ametop. In another study with the same model [70], saturated solution of tetracaine, lidocaine, and benzocaine are compared showing that tetracaine is superior to lidocaine, which is superior to benzocaine
Corticosteroids (Hydrocortisione, bethametasone valerate)	Hydrocortisone is classified as "mild" in the topical corticosteroids clinical ranking rating, while betamethasone valerate is classified as "potent" [71]
Immunossupressors (Cyclosporine, tacrolimus)	Topical tacrolimus is effective in the treatment of atopic dermatitis [72], while Cyclosporin A has shown limited or no activity in atopic dermatitis [34, 35]

Source: Trottet and Maibach [15].

Table 4.7 *In vitro* potency vs. clinical efficacy ranking of some anesthetics, corticosteroids, and immunossupressors.

Target class/drugs	*In vitro* potency ranking	Clinical efficacy ranking
Anaesthetics (Benzocaine, lidocaine, prilocaine, tetracaine)	Tetracaine > Prilocaine ~ Lidocaine > Benzocaine	Tetracaine > Prilocaine + Lidocaine > Benzocaine
Corticosteroids (Hydrocortisione, bethametasone valerate)	Bethametasone valerate > Hydrocortisone	Bethametasone valerate > Hydrocortisone
Immunossupressors (Cyclosporine, tacrolimus)	Tacrolimus > Cyclosporin A	Tacrolimus > Cyclosporin A

Source: Trottet and Maibach [15].

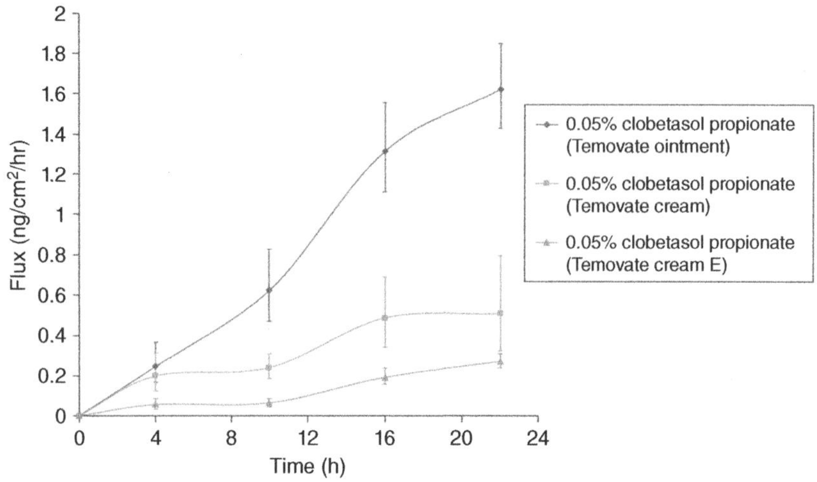

Figure 4.4 Percutaneous flux of clobetasol propionate from three different formulations. Source: Trottet [12].

4.5 Dermal Pharmacokinetics/Pharmacodynamics

In oral drug discovery, the triad solubility, permeability, and potency are considered to progress compounds [75]. This should apply to topical compounds too. The percutaneous flux is the combination of solubility and permeability, so one could hypothesize that using a combination of the percutaneous flux and potency of the compound could help to predict efficacy.

Pulling information from the literature, an empirical correlation can be established between the ratio of percutaneous flux (in ng/cm^2/h) and potency (in ng/ml) with clinical efficacy for a compound acting in the bottom of the epidermis or in the dermis (Figure 4.5) [76].

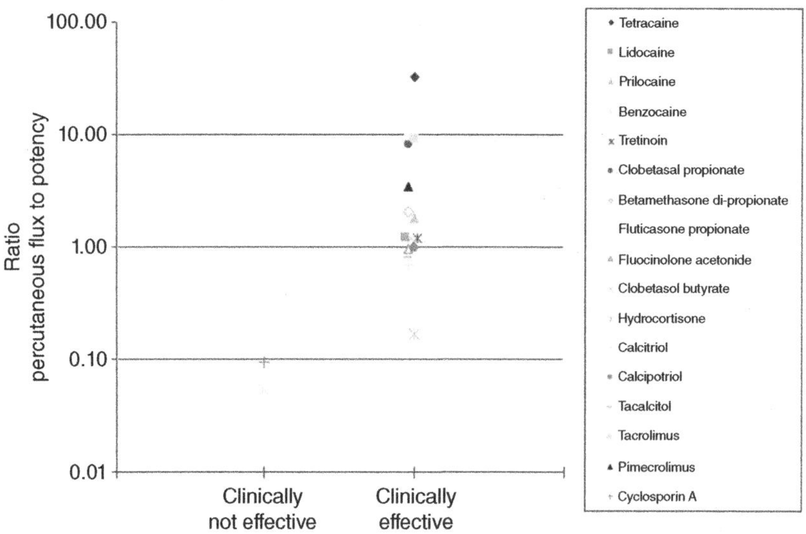

Figure 4.5 Clinical efficacy vs. percutaneous flux/*in vitro* potency. Source: Trottet and Maibach [76, 77].

More data would be required to validate this simple correlation, especially by populating it with compounds having failed to show clinical efficacy. The finding above suggests that if the percutaneous flux (in ng/cm^2/h) is superior to the potency of the compound for its target (in ng/ml), one would expect topical efficacy for a target.

When making a pharmacokinetics/pharmacodynamics (PK/PD) prediction, the general approach states that 50% of target engagement would be achieved when the free target site concentration is equal to the free EC_{50} (EC = Effective Concentration). This fits well with Eq. (4.5):

$$C_{free\ upper\ dermis\ in\ vivo}\ [\text{ng/ml}] = \text{Flux}\ [\text{ng/cm}^2/\text{h}] * 1\ [\text{cm/h}]$$

Overall, the empirical correlation linking flux/potency vs. clinical efficacy (Figure 4.5) can be reasonably well explained by the theoretical Eq. (4.5) (even though this equation uses several approximations).

4.6 Assessment of Systemic Exposure

One key advantage of a topical administration is the decreased risk of toxicological findings because the plasma concentration levels are expected to be low. This perceived element is generally the driving rationale for developing a topical medicine. This has proven to be true when looking at key topical target classes such as corticosteroids, immunomodulators, retinoids, and vitamin D3 derivatives which have side effects difficult to manage if given systemically for long periods, but which are effective and mostly safe if given topically.

The principle behind assessing circulating plasma concentration is well described by the classic pharmacokinetic Eq. (4.6), which allows calculation of the steady-state concentration in plasma based on input rate and systemic clearance.

$$\text{Total Plasma Concentration}_{steady\ state}\ (\mu\text{g/ml}) = \frac{\text{input rate}\ (\mu\text{g/h})}{\text{clearance}\ (\text{ml/h})} \tag{4.6}$$

The input rate for a topical is dependent on body surface area, the topical bioavailability (which is near 100% for most solutes as skin first pass only plays a minor role), and the input flux (percutaneous flux):

$$\text{Input Rate}\ (\mu\text{g/h}) = \text{surface area}\ (\text{cm}^2) * \text{bioavailability} * \text{input flux}\ (\mu\text{g/cm}^2/\text{h}) \tag{4.7}$$

From Eqs. (4.6) and (4.7):

$$\text{Total Plasma Concentration}_{\text{steady state}}\ (\mu\text{g/ml}) = \frac{\text{surface area}\ (\text{cm}^2) * \text{bioavailability} * \text{input flux}\ (\mu\text{g/cm}^2/\text{h})}{\text{clearance}\ (\text{ml/h})} \tag{4.8}$$

Transforming total plasma concentration into free plasma concentration from Eq. (4.8):

$$\text{Free Plasma Concentration}_{steady\ state}\ (\mu\text{g/ml}) = \frac{\text{surface area}\ (\text{cm}^2) * \text{bioavailability} * \text{input flux}\ (\mu\text{g/cm}^2/\text{h}) * (100 - \text{plasma protein binding})}{\text{clearance}\ (\text{ml/h}) * 100} \tag{4.9}$$

The risk of systemic exposure vs. the local target engagment (in other words what people sometimes called the "therapeutic index" or "therapeutic window") will be driven by the ratio of free skin concentration (where Eq. (4.5) can be used) and free plasma concentration [77]. This gives, therefore,

$$\text{Therapeutic Index} = \frac{\text{input flux }(\mu g/cm^2/h) * 1\ (h/cm) * \text{clearance}(ml/h) * 100}{\text{surface area }(cm^2) * \text{bioavailability} * \text{input flux }(\mu g/cm^2/h) * (100 - \text{plasma protein binding})}$$

$$\text{Therapeutic Index} = \frac{1\ (h/cm) * \text{clearance}(ml/h) * 100}{\text{surface area }(cm^2) * \text{bioavailability} * (100 - \text{plasma protein binding})} \quad (4.10)$$

Equation (4.10) can therefore be used as a guiding element to assess the potential of a compound to give systemic side effects as well as to rank a series of compounds [77].

4.7 Screening Cascade Approach to Select a Dermal Drug Candidate

During a drug discovery project, the project team will try to discharge as much risk as possible for the selected candidate. The challenge is to discharge such risks effectively such that the cost (money and ressources) used is well balanced vs. the occurrence and importance of the risk. It is, therefore, good to understand the type of reason for failure that can occur in a dermal drug discovery project. Table 4.8 lists such reasons and their occurrence, according to the experience of the author of this chapter.

Using Table 4.8, one can then focus on the main elements to consider. Seven parameters covering four areas appear to be key to address.

4.7.1 Efficacy (Lack of Target Engagement)

As seen previously, the ratio of **percutaneous flux** and **potency** may be a good element to predict efficacy and so both should be considered early in the screening cascade of a topical. One would be tempted to say that the more potent the compound, the better, and the higher the percutaneous flux, the better; however, it is in fact the higher the ratio of flux vs. potency that

Table 4.8 List of reasons of possible failures during a topical drug development project and their occurrence.

Reason of failure in topical drug development	Occurrence
Wrong pharmacology	Same as in any drug discovery program
Lack of target engagement after topical application	+++
Drug candidate chemical instability in formulation	+
Local topical irritation	+
Photo toxicity	+
Systemic exposure	Not usual but could be considered
Formulation physical or microbiological instability	Unlikely (with known formulation systems)

would need to be considered. If potency is part of the standard package of test performed early in any lead optimization project, this is not the case for the assessment of the percutaneous flux, which can be long and costly to perform. There are however ways to make the generation of such data more cost-effective by generating more data with the use of cassette dosing and with limited compound material (down to 5–10 mg) [78].

4.7.2 Developability

Stability of the compound in the formulation commonly drives failures or pushes the project teams to consider anhydrous formulations because of instability in water. Creams, gels, and lotions all contain water and hydrolysis is quite common. A potential solution could be to switch to anhydrous formulations; however, such formulations often lead to aesthetic issues and then compliance issues for the patients and should be avoided if possible, which would be even more true for the cosmetic industry. **Aqueous solution stability** would be of value to test early in the screening cascade.

4.7.3 Local Safety

Because local skin concentration can be quite high after a topical application, especially in the upper part of the skin tissues, off target (but as well on target) can occur, and one may want to check that such events do not occur. Such **local irritation tests** could be performed *in vitro* or *in vivo*. For a cosmeceutical mindset one would want to consider *in vitro* assays. One could, for example consider the use of Reconstituted Human Epidermis following various biomarkers representative of irritation [79–81]. Such methods have even been shown to be superior (on specificity) to the *in vivo* irritation studies performed in rabbit [79].

Photosafety will be a second element to consider for local safety. For that, **UV absorption** is a parameter that can be assessed. ICH guidelines – S10 – Photosafety Evaluation of Pharmaceuticals [82] gives the cascade to follow for Photosafety evaluation. An interesting element is the fact that if the Molar Extinction Coefficient (MEC) is inferior to 1000 l/mol/cm for a peak or tail above 290 nm, there is no concern for phototoxicity and no more studies are requested. Checking early for MEC above 290 nm is, therefore, an easy and cheap way to discriminate/rank compounds if a program team is in a lead optimization program or has many compounds to chose from.

4.7.4 Systemic Safety

When moving to the dermal route, the project team expects not to have systemic side effects. This is generally the case and not often an issue as shown by the small number of reported cases for a local corticosteroid, immunosuppressors, retinoid, or vitamin-D3 derivative, even though such target classes are known to give quite concerning side effects when giving orally long term. This can, however, occur, but as seen previously, it is possible to rationalize and estimate the risk of systemic exposure using Eq. (4.10), which requires as input data **systemic clearance** and **plasma protein binding**. Such data are part of the screening package used in an oral lead optimization project. Plasma protein binding is nowadays generated most often using an equilibrium dialysis approach [83]. Clearance can be assessed either by performing *in vivo* PK studies in rat or/and other species and scaling the clearance observed to the one expected in man [84, 85] or alternatively by determining the *in vitro* intrinsic clearance in human liver microsomes or in human hepatocytes [86].

4.8 Opportunities for Repurposing Molecules into Dermally Active Treatments for Cosmeceutical or Pharmaceutical Approaches

Discovering and developing a new chemical entity from the start to get to an approved drug is a long and cumbersome path as there are many pitfalls and many properties that the molecule will need to meet. Such an effort will need substantial resources that few companies will be prepared to make. There is, however, a good rationale to consider the dermal route opportunistically, as the costs and the risks could be decreased substantially.

On the cost side, one has to realize that there are nowadays a large number of asset molecules that exist in the pharma industry that are not being exploited. Indeed, the pharma industry nowadays works on more therapeutic targets than it used to, thanks to the advances in the understanding of the molecular basis of diseases, as well as the push from payors to deliver better medicines compared to standard of care. This means that there are now a very large number of new targets that could be investigated for existing molecules. Moreover, because of the high risk of target-related toxicity, a large number of asset molecules for various targets end up being abandoned as no viable indication for a systemic use can be found. Some of these molecules could be of interest for a skin disease as systemic toxicology would not be an issue in most cases. There is, therefore, the potential to pick these molecules without the need to invest in long and costly lead optimization effort. The difficulty in this exercise will be to put in place the right screening cascade such that the right and best molecules are selected. There will be a tendency to look only at the most advanced molecules, the ones which will have reached the candidate selection stage. This may be very tempting as part of the package will exist and some large quantity of the molecule will have been already synthesized. Tempting but probably wrong as in most cases such candidate molecules will have a low clearance and low protein binding. This will be the exact opposite of what is desired for a dermal molecule where one wants to limit the free plasma concentration. Therefore, the chances to get to a better and safer dermal molecule will likely lie in less advanced molecules. Still, with the right mindset and a good screening cascade in place, one could harvest potentially good dermal molecules and at low cost as such molecules would have already been synthesized and partly assessed.

On the risk side, as already discussed, the dermal route allows to limit the exposure of the applied molecule to the whole body. By doing so, local pharmacological effect can be obtained in the skin while preventing the molecule exerting its pharmacological effect systemically. As described, this is well proven by key topical target classes such as corticosteroids, immunosuppressors, retinoids, and vitamin D3 derivatives. Furthermore, now, with the theoretical possibility to define the systemic safety window *in silico* such as with Eq. (4.10), one can assess better the risk of systemic side effects and make a better selection of the molecule to progress. This element will be of particular help for a cosmeceutical approach, where the lack of systemic side effects will be paramount.

As an example of such an evaluation (Table 4.9), the therapeutic index is assessed for a target located in the upper dermis with a molecule with a systemic clearance equivalent to 50% of the hepatic blood flow and 98% plasma protein binding.

4.9 Conclusion

The selection of a dermal drug presents more risks than the selection of an oral drug, as the knowledge of the key factors affecting failure and success of a topically applied molecule are not well understood. However, with the right assessment of a few criteria, such risks could be better assessed and improve the quality of dermal drug candidates. Interestingly, most of the assessment can be

Table 4.9 Example of calculated therapeutic index evaluation.

% of surface area treated	Surface area treated (cm^2)[a]	Systemic clearance (l/h)	Plasma protein binding (%)	Therapeutic index[b]
30	5400	40	98	370
10	1800	40	98	1111
3	540	40	98	3704
1	180	40	98	11 111

a) Assuming total skin area for 70 kg adult = 18 000 cm^2.
b) Using Eq. (4.10) and assuming topical bioavailability is 100%.

obtained with *in vitro* assays, which should make such an evaluation reasonable in terms of timelines and costs. Moreover, the pharma industry has a rich unused pipeline of potential molecules, many from failed research projects, that are just waiting to be picked up by opportunistic project teams willing to put such compounds through an appropriate screening cascade.

References

1 Hughes, J.P., Rees, S., Kalindjian, S.B., and Philpott, K.L. (2011). Principles of early drug discovery. *Br. J. Pharmacol.* 162 (6): 1239–1249.

2 Morgan, P., Van Der Graaf, P.H., Arrowsmith, J. et al. (2012). Can the flow of medicines be improved? Fundamental pharmacokinetic and pharmacological principles toward improving Phase II survival. *Drug Discovery Today* 17 (9–10): 419–424.

3 Paul, S.M., Mytelka, D.S., Dunwiddie, C.T. et al. (2010). How to improve R&D productivity: the pharmaceutical industry's grand challenge. *Nat. Rev. Drug Discovery* 9 (3): 203–214.

4 Smith, D.A., Di, L., and Kerns, E.H. (2010). The effect of plasma protein binding on in vivo efficacy: misconceptions in drug discovery. *Nat. Rev. Drug Discovery* 9 (12): 929–939.

5 N. Atrux-Tallau, J.M. Linget, C. Wilson, et al. (2018). Poster–binding of drugs to dermis tissue: implication of physicochemistry and inter-species comparison. Perspectives in Percutaneous Penetration Conférence. La Grande Motte (France).

6 Trottet, L. and Maibach, H.I. (2017). Topical vehicle selection: myths and reality. In: *Dermal Drug Selection and Development: An Industrial Perspective*, 95–107. Springer.

7 Food & Drug Administration (2016). *Draft guidance on acyclovir*. Food & Drug Administration.

8 Scheuplein, R.J. and Blank, I.H. (1971). Permeability of the skin. *Physiol. Rev.* 51 (4): 702–747.

9 Siddiqui, O., Roberts, M.S., and Polack, A.E. (1989). Percutaneous absorption of steroids: relative contributions of epidermal penetration and dermal clearance. *J. Pharmacokinet. Biopharm.* 17 (4): 405–424.

10 Anissimov, Y.G. (2014). Mathematical models for skin toxicology. *Expert Opin. Drug Metab. Toxicol.* 10 (4): 551–560.

11 Cordero, J.A., Camacho, M., Obach, R. et al. (2001). In vitro based index of topical anti-inflammatory activity to compare a series of NSAIDs. *Eur. J. Pharm. Biopharm.* 51 (2): 135–142.

12 Trottet, L. (2004). Topical pharmacokinetics for a rational and effective topical drug development process. PhD thesis. University of Greenwich.

13 Bunge, A.L. and Cleek, R.L. (1995). A new method for estimating dermal absorption from chemical exposure: 2. Effect of molecular weight and octanol-water partitioning. *Pharm. Res.* 12 (1): 88–95.

14 Ibrahim, R. and Kasting, G.B. (2010). Improved method for determining partition and diffusion coefficients in human dermis. *J. Pharm. Sci.* 99 (12): 4928–4939.

15 Trottet, L. and Maibach, H.I. (2017). Key factors affecting the efficacy of a topical drug candidate: learnings from past topical drug development. In: *Dermal Drug Selection and Development: An Industrial Perspective*, 29–38. Springer.

16 Vanbreuseghem, R., Van Cutsem, J., and Thienpont, D. (1967). Experimental study of a new fungistatic agent: R 10100 1-(1,2,3,4-tetrahydro-1-naphthyl)-5-(ethoxycarbonyl)-imidazole nitrate. *Chemotherapy* 12 (2): 107–122.

17 Brugmans, J. and Van Cutsem, J. (1969). A three-stage clinical evaluation of etonam (R 10100) in the treatment of chronic athlete's foot. *Dermatologica* 138: 403–417.

18 Greer, D. and Jolly, H.W. (1987). Comparative trial of a two-dosage schedule of ketoconazole 2% cream for the treatment of tinea pedis. *J. Am. Acad. Dermatol.* 17 (1): 53–56.

19 Villars, V. and Jones, T.C. (1989). Clinical efficacy and tolerability of terbinafine (Lamisil) – a new topical and systemic fungicidal drug for treatment of dermatomycoses. *Clin. Exp. Dermatol.* 14 (2): 124–127.

20 Nilsen, A.E., Aasen, T., Halsos, A.M. et al. (1982). Efficacy of oral acyclovir in the treatment of initial and recurrent genital herpes. *Lancet* 2 (8298): 571–573.

21 Fiddian, A.P. and Ivanyi, L. (1983). Topical acyclovir in the management of recurrent herpes labialis. *Br. J. Dermatol.* 109 (3): 321–326.

22 Van Vloten, W.A., Swart, R.N., and Pot, F. (1983). Topical acyclovir therapy in patients with recurrent orofacial herpes simplex infections. *J. Antimicrob. Chemother.* 12 (Suppl B): 89–93.

23 Degreef, H. and Famciclovir Herpes Zoster Clinical Study Group (1994). Famciclovir, a new oral antiherpes drug: results of the first controlled clinical study demonstrating its efficacy and safety in the treatment of uncomplicated herpes zoster in immunocompetent patients. *Int. J. Antimicrob. Agents* 4 (4): 241–246.

24 Shafran, S.D., Aoki, F.Y., Lassonde, M. et al. A randomized, multicentre, double-blind, placebo-controlled, dose ranging, clinic-initiated study of oral famciclovir in the treatment of recurrent episodes of genital herpes. Annual Meeting of the Canadian Society for Clinical Investigation and the Royal College of Physicians and Surgeons of Canada, Toronto, Ontario, Canada, September 14–19, 1994. *Clin. Invest. Med.* 17, 4 SUPPL: B76–B1994.

25 Spruance, S.L., Rea, T.L., Thoming, C. et al. (1997). Penciclovir cream for the treatment of herpes simplex labialis. A randomized, multicenter, double-blind, placebo-controlled trial. Topical Penciclovir Collaborative Study Group. *JAMA* 277 (17): 1374–1379.

26 Reiss, F. and Campbell, R.M. (1954). The effect of topical application of vitamin A with special reference to the senile skin. *Dermatologica* 108: 121–128.

27 Bourke, J.F., Iqbal, S.J., and Hutchinson, P.E. (1996). Vitamin D analogues in psoriasis: effects on systemic calcium homeostasis. *Br. J. Dermatol.* 135 (3): 347–354.

28 Gilleaudeau, P., Mclane, J.A., Heftler, N. et al. (1996). Modulation of epidermal differentiation, tissue inflammation, and T-lymphocyte infiltration in psoriatic plaques by topical calcitriol. *J. Cutaneous Pathol.* 23: 419–430.

29 Griffiths, C.E., Powles, A.V., Baker, B.S. et al. (1987). Topical cyclosporin and psoriasis. *Lancet* 1 (8536): 806.

30 Hermann, R.C., Taylor, R.S., Ellis, C.N. et al. (1988). Topical ciclosporin for psoriasis: in vitro skin penetration and clinical study. *Skin Pharmacol. Off. J. Skin Pharmacol. Soc.* 1 (4): 246–249.

31 Schulze, H.J., Mahrle, G., and Steigleder, G.K. (1990). Topical cyclosporin A in psoriasis. *Br. J. Dermatol.* 122 (1): 113–114.

32 Gilhar, A., Winterstein, G., and Golan, D.T. (1988). Topical cyclosporine in psoriasis. *J. Am. Acad. Dermatol.* 18 (2 Pt 1): 378–379.

33 Schauder, C.S. and Gorsulowski, D.C. (1986). Topical cyclosporine A in the treatment of psoriasis. *Clin. Res.* 34: 1007 A.

34 de Prost, Y., Bodemer, C., and Teillac, D. (1989). Double-blind randomized placebo-controlled trial of local cyclosporine in atopic dermatitis. *Arch. Dermatol.* 125 (4): 570.

35 Reitamo, S., Käyhkö, K., Lauerma, A.I., and Mustakallio, K.K. (1989). Topical cyclosporine and contact dermatitis in guinea pig and man. *Arch. Dermatol.* 125 (4): 568.

36 Freeman, A.K., Linowski, G.J., Brady, C. et al. (2003). Tacrolimus ointment for the treatment of psoriasis on the face and intertriginous areas. *J. Am. Acad. Dermatol.* 48 (4): 564–568.

37 Hanifin, J. and Sober, A. (2000). Comparison of topically applied tacrolimus ointment vs. vehicle in adult patients wirth atopic dermatitis. Am. Acad. Dermatol. Conf., p. Poster 16.

38 Nasr, I.S. (2000). Topical tacrolimus in dermatology. *Clin. Exp. Dermatol.* 25 (3): 250–254.

39 Paller, A., Eichenfield, L.F., Leung, D. et al. (2000). Topically applied tacrolimus ointment vs. vehicle ointment in pediatric patients with atopic dermatitis. Am. Acad. Dermatol. Conf., p. Poster 111.

40 Zonneveld, I.M., Rubins, A., Jablonska, S. et al. (1998). Topical tacrolimus is not effective in chronic plaque psoriasis. A pilot study. *Arch. Dermatol.* 134 (9): 1101–1102.

41 Goldman, L., Thompson, R.G., and Trice, E.R. (1952). Cortisone acetate in skin disease; local effect in the skin from topical application and local injection. *AMA Arch. Dermatol. Syphilol.* 65 (2): 177–186.

42 Newman, B.A. and Feldman, F.F. (1951). Effects of topical cortisone ointment on chronic discoid lupus erythematosus and necrobiosis lipoidic diabeticorum. *J. Invest. Dermatol.* 17: 3–6.

43 Spies, T.D. and Stone, R.E. (1950). Effect of local application of synthetic cortisone acetate on lesions or iritis and uveitis, of allergic dermatitis, and psoriasis. *South. Med. J.* 43: 871–874.

44 Sulzberger, M.B. and Witten, V.H. (1952). Effect of topically applied compound F in selected dermatoses. *J. Invest. Dermatol.* 19: 101–102.

45 Aalto-Korte, K. (1995). Improvement of skin barrier function during treatment of atopic dermatitis. *J. Am. Acad. Dermatol.* 33 (6): 969–972.

46 Aalto-Korte, K. and Turpeinen, M. (1995). Quantifying systemic absorption of topical hydrocortisone in erythroderma. *Br. J. Dermatol.* 133: 403–408.

47 Aalto-Korte, K. and Turpeinen, M. (1996). Transepidermal water loss predicts systemic absorption of topical hydrocortisone in atopic dermatitis. *Br. J. Dermatol.* 135: 489–504.

48 Aalto-Korte, K. and Turpeinen, M. (1997). Weak barrier is associated with weak reservoir of topical hydrocortisone in patients with dermatitis. *J. Invest. Dermatol.* 109: 459 (Abstract 332).

49 Maibach, H.I. (1976). In vivo percutaneous penetration of corticoids in man and unresolved problems in their efficacy. *Dermatologica* 152 (Suppl 1): 11–25.

50 Matsumoto, M., Sugiura, H., and Uehara, M. (2000). Skin barrier function in patients with completely healed atopic dermatitis. *J. Dermatol. Sci.* 23 (3): 178–182.

51 Turpeinen, M., Lehtokoski-Lehtiniemi, E., Leisti, S., and Salo, O.P. (1988). Percutaneous absorption of hydrocortisone during and after the acute phase of dermatitis in children. *Pediatr. Dermatol.* 5 (4): 276–279.

52 Turpeinen, M., Mashkilleyson, N., Björkstén, F., and Salo, O.P. (1988). Percutaneous absorption of hydrocortisone during exacerbation and remission of atopic dermatitis in adults. *Acta Derm. Venereol.* 68 (4): 331–335.

53 Turpeinen, M. (1988). Influence of age and severity of dermatitis on the percutaneous absorption of hydrocortisone in children. *Br. J. Dermatol.* 118 (4): 517–522.

54 Wester, R.C., Bucks, D.A., and Maibach, H.I. (1983). In vivo percutaneous absorption of hydrocortisone in psoriatic patients and normal volunteers. *J. Am. Acad. Dermatol.* 8 (5): 645–647.

55 McNeal, E.T., Lewandowski, G.A., Daly, J.W., and Creveling, C.R. (1985). [3H]Batrachotoxinin A 20 alpha-benzoate binding to voltage-sensitive sodium channels: a rapid and quantitative assay for local anesthetic activity in a variety of drugs. *J. Med. Chem.* 28 (3): 381–388.

56 Gupta, S.P., Gupta, J.K., and Saha, R.N. (1990). A quantitative structure-activity relationship study on the inhibitory effects of local anesthetics on sodium flux, phosphoinositide breakdown, and binding to sodium channels. *Drug Des. Delivery* 6 (2): 131–135.

57 Bonsmann, U. and Rohdewald, P. (1998). Metabolisierung und rezeptoraffinitat topisch angewandter glucocorticoidester. *Z. Hautkr.* 73: 581–590.

58 Lutsky, B.N., Millonig, R.C., Wojnar, R.J. et al. (1986). Androstene-17-thioketals. 2nd communication: pharmacological profiles of tipredane and (11 beta, 17 alpha)-17-(ethylthio)-9 alpha-fluoro-17-[2-(fluoroethyl)thio]-11 beta-hydroxy-androsta-1,4-dien-3-one, structurally novel 20-thiasteroids possessing potent and selective topical antiinflammatory activity. *Arzneimittelforschung* 36 (12): 1787–1795.

59 Muramatsu, M., Tanaka, M., Otomo, S. et al. (1985). Characteristics of binding of a new anti-inflammatory glucocorticoid, hydrocortisone 17-butyrate 21-propionate (HBP), to glucocorticoid receptors of rat liver. *Jpn. J. Pharmacol.* 37 (2): 143–150.

60 Ponec, M. (1982). Glucocorticoids and cultured human skin cells: specific intracellular binding and structure-activity relationships. *Br. J. Dermatol.* 107 (Suppl 23): 24–29.

61 Smith, K. and Shuster, S. (1987). A comparison of the glucocorticoid receptors in the dermis and epidermis of the rat. *Clin. Exp. Dermatol.* 12 (2): 89–92.

62 Wurthwein, G., Rehder, S., and Rohdewald, P. (1992). Lipophilicity and receptor affinity of glucocorticoids. *Pharm Ztg Wiss* 137: 161–167.

63 T. Micromedex, *Cyclosporine.* 2003.

64 Furlanut, M., Baraldo, M., Pea, F. et al. (1996). Blood concentrations and clinical effect of cyclosporin in psoriasis. *Ther. Drug Monit.* 18 (5): 544–548.

65 Koch, K., Newborg, M.F., Hanson, D.C. et al. (1995). The C-32 triacetyl-L-rhamnose derivative of ascomycin: a potent, orally active macrolactone immunosuppressant. *J. Med. Chem.* 38 (8): 1255–1258.

66 Stuetz, A., Grassberger, M., and Meingassner, J.G. (2001). Pimecrolimus (Elidel, SDZ ASM 981)–preclinical pharmacologic profile and skin selectivity. *Semin. Cutaneous Med. Surg.* 20 (4): 233–241.

67 G. The European FK 506 Multicentre Psoriasis Study Group (1996). Systemic tacrolimus (FK 506) is effective for the treatment of psoriasis in a double-blind, placebo-controlled study. *Arch Derm* 132: 419–423.

68 Winiski, A., Wang, S., Schwendinger, B., and Stuetz, A. (2002). Inhibitory activity of pimecrolimus and tacrolimus on induced cytokine mRNA and protein expression in a human T cell line (Jurkat) measured via RT-PCR and Elisa. *J. Invest. Dermatol.* 119: 347.

69 McCafferty, D.F., Woolfson, A.D., and Boston, V. (1989). In vivo assessment of percutaneous local anaesthetic preparations. *Br. J. Anaesth.* 62 (1): 17–21.

70 McCafferty, D.F., Woolfson, A.D., McClelland, K.H., and Boston, V. (1988). Comparative in vivo and in vitro assessment of the percutaneous absorption of local anaesthetics. *Br. J. Anaesth.* 60 (1): 64–69.

71 Parfitt, K. (ed.) (1999). *Martindale: The Complete Drug Reference*, 32e. Pharmaceutical Press.

72 Lauerma, A.I., Maibach, H.I., Granlund, H. et al. (1992). Inhibition of contact allergy reactions by topical FK506. *Lancet.* 340 (8818): 556.

73 Pershing, L.K. and Corlett, J.L. (1998). Potency assessment of clobetasol propionate in four commercial formulations using a colorimeter and population based Emax modeling. *J. Invest. Dermatol.*, Abstract 1247: 680.

74 Weigmann, H.J., Lademann, J., Schanzer, S. et al. (2001). Correlation of the local distribution of topically applied substances inside the stratum corneum determined by tape-stripping to differences in bioavailability. *Skin Pharmacol. Appl. Skin Physiol.* 14 (Suppl 1): 98–102.

75 Lipinski, C.A., Lombardo, F., Dominy, B.W., and Feeney, P.J. (2001). Experimental and computational approaches to estimate solubility and permeability in drug discovery and development settings. *Adv. Drug Delivery Rev.* 46 (1–3): 3–26.

76 Trottet, L. and Maibach, H.I. (2017). Assessing topical efficacy. In: *Dermal Drug Selection and Development: An Industrial Perspective*, 69–86. Springer.

77 Trottet, L. and Maibach, H.I. (2017). Assessing the therapeutic index. In: *Dermal Drug Selection and Development: An Industrial Perspective*, 87–93. Springer.

78 Trottet, L. and Maibach, H.I. (2017). Topical drug candidate selection criteria and cascade. In: *Dermal Drug Selection and Development: An Industrial Perspective*, 115–131. Springer.

79 Jírová, D., Basketter, D., Liebsch, M. et al. (2010). Comparison of human skin irritation patch test data with in vitro skin irritation assays and animal data. *Contact Dermatitis* 62 (2): 109–116.

80 Spielmann, H., Hoffmann, S., Liebsch, M. et al. (2007). The ECVAM international validation study on in vitro tests for acute skin irritation: report on the validity of the EPISKIN and EpiDerm assays and on the Skin Integrity Function Test. *Altern. Lab. Anim.* 35 (6): 559–601.

81 Tornier, C., Rosdy, M., and Maibach, H.I. (2006). In vitro skin irritation testing on reconstituted human epidermis: reproducibility for 50 chemicals tested with two protocols. *Toxicol. Vitro Int. J. Publ. Assoc. BIBRA* 20 (4): 401–416.

82 ICH. (2015). PhotoSafety evaluation of pharmaceuticals – S10 – ICH Harmonised Tripartite Guideline.

83 Waters, N.J., Jones, R., Williams, G., and Sohal, B. (2008). Validation of a rapid equilibrium dialysis approach for the measurement of plasma protein binding. *J. Pharm. Sci.* 97 (10): 4586–4595.

84 Mahmood, I. and Balian, J.D. (1996). Interspecies scaling: predicting clearance of drugs in humans. Three different approaches. *Xenobiotica Fate Foreign Compd. Biol. Syst.* 26 (9): 887–895.

85 Jolivette, L.J. and Ward, K.W. (2005). Extrapolation of human pharmacokinetic parameters from rat, dog, and monkey data: molecular properties associated with extrapolative success or failure. *J. Pharm. Sci.* 94 (7): 1467–1483.

86 Riede, J., Poller, B., Umehara, K. et al. (2016). New IVIVE method for the prediction of total human clearance and relative elimination pathway contributions from in vitro hepatocyte and microsome data. *Eur. J. Pharm. Sci. Off. J. Eur. Fed. Pharm. Sci.* 86: 96–102.

5

Preformulation and Physicochemical Characterization Underpinning the Development of Controlled Drug Delivery Systems

Ronak Savla[1] *and Julien Meissonnier*[2]

[1] *Science and Technology, Catalent Pharma Solutions, Somerset, NJ, USA*
[2] *Science and Technology, Catalent Pharma Solutions, Beinheim, France*

Controlled drug delivery (CDD) systems play an important role in clinical treatment of diseases. CDD systems aim to solve three main challenges facing drug developers and clinicians: unmet clinical needs in terms of multiple daily doses and poorer adherence; nonoptimal drug pharmacokinetics; and loss of patent protection and exclusivity to immediate release (IR) dosage form. Development of CDD systems is often considered later in the drug product development timeline because of the relatively high resource, particularly the amount of active pharmaceutical ingredient (API), required to develop formulations and test them *in vivo*. Before embarking on creating prototypes, drug developers can save time, money, and resources by defining the target product profile (dissolution profile, drug release profile, and patient factors) and conducting a thorough preformulation assessment. This chapter covers the fundamental preformulation tests to quantitate the physciochemical characteristics, solid-state chemistry, stability, and bulk properties of APIs that are critical during the formulation development of CDD systems.

5.1 When Is a Controlled Drug Delivery System Needed?

Historically, CDD systems have been and continue to be used as the preferred life cycle management (LCM) tool. The introductory IR drug product whose dosing regimen calls for multiple dosing per day can be simplified to be taken once daily while preserving the efficacy and perhaps improving the safety. CDD are a natural extension of a drug product life and have been used as a LCM tool for many IR products that were dosed multiple times per day.

There may be cases where assessment of CDD approaches during compound discovery may be warranted by critical attributes such as the need for once daily dosing or specific plasma profiles, or to overcome molecule challenges such as reducing the "peak-and-trough" concentrations to minimize adverse effects [1, 2]. CDD may provide commercial advantage for a drug that is not first to market within a class of drugs that require multiple daily doses. In such cases, the drug developability assessment and CDD feasibility are highly linked. A company taking this approach would be wise to promote cross-functional collaboration between the various groups of discovery and development scientists [3]. Screening compounds with properties suitable for CDD, which may be more stringent than compounds aimed for IR delivery, can de-risk future CDD formulation development. The more stringent requirements of molecules for CDD development are covered in later sections of this chapter.

Fundamentals of Drug Delivery, First Edition.
Edited by Heather A. E. Benson, Michael S. Roberts, Adrian C. Williams, and Xiaowen Liang.

The increasing percentage of more potent or more insoluble drugs has necessitated the CDD systems also address other shortcomings in addition to multiple daily doses. These areas of address can be broadly divided in unmet clinical needs, less than optimal drug characteristics, and commercial interests. Most times that CDD is needed will contain elements of the three areas.

5.2 Optimizing Drug Characteristics

The drug dose and regional absorption are important drug characteristics that can either limit CDD formulation options or guide the formulator to preferred development pathways. Achieving once daily administration requires a reasonable dose to avoid large tablets and capsules and the drug being absorbed in both the small and large intestines. Bioavailability enhancement technologies can be combined with CDD technologies to improve solubility of lipophilic drugs to achieve lower dose requirements and promote absorption. Drugs with dosing dependent on weight (i.e. pediatrics), renal or hepatic function (commonly seen in geriatrics), and titration to reach efficacy or minimize safety, require a range of doses. A multiparticulate approach in which the same beads are filled in different size capsules would offer a more streamlined approach compared with development of different size matrix tablets. There are several molecule specific characteristics that can be improved using CDD:

- Enable longer duration of action
- Maintain plasma concentration within therapeutic range
- Target drug release in specific regions of the gastrointestinal tract (GIT)

The most common reason for the use of CDD systems is to increase the duration of action (effect) of a drug. This is especially for drugs that have short half-lives and require multiple doses per day. It is a well-accepted fact, drugs that require frequent dosing decrease patient convenience and adherence [4–6]. The conversion of memantine IR formulation to extended release (ER) formulation (Namenda XR®) is a more convenient dosage form for Alzheimer's disease patients [7], whose underlying disease may be a big contributor for a patient missing to take their dose.

Modifying the pharmacokinetics of a drug can improve its efficacy and safety. The therapeutic index (ratio of a drug dose that causes the therapeutic effect to the dose that causes toxicity) varies for drugs. To ensure optimal safety and efficacy, the peak (and trough) must remain within the therapeutic index. Vitamin D insufficiency and secondary hyperparathyroidism (SHPT) are associated with loss of bone density and higher rates of fractures in patients with chronic kidney disease (CKD) [8]. Intermittent, high-dose regimens of IR calcifediol formulations fail to achieve meaningful reduction in parathyroid hormone (PTH) due to upregulation of CYP241 caused by spikes in serum 25-hydroxyvitamin D concentration. The ER calcifediol formulation achieves a gradual increase in serum 25-hydroxyvitamin D producing more greater and more prolonged suppression of plasma PTH in CKD patients without upregulated CYP24A1 expression [9]. The delivery of potent and poorly soluble ER calcifediol was enabled through the combination of a semisolid waxy matrix [10, 11] in a novel type polysaccharide-based soft capsule shell that can encapsulate fill materials that have a high melting temperature [12, 13].

CDD systems, especially delayed release formulations, are ideal for drugs whose release needs to be targeted at a specific region of the GIT. Different enteric coatings have been used to specifically deliver mesalazine at different regions of the GIT as shown in Table 5.1 [7, 14, 15]. Because of the heterogeneity in the pathological regions seen in patients with inflammatory bowel disease, the choice of mesalazine formulation is individualized for the patient.

Table 5.1 Target sites of enteric-coated marketed mesalazine drugs.

Brand name	Enteric coating polymer	Target site release			
		Duodenum	Ileum	Colon	Rectum
Asacol®	Eudragit® S		X	X	
Asacol HD	Eudragit S, Eudragit L		X	X	
Salofalk®	Eudragit L®, Eudragit E		X	X	
Salofalk Granu-Stix®	Eudragit L100, Eudragit E40D			X	X
Claversal®	Eudragit S		X	X	
Pentasa®	Ethylcellulose microgranule	X	X	X	

Source: Refs. [7, 14, 15].

Many drugs are unstable in the acidic pH of the stomach. Studying the stability of a drug in acidic conditions (simulated gastric fluid or pH 1.2) should be standard in every preformulation program. These drugs are excellent candidates to be formulated as CDD systems. Omeprazole, a proton pump inhibitor, is acid labile. The marketed dosage form is an enteric coated tablet that releases drug in the small intestine. As with several other antibiotics, the macrolide erythromycin is acid-labile, and in low pH conditions undergoes intramolecular dehydration reaction resulting in an inactive and more toxic form (anhydroerythromycin A) [16]. The importance of macrolide antibiotics spurred the development of the analogues, azithromycin and clarithromycin, which have improved acidic stability compared to erythromycin.

Certain indications may require that the CDD dosage form has an IR feature followed by an increased duration of action. The IR feature is usually achieved using a loading dose to reach therapeutic concentrations. If the dosing requires multiple daily doses, the subsequent doses do not need a loading dose because they are simply maintaining the steady-state therapeutic drug concentration. A once-daily formulation such as zolpidem tartrate ER tablets (Ambien® CR) is designed for a biphasic absorption profile: an IR component induces rapid onset of sleep and the ER component provides the maintenance dose for longer duration of action. The elimination half-life of zolpidem is 2.8 hours; therefore, the drug does not accumulate [17, 18] and lead to excessive sleepiness during the day. Methylphenidate HCl ER tablets (Concerta®) use an osmotically active CR core with an IR drug overcoat to also achieve rapid onset followed by complete elimination to have a 12-hour duration of action [15]. A unique approach is the fixed-dose combination of memantine hydrochloride ER and donepezil hydrochloride (Namzaric®) [19]. Similar principles are achieved for some diltiazem formulations, including a combination of IR and ER beadlets (Tildiem SPC) [20].

5.3 Defining the Product Profile

The quality target product profiles (QTPP) are essential elements for product development because they describe the design criteria for the drug product (patient needs) and form the basis for development (CDD types, technologies, and dosage forms) (Figure 5.1) [22]. In addition, QTPP provide an understanding of factors that will ensure quality, safety, and efficacy of the drug product for the patient.

The need for a robust QTPP is amplified by the fact that one of the biggest challenges in the development of not just a CDD drug product, but also any drug product, is reproduction of dissolution

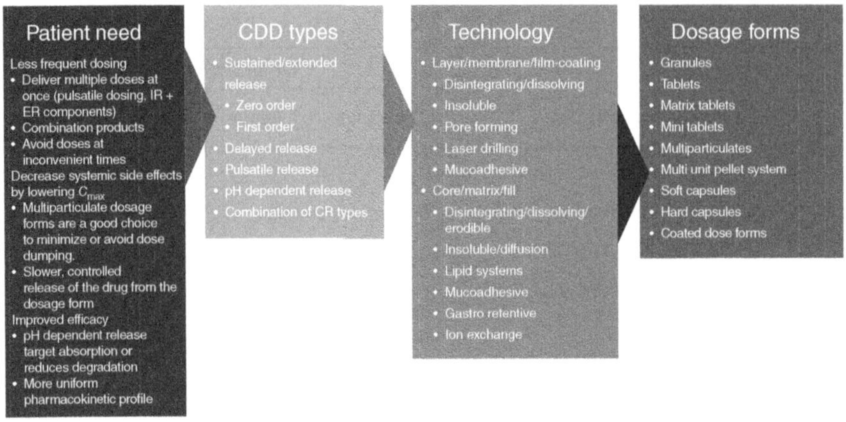

Figure 5.1 Decision tree process for selection of a CDD technology. Source: Modified from Thombre [21].

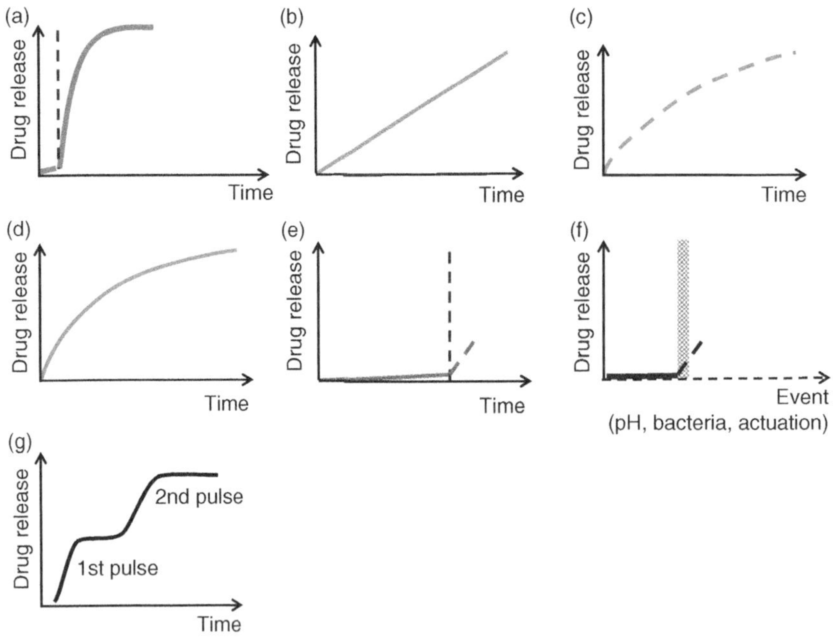

Figure 5.2 Dissolution profiles for CDD systems. (a) Immediate release, (b) zero-order, (c) pseudo zero-order, (d) first-order, (e) delayed release, (f) stimuli-dependent release, and (g) pulsatile release.

profiles, which can be affected by changing manufacturing sites and small changes in performance of release controlling polymers. Dissolution methods are an integral part of the FDA's Scale Up and Post-Approval Changes (SUPAC) Modified Release Guidance [23]. Every regulatory authority will request details of dissolution method development, dissolution specification, and clinical relevance. Some of the CDD dissolution profiles are shown in Figure 5.2.

Addressing unmet patient needs or improving patient therapy is the most commonly communicated reason for pursuing CDD formulations. A list of patient benefits along with CDD dosage form options are highlighted in Figure 5.1. The once or twice daily dosing regimen offered by CDD dosage forms can improve patient acceptance and adherence across all patient groups [5, 24]. The major patient groups who stand to benefit most from CDD are geriatrics and pediatrics [25–32]. There are several pediatric and geriatric considerations that need to be designed into CDD dosage forms including dosage form and dose flexibility, ease of dosing, and palatability. CDD dosage forms are typically larger to contain a larger dose, controlled release excipients, and other excipients and may present a more significant swallowability challenge in these groups affecting the adherence benefit of a reduced dosing regimen. In addition, there may be certain population-specific excipient safety concerns with pediatrics. It would save both time and resources to avoid studying drug-excipient compatibility (discussed later in the chapter) with those excipients not likely to make it into the final dosage form. Further considerations about building in patient factors and medical needs into CDD dosage forms are discussed in the following paragraphs.

Nearly 1 in 10 adults in the United States suffers from chronic pain, which can limit daily function and negatively impact quality of life [33]. There is insufficient/low-quality evidence to guide clinicians to select a short-acting vs. a long-acting opioid [34, 35]. However, their use is associated with increased risk of fatal overdose and misuse and abuse leading to addiction. With the recent rise in opioid abuse [36], there is increasing interest to combine abuse deterrent and sustained release technologies [37, 38]. Special considerations for this application include limiting the release of the high drug load if the patient chews, grinds, or tried to extract drug using solvents [39]. Multiple approaches have been developed to create abuse deterrent properties: physical and chemical barriers, agonist/antagonist combination, aversive agents, prodrugs, and sustained release injections and implants [40–42]. These requirements create incremental complexity to the design and development of an efficient and approvable opioid CDD.

An important consideration in the design of a CDD dosage form is the FDA required pediatric studies under the Pediatric Research Equity Act (PREA). Before the PREA was enacted, more than 80% of drugs approved for use in adults were being used in children without fully understanding the efficacy, safety, and dosing requirements for children. A large tablet size, seen with many CDD dosage forms, would not be appropriate for pediatric patients. Pediatric patients are dosed based on weight, so there is a greater need for dosing flexibility. For oral solid dosage forms, this is usually accomplished by manipulating the dosage form. CDD tablets and capsules may lose their release properties if broken, crushed, or opened. Physiological and metabolic differences in children may have significant impact on the biopharmaceutical classification of the drug and their pharmacokinetics [43–45]. A multiparticulate CDD formulation would be suitable for infants as they can be sprinkled on applesauce or mixed with liquids.

5.4 Preformulation and Physicochemical Characterization Underpinning Development of CDD

Essentially, preformulation is preparative work for formulation and dosage form development. The goal of formulation prescreening or preformulation is minimization of risk, costs, and time during formulation development. These goals are accomplished by studying fundamental physical and chemical characteristics of a drug, alone and in combination with excipients, and assessing the feasibility of creating a CDD system. Some common preformulation studies that are critical in development of CDD systems are listed in Table 5.2. In the previous sections, we described clinical and pharmaceutical situations that are likely to gain the most benefit from CDD and the importance and strategies to define the product profile. The physicochemical characterization techniques should be quick, focused on expected dosage forms, predict processing problems, require minimal quantities of API, and use easily available analytical instruments [48].

Table 5.2 List of common preformulation activities.

Physicochemical properties
Solubility • Physiological pH range • Pharmaceutical and chemical solvents Solid-state chemistry • Polymorphs and crystallinity • Salt forms Excipient compatibility
Stability • Solid state (temperature, temperature/relative humidity, oxygen, light, acid, and base) • Solution state (pH, temperature, oxygen, light)
Ionization properties, p*K*a, partitioning
Wettability
Particle size distribution
Dissolution • Dosage form method development • Intrinsic Permeability
Particulate and mechanical properties
Particle size and shape analysis
Bulk powder properties • Powder density • Angle of repose • Powder flow Compression

Source: Refs. [46, 47].

Development of a CDD is typically undertaken during late development stages or postapproval. One of the reasons is that CDD development requires a lot of API (5–7 g), which may not be available during the early stages. The positive aspect of this re-formulation approach is that formulators already have access to robust physicochemical, biopharmaceutical, and pharmacokinetic information about the drug. There should be minimal, if any, additional preformulation experiments needed to assess a molecule's feasibility of being formulated as a CDD dosage form. The formulator's role in this instance is not necessarily to collect additional data, but to utilize the already collected data to assess the feasibility and plan the development of the CDD. The formulator's role at a generic company or one that is developing a 505(b)2 reformulation will involve more preformulation experiments. Besides physicochemical and biopharmaceutical data, the formulator needs to know the drug's efficacy, potency, and safety. Figure 5.3 shows the simplified decision tree that a formulator may use when selecting a CDD system.

5.4.1 Feasibility and Risk Assessment

Although there are many factors that contribute to the success or failure of CDD systems, the most prominent ones are dose, dose/solubility ratio, stability, regional permeability, elimination half-life, and metabolism [21]. These factors are comprehensively listed in Table 5.3. This chapter will cover some of the most important factors in detail. CDD drug products with doses in the range

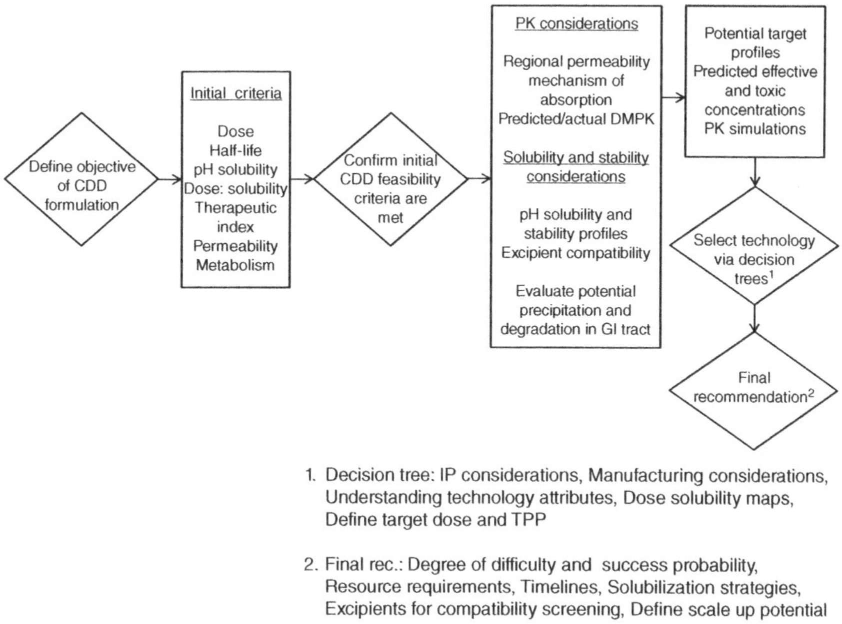

Figure 5.3 Linking patient needs to CDD technology and dosage forms.

of 10–250 mg are ideal as they do not present uniformity challenges and can likely be formulated into a single, swallowable dosage unit. Large doses are indicative of poor molecule solubility (high dose/solubility ratio, poor absorption, etc.). Drugs with low dose/solubility ratios are more feasible to formulate as CDD formulations and have more technology options. Earlier in the chapter, it was mentioned that once and twice daily doses improve patient adherence. Dosage unit size and number of dosage units per dose are other factors that impact patient adherence. The large amounts of excipients that are required to make a CDD increase the dosage unit size and may present swallowability challenges [50, 51].

Because of the ER and absorption profile of CDD formulations, drugs are more likely to interact with food. Drugs that experience significant food effect, either positive or negative, present higher risk for CDD development. One classic therapeutic category would be targeted anticancer drugs [52, 53]. The food effect would introduce unpredictability of pharmacokinetics and increase the challenges of obtaining the desired plasma profile for the drug product. CDD formulations that rely on pH for drug release may not behave as predicted with co-administration with food because food increases stomach pH from 1–2 to pH 5–7. The pH increase is followed by increased gastric secretion and gastric emptying lowers the duodenal pH. Mesalamine delayed release tablets [15] and fixed dose combination of doxylamine succinate and pyridoxine HCl delayed release tablets [35] are labeled to be taken on an empty stomach.

5.4.2 Solubility and Dissolution Rate

Solubility is an inexact scientific value and solubility measurement is an inexact science. A soluble compound is described differently depending on the source. According to the United States Pharmacopeia (USP) definition, solubility is characterized by parts of solvent required for one part of solute. The biopharmaceutical classification system (BCS) definition looks to answer if the highest

Table 5.3 Factors influencing feasibility of CDD systems development.

Physicochemical factors		
Dose	<1 mg	Content uniformity issue
	10–250 mg	Average degree of difficulty
	≫250–300 mg	Multiple dosage units may be needed
Dose/solubility ratio (highest dose:lowest solubility in the pH range 1–7.5)	<1 ml	• Straightforward CDD development • Multiple technology options exist • May need a more lipophilic salt form if high solubility causes very fast dissolution
	1–100 ml	Average degree of difficulty
	100–1000 ml	CDD is challenging but feasible
	>1000 ml	Need solubilization
	>10 000 ml	CDD is not recommended
Stability	Generally stable as a solid or solution and with common CR excipients	Predict average degree of difficulty
	Compound shows or is predicted to have significant degradation	Predict higher degree of difficulty
Biopharmaceutical factors		
Regional colonic absorption	Good absorption across entire GIT • $P_{app, Caco-2} > 10^{-5}$ cm/s, $k_a > 0.01$ min^{-1} • Frel$_{colon} > 60\%$ • BCS Class I and stable in colon chime (<1% degraded in 1 h)	CDD development likely to be feasible
	Risk for poor colon absorption • $P_{app, Caco-2} = 10^{-6}$–$10^{-5}$ cm/s • Frel$_{colon}$ 30–60% • High permeability after efflux inhibition or intermediate solubility or rapid degradation in colon chime (>10% degraded in 1 h)	CDD development is challenging but feasible
	Poor colonic absorption • $P_{app, Caco-2} < 10^{-6}$ cm/s, $k_a < 0.01$ min^{-1} • Frel$_{colon} < 30\%$ • Low permeability according to BCS or volume to dissolve highest dose at pH 5.5–7.5 is 0.10 l	CDD development is not recommended Nontraditional approaches may be helpful
Pharmacokinetic factors		
PK or PD half-life	<1–2 h	Half-life too short
	2–10 h	Acceptable half-life
	≫10 h	Compound might not need CDD
Metabolism and efflux	High presystemic metabolism	Relative BA might be low
	P-gp or CYP3A4 substrate	CDD performance is difficult to predict

Source: Refs. [21, 49].

Table 5.4 Solubility predictions using different *in silico* methods.

Models	Descriptors	Method	Huuskonen data (r^2)	Solubility Challenge test (r^2)
Pipeline Pilot v7.5: solubility	Electrotopological indices	ANN	0.86	0.29
Pipeline Pilot v7.5: ADMET solubility	Atomic, topological	Genetic algorithm, multiple linear regression (MLR)	0.83	0.28
MOE v2010.10	Fragments, 76 atom types	Fragment-based, MLR	0.91	0.44
ACD lab v 12.0	Atomic	Experiments w/QSPR (quantitative structure activity relationship)	0.92	0.53
QikProp v3.4	2D and 3D	Experiments w/QSPR	0.88	0.34
ADMET Predictor v6	Atomic	Three components partial least square regression (PLS)	0.94	0.32
volsurf+ 1.0.6	3D	QSPR w/3D	0.85	0.44
FAF-Drugs2	Atomic, topological	MLR	0.82	0.18
VCC lab	Electrotopological indices	ANN	0.79	0.39
ISIDA	Fragments	QSPR, clustering	0.93	0.35

dose can be solubilized in 250 ml of aqueous buffers ranging from pH 1.0 to 7.5 [54]. The Developability Classification System looks at the dose/solubility ratio in biorelevant media [55]. Solubility depends on several molecular traits: chemical structure, functional groups, ionizable groups, stereochemistry, and solid-state properties (crystallinity, purity, etc.). As such, there is no single solubility value or measurement method that sufficiently characterizes a drug molecule to make a feasibility decision and determine the degree of difficulty to formulate a drug in a CDD approach. The formulator must use a combination of predictions and measurements to fully understand the solubility behavior of a molecule.

The two most commonly used solubility assessment methods are *in silico* prediction and *in vitro* measurements. There are numerous commercially available solubility prediction tools. Each tool uses a different class of chemical descriptors and analyzes their impact on solubility using different methods. Table 5.4 shows the prediction accuracy for several publicly available data sets. The Huuskonen Data is a set of 1033 compounds with experimentally measured solubilities [56]. The results of the Huuskonen data set are satisfactory for all models due to the likelihood that these data were used for training the model [57]. The Solubility Challenge was a collaborative competition to develop the best model to predict solubility using a high-quality experimental data set [58]. Compared to the Huuskonen data, all 100 models performed poorly to predict the 50 test molecules in the Solubility Challenge [57].

Adequate solubility across the physiological pH range (pH 1–8) is perhaps the most important parameter for feasibility assessment and CDD development. Unlike IR formulations, which are mainly absorbed in the upper GIT, CDD usually depend on significant absorption in the distal small intestine and colon. Due to the broad range of GI residence times, t_{max} for CDD formulations may be reached more than 18 hours after dosing. In order to achieve once daily dosing, the drugs will most likely need to be absorbed at all points along the GIT including the colon [49]. Therefore, the drug should have adequate solubility and permeability and not precipitate along the physiological pH range. Further, the dose/solubility ratio provides more drug-specific insights. Drugs with pH-dependent solubility tend to be poor candidates for matrix based oral CDD dosage forms. A multiparticulate dosage form may be advantageous to maintain constant release of drug throughout the pH gradient of the GI tract [46].

The Noyes–Whitney equation (Eq. (5.1)) relates the impact of drug solubility (C_s) and drug particle surface area (A) on dissolution rate (dC/dt) wherein D is the diffusion coefficient, C is the concentration of drug in the bulk medium, and h is the height of the boundary layer. Regardless of the CDD format, the drug must display appropriate dissolution rate and solubility throughout the physiological pH range. A drug with very rapid dissolution has low feasibility to be formulated as a CDD. A drug that inherently has very slow dissolution may not need to be formulated as a CDD. For matrix systems, the dissolution rate of the drug in the matrix must be greater than the dissolution rate of the drug in the GI fluid to maintain controlled release [59, 60].

$$\frac{dC}{dt} = \frac{DA(C_s - C)}{h} \tag{5.1}$$

Historically, CDD was focused on decreasing the dissolution rate of soluble and highly soluble drugs to attain constant plasma levels using techniques such as creating salt forms with lower solubility [47]. Another approach is to slow the drug's dissolution through granulation with fatty or waxy lipophilic excipients such as stearic acid, castor wax, high molecular weight polyethylene glycol, glycerylmonosterate, or white wax. These oily materials may introduce interpatient variability because of the differences in digestion, temperature, and mechanical stress after administration. The inflation of molecular properties of new molecular entities (NMEs) especially lipophilicity and molecular weight has been well documented [61]. Multiple reports estimate that 60–90% of pipeline NMEs have poor solubility or dissolution rate [61, 62]. These challenging NMEs will require multiple enabling technologies: combination of solubility enhancement and CDD or targeted delivery to achieve adequate bioavailability and pharmacokinetics.

Both the FDA and USP have guidance documents focused on development of dissolution methods for CDD systems. Both specify that dissolution methods include at least three sampling time points and lower and higher limits at each time point. Multistage approaches have been developed to simulate the pH changes as a drug passes through the GI tract. The first step for developing a dissolution method is drug solubility determination in water, 0.1 N HCl, simulated intestinal fluid, and pH 1–8. Solubility in these media provides clues on *in vivo* performance and necessary sink conditions for *in vitro* dissolution testing.

Improving solubility is necessary. Otherwise, there would be more difficulty in CDD development and high doses/bigger or more dosage units would be required. CDD dosage forms contain high levels (30%) of matrix-forming polymers and conventional excipients to make tablets or capsules. Because the rate-controlling mechanisms can be destroyed, tablet splitting, or crushing are contraindicated for CDD dosage forms.

5.4.3 Permeability

After dissolving in the GIT fluid, drugs need to have high permeability to be absorbed in the bloodstream by paracellular, transcellular, or both routes. The paracellular route is the main route for highly hydrophilic molecules since they cannot partition into the cell membrane. The transcellular route can be passive or carrier-mediated. Molecules with a passive transcellular absorption mechanism present an average level of difficulty for CDD formulation. The performance of CDD is difficult to predict with other mechanisms or if the drug is subject to efflux.

Unlike IR formulations, in which drug absorption primarily occurs in the upper GIT, CDD formulations must display adequate permeability throughout the GIT including the colon. The colonic permeability is a factor during feasibility assessment. A gabapentin prodrug, gabapentin enecarbil (XP13512; Horizant®) demonstrated significantly higher intracolonic absorption compared to gabapentin; thus, allowing the development of a sustained release formulation [63]. Gabapentin absorption is limited by solute transporters that have low capacity, limited distribution, and variable expression. Gabapentin enacarbil is absorbed by high-capacity nutrient transporters that are located throughout the intestine. The new prodrug formulation decreases daily doses from three or more to one or two depending on the specific indication [21].

The apparent permeability in Caco-2 cells ($P_{\text{app, Caco-2}}$) and relative colonic bioavailability compared to that of the small intestine ($\text{Frel}_{\text{colon}}$) are listed in Table 5.4. Caco-2 permeability studies are routinely performed, and some degree of correlation has been established for human colonic permeability [23, 64, 65]. There have been investigations to study other cell lines as alternatives to Caco-2 [66, 67].The colon presents a unique environment compared to the upper GIT. It contains a higher number of microbes, most of which are anaerobes. This presents a challenge to study colonic stability of drug *in vitro* and correlate with *in vivo* behavior. There are several approaches to coculture epithelial cells with microbes including on microfluidic chips to better mimic the *in vivo* environment of the GIT [68, 69].

5.4.4 Drug and Drug Product Particle Sizes

In addition to influencing the dissolution rate and bioavailability, the drug particle size can predict drug release rate in CDD, and challenges with content uniformity, stability, and processing [70]. Micronization is a common approach for drug particle size reduction to improve content uniformity. This is reflected in the International Conference on Harmonization (ICH) guideline Q6A [71], which states that a particle size specification is required when the drug particle size is critical to product performance or manufacturability. However, micronized drug particles, especially those with high surface energy, tend to agglomerate. Setting a particle size specification to ensure content uniformity and avoid cohesive agglomeration or a deagglomeration step such as adding excipients (e.g. silicon dioxide), which result in drug product particles, are beneficial approaches [18].

For solid oral dosage forms, the impact of particle size of drug and excipients particles is significant because many of the unit operations involved in the pharmaceutical manufacturing process manipulate particle size. Unit operations such as granulation, roller compaction, and coating are particle size enlargement techniques, whereas milling and grinding are particle size reduction techniques. Homogeneity becomes a bigger challenge when blending or mixing powders of drastically different particle size ranges, and control of particle size distribution may be necessary for raw and in-process materials at the level of different unit operations.

Many CDD products are capsules filled with coated sugar beads/spheres. The manufacture of CDD beads involves coating sugar spheres with layers of drug and CDD polymer. This process essentially increases the particle size of the initial sugar sphere. Monitoring the growth of the particles is directly correlated with coating thickness and may be an effective quality control approach. Because of the small size of the sugar spheres, measuring weight gain is not meaningful. Rather different such as dynamic imaging analysis or simple sieving analysis can offer insights on particle size distribution and size increase. Like with any drug or drug product particle, a wide variance in size distribution will introduce challenges in content uniformity and manufacturing. Setting a particle size specification with acceptance criteria for starting materials and after each step would minimize intra- and inter-batch variability: uncoated sugar spheres (baseline); drug-coated spheres (dose uniformity and precision); CDD polymer coated spheres (drug release); and final blended spheres (drug product performance). The importance of measuring size of drug particles depends on how the drug will be applied to the sugar sphere. If the drug is dissolved in the coating solution, particle size is not required. If the drug is suspended in the coating solvent, drug particle size should be monitored.

There are many techniques to determine particle size and shape [72, 73]. Microscopy and laser diffraction are the most commonly used techniques. The biggest challenge with studying drug particle size is precise and repeatable measurements. Each technique has its own limitations and it would be advisable to employ multiple imaging or measurement techniques. Similar to solubility, precise drug particle size cannot be represented as a single number.

Laser diffraction cannot discriminate between discreet drug particles and agglomerates. Wet and dry dispersion methods can be employed during drug particle size measurements to obtain data on discreet particle size [73]. There may be cases where the formulator may intend to assess particle agglomeration and choose to use reduced or no dispersion to preserve agglomerates.

It is also important to note whether the measurement technique represents data as volume-based distribution vs. in numerical terms. The largest particles primarily influence the size distribution in volume-based methods. The volume-based number below which 90% of particles are measured (D_{90}) is sensitive to subtle changes in milling parameters and has the greatest impact on dissolution rate. On the other hand, a number-based distribution is influenced by the smallest, most abundant particles.

Laser diffraction methods are highly influenced by drug particle shape. The results of laser diffraction are expressed in terms of equivalent sphere to allow a single value for particle size expression [72]. A needle-like drug particle with a length of 100 μm may be converted to an equivalent sphere of 5 μm. A secondary method such as optical microscopy or scanning electron microscopy (SEM) allows particle size measurement in the context of nonspherical drug particles.

Yu et al. [74] studied the particle shapes of over 1000 API. For the data set, they reported the aspect ratio, which is the ratio between the longest and shortest Feret diameters. A sphere would have an aspect ratio of 1.0, whereas a cube would have an aspect ratio of 1.6. The median aspect ratio was between 0.6 and 0.8, which represents a length that is 1.3 times greater than the breadth. The group also studied the sphericity of 282 excipients and found that the excipients had a higher median sphericity value compared with API. This is expected because excipients are designed to have optimal handling properties such as powder flow.

5.4.5 Solid-State Chemistry

Solid-state chemistry, characterization, and research are the crucial backbone in the development of NMEs. Understanding the physical characteristics of a NME and how to manipulate them allow us to optimize product development (e.g. hygroscopicity, chemical stability, compressibility, and flowability) and address biopharmaceutical challenges (solubility and bioavailability). Discovering and patenting different solid forms of a drug molecule offers strategies to maintain product exclusivity or extend patent protection. Although these aspects of solid-state chemistry may not be the most pressing during CDD development, knowledge of solid-state chemistry will focus other preformulation activities such as excipient compatibility.

5.4.5.1 Crystallinity and Polymorphism

Polymorphism occurs when a molecule exists in multiple crystalline forms. An example is how carbon can exist as graphene and diamond. The prevalence in the pharmaceutical space makes polymorph screening an important aspect of any preformulation program. It is estimated that 32–57% of small organic molecules with molecular weight less than 600 Da have true polymorphs. If solvates are included, the prevalence jumps to 56–87% [75].

Polymorphism is a challenge in pharmaceutical development and can greatly impact the solubility and dissolution rate, which are important parameters for controlling drug release rate from CDD systems and drug absorption. Other properties influenced by polymorphism include color, hardness, thermal behavior, physical and chemical stability, and bioavailability. A polymorph screening approach deliberately attempts to create new polymorphs by recrystallizing a drug in different solvents. Understanding all potential polymorphs is important in case new processing approaches or solvents used during manufacturing cause polymorphic changes.

Powder X-ray diffraction (PXRD) is the most commonly used preformulation technique to screen for polymorphs. The typical polymorph screening approach involves placing a drug in different solvents to find the most stable polymorphs. A small-scale screening will look at 8–10 solvents, whereas a large-scale screening will employ more than 50 solvents. Although it is the most common preformulation technique, PXRD cannot provide information on solvation state, melting point, water uptake, solubility, physical stability, and form changes upon drying. A complete solid-state characterization protocol would also include techniques such as differential scanning calorimetry, thermogravimetry, and hot-stage microscopy.

5.4.5.2 Salts

Salts of basic drugs are ionized at the lower pH of the stomach and are not absorbed at a great extent until reaching the neutral or mildly alkaline region of the intestine, where there is a sufficient percentage of unionized drugs.

Different salt forms of a drug have different solubilities. With CDD systems, the intent is to control the solubility and dissolution to allow for prolonged drug release. Highly soluble drugs, such as those in DCS Class I, may benefit from creating a salt form with lower solubility. Recent chemistry techniques and the drive to develop highly targeted drugs have introduced drugs and drug candidates with very poor solubility profiles. These drugs still require solubility enhancement to be made into CDD systems, but the magnitude of solubility increase may not need to be as high as if an IR formulation was being developed. Metoprolol is a classic example [76, 77]. The tartrate salt has water solubility greater than 700 mg/ml and is used in the IR formulation, which is dosed twice daily. The succinate salt is used in the once daily ER formulation and has water solubility around 200 mg/ml.

5.4.6 Stability

A drug's stability decreases when it is combined with excipients, and the stability continues to decline as the drug concentration decreases. Studying a drug's chemical liability in both solution and solid states followed by compatibility with excipients forms the basis of formulation quality by design (QbD). Table 5.5 summarizes solution- and solid-state forced degradation studies. These studies are extremely valuable to design manufacturing and processing conditions. Drugs that degrade in response to high temperatures should not be formulated using hot melt extrusion. Wet granulation is a common manufacturing process for high dose drugs; however, for moisture sensitive drugs, fluid bed granulation offers an option to minimize time the drug is in contact with water.

Table 5.5 Forced degradation studies.

Solution state forced degradation studies		
Study type	**Medium and conditions**	**Duration**
Acid	0.1 N HCl, 50–121 °C (10 °C increments)	Daily up to 1 wk
Base	0.1 N NaOH, 50–121 °C (10 °C increments)	Daily up to 1 wk
Heat	Water, 50–121 °C (10 °C increments)	Daily up to 1 wk
UV light	Water, 1.2 million lux hours at 25 °C	22 h and 15 min (ICH option 2)
Fluorescence light	Water, 1000 Wh/m^2 at 25 °C	10 d and 8 h (ICH option 2)
Oxidation 1	3–30% H_2O_2 at 25 °C	Daily up to 1 wk
Oxidation 2	Water, pressurized O_2 with free radical initiator at 25 °C	Daily up to 1 wk

Solid state forced degradation studies		
Study type	**Conditions**	**Duration**
Dry heat	56 °C open and closed bottles	Weekly up to 1 mo
Moist heat	40 °C/75% RH open and closed bottles	Weekly up to 1 mo
UV light	1.2 million lux hours at 25 °C, open dish	22 h and 15 min (ICH option 2)
Fluorescence light	1000 Wh/m^2 at 25 °C, open dish	10 d and 8 h (ICH option 2)
Oxidation	Pressurized O_2 with free radical initiator at 25 °C	Daily up to 1 wk

Source: Ku [46]. © 2010 John Wiley & Sons.

Spectroscopy, chromatography, and thermal analysis are among the variety of analytical technologies used during stability testing. Chromatography techniques (HPLC, GC, and HP-TLC) are the most sensitive techniques with typical detection limits for impurities in the range of 0.01–0.1% [46]. The least sensitive methods include PXRD, IR/Raman, and nuclear magnetic resonance (NMR).

The ICH Q1A guideline defines the stability data package required for a new drug substance or drug product during the registration application [78]. The guideline is aimed to provide evidence of how environmental factors (temperature, humidity, and light) influence quality of drug substance or drug product, establish shelf life for a drug product, and recommend drug product storage conditions. In regard to CDD approaches of re-formulation, the drug substance stability data should be available from the IR formulation. According to ICH Q1C [79], the parent stability guideline should be followed for CDD dosage form; however, a reduced stability data package may be acceptable on a case-by-case basis.

5.4.7 Excipient Compatibility

Despite the ubiquity of inactive ingredients/excipients in oral dosage forms, there is no standard method for testing drug excipient compatability and most reported test methods have poor predictive value for the stability of the final dosage form. Excipient compatibility testing protocols suffer from multiple shortcomings that limit their full potential: poor sample uniformity, unrealistic drug-excipient ratio, exclusion of key excipients, testing compatibility with a single or binary mixture of excipients, etc. To limit the number of samples and resources, compatability studies do not test excipient blends that fully replicate the final dosage form. Excipient compatibility studies are time- and resource-consuming. It is common that compatibility testing will be performed using a dozen or more excipients and require preparation of several dozen samples.

The general approach is to begin by conducting a literature search to find if there are any known incompatibilities between functional groups of the drug and excipient. Next step would be compatibility screening with a single excipient at ratios that simulate finished dosage forms. The binary mixture (drug and excipient) can be placed in open bottles at 40 °C/75% RH for two and four weeks for study and at 5 °C for troubleshooting. High-performance liquid chromatography (HPLC) is the most studied method to get the baseline and assess drug recovery at the study time points. The relative retention times on the HPLC column are used to quantify percent potency and impurity. Excipients demonstrating any incompatibilities can be eliminated. The drug alone would serve as the control and ideally, the drug would be stable by itself. Some protocols may call for addition of 5–20% water since moisture is a trigger for many degradation pathways.

5.4.8 Bulk Powder Properties

Assessing bulk powder properties is important to predict any manufacturing challenges that may arise due to poor powder flow. Bulk density is also an important parameter for determining the size of the final dosage form. Bulk powders are complex systems because they contain millions of nonidentical particles that undergo unique arrangements. The surface characteristics of drug and excipient particles are influenced by physicochemical properties, spatial arrangement, and orientation.

There are several manufacturing processes that are influenced by powder flow and bulk density. Direct compression is a simple and low-cost processing method for manufacturing matrix tablets. At a commercial scale, it is vital that the API and excipients are homogeneous, free flowing, and compressible. Often, wet and dry granulation methods need to be employed to reach necessary homogeneity and flow properties of the powder blend prior to direct compression. Other manufacturing steps such as blending, conveying, and encapsulation are impacted by bulk powder properties.

The Carr's (compressibility) index [80] and Hausner ratio [81] are commonly used calculations to describe flowability by comparing taped density and pour density [82, 83]. Interpretation of values for these tests are listed in Table 5.6. Despite being also called compressibility index, the evaluation of the Carr's index using tapped density tester does not apply compression forces. The particles are repacked under a tapping motion (some testers also have a rotary motion). Powders with a Carr's Index >33% have poor powder flow, whereas those with Carr's Index <20% have good powder flow (Table 5.5). There are caveats to using Carr's index in the cases of powders with extreme flow properties. The Carr's index may be zero for both powders with perfect flow and those with very poor flow that fail to consolidate. Therefore, it is important to use Carr's index in conjunction with other tests.

$$\text{Carr's index} = 100\% \times \frac{\text{tapped density} - \text{pour density}}{\text{tapped density}}$$

$$\text{Hausner ratio} = \frac{\text{tapped density}}{\text{bulk density}}$$

The angle of repose is another measure of powder flow. It provides insights into how a powder bed responds to stresses such as agitation or air force. Typically, a fixed amount of powder is poured onto a flat surface into a conical heap. The measured incline angle of the heap is the angle of repose. The interpretation of values is provided in Table 5.5.

Table 5.6 Bulk powder test results.

Test	Results interpretation	Extreme cases
Carr's index	>33% – poor powder flow <20% – good powder flow	• Powders with perfect flow and those with very poor flow may have Carr's index of zero
Angle of repose	>40% – poor powder flow <25% – good powder flow	• Powder with perfect flow have an angle of repose near 0° • Powders with poor flow may have an angle of repose >90°
Flow meter	$\alpha/\alpha_{ref} > 0.9$ – very good α/α_{ref} 0.8–0.9 – good α/α_{ref} 0.7–0.8 – satisfactory α/α_{ref} 0.6–0.7 – medium α/α_{ref} 0.5–0.6 – unsatisfactory α/α_{ref} 0.4–0.5 – poor $\alpha/\alpha_{ref} < 0.4$ – very poor	

The European Pharmacopeia [84] has a monograph for measuring powder flowability using a flow meter. This is a direct way to measure the rate of powder flow. It is important to note that the powder must be highly flowable to give a measurable flow rate using the traditional glass funnel approach. Commercial flow testers use vibration and tapping to induce flow.

Newer CDD approaches such as semisolid lipid matrices encapsulated in soft gelatin capsules do not face powder flow challenges, but the flow of the fill formulation during softgel encapsulation is a major factor. Compared with liquid fill formulations, semisolid fill formulations have higher melting temperatures. The semisolid formulation must be heated to make it free flowing to be encapsulated into soft gelatin capsules using a rotary die process.

5.4.9 Drug Metabolism and Pharmacokinetic Modeling

5.4.9.1 Guiding the Design of CDD Dosage Forms

The power of pharmacokinetic modeling is greatly influenced by the amount and quality of data. In the early development setting for the NME, a lot of the data may not be available, available in only preclinical animal species, or may need to be estimated based on similar molecules. Some of the factors needed for modeling include dose, solubility, permeability, half-life, site of absorption, therapeutic window, pH sensitivity, food effect, log P, pK_a, solubility in fasting and fed state simulated intestinal fluids (FaSSIF and FeSSIF), unbound fraction in human plasma, ratio of blood to plasma, intrinsic clearance (Cl_{int}), and fraction absorbed. Since CDD dosage forms are usually an evolution of IR drug product, essentially these factors will be known by the innovator company. Because of the ER nature of the product, there will be slower absorption of the drug (k_a), a longer time to maximum concentration (t_{max}) and the peak plasma concentration (C_{max}) will be lower. However, the area under the plasma drug concentration curve (AUC) should be the same for a properly formulated CDD product as the IR product. The t_{max} is important for a CDD product because it influences dosing interval. A CDD product with a t_{max} that is reached within a couple of hours after dosing will not suffice for a once or twice daily dosing regimen. A high C_{max} value may indicate inadequate formulation or alcohol induced dose dumping [38].

5.4.9.2 Establishing *In Vitro–In Vivo* Correlation (IVIVC)

In vitro dissolution tests and preclinical animal studies are the traditional preclinical tests to evaluate the performance of CDD. The main goal of these studies is to establish *in vitro–in vivo* correlation (IVIVC) for the dissolution test, study the mechanism of release, and select the prototype for human studies. Advances in physiologically based pharmacokinetic (PBPK) modeling have resulted in greater incorporation of these tools to guide CDD formulation development [85].

5.4.9.3 Physiologically Based Pharmacokinetic (PBPK) Modeling Tools

Given the increased complexity of CDD formulations compared with IR formulations, additional details such as simulating absorption in the lower intestines and region-dependent absorption must be incorporated. There are numerous modeling tools commercially available such as GastroPlus™ (Simulations-Plus, Inc.) and Phoenix WinNonlin (Certara, L.P.). Simulation tools can be particularly useful to predict plasma profiles based on *in vitro* dissolution data, dose, and clearance and to deconvolute data from pharmacokinetic studies.

The advanced compartmental absorption and transit (ACAT) model in GastroPlus explicitly models every relevant process involved in drug absorption such as drug release from formulation, first-pass effect in the gut and liver, and saturable carrier-mediated transport [86]. The ACAT was originally designed by Yu et al. [87] and enhanced to include pH-dependent solubility and effect on dissolution rate, gut and liver metabolism, absorption from stomach and colon, and other regional factors.

For 505(b)(2) applications, using PK modeling tools to see if a CDD is a viable option, if it offers further PK/PD advantages, and to predict if the CDD meets the 505(b)(2) bridging bioequivalence requirements using results from a single dose study provide both cost and time savings. The latter advantages would be beneficial for companies seeking a 505(b)2 approach for the CDD dosage form.

5.5 Conclusion

CDD is a valuable tool to develop drugs to address unmet clinical needs, overcome less than optimal drug characteristics, and open more commercial opportunities. The successful development of CDD is based on a strong preformulation strategy to characterize drug molecules and use the information to assess feasibility and minimize risks, costs, and time.

References

1 Hu, M. (2014). Early investment in controlled release: discovery controlled release assessment and formulation strategies. *Controlled & Modified Drug Release*, Philadelphia, PA (7–8 May 2014).

2 Rovner, E.S. and Wein, A.J. (2002). Once-daily, extended-release formulations of antimuscarinic agents in the treatment of overactive bladder: a review. *Eur. Urol.* 41 (1): 6–14.

3 Saxena, V., Panicucci, R., Joshi, Y., and Garad, S. (2009). Developability assessment in pharmaceutical industry: an integrated group approach for selecting developable candidates. *J. Pharm. Sci.* 98 (6): 1962–1979.

4 Osterberg, L. and Blaschke, T. (2005). Adherence to medication. *N. Engl. J. Med.* 353 (5): 487–497.

5 Coleman, C.I., Limone, B., Sobieraj, D.M. et al. (2012). Dosing frequency and medication adherence in chronic disease. *J. Manag. Care Pharm.* 18 (7): 527–539.

6 Srivastava, K., Arora, A., Kataria, A. et al. (2013). Impact of reducing dosing frequency on adherence to oral therapies: a literature review and meta-analysis. *Patient Prefer. Adherence* 7: 419–434.

7 Plosker, G.L. (2015). Memantine extended release (28 mg once daily): a review of its use in Alzheimer's disease. *Drugs* 75 (8): 887–897.

8 Dusso, A.S. (2011). Kidney disease and vitamin D levels: 25-hydroxyvitamin D, 1,25-dihydroxyvitamin D, and VDR activation. *Kidney Int. Suppl.* 1 (4): 136–141.

9 Petkovich, M., Melnick, J., White, J. et al. (2015). Modified-release oral calcifediol corrects vitamin D insufficiency with minimal CYP24A1 upregulation. *J. Steroid Biochem. Mol. Biol.* 148: 283–289.

10 Bishop, C.W., Tabash, S.P., Agudoawu, S.A. et al. (2013). Methods and compositions for controlled release oral dosage of a vitamin D compound. Google Patents US8778373B2.

11 White, J.A., Tabash, S.P., Agudoawu, S.A., and Melnick, J.Z. (2018). Stabilized modified release vitamin D formulation and method of administering same. Google Patents US9861644B2.

12 Tanner, K.E., Draper, P.R., Getz, J.J. et al. (2002). Film forming compositions comprising modified starches and iota-carrageenan and methods for manufacturing soft capsules using same. Google Patents US6582727.

13 Tanner, K.E., Shelley, R.S., Stroud, N.S., and Youngblood, E. (2012). Non-gelatin soft capsule system. Google Patents US8231896B2.

14 Lautenschlager, C., Schmidt, C., Fischer, D., and Stallmach, A. (2014). Drug delivery strategies in the therapy of inflammatory bowel disease. *Adv. Drug Delivery Rev.* 71: 58–76.

15 Ding, H. (2016). Modified-release drug products and drug devices. In: *Applied Biopharmaceutics & Pharmacokinetics*, 7e (eds. L. Shargel and A.B.C. Yu). New York, NY: McGraw-Hill Education.

16 Cyphert, E.L., Wallat, J.D., Pokorski, J.K., and von Recum, H.A. (2017). Erythromycin modification that improves its acidic stability while optimizing it for local drug delivery. *Antibiotics* 6 (2).

17 Kirkwood, C., Neill, J., and Breden, E. (2007). Zolpidem modified-release in insomnia. *Neuropsychiatr. Dis. Treat.* 3 (5): 521–526.

18 Huang, C.Y. and Sherry, K.M. (2010). Prediction of drug particle size and content uniformity in low-dose solid dosage forms. *Int. J. Pharm.* 383 (1–2): 70–80.

19 Boinpally, R., Chen, L., Zukin, S.R. et al. (2015). A novel once-daily fixed-dose combination of memantine extended release and donepezil for the treatment of moderate to severe Alzheimer's disease: two phase I studies in healthy volunteers. *Clin. Drug Invest.* 35 (7): 427–435.

20 Tildiem, L.A. (2020). 200mg prolonged-release capsules. electronic Medicines Compendium (eMC). www.medicines.org.uk/emc/product/4971/smpc#PHARMACOKINETIC_PROPS (accessed 20 November 2020).

21 Thombre, A.G. (2005). Assessment of the feasibility of oral controlled release in an exploratory development setting. *Drug Discovery Today* 10 (17): 1159–1166.

22 Maguire J, Peng D.. How to identify critical quality attributes and critical process parameters. *FDA/PQRI 2nd Conference,* North Bethesda, MD (6 October 2015). http://pqri.org/wp-content/uploads/2015/10/01-How-to-identify-CQA-CPP-CMA-Final.pdf.

23 Yee, S. (1997). In vitro permeability across Caco-2 cells (colonic) can predict in vivo (small intestinal) absorption in man--fact or myth. *Pharm. Res.* 14 (6): 763–766.

24 Weeda, E.R., Coleman, C.I., McHorney, C.A. et al. (2016). Impact of once- or twice-daily dosing frequency on adherence to chronic cardiovascular disease medications: a meta-regression analysis. *Int. J. Cardiol.* 216: 104–109.

25 Lopez, F.A. and Leroux, J.R. (2013). Long-acting stimulants for treatment of attention-deficit/hyperactivity disorder: a focus on extended-release formulations and the prodrug lisdexamfetamine dimesylate to address continuing clinical challenges. *Atten. Defic. Hyperact. Disord.* 5 (3): 249–265.

26 Caballero, J., Darsey, E.H., Walters, F., and Belden, H.W. (2017). Methylphenidate extended-release oral suspension for the treatment of attention-deficit/hyperactivity disorder: a practical guide for pharmacists. *Integr. Pharm. Res. Pract.* 6: 163–171.

27 Craig, W.A. (2004). Overview of newer antimicrobial formulations for overcoming pneumococcal resistance. *Am. J. Med.* 117 (Suppl 3A): 16S–22S.

28 Arguedas, A., Soley, C., Kamicker, B.J., and Jorgensen, D.M. (2011). Single-dose extended-release azithromycin versus a 10-day regimen of amoxicillin/clavulanate for the treatment of children with acute otitis media. *Int. J. Infect. Dis.* 15 (4): e240–e248.

29 Falagas, M.E., Karagiannis, A.K., Nakouti, T., and Tansarli, G.S. (2015). Compliance with once-daily versus twice or thrice-daily administration of antibiotic regimens: a meta-analysis of randomized controlled trials. *PLoS One* 10 (1): e0116207.

30 Perucca, E. (2009). Extended-release formulations of antiepileptic drugs: rationale and comparative value. *Epilepsy Curr.* 9 (6): 153–157.

31 Herranz, J.L., Arteaga, R., Adin, J., and Armijo, J.A. (2006). Conventional and sustained-release valproate in children with newly diagnosed epilepsy: a randomized and crossover study comparing clinical effects, patient preference and pharmacokinetics. *Eur. J. Clin. Pharmacol.* 62 (10): 805–815.

32 Eeg-Olofsson, O., Nilsson, H.L., Tonnby, B. et al. (1990). Diurnal variation of carbamazepine and carbamazepine-10,11-epoxide in plasma and saliva in children with epilepsy: a comparison between conventional and slow-release formulations. *J. Child Neurol.* 5 (2): 159–165.

33 Nahin, R.L. (2015). Estimates of pain prevalence and severity in adults: United States, 2012. *J. Pain* 16 (8): 769–780.

34 Chou, R., Clark, E., and Helfand, M. (2003). Comparative efficacy and safety of long-acting oral opioids for chronic non-cancer pain: a systematic review. *J. Pain Symptom Manage.* 26 (5): 1026–1048.

35 Chou, R., Fanciullo, G.J., Fine, P.G. et al. (2009). Clinical guidelines for the use of chronic opioid therapy in chronic noncancer pain. *J. Pain* 10 (2): 113–130.

36 Skolnick, P. (2018). The opioid epidemic: crisis and solutions. *Annu. Rev. Pharmacol. Toxicol.* 58 (1): 143–159.

37 Michna, E., Kirson, N.Y., Shei, A. et al. (2014). Use of prescription opioids with abuse-deterrent technology to address opioid abuse. *Curr. Med. Res. Opin.* 30 (8): 1589–1598.

38 Jedinger, N., Khinast, J., and Roblegg, E. (2014). The design of controlled-release formulations resistant to alcohol-induced dose dumping--a review. *Eur. J. Pharm. Biopharm.* 87 (2): 217–226.

39 U.S. Food and Drug Administration (2015). *Guidance for Industry: Abuse-Deterrent Opioids—Evaluation and Labeling*. Silver Spring, MD: US Department of Health and Human Services.

40 Litman, R.S., Pagan, O.H., and Cicero, T.J. (2018). Abuse-deterrent opioid formulations. *Anesthesiology* 128 (5): 1015–1026.

41 Maincent, J. and Zhang, F. (2016). Recent advances in abuse-deterrent technologies for the delivery of opioids. *Int. J. Pharm.* 510 (1): 57–72.

42 Jones, M.R., Carney, M.J., Kaye, R.J. et al. (2016). Drug formulation advances in extended-release medications for pain control. *Curr. Pain Headache Rep.* 20 (6): 36.

43 Gandhi, S.V., Rodriguez, W., Khan, M., and Polli, J.E. (2014). Considerations for a Pediatric Biopharmaceutics Classification System (BCS): application to five drugs. *AAPS PharmSciTech.* 15 (3): 601–611.

44 Norris, R.E. and Adamson, P.C. (2012). Challenges and opportunities in childhood cancer drug development. *Nat. Rev. Cancer* 12 (11): 776–782.

45 Batchelor, H.K., Fotaki, N., and Klein, S. (2014). Paediatric oral biopharmaceutics: key considerations and current challenges. *Adv. Drug Delivery Rev.* 73: 102–126.

46 Ku, S. (2010). Preformulation consideration for drugs in oral CR formulation. In: *Oral Controlled Release Formulation Design and Drug Delivery: Theory to Practice* (eds. H. Wen and K. Park), 47–70. John Wiley & Sons, Inc.

47 Chrzanowski, F.A. and Ahmad, K. (2017). The preparation and evaluation of salt forms of linogliride with reduced solubilities as candidates for extended release. *Drug Dev. Ind. Pharm.* 43 (3): 421–431.

48 Chrzanowski, F. (2008). Preformulation considerations for controlled release dosage forms. Part I. Selecting candidates. *AAPS PharmSciTech.* 9 (2): 635–638.

49 Zhang, H.S. and J.M. (2010). Biopharmaceutic consideration and assessment for Oral controlled release formulations. In: *Oral Controlled Release Formulation Design and Drug Delivery: Theory to Practice* (ed. H.P.K. Wen), 33–45. Hoboken, NJ: Wiley.

50 Fields, J., Go, J.T., and Schulze, K.S. (2015). Pill properties that cause dysphagia and treatment failure. *Curr. Ther. Res. Clin. Exp.* 77: 79–82.

51 Schiele, J.T., Quinzler, R., Klimm, H.D. et al. (2013). Difficulties swallowing solid oral dosage forms in a general practice population: prevalence, causes, and relationship to dosage forms. *Eur. J. Clin. Pharmacol.* 69 (4): 937–948.

52 Kang, S.P. and Ratain, M.J. (2010). Inconsistent labeling of food effect for oral agents across therapeutic areas: differences between oncology and non-oncology products. *Clin. Cancer Res.* 16 (17): 4446–4451.

53 Ratain, M.J. (2012). Importance of food effects for oral oncology drugs. *Clin. Adv. Hematol. Oncol.* 10 (6): 397–398.

54 Amidon, G.L., Lennernas, H., Shah, V.P., and Crison, J.R. (1995). A theoretical basis for a biopharmaceutic drug classification: the correlation of in vitro drug product dissolution and in vivo bioavailability. *Pharm. Res.* 12 (3): 413–420.

55 Butler, J.M. and Dressman, J.B. (2010). The developability classification system: application of biopharmaceutics concepts to formulation development. *J. Pharm. Sci.* 99 (12): 4940–4954.

56 Huuskonen, J. (2000). Estimation of aqueous solubility for a diverse set of organic compounds based on molecular topology. *J. Chem. Inf. Comput. Sci.* 40 (3): 773–777.

57 Chevillard, F., Lagorce, D., Reynes, C. et al. (2012). In silico prediction of aqueous solubility: a multimodel protocol based on chemical similarity. *Mol. Pharmaceutics* 9 (11): 3127–3135.

58 Hewitt, M., Cronin, M.T., Enoch, S.J. et al. (2009). In silico prediction of aqueous solubility: the solubility challenge. *J. Chem. Inf. Model.* 49 (11): 2572–2587.

59 Siepmann, J. and Siepmann, F. (2012). Modeling of diffusion controlled drug delivery. *J. Controlled Release* 161 (2): 351–362.

60 Bruschi, M. (2015). *Strategies to Modify the Drug Release from Pharmaceutical Systems*, 63–86. Woodhead Publishing Limited.

61 Leeson, P.D. (2016). Molecular inflation, attrition and the rule of five. *Adv. Drug Delivery Rev.* 101: 22–33.

62 Bayliss, M.K., Butler, J., Feldman, P.L. et al. (2016). Quality guidelines for oral drug candidates: dose, solubility and lipophilicity. *Drug Discovery Today* 21 (10): 1719–1727.
63 Cundy, K.C., Annamalai, T., Bu, L. et al. (2004). XP13512 [(+/−)-1-([(alpha-isobutanoyloxyethoxy)carbonyl] aminomethyl)-1-cyclohexane acetic acid], a novel gabapentin prodrug: II. Improved oral bioavailability, dose proportionality, and colonic absorption compared with gabapentin in rats and monkeys. *J. Pharmacol. Exp. Ther.* 311 (1): 324–333.
64 Meunier, V., Bourrie, M., Berger, Y., and Fabre, G. (1995). The human intestinal epithelial cell line Caco-2; pharmacological and pharmacokinetic applications. *Cell. Biol. Toxicol.* 11 (3–4): 187–194.
65 Rubas, W., Cromwell, M.E., Shahrokh, Z. et al. (1996). Flux measurements across Caco-2 monolayers may predict transport in human large intestinal tissue. *J. Pharm. Sci.* 85 (2): 165–169.
66 Gres, M.C., Julian, B., Bourrie, M. et al. (1998). Correlation between oral drug absorption in humans, and apparent drug permeability in TC-7 cells, a human epithelial intestinal cell line: comparison with the parental Caco-2 cell line. *Pharm. Res.* 15 (5): 726–733.
67 Collett, A., Sims, E., Walker, D. et al. (1996). Comparison of HT29-18-C1 and Caco-2 cell lines as models for studying intestinal paracellular drug absorption. *Pharm. Res.* 13 (2): 216–221.
68 Kimura, H., Sakai, Y., and Fujii, T. (2017). Organ/body-on-a-chip based on microfluidic technology for drug discovery. *Drug Metab. Pharmacokinet.*
69 von Martels, J.Z.H., Sadaghian Sadabad, M., Bourgonje, A.R. et al. (2017). The role of gut microbiota in health and disease: in vitro modeling of host-microbe interactions at the aerobe-anaerobe interphase of the human gut. *Anaerobe* 44: 3–12.
70 Shekunov, B.Y., Chattopadhyay, P., Tong, H.H., and Chow, A.H. (2007). Particle size analysis in pharmaceutics: principles, methods and applications. *Pharm. Res.* 24 (2): 203–227.
71 International Conference on Harmonization (ICH) (1999). Q6A: specifications: test procedures and acceptance criteria for new drug substances and new drug products: chemical substances.
72 Snorek, S.M., Bauer, J.F., Chidambaram, N. et al. (2007). PQRI recommendations on particle-size analysis of drug substances used in oral dosage forms. *J. Pharm. Sci.* 96 (6): 1451–1467.
73 Gamble, J.F., Tobyn, M., and Hamey, R. (2015). Application of image-based particle size and shape characterization systems in the development of small molecule pharmaceuticals. *J. Pharm. Sci.* 104 (5): 1563–1574.
74 Yu, W., Liao, L., Bharadwaj, R., and Hancock, B.C. (2017). What is the "typical" particle shape of active pharmaceutical ingredients? *Powder Technol.* 313: 1–8.
75 Hilfiker, R. (2006). *Polymorphism: In the Pharmaceutical Industry*. Wiley.
76 Sandberg, A., Abrahamsson, B., Regardh, C.G. et al. (1990). Pharmacokinetic and biopharmaceutic aspects of once daily treatment with metoprolol CR/ZOK: a review article. *J. Clin. Pharmacol.* 30 (2 Suppl): S2–S16.
77 Dahdal, W.Y.B.J.L. (2007). Cardiovascular disorders. In: *Gibaldi's Drug Delivery Systems in Pharmaceutical Care* (ed. A.L.M. Desai), 197–232. Bethesda, MD: American Society of Health-System Pharmacists.
78 U.S. Food and Drug Administration (2001). *Stability Testing of New Drug Substances and Products, ICH Q1A*. Washington, DC: FDA.
79 ICH F (1997). *Q1C Stability Testing for New Dosage Forms. Guidance for Industry*. US Department of Health and Human Services Food and Drug Administration Center for Drug Evaluation and Research. Center for Biologics Evaluation and Research, ICH.
80 Carr, R.L. (1965). Evaluating flow properties of solids. *Chem. Eng.* 72 (69).

81 Hausner, H.H. (1967). Friction conditions in a mass of metal powder. *Int. J. Powder Metall* 3: 7–13.

82 USP (2007). <1174> Powder flow. USP30 NF 25.

83 USP (2007). <616> Bulk density and tapped density. USP30 NF 25.

84 EP (2008). <2.9.36> Powder flow. European Pharmacopeia 6.0.

85 Kesisoglou, F., Balakrishnan, A., and Manser, K. (2016). Utility of PBPK absorption modeling to guide modified release formulation development of gaboxadol, a highly soluble compound with region-dependent absorption. *J. Pharm. Sci.* 105 (2): 722–728.

86 Lukacova, V., Woltosz, W.S., and Bolger, M.B. (2009). Prediction of modified release pharmacokinetics and pharmacodynamics from in vitro, immediate release, and intravenous data. *AAPS J.* 11 (2): 323–334.

87 Yu, L.X., Lipka, E., Crison, J.R., and Amidon, G.L. (1996). Transport approaches to the biopharmaceutical design of oral drug delivery systems: prediction of intestinal absorption. *Adv. Drug Delivery Rev.* 19 (3): 359–376.

6

Mathematical Models Describing Kinetics Associated with Controlled Drug Delivery Across Membranes

Annette L. Bunge

Chemical and Biological Engineering Department, Colorado School of Mines, Golden, CO, USA

6.1 Introduction

6.1.1 General Description

Transport by passive diffusion into and through membranes is important in a variety of research, industrial, and biological processes. In medicine, artificial membranes are used in many applications including drug delivery, hemodialysis, artificial organs, tissue regeneration, and diagnostic devices. Except for parenteral administration, drug delivery to the target site frequently involves transport through a membrane that affects the therapeutic response. For topical dermatologic and transdermal drug products, the time course of delivery depends on the release rate from the product, which might include an artificial membrane, combined with penetration through and distribution in the skin, which behaves as a membrane. Mathematical models of membrane processes in drug delivery are useful therefore for understanding and predicting drug absorption and distribution, interpreting experimental permeation data, and designing new improved delivery methods.

We present here a summary of membrane model solutions derived over many years along with a description of the solution methods. The cases considered were driven by specific research questions and not by a systematic or comprehensive examination of membrane diffusion models. Some of these models have been solved by others. However, because we derived our own solutions, we did not actively look for these other solutions. Where we are aware of other solutions, citations have been provided.

The models described in this chapter involve passive one-dimensional Fickian diffusion of a solute through one or two membranes of thickness L_i and diffusivity D_i ($i = 1$ or 2), separating vehicle (V) and receptor (R) solutions (Figure 6.1). The membranes are initially free of solute. Typically, the absorbing solute transfers from the vehicle into the membrane(s), and then diffuses through the membrane(s) to enter the receptor solution. In a few models, the solute transfers from the membrane to the vehicle. Initially, the solute concentration in the vehicle (C_V) is constant everywhere in the vehicle as is the solute concentration in the receptor (C_R), which is often (although not always) zero.

The models are classified into one of three types. Type A and B models consider one membrane; type C models consider two membranes. In the type A and C models, the solute concentration in the vehicle is independent of position; thus, functionally, the vehicle is well-stirred. In the type B models, the vehicle is unstirred and the solute concentration varies with position x. In all the models presented here, the solute concentration in the receptor solution is assumed to be independent of

Fundamentals of Drug Delivery, First Edition.
Edited by Heather A. E. Benson, Michael S. Roberts, Adrian C. Williams, and Xiaowen Liang.

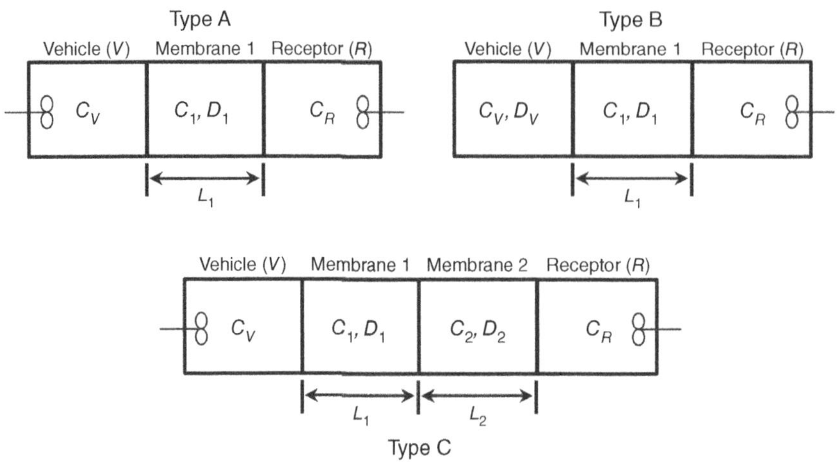

Figure 6.1 Schematic diagrams of the three types of membrane models.

time and position; i.e. the receptor solution has a large volume and is well-stirred. Conditions for the models included in this chapter are listed in Tables 6.1–6.3.

In the simplest model, there is one membrane, the solute concentrations in the vehicle and receptor solutions do not change in time or position (i.e. the vehicle and receptor both have large volumes and are well-stirred), and there are no mass transfer limitations between the membrane and the vehicle or the receptor solution. Other models describe scenarios with (i) no receptor solution (i.e. there is no flux from the membrane to the receptor solution); (ii) mass transfer resistances between the membrane and the vehicle and/or receptor solutions; (iii) time variation of the solute concentrations in a finite volume vehicle solution; (iv) an infinitely thick membrane; and (v) position-dependent solute concentrations in an semi-infinite unstirred vehicle.

Depending on the nature of the model, different methods were used to solve the governing equations describing the solute concentrations in the membrane(s), and vehicle and receptor solutions. For example, diffusion in one membrane with constant solute concentrations in the vehicle and receptor can be solved using separation of variables or Laplace transforms. However, models requiring the simultaneous solution of two or more differential equations (e.g. when the solute concentration in the vehicle changes or when there are two membranes) cannot be solved by separation of variables, and Laplace transform methods are used instead. The model solutions listed in Section 6.2 include equations describing the solute concentration in each compartment along with expressions for the average concentration, flux, and cumulative mass of solute entering or leaving the membrane(s).

The separation of variable and Laplace transform solution methods are described in Section 6.3, along with examples, to assist readers in developing solutions to other membrane models. In the Laplace transform method, the transformed equations are solved in the Laplace domain, which are then inverted back to the time domain. Analytical inversions from the Laplace domain generate equations in time and position. Laplace solutions can be inverted numerically to the time domain using various software packages; therefore, when available, Laplace domain solutions are included. For one model (A4), only Laplace domain equations are provided for the time period after the vehicle was replaced with fresh solution because inversion to the time domain is impractically complex.

Table 6.1 Type A models: well-stirred vehicle on one membrane.

Model number	Membrane characteristics		Applicable time period	Vehicle characteristics				Receptor characteristics					Vehicle/membrane boundary			Membrane/receptor boundary		
	Finite	Semi-infinite		Constant C_V	Volume	t_{ref}	C_V at t_{ref}	None	Well stirred	Constant C_R	Volume	C_R at t_{ref}	Local equilibrium	Mass transfer limited	No flux	Local equilibrium	Mass transfer limited	No flux
A1	x		$t \geq 0$	x	Infinite	0	C_V^o		x	x	Infinite	0	x			x		
A2	x		$0 \leq t \leq t_c$	x	Infinite	0	C_V^o		x	x	Infinite	0	x			x		
			$t \geq t_c$	na[a]	na	t_c	na		x	x	Infinite	0			x	x		
A3	x		$0 \leq t \leq t_c$	x	Infinite	0	C_V^o		x	x	Infinite	0	x			x		
			$t \geq t_c$	x	Infinite	t_c	0		x	x	Infinite	0	x			x		
A4	x		$0 \leq t \leq t_c$	x	Infinite	0	C_V^o		x	x	Infinite	0	x			x		
			$t \geq t_c$	x	Infinite	t_c	0		x	x	Infinite	0		x		x		
A5	x		$t \geq 0$	x	Infinite	0	C_V^o		x	x	Infinite	0		x		x		
A6	x		$t \geq 0$	x	Infinite	0	C_V^o	x	na	na	na	na		x				x
A7	x		$t \geq 0$	x	Infinite	0	C_V^o	x	na	na	na	na	x					x
A8	x		$t \geq 0$	x	Infinite	0	C_V^o		x	x	Infinite	0	x				x	
A9	x		$t \geq 0$	x	Infinite	0	≥ 0		x	x	Infinite	≥ 0		x			x	
A9a	x		$t \geq 0$	x	Infinite	0	C_V^o		x	x	Infinite	0		x			x	
A9b	x		$t \geq 0$	x	Infinite	0	0		x	x	Infinite			x			x	
A10	x		$t \geq 0$		Finite	0	C_V^o		x	x	Infinite	0	x			x		
A11	x		$t \geq 0$		Finite	0	C_V^o	x	na	na	na	na	x					x
A12		x	$t \geq 0$	x	Infinite	0	C_V^o	x	na	na	na	na	x					
A13		x	$t \geq 0$		Finite	0	C_V^o	x	na	na	na	na	x					
A14	x		$t \geq 0$		Finite	0	C_V^o		x		Finite	0	x			x		

a) Not applicable.

Table 6.2 Type B models: unstirred semi-infinite vehicle on one membrane.[a]

Model number	Vehicle characteristics		Receptor characteristics				
	t_{ref}	C_V at t_{ref}	None	Well stirred	Constant C_R	Volume	C_R at t_{ref}
B1	0	C_V^o		x	x	Infinite	0
B2	0	C_V^o	x			na[b]	na

a) The vehicle/membrane and membrane/receptor boundaries are at local equilibrium.
b) Not applicable.

Table 6.3 Type C models: well-stirred vehicle on two membranes in series.[a]

	Vehicle characteristics		Receptor characteristics				
Model number	t_{ref}	C_V at t_{ref}	**None**	**Well stirred**	**Constant C_R**	**Volume**	C_R at t_{ref}
C1	0	C_V^o		x	x	Infinite	0

a) Not applicable.

6.1.2 Governing Equations

6.1.2.1 Differential Equations

The solute concentrations in the membrane(s), vehicle, and receptor solutions must satisfy differential mass balances describing diffusion within the membrane(s) and vehicle and between the membrane and vehicle as appropriate for the particular model. The following equations describe the solute concentration as a function of time (t) and position (x) for the specified compartments and conditions assuming the time period of interest begins at $t = 0$.

Membrane 1:

$$\frac{\partial C_1}{\partial t} = D_1 \frac{\partial^2 C_1}{\partial x^2} \qquad \text{for } 0 \le x \le L_1, t > 0$$

Membrane 2:

$$\frac{\partial C_2}{\partial t} = D_2 \frac{\partial^2 C_2}{\partial x^2} \qquad \text{for } L_1 \le x \le L_1 + L_2, t > 0$$

Unstirred vehicle:

$$\frac{\partial C_V}{\partial t} = D_V \frac{\partial^2 C_V}{\partial x^2} \qquad \text{for} - L_V \le x \le 0, t > 0$$

Well-stirred vehicle with finite volume:

$$V_V \frac{dC_V}{dt} = A\, D_1 \left.\frac{\partial C_1}{\partial x}\right|_{x=0} \qquad \text{for } t > 0$$

6.1.2.2 Dimensionless Differential Equations

It is convenient to write the differential equations listed above in terms of the dimensionless concentration in each compartment i ($\widetilde{C}_i$) as a function of the dimensionless time ($\tau = tD_1/L_1$) and position ($\xi = x/L_1$). Definitions of $\widetilde{C}_i$ and the other dimensionless parameters that appear in these equations are defined in Section 6.1.4.

Membrane 1:

$$\frac{\partial \widetilde{C}_1}{\partial \tau} = \frac{\partial^2 \widetilde{C}_1}{\partial \xi^2} \qquad \text{for } 0 \le \xi \le 1, \tau > 0$$

Membrane 2:

$$\frac{\partial \widetilde{C}_2}{\partial \tau} = G_{1,2}\, \eta_2^2 \frac{\partial^2 \widetilde{C}_2}{\partial \xi^2} \qquad \text{for } 1 \le \xi \le 1 + \eta_2, \tau > 0$$

Unstirred vehicle:

$$\frac{\partial \widetilde{C}_V}{\partial \tau} = K_{1/V}^2 \gamma^2 \frac{\partial^2 \widetilde{C}_V}{\partial \xi^2} \qquad \text{for} - \eta_V \leq \xi \leq 0, \tau > 0$$

Well-stirred vehicle with finite volume:

$$\frac{d\widetilde{C}_V}{d\tau} = \frac{1}{\Lambda_V} \left. \frac{\partial \widetilde{C}_1}{\partial \xi} \right|_{\xi=0} \qquad \text{for } \tau > 0$$

6.1.2.3 Initial and Boundary Conditions

This is a summary of the initial and boundary conditions that arise in the various models. These are presented in dimensional form on the left and in dimensionless form on the right using the dimensionless variables defined in Section 6.1.4.

Zero solute concentration in compartment i:

$$C_i = 0 \qquad \widetilde{C}_i = 0 \qquad \text{where } i = 1, 2, \text{or } R$$

Constant solute concentration in the vehicle:

$$C_V = C_V^o \qquad \widetilde{C}_V = 1$$

Constant solute concentration in the receptor:

$$C_R = C_R^o \qquad \widetilde{C}_R = \widetilde{C}_R^o$$

Local equilibrium at the interface:

$$C_i = K_{i/j} C_j \qquad \widetilde{C}_i = \widetilde{C}_j$$

where $i = 1, 2,$ or R and $j = 1, 2, V,$ or R and $j \neq i$. The equilibrium partition coefficient between compartment i and j ($K_{i/j}$) is assumed to be constant.

No accumulation of mass at the interface (i.e. continuity of flux at the interface):

Between membranes 1 and 2:

$$D_1 \frac{\partial C_1}{\partial x} = D_2 \frac{\partial C_2}{\partial x} \qquad \frac{\partial \widetilde{C}_1}{\partial \xi} = \Gamma_2 \frac{\partial \widetilde{C}_2}{\partial \xi} = \frac{\eta_2}{B_{1,2}} \frac{\partial \widetilde{C}_2}{\partial \xi}$$

Between membrane 1 and vehicle:

$$D_1 \frac{\partial C_1}{\partial x} = D_V \frac{\partial C_V}{\partial x} \qquad \frac{\partial \widetilde{C}_1}{\partial \xi} = \Gamma_V \frac{\partial \widetilde{C}_V}{\partial \xi} = \frac{\eta_V}{B_{1,V}} \frac{\partial \widetilde{C}_V}{\partial \xi}$$

Vehicle-side mass transfer resistance:

General Case ($C_V \leq C_V^o$):

$$\frac{\partial C_1}{\partial x} = -\frac{k_V}{D_1}(K_{1/V} C_V - C_1) \qquad \frac{\partial \widetilde{C}_1}{\partial \xi} = -\alpha_V (\widetilde{C}_V - \widetilde{C}_1)$$

Special Case ($C_V = C_V^o$):

$$\frac{\partial C_1}{\partial x} = -\frac{k_V}{D_1}(K_{1/V} C_V^o - C_1) \qquad \frac{\partial \widetilde{C}_1}{\partial \xi} = -\alpha_V (1 - \widetilde{C}_1)$$

In these expressions, k_V is the mass transfer coefficient describing the resistance in transfer from the vehicle solution to membrane 1, and $K_{1/V}$ is the equilibrium partition coefficient between membrane 1 and the vehicle solution. Both k_V and $K_{1/V}$ are assumed to be constant.

Receptor-side mass transfer resistance:

General Case ($C_R \neq 0$)

One membrane models:

$$\frac{\partial C_1}{\partial x} = -\frac{k_R}{D_1}\left(\frac{C_1}{K_{1/R}} - C_R\right) \qquad \frac{\partial \widetilde{C}_1}{\partial \xi} = -\alpha_R(\widetilde{C}_1 - \widetilde{C}_R)$$

Two membrane models:

$$\frac{\partial C_2}{\partial x} = \frac{-k_R}{D_2}\left(\frac{C_2}{K_{2/R}} - C_R\right) \qquad \frac{\partial \widetilde{C}_2}{\partial \xi} = -\frac{\alpha_R}{\eta_2}(\widetilde{C}_2 - \widetilde{C}_R)$$

Special Case ($C_R = 0$):

One membrane models:

$$\frac{\partial C_1}{\partial x} = \frac{-k_R}{D_1}\left(\frac{C_1}{K_{1/R}}\right) \qquad \frac{\partial \widetilde{C}_1}{\partial \xi} = -\alpha_R \widetilde{C}_1$$

Two membrane models:

$$\frac{\partial C_2}{\partial x} = \frac{-k_R}{D_2}\left(\frac{C_2}{K_{2/R}}\right) \qquad \frac{\partial \widetilde{C}_2}{\partial \xi} = -\frac{\alpha_R \widetilde{C}_2}{\eta_2}$$

In the above expressions, k_R is the mass transfer coefficient describing the resistance in transfer from the membrane i to the receptor solution, and $K_{i/R}$ is the equilibrium partition coefficient between membrane i and the receptor solution (where $i = 1$ in the one-membrane model and 2 in the two-membrane model). Both k_R and $K_{i/R}$ are assumed to be constant.

No flux:

$$\frac{\partial C_i}{\partial x} = 0 \qquad \frac{\partial \widetilde{C}_i}{\partial \xi} = 0 \qquad \text{where } i = 1, 2, \text{ or } V$$

6.1.3 Other Derived Quantities

For each model, the differential equations and initial and boundary conditions are solved to give the solute concentrations in the membrane(s), vehicle and receptor compartments as appropriate. In experiments, other quantities are measured such as the solute flux or cumulative mass of solute that crosses a specified boundary (e.g. from the vehicle to the membrane or from the membrane to the vehicle), or the average solute concentration or solute mass in a membrane or compartment. Therefore, equations for the flux, cumulative mass crossing a specified boundary, and the average concentration are also presented. These are derived from the equations for the solute concentration using the expressions presented below, which are written in both dimensional and dimensionless form. The dimensionless form of a given quantity is designated by the tilde (~) above the symbol representing that quantity.

Flux of solute across membrane i: J_i

$$J_i = -D_i \frac{\partial C_i}{\partial x}$$

$$\widetilde{J}_i = \frac{J_i L_1}{K_{1/V} D_1 C_V^{\text{o}}} = -\Gamma_i \frac{\partial \widetilde{C}_i}{\partial \xi}$$

For membrane 1,

$$\widetilde{J}_1 = \frac{J_1 L_1}{K_{1/V} D_1 C_V^o} = -\frac{\partial \widetilde{C}_1}{\partial \xi}$$

For membrane 2,

$$\widetilde{J}_2 = \frac{J_2 L_1}{K_{1/V} D_1 C_V^o} = -\Gamma_2 \frac{\partial \widetilde{C}_2}{\partial \xi} = -\frac{\eta_2}{B_{1,2}} \frac{\partial \widetilde{C}_2}{\partial \xi}$$

For an unstirred vehicle,

$$\widetilde{J}_V = \frac{J_V L_1}{K_{1/V} D_1 C_V^o} = -\Gamma_V \frac{\partial \widetilde{C}_V}{\partial \xi} = -K_{1/V}\, \gamma^2 \frac{\partial \widetilde{C}_V}{\partial \xi}$$

Cumulative mass transfer from the vehicle to membrane 1: $M_{V,1}$

$$M_{V,1} = A \int_0^t -D_1 \left.\frac{\partial C_1}{\partial x}\right|_{x=0} dt = A \int_0^t -D_V \left.\frac{\partial C_V}{\partial x}\right|_{x=0} dt$$

$$\widetilde{M}_{V,1} = \frac{M_{V,1}}{K_{1/V} L_1 A C_V^o} = \int_0^\tau \left.\widetilde{J}_1\right|_{\xi=0} d\tau = \int_0^\tau \left.\widetilde{J}_V\right|_{\xi=0} d\tau$$

where the right-hand term in these equations, which describes diffusion in the vehicle, only applies when the vehicle is unstirred.

Cumulative mass transfer from membrane 1 to the receptor (one-membrane models): $M_{1,R}$

$$M_{1,R} = A \int_0^t -D_1 \left.\frac{\partial C_1}{\partial x}\right|_{x=L_1} dt = A \int_0^t -D_R \left.\frac{\partial C_R}{\partial x}\right|_{x=L_1} dt$$

$$\widetilde{M}_{1,R} = \frac{M_{1,R}}{K_{1/V} L_1 A C_V^o} = \int_0^\tau \left.\widetilde{J}_1\right|_{\xi=1} d\tau = \int_0^\tau \left.\widetilde{J}_R\right|_{\xi=1} d\tau$$

where the right-hand term in these equations, which describes diffusion in the receptor, only applies when the receptor is unstirred. This situation is not considered in any of the models presented here.

Cumulative mass transfer from membrane 2 to *the receptor (two-membrane models): $M_{2,R}$*

$$M_{2,R} = A \int_0^t -D_2 \left.\frac{\partial C_2}{\partial x}\right|_{x=L_1+L_2} dt = A \int_0^t -D_R \left.\frac{\partial C_R}{\partial x}\right|_{x=L_1+L_2} dt$$

$$\widetilde{M}_{2,R} = \frac{M_{2,R}}{K_{1/V} L_1 A C_V^o} = \int_0^\tau \left.\widetilde{J}_2\right|_{\xi=1+\eta_2} d\tau = \int_0^\tau \left.\widetilde{J}_R\right|_{\xi=1+\eta_2} d\tau$$

where the right-hand term in these equations, which describes diffusion in the receptor, only applies when the receptor is unstirred. This situation is not considered in any of the models presented here.

Cumulative mass transfer between membranes 1 and 2 (two-membrane models): $M_{1,2}$

$$M_{1,2} = A \int_0^t -D_1 \left.\frac{\partial C_1}{\partial x}\right|_{x=L_1} dt = A \int_0^t -D_2 \left.\frac{\partial C_2}{\partial x}\right|_{x=L_1} dt$$

$$\widetilde{M}_{1,2} = \frac{M_{1,2}}{K_{1/V} L_1 A C_V^o} = \int_0^\tau \left.\widetilde{J}_1\right|_{\xi=1} d\tau = \int_0^\tau \left.\widetilde{J}_2\right|_{\xi=1} d\tau$$

Average concentration in compartment i: $\langle C_i \rangle$ where i = 1, 2, or V:

$$\langle C_1 \rangle = \frac{1}{L_1} \int_0^{L_1} C_1 \, dx \qquad \langle \widetilde{C}_1 \rangle = \int_0^1 \widetilde{C}_1 \, d\xi$$

$$\langle C_2 \rangle = \frac{1}{L_2} \int_{L_1}^{L_1+L_2} C_2 \, dx \qquad \langle \widetilde{C}_2 \rangle = \frac{1}{\eta_2} \int_1^{1+\eta_2} \widetilde{C}_2 \, d\xi$$

$$\langle C_V \rangle = \frac{1}{L_V} \int_{-L_V}^{0} C_V \, dx \qquad \langle \widetilde{C}_V \rangle = \frac{1}{\eta_V} \int_{-\eta_V}^{0} \widetilde{C}_V \, d\xi$$

The expressions for average concentration in the vehicle, only apply to unstirred vehicles with finite volume. This does not occur in any of the models presented in this chapter, which only consider vehicles of infinite volume or vehicles of finite volume that are well stirred.

6.1.4 Dimensionless Variables and Groups

This section provides an alphabetical listing of the dimensionless variables and groups that appear in the dimensionless equations representing the various models. The definitions of a few dimensionless parameters are written in terms of the permeability coefficient ($P_{i,V}$), as defined in the skin literature, which quantifies the steady-state rate of solute mass transfer through membrane i of thickness L_i from a vehicle V:

$$P_{i,V} = \frac{D_i \, K_{i/V}}{L_i}$$

where D_i is the diffusivity of membrane i, and $K_{i/V}$ is the equilibrium partition coefficient between membrane i and the vehicle, which are assumed to be constant. In the synthetic membrane literature, permeability is defined as $L_i P_{i,V}$, which is independent of the membrane thickness L_i, and $P_{i,V}$ is called the permeance.

$B_{i,j}$ = Ratio of the permeability coefficients for solute diffusion into membranes i and j from the vehicle V. This group only arises in models with two membranes.

$$B_{i,j} = \frac{P_{i,V}}{P_{j,V}} = \frac{K_{i/V} D_i / L_i}{K_{j/V} D_j / L_j} = \frac{K_{i/j} D_i L_j}{D_j L_i}$$

$\widetilde{C}_i$ = Dimensionless concentration in the vehicle V, membrane i (where $i = 1$ in the one-membrane models and $i = 1$ or 2 in the two-membrane models) or receptor R

$$\widetilde{C}_i = \frac{K_{1/i} \, C_i}{K_{1/V} \, C_V^o} = \frac{C_i}{K_{i/V} \, C_V^o}$$

$G_{i,j}$ = Ratio of the lag times in membranes i and j. This group only arises in models with two membranes.

$$G_{i,j} = \frac{L_i^2 \, D_j}{L_j^2 \, D_i} = \frac{t_{lag,i}}{t_{lag,j}}$$

$\widetilde{J}_i$ = Dimensionless flux in membrane i

$$\widetilde{J}_i = \frac{J_i \, L_1}{K_{1/V} \, D_1 \, C_V^0}$$

$\widetilde{M}_{j,k}$ = Dimensionless mass transfer from j to k, where j is vehicle V or membrane 1 or 2, and k is membrane 1 or 2 or the receptor R and $k \neq j$.

$$\widetilde{M}_{j,k} = \frac{M_{j,k}}{K_{1/V}\, L_1\, A\, C_V^0}$$

α_R = Ratio of the mass transfer coefficient in (or clearance rate into) the receptor (k_R) and the chemical penetration rate across membrane i from the vehicle to the receptor ($P_{i,V}/K_{R/V}$). For models with one membrane, $i = 1$. For models with two membranes, $i = 2$.

$$\alpha_R = \frac{k_R\, K_{R/V}}{P_{i,V}} = \frac{k_R\, K_{R/V}}{K_{i/V} D_i / L_i}$$

α_V = Ratio of the mass transfer coefficient from the vehicle to membrane 1 (k_V) and the chemical penetration rate across membrane 1 from the vehicle as represented by the permeability coefficient for membrane 1 from vehicle V ($P_{1,V}$)

$$\alpha_V = \frac{k_V\, K_{1/V}}{P_{1,V}} = \frac{k_V}{D_1/L_1}$$

γ = Ratio of solution-diffusion through the vehicle compared to membrane 1. This group only arises for models with an unstirred vehicle.

$$\gamma = \frac{1}{K_{1/V}} \sqrt{\frac{D_V}{D_1}}$$

Γ_i = Ratio of the solution-diffusion through membrane i compared to membrane 1. This group only arises in models with an unstirred vehicle or with two membranes.

$$\Gamma_i = \frac{1}{K_{1/i}} \frac{D_i}{D_1}$$

For models with an unstirred vehicle

$$\Gamma_V = \frac{1}{K_{1/V}} \frac{D_V}{D_1} = K_{1/V}\, \gamma^2$$

For models with two membranes

$$\Gamma_2 = \frac{1}{K_{1/2}} \frac{D_2}{D_1} = \frac{\eta_2}{B_{1,2}}$$

η_i = Ratio of the thickness of the vehicle or another membrane relative to the thickness of membrane 1 where L_i is equal to L_V for a finite vehicle or to L_2 for a second membrane.

$$\eta_i = \frac{L_i}{L_1}$$

Λ_V = Ratio of the solute holding capacities of the vehicle and membrane 1. This group only arises in models with a finite vehicle.

$$\Lambda_V = \frac{V_V}{L_1\, K_{1/V}\, A}$$

ξ = Dimensionless position in the membranes relative to the thickness of membrane 1.

$$\xi = \frac{x}{L_1}$$

τ = Dimensionless time relative to the time to diffuse through membrane 1.

$$\tau = \frac{D_1\, t}{L_1^2}$$

6.2 Model Solutions

The solutions to the models are organized by type: (A) well-stirred vehicle on one membrane, (B) unstirred vehicle on one membrane, and (C) well-stirred vehicle on two membranes. The solutions for each model include a schematic diagram, the defining boundary, and initial conditions, and equations for the concentration, flux, and cumulative mass transfer in the time domain (and sometimes Laplace domain).

6.2.1 Type A Models – Well-Stirred Vehicle on One Membrane

Solutions to 13 type A models are presented. Conditions for each of these are summarized in Table 6.1.

6.2.1.1 Model A1

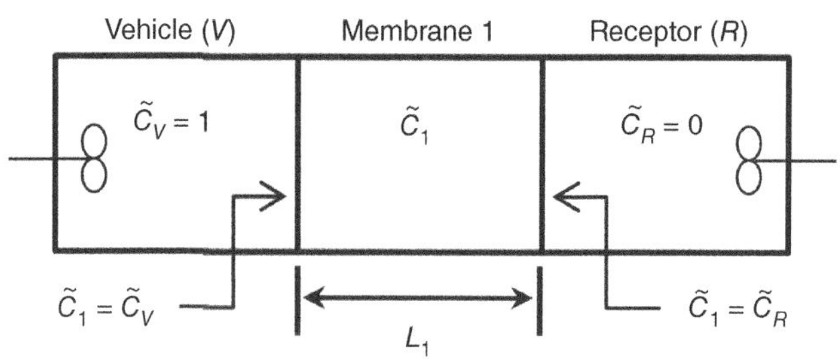

Description:

Vehicle: Infinite volume, well-stirred, constant solute concentration
Membranes: One, finite thickness, solute concentration initially zero
Receptor: Infinite volume, well-stirred, negligible concentration (sink conditions)
Vehicle/Membrane: Local equilibrium
Membrane/Receptor: Local equilibrium

Differential equation:

$$\frac{\partial \tilde{C}_1}{\partial \tau} = \frac{\partial^2 \tilde{C}_1}{\partial \xi^2} \qquad \text{for } 0 \le \xi \le 1, \tau \ge 0$$

Boundary conditions:

$$\tilde{C}_1 = \tilde{C}_V = 1 \qquad \text{at } \xi = 0, \tau > 0$$

$$\tilde{C}_1 = \tilde{C}_R = 0 \qquad \text{at } \xi = 1, \tau \ge 0$$

Initial condition:

$$\tilde{C}_1 = 0 \qquad \text{at } \tau = 0, 0 \le \xi \le 1$$

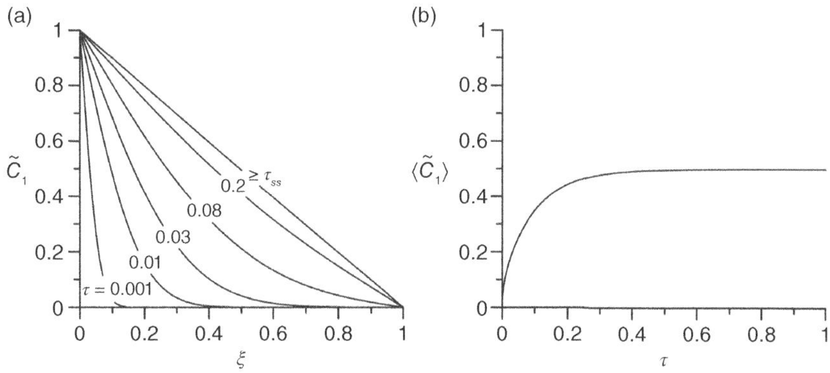

Figure 6.2 Distribution of the dimensionless concentration in the membrane $\tilde{C}_1$ at various values of τ that are less than or greater than the time to reach steady state τ_{ss} (a), and the dimensionless average concentration $\langle \tilde{C}_1 \rangle$ plotted as a function of τ (b).

6.2.1.1.1 Time Domain Solutions

Concentration Membrane 1 ($0 \leq \xi \leq 1$):

$$\tilde{C}_1 = 1 - \xi - \frac{2}{\pi}\sum_{n=1}^{\infty}\frac{\sin(n\pi\xi)}{n}\exp[-n^2\pi^2\tau] \tag{A1.1}$$

$$\langle \tilde{C}_1 \rangle = \frac{1}{2} - \frac{2}{\pi^2}\sum_{n=1}^{\infty}\frac{[1-(-1)^n]}{n^2}\exp[-n^2\pi^2\tau]$$

$$= \frac{1}{2} - \frac{4}{\pi^2}\sum_{n=0}^{\infty}\frac{\exp[-(2n+1)^2\pi^2\tau]}{(2n+1)^2} \tag{A1.2}$$

Flux Membrane 1 ($0 \leq \xi \leq 1$):

$$\tilde{J}_1 = 1 + 2\sum_{n=1}^{\infty}\cos(n\pi\xi)\exp[-n^2\pi^2\tau] \tag{A1.3}$$

Cumulative Mass Transfer

From vehicle to membrane 1:

$$\tilde{M}_{V,1} = \tau + \frac{1}{3} - \frac{2}{\pi^2}\sum_{n=1}^{\infty}\frac{\exp[-n^2\pi^2\tau]}{n^2} \tag{A1.4}$$

From membrane 1 to receptor:

$$\tilde{M}_{1,R} = \tau - \frac{1}{6} - \frac{2}{\pi^2}\sum_{n=1}^{\infty}\frac{(-1)^n\exp[-n^2\pi^2\tau]}{n^2} \tag{A1.5}$$

Plots and Calculations Figures 6.2 and 6.3 show the concentration ($\tilde{C}_1$ and $\langle \tilde{C}_1 \rangle$), flux in and out of the membrane ($\tilde{J}_1$ at $\xi = 0$ and 1), and the cumulative mass transfer ($\tilde{M}_{V,1}$ and $\tilde{M}_{1,R}$) for time τ increasing from zero until after the time to reach steady state τ_{ss}. Table 6.4 summarizes the time τ required to reach a designated percentage of the steady-state value for each of these quantities as well as the percentage of the steady-state value that is achieved at $\tau = 0.4$, 0.5, and 0.6.

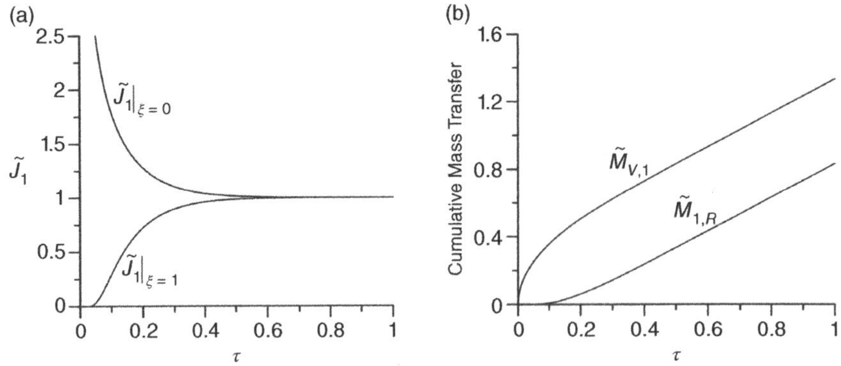

Figure 6.3 Dimensionless flux and mass transfer from the vehicle to the membrane ($\tilde{J}_1|_{\xi=0}$ and $\tilde{M}_{V,1}$, respectively) and from the membrane to the receptor ($\tilde{J}_1|_{\xi=1}$ and $\tilde{M}_{1,R}$, respectively) plotted as a function of τ.

Table 6.4 Estimates of the time τ required to approach steady state for Model A1.

Quantity	**Maximum value of τ for P to satisfy the designated criterion**[a)]			**P (%) at the designated τ**[a)]		
	$P \leq 1\%$	$P \leq 0.5\%$	$P \leq 0.1\%$	$\tau = 0.4$	$\tau = 0.50$	$\tau = 0.60$
$\langle \tilde{C}_1 \rangle$	0.45	0.52	0.68	1.6	0.58	0.22
$\tilde{J}_1$ at $\xi = 0$ and 1	0.54	0.61	0.77	3.9	1.4	0.54
$\tilde{M}_{V,1}$	0.35	0.41	0.55	0.53	0.17	0.058
$\tilde{M}_{1,R}$	0.44	0.49	0.62	1.7	0.43	0.13

a) P is the relative deviation from steady state defined as the absolute value of the difference in the designated quantity relative to its steady-state value divided by its steady-state value.

Notes The dimensionless time-axis intercepts for the steady-state cumulative mass transfer from vehicle to membrane 1 (at $\tau = -1/3$) and from membrane 1 to the receptor (at $\tau = 1/6$) can be found by recognizing the identities (see Section 6.3.3)

$$\sum_{n=1}^{\infty} \frac{1}{(n\pi)^2} = \frac{1}{6} \tag{A1.6}$$

$$\sum_{n=1}^{\infty} \frac{(-1)^n}{(n\pi)^2} = -\frac{1}{12} \tag{A1.7}$$

or by a more general method using expressions in the Laplace domain, which is described in Example A in Section 6.3.2.3.

6.2.1.1.2 Laplace Domain Solutions

Concentration Membrane 1 ($0 \leq \xi \leq 1$):

$$\mathcal{L}\{\tilde{C}_1\} = \frac{\cosh(\sqrt{s}\xi)}{s} - \frac{\cosh(\sqrt{s})\sinh(\sqrt{s}\xi)}{s\sinh(\sqrt{s})} \tag{A1.8}$$

Flux Membrane 1 ($0 \le \xi \le 1$):

$$\mathcal{L}\{\tilde{J}_1\} = \sqrt{s}\left[-\frac{\sinh(\sqrt{s}\xi)}{s} + \frac{\cosh(\sqrt{s})\cosh(\sqrt{s}\xi)}{s\sinh(\sqrt{s})}\right] \quad \text{(A1.9)}$$

Notes The expression for concentration in the Laplace domain can be inverted using the method of residues as described in Example A in Section 6.3.2.3.

6.2.1.2 Model A2

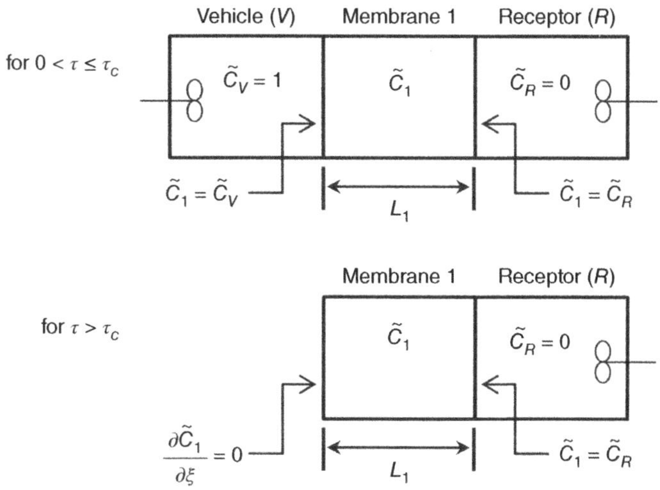

Description:

Vehicle: Infinite volume, well-stirred, constant solute concentration until removed at $\tau = \tau_c$, and then no flux from the vehicle side of the membrane
Membranes: One, finite thickness, solute concentration initially zero
Receptor: Well-stirred, negligible concentration (sink conditions)
Vehicle/Membrane: Local equilibrium
Membrane/Receptor: Local equilibrium

Differential equation:

$$\frac{\partial \tilde{C}_1}{\partial \tau} = \frac{\partial^2 \tilde{C}_1}{\partial \xi^2} \qquad \text{for } 0 \le \xi \le 1, \tau \ge 0$$

Boundary conditions:

$$\tilde{C}_1 = \tilde{C}_V = 1 \qquad \text{at } \xi = 0, 0 < \tau \le \tau_c$$

$$\frac{\partial \tilde{C}_1}{\partial \xi} = 0 \qquad \text{at } \xi = 0, \tau > \tau_c$$

$$\tilde{C}_1 = \tilde{C}_R = 0 \qquad \text{at } \xi = 1, \tau \ge 0$$

Initial condition:

$$\tilde{C}_1 = 0 \qquad \text{at } \tau = 0, 0 \le \xi \le 1$$

6.2.1.2.1 Time Domain Solutions

Eigenvalues

$$\lambda_n = (2n+1)\pi/2 \tag{A2.1}$$

Concentration Membrane 1 ($0 \leq \xi \leq 1$):

for $0 < \tau \leq \tau_c$ (from Model A1)

$$\widetilde{C}_1 = 1 - \xi - 2\sum_{m=1}^{\infty} \frac{\sin(m\pi\xi)}{m\pi} \exp[-m^2\pi^2\tau] \tag{A2.2}$$

$$\langle \widetilde{C}_1 \rangle = \frac{1}{2} - \frac{2}{\pi^2}\sum_{m=1}^{\infty} \frac{[1-(-1)^m]}{m^2} \exp[-m^2\pi^2\tau]$$

$$= \frac{1}{2} - \frac{4}{\pi^2}\sum_{m=0}^{\infty} \frac{\exp[-(2m+1)^2\pi^2\tau]}{(2m+1)^2} \tag{A2.3}$$

for $\tau \geq \tau_c$

$$\widetilde{C}_1 = 2\sum_{n=0}^{\infty} \left(\frac{1}{\lambda_n^2} - 2\sum_{m=1}^{\infty} \frac{\exp[-m^2\pi^2\tau_c]}{m^2\pi^2 - \lambda_n^2} \right) \cos(\lambda_n\xi) \exp[-\lambda_n^2(\tau - \tau_c)] \tag{A2.4}$$

$$\langle \widetilde{C}_1 \rangle = 2\sum_{n=0}^{\infty} \frac{(-1)^n \exp[-\lambda_n^2(\tau - \tau_c)]}{\lambda_n} \left(\frac{1}{\lambda_n^2} - 2\sum_{m=1}^{\infty} \frac{\exp[-m^2\pi^2\tau_c]}{m^2\pi^2 - \lambda_n^2} \right) \tag{A2.5}$$

Flux Membrane 1 ($0 \leq \xi \leq 1$):

for $0 < \tau \leq \tau_c$ (from Model A1)

$$\widetilde{J}_1 = 1 + 2\sum_{m=1}^{\infty} \cos(m\pi\xi) \exp[-m^2\pi^2\tau] \tag{A2.6}$$

for $\tau \geq \tau_c$

$$\widetilde{J}_1 = 2\sum_{n=0}^{\infty} \left(\frac{1}{\lambda_n} - 2\lambda_n \sum_{m=1}^{\infty} \frac{\exp[-m^2\pi^2\tau_c]}{m^2\pi^2 - \lambda_n^2} \right) \sin(\lambda_n\xi) \exp[-\lambda_n^2(\tau - \tau_c)] \tag{A2.7}$$

Cumulative Mass Transfer

From vehicle to membrane 1:

for $0 < \tau \leq \tau_c$ (from Model A1)

$$\widetilde{M}_{V,1} = \tau + \frac{1}{3} - 2\sum_{m=1}^{\infty} \frac{\exp[-m^2\pi^2\tau]}{m^2\pi^2} \tag{A2.8}$$

for $\tau \geq \tau_c$

$$\widetilde{M}_{V,1} = \tau_c + \frac{1}{3} - 2\sum_{m=1}^{\infty} \frac{\exp[-m^2\pi^2\tau_c]}{m^2\pi^2} \tag{A2.9}$$

From membrane 1 to receptor:

for $0 < \tau \leq \tau_c$ (from Model A1)

$$\widetilde{M}_{1,R} = \tau - \frac{1}{6} - 2\sum_{m=1}^{\infty} \frac{(-1)^m \exp[-m^2\pi^2\tau]}{m^2\pi^2} \tag{A2.10}$$

for $\tau \geq \tau_c$

$$\widetilde{M}_{1,R} = \tau_c + \frac{1}{3} - 2\sum_{m=1}^{\infty}\frac{(-1)^m \exp[-m^2\pi^2\tau_c]}{m^2\pi^2} - 4\sum_{n=0}^{\infty}\frac{(-1)^n(1-\exp[-\lambda_n^2(\tau-\tau_c)])}{\lambda_n}\left\{\sum_{m=1}^{\infty}\frac{\exp[-m^2\pi^2\tau_c]}{m^2\pi^2-\lambda_n^2}\right\} - 2\sum_{n=0}^{\infty}\frac{(-1)^n \exp[-\lambda_n^2(\tau-\tau_c)]}{\lambda_n^3} \tag{A2.11}$$

or equivalently

$$\widetilde{M}_{1,R} = \tau_c + \frac{1}{3} - 2\sum_{m=1}^{\infty}\frac{\exp[-m^2\pi^2\tau_c]}{m^2\pi^2} + 4\sum_{n=0}^{\infty}\frac{(-1)^n \exp[-\lambda_n^2(\tau-\tau_c)]}{\lambda_n}\left\{\sum_{m=1}^{\infty}\frac{\exp[-m^2\pi^2\tau_c]}{m^2\pi^2-\lambda_n^2}\right\} - 2\sum_{n=0}^{\infty}\frac{(-1)^n \exp[-\lambda_n^2(\tau-\tau_c)]}{\lambda_n^3} \tag{A2.12}$$

Plots Figures 6.4 and 6.5 show the time variation of concentration $\widetilde{C}_1$ and the average concentration $\langle\widetilde{C}_1\rangle$, respectively, for selected values of τ_c. Figure 6.6 shows the time variation of the flux and cumulative mass transfer from the membrane to the receptor to the membrane ($\widetilde{J}_1\big|_{\xi=1}$ and $\widetilde{M}_{1,R}$, respectively) for selected values of τ_c. After the solute is removed, the gradient of $\widetilde{C}_1$ at the vehicle-membrane interface ($\xi = 1$) is zero as required by the no-flux boundary condition.

Notes The second equation listed for $\widetilde{M}_{1,R}$ when $\tau > \tau_c$ can be derived by equating the equations describing $\langle\widetilde{C}_1\rangle$ for $0 < \tau \leq \tau_c$ and $\langle\widetilde{C}_1\rangle$ for $\tau > \tau_c$ at $\tau = \tau_c$ as follows:

$$\langle\widetilde{C}_1\rangle = \frac{1}{2} - 2\sum_{m=1}^{\infty}\frac{[1-(-1)^m]\exp[-m^2\pi^2\tau_c]}{m^2\pi^2} = 2\sum_{n=0}^{\infty}\frac{(-1)^n}{\lambda_n}\left(\frac{1}{\lambda_n^2} - 2\sum_{m=1}^{\infty}\frac{\exp[-m^2\pi^2\tau_c]}{m^2\pi^2-\lambda_n^2}\right) \tag{A2.13}$$

which after expanding the right-hand side gives:

$$\langle\widetilde{C}_1\rangle = \frac{1}{2} - 2\sum_{m=1}^{\infty}\frac{[1-(-1)^m]\exp[-m^2\pi^2\tau_c]}{m^2\pi^2} = 2\sum_{n=0}^{\infty}\frac{(-1)^n}{\lambda_n^3} - 4\sum_{n=0}^{\infty}\frac{(-1)^n}{\lambda_n}\sum_{m=1}^{\infty}\frac{\exp[-m^2\pi^2\tau_c]}{m^2\pi^2-\lambda_n^2} \tag{A2.14}$$

This can be simplified further using the identity (see Section 6.3.3)

$$\sum_{n=0}^{\infty}\frac{(-1)^n}{\lambda_n^3} = \sum_{n=0}^{\infty}(-1)^n\left(\frac{2}{(2n+1)\pi}\right)^3 = \frac{1}{4} \tag{A2.15}$$

to obtain:

$$\sum_{m=1}^{\infty}\frac{[1-(-1)^m]\exp[-m^2\pi^2\tau_c]}{m^2\pi^2} = 2\sum_{n=0}^{\infty}\frac{(-1)^n}{\lambda_n}\left\{\sum_{m=1}^{\infty}\frac{\exp[-m^2\pi^2\tau_c]}{m^2\pi^2-\lambda_n^2}\right\} \tag{A2.16}$$

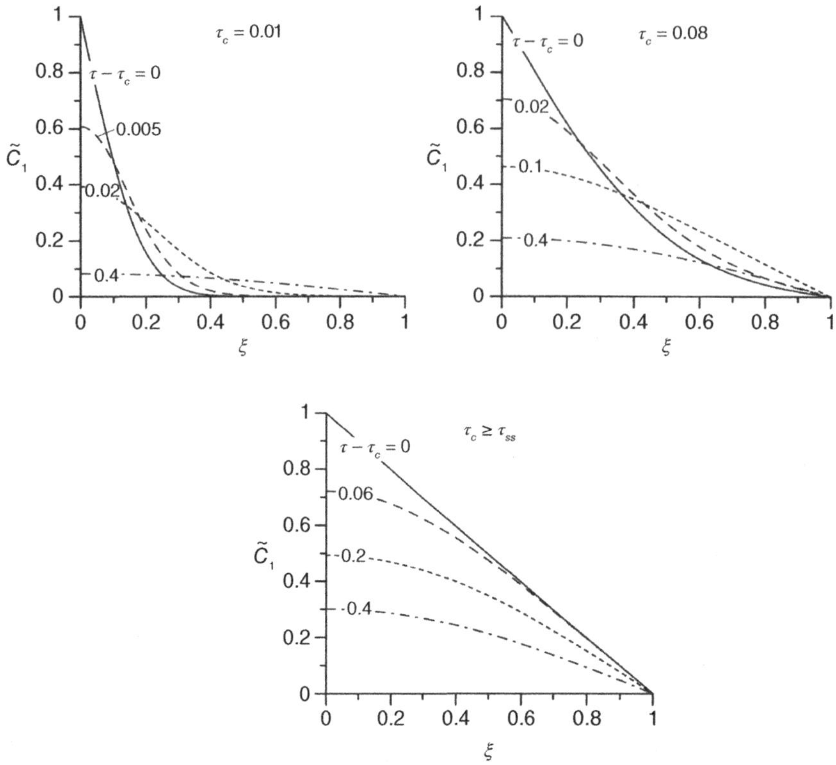

Figure 6.4 Distribution of the dimensionless concentration in the membrane $\tilde{C}_1$ at various values of τ after the solute is removed at $\tau = \tau_c$ for values of τ_c that are less than (0.01 and 0.08) or greater than the time required to establish steady state τ_{ss}.

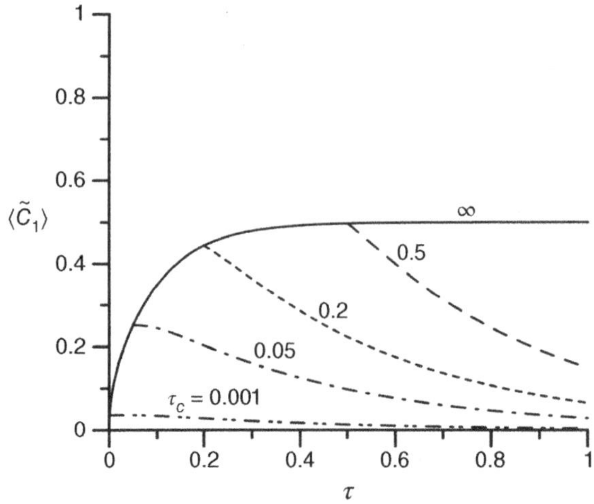

Figure 6.5 Dimensionless average concentration in the membrane $\langle \tilde{C}_1 \rangle$ plotted as a function of τ for selected values of τ_c.

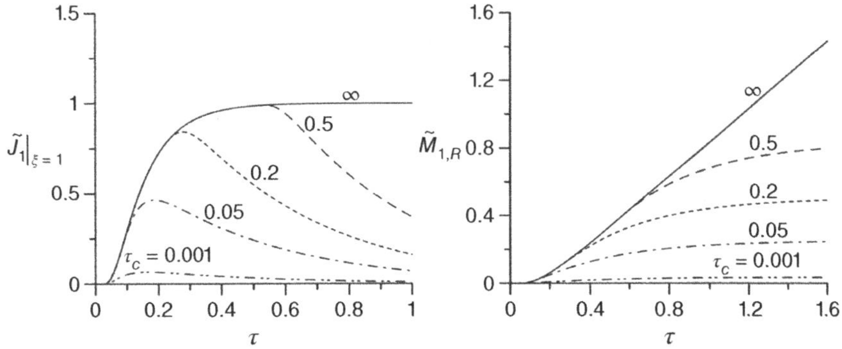

Figure 6.6 Dimensionless flux and cumulative mass transfer from the membrane to the receptor ($\widetilde{J}_1\big|_{\xi=1}$ and $\widetilde{M}_{1,R}$, respectively) plotted as a function of τ for selected values of τ_c.

which can be expanded to yield:

$$\sum_{m=1}^{\infty}\frac{(-1)^m \exp[-m^2\pi^2\tau_c]}{m^2\pi^2} = -2\sum_{n=0}^{\infty}\frac{(-1)^n}{\lambda_n}\left\{\sum_{m=1}^{\infty}\frac{\exp[-m^2\pi^2\tau_c]}{m^2\pi^2-\lambda_n^2}\right\} + \sum_{m=1}^{\infty}\frac{\exp[-m^2\pi^2\tau_c]}{m^2\pi^2} \tag{A2.17}$$

This expression is then substituted into the first equation listed for $\widetilde{M}_{1,R}$ to obtain:

$$\begin{aligned}\widetilde{M}_{1,R} = \tau_c + \frac{1}{3} + 4\sum_{n=0}^{\infty}\frac{(-1)^n}{\lambda_n}\left\{\sum_{m=1}^{\infty}\frac{\exp[-m^2\pi^2\tau_c]}{m^2\pi^2-\lambda_n^2}\right\} \\ - 2\sum_{m=1}^{\infty}\frac{\exp[-m^2\pi^2\tau_c]}{m^2\pi^2} - 4\sum_{n=0}^{\infty}\frac{(-1)^n}{\lambda_n}\left\{\sum_{m=1}^{\infty}\frac{\exp[-m^2\pi^2\tau_c]}{m^2\pi^2-\lambda_n^2}\right\} \\ + 4\sum_{n=0}^{\infty}\frac{(-1)^n \exp[-\lambda_n^2(\tau-\tau_c)]}{\lambda_n}\left\{\sum_{m=1}^{\infty}\frac{\exp[-m^2\pi^2\tau_c]}{m^2\pi^2-\lambda_n^2}\right\} \\ - 2\sum_{n=0}^{\infty}\frac{(-1)^n \exp[-\lambda_n^2(\tau-\tau_c)]}{\lambda_n^3}\end{aligned} \tag{A2.18}$$

which simplifies as:

$$\begin{aligned}\widetilde{M}_{1,R} = \tau_c + \frac{1}{3} - 2\sum_{m=1}^{\infty}\frac{\exp[-m^2\pi^2\tau_c]}{m^2\pi^2} \\ + 4\sum_{n=0}^{\infty}\frac{(-1)^n \exp[-\lambda_n^2(\tau-\tau_c)]}{\lambda_n}\left\{\sum_{m=1}^{\infty}\frac{\exp[-m^2\pi^2\tau_c]}{m^2\pi^2-\lambda_n^2}\right\} \\ - 2\sum_{n=0}^{\infty}\frac{(-1)^n \exp[-\lambda_n^2(\tau-\tau_c)]}{\lambda_n^3}\end{aligned} \tag{A2.19}$$

At $\tau = \tau_c$, this expression collapses to give the expected result:

$$\widetilde{M}_{1,R} = \tau_c - \frac{1}{6} - 2\sum_{m=1}^{\infty}\frac{(-1)^m \exp[-m^2\pi^2\tau_c]}{m^2\pi^2} \tag{A2.20}$$

6.2.1.3 Model A3

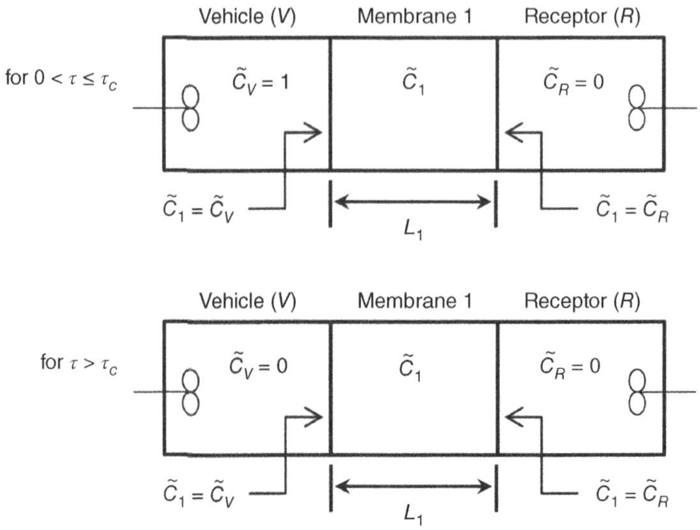

Description:

Vehicle: Infinite volume, well-stirred, constant solute concentration until replaced at $\tau = \tau_c$ with an infinite volume, well-stirred solute-free solution (sink conditions)

Membranes: One, finite thickness, solute concentration initially zero

Receptor: Well-stirred, negligible concentration (sink conditions)

Vehicle/Membrane: Local equilibrium

Membrane/Receptor: Local equilibrium

Differential equation:

$$\frac{\partial \tilde{C}_1}{\partial \tau} = \frac{\partial^2 \tilde{C}_1}{\partial \xi^2} \qquad \text{for } 0 \leq \xi \leq 1, \tau \geq 0$$

Boundary conditions:

$$\tilde{C}_1 = \tilde{C}_V = 1 \qquad \text{at } \xi = 0, 0 < \tau \leq \tau_c$$

$$\tilde{C}_1 = \tilde{C}_V = 0 \qquad \text{at } \xi = 0, \tau > \tau_c$$

$$\tilde{C}_1 = \tilde{C}_R = 0 \qquad \text{at } \xi = 1, \tau \geq 0$$

Initial condition:

$$\tilde{C}_1 = 0 \qquad \text{at } \tau = 0, 0 \leq \xi \leq 1$$

6.2.1.3.1 Time Domain Solutions

Concentration Membrane 1 ($0 \leq \xi \leq 1$):

for $0 < \tau \leq \tau_c$ (from Model A1)

$$\tilde{C}_1 = 1 - \xi - 2\sum_{m=1}^{\infty} \frac{\sin(n\pi\xi)}{n\pi} \exp[-n^2\pi^2\tau] \tag{A3.1}$$

$$\langle \widetilde{C}_1 \rangle = \frac{1}{2} - \frac{2}{\pi^2}\sum_{n=1}^{\infty}\frac{[1-(-1)^n]}{n^2}\exp[-n^2\pi^2\tau] = \frac{1}{2} - \frac{4}{\pi^2}\sum_{n=0}^{\infty}\frac{\exp[-(2n+1)^2\pi^2\tau]}{(2n+1)^2} \tag{A3.2}$$

for $\tau \geq \tau_c$

$$\widetilde{C}_1 = \frac{2}{\pi}\sum_{n=1}^{\infty}\frac{\sin(n\pi\xi)}{n}\{\exp[n^2\pi^2\tau_c] - 1\}\exp[-n^2\pi^2\tau] \tag{A3.3}$$

$$\langle \widetilde{C}_1 \rangle = \frac{2}{\pi^2}\sum_{n=1}^{\infty}\frac{[1-(-1)^n]}{n^2}\{\exp[n^2\pi^2\tau_c] - 1\}\exp[-n^2\pi^2\tau] \tag{A3.4}$$

$$\langle \widetilde{C}_1 \rangle = \frac{4}{\pi^2}\sum_{n=0}^{\infty}\frac{1}{(2n+1)^2}\{\exp[(2n+1)^2\pi^2\tau_c] - 1\}\exp[-(2n+1)^2\pi^2\tau] \tag{A3.5}$$

Flux Membrane 1 ($0 \leq \xi \leq 1$):

for $0 < \tau \leq \tau_c$

$$\widetilde{J}_1 = 1 + 2\sum_{n=1}^{\infty}\cos(n\pi\xi)\exp[-n^2\pi^2\tau] \tag{A3.6}$$

for $\tau \geq \tau_c$

$$\widetilde{J}_1 = 2\sum_{n=1}^{\infty}\cos(n\pi\xi)\{1 - \exp[n^2\pi^2\tau_c]\}\exp[-n^2\pi^2\tau] \tag{A3.7}$$

Cumulative Mass Transfer

From vehicle to membrane 1:

for $0 < \tau \leq \tau_c$

$$\widetilde{M}_{V,1} = \tau + \frac{1}{3} - \frac{2}{\pi^2}\sum_{n=1}^{\infty}\frac{\exp[-n^2\pi^2\tau]}{n^2} \tag{A3.8}$$

for $\tau \geq \tau_c$

$$\widetilde{M}_{V,1} = \tau_c - \frac{2}{\pi^2}\sum_{n=1}^{\infty}\frac{1}{n^2}\{1 - \exp[n^2\pi^2\tau_c]\}\exp[-n^2\pi^2\tau] \tag{A3.9}$$

From membrane 1 to receptor:

for $0 < \tau \leq \tau_c$

$$\widetilde{M}_{1,R} = \tau - \frac{1}{6} - \frac{2}{\pi^2}\sum_{n=1}^{\infty}\frac{(-1)^n\exp[-n^2\pi^2\tau]}{n^2} \tag{A3.10}$$

for $\tau \geq \tau_c$

$$\widetilde{M}_{1,R} = \tau_c - \frac{2}{\pi^2}\sum_{n=1}^{\infty}\frac{(-1)^n}{n^2}\{1 - \exp[n^2\pi^2\tau_c]\}\exp[-n^2\pi^2\tau] \tag{A3.11}$$

Plots Figures 6.7 and 6.8 show the time variation of concentration $\widetilde{C}_1$ and the average concentration $\langle \widetilde{C}_1 \rangle$, respectively, for selected values of τ_c. Figures 6.9 and 6.10, respectively, show the time variation of the flux and the cumulative mass transfer for selected values of τ_c. After the solute is removed at $\tau = \tau_c$, $\widetilde{C}_1 = 0$ at the vehicle-membrane interface ($\xi = 0$), which causes $\widetilde{J}_1\big|_{\xi=0} < 0$ with

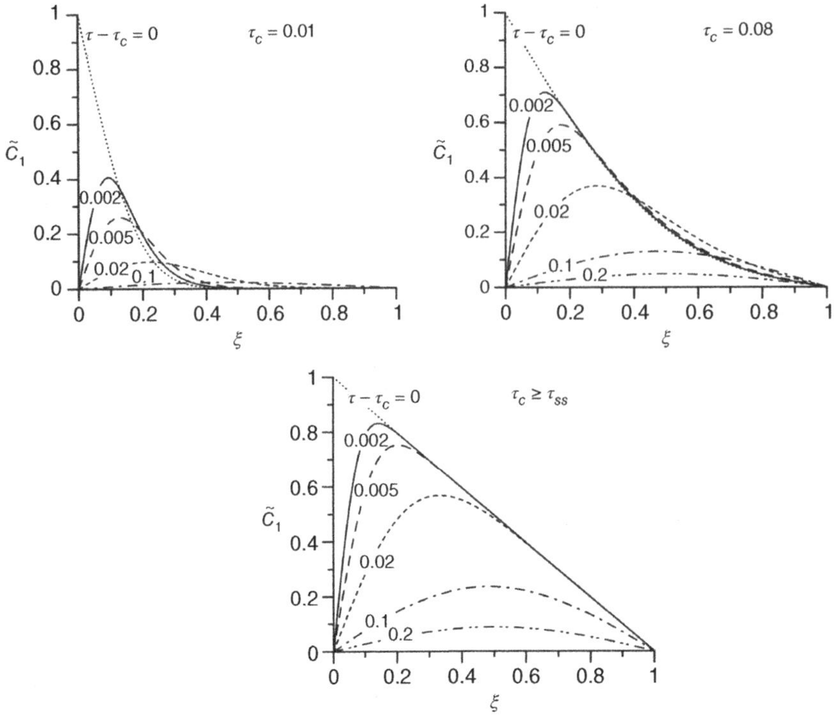

Figure 6.7 Distribution of the dimensionless concentration in the membrane $\tilde{C}_1$ at various values of τ after the solute is removed at $\tau = \tau_c$ for values of τ_c that are less than (0.01 and 0.08) or greater than the time required to establish steady state τ_{ss}.

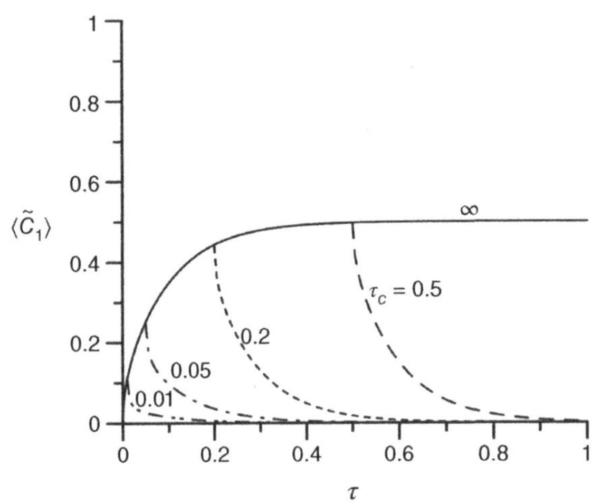

Figure 6.8 Dimensionless average concentration in the membrane $\langle \tilde{C}_1 \rangle$ plotted as a function of τ for selected values of τ_c.

a corresponding decrease in the cumulative mass that transfers from the vehicle to the membrane $\tilde{M}_{V,1}$. Because mass transfers from the membrane to the vehicle, the time required for the average concentration and flux from the membrane to the receptor to reach zero and for the cumulative mass transfer to the receptor to reach its final value is shorter than for Model A2, which assumes no flux at $\xi = 0$ (compare Figures 6.8, 6.0 and 6.10 to Figures 6.5 and 6.6.).

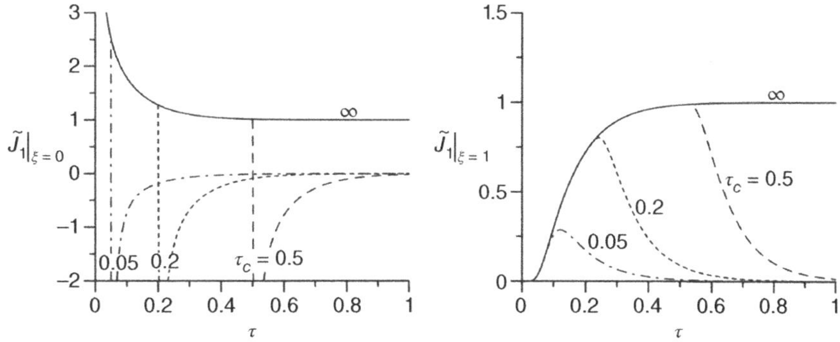

Figure 6.9 Dimensionless flux from the vehicle to the membrane ($\tilde{J}_1|_{\xi=0}$) and from the membrane to the receptor ($\tilde{J}_1|_{\xi=1}$) plotted as a function of τ for selected values of τ_c.

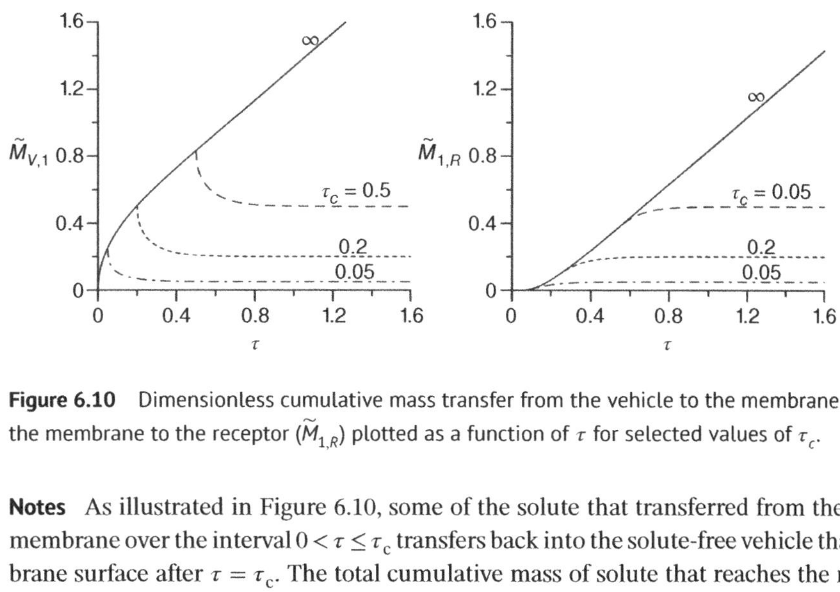

Figure 6.10 Dimensionless cumulative mass transfer from the vehicle to the membrane ($\tilde{M}_{V,1}$) and from the membrane to the receptor ($\tilde{M}_{1,R}$) plotted as a function of τ for selected values of τ_c.

Notes As illustrated in Figure 6.10, some of the solute that transferred from the vehicle into the membrane over the interval $0 < \tau \leq \tau_c$ transfers back into the solute-free vehicle that is on the membrane surface after $\tau = \tau_c$. The total cumulative mass of solute that reaches the receptor solution can be determined from the infinite time limit of the cumulative mass transfer from the vehicle to membrane 1 or from membrane 1 to the receptor, which gives the result:

$$\lim_{\tau\to\infty} \tilde{M}_{V,1} = \lim_{\tau\to\infty} \tilde{M}_{1,R} = \tau_c \tag{A3.12}$$

By applying the definition of the dimensionless cumulative mass, the total cumulative mass of solute that penetrates the membrane to reach the receptor solution is determined to be

$$\frac{M_{V,1}}{A} = \frac{K_{1/V} D_1}{L_1} C_V^o t_c = P_{i,V} C_V^o t_c \tag{A3.13}$$

where D_1 (length^2/time) and L_1 (length) are the diffusion coefficient and thickness of membrane 1, respectively, $K_{1/V}$ (unitless) is the equilibrium partition coefficient between membrane 1 and the vehicle, $P_{i,V}$ (length/time) is the steady-state permeability coefficient for solute diffusion through membrane i from vehicle V, and t_c is the total time that the vehicle at constant solute concentration is in contact with the membrane surface.

The fraction of solute absorbed into membrane 1 at the end of the exposure (i.e. at $\tau = \tau_c$) that eventually diffuses to the receptor solution (F) can be determined from the following mass balance

$$F = \frac{\lim_{\tau\to\infty} \widetilde{M}_{1,R} - \widetilde{M}_{1,R}\Big|_{\tau=\tau_c}}{\langle \widetilde{C}_1 \rangle\Big|_{\tau=\tau_c}} \tag{A3.14}$$

to be

$$F = \frac{\frac{1}{3} + 4\sum_{n=1}^{\infty} \frac{(-1)^n \exp[-n^2\pi^2\tau_c]}{n^2\pi^2}}{1 - \frac{8}{\pi^2}\sum_{n=0}^{\infty} \frac{\exp[-(2n+1)^2\pi^2\tau_c]}{(2n+1)^2}} \tag{A3.15}$$

If τ_c is large enough that steady state is reached before the solute containing vehicle is removed, then the summations of the exponentials in Eq. (A3.15) go to zero, and $F = 1/3$. If τ_c is smaller than the time required to establish steady state, then F is smaller than 1/3. In the limit of τ_c approaching zero, F is zero (see identities in Section 6.3.3).

6.2.1.4 Model A4

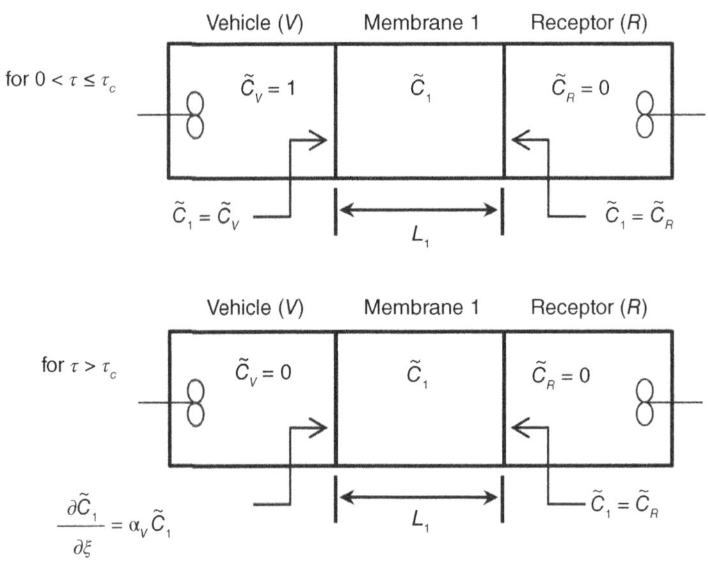

Source: Based on Frasch and Bunge [1].

Description:

Vehicle: Infinite volume, well-stirred, constant solute concentration until replaced at $\tau = \tau_c$ with an infinite volume, well-stirred solute-free solution (sink conditions)

Membranes: One, finite thickness, solute concentration initially zero

Receptor: Well-stirred, negligible concentration (sink conditions)

Vehicle/Membrane: Local equilibrium for $0 \le \tau < \tau_c$ and mass transfer limited for $\tau \ge \tau_c$

Membrane/Receptor: Local equilibrium

Differential equation:

$$\frac{\partial \widetilde{C}_1}{\partial \tau} = \frac{\partial^2 \widetilde{C}_1}{\partial \xi^2} \qquad \text{for } 0 \le \xi \le 1, \tau \ge 0$$

Boundary conditions:

$$\widetilde{C}_1 = \widetilde{C}_V = 1 \qquad \text{at } \xi = 0, 0 < \tau \le \tau_c$$

$$\frac{\partial \widetilde{C}_1}{\partial \xi} = \alpha_V \widetilde{C}_1 \qquad \text{at } \xi = 0, \tau > \tau_c$$

$$\widetilde{C}_1 = \widetilde{C}_R = 0 \qquad \text{at } \xi = 1, \tau \ge 0$$

Initial condition:

$$\widetilde{C}_1 = 0 \qquad \text{at } \tau = 0, 0 \le \xi \le 1$$

6.2.1.4.1 Time Domain Solutions

Concentration Membrane 1 ($0 \le \xi \le 1$):

for $0 < \tau \le \tau_c$ (from Model A1)

$$\widetilde{C}_1 = 1 - \xi - 2\sum_{n=1}^{\infty} \frac{\sin(n\pi\xi)}{n\pi} \exp[-n^2\pi^2\tau] \tag{A4.1}$$

$$\langle \widetilde{C}_1 \rangle = \frac{1}{2} - \frac{2}{\pi^2}\sum_{n=1}^{\infty} \frac{[1-(-1)^n]}{n^2} \exp[-n^2\pi^2\tau] = \frac{1}{2} - \frac{4}{\pi^2}\sum_{n=0}^{\infty} \frac{\exp[-(2n+1)^2\pi^2\tau]}{(2n+1)^2} \tag{A4.2}$$

for $\tau \ge \tau_c$ (see Laplace domain solutions)

Flux Membrane 1 ($0 \le \xi \le 1$):

for $0 < \tau \le \tau_c$ (from Model A1)

$$\widetilde{J}_1 = 1 + 2\sum_{n=1}^{\infty} \cos(n\pi\xi) \exp[-n^2\pi^2\tau] \tag{A4.3}$$

for $\tau \ge \tau_c$ (see Laplace domain solutions)

Cumulative Mass Transfer

From vehicle to membrane 1:

for $0 < \tau \le \tau_c$ (from Model A1)

$$\widetilde{M}_{V,1} = \tau + \frac{1}{3} - 2\sum_{n=1}^{\infty} \frac{\exp[-n^2\pi^2\tau]}{n^2\pi^2} \tag{A4.4}$$

for $\tau \ge \tau_c$ (see Laplace domain solutions)

From membrane 1 to receptor:

for $0 < \tau \le \tau_c$ (from Model A1)

$$\widetilde{M}_{1,R} = \tau - \frac{1}{6} - 2\sum_{n=1}^{\infty} \frac{(-1)^n \exp[-n^2\pi^2\tau]}{n^2\pi^2} \tag{A4.5}$$

for $\tau \ge \tau_c$ (see Laplace domain solutions)

Notes Some of the solute that is transferred from the vehicle into the membrane over the interval $0 < \tau \leq \tau_c$ transfers back into the solute-free vehicle that is on the membrane surface after $\tau = \tau_c$. This model describes the scenario in which solute absorbed into the membrane (from a vehicle without a mass transfer resistance) evaporates with a rate that depends on the gas phase mass transfer resistance and the solute saturation pressure [1]. The equation defining the fraction (F) of solute in membrane 1 at $\tau = \tau_c$ that eventually diffuses to the receptor solution, Eq. (A4.20) repeated here, is derived from the Laplace domain solution as described below:

$$F = \frac{3+\alpha_V}{3(1+\alpha_V)} \frac{\left[1 + \frac{12}{\pi^2}\sum_{n=1}^{\infty}\frac{(-1)^n}{n^2}\exp[-n^2\pi^2\tau_c]\right]}{\left[1 - \frac{8}{\pi^2}\sum_{n=0}^{\infty}\frac{\exp[-(2n+1)^2\pi^2\tau_c]}{(2n+1)^2}\right]} \tag{A4.20}$$

If τ_c is large enough that steady state is reached before the solute containing vehicle is removed, then the summations of the exponentials go to zero, and

$$F = \frac{3+\alpha_V}{3(1+\alpha_V)} \tag{A4.6}$$

If τ_c is smaller than the time required to establish steady state, then F is smaller than $(1+\alpha_V)/(1+\alpha_V)/3$. In the limit of τ_c approaching zero, F is zero (see identities in Section 6.3.3). If the resistance to mass transfer from the membrane to the vehicle is very large for $\tau \geq \tau_c$, then $\alpha_V = 0$ and $F = 1$, as it should. If mass transfer from the membrane to the vehicle is not limited for $\tau \geq \tau_c$, then $\alpha_V \to \infty$ and $F = 1/3$, consistent with Model A3.

6.2.1.4.2 Laplace Domain Solutions

Other Definitions

$$Z = \frac{\alpha_V + 1}{s} + 2\sum_{n=1}^{\infty}\frac{1}{(n^2\pi^2+s)}\exp[-n^2\pi^2\tau_c] \tag{A4.7}$$

Concentration Membrane 1 ($0 \leq \xi \leq 1$):

for $0 < \tau \leq \tau_c$ (from Model A1)

$$\mathcal{L}\{\widetilde{C}_1\} = \frac{\cosh(\sqrt{s}\xi)}{s} - \frac{\cosh(\sqrt{s})\sinh(\sqrt{s}\xi)}{s\sinh(\sqrt{s})} \tag{A4.8}$$

for $\tau \geq \tau_c$

$$\mathcal{L}\{\widetilde{C}_1(\tau-\tau_c,\xi)\} = \frac{Z\sinh(\sqrt{s}[1-\xi])}{-\sqrt{s}\cosh(\sqrt{s}) - \alpha_V\sinh(\sqrt{s})} + \frac{1-\xi}{s} - \frac{2}{\pi}\sum_{n=1}^{\infty}\frac{\sin(n\pi\xi)}{n(n^2\pi^2+s)}\exp[-n^2\pi^2\tau_c] \tag{A4.9}$$

$$\mathcal{L}\{\langle\widetilde{C}_1(\tau-\tau_c)\rangle\} = \frac{Z[1-\cosh(\sqrt{s})]}{s\cosh(\sqrt{s}) - \alpha_V\sqrt{s}\sinh(\sqrt{s})} + \frac{1}{2s} - 2\sum_{n=1}^{\infty}\frac{[1-(-1)^n]}{(n^2\pi^2+s)}\exp[-n^2\pi^2\tau_c] \tag{A4.10}$$

Flux Membrane 1 ($0 \leq \xi \leq 1$):

for $0 < \tau \leq \tau_c$ (from Model A1)

$$\mathcal{L}\{\widetilde{J}_1\} = \sqrt{s}\left(-\frac{\sinh(\sqrt{s}\xi)}{s} + \frac{\cosh(\sqrt{s})\cosh(\sqrt{s}\xi)}{s\sinh(\sqrt{s})}\right) \tag{A4.11}$$

for $\tau \geq \tau_c$

$$\mathcal{L}\{\widetilde{J}_1(\tau - \tau_c, \xi)\} = \frac{-\sqrt{s}\,Z\cosh[\sqrt{s}(1-\xi)]}{\sqrt{s}\cosh(\sqrt{s}) + \alpha_V\sinh(\sqrt{s})} + \frac{1}{s} + 2\sum_{n=1}^{\infty}\frac{\cos(n\pi\xi)}{(n^2\pi^2 + s)}\exp[-n^2\pi^2\tau_c] \tag{A4.12}$$

Cumulative Mass Transfer

for $\tau \geq \tau_c$

From vehicle to membrane 1:

$$\mathcal{L}\{\widetilde{M}_{V,1}(\tau - \tau_c, \xi) - \widetilde{M}_{V,1}(\tau_c, \xi)\} = \mathcal{L}\{\widetilde{J}_1((\tau - \tau_c, \xi), \xi)\big|_{\xi=0}/s\} = \frac{-Z\cosh[\sqrt{s}]}{s\cosh(\sqrt{s}) + \alpha_V\sqrt{s}\sinh(\sqrt{s})} + \frac{1}{s^2} + \frac{2}{s}\sum_{n=1}^{\infty}\frac{1}{(n^2\pi^2 + s)}\exp[-n^2\pi^2\tau_c] \tag{A4.13}$$

From membrane 1 to receptor:

$$\mathcal{L}\{\widetilde{M}_{1,R}(\tau - \tau_c, \xi) - \widetilde{M}_{1,R}(\tau_c, \xi)\} = \mathcal{L}\left\{\widetilde{J}_1(\tau - \tau_c, \xi)\big|_{\xi=1}/s\right\} = \frac{-Z}{s\cosh(\sqrt{s}) + \alpha_V\sqrt{s}\sinh(\sqrt{s})} + \frac{1}{s^2} + \frac{2}{s}\sum_{n=1}^{\infty}\frac{(-1)^n}{(n^2\pi^2 + s)}\exp[-n^2\pi^2\tau_c] \tag{A4.14}$$

Notes Some of the solute that transferred from the vehicle to the membrane over the interval $0 < \tau \leq \tau_c$ transfers from the membrane into the solute-free vehicle that is on the membrane surface after $\tau = \tau_c$. A resistance in the solute-free vehicle represents the scenario in which solute absorbed into the membrane evaporates with a rate that depends on the gas phase mass transfer resistance and the solute saturation pressure [1].

The maximum total cumulative mass of solute that transfers into or out of membrane 1 can be determined by applying the final value theorem to the cumulative mass transfer equations written in the Laplace domain; see Eq. (6.99). We determine thereby that the total cumulative mass transfer from the vehicle to membrane 1 after the exposure to solute ends at $\tau = \tau_c$ is described by Eqs. (A4.15) and (A4.16):

$$\lim_{\tau\to\infty}[\widetilde{M}_{V,1}(\tau - \tau_c) - \widetilde{M}_{V,1}(\tau_c)] = \lim_{s\to 0}\mathcal{L}\{s[\widetilde{M}_{V,1}(\tau - \tau_c) - \widetilde{M}_{V,1}(\tau_c)]\} = \lim_{s\to 0}\mathcal{L}\left\{\widetilde{J}_1(\tau - \tau_c, \xi)\big|_{\xi=0}\right\} \tag{A4.15}$$

$$\lim_{\tau\to\infty}[\widetilde{M}_{V,1}(\tau - \tau_c) - \widetilde{M}_{V,1}(\tau_c)] = -\frac{2\alpha_V}{6(1+\alpha_V)}\left[1 + \frac{6}{\pi^2}\sum_{n=1}^{\infty}\frac{1}{n^2}\exp[-n^2\pi^2\tau_c]\right] \tag{A4.16}$$

and by Eqs. (A4.17) and (A4.18) for transfer from membrane 1 to the receptor for $\tau > \tau_c$:

$$\lim_{\tau\to\infty}[\widetilde{M}_{1,R}(\tau-\tau_c)-\widetilde{M}_{1,R}(\tau_c)] = \lim_{s\to 0}\mathcal{L}\{s[\widetilde{M}_{1,R}(\tau-\tau_c)-\widetilde{M}_{1,R}(\tau_c)]\}$$
$$= \lim_{s\to 0}\mathcal{L}\left\{\widetilde{J}_1(\tau-\tau_c,\xi)\Big|_{\xi=1}\right\} \quad \text{(A4.17)}$$

$$\lim_{\tau\to\infty}[\widetilde{M}_{1,R}(\tau-\tau_c)-\widetilde{M}_{1,R}(\tau_c)] = \frac{3+\alpha_V}{6(1+\alpha_V)}\left[1+\frac{12}{\pi^2}\sum_{n=1}^{\infty}\frac{(-1)^n}{n^2}\exp[-n^2\pi^2\tau_c]\right] \quad \text{(A4.18)}$$

The fraction of solute in membrane 1 at $\tau = \tau_c$ that eventually transfers into the receptor solution (F) is specified by the following mass balance:

$$F = \frac{\lim_{\tau\to\infty}[\widetilde{M}_{1,R}(\tau-\tau_c)-\widetilde{M}_{1,R}(\tau_c)]}{\langle\widetilde{C}_1\rangle\Big|_{\tau=\tau_c}} \quad \text{(A4.19)}$$

which combined with Eqs. (A4.2) and (A4.18) gives

$$F = \frac{3+\alpha_V}{3(1+\alpha_V)}\frac{\left[1+\frac{12}{\pi^2}\sum_{n=1}^{\infty}\frac{(-1)^n}{n^2}\exp[-n^2\pi^2\tau_c]\right]}{\left[1-\frac{8}{\pi^2}\sum_{n=0}^{\infty}\frac{\exp[-(2n+1)^2\pi^2\tau_c]}{(2n+1)^2}\right]} \quad \text{(A4.20)}$$

6.2.1.5 Model A5

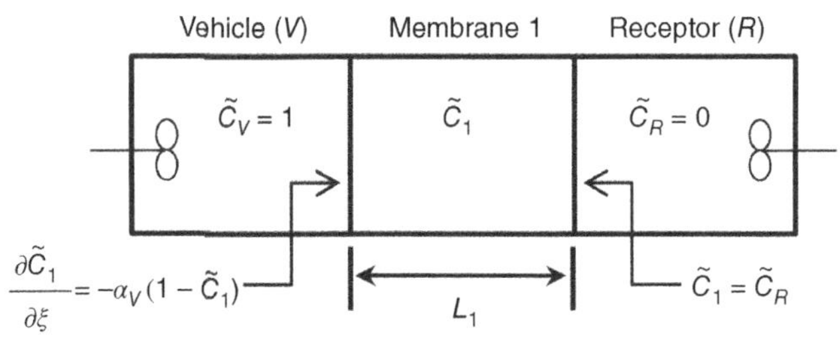

Description:

Vehicle: Infinite volume, well-stirred, constant solute concentration
Membranes: One, finite thickness, solute concentration initially zero
Receptor: Infinite volume, well-stirred, negligible concentration (sink conditions)
Vehicle/Membrane: Mass transfer limited
Membrane/Receptor: Local equilibrium

Differential equation:

$$\frac{\partial\widetilde{C}_1}{\partial\tau} = \frac{\partial^2\widetilde{C}_1}{\partial\xi^2} \qquad \text{for } 0\le\xi\le 1, \tau\ge 0$$

Boundary conditions:

$$\frac{\partial\widetilde{C}_1}{\partial\xi} = -\alpha_V(1-\widetilde{C}_1) \qquad \text{at } \xi = 0, \tau > 0$$

$$\widetilde{C}_1 = \widetilde{C}_R = 0 \qquad \text{at } \xi = 1, \tau \ge 0$$

Initial condition:

$$\widetilde{C}_1 = 0 \qquad \text{at } \tau = 0, 0 \le \xi \le 1$$

6.2.1.5.1 *Time Domain Solutions*

Characteristic Equation for Eigenvalues

$$\lambda_n \cot \lambda_n = -\alpha_V \tag{A5.1}$$

Concentration Membrane 1 ($0 \le \xi \le 1$):

$$\widetilde{C}_1 = \frac{\alpha_V}{\alpha_V + 1}(1 - \xi) - 2\alpha_V \sum_{n=1}^{\infty} \frac{\alpha_V \sin(\lambda_n \xi) + \lambda_n \cos(\lambda_n \xi)}{\lambda_n(\alpha_V^2 + \alpha_V + \lambda_n^2)} \exp[-\lambda_n^2 \tau] \tag{A5.2}$$

$$\langle \widetilde{C}_1 \rangle = \frac{\alpha_V}{2(\alpha_V + 1)} - 2\sum_{n=1}^{\infty} \frac{\alpha_V^2 - (\alpha_V^2 + \lambda_n^2)\cos(\lambda_n)}{\lambda_n^2(\alpha_V^2 + \alpha_V + \lambda_n^2)} \exp[-\lambda_n^2 \tau] \tag{A5.3}$$

Flux Membrane 1 ($0 \le \xi \le 1$):

$$\widetilde{J}_1 = \frac{\alpha_V}{\alpha_V + 1} - 2\alpha_V \sum_{n=1}^{\infty} \frac{\lambda_n \sin(\lambda_n \xi) - \alpha_V \cos(\lambda_n \xi)}{\alpha_V^2 + \alpha_V + \lambda_n^2} \exp[-\lambda_n^2 \tau] \tag{A5.4}$$

Cumulative Mass Transfer

From vehicle to membrane 1:

$$\widetilde{M}_{V,1} = \frac{\alpha_V}{\alpha_V + 1}\tau + \frac{\alpha_V^2}{3(\alpha_V + 1)^2} - 2\alpha_V^2 \sum_{n=1}^{\infty} \frac{\exp[-\lambda_n^2 \tau]}{\lambda_n^2(\alpha_V^2 + \alpha_V + \lambda_n^2)} \tag{A5.5}$$

From membrane 1 to receptor:

$$\widetilde{M}_{1,R} = \frac{\alpha_V}{\alpha_V + 1}\tau - \frac{\alpha_V(3 + \alpha_V)}{6(\alpha_V + 1)^2} - 2\sum_{n=1}^{\infty} \frac{(\alpha_V^2 + \lambda_n^2)\cos(\lambda_n)}{\lambda_n^2(\alpha_V^2 + \alpha_V + \lambda_n^2)} \exp[-\lambda_n^2 \tau] \tag{A5.6}$$

Notes Equations (A5.3), (A5.5), and (A5.6) are consistent with the mass balance requirement that $\langle \widetilde{C}_1 \rangle = \widetilde{M}_{V,1} - \widetilde{M}_{1,R}$. Carslaw and Jaeger [2] list the first six roots, λ_n, of Eq. (A5.1) for various values of α_V.

6.2.1.6 Model A6

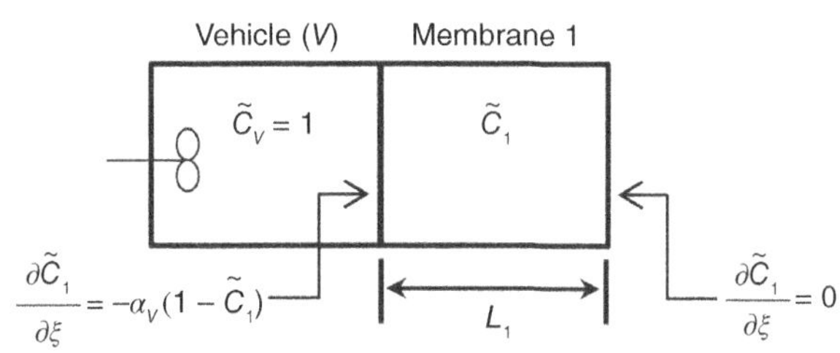

Description:

Vehicle: Infinite volume, well-stirred, constant solute concentration
Membranes: One, finite thickness, solute concentration initially zero
Receptor: None

Vehicle/Membrane: Mass transfer limited
Membrane/Receptor: No flux

Differential equation:

$$\frac{\partial \widetilde{C}_1}{\partial \tau} = \frac{\partial^2 \widetilde{C}_1}{\partial \xi^2} \qquad \text{for } 0 \le \xi \le 1, \tau \ge 0$$

Boundary conditions:

$$\frac{\partial \widetilde{C}_1}{\partial \xi} = -\alpha_V(1 - \widetilde{C}_1) \qquad \text{at } \xi = 0, \tau > 0$$

$$\frac{\partial \widetilde{C}_1}{\partial \xi} = 0 \qquad \text{at } \xi = 1, \tau > 0$$

Initial condition:

$$\widetilde{C}_1 = 0 \qquad \text{at } \tau = 0, 0 \le \xi \le 1$$

6.2.1.6.1 Time Domain Solutions

Characteristic Equation for Eigenvalues

$$\lambda_n \tan \lambda_n = \alpha_V \tag{A6.1}$$

Concentration Membrane 1 ($0 \le \xi \le 1$):

$$\widetilde{C}_1 = 1 - 2\alpha_V \sum_{n=1}^{\infty} \frac{\alpha_V \sin(\lambda_n \xi) + \lambda_n \cos(\lambda_n \xi)}{\lambda_n(\alpha_V^2 + \alpha_V + \lambda_n^2)} \exp[-\lambda_n^2 \tau] \tag{A6.2}$$

$$\langle \widetilde{C}_1 \rangle = 1 - 2\alpha_V^2 \sum_{n=1}^{\infty} \frac{\exp[-\lambda_n^2 \tau]}{\lambda_n^2(\alpha_V^2 + \alpha_V + \lambda_n^2)} \tag{A6.3}$$

Flux Membrane 1 ($0 \le \xi \le 1$):

$$\widetilde{J}_1 = 2\alpha_V \sum_{n=1}^{\infty} \frac{\alpha_V \cos(\lambda_n \xi) - \lambda_n \sin(\lambda_n \xi)}{\alpha_V^2 + \alpha_V + \lambda_n^2} \exp[-\lambda_n^2 \tau] \tag{A6.4}$$

Cumulative Mass Transfer

From vehicle to membrane 1:

$$\widetilde{M}_{V,1} = 1 - 2\alpha_V^2 \sum_{n=1}^{\infty} \frac{\exp[-\lambda_n^2 \tau]}{\lambda_n^2(\alpha_V^2 + \alpha_V + \lambda_n^2)} \tag{A6.5}$$

From membrane 1 to receptor:

$$\widetilde{M}_{1,R} = 0 \tag{A6.6}$$

Notes Because there is no flux out of membrane 1, $\widetilde{M}_{V,1} = \langle \widetilde{C}_1 \rangle$. In addition, because $\langle \widetilde{C}_1 \rangle = 0$ at $\tau = 0$, it follows from Eq. (A6.3) that

$$\sum_{n=1}^{\infty} \frac{1}{\lambda_n^2(\alpha_V^2 + \alpha_V + \lambda_n^2)} = \frac{1}{2\alpha_V^2} \tag{A6.7}$$

Carslaw and Jaeger [2] list the first six roots, λ_n, of Eq. (A6.1) for various values of α_V.

6.2.1.7 Model A7

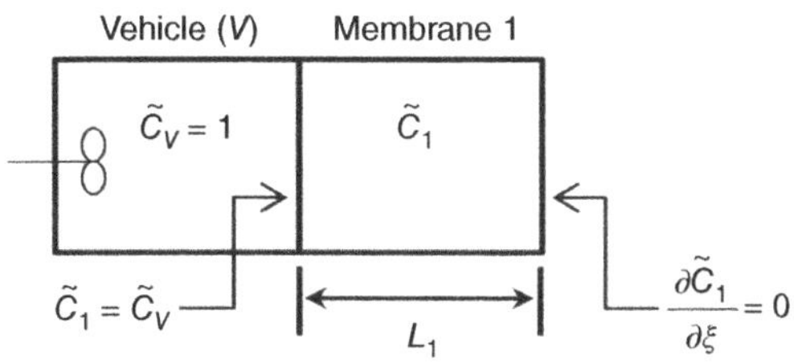

Description:

Vehicle: Infinite volume, well-stirred, constant solute concentration
Membranes: One, finite thickness, solute concentration initially zero
Receptor: None
Vehicle/Membrane: Local equilibrium
Membrane/Receptor: No flux

Differential equation:

$$\frac{\partial \tilde{C}_1}{\partial \tau} = \frac{\partial^2 \tilde{C}_1}{\partial \xi^2} \qquad \text{for } 0 \le \xi \le 1, \tau \ge 0$$

Boundary conditions:

$$\tilde{C}_1 = \tilde{C}_V = 1 \qquad \text{at } \xi = 0, \tau > 0$$

$$\frac{\partial \tilde{C}_1}{\partial \xi} = 0 \qquad \text{at } \xi = 1, \tau > 0$$

Initial condition:

$$\tilde{C}_1 = 0 \qquad \text{at } \tau = 0, 0 \le \xi \le 1$$

6.2.1.7.1 Time Domain Solutions

Eigenvalues

$$\lambda_n = (2n+1)\pi/2 \tag{A7.1}$$

Concentration Membrane 1 ($0 \le \xi \le 1$):

$$\tilde{C}_1 = 1 - 2\sum_{n=0}^{\infty} \frac{\sin(\lambda_n \xi)}{\lambda_n} \exp[-\lambda_n^2 \tau] \tag{A7.2}$$

$$\langle \tilde{C}_1 \rangle = 1 - 2\sum_{n=0}^{\infty} \frac{\exp[-\lambda_n^2 \tau]}{\lambda_n^2} \tag{A7.3}$$

Flux Membrane 1 ($0 \le \xi \le 1$):

$$\tilde{J}_1 = 2\sum_{n=1}^{\infty} \cos(\lambda_n \xi) \exp[-\lambda_n^2 \tau] \tag{A7.4}$$

Cumulative Mass Transfer

From vehicle to membrane 1:

$$\widetilde{M}_{V,1} = 2\sum_{n=1}^{\infty}\frac{(1 - \exp[-\lambda_n^2\tau])}{\lambda_n^2} \tag{A7.5}$$

From membrane 1 to receptor:

$$\widetilde{M}_{1,R} = 0 \tag{A7.6}$$

Notes Because there is no flux out of membrane 1, $\widetilde{M}_{V,1} = \langle\widetilde{C}_1\rangle$.

6.2.1.8 Model A8

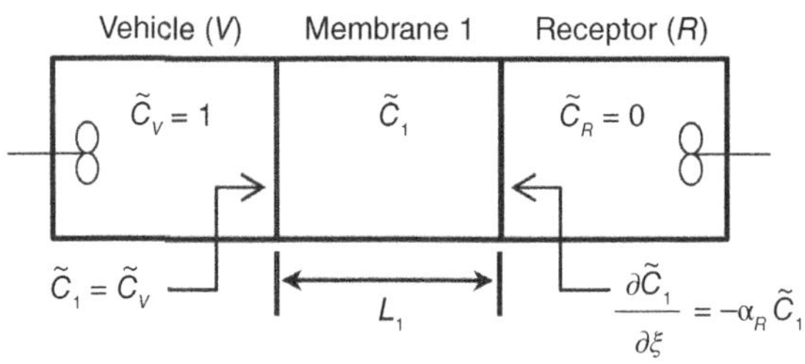

Description:

Vehicle: Infinite volume, well-stirred, constant solute concentration
Membranes: One, finite thickness, solute concentration initially zero
Receptor: Infinite volume, well-stirred, negligible concentration (sink conditions)
Vehicle/Membrane: Local equilibrium
Membrane/Receptor: Mass transfer limited

Differential equation:

$$\frac{\partial\widetilde{C}_1}{\partial\tau} = \frac{\partial^2\widetilde{C}_1}{\partial\xi^2} \qquad \text{for } 0 \le \xi \le 1, \tau \ge 0$$

Boundary conditions:

$$\widetilde{C}_1 = \widetilde{C}_V = 1 \qquad \text{at } \xi = 0, \tau > 0$$

$$\frac{\partial\widetilde{C}_1}{\partial\xi} = -\alpha_R\widetilde{C}_1 \qquad \text{at } \xi = 1, \tau > 0$$

$$\widetilde{C}_R = 0 \qquad \text{at } \tau > 0$$

Initial condition:

$$\widetilde{C}_1 = 0 \qquad \text{at } \tau = 0, 0 \le \xi \le 1$$

6.2.1.8.1 Time Domain Solutions

Characteristic Equation for Eigenvalues

$$\lambda_n \cot\lambda_n = -\alpha_R \tag{A8.1}$$

Concentration Membrane 1 ($0 \le \xi \le 1$):

$$\widetilde{C}_1 = 1 - \frac{\alpha_R}{\alpha_R + 1}\xi - 2\sum_{n=1}^{\infty}\frac{(\alpha_R^2 + \lambda_n^2)\sin(\lambda_n\xi)}{\lambda_n(\alpha_R^2 + \alpha_R + \lambda_n^2)}\exp[-\lambda_n^2\tau] \tag{A8.2}$$

$$\langle\widetilde{C}_1\rangle = \frac{2 + \alpha_R}{2(\alpha_R + 1)} - 2\sum_{n=1}^{\infty}\frac{(\alpha_R^2 + \lambda_n^2)(1 - \cos(\lambda_n))}{\lambda_n^2(\alpha_R^2 + \alpha_R + \lambda_n^2)}\exp[-\lambda_n^2\tau] \tag{A8.3}$$

Flux Membrane 1 ($0 \le \xi \le 1$):

$$\widetilde{J}_1 = \frac{\alpha_R}{\alpha_R + 1} + 2\sum_{n=1}^{\infty}\frac{(\alpha_R^2 + \lambda_n^2)\cos(\lambda_n\xi)}{\alpha_R^2 + \alpha_R + \lambda_n^2}\exp[-\lambda_n^2\tau] \tag{A8.4}$$

Cumulative Mass Transfer From vehicle to membrane 1:

$$\widetilde{M}_{V,1} = \frac{\alpha_R}{\alpha_R + 1}\tau + \frac{\alpha_R^2 + 3\alpha_R + 3}{3(\alpha_R + 1)^2} - 2\sum_{n=1}^{\infty}\frac{(\alpha_R^2 + \lambda_n^2)}{\lambda_n^2(\alpha_R^2 + \alpha_R + \lambda_n^2)}\exp[-\lambda_n^2\tau] \tag{A8.5}$$

From membrane 1 to receptor:

$$\widetilde{M}_{1,R} = \frac{\alpha_R}{\alpha_R + 1}\tau - \frac{\alpha_R(3 + \alpha_R)}{6(\alpha_R + 1)^2} - 2\sum_{n=1}^{\infty}\frac{(\alpha_R^2 + \lambda_n^2)\cos(\lambda_n)}{\lambda_n^2(\alpha_R^2 + \alpha_R + \lambda_n^2)}\exp[-\lambda_n^2\tau] \tag{A8.6}$$

Notes Equations (A5.3), (A5.5), and (A5.6) are consistent with the mass balance requirement that $\langle\widetilde{C}_1\rangle = \widetilde{M}_{V,1} - \widetilde{M}_{1,R}$. Carslaw and Jaeger [2] list the first six roots, λ_n, of Eq. (A8.1) for various values of α_R.

6.2.1.9 Model A9

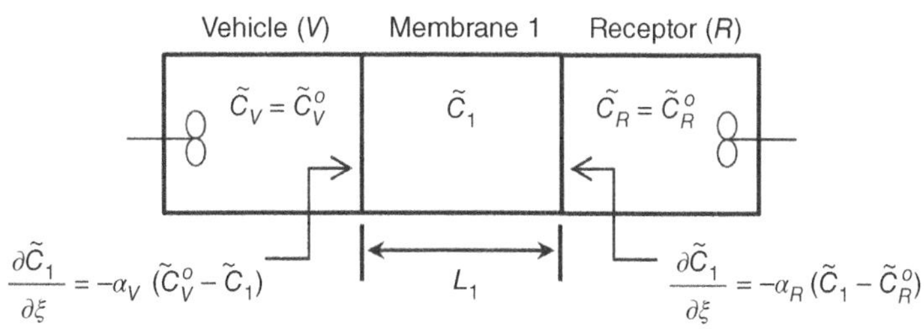

Source: Based on McCarley and Bunge [3, 4].

Description:

Vehicle: Infinite volume, well-stirred, constant solute concentration
Membranes: One, finite thickness, solute concentration initially zero
Receptor: Infinite volume, well-stirred, constant solute concentration
Vehicle/Membrane: Mass transfer limited
Membrane/Receptor: Mass transfer limited
Notes: Vehicle and receptor concentrations are restricted as follows:
if $C_V^o \neq 0$, then $\widetilde{C}_V^o = 1$, $\widetilde{C}_1 = \frac{C_1}{K_{1/V}C_V^o}$, and $\widetilde{C}_R^o = \frac{C_R^o}{K_{R/V}C_V^o}$
if $C_V^o = 0$, then $\widetilde{C}_V^o = 0$, $\widetilde{C}_1 = \frac{C_1}{K_{1/R}C_R^o}$, and $\widetilde{C}_R^o = 1$

Model A9a represents the limiting case of $\widetilde{C}_V^o = 1$ and $\widetilde{C}_R^o = 0$. Model A9b represents the complementary limiting case of $\widetilde{C}_V^o = 0$ and $\widetilde{C}_R^o = 1$.

Differential equation:

$$\frac{\partial \widetilde{C}_1}{\partial \tau} = \frac{\partial^2 \widetilde{C}_1}{\partial \xi^2} \qquad \text{for } 0 \le \xi \le 1,\ \ \tau \ge 0$$

Boundary conditions:

$$\frac{\partial \widetilde{C}_1}{\partial \xi} = -\alpha_V(\widetilde{C}_V^o - \widetilde{C}_1) \qquad \text{at } \xi = 0,\ \ \tau > 0$$

$$\frac{\partial \widetilde{C}_1}{\partial \xi} = -\alpha_R(\widetilde{C}_1 - \widetilde{C}_R^o) \qquad \text{at } \xi = 1,\ \ \tau > 0$$

Initial condition:

$$\widetilde{C}_1 = 0 \qquad \text{at } \tau = 0,\ \ 0 \le \xi \le 1$$

6.2.1.9.1 Time Domain Solutions

Characteristic Equation for Eigenvalues

$$\lambda_n \cot \lambda_n = \frac{\lambda_n^2 - \alpha_R\alpha_V}{\alpha_R + \alpha_V} \tag{A9.1}$$

Concentration Membrane 1 ($0 \le \xi \le 1$):

$$\widetilde{C}_1 = \frac{\alpha_V(1 + \alpha_R - \alpha_R\xi)\widetilde{C}_V^o + \alpha_R(1 + \alpha_V\xi)\widetilde{C}_R^o}{\alpha_R + \alpha_V + \alpha_R\alpha_V} - 2\sum_{n=1}^{\infty} \exp[-\lambda_n^2\tau] \times \frac{[(\alpha_R^2 + \lambda_n^2)\alpha_V\widetilde{C}_V^o + (\lambda_n^2 - \alpha_R\alpha_V)\alpha_R\widetilde{C}_R^o/\cos(\lambda_n)]}{\lambda_n[(\alpha_R^2 + \lambda_n^2)(\alpha_V^2 + \lambda_n^2) + (\alpha_R\alpha_V + \lambda_n^2)(\alpha_R + \alpha_V)]} \times [\lambda_n \cos(\lambda_n\xi) + \alpha_V \sin(\lambda_n\xi)] \tag{A9.2}$$

$$\langle\widetilde{C}_1\rangle = \frac{\alpha_V(2 + \alpha_R)\widetilde{C}_V^o + \alpha_R(2 + \alpha_V)\widetilde{C}_R^o}{2(\alpha_R + \alpha_V + \alpha_R\alpha_V)} - 2\sum_{n=1}^{\infty} \exp[-\lambda_n^2\tau] \times \frac{[(\alpha_R^2 + \lambda_n^2)\alpha_V\widetilde{C}_V^o + (\lambda_n^2 - \alpha_R\alpha_V)\alpha_R\widetilde{C}_R^o/\cos(\lambda_n)]}{\lambda_n^2[(\alpha_R^2 + \lambda_n^2)(\alpha_V^2 + \lambda_n^2) + (\alpha_R\alpha_V + \lambda_n^2)(\alpha_R + \alpha_V)]} \times [\alpha_V + \lambda_n \sin(\lambda_n) - \alpha_V \cos(\lambda_n)] \tag{A9.3}$$

or equivalently

$$\langle\widetilde{C}_1\rangle = \frac{\alpha_V(2 + \alpha_R)\widetilde{C}_V^o + \alpha_R(2 + \alpha_V)\widetilde{C}_R^o}{2(\alpha_R + \alpha_V + \alpha_R\alpha_V)} - 2\sum_{n=1}^{\infty} \exp[-\lambda_n^2\tau] \times \frac{[(\alpha_R^2 + \lambda_n^2)\alpha_V\widetilde{C}_V^o + (\lambda_n^2 - \alpha_R\alpha_V)\alpha_R\widetilde{C}_R^o/\cos(\lambda_n)]}{\alpha_V\lambda_n^2[(\alpha_R^2 + \lambda_n^2)(\alpha_V^2 + \lambda_n^2) + (\alpha_R\alpha_V + \lambda_n^2)(\alpha_R + \alpha_V)]} \times \left[\alpha_V + \frac{\alpha_R(\alpha_V^2 + \lambda_n^2)}{(\lambda_n^2 - \alpha_R\alpha_V)}\cos(\lambda_n)\right] \tag{A9.4}$$

Flux Membrane 1 ($0 \le \xi \le 1$):

$$\widetilde{J}_1 = \frac{\alpha_V \alpha_R (\widetilde{C}_V^o - \widetilde{C}_R^o)}{\alpha_R + \alpha_V + \alpha_R \alpha_V} - 2\sum_{n=1}^{\infty} \exp[-\lambda_n^2 \tau] \times \frac{[(\alpha_R^2 + \lambda_n^2)\alpha_V \widetilde{C}_V^o + (\lambda_n^2 - \alpha_R \alpha_V)\alpha_R \widetilde{C}_R^o / \cos(\lambda_n)]}{[(\alpha_R^2 + \lambda_n^2)(\alpha_V^2 + \lambda_n^2) + (\alpha_R \alpha_V + \lambda_n^2)(\alpha_R + \alpha_V)]} [\lambda_n \sin(\lambda_n \xi) - \alpha_V \cos(\lambda_n \xi)] \tag{A9.5}$$

Cumulative Mass Transfer

From vehicle to membrane 1:

$$\begin{aligned}\widetilde{M}_{V,1} = {} & \frac{\alpha_V \alpha_R (\widetilde{C}_V^o - \widetilde{C}_R^o)}{\alpha_R + \alpha_V + \alpha_R \alpha_V} \tau \\ & + \frac{2\alpha_V^2 (3 + 3\alpha_R + \alpha_R^2) \widetilde{C}_V^o + \alpha_V \alpha_R (6 + 3\alpha_R + 3\alpha_V + \alpha_R \alpha_V) \widetilde{C}_R^o}{6(\alpha_R + \alpha_V + \alpha_R \alpha_V)^2} \\ & - 2\alpha_V \sum_{n=1}^{\infty} \frac{[(\alpha_R^2 + \lambda_n^2)\alpha_V \widetilde{C}_V^o + (\lambda_n^2 - \alpha_R \alpha_V)\, \alpha_R \widetilde{C}_R^o / \cos(\lambda_n)]}{\lambda_n^2 [(\alpha_R^2 + \lambda_n^2)(\alpha_V^2 + \lambda_n^2) + (\alpha_R \alpha_V + \lambda_n^2)(\alpha_R + \alpha_V)]} \exp[-\lambda_n^2 \tau]\end{aligned} \tag{A9.6}$$

From membrane 1 to receptor:

$$\begin{aligned}\widetilde{M}_{1,R} = {} & \frac{\alpha_V \alpha_R (\widetilde{C}_V^o - \widetilde{C}_R^o)}{\alpha_R + \alpha_V + \alpha_R \alpha_V} \tau \\ & - \frac{\alpha_V \alpha_R (6 + 3\alpha_R + 3\alpha_V + \alpha_R \alpha_V) \widetilde{C}_V^o + 2\alpha_R^2 (3 + 3\alpha_V + \alpha_V^2) \widetilde{C}_R^o}{6(\alpha_R + \alpha_V + \alpha_R \alpha_V)^2} \\ & - 2\sum_{n=1}^{\infty} \exp[-\lambda_n^2 \tau] \frac{[(\alpha_R^2 + \lambda_n^2)\alpha_V \widetilde{C}_V^o + (\lambda_n^2 - \alpha_R \alpha_V)\, \alpha_R \widetilde{C}_R^o / \cos(\lambda_n)]}{\lambda_n^2 [(\alpha_R^2 + \lambda_n^2)(\alpha_V^2 + \lambda_n^2) + (\alpha_R \alpha_V + \lambda_n^2)(\alpha_R + \alpha_V)]} \\ & \times [\alpha_V \cos(\lambda_n) - \lambda_n \sin(\lambda_n)]\end{aligned} \tag{A9.7}$$

or equivalently

$$\begin{aligned}\widetilde{M}_{1,R} = {} & \frac{\alpha_V \alpha_R (\widetilde{C}_V^o - \widetilde{C}_R^o)}{\alpha_R + \alpha_V + \alpha_R \alpha_V} \tau \\ & - \frac{\alpha_V \alpha_R (6 + 3\alpha_R + 3\alpha_V + \alpha_R \alpha_V) \widetilde{C}_V^o + 2\alpha_R^2 (3 + 3\alpha_V + \alpha_V^2) \widetilde{C}_R^o}{6(\alpha_R + \alpha_V + \alpha_R \alpha_V)^2} \\ & + 2\sum_{n=1}^{\infty} \exp[-\lambda_n^2 \tau] \frac{[(\alpha_R^2 + \lambda_n^2)\alpha_V \widetilde{C}_V^o + (\lambda_n^2 - \alpha_R \alpha_V)\alpha_R \widetilde{C}_R^o / \cos(\lambda_n)]}{\lambda_n^2 [(\alpha_R^2 + \lambda_n^2)(\alpha_V^2 + \lambda_n^2) + (\alpha_R \alpha_V + \lambda_n^2)(\alpha_R + \alpha_V)]} \\ & \times \left[\frac{\alpha_R (\alpha_V^2 + \lambda_n^2)}{(\lambda_n^2 - \alpha_R \alpha_V)} \cos(\lambda_n) \right]\end{aligned} \tag{A9.8}$$

Notes Equation (A9.3) for $\langle \widetilde{C}_1 \rangle$ includes a correction to the equation listed in McCarley et al. [3]. The second equation listed for $\langle \widetilde{C}_1 \rangle$, Eq. (A9.4), is derived by eliminating $\sin(\lambda_n)$ from Eq. (A9.3) using the characteristic equation for the eigenvalues. Specifically, the relationship

$$\lambda_n \sin(\lambda_n) - \alpha_V \cos(\lambda_n) = \alpha_R \left(\frac{\alpha_v^2 + \lambda_n^2}{\lambda_n^2 - \alpha_R \alpha_V} \right) \cos(\lambda_n) \tag{A9.9}$$

is derived using the following rearrangement of the characteristic equation, Eq. (A9.1)

$$\lambda_n \sin(\lambda_n) = \left(\frac{\alpha_R + \alpha_V}{\lambda_n^2 - \alpha_R \alpha_V} \right) \lambda_n^2 \cos(\lambda_n) \tag{A9.10}$$

Table 6.5 First 10 eigenvalues (λ_n) defined by Eq. (A9.1) for selected values of α_V and α_R.[a)]

	α_i	0.1	0.1	0.1	1	1	10
n	α_j	0.1	1	10	1	10	10
1		0.44352	0.92925	1.48991	1.30654	1.87531	2.62768
2		3.20399	3.45248	4.32711	3.67319	4.50726	5.30732
3		6.31485	6.45244	7.24107	6.58462	7.35496	8.06714
4		9.44600	9.53970	10.20960	9.63168	10.29261	10.9087
5		12.58227	12.65314	13.22148	12.72324	13.28668	13.81919
6		15.72069	15.77760	16.26535	15.83411	16.31893	16.78269
7		18.86016	18.90768	19.33210	18.95497	19.37753	19.78551
8		22.00024	22.04102	22.41524	22.08166	22.45463	22.81725
9		25.14070	25.17641	25.51025	25.21203	25.54500	25.87044
10		28.28141	28.31317	28.61404	28.34486	28.64510	28.93974

a) If $i = V, j = R$ and if $i = R, j = V$

Equation (A9.4) was provided in the published correction [4] to McCarley et al. [3]. Equation (A9.8) is derived by substituting the expression from Eq. (A9.9) into Eq. (A9.7). Table 6.5 lists the first 10 eigenvalues defined by Eq. (A9.1) for selected values of α_V and α_R.

6.2.1.10 Model A9a

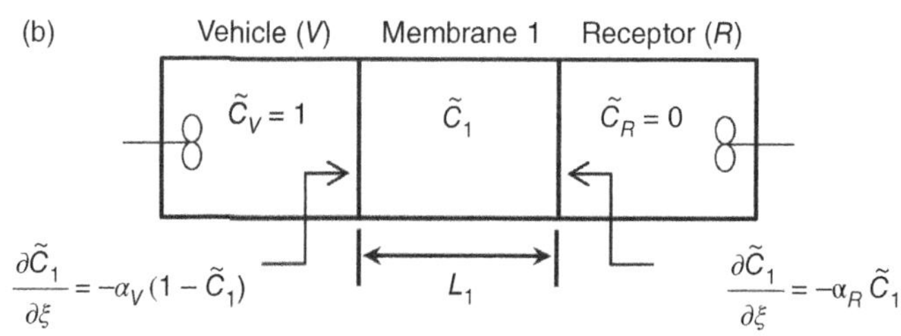

Source: Based on McCarley and Bunge [3, 4].

Description:

Vehicle: Infinite volume, well-stirred, constant solute concentration
Membranes: One, finite thickness, solute concentration initially zero
Receptor: Infinite volume, well-stirred, negligible concentration (sink conditions)
Vehicle/Membrane: Mass transfer limited
Membrane/Receptor: Mass transfer limited

Differential equation:

$$\frac{\partial \tilde{C}_1}{\partial \tau} = \frac{\partial^2 \tilde{C}_1}{\partial \xi^2} \qquad \text{for } 0 \leq \xi \leq 1, \ \tau \geq 0$$

Boundary conditions:

$$\frac{\partial \widetilde{C}_1}{\partial \xi} = -\alpha_V(1-\widetilde{C}_1) \qquad \text{at } \xi = 0, \quad \tau > 0$$

$$\frac{\partial \widetilde{C}_1}{\partial \xi} = -\alpha_R \widetilde{C}_1 \qquad \text{at } \xi = 1, \quad \tau > 0$$

Initial condition:

$$\widetilde{C}_1 = 0 \qquad \text{at } \tau = 0, \quad 0 \le \xi \le 1$$

6.2.1.10.1 Time Domain Solutions

Characteristic Equation for Eigenvalues

$$\lambda_n \cot \lambda_n = \frac{\lambda_n^2 - \alpha_R\alpha_V}{\alpha_R + \alpha_V} \tag{A9.1}$$

Concentration Membrane 1 ($0 \le \xi \le 1$):

$$\widetilde{C}_1 = \frac{\alpha_V(1+\alpha_R-\alpha_R\xi)}{\alpha_R+\alpha_V+\alpha_R\alpha_V} - 2\alpha_V \sum_{n=1}^{\infty} \frac{(\alpha_R^2+\lambda_n^2)[\lambda_n \cos(\lambda_n\xi) + \alpha_V \sin(\lambda_n\xi)]}{\lambda_n[(\alpha_R^2+\lambda_n^2)(\alpha_V^2+\lambda_n^2)+(\alpha_R\alpha_V+\lambda_n^2)(\alpha_R+\alpha_V)]} \exp[-\lambda_n^2\tau] \tag{A9.11}$$

$$\langle \widetilde{C}_1 \rangle = \frac{\alpha_V(2+\alpha_R)}{2(\alpha_R+\alpha_V+\alpha_R\alpha_V)} - 2\alpha_V \sum_{n=1}^{\infty} \exp[-\lambda_n^2\tau] \times \frac{[\alpha_V + \lambda_n \sin(\lambda_n) - \alpha_V \cos(\lambda_n)](\alpha_R^2+\lambda_n^2)}{\lambda_n^2[(\alpha_R^2+\lambda_n^2)(\alpha_V^2+\lambda_n^2)+(\alpha_R\alpha_V+\lambda_n^2)(\alpha_R+\alpha_V)]} \tag{A9.12}$$

or equivalently

$$\langle \widetilde{C}_1 \rangle = \frac{\alpha_V(2+\alpha_R)}{2(\alpha_R+\alpha_V+\alpha_R\alpha_V)} - 2\alpha_V \sum_{n=1}^{\infty} \left[\alpha_V + \frac{\alpha_R(\alpha_V^2+\lambda_n^2)}{(\lambda_n^2-\alpha_R\alpha_V)}\cos(\lambda_n)\right] \times \frac{(\alpha_R^2+\lambda_n^2)}{\lambda_n^2[(\alpha_R^2+\lambda_n^2)(\alpha_V^2+\lambda_n^2)+(\alpha_R\alpha_V+\lambda_n^2)(\alpha_R+\alpha_V)]} \exp[-\lambda_n^2\tau] \tag{A9.13}$$

Flux Membrane 1 ($0 \le \xi \le 1$):

$$\widetilde{J}_1 = \frac{\alpha_V\alpha_R}{\alpha_R+\alpha_V+\alpha_R\alpha_V} - 2\alpha_V \sum_{n=1}^{\infty} \frac{(\alpha_R^2+\lambda_n^2)(\lambda_n \sin(\lambda_n\xi) - \alpha_V \cos(\lambda_n\xi))}{[(\alpha_R^2+\lambda_n^2)(\alpha_V^2+\lambda_n^2)+(\alpha_R\alpha_V+\lambda_n^2)(\alpha_R+\alpha_V)]} \exp[-\lambda_n^2\tau] \tag{A9.14}$$

Cumulative Mass Transfer

From vehicle to membrane 1:

$$\widetilde{M}_{V,1} = \frac{\alpha_V\alpha_R}{\alpha_R+\alpha_V+\alpha_R\alpha_V}\tau + \frac{(3+3\alpha_R+\alpha_R^2)\alpha_V^2}{3(\alpha_R+\alpha_V+\alpha_R\alpha_V)^2} - 2\alpha_V^2 \sum_{n=1}^{\infty} \frac{(\alpha_R^2+\lambda_n^2)}{\lambda_n^2[(\alpha_R^2+\lambda_n^2)(\alpha_V^2+\lambda_n^2)+(\alpha_R\alpha_V+\lambda_n^2)(\alpha_R+\alpha_V)]} \exp[-\lambda_n^2\tau] \tag{A9.15}$$

From membrane 1 to receptor:

$$\widetilde{M}_{1,R} = \frac{\alpha_V\alpha_R}{\alpha_R + \alpha_V + \alpha_R\alpha_V}\tau - \frac{(6 + 3\alpha_R + 3\alpha_V + \alpha_R\alpha_V)\alpha_V\alpha_R}{6(\alpha_R + \alpha_V + \alpha_R\alpha_V)^2} - 2\alpha_V\sum_{n=1}^{\infty}\frac{(\alpha_R^2 + \lambda_n^2)(\alpha_V\cos(\lambda_n) - \lambda_n\sin(\lambda_n))}{\lambda_n^2[(\alpha_R^2 + \lambda_n^2)(\alpha_V^2 + \lambda_n^2) + (\alpha_R\alpha_V + \lambda_n^2)(\alpha_R + \alpha_V)]}\exp[-\lambda_n^2\tau] \quad \text{(A9.16)}$$

6.2.1.11 Model A9b

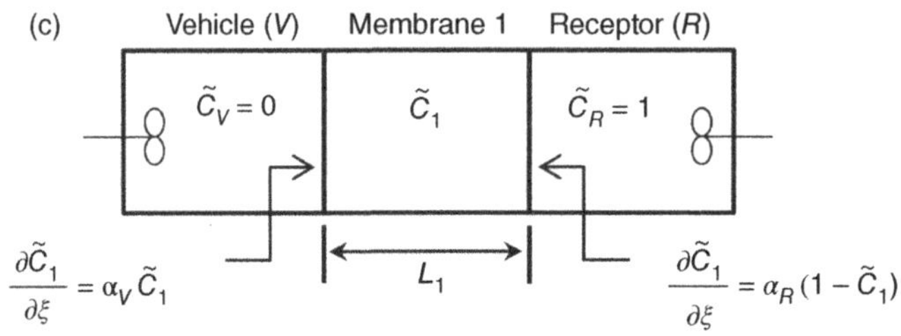

Source: Based on McCarley and Bunge [3, 4].

Description:

Vehicle: Infinite volume, well-stirred, negligible concentration (sink conditions)
Membranes: One, finite thickness, solute concentration initially zero
Receptor: Infinite volume, well-stirred, constant solute concentration
Vehicle/Membrane: Mass transfer limited
Membrane/Receptor: Mass transfer limited

Differential equation:

$$\frac{\partial\widetilde{C}_1}{\partial\tau} = \frac{\partial^2\widetilde{C}_1}{\partial\xi^2} \qquad \text{for } 0 \le \xi \le 1, \ \tau \ge 0$$

Boundary conditions:

$$\frac{\partial\widetilde{C}_1}{\partial\xi} = \alpha_V\widetilde{C}_1 \quad \text{at } \xi = 0, \ \tau > 0$$

$$\frac{\partial\widetilde{C}_1}{\partial\xi} = \alpha_R(1 - \widetilde{C}_1) \quad \text{at } \xi = 1, \ \tau > 0$$

Initial condition:

$$\widetilde{C}_1 = 0 \qquad \text{at } \tau = 0, \ 0 \le \xi \le 1$$

6.2.1.11.1 Time Domain Solutions

Characteristic Equation for Eigenvalues

$$\lambda_n\cot\lambda_n = \frac{\lambda_n^2 - \alpha_R\alpha_V}{\alpha_R + \alpha_V} \quad \text{(A9.1)}$$

Concentration Membrane 1 ($0 \le \xi \le 1$):

$$\widetilde{C}_1 = \frac{\alpha_R(1+\alpha_V\xi)}{\alpha_R+\alpha_V+\alpha_R\alpha_V} - 2\sum_{n=1}^{\infty}\exp[-\lambda_n^2\tau]\,\frac{[\lambda_n\cos(\lambda_n\xi)+\alpha_V\sin(\lambda_n\xi)](\lambda_n^2-\alpha_R\alpha_V)\alpha_R/\cos(\lambda_n)}{\lambda_n[(\alpha_R^2+\lambda_n^2)(\alpha_V^2+\lambda_n^2)+(\alpha_R\alpha_V+\lambda_n^2)(\alpha_R+\alpha_V)]} \tag{A9.17}$$

$$\langle\widetilde{C}_1\rangle = \frac{\alpha_R(2+\alpha_V)}{2(\alpha_R+\alpha_V+\alpha_R\alpha_V)} - 2\sum_{n=1}^{\infty}\exp[-\lambda_n^2\tau]\frac{[\alpha_V+\lambda_n\sin(\lambda_n)-\alpha_V\cos(\lambda_n)](\lambda_n^2-\alpha_R\alpha_V)\alpha_R/\cos(\lambda_n)}{\lambda_n^2[(\alpha_R^2+\lambda_n^2)(\alpha_V^2+\lambda_n^2)+(\alpha_R\alpha_V+\lambda_n^2)(\alpha_R+\alpha_V)]} \tag{A9.18}$$

or equivalently

$$\langle\widetilde{C}_1\rangle = \frac{\alpha_R(2+\alpha_V)}{2(\alpha_R+\alpha_V+\alpha_R\alpha_V)} - 2\sum_{n=1}^{\infty}\left[\alpha_V+\frac{\alpha_R(\alpha_V^2+\lambda_n^2)}{(\lambda_n^2-\alpha_R\alpha_V)}\cos(\lambda_n)\right]\exp[-\lambda_n^2\tau] \times \frac{(\lambda_n^2-\alpha_R\alpha_V)\alpha_R/\cos(\lambda_n)}{\lambda_n^2[(\alpha_R^2+\lambda_n^2)(\alpha_V^2+\lambda_n^2)+(\alpha_R\alpha_V+\lambda_n^2)(\alpha_R+\alpha_V)]} \tag{A9.19}$$

Flux Membrane 1 ($0 \le \xi \le 1$):

$$\widetilde{J}_1 = -\frac{\alpha_V\alpha_R}{\alpha_R+\alpha_V+\alpha_R\alpha_V} - 2\sum_{n=1}^{\infty}\exp[-\lambda_n^2\tau] \times \frac{[\lambda_n\sin(\lambda_n\xi)-\alpha_V\cos(\lambda_n\xi)](\lambda_n^2-\alpha_R\alpha_V)\alpha_R/\cos(\lambda_n)}{(\alpha_R^2+\lambda_n^2)(\alpha_V^2+\lambda_n^2)+(\alpha_R\alpha_V+\lambda_n^2)(\alpha_R+\alpha_V)} \tag{A9.20}$$

Cumulative Mass Transfer

From vehicle to membrane 1:

$$\widetilde{M}_{V,1} = -\frac{\alpha_V\alpha_R}{\alpha_R+\alpha_V+\alpha_R\alpha_V}\tau + \frac{\alpha_V\alpha_R(6+3\alpha_R+3\alpha_V+\alpha_R\alpha_V)}{6(\alpha_R+\alpha_V+\alpha_R\alpha_V)^2} - 2\alpha_V\sum_{n=1}^{\infty}\frac{(\lambda_n^2-\alpha_R\alpha_V)\alpha_R/\cos(\lambda_n)}{\lambda_n^2[(\alpha_R^2+\lambda_n^2)(\alpha_V^2+\lambda_n^2)+(\alpha_R\alpha_V+\lambda_n^2)(\alpha_R+\alpha_V)]}\exp[-\lambda_n^2\tau] \tag{A9.21}$$

From membrane 1 to receptor:

$$\widetilde{M}_{1,R} = -\frac{\alpha_V\alpha_R}{\alpha_R+\alpha_V+\alpha_R\alpha_V}\tau - \frac{2\alpha_R^2(3+3\alpha_V+\alpha_V^2)}{6(\alpha_R+\alpha_V+\alpha_R\alpha_V)^2} - 2\sum_{n=1}^{\infty}\exp[-\lambda_n^2\tau] \times \frac{[\alpha_V\cos(\lambda_n)-\lambda_n\sin(\lambda_n)][(\lambda_n^2-\alpha_R\alpha_V)\alpha_R/\cos(\lambda_n)]}{\lambda_n^2[(\alpha_R^2+\lambda_n^2)(\alpha_V^2+\lambda_n^2)+(\alpha_R\alpha_V+\lambda_n^2)(\alpha_R+\alpha_V)]} \tag{A9.22}$$

or equivalently

$$\widetilde{M}_{1,R} = -\frac{\alpha_V\alpha_R}{\alpha_R+\alpha_V+\alpha_R\alpha_V}\tau - \frac{2\alpha_R^2(3+3\alpha_V+\alpha_V^2)}{6(\alpha_R+\alpha_V+\alpha_R\alpha_V)^2} + 2\sum_{n=1}^{\infty}\left[\frac{\alpha_R(\alpha_V^2+\lambda_n^2)}{(\lambda_n^2-\alpha_R\alpha_V)}\cos(\lambda_n)\right]\exp[-\lambda_n^2\tau] \times \frac{[(\lambda_n^2-\alpha_R\alpha_V)\,\alpha_R/\cos(\lambda_n)]}{\lambda_n^2[(\alpha_R^2+\lambda_n^2)(\alpha_V^2+\lambda_n^2)+(\alpha_R\alpha_V+\lambda_n^2)(\alpha_R+\alpha_V)]} \tag{A9.23}$$

6.2.1.12 Model A10

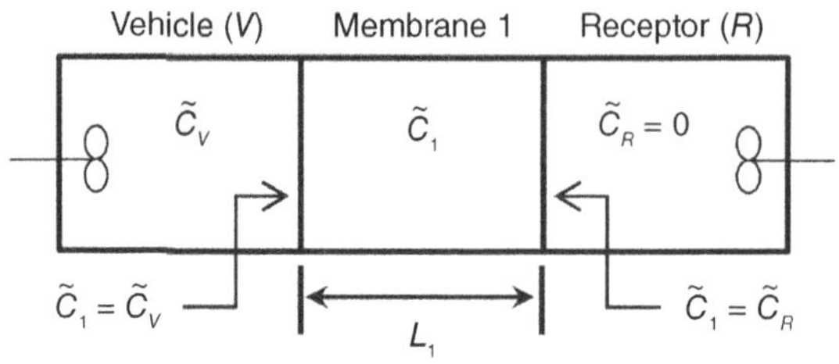

Description:

Vehicle: Finite volume, well-stirred
Membranes: One, finite thickness, solute concentration initially zero
Receptor: Infinite volume, well-stirred, negligible concentration (sink conditions)
Vehicle/Membrane: Local equilibrium
Membrane/Receptor: Local equilibrium

Differential equations:

$$\frac{\partial \widetilde{C}_1}{\partial \tau} = \frac{\partial^2 \widetilde{C}_1}{\partial \xi^2} \qquad \text{for } 0 \le \xi \le 1, \quad \tau \ge 0$$

$$\frac{d\widetilde{C}_V}{d\tau} = \frac{1}{\Lambda_V} \left.\frac{\partial \widetilde{C}_1}{\partial \xi}\right|_{\xi=0} \qquad \text{for } \tau \ge 0$$

Boundary conditions:

$$\widetilde{C}_1 = \widetilde{C}_V \qquad \text{at } \xi = 0, \quad \tau > 0$$

$$\widetilde{C}_1 = \widetilde{C}_R = 0 \qquad \text{at } \xi = 1, \quad \tau > 0$$

Initial condition:

$$\widetilde{C}_V = 1 \qquad \text{at } \tau = 0$$

$$\widetilde{C}_1 = 0 \qquad \text{at } \tau = 0, \quad 0 \le \xi \le 1$$

6.2.1.12.1 Time Domain Solutions

Characteristic Equation for Eigenvalues

$$\lambda_n \tan \lambda_n = 1/\Lambda_V \tag{A10.1}$$

Concentration

Membrane 1 ($0 \le \xi \le 1$):

$$\widetilde{C}_1 = 2\Lambda_V \sum_{n=1}^{\infty} \frac{\sin[\lambda_n(1-\xi)]}{(1 + \Lambda_V + \Lambda_V^2 \lambda_n^2)\sin(\lambda_n)} \exp[-\lambda_n^2 \tau] \tag{A10.2}$$

$$\langle \widetilde{C}_1 \rangle = 2\Lambda_V \sum_{n=1}^{\infty} \frac{1 - \cos(\lambda_n)}{(1 + \Lambda_V + \Lambda_V^2 \lambda_n^2)\lambda_n \sin(\lambda_n)} \exp[-\lambda_n^2 \tau] \tag{A10.3}$$

Vehicle

$$\widetilde{C}_V = 2\Lambda_V \sum_{n=1}^{\infty} \frac{\exp[-\lambda_n^2 \tau]}{1 + \Lambda_V + \Lambda_V^2 \lambda_n^2} \tag{A10.4}$$

Flux Membrane 1 ($0 \leq \xi \leq 1$):

$$\widetilde{J}_1 = 2\Lambda_V \sum_{n=1}^{\infty} \left[\frac{\lambda_n \cos[\lambda_n(1-\xi)]}{(1 + \Lambda_V + \Lambda_V^2 \lambda_n^2)\sin(\lambda_n)} \right] \exp[-\lambda_n^2 \tau] \tag{A10.5}$$

Cumulative Mass Transfer

From vehicle to membrane 1:

$$\widetilde{M}_{V,1} = \Lambda_V - 2\Lambda_V^2 \sum_{n=1}^{\infty} \frac{\exp[-\lambda_n^2 \tau]}{1 + \Lambda_V + \Lambda_V^2 \lambda_n^2} \tag{A10.6}$$

From membrane 1 to receptor:

$$\widetilde{M}_{1,R} = \Lambda_V - 2\Lambda_V^2 \sum_{n=1}^{\infty} \frac{\exp[-\lambda_n^2 \tau]}{\cos(\lambda_n)(1 + \Lambda_V + \Lambda_V^2 \lambda_n^2)} \tag{A10.7}$$

Plots Figures 6.11 and 6.12 show, respectively, the time variation of the concentration $\widetilde{C}_1$ and the average concentration $\langle \widetilde{C}_1 \rangle$ for selected values of Λ_V. As τ increases, $\widetilde{C}_1$ at $\xi = 0$ decreases by smaller amounts as the holding capacity of the vehicle relative the membrane Λ_V increases. If Λ_V is large enough, then, at larger τ, $\widetilde{C}_1$ varies linearly with position ξ, indicating a pseudo-steady state with

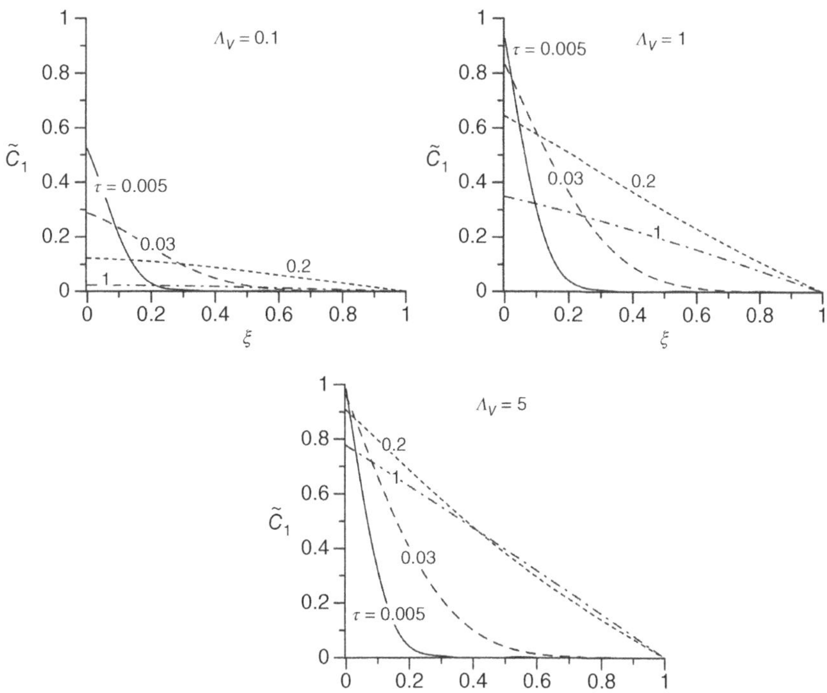

Figure 6.11 Distribution of the dimensionless concentration in the membrane $\widetilde{C}_1$ at various values of τ for selected values of Λ_V.

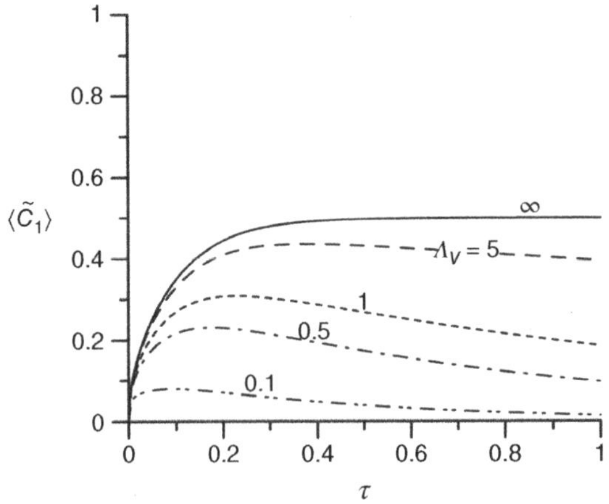

Figure 6.12 Dimensionless average concentration in the membrane $\langle \widetilde{C}_1 \rangle$ plotted as a function of τ for selected values of Λ_V.

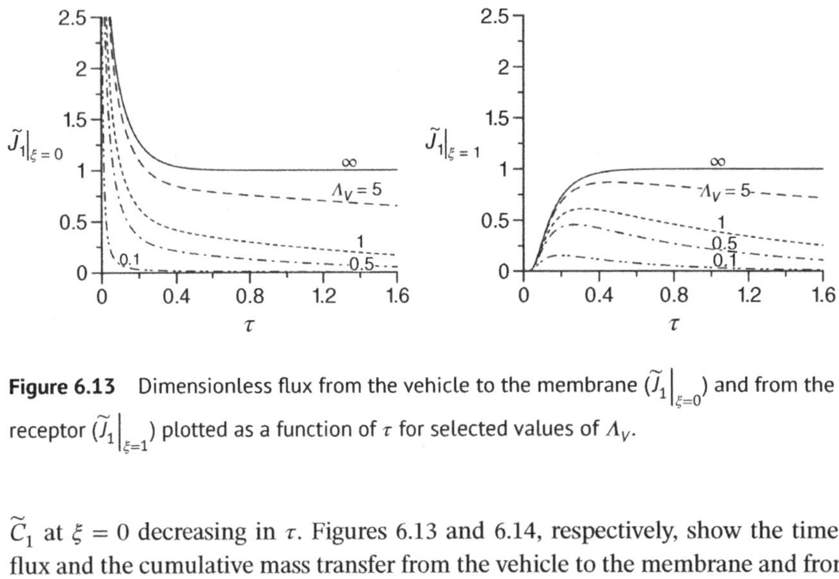

Figure 6.13 Dimensionless flux from the vehicle to the membrane ($\widetilde{J}_1\big|_{\xi=0}$) and from the membrane to the receptor ($\widetilde{J}_1\big|_{\xi=1}$) plotted as a function of τ for selected values of Λ_V.

$\widetilde{C}_1$ at $\xi = 0$ decreasing in τ. Figures 6.13 and 6.14, respectively, show the time variation of the flux and the cumulative mass transfer from the vehicle to the membrane and from the membrane to the receptor for selected values of Λ_V. Figure 6.15 shows the time variation of the cumulative mass transfer from the vehicle to the membrane $M_{V,1}$ and from the membrane to the receptor $M_{1,R}$ relative to the initial mass in the vehicle ($V_V C_V^o$). These cumulative mass ratios correspond to the dimensionless cumulative mass transfer divided by Λ_V (i.e. $\widetilde{M}_{V,1}/\Lambda_V$ and $\widetilde{M}_{1,R}/\Lambda_V$, respectively).

Notes The equation listed for $\widetilde{C}_1$, Eq. (A10.2), is the dimensionless representation of eq. (8) on p. 128 in Carslaw and Jaeger [2]. The equation listed for $\widetilde{J}_1$, Eq. (A10.5) is presented in dimensional form in Franz et al. [5] for $\xi = 1$. From the mass balance on the vehicle $\widetilde{M}_{V,1} = \Lambda_V(1 - \widetilde{C}_V)$. In addition, because $\widetilde{C}_V = 1$ at $\tau = 0$, it follows from Eq. (A10.4) that

$$\sum_{n=1}^{\infty} \frac{1}{1 + \Lambda_V + \lambda_n^2 \Lambda} = \frac{1}{2\Lambda_V} \tag{A10.8}$$

Carslaw and Jaeger [2] list the first six roots, λ_n, of Eq. (A10.1) for various values of Λ_V.

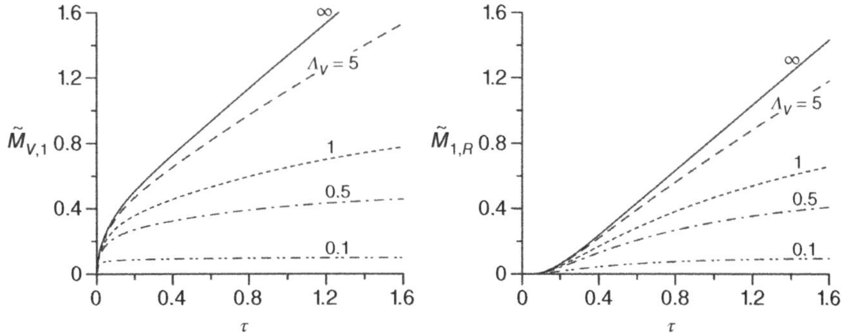

Figure 6.14 Dimensionless cumulative mass transfer from the vehicle to the membrane ($\widetilde{M}_{V,1}$) and from the membrane to the receptor ($\widetilde{M}_{1,R}$) plotted as a function of τ for selected values of Λ_V.

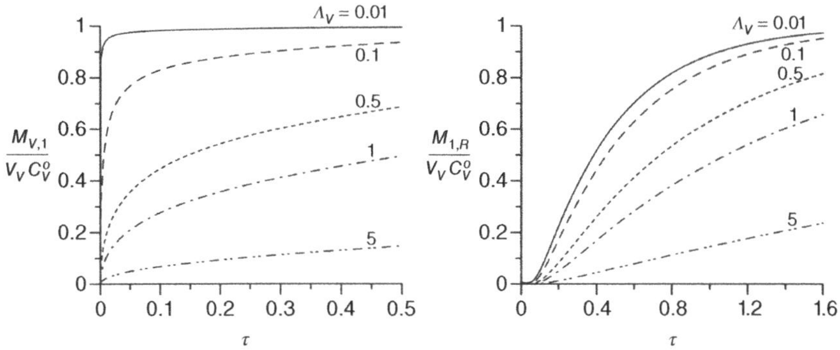

Figure 6.15 Cumulative mass transfer from the vehicle to the membrane ($M_{V,1}$) and from the membrane to the receptor ($M_{1,R}$) relative to the initial mass in the vehicle ($V_V C_V^o$) plotted as a function of τ for selected values of Λ_V.

6.2.1.12.2 Laplace Domain Solutions

Concentration

Membrane 1 ($0 \leq \xi \leq 1$):

$$\mathcal{L}\{\widetilde{C}_1\} = \frac{\Lambda_V \sinh[\sqrt{s}(1-\xi)]}{\sqrt{s}\cosh\sqrt{s} + \Lambda_V \sinh\sqrt{s}} \tag{A10.9}$$

Vehicle:

$$\mathcal{L}\{\widetilde{C}_V\} = \frac{1}{s}\left[1 - \frac{\cosh(\sqrt{s})}{\cosh(\sqrt{s}) + \Lambda_V\sqrt{s}\sinh(\sqrt{s})}\right] \tag{A10.10}$$

or equivalently

$$\mathcal{L}\{\widetilde{C}_V\} = \frac{\Lambda_V \sinh(\sqrt{s})}{\sqrt{s}\cosh(\sqrt{s}) + \Lambda_V s\ \sinh(\sqrt{s})} \tag{A10.11}$$

Notes These expressions for concentration can be inverted using the method of residues as described for Example B in Section 6.3.2.4.

6.2.1.13 Model A11

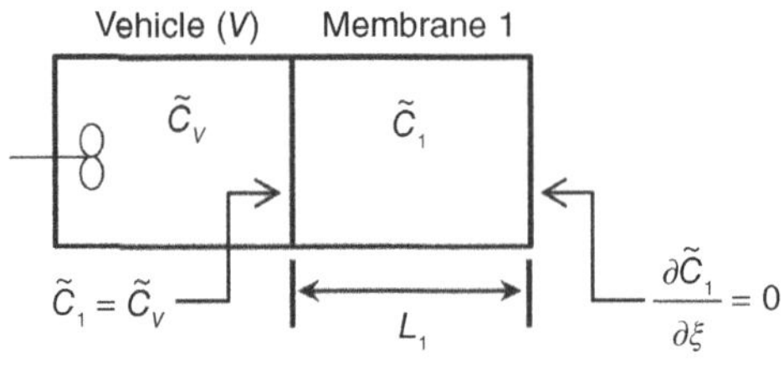

Description:

Vehicle: Finite volume, well-stirred
Membranes: One, finite thickness, solute concentration initially zero
Receptor: None
Vehicle/Membrane: Local equilibrium
Membrane/Receptor: No flux

Differential equations:

$$\frac{\partial \tilde{C}_1}{\partial \tau} = \frac{\partial^2 \tilde{C}_1}{\partial \xi^2} \qquad \text{for } 0 \le \xi \le 1, \ \ \tau \ge 0$$

$$\frac{d\tilde{C}_V}{d\tau} = \frac{1}{\Lambda_V} \left.\frac{\partial \tilde{C}_1}{\partial \xi}\right|_{\xi=0} \qquad \text{for } \tau \ge 0$$

Boundary conditions:

$$\tilde{C}_1 = \tilde{C}_V \qquad \text{at } \xi = 0, \ \ \tau > 0$$

$$\frac{\partial \tilde{C}_1}{\partial \xi} = 0 \qquad \text{at } \xi = 1, \ \ \tau > 0$$

Initial condition:

$$\tilde{C}_V = 1 \qquad \text{at } \tau = 0$$

$$\tilde{C}_1 = 0 \qquad \text{at } \tau = 0, \ \ 0 \le \xi \le 1$$

6.2.1.13.1 Time Domain Solutions

Characteristic Equation for Eigenvalues

$$\lambda_n \cot \lambda_n = -1/\Lambda_V \tag{A11.1}$$

Concentration

Membrane 1 ($0 \le \xi \le 1$):

$$\tilde{C}_1 = \frac{\Lambda_V}{\Lambda_V + 1} + 2\sum_{n=1}^{\infty} \frac{\Lambda_V \lambda_n \cos(\lambda_n(1-\xi))}{(2\Lambda_V + 1)\lambda_n \cos \lambda_n + (1 - \lambda_n^2 \Lambda_V)\sin \lambda_n} \exp[-\lambda_n^2 \tau] \tag{A11.2}$$

or equivalently

$$\tilde{C}_1 = \frac{\Lambda_V}{\Lambda_V + 1} + 2\Lambda_V \sum_{n=1}^{\infty} \frac{\cos[\lambda_n(1-\xi)]}{\cos \lambda_n (1 + \Lambda_V + \lambda_n^2 \Lambda_V^2)} \exp[-\lambda_n^2 \tau] \tag{A11.3}$$

$$\langle \tilde{C}_1 \rangle = \frac{\Lambda_V}{\Lambda_V + 1} - 2\Lambda_V^2 \sum_{n=1}^{\infty} \frac{\exp[-\lambda_n^2 \tau]}{1 + \Lambda_V + \lambda_n^2 \Lambda_V^2} \tag{A11.4}$$

Vehicle

$$\widetilde{C}_V = \frac{\Lambda_V}{\Lambda_V + 1} + 2\Lambda_V \sum_{n=1}^{\infty} \frac{\exp[-\lambda_n^2 \tau]}{1 + \Lambda_V + \lambda_n^2 \Lambda_V^2} \tag{A11.5}$$

Flux Membrane 1 ($0 \le \xi \le 1$):

$$\widetilde{J}_1 = 2\Lambda_V^2 \sum_{n=1}^{\infty} \frac{\lambda_n^2 \sin[\lambda_n(1-\xi)]}{\sin(\lambda_n)(1 + \Lambda_V + \lambda_n^2 \Lambda_V^2)} \exp[-\lambda_n^2 \tau] \tag{A11.6}$$

Cumulative Mass Transfer

From vehicle to membrane 1:

$$\widetilde{M}_{V,1} = \frac{\Lambda_V}{\Lambda_V + 1} - 2\Lambda_V^2 \sum_{n=1}^{\infty} \frac{\exp[-\lambda_n^2 \tau]}{1 + \Lambda_V + \lambda_n^2 \Lambda_V^2} \tag{A11.7}$$

From membrane 1 to receptor:

$$\widetilde{M}_{1,R} = 0 \tag{A11.8}$$

Notes Equation (A11.3) for $\widetilde{C}_1$ is presented in dimensional form in Carslaw and Jaeger [2] (p. 128, eq. (6)). It is derived by using the characteristic equation for the eigenvalues to eliminate $\sin(\lambda_n)$ from Eq. (A11.2).

Additional relationships can be derived. Because there is no flux to the receptor solution, the solute mass in membrane 1 and the vehicle is equal to the initial mass of solute in the vehicle:

$$\langle C_1 \rangle L_1 A + C_V V_V = C_V^o V_V$$

which in terms of the dimensionless concentrations reduces to

$$\langle \widetilde{C}_1 \rangle + \Lambda_V \widetilde{C}_V = \Lambda_V$$

Also, by mass balance $\widetilde{M}_{V,1} = \langle \widetilde{C}_1 \rangle$. In addition, because $\langle \widetilde{C}_1 \rangle = 0$ at $\tau = 0$ (and also because $\widetilde{C}_V = 1$ at $\tau = 0$), it follows that

$$\sum_{n=1}^{\infty} \frac{1}{1 + \Lambda_V + \lambda_n^2 \Lambda_V^2} = \frac{1}{2\Lambda_V(\Lambda_V + 1)} \tag{A11.9}$$

Carslaw and Jaeger [2] list the first six roots, λ_n, of Eq. (A11.1) for various values of Λ_V.

6.2.1.14 Model A12

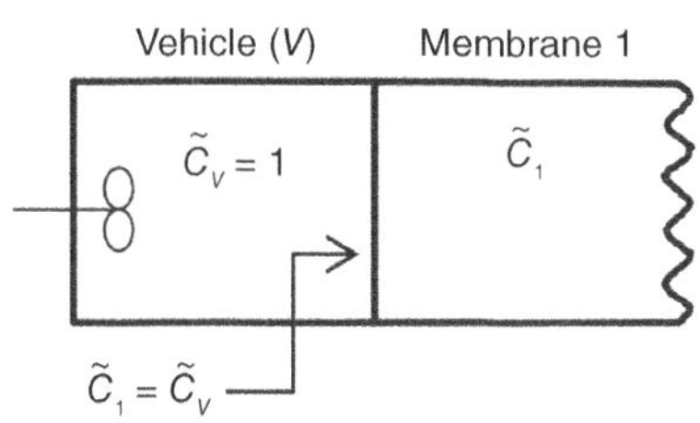

Description:

Vehicle: Infinite volume, well-stirred, constant solute concentration
Membranes: One, semi-infinitely thick, solute concentration initially zero
Receptor: None
Vehicle/Membrane: Local equilibrium

Differential equation:

$$\frac{\partial \widetilde{C}_1}{\partial \tau} = \frac{\partial^2 \widetilde{C}_1}{\partial \xi^2} \qquad \text{for } \xi \geq 0, \quad \tau \geq 0$$

Boundary conditions:

$$\widetilde{C}_1 = \widetilde{C}_V = 1 \qquad \text{at } \xi = 0, \quad \tau > 0$$

$$\widetilde{C}_1 = 0 \qquad \text{at } \xi \to \infty, \quad \tau > 0$$

Initial conditions:

$$\widetilde{C}_1 = 0 \qquad \text{at } \tau = 0, \quad \xi \geq 0$$

6.2.1.14.1 *Time Domain Solutions*

Concentration Membrane 1 ($\xi \geq 0$):

$$\widetilde{C}_1 = \operatorname{erfc}\left(\frac{\xi}{2\sqrt{\tau}}\right) \tag{A12.1}$$

Flux Membrane 1 ($\xi > 0$):

$$\widetilde{J}_1 = \frac{1}{\sqrt{\pi\tau}} \exp\left[-\frac{\xi^2}{4\tau}\right] \tag{A12.2}$$

Cumulative Mass Transfer From vehicle to membrane 1:

$$\widetilde{M}_{V,1} = 2\sqrt{\tau/\pi} \tag{A12.3}$$

Notes Equations describing concentration, flux, and cumulative mass transfer have been reported by many authors, including Crank [6] and Carslaw and Jaeger [2].

6.2.1.15 Model A13

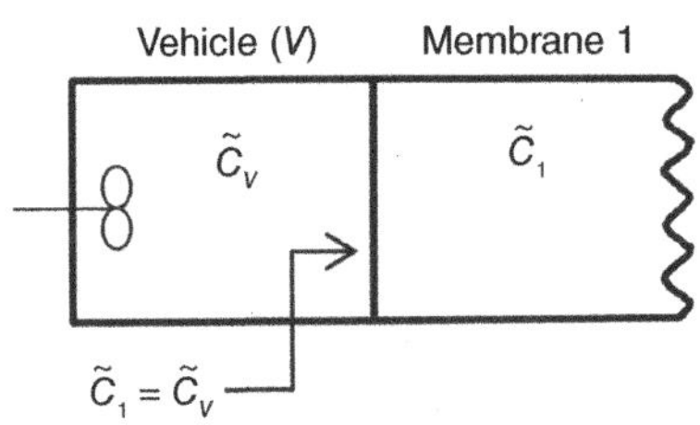

Description:

Vehicle: Finite volume, well-stirred
Membranes: One, semi-infinitely thick, solute concentration initially zero
Receptor: None
Vehicle/Membrane: Local equilibrium

Differential equation:

$$\frac{\partial \widetilde{C}_1}{\partial \tau} = \frac{\partial^2 \widetilde{C}_1}{\partial \xi^2} \qquad \text{for } \xi \geq 0, \quad \tau \geq 0$$

$$\frac{d\widetilde{C}_V}{d\tau} = \frac{1}{\Lambda_V} \left.\frac{\partial \widetilde{C}_1}{\partial \xi}\right|_{\xi=0} \qquad \text{for } \tau \geq 0$$

Boundary conditions:

$$\widetilde{C}_1 = \widetilde{C}_V \qquad \text{at } \xi = 0, \quad \tau > 0$$

$$\widetilde{C}_1 = 0 \qquad \text{at } \xi \to \infty, \quad \tau > 0$$

Initial conditions:

$$\widetilde{C}_V = 1 \qquad \text{at } \tau = 0$$

$$\widetilde{C}_1 = 0 \qquad \text{at } \tau = 0, \quad \xi \geq 0$$

6.2.1.15.1 Time Domain Solutions

Concentration

Membrane 1 ($\xi \geq 0$):

$$\widetilde{C}_1 = \exp\left[\frac{\tau}{\Lambda_V^2} + \frac{\xi}{\Lambda_V}\right] \text{erfc}\left(\frac{\sqrt{\tau}}{\Lambda_V} + \frac{\xi}{2\sqrt{\tau}}\right) \tag{A13.1}$$

Vehicle:

$$\widetilde{C}_V = \exp\left[\frac{\tau}{\Lambda_V^2}\right] \text{erfc}\left(\frac{\sqrt{\tau}}{\Lambda_V}\right) \tag{A13.2}$$

Flux Membrane 1 ($\xi \geq 0$):

$$\widetilde{J}_1 = \frac{1}{\sqrt{\pi\tau}} \exp\left[-\frac{\xi^2}{4\tau}\right] - \frac{1}{\Lambda_V} \exp\left[\frac{\tau}{\Lambda_V^2} + \frac{\xi}{\Lambda_V}\right] \text{erfc}\left(\frac{\sqrt{\tau}}{\Lambda_V} + \frac{\xi}{2\sqrt{\tau}}\right) \tag{A13.3}$$

Cumulative Mass Transfer From vehicle to membrane 1:

$$\widetilde{M}_{V,1} = \Lambda_V \left(1 - \exp\left[\frac{\tau}{\Lambda_V^2}\right] \text{erfc}\left(\frac{\sqrt{\tau}}{\Lambda_V}\right)\right) \tag{A13.4}$$

Notes By mass balance $\widetilde{M}_{V,1} = \Lambda_V(1 - \widetilde{C}_V)$. In the limit of large Λ_V, the equations describing this model match those of the well-stirred vehicle with infinite vehicle volume (Model A12).

6.2.1.15.2 Laplace Domain Solutions

Concentration

Membrane 1 ($0 \leq \xi \leq 1$):

$$\mathcal{L}\{\widetilde{C}_1\} = \frac{\Lambda_V}{\sqrt{s}(1 + \sqrt{s}\Lambda_V)} \exp[-\sqrt{s}\,\xi] \tag{A13.5}$$

Vehicle:

$$\mathcal{L}\{\widetilde{C}_V\} = \frac{\Lambda_V}{\sqrt{s}(1 + \sqrt{s}\Lambda_V)} \tag{A13.6}$$

6.2.2 Type B Models – Unstirred Semi-infinite Vehicle on One Membrane

Solutions to two models are presented. In both models, the membrane is finite and in local equilibrium with the vehicle at the vehicle-membrane interface.

6.2.2.1 Model B1

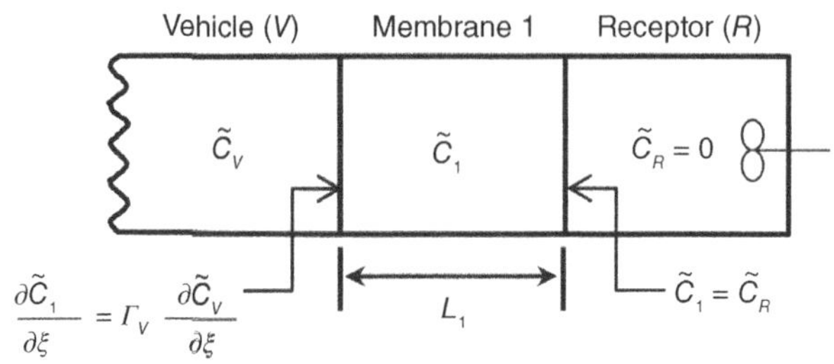

Description:

Vehicle: Semi-infinite volume, unstirred
Membranes: One, finite thickness, solute concentration initially zero
Receptor: Infinite volume, well-stirred, negligible concentration (sink conditions)
Vehicle/Membrane: Local equilibrium
Membrane/Receptor: Local equilibrium

Differential equations:

$$\frac{\partial \tilde{C}_1}{\partial \tau} = \frac{\partial^2 \tilde{C}_1}{\partial \xi^2} \qquad \text{for } 0 \le \xi \le 1, \quad \tau \ge 0$$

$$\frac{\partial \tilde{C}_V}{\partial \tau} = K_{1/V}^2 \gamma^2 \frac{\partial^2 \tilde{C}_V}{\partial \xi^2} \qquad \text{for } -\eta_V \le \xi \le 0, \quad \tau \ge 0$$

Boundary conditions:

$$\tilde{C}_V = 1 \qquad \text{at } \xi \to -\infty, \quad \tau > 0$$

$$\tilde{C}_1 = \tilde{C}_V \qquad \text{at } \xi = 0, \quad \tau > 0$$

$$\frac{\partial \tilde{C}_1}{\partial \xi} = \Gamma_V \frac{\partial \tilde{C}_V}{\partial \xi} \qquad \text{at } \xi = 0, \quad \tau \ge 0$$

$$\tilde{C}_1 = \tilde{C}_R = 0 \qquad \text{at } \xi = 1, \quad \tau \ge 0$$

Initial condition:

$$\tilde{C}_V = 1 \qquad \text{at } \tau = 0, \quad -\infty < \xi < 0$$

$$\tilde{C}_1 = 0 \qquad \text{at } \tau = 0, \quad 0 \le \xi \le 1$$

6.2.2.1.1 Time Domain Solutions

Concentration

Membrane 1 ($0 \le \xi \le 1$):

$$\widetilde{C}_1 = \frac{\gamma}{\gamma+1}\sum_{n=0}^{\infty}\left(\frac{\gamma-1}{\gamma+1}\right)^n\left[\operatorname{erf}\left(\frac{2n+2-\xi}{2\sqrt{\tau}}\right)-\operatorname{erf}\left(\frac{2n+\xi}{2\sqrt{\tau}}\right)\right] \tag{B1.1}$$

or equivalently

$$\widetilde{C}_1 = \frac{\gamma}{\gamma+1}\sum_{n=0}^{\infty}\left(\frac{\gamma-1}{\gamma+1}\right)^n\left[\operatorname{erfc}\left(\frac{2n+\xi}{2\sqrt{\tau}}\right)-\operatorname{erfc}\left(\frac{2n+2-\xi}{2\sqrt{\tau}}\right)\right] \tag{B1.2}$$

$$\begin{aligned}\langle\widetilde{C}_1\rangle = \frac{2\gamma}{\gamma+1}\sum_{n=0}^{\infty}\left(\frac{\gamma-1}{\gamma+1}\right)^n&\left\{\sqrt{\frac{\tau}{\pi}}\left(\exp\left[-\frac{n^2}{4\tau}\right]+\exp\left[-\frac{(1+n)^2}{4\tau}\right]-2\exp\left[-\frac{(1+2n)^2}{4\tau}\right]\right)\right.\\ &\left.+n\operatorname{erf}\left(\frac{n}{\sqrt{\tau}}\right)+(1+n)\operatorname{erf}\left(\frac{1+n}{\sqrt{\tau}}\right)-(1+2n)\operatorname{erf}\left(\frac{1+2n}{2\sqrt{\tau}}\right)\right\}\end{aligned} \tag{B1.3}$$

or equivalently

$$\begin{aligned}\langle\widetilde{C}_1\rangle = \frac{2\gamma}{\gamma+1}\sum_{n=0}^{\infty}\left(\frac{\gamma-1}{\gamma+1}\right)^n&\left\{\sqrt{\frac{\tau}{\pi}}\left(\exp\left[-\frac{n^2}{4\tau}\right]+\exp\left[-\frac{(1+n)^2}{4\tau}\right]-2\exp\left[-\frac{(1+2n)^2}{4\tau}\right]\right)\right.\\ &\left.-n\operatorname{erfc}\left(\frac{n}{\sqrt{\tau}}\right)-(1+n)\operatorname{erfc}\left(\frac{1+n}{\sqrt{\tau}}\right)+(1+2n)\operatorname{erfc}\left(\frac{1+2n}{2\sqrt{\tau}}\right)\right\}\end{aligned} \tag{B1.4}$$

Vehicle ($-\infty < \xi \le 0$):

$$\begin{aligned}\widetilde{C}_V = 1 - \frac{1}{\gamma+1}\sum_{n=0}^{\infty}&\left(\frac{\gamma-1}{\gamma+1}\right)^n\\ &\times\left\{\operatorname{erfc}\left[\left(2n+2-\frac{\xi}{\gamma K_{1/V}}\right)\Big/2\sqrt{\tau}\right]+\operatorname{erfc}\left[\left(2n-\frac{\xi}{\gamma K_{1/V}}\right)\Big/2\sqrt{\tau}\right]\right\}\end{aligned} \tag{B1.5}$$

Flux Membrane 1 ($0 \le \xi \le 1$):

$$\widetilde{J}_1 = \frac{\gamma}{\sqrt{\pi\tau}(\gamma+1)}\sum_{n=0}^{\infty}\left(\frac{\gamma-1}{\gamma+1}\right)^n\left(\exp\left[-\frac{(2n+\xi)^2}{4\tau}\right]+\exp\left[-\frac{(2n+2-\xi)^2}{4\tau}\right]\right) \tag{B1.6}$$

Vehicle ($-\infty < \xi \le 0$):

$$\begin{aligned}\widetilde{J}_V = \frac{\gamma}{\sqrt{\pi\tau}(\gamma+1)}\sum_{n=0}^{\infty}&\left(\frac{\gamma-1}{\gamma+1}\right)^n\\ &\times\left\{\exp\left[-\left(2n-\frac{\xi}{\gamma K_{1/V}}\right)^2\Big/4\tau\right]+\exp\left[-\left(2n+2-\frac{\xi}{\gamma K_{1/V}}\right)^2\Big/4\tau\right]\right\}\end{aligned} \tag{B1.7}$$

Cumulative Mass Transfer

From vehicle to membrane 1:

$$\begin{aligned}\widetilde{M}_{V,1} = \frac{2\gamma}{\gamma+1}\sum_{n=0}^{\infty}\left(\frac{\gamma-1}{\gamma+1}\right)^n&\left\{\sqrt{\frac{\tau}{\pi}}\left(\exp\left[-\left(\frac{(n+1)^2}{\tau}\right)\right]+\exp\left[-\left(\frac{n^2}{\tau}\right)\right]\right)\right.\\ &\left.-(n+1)\operatorname{erfc}\left(\frac{n+1}{\sqrt{\tau}}\right)-n\operatorname{erfc}\left(\frac{n}{\sqrt{\tau}}\right)\right\}\end{aligned} \tag{B1.8}$$

From membrane 1 to receptor:

$$\widetilde{M}_{1,R} = \frac{2\gamma}{\gamma+1}\sum_{n=0}^{\infty}\left(\frac{\gamma-1}{\gamma+1}\right)^n\left\{2\sqrt{\frac{\tau}{\pi}}\exp\left[-\frac{(2n+1)^2}{4\tau}\right] - (2n+1)\mathrm{erfc}\left(\frac{2n+1}{2\sqrt{\tau}}\right)\right\} \quad \text{(B1.9)}$$

Notes By mass balance, $\langle\widetilde{C}_1\rangle = \widetilde{M}_{V,1} - \widetilde{M}_{1,R}$. Parks et al. [7] present equations for $\widetilde{C}_1$, $\widetilde{C}_V$, and $\widetilde{M}_{1,R}$. The equations for $\widetilde{C}_1$ and $\widetilde{C}_V$ in Parks et al. have an incorrect sign on ξ (i.e. x/L_m in Parks et al.), which is corrected here. At long times,

$$\lim_{\tau\to\infty}\widetilde{M}_{1,R} = \lim_{\tau\to\infty}\widetilde{M}_{V,1} = \frac{2\gamma}{\sqrt{\pi}}(\sqrt{\tau} - \sqrt{\tau_{lag}}) \qquad \text{where } \sqrt{\tau_{lag}} = \frac{\gamma\sqrt{\pi}}{2} \quad \text{(B1.10)}$$

6.2.2.1.2 Laplace Domain Solutions

Concentration

Membrane 1 ($0 \le \xi \le 1$):

$$\mathcal{L}\{\widetilde{C}_1\} = \frac{\gamma}{s(\gamma+1)}\frac{(\exp[-\sqrt{s}\,\xi] - \exp[-\sqrt{s}\,(2-\xi)])}{[1-\exp[-2\sqrt{s}](\gamma-1)/(\gamma+1)]} \quad \text{(B1.11)}$$

or equivalently

$$\mathcal{L}\{\widetilde{C}_1\} = \frac{\gamma}{s(\gamma+1)}\sum_{n=0}^{\infty}\left(\frac{\gamma-1}{\gamma+1}\right)^n(\exp[-\sqrt{s}\,(2n+\xi)] - \exp[-\sqrt{s}\,(2n+2-\xi)]) \quad \text{(B1.12)}$$

Vehicle ($-\infty < \xi \le 0$):

$$\mathcal{L}\{\widetilde{C}_V\} = \frac{1}{s} - \frac{\exp\left[\sqrt{s}\frac{\xi}{\gamma K_{1/V}}\right] + \exp\left[-\sqrt{s}\left(2-\frac{\xi}{\gamma K_{1/V}}\right)\right]}{s(\gamma+1)[1-\exp[-2\sqrt{s}](\gamma-1)/(\gamma+1)]} \quad \text{(B1.13)}$$

or equivalently

$$\mathcal{L}\{\widetilde{C}_V\} = \frac{1}{s} - \frac{1}{s(\gamma+1)}\sum_{n=0}^{\infty}\left(\frac{\gamma-1}{\gamma+1}\right)^n\left\{\exp\left[-\sqrt{s}\left(2n-\frac{\xi}{\gamma K_{1/V}}\right)\right] + \exp\left[-\sqrt{s}\left(2n+2-\frac{\xi}{\gamma K_{1/V}}\right)\right]\right\} \quad \text{(B1.14)}$$

Notes Equations (B1.12) and (B1.14) are derived from Eqs. (B1.11) and (B1.13), respectively, using the series expansion for $a^2 < 1$

$$\frac{1}{1-a} = \sum_{n=0}^{\infty}a^n$$

where $a = \left(\frac{\gamma-1}{\gamma+1}\right)e^{-2\sqrt{s}}$.

6.2.2.2 Model B2

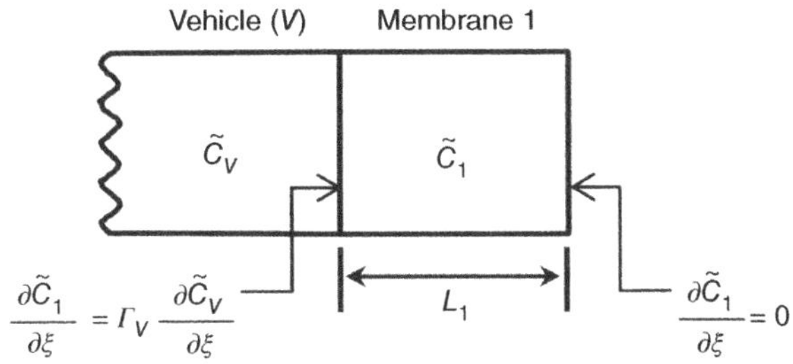

Description:

Vehicle: Semi-infinite volume, unstirred
Membranes: One, finite thickness, solute concentration initially zero
Receptor: None
Vehicle/Membrane: Local equilibrium
Membrane/Receptor: No flux

Differential equations:

$$\frac{\partial \tilde{C}_1}{\partial \tau} = \frac{\partial^2 \tilde{C}_1}{\partial \xi^2} \qquad \text{for } 0 \le \xi \le 1, \quad \tau \ge 0$$

$$\frac{\partial \tilde{C}_V}{\partial \tau} = K_{1/V}^2 \gamma^2 \frac{\partial^2 \tilde{C}_V}{\partial \xi^2} \qquad \text{for } -\eta_V \le \xi \le 0, \quad \tau > 0$$

Boundary conditions:

$$\tilde{C}_V = 1 \qquad \text{at } \xi \to -\infty, \quad \tau > 0$$

$$\tilde{C}_1 = \tilde{C}_V \qquad \text{at } \xi = 0, \quad \tau > 0$$

$$\frac{\partial \tilde{C}_1}{\partial \xi} = \Gamma_V \frac{\partial \tilde{C}_V}{\partial \xi} \qquad \text{at } \xi = 0, \quad \tau > 0$$

$$\frac{\partial \tilde{C}_1}{\partial \xi} = 0 \qquad \text{at } \xi = 1, \quad \tau > 0$$

Initial condition:

$$\tilde{C}_V = 1 \qquad \text{at } \tau = 0, \quad -\infty < \xi < 0$$

$$\tilde{C}_1 = 0 \qquad \text{at } \tau = 0, \quad 0 \le \xi \le 1$$

6.2.2.2.1 Time Domain Solutions

Concentration

Membrane 1 ($0 \le \xi \le 1$):

$$\tilde{C}_1 = \frac{\gamma}{\gamma + 1} \sum_{n=0}^{\infty} \left(\frac{1-\gamma}{1+\gamma} \right)^n \left[\operatorname{erfc}\left(\frac{2n+\xi}{2\sqrt{\tau}} \right) + \operatorname{erfc}\left(\frac{2n+2-\xi}{2\sqrt{\tau}} \right) \right] \tag{B2.1}$$

$$\langle \tilde{C}_1 \rangle = \frac{2\gamma}{\gamma + 1} \sum_{n=0}^{\infty} \left(\frac{1-\gamma}{1+\gamma} \right)^n \left\{ \sqrt{\frac{\tau}{\pi}} \left(\exp\left[-\left(\frac{n^2}{\tau} \right) \right] - \exp\left[-\left(\frac{(n+1)^2}{\tau} \right) \right] \right) + (n+1) \operatorname{erfc}\left(\frac{n+1}{\sqrt{\tau}} \right) - n \operatorname{erfc}\left(\frac{n}{\sqrt{\tau}} \right) \right\} \tag{B2.2}$$

Vehicle ($-\infty < \xi \leq 0$):

$$\widetilde{C}_V = 1 - \frac{1}{\gamma+1}\sum_{n=0}^{\infty}\left(\frac{1-\gamma}{1+\gamma}\right)^n \times \left\{\text{erfc}\left[\left(2n - \frac{\xi}{\gamma K_{1/V}}\right)\Big/2\sqrt{\tau}\right] - \text{erfc}\left[\left(2n+2-\frac{\xi}{\gamma K_{1/V}}\right)\Big/2\sqrt{\tau}\right]\right\} \tag{B2.3}$$

Flux

Membrane 1 ($0 \leq \xi \leq 1$):

$$\widetilde{J}_1 = \frac{\gamma}{\sqrt{\pi\tau}(\gamma+1)}\sum_{n=0}^{\infty}\left(\frac{1-\gamma}{1+\gamma}\right)^n\left(\exp\left[-\frac{(2n+\xi)^2}{4\tau}\right] - \exp\left[-\frac{(2n+2-\xi)^2}{4\tau}\right]\right) \tag{B2.4}$$

Vehicle ($-\infty < \xi \leq 0$):

$$\widetilde{J}_V = \frac{\gamma}{\sqrt{\pi\tau}(\gamma+1)}\sum_{n=0}^{\infty}\left(\frac{1-\gamma}{1+\gamma}\right)^n \times \left\{\exp\left[-\left(2n-\frac{\xi}{\gamma K_{1/V}}\right)^2\Big/4\tau\right] - \exp\left[-\left(2n+2-\frac{\xi}{\gamma K_{1/V}}\right)^2\Big/4\tau\right]\right\} \tag{B2.5}$$

Cumulative Mass Transfer

From vehicle to membrane 1:

$$\widetilde{M}_{V,1} = \frac{2\gamma}{\gamma+1}\sum_{n=0}^{\infty}\left(\frac{1-\gamma}{1+\gamma}\right)^n\left\{\sqrt{\frac{\tau}{\pi}}\left(\exp\left[-\left(\frac{n^2}{\tau}\right)\right] - \exp\left[-\left(\frac{(n+1)^2}{\tau}\right)\right]\right) + (n+1)\text{erfc}\left(\frac{n+1}{\sqrt{\tau}}\right) - n\ \text{erfc}\left(\frac{n}{\sqrt{\tau}}\right)\right\} \tag{B2.6}$$

From membrane 1 to receptor:

$$\widetilde{M}_{1,R} = 0 \tag{B2.7}$$

Notes Because there is no flux from the membrane at $\xi = 1$, $\langle \widetilde{C}_1 \rangle = \widetilde{M}_{V,1}$. Parks [8] presents equations for $\widetilde{C}_1$ and $\widetilde{M}_{1,R}$ in dimensional form (identified, respectively, as C_{sink} in Eq. (3.24) and M_A in Eq. (3.27)). The equation for M_A in Parks has an incorrect sign on the exponential of $(n+1)^2/\tau$, which is corrected here.

6.2.2.2.2 Laplace Domain Solutions

Concentration Membrane 1 ($0 \leq \xi \leq 1$):

$$\mathcal{L}\{\widetilde{C}_1\} = \frac{\gamma}{s(\gamma+1)}\frac{(\exp[-\sqrt{s}\,\xi] + \exp[-\sqrt{s}\,(2-\xi)])}{[1+\exp[-2\sqrt{s}](\gamma-1)/(\gamma+1)]} \tag{B2.8}$$

or equivalently

$$\mathcal{L}\{\widetilde{C}_1\} = \frac{\gamma}{s(\gamma+1)}\sum_{n=0}^{\infty}\left(\frac{1-\gamma}{1+\gamma}\right)^n(\exp[-\sqrt{s}\,(2n+\xi)] + \exp[-\sqrt{s}\,(2n+2-\xi)]) \tag{B2.9}$$

Vehicle ($-\infty < \xi \leq 0$):

$$\mathcal{L}\{\widetilde{C}_V\} = \frac{1}{s} - \frac{1}{s(\gamma+1)}\frac{\exp[\sqrt{s}\,\xi/\beta](1-\exp[-2\sqrt{s}])}{[1+\exp[-2\sqrt{s}](\gamma-1)/(\gamma+1)]} \tag{B2.10}$$

or equivalently

$$\mathcal{L}\{\widetilde{C}_V\} = \frac{1}{s} - \frac{1}{s(\gamma+1)}\sum_{n=0}^{\infty}\left(\frac{1-\gamma}{1+\gamma}\right)^n \left\{\exp\left[-\sqrt{s}\left(2n - \frac{\xi}{\gamma K_{1/V}}\right)\right] - \exp\left[-\sqrt{s}\left(2n+2-\frac{\xi}{\gamma K_{1/V}}\right)\right]\right\} \tag{B2.11}$$

Notes Equations (B2.9) and (B2.11) are derived from Eqs. (B2.8) and (B2.10), respectively, using the series expansion for $a^2 < 1$

$$\frac{1}{1-a} = \sum_{n=0}^{\infty} a^n$$

where $a = \left(\frac{1-\gamma}{1+\gamma}\right) e^{-2\sqrt{s}}$

6.2.3 Type C – Well Stirred Vehicle on Two Membranes in Series

The solution of the one model containing two finite membranes in series is presented.

6.2.3.1 Model C1

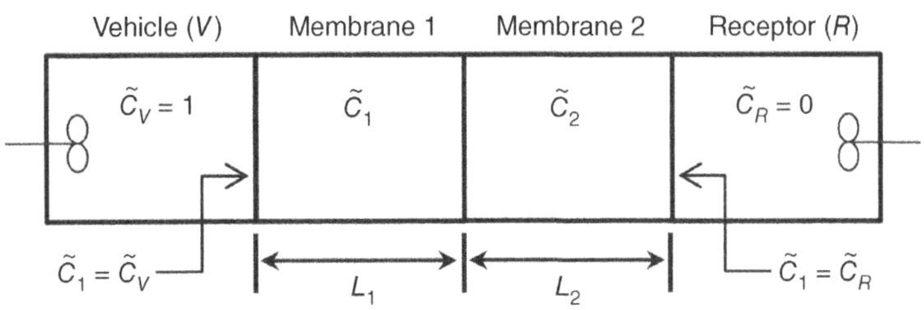

Source: Based on Cleek and Bunge [9].

Description:

Vehicle: Infinite volume, well-stirred, constant solute concentration
Membranes: Two, finite thickness, solute concentration initially zero
Receptor: Infinite volume, well-stirred, negligible concentration (sink conditions)
Vehicle/Membrane: Local equilibrium
Membrane/Receptor: Local equilibrium

Differential equations:

$$\frac{\partial \widetilde{C}_1}{\partial \tau} = \frac{\partial^2 \widetilde{C}_1}{\partial \xi^2} \qquad \text{for } 0 \le \xi \le 1, \quad \tau \ge 0$$

$$\frac{\partial \widetilde{C}_2}{\partial \tau} = G_{1,2}\,\eta_2^2 \frac{\partial^2 \widetilde{C}_2}{\partial \xi^2} \qquad \text{for } 1 \le \xi \le 1+\eta_2, \quad \tau \ge 0$$

Boundary conditions:

$$\widetilde{C}_1 = \widetilde{C}_V = 1 \qquad \text{at } \xi = 0, \quad \tau > 0$$

$$\widetilde{C}_1 = \widetilde{C}_2 \qquad \text{at } \xi = 1, \quad \tau > 0$$

$$\frac{\partial \widetilde{C}_1}{\partial \xi} = \Gamma_2 \frac{\partial \widetilde{C}_2}{\partial \xi} = \frac{\eta_2}{B_{1,2}} \frac{\partial \widetilde{C}_2}{\partial \xi} \qquad \text{at } \xi = 1, \quad \tau > 0$$

$$\widetilde{C}_2 = \widetilde{C}_R = 0 \qquad \text{at } \xi = 1 + \eta_2, \quad \tau \geq 0$$

Initial condition:

$$\widetilde{C}_1 = 0 \quad \text{at } \tau = 0, \quad 0 \leq \xi \leq 1$$

$$\widetilde{C}_2 = 0 \quad \text{at } \tau = 0, \quad 1 \leq \xi \leq 1 + \eta_2$$

6.2.3.1.1 Time Domain Solutions

Characteristic Equation for Eigenvalues:

$$\frac{\tan(\lambda_n/\sqrt{G_{1,2}})}{\tan(\lambda_n)} = -\frac{1}{B_{1,2}\sqrt{G_{1,2}}} \tag{C1.1}$$

Other Definitions

$$\sigma_n = \left[\frac{B_{1,2}+1}{B_{1,2}\sqrt{G_{1,2}}}\right] \cos(\lambda_n/\sqrt{G_{1,2}}) \cos(\lambda_n) - \left[\frac{G_{1,2}\,B_{1,2}+1}{B_{1,2}\,G_{1,2}}\right] \sin(\lambda_n/\sqrt{G_{1,2}}) \sin(\lambda_n) \tag{C1.2}$$

Concentration

Membrane 1 ($0 \leq \xi \leq 1$):

$$\widetilde{C}_1 = \frac{B_{1,2}+1-\xi}{B_{1,2}+1} + \frac{2}{B_{1,2}\sqrt{G_{1,2}}} \sum_{n=1}^{\infty} \frac{\exp[-\lambda_n^2 \tau]}{\lambda_n \sigma_n} \left\{ \cos(\lambda_n/\sqrt{G_{1,2}}) \sin(\lambda_n(1-\xi)) + \right.$$
$$\left. + B_{1,2}\sqrt{G_{1,2}} \sin(\lambda_n/\sqrt{G_{1,2}}) \cos(\lambda_n(1-\xi)) \right\} \tag{C1.3}$$

or equivalently

$$\widetilde{C}_1 = \frac{B_{1,2}+1-\xi}{B_{1,2}+1} - \frac{2}{B_{1,2}\sqrt{G_{1,2}}} \sum_{n=1}^{\infty} \frac{\cos(\lambda_n/\sqrt{G_{1,2}}) \sin(\lambda_n \xi)}{\lambda_n \sigma_n \cos(\lambda_n)} \exp[-\lambda_n^2 \tau] \tag{C1.4}$$

or equivalently

$$\widetilde{C}_1 = \frac{B_{1,2}+1-\xi}{B_{1,2}+1} + 2\sum_{n=1}^{\infty} \frac{\sin(\lambda_n/\sqrt{G_{1,2}}) \sin(\lambda_n \xi)}{\lambda_n \sigma_n \sin(\lambda_n)} \exp[-\lambda_n^2 \tau] \tag{C1.5}$$

$$\langle \widetilde{C}_1 \rangle = \frac{1+2B_{1,2}}{2(1+B_{1,2})} + \frac{2}{B_{1,2}\sqrt{G_{1,2}}} \sum_{n=1}^{\infty} \frac{\exp[-\lambda_n^2 \tau]}{\lambda_n^2 \sigma_n} \left\{ \cos(\lambda_n/\sqrt{G_{1,2}})[1-\cos(\lambda_n)] + \right.$$
$$\left. + B_{1,2}\sqrt{G_{1,2}} \sin(\lambda_n/\sqrt{G_{1,2}}) \sin(\lambda_n) \right\} \tag{C1.6}$$

or equivalently

$$\langle \widetilde{C}_1 \rangle = \frac{1 + 2B_{1,2}}{2(1 + B_{1,2})} - \frac{2}{B_{1,2}\sqrt{G_{1,2}}} \sum_{n=1}^{\infty} \frac{\cos(\lambda_n/\sqrt{G_{1,2}})[1 - \cos(\lambda_n)]}{\lambda_n^2 \sigma_n \cos(\lambda_n)} \exp[-\lambda_n^2 \tau] \tag{C1.7}$$

or equivalently

$$\langle \widetilde{C}_1 \rangle = \frac{1 + 2B_{1,2}}{2(1 + B_{1,2})} + 2\sum_{n=1}^{\infty} \frac{\sin(\lambda_n/\sqrt{G_{1,2}})[1 - \cos(\lambda_n)]}{\lambda_n^2 \sigma_n \sin(\lambda_n)} \exp[-\lambda_n^2 \tau] \tag{C1.8}$$

Membrane 2 ($1 < \xi < 1 + \eta_2$):

$$\widetilde{C}_2 = \frac{B_{1,2}(\eta_2 + 1 - \xi)}{\eta_2(B_{1,2} + 1)} + 2\sum_{n=1}^{\infty} \frac{\exp[-\lambda_n^2 \tau]}{\lambda_n \sigma_n} \sin\left[\frac{\lambda_n}{\sqrt{G_{1,2}}}\left(\frac{\eta_2 + 1 - \xi}{\eta_2}\right)\right] \tag{C1.9}$$

$$\langle \widetilde{C}_2 \rangle = \frac{B_{1,2}}{2(1 + B_{1,2})} + 2\sqrt{G_{1,2}} \sum_{n=1}^{\infty} \frac{[1 - \cos(\lambda_n/\sqrt{G_{1,2}})]}{\lambda_n^2 \sigma_n} \exp[-\lambda_n^2 \tau] \tag{C1.10}$$

Flux

Membrane 1 ($0 \le \xi \le 1$):

$$\widetilde{J}_1 = \frac{1}{1 + B_{1,2}} + \frac{2}{B_{1,2}\sqrt{G_{1,2}}} \sum_{n=1}^{\infty} \frac{\exp[-\lambda_n^2 \tau]}{\sigma_n}$$
$$\times [\cos(\lambda_n/\sqrt{G_{1,2}}) \cos(\lambda_n(1 - \xi)) - B_{1,2}\sqrt{G_{1,2}} \sin(\lambda_n/\sqrt{G_{1,2}}) \sin(\lambda_n(1 - \xi))] \tag{C1.11}$$

or equivalently

$$\widetilde{J}_1 = \frac{1}{1 + B_{1,2}} + \frac{2}{B_{1,2}\sqrt{G_{1,2}}} \sum_{n=1}^{\infty} \frac{\cos(\lambda_n/\sqrt{G_{1,2}}) \cos(\lambda_n \xi)}{\sigma_n \cos(\lambda_n)} \exp[-\lambda_n^2 \tau] \tag{C1.12}$$

or equivalently

$$\widetilde{J}_1 = \frac{1}{1 + B_{1,2}} - 2\sum_{n=1}^{\infty} \frac{\sin(\lambda_n/\sqrt{G_{1,2}}) \cos(\lambda_n \xi)}{\sigma_n \sin(\lambda_n)} \exp[-\lambda_n^2 \tau] \tag{C1.13}$$

Membrane 2 ($1 < \xi < 1 + \eta_2$):

$$\widetilde{J}_2 = \frac{1}{1 + B_{1,2}} + \frac{2}{B_{1,2}\sqrt{G_{1,2}}} \sum_{n=1}^{\infty} \frac{\exp[-\lambda_n^2 \tau]}{\sigma_n} \cos\left[\frac{\lambda_n}{\sqrt{G_{1,2}}}\left(\frac{\eta_2 + 1 - \xi}{\eta_2}\right)\right] \tag{C1.14}$$

Cumulative Mass Transfer

From vehicle to membrane 1:

$$\widetilde{M}_{V,1} = \frac{1}{1 + B_{1,2}}\left[\tau + \frac{B_{1,2} + G_{1,2} + 3B_{1,2}G_{1,2} + 3G_{1,2}B_{1,2}^2}{3G_{1,2}(1 + B_{1,2})}\right]$$
$$- \frac{2}{B_{1,2}\sqrt{G_{1,2}}} \sum_{n=1}^{\infty} \frac{\exp[-\lambda_n^2 \tau]}{\lambda_n^2 \sigma_n} \left\{\cos(\lambda_n/\sqrt{G_{1,2}}) \cos(\lambda_n)\right.$$
$$\left. -B_{1,2}\sqrt{G_{1,2}} \sin(\lambda_n/\sqrt{G_{1,2}}) \sin(\lambda_n)\right\} \tag{C1.15}$$

or equivalently

$$\widetilde{M}_{V,1} = \frac{1}{1+B_{1,2}}\left[\tau + \frac{B_{1,2}+G_{1,2}+3B_{1,2}G_{1,2}+3G_{1,2}B_{1,2}^2}{3G_{1,2}(1+B_{1,2})}\right] + 2\sum_{n=1}^{\infty}\frac{\sin(\lambda_n/\sqrt{G_{1,2}})}{\lambda_n^2\sigma_n\sin(\lambda_n)}\exp[-\lambda_n^2\tau] \tag{C1.16}$$

or equivalently

$$\widetilde{M}_{V,1} = \frac{1}{1+B_{1,2}}\left[\tau + \frac{B_{1,2}+G_{1,2}+3B_{1,2}G_{1,2}+3B_{1,2}G_{1,2}^2}{3G_{1,2}(1+B_{1,2})}\right] - \frac{2}{B_{1,2}\sqrt{G_{1,2}}}\sum_{n=1}^{\infty}\frac{\cos(\lambda_n/\sqrt{G_{1,2}})}{\lambda_n^2\sigma_n\cos(\lambda_n)}\exp[-\lambda_n^2\tau] \tag{C1.17}$$

From membrane 1 to membrane 2:

$$\widetilde{M}_{1,2} = \frac{1}{1+B_{1,2}}\left[\tau - \frac{G_{1,2}-2B_{1,2}+3B_{1,2}G_{1,2}}{6G_{1,2}(1+B_{1,2})}\right] - \frac{2}{B_{1,2}\sqrt{G_{1,2}}}\sum_{n=1}^{\infty}\frac{\cos(\lambda_n/\sqrt{G_{1,2}})}{\lambda_n^2\sigma_n}\exp[-\lambda_n^2\tau] \tag{C1.18}$$

From membrane 2 to receptor:

$$\widetilde{M}_{2,R} = \frac{1}{1+B_{1,2}}\left[\tau - \frac{3+G_{1,2}+B_{1,2}+3B_{1,2}G_{1,2}}{6G_{1,2}(1+B_{1,2})}\right] - \frac{2}{B_{1,2}\sqrt{G_{1,2}}}\sum_{n=1}^{\infty}\frac{\exp[-\lambda_n^2\tau]}{\lambda_n^2\sigma_n} \tag{C1.19}$$

Notes The algebraically equivalent forms of Eq. (C1.3) for $\widetilde{C}_1$, Eqs. (C1.4) and (C1.5), are derived using Eq. (C1.1) defining the eigenvalues λ_n rewritten as follows:

$$B_{1,2}\sqrt{G_{1,2}}\frac{\sin(\lambda_n/\sqrt{G_{1,2}})}{\sin(\lambda_n)} = -\frac{\cos(\lambda_n/\sqrt{G_{1,2}})}{\cos(\lambda_n)}$$

combined with the following trigonometric identities:

$$\sin(\lambda_n(1-\xi)) = \sin(\lambda_n)\cos(\lambda_n\xi) - \cos(\lambda_n)\sin(\lambda_n\xi)$$
$$\cos(\lambda_n(1-\xi)) = \cos(\lambda_n)\cos(\lambda_n\xi) + \sin(\lambda_n)\sin(\lambda_n\xi)$$

Algebraically equivalent forms of the average concentration in the membranes, flux, and cumulative mass transfer are derived from Eqs. (C1.3)–(C1.5) for $\widetilde{C}_1$. Note that at the boundary between the two membranes the dimensionless concentrations and dimensionless fluxes are equal as required by the boundary conditions:

$$\left.\widetilde{C}_1\right|_{\xi=1} = \left.\widetilde{C}_2\right|_{\xi=1} = \frac{B_{1,2}}{1+B_{1,2}} + 2\sum_{n=1}^{\infty}\frac{\sin(\lambda_n/\sqrt{G_{1,2}})}{\lambda_n\sigma_n}\exp[-\lambda_n^2\tau] \tag{C1.20}$$

$$\left.\widetilde{J}_1\right|_{\xi=1} = \left.\widetilde{J}_2\right|_{\xi=1} = \frac{1}{1+B_{1,2}} + \frac{2}{B_{1,2}\sqrt{G_{1,2}}}\sum_{n=1}^{\infty}\frac{\cos(\lambda_n/\sqrt{G_{1,2}})}{\sigma_n}\exp[-\lambda_n^2\tau] \tag{C1.21}$$

The first 10 eigenvalues defined by Eq. (C1.1) for selected values of $B_{1,2}$ and $G_{1,2}$ are presented in Table 6.6.

Table 6.6 First 10 eigenvalues (λ_n) defined by Eq. (C1.1) for $G_{1,2} = 10$.

n	$B_{1,2}$ 1	10
1	1.982414	1.626692
2	4.735596	4.714918
3	7.532544	7.814132
4	9.677216	9.885017
5	11.519176	11.078920
6	14.207614	14.144885
7	17.031594	17.249751
8	19.340093	19.744521
9	21.084129	20.561527
10	23.682204	23.575270

6.2.3.1.2 Laplace Domain Solutions

Concentration

Membrane 1 ($0 \leq \xi \leq 1$):

$$\mathcal{L}\{\widetilde{C}_1\} = \frac{B_{1,2}\sqrt{G_{1,2}}\cosh[\sqrt{s}(1-\xi)]\sinh(\sqrt{s/G_{1,2}}) + \sinh[\sqrt{s}(1-\xi)]\cosh(\sqrt{s/G_{1,2}})}{s[B_{1,2}\sqrt{G_{1,2}}\cosh(\sqrt{s})\sinh(\sqrt{s/G_{1,2}}) + \sinh(\sqrt{s})\cosh(\sqrt{s/G_{1,2}})]} \tag{C1.22}$$

Membrane 2 ($1 \leq \xi \leq 1 + \eta_2$):

$$\mathcal{L}\{\widetilde{C}_2\} = \frac{B_{1,2}\sqrt{G_{1,2}}\sinh[\sqrt{s}(1+\eta_2-\xi)/\eta_2]}{s[\cosh(\sqrt{s/G_{1,2}})\sinh(\sqrt{s}) + B_{1,2}\sqrt{G_{1,2}}\sinh(\sqrt{s/G_{1,2}})\cosh(\sqrt{s})]} \tag{C1.23}$$

6.3 Solution Methods

The models presented in this chapter were solved by separation of variables or Laplace transform methods. Laplace transform methods were required for models with varying concentrations in more than one compartment (e.g. the vehicle and membrane, or two membranes). Some problems are amenable to solution by both methods (e.g. see Example A below). This section describes the application of these two solution methods to membrane models, starting with separation of variables. Descriptions of the two methods applied to membrane models can also be found in Chapter 2 of Crank [6], Chapter 12 of Carslaw and Jaeger [2] along with other sources (e.g. [10, 11]).

For membrane models that exhibit a nonzero steady-state flux ($\widetilde{J}^{ss}$), the cumulative mass transfer across the boundary between compartments i and j will reach steady state ($\widetilde{M}_{i,j}^{ss}$) at long times, which is represented as

$$\widetilde{M}_{i,j}^{ss} = \widetilde{J}^{ss}(\tau - \tau_o) \tag{6.1}$$

where τ_o is the intercept on the τ-axis (i.e. the dimensionless time axis). For transfer out of a membrane, τ_o is a positive expression or number representing the dimensionless lag time; for transfer into a membrane, τ_0 is a negative expression or number representing the dimensionless lead time [12, 13]. Thus, in the equations describing the cumulative mass transfer across one membrane from the vehicle to the receptor, τ_0 will be negative for $\widetilde{M}_{V,1}$ and positive for $\widetilde{M}_{1,R}$. It is often convenient to derive expressions for τ_0 using equations in the Laplace domain as described in Section 6.3.2 and illustrated in Example A. Laplace domain derivations for τ_0 are sometimes helpful even when the problem has been solved using separation of variables.

6.3.1 Separation of Variables Solutions

In developing a solution to a partial differential equation by separation of variables, one assumes that it is possible to separate the contributions of the independent variables into separate functions of only one independent variable. For example, consider the dependent variable θ that depends on the independent variables time (t) and position (x) as specified in the following partial differential equation and restricting conditions; i.e. one initial condition (IC) and two boundary conditions (BCs):

$$\frac{\partial \theta}{\partial t} = \frac{\partial^2 \theta}{\partial x^2} \qquad \text{for } t > 0 \quad \text{for } 0 \leq x \leq L \tag{6.2}$$

$$\text{IC}: \theta = \theta_0 \qquad \text{at } t = 0 \quad \text{for } 0 \leq x \leq L \tag{6.3}$$

$$\text{BC 1}: \frac{\partial \theta}{\partial x} = 0 \qquad \text{at } x = 0 \quad \text{for } t > 0 \tag{6.4}$$

$$\text{BC 2}: \theta = 0 \qquad \text{at } x = L \quad \text{for } t > 0 \tag{6.5}$$

in which θ_0 is assumed to be a constant. To solve this problem by separation of variables, we assume that the dependence on t and x can be separated into the product of two functions T and X:

$$\theta(t, x) = T(t)X(x) \tag{6.6}$$

which depend only on t and x, respectively. If this assumption is incorrect, then the solution will fail during the procedure. The procedure for solving a partial differential equation using separation of variables will be presented by solving Eq. (6.2) for the initial and boundary conditions listed as Eqs. (6.3)–(6.5).

6.3.1.1 Separating the Partial Differential Equation of *N* Independent Variables into *N* Ordinary Differential Equations

Begin by differentiating Eq. (6.6) to obtain the terms required in Eqs. (6.2)–(6.5):

$$\frac{\partial \theta}{\partial t} = X\frac{dT}{dt} \tag{6.7}$$

$$\frac{\partial \theta}{\partial x} = T\frac{dX}{dx} \tag{6.8}$$

$$\frac{\partial^2 \theta}{\partial x^2} = T\frac{d^2X}{dx^2} \tag{6.9}$$

These are substituted into the partial differential equation and the restricting conditions. for the problem specified by Eqs. (6.2)–(6.5):

$$X\frac{dT}{dt} = T\frac{d^2X}{dx^2} \qquad \text{for } t = 0 \text{ for } 0 \leq x \leq L \tag{6.10}$$

$$\text{IC}: \theta = TX = \theta_0 \qquad \text{at } t = 0 \quad \text{for } 0 \leq x \leq L \tag{6.11}$$

$$\text{BC 1}: \frac{\partial \theta}{\partial x} = T\frac{dX}{dx} = 0 \quad \text{or} \quad \frac{dX}{dx} = 0 \quad \text{at}\, x = 0 \quad \text{for } t > 0 \tag{6.12}$$

$$\text{BC 2}: \theta = TX = 0 \quad \text{or} \quad X = 0 \quad \text{at}\, x = L \quad \text{for } t > 0 \tag{6.13}$$

Equation (6.10) can be rearranged as follows:

$$\frac{1}{T}\frac{dT}{dt} = \frac{1}{X}\frac{d^2X}{dx^2} \tag{6.14}$$

so that the left-hand side of Eq. (6.14) is a function of only t, and the right-hand side of Eq. (6.14) is a function of only x; i.e. the partial differential equation, Eq. (6.2), has been successfully separated. Because Eq. (6.14) must be satisfied for all values of x and t, one must conclude that each side of Eq. (6.14) must be equal to the same constant value, called the separation constant. In order to define this separation constant with an unambiguous sign, we specify it as $\pm\lambda^2$, where the sign (either plus or minus) will be chosen consistent with the physics of the problem. Thus, Eq. (6.14) gives:

$$\frac{1}{T}\frac{dT}{dt} = \frac{1}{X}\frac{d^2X}{dx^2} = \pm\lambda^2 \tag{6.15}$$

which can be divided into two separate ordinary differential equations:

$$\frac{1}{T}\frac{dT}{dt} = \pm\lambda^2 \tag{6.16}$$

$$\frac{1}{X}\frac{d^2X}{dx^2} = \pm\lambda^2 \tag{6.17}$$

in which the sign on λ^2 (either positive or negative) must be the same in both equations.

6.3.1.2 Choosing the Sign (Positive or Negative) on the Separation Constant

If the sign on the separation constant λ^2 is chosen to be negative, then Eqs. (6.16) and (6.17) become

$$\frac{1}{T}\frac{dT}{dt} = -\lambda^2 \tag{6.18}$$

$$\frac{d^2X}{dx^2} + \lambda^2 X = 0 \tag{6.19}$$

which have the solutions:

$$T = I\, \exp(-\lambda^2 t) \tag{6.20}$$

$$X = A\sin(\lambda x) + B\cos(\lambda x) \tag{6.21}$$

where I, A, and B are constants of integration. For the choice of a negative sign on λ^2, X is periodic in x and T decreases exponentially in time. For many physical problems, the time dependence is most significant at short times and it decreases in importance as time becomes large. Thus, an exponentially decreasing time dependence is often physically reasonable. In addition, for the separation of variable solution to work, at least one of the two independent variable functions (i.e. either T or X) must produce periodic functions. In this case, the function $X(x)$ is periodic (i.e. it consists of sine and cosine functions). If the dependent variable θ is a function of more than two independent variables (e.g. $\theta(x, y, z)$ or $\theta(t, x, y, z)$, where x, y, and z represent the three-dimensional position in Cartesian coordinates), then at most one of the independent variable functions can be nonperiodic.

If the sign on λ^2 is chosen to be positive, then Eqs. (6.16) and (6.17) become

$$\frac{1}{T}\frac{dT}{dt} = \lambda^2 \tag{6.22}$$

$$\frac{d^2X}{dx^2} - \lambda^2 X = 0 \tag{6.23}$$

which have solutions:

$$T = I\ \exp(\lambda^2 t) \tag{6.24}$$

$$X = A\sinh(\lambda x) + B\cosh(\lambda x) \tag{6.25}$$

where I, A, and B are constants of integration. For the choice a positive sign on λ^2, X is not periodic in x and T increases exponentially in time. For many physical problems, this time dependence is unreasonable. Equally important, a solution by separation of variables is not possible for this choice of sign on λ^2 because neither the $X(x)$ or $T(t)$ functions are periodic. Based on these results, we will choose the sign on λ^2 to be negative and Eqs. (6.20) and (6.21) define the functions for $X(x)$ or $T(t)$, respectively.

6.3.1.3 Finding the Constants of Integration and the Eigenvalues

To find the constants of integration (I, A, and B), we start with the restrictions on the independent variable that generated the periodic functions. In this example, that is $X(x)$:

$$X = A\sin(\lambda x) + B\cos(\lambda x) \tag{6.26}$$

which is solved for the boundary conditions:

$$\frac{dX}{dx} = 0 \qquad \text{at } x = 0 \quad \text{for } t > 0 \tag{6.27}$$

$$X = 0 \qquad \text{at } x = L \quad \text{for } t > 0 \tag{6.28}$$

Differentiating Eq. (6.26) produces:

$$\frac{dX}{dx} = A\lambda\cos(\lambda x) - B\lambda\sin(\lambda x) \tag{6.29}$$

which when evaluated at $x = 0$ for the condition that $dX/dx = 0$ gives the result that $A = 0$ and Eq. (6.26) reduces to

$$X = B\cos(\lambda x) \tag{6.30}$$

Applying the boundary condition specified in Eq. (6.28), gives the following:

$$X = B\cos(\lambda L) = 0 \tag{6.31}$$

which is satisfied for two cases. In the first case, $B = 0$. This solution states that $\theta = 0$ for all values of t and x. While this satisfies the differential equation, it is ignored as a *trivial solution*.

In the second case, Eq. (6.31) will be satisfied, if λL is equal to $\pi/2$, $3\pi/2$, $5\pi/2$, and so on. That is, Eq. (6.31) is satisfied if $\lambda_n = (2n+1)\pi/2L$, where n is any positive integer greater than or equal to zero (the negative integers are not included because they duplicate the results from the positive integers; i.e. $\cos(-\lambda_n L) = \cos(\lambda_n L)$). As a consequence, we are not able to determine the value of B (yet, we will determine it later), and we have discovered that X has an infinite number of solutions. We will designate the distinctive solutions to X as follows:

$$X_n = B_n\cos(\lambda_n x) \tag{6.32}$$

where

$$\lambda_n = \frac{(2n+1)\pi}{2L} \qquad \text{for integer values of } n \geq 0 \tag{6.33}$$

As a result, the solution to the time equation also has an infinite number of solutions, represented as

$$T_n = I_n \exp(-\lambda_n^2 t) \tag{6.34}$$

where B_n and I_n in Eqs. (6.32) and (6.34) designate the constants of integration corresponding to the solution for a given λ_n.

The periodic function (in this case, $\cos(\lambda_n x) = 0$) is called the *eigenfunction*. The values of λ_n that cause the eigenfunction to be satisfied are called *eigenvalues* (i.e. literally values of the zeroes). Before we find the remaining constants of integration (i.e. B_n and I_n), we first combine T_n and X_n to obtain the complete solution of θ in a procedure called superposition described next.

6.3.1.4 Superposition

The complete solution to a partial differential equation is the sum of all possible solutions. The process of summing all possible solutions is called *superposition*. For this example problem, there are an infinite number of solutions because there are an infinite number of eigenvalues, λ_n. Thus, the complete solution to the partial differential equation of θ is:

$$\theta = \sum_{n=0}^{\infty} T_n(t) X_n(x) = \sum_{n=0}^{\infty} I_n B_n \cos(\lambda_n x) \exp(-\lambda_n^2 t) \tag{6.35}$$

As noted above, summing over values of $n \geq 0$ includes all possible solutions of the partial differential equation because the solutions with negative λ_n duplicate (i.e. are not linearly independent of) the solutions with positive λ_n; i.e. $\cos(-\lambda_n x) = \cos(\lambda_n x)$. Since the product of two constants is still a constant, we will replace $I_n\ B_n$ in Eq. (6.35) with simply B_n.

$$\theta = \sum_{n=0}^{\infty} B_n \cos(\lambda_n x) \exp(-\lambda_n^2 t) \tag{6.36}$$

Next, we will use the final remaining restricting condition, the initial condition, to determine the constants of integration B_n.

6.3.1.5 Finding the Remaining Constants of Integration

To find B_n we apply the initial condition, Eq. (6.11), which states that $\theta = \theta_o$ at $t = 0$ to Eq. (6.36):

$$\theta = \theta_o = \sum_{n=0}^{\infty} B_n \cos(\lambda_n x) \qquad \text{at} \quad t = 0 \tag{6.37}$$

To find the infinite number of B_n constants, we use a special property of eigenfunctions called *orthogonality*. Mathematically, orthogonal functions like X_n have the property:

$$\int_a^b X_n(\lambda_n x)\, X_m(\lambda_m x)\, r(x) dx = 0 \qquad \text{for } n \neq m$$

$$= \int_a^b X_n^2(\lambda_n x)\, r(x)\, dx \qquad \text{for } n = m \tag{6.38}$$

where $r(x)$ is a weighting factor, and a and b are the lower and upper bounds on x as specified by the bounds on the differential equation that generated the eigenfunction $X_n(x)$. For this example, $a = 0$ and $b = L$. The weighting factor $r(x)$ depends on the differential equation that generated the eigenfunction. For this example, the cosine eigenfunction was generated by the harmonic equation (Eq. (6.19)), for which $r(x) = 1$. Bessel's equation, which arises in problems with cylindrical geometry, also can generate eigenfunctions. For eigenfunctions generated from Bessel's equation $r(x) = x$. More information on weighting functions of other differential equations can be found in textbooks describing solutions to ordinary and partial differential equations.

We use the property of orthogonality to find each of the constants B_n separately. To do this, we multiply both sides of Eq. (6.38) by the eigenfunction X_m and then integrate over the bounds of the problem (i.e. between $a = 0$ and $b = L$) and recognizing that the integral of the summation can be written as the summation of the integrals as follows:

$$\int_0^L \theta_o[\cos(\lambda_m x)]dx = \int_0^L \sum_{n=0}^{\infty} B_n \cos(\lambda_n x)[\cos(\lambda_m x)]dx$$
$$= \sum_{n=0}^{\infty} \int_0^L B_n \cos(\lambda_n x)[\cos(\lambda_m x)]dx \tag{6.39}$$

Since the right-hand side of Eq. (6.39) is only nonzero when $m = n$, it follows that

$$\int_0^L \theta_o \cos(\lambda_n x)dx = \int_0^L B_n \cos^2(\lambda_n x)dx \tag{6.40}$$

where the constant B_n is not a function of x and can be taken out of the integration on x and solved for B_n, which gives the result

$$B_n = \int_0^L \theta_o[\cos(\lambda_n x)]dx \Big/ \int_0^L \cos^2(\lambda_n x)dx \tag{6.41}$$

The integrals indicated in Eq. (6.41) are then evaluated to obtain numerical values for B_n. For this problem, the integral in the numerator is

$$\int_0^L \theta_o[\cos(\lambda_n x)]dx = \frac{\theta_o}{\lambda_n}\sin(\lambda_n x)\Big|_{x=0}^{x=L} = \frac{\theta_o}{\lambda_n}\sin(\lambda_n L) - 0 = \frac{\theta_o}{\lambda_n}(-1)^n \tag{6.42}$$

which includes the identity $\sin(\lambda_n L) = \sin((2n+1)\pi/2) = (-1)^n$. Evaluating the integral in the denominator of Eq. (6.41) yields:

$$\int_0^L \cos^2(\lambda_n x)dx = \frac{x}{2} + \frac{1}{4\lambda_n}\sin(2\lambda_n x)\Big|_0^L = \frac{L}{2} \tag{6.43}$$

Combining the results from Eqs. (6.42) and (6.43), we obtain the following expression for B_n:

$$B_n = \frac{2\theta_o(-1)^n}{\lambda_n L} = \frac{4\theta_o(-1)^n}{(2n+1)\pi} \tag{6.44}$$

This equation is introduced into Eq. (6.36) to give the final solution:

$$\theta = 2\theta_o \sum_{n=0}^{\infty} \frac{(-1)^n}{\lambda_n L}\cos(\lambda_n x)\exp(-\lambda_n^2 t) \quad \text{where} \quad \lambda_n = \frac{(2n+1)\pi}{2L} \tag{6.45}$$

The integrals that arise in the determination of B_n depend on the eigenfunction $X(x)$ and the restricting condition at the nonzero boundary. Diffusion in linear, rectangular, or cubic geometry will produce sine and cosine eigenfunctions, which for common boundary conditions at $x = L$ give eigenvalues of $\lambda_n = (2n+1)\pi/2L$ or $\lambda_n = n\pi/L$. A list of integrals that arise in the evaluation of B_n for sine and cosine eigenfunctions with these eigenvalues are presented in Table 6.7.

Table 6.7 Common integrals in separation of variable problems generating sines and cosines.[a)]

$\int_0^1 \sin(n\pi\xi)d\xi = 0 \quad \text{if } n = 0$ $= \dfrac{1-(-1)^n}{n\pi} \quad \text{if } n \neq 0$	$\int_0^1 \cos(n\pi\xi)d\xi = 1 \quad \text{if } n = 0$ $= 0 \quad \text{if } n \neq 0$
$\int_0^1 \sin\left[\dfrac{(2n+1)\pi\xi}{2}\right] d\xi = \dfrac{2}{(2n+1)\pi}$	$\int_0^1 \cos\left[\dfrac{(2n+1)\pi\xi}{2}\right] d\xi = \dfrac{2(-1)^n}{(2n+1)\pi}$
$\int_0^1 \sin^2(n\pi\xi)d\xi = 0 \text{ if } n = 0$ $= \dfrac{1}{2} \text{if } n \neq 0$	$\int_0^1 \cos^2(n\pi\xi)d\xi = 1 \text{ if } n = 0$ $= \dfrac{1}{2} \text{if } n \neq 0$
$\int_0^1 \sin^2\left[\dfrac{(2n+1)\pi\xi}{2}\right] d\xi = \dfrac{1}{2}$	$\int_0^1 \cos^2\left[\dfrac{(2n+1)\pi\xi}{2}\right] d\xi = \dfrac{1}{2}$
$\int_0^1 \xi \sin(n\pi\xi)d\xi = 0 \text{ if } n = 0$ $= \dfrac{-(-1)^n}{n\pi} \text{if } n \neq 0$	$\int_0^1 \xi \cos(n\pi\xi)d\xi = \dfrac{1}{2} \text{ if } n = 0$ $= \dfrac{(-1)^n - 1}{n^2\pi^2} \text{ if } n > 0$
$\int_0^1 \xi \sin\left[\dfrac{(2n+1)\pi\xi}{2}\right] d\xi = \dfrac{4(-1)^n}{(2n+1)^2\pi^2}$	$\int_0^1 \xi \cos\left[\dfrac{(2n+1)\pi\xi}{2}\right] d\xi$ $= \dfrac{2(-1)^n}{(2n+1)\pi} - \dfrac{4}{(2n+1)^2\pi^2}$
$\int_0^1 \xi^2 \sin(n\pi\xi)d\xi = 0 \text{ if } n = 0$ $= \dfrac{(2-n^2\pi^2)(-1)^n - 2}{n^3\pi^3} \text{ if } n \neq 0$	$\int_0^1 \xi^2 \cos(n\pi\xi)d\xi = \dfrac{1}{3} \text{ if } n = 0$ $= \dfrac{2(-1)^n}{n^2\pi^2} \text{if } n \neq 0$
$\int_0^1 \xi^2 \sin\left[\dfrac{(2n+1)\pi\xi}{2}\right] d\xi$ $= \dfrac{8(-1)^n}{(2n+1)^2\pi^2} - \dfrac{16}{(2n+1)^3\pi^3}$	$\int_0^1 \xi^2 \cos\left[\dfrac{(2n+1)\pi\xi}{2}\right] d\xi$ $= 2(-1)^n \left[\dfrac{(2n+1)^2\pi^2 - 8}{(2n+1)^3\pi^3}\right]$

a) Trigonometric identities used in developing these equations.
$\sin(n\pi) = 0$ $\quad$ $\cos(n\pi) = (-1)^n$
$\sin[(2n+1)\pi/2] = (-1)^n$ $\quad$ $\cos[(2n+1)\pi/2] = 0$

6.3.1.6 Guidelines for Using Separation of Variable Methods to Solve Partial Differential Equations

1. *The differential equation must be linear*. This means that no terms in the differential equation involve products of the dependent variable (e.g. θ^2) or derivatives of the dependent variables (e.g. $\theta\, \partial\theta/\partial x$).
2. *The differential equation must be homogeneous*. This means there are no terms in the differential equation that do not involve the dependent variable; e.g. θ).
3. *The differential equation and restricting conditions must be separable*. This means that the independent variables can be separated by representing the dependent variable as the product of separate functions that each contain only one independent variable; e.g. $\theta(t, x) = T(t)X(x)$.
4. The separated parts of the partial differential equation each produce an ordinary differential equation involving the independent variable in that part. *At most one of the ordinary differential equations generated from the separation process can produce solutions that are not periodic functions* (e.g. differential equations that generate solutions that are periodic functions include the harmonic equation or Bessel's equation). This requirement should be considered in choosing the sign of the separation constant; i.e. the positive or negative sign on λ^2. In some problems, all of the ordinary differential equations generated from the separation process will produce solutions that are periodic functions.

5. *Boundary conditions for the ordinary differential equation generating periodic functions must be homogeneous.* This means that at each of the boundaries of the periodic function (e.g. at $x = 0$ and at $x = L$ in the previous example), either $\theta = 0$, or $\partial\theta/\partial x = 0$, or $\theta + \alpha\, \partial\theta/\partial x = 0$, where α is a constant. (Note that the boundary conditions on x in the previous example met this requirement.)

6.3.1.7 Methods for Making a Nonhomogeneous Partial Differential Equation or Nonhomogeneous Boundary Conditions Homogeneous

Separation of variable solutions can sometimes be used for problems with a nonhomogeneous partial differential equation or with nonhomogeneous boundary conditions. The procedure is to divide the dependent variable θ into the sum of two functions. One function is chosen to satisfy the nonhomogeneous parts of the differential equation or boundary conditions, so that the other function does not need to do this.

For example, consider the dependent variable θ that depends on independent variables t and x as specified in the following partial differential equation and restricting conditions (i.e. one initial and two boundary conditions):

$$\frac{\partial\theta}{\partial t} = \frac{\partial^2\theta}{\partial x^2} \qquad \text{for } t > 0 \quad \text{for } 0 \le x \le L \tag{6.46}$$

$$\text{IC } \theta = \theta_o \qquad \text{at } t = 0 \quad \text{for } 0 \le x \le L \tag{6.47}$$

$$\text{BC 1}: \theta = \theta_1 \qquad \text{at } x = 0 \quad \text{for } t > 0 \tag{6.48}$$

$$\text{BC 2}: \theta = 0 \qquad \text{at } x = L \quad \text{for } t > 0 \tag{6.49}$$

In this example, the boundary condition at $x = 0$ is not homogeneous (i.e. $\theta \neq 0$). This problem can still be solved by separation of variables, if we define θ as follows:

$$\theta(t, x) = \theta_\infty(x) - \varphi(t, x) \tag{6.50}$$

where θ_∞ is the steady-state solution of Eq. (6.46), which depends on x but not t, and satisfies the boundary conditions of the unsteady-state problem, Eqs. (6.48) and (6.49). That is,

$$\frac{d^2\theta_\infty}{dx^2} = 0 \qquad \text{for } 0 \le x \le L \tag{6.51}$$

$$\text{BC 1}: \theta_\infty = \theta_1 \qquad \text{at } x = 0 \tag{6.52}$$

$$\text{BC 2}: \theta_\infty = 0 \qquad \text{at } x = L \tag{6.53}$$

As specified by Eq. (6.50), $\varphi(t, x)$ is the time-dependent function that must be added to the steady-state solution θ_∞ to satisfy the complete solution θ. Solving Eq. (6.51) for the boundary conditions, Eqs. (6.52) and (6.53) yields

$$\theta_\infty = \theta_1(1 - x/L) \tag{6.54}$$

Substituting Eq. (6.50) into Eqs. (6.46)–(6.49) the following:

$$\frac{\partial\varphi}{\partial t} = \frac{d^2\theta_\infty}{dx^2} - \frac{\partial^2\varphi}{\partial x^2} \qquad \text{for } t > 0 \quad \text{for } 0 \le x \le L \tag{6.55}$$

$$\text{IC}: \theta = \theta_\infty - \varphi = \theta_o \qquad \text{at } t = 0 \quad \text{for } 0 \le x \le L \tag{6.56}$$

$$\text{BC 1}: \theta = \theta_\infty - \varphi = \theta_1 \qquad \text{at } x = 0 \quad \text{for } t > 0 \tag{6.57}$$

$$\text{BC 2}: \theta = \theta_\infty - \varphi = 0 \qquad \text{at } x = L \quad \text{for } t > 0 \tag{6.58}$$

Since θ_∞ satisfies Eqs. (6.51)–(6.53), Eqs. (6.55)–(6.58) simplify to

$$\frac{\partial \varphi}{\partial t} = \frac{\partial^2 \varphi}{\partial x^2} \qquad \text{for } t > 0 \quad \text{for } 0 \leq x \leq L \tag{6.59}$$

$$\text{IC} : \varphi = \theta_1(1 - x/L) - \theta_o \qquad \text{at } t = 0 \quad \text{for } 0 \leq x \leq L \tag{6.60}$$

$$\text{BC 1} : \varphi = 0 \qquad \text{at } x = 0 \quad \text{for } t > 0 \tag{6.61}$$

$$\text{BC 2} : \varphi = 0 \qquad \text{at } x = L \quad \text{for } t > 0 \tag{6.62}$$

Notice that the differential equation and restricting conditions on φ meet the requirements of a homogenous partial differential equation and boundary conditions for the independent variable that would produce periodic functions after applying the separation of variable method (i.e. on x). Consequently, Eqs. (6.59)–(6.62) can be solved using separation of variables as described above.

This same technique can also be used to solve a nonhomogeneous partial differential equation such as

$$\frac{\partial \theta}{\partial t} = \frac{\partial^2 \theta}{\partial x^2} + H \qquad \text{for } t > 0 \, \text{for } 0 \leq x \leq L \tag{6.63}$$

where H is a constant. For this example, we keep the same initial and boundary conditions of the previous example:

$$\text{IC} : \theta = \theta_o \qquad \text{at } t = 0 \quad \text{for } 0 \leq x \leq L \tag{6.64}$$

$$\text{BC 1} : \theta = \theta_1 \qquad \text{at } x = 0 \quad \text{for } t > 0 \tag{6.65}$$

$$\text{BC 2} : \theta = 0 \qquad \text{at } x = L \quad \text{for } t > 0 \tag{6.66}$$

Following the same procedure described above for problems with nonhomogeneous boundary conditions, we choose $\theta(t, x)$ to be defined by Eq. (6.50), where θ_∞ is required to satisfy the following equations:

$$\frac{d^2 \theta_\infty}{dx^2} = -H \qquad \text{for } 0 \leq x \leq L \tag{6.67}$$

$$\text{BC 1} : \theta_\infty = \theta_1 \qquad \text{at } x = 0 \tag{6.68}$$

$$\text{BC 2} : \theta_\infty = 0 \qquad \text{at } x = L \tag{6.69}$$

Equation (6.67) can be solved by integrating twice, which gives:

$$\theta_\infty = -H\frac{x^2}{2} + I_1 x + I_2 \tag{6.70}$$

where the two constants of integration I_1 and I_2 are determined from the boundary conditions, Eqs. (6.68) and (6.69), to give the following solution for θ_∞:

$$\theta_\infty = \left(1 - \frac{x}{L}\right)\left(\theta_1 + \frac{HLx}{2}\right) \tag{6.71}$$

Substituting Eq. (6.50) into Eqs. (6.63)–(6.66) yields:

$$-\frac{\partial \varphi}{\partial t} = \frac{d^2 \theta_\infty}{dx^2} - \frac{\partial^2 \varphi}{\partial x^2} + H \qquad \text{for } t > 0 \quad \text{for } 0 \leq x \leq L \tag{6.72}$$

$$\text{IC} : \theta = \theta_\infty - \varphi = \theta_o \qquad \text{at } t = 0 \quad \text{for } 0 \leq x \leq L \tag{6.73}$$

$$\text{BC 1}: \theta = \theta_\infty - \varphi = \theta_1 \qquad \text{at } x = 0 \tag{6.74}$$

$$\text{BC 2}: \theta = \theta_\infty - \varphi = 0 \qquad \text{at } x = L \tag{6.75}$$

which after substituting Eq. (6.71) for θ_∞ (or equivalently, including Eqs. (6.67)–(6.69)) produces:

$$\frac{\partial \varphi}{\partial t} = \frac{\partial^2 \varphi}{\partial x^2} \qquad \text{for } t > 0 \quad \text{for } 0 \le x \le L \tag{6.76}$$

$$\text{IC}: \varphi = \left(1 - \frac{x}{L}\right)\left(\theta_1 + \frac{HLx}{2}\right) - \theta_o \qquad \text{at } t = 0 \quad \text{for } 0 \le x \le L \tag{6.77}$$

$$\text{BC 1}: \varphi = 0 \qquad \text{at } x = 0 \quad \text{for } t > 0 \tag{6.78}$$

$$\text{BC 2}: \varphi = 0 \qquad \text{at } x = L \quad \text{for } t > 0 \tag{6.79}$$

which can be solved using separation of variables as described above.

6.3.1.8 Choosing the Index Starting Value on the Sum of All Solutions (i.e. should $n = 1$ or 0?)

Remember that the complete solution to a partial differential equation is the sum of all possible solutions. The starting index on the summation of all solutions must be chosen to include all solutions. In the first example in this section, eigenvalues satisfied the eigenfunction $\cos(\lambda_n L) = 0$. Thus, $\lambda_n = \pi/2L, 3\pi/2L, 5\pi/2L$, and so on. Note that eigenvalues also occur at negative values (e.g. $\lambda_n = -\pi/2L, -3\pi/2L, -5\pi/2L$). However, because $\cos(-\lambda_n x) = \cos(\lambda_n x)$, solutions written with the negative values of λ_n are not independent from the solutions written with the positive λ_n. If λ_n is written as $(2n+1)\pi/2L$, then the index on the sum of all solutions should begin with $n = 0$.

Other situations can occur. For example, if the separation of variable solution generates the harmonic equation:

$$\frac{d^2X}{dx^2} + \lambda^2 X = 0 \tag{6.80}$$

$$X = A\sin(\lambda x) + B\cos(\lambda x) \tag{6.81}$$

with the following boundary conditions

$$\text{BC 1}: X = 0 \qquad \text{at } x = 0 \tag{6.82}$$

$$\text{BC 2}: X = 0 \qquad \text{at } x = L \tag{6.83}$$

then

$$X_n = A_n \sin(\lambda_n x) \quad \text{where} \quad \lambda_n = n\pi/L \tag{6.84}$$

In this case, $\sin(\lambda_n x) = 0$ for all values of x when $n = 0$, and consequently, the index on the sum of all solutions should start at $n = 1$ (i.e. because the solution at $n = 0$ is zero).

However, if the following boundary conditions were applied to Eq. (6.81):

$$\text{BC 1}: \frac{dX}{dx} = 0 \qquad \text{at } x = 0 \tag{6.85}$$

$$\text{BC 1}: \frac{dX}{dx} = 0 \qquad \text{at } x = L \tag{6.86}$$

then

$$X_n = A_n \cos(\lambda_n x) \quad \text{where} \quad \lambda_n = n\pi/L \tag{6.87}$$

In this case, $\cos(\lambda_n x) = 1$ (i.e. not zero) when $n = 0$, and thus, the index on the sum of all solutions should begin with $n = 0$. This example illustrates an additional point. When finding the constants in the summation, the denominator involves the integral listed in Eq. (6.88), which exhibit different values depending on whether n is or is not equal to zero.

$$\int_0^L \cos^2(\lambda_n x)dx = \int_0^L dx = x|_0^L = L \text{ for } n = 0$$
$$= \int_0^L \cos^2(\lambda_n x)dx = \frac{x}{2} + \frac{1}{4\lambda_n}\sin(2\lambda_n x)\Big|_0^L = \frac{L}{2} \text{ for } n \neq 0 \tag{6.88}$$

A list of integrals that commonly arise in the evaluation of constants in separation of variable problems generating sines and cosines with $\lambda_n = (2n+1)\pi/2L$ or $\lambda_n = n\pi/L$ are provided in Table 6.7.

6.3.2 Laplace Transform Solutions

Laplace transform solutions to the partial differential equation(s) arising in the membrane diffusion problems described here involve taking the Laplace transform on time t or its dimensionless form τ. The Laplace transform of a function of τ, $f(\tau)$, denoted as $\mathcal{L}\{f(\tau)\}$, is defined by the following integration:

$$\mathcal{L}\{f(\tau)\} = \int_0^\infty f(\tau)e^{-s\tau}d\tau \tag{6.89}$$

This operation transforms the τ-dependent function $f(\tau)$ into a new function $\mathcal{L}\{f(\tau)\}$, which depends on s. Laplace transform solutions involve converting the partial differential equation and restricting conditions involving τ and ξ (e.g. $\widetilde{C}_1(\tau,\xi)$) into new equation(s), usually algebraic, of the function $\mathcal{L}\{\widetilde{C}_1(\tau,\xi)\}$, which depend on s and ξ. Once the transformed equations are solved in the "s-domain," $\mathcal{L}\{\widetilde{C}_1(\tau,\xi)\}$ is inverted back to the τ-domain through the following operation:

$$\widetilde{C}_1(\tau,\xi) = \frac{1}{2\pi i}\int_{\alpha-i\infty}^{\alpha+i\infty} e^{st}\mathcal{L}\{\widetilde{C}_1(\tau,\xi)\}ds \tag{6.90}$$

which can be performed analytically to obtain algebraic equations for $\widetilde{C}_1(\tau,\xi)$ or by numerical integration to generate values for $\widetilde{C}_1(\tau,\xi)$ at specific values of τ and ξ. Several of the models considered here produce more complicated equations for $\mathcal{L}\{\widetilde{C}_1(\tau,\xi)\}$ that were inverted to $\widetilde{C}_1(\tau,\xi)$ using the method of residues, which is described below. In a few models, $\mathcal{L}\{\widetilde{C}_1(\tau,\xi)\}$ is given and in one case (Model A4), $\mathcal{L}\{\widetilde{C}_1(\tau,\xi)\}$ is listed without the inversion to $\widetilde{C}_1(\tau,\xi)$. Examples of Laplace transform solutions are provided for Model A1 (which can also be solved by separation of integrals) and Model A10. Carslaw and Jaeger [2] also present examples of Laplace transform solutions including inversion of the Laplace domain solutions by method of residues.

6.3.2.1 Using Laplace Transforms to Determine Lag Times, Steady-state Values and Other Derived Quantities

Laplace domain equations for the dimensionless flux and average concentrations in membranes 1 and 2 and flux in the vehicle can be derived from the Laplace transform of the dimensionless solute concentration as specified below.

$$\mathcal{L}\{\widetilde{J}_1\} = -\frac{d}{d\xi}\mathcal{L}\{\widetilde{C}_1\} \tag{6.91}$$

$$\mathcal{L}\{\widetilde{J}_2\} = -\Gamma_2\frac{d}{d\xi}\mathcal{L}\{\widetilde{C}_2\} = -\frac{\eta_2}{B_{1,2}}\frac{d}{d\xi}\mathcal{L}\{\widetilde{C}_2\} \tag{6.92}$$

$$\mathcal{L}\{\widetilde{J}_V\} = -\Gamma_V \frac{d}{d\xi}\mathcal{L}\{\widetilde{C}_V\} = -K_{1/V}\,\gamma^2 \frac{d}{d\xi}\mathcal{L}\{\widetilde{C}_V\} \tag{6.93}$$

$$\mathcal{L}\{\langle\widetilde{C}_1\rangle\} = \int_0^1 \mathcal{L}\{\widetilde{C}_1\}d\xi \tag{6.94}$$

$$\mathcal{L}\{\langle\widetilde{C}_2\rangle\} = \frac{1}{\eta_2}\int_1^{1+\eta_2} \mathcal{L}\{\widetilde{C}_2\}d\xi \tag{6.95}$$

The cumulative mass transfer across the boundary between compartments i and j is determined from the time integral of the flux at that boundary:

$$\widetilde{M}_{i,j} = \int_0^\tau \widetilde{J}_i\Big|_{\xi=i-j\text{ boundary}} d\tau = \int_0^\tau \widetilde{J}_j\Big|_{\xi=i-j\text{ boundary}} d\tau \tag{6.96}$$

In the Laplace domain, the cumulative mass transfer across a boundary $\mathcal{L}\{\widetilde{M}_{i,j}\}$ is

$$\mathcal{L}\{\widetilde{M}_{i,j}\} = \frac{1}{s}\mathcal{L}\left\{\widetilde{J}_i\Big|_{\xi=i-j\text{ boundary}}\right\} = \frac{1}{s}\mathcal{L}\left\{\widetilde{J}_j\Big|_{\xi=i-j\text{ boundary}}\right\} \tag{6.97}$$

which is derived from the Laplace transform of the time integral of the flux as described in Eq. (6.98) for the generic function of $\tau, f(\tau)$:

$$\mathcal{L}\left\{\int_0^\tau f(\hat{\tau})d\hat{\tau}\right\} = \int_0^\infty \left[\int_0^\tau f(\hat{\tau})d\hat{\tau}\right] e^{-s\tau}d\tau = \frac{1}{s}\mathcal{L}\{f(\tau)\} \tag{6.98}$$

where $\hat{\tau}$ is a dummy variable of integration.

For models that exhibit steady-state conditions, steady-state expressions for concentrations and flux can be derived from equations in the Laplace domain using the final value theorem:

$$\lim_{\tau\to\infty} f(\tau) = \lim_{s\to 0} s\,\mathcal{L}\{f(\tau)\} \tag{6.99}$$

which applies if the time domain function $f(\tau)$ is bounded on the time interval $(0,\infty)$. Thus, the steady-state value of the dimensionless concentration for membrane 1 ($\widetilde{C}_1^{ss}$) can be derived by performing the operation:

$$\widetilde{C}_1^{ss} = \lim_{s\to 0} s\,\mathcal{L}\{\widetilde{C}_1\} \tag{6.100}$$

Similarly, the steady-state flux $\widetilde{J}^{ss}$ is

$$\widetilde{J}^{ss} = \lim_{s\to 0} s\,\mathcal{L}\{\widetilde{J}_k\} \tag{6.101}$$

where $\widetilde{J}_k$ is the flux in membrane k. The dimensionless steady-state cumulative mass across the boundary between compartments i and j ($\widetilde{M}_{i,j}^{ss}$) is represented by

$$\widetilde{M}_{i,j}^{ss} = \widetilde{J}^{ss}(\tau - \tau_o) \tag{6.102}$$

where τ_o is the τ-axis intercept. Expressions for τ_o at the i–j boundary can be derived by performing the operation specified in Eq. (6.103) from Siegel [14]:

$$\tau_o = \lim_{s\to 0}\left[\frac{1}{s}\left(1 - s\,\mathcal{L}\left\{\widetilde{J}_k\Big|_{\xi=i-j\text{ boundary}}\right\}\Big/\widetilde{J}^{ss}\right)\right] \tag{6.103}$$

where $\widetilde{J}_k$ is the flux in membrane k, which is either compartment i or j depending on the boundary location. The derivation of τ_o from the Laplace domain solution is illustrated in Example A below.

6.3.2.2 Inversion of Laplace Transformed Functions to Time Domain Functions by Method of Residues

The Laplace transformed function $\mathcal{L}\{f(\tau)\}$ can be inverted to the time domain function $f(\tau)$ using the method of residues provided the singularities in $\mathcal{L}\{f(\tau)\}$ are restricted to mth-order poles and $\mathcal{L}\{f(\tau)\}$ is single-valued at all other points. If this is the case, then the function $f(\tau)$ can be obtained from $\mathcal{L}\{f(\tau)\}$ by performing the following operation

$$\begin{aligned} f(\tau) &= \frac{1}{2\pi i}\int_{\alpha-i\infty}^{\alpha+i\infty} e^{s\tau}\mathcal{L}\{f(\tau)\}ds \\ &= \sum_n \text{Res}(e^{s\tau}\mathcal{L}\{f(\tau)\}; s_n) \end{aligned} \tag{6.104}$$

where Res $(e^{st}\mathcal{L}\{f(\tau)\}\,; s_n)$ is the residue for an mth-order pole at $s = s_n$, which is defined as follows:

$$\text{Res}\,(e^{s\tau}\mathcal{L}\{f(\tau)\}\,; s_n) = \lim_{s\to s_n}\left[\frac{1}{(m-1)!}\frac{d^{m-1}}{ds^{m-1}}\{(s-s_n)^m e^{s\tau}\mathcal{L}\{f(\tau)\}\}\right] \tag{6.105}$$

As a result, the residue for the first-order pole is defined as follows:

$$\text{Res}\,(e^{s\tau}\mathcal{L}\{f(\tau)\}\,; s_n) = \lim_{s\to s_n}[(s-s_n)e^{s\tau}\mathcal{L}\{f(\tau)\}] \tag{6.106}$$

Inversion involves two steps: (i) identifying the singularities in $\mathcal{L}\{f(\tau)\}$, and (ii) determining the residue at each singularity. How to do this is illustrated below for two examples.

6.3.2.3 Example A – Model A1

The dimensionless differential equation and the initial and boundary conditions describing the one-dimensional diffusion through one finite membrane, initially solute-free, that is exposed at time zero to a vehicle with a constant solute concentration and a receptor solution containing no solute are the following:

$$\frac{\partial \widetilde{C}_1}{\partial \tau} = \frac{\partial^2 \widetilde{C}_1}{\partial \xi^2} \qquad \text{for } 0 \le \xi \le 1,\ \tau > 0 \tag{6.107}$$

$$\text{IC}: \widetilde{C}_1 = 0 \qquad \text{at } \tau = 0,\ 0 \le \xi \le 1 \tag{6.108}$$

$$\text{BC 1}: \widetilde{C}_1 = \widetilde{C}_V = 1 \qquad \text{at } \xi = 0,\ \tau > 0 \tag{6.109}$$

$$\text{BC 2}: \widetilde{C}_1 = \widetilde{C}_R = 0 \qquad \text{at } \xi = 1,\ \tau \ge 0 \tag{6.110}$$

This problem is in the form described by Eqs. (6.46) and (6.49), which can be solved by separation of variables after replacing $\widetilde{C}_1$ with the linear combination of the steady-state solution ($\widetilde{C}_1^{SS} = 1 - \xi$) and an unsteady-state function as described in the general case by Eq. (6.50).

This problem can also be solved by Laplace transforms and then inverted to the time domain using the method of residues as illustrated below. Derivation of the lag and lead times (i.e. τ_0 in Eq. (6.102)) from equations in the Laplace domain is also demonstrated (starting after Eq. (6.130)).

The Laplace transform of Eq. (6.107) is

$$s\mathcal{L}\{\widetilde{C}_1\} - \widetilde{C}_1\Big|_{\tau=0} = \frac{d^2\mathcal{L}\{\widetilde{C}_1\}}{d\xi^2} \tag{6.111}$$

which after applying the initial condition Eq. (6.108), $\widetilde{C}_1\Big|_{\tau=0} = 0$, simplifies to

$$\frac{d^2\mathcal{L}\{\widetilde{C}_1\}}{d\xi^2} - s\mathcal{L}\{\widetilde{C}_1\} = 0 \tag{6.112}$$

The solution to this ordinary differential equation in ξ is

$$\mathcal{L}\{\widetilde{C}_1\} = A\sinh(\sqrt{s}\xi) + B\cosh(\sqrt{s}\xi) \tag{6.113}$$

where the integration constants

$$A = -\frac{\cosh(\sqrt{s})}{s\sinh(\sqrt{s})} \quad \text{and} \quad B = \frac{1}{s} \tag{6.114}$$

are determined by applying the Laplace transform of the boundary conditions, Eqs. (6.109) and (6.110):

$$\mathcal{L}\{\widetilde{C}_1\} = \mathcal{L}\{1\} = \frac{1}{s} \qquad \text{at } \xi = 0, \ \ \tau > 0 \tag{6.115}$$

$$\mathcal{L}\{\widetilde{C}_1\} = \mathcal{L}\{0\} = 0 \qquad \text{at } \xi = 1, \ \ \tau > 0 \tag{6.116}$$

Thus, the solution for $\widetilde{C}_1$ in the Laplace domain is:

$$\mathcal{L}\{\widetilde{C}_1\} = \frac{\cosh(\sqrt{s}\xi)}{s} - \frac{\cosh(\sqrt{s})\sinh(\sqrt{s}\xi)}{s\sinh(\sqrt{s})} \tag{6.117}$$

Without inverting $\mathcal{L}\{\widetilde{C}_1\}$, we can determine the dimensionless steady-state concentration, $\widetilde{C}_1^{ss} = 1$, by performing the operation specified in Eq. (6.100).

$$\widetilde{C}_1^{ss} = \lim_{s\to 0} s\,\mathcal{L}\{\widetilde{C}_1\} = \lim_{s\to 0}\left[\cosh(\sqrt{s}\xi) - \frac{\cosh(\sqrt{s})\sinh(\sqrt{s}\xi)}{\sinh(\sqrt{s})}\right] = 1 - \xi \tag{6.118}$$

Equation (6.117) can be inverted by the method of residues, provided all points are single-valued except at the poles (i.e. values of s which are singular). Although the function $\sqrt{s}$ is not single valued, $\mathcal{L}\{\widetilde{C}_1\}$ is single valued except at the poles as illustrated by writing Eq. (6.117) as follows:

$$\mathcal{L}\{\widetilde{C}_1\} = \frac{\cosh(\sqrt{s}\xi)}{s} - \frac{\cosh(\sqrt{s})}{s}\frac{\sinh(\sqrt{s}\xi)/\sqrt{s}}{\sinh(\sqrt{s})/\sqrt{s}} \tag{6.119}$$

and the polynomial expansions of $\sinh(\sqrt{s}\xi)/\sqrt{s}$ and $\cosh(\sqrt{s}\xi)$

$$\begin{aligned}\frac{\sinh(\sqrt{s}\xi)}{\sqrt{s}} &= \frac{\sqrt{s}}{\sqrt{s}}\xi + \frac{\sqrt{s}}{\sqrt{s}}\frac{s\xi^3}{3!} + \frac{\sqrt{s}}{\sqrt{s}}\frac{s^2\xi^5}{5!} + \frac{\sqrt{s}}{\sqrt{s}}\frac{s^3\xi^7}{7!} + \cdots \\ &= \xi + \frac{s\xi^3}{3!} + \frac{s^2\xi^5}{5!} + \frac{s^3\xi^7}{7!} + \cdots \end{aligned} \tag{6.120}$$

$$\begin{aligned}\cosh(\sqrt{s}\xi) &= 1 + \frac{(\sqrt{s}\xi)^2}{2!} + \frac{(\sqrt{s}\xi)^4}{4!} + \frac{(\sqrt{s}\xi)^6}{6!} + \cdots \\ &= 1 + \frac{s\xi^2}{2!} + \frac{s^2\xi^4}{4!} + \frac{s^3\xi^6}{6!} + \cdots \end{aligned} \tag{6.121}$$

which do not involve $\sqrt{s}$. Poles (singularities) in Eq. (6.117) occur at $s = 0$ and at $\sqrt{s} = in\pi$ (equivalent to $s = -n^2\pi^2$), which satisfies the condition that $\sinh(i\sqrt{s}/i) = i\sin(\sqrt{s}/i) = 0$, where $i^2 = -1$.

Therefore, the inversion of Eq. (6.117) is the summation of the residues, defined as follows:

$$\widetilde{C}_1(\tau,\xi) = \lim_{s\to 0}\left[s\exp[s\tau]\left(\frac{\cosh(\sqrt{s}\xi)}{s} - \frac{\cosh(\sqrt{s})\sinh(\sqrt{s}\xi)}{s\sinh(\sqrt{s})}\right)\right] + \lim_{\substack{\sqrt{s}\to in\pi\\ s\to -n^2\pi^2}}\left[(s+n^2\pi^2)\exp[s\tau]\left(\frac{\cosh(\sqrt{s}\xi)}{s} - \frac{\cosh(\sqrt{s})\sinh(\sqrt{s}\xi)}{s\sinh(\sqrt{s})}\right)\right] \tag{6.122}$$

The first term in Eq. (6.122) has the solution:

$$\lim_{s\to 0}\left[se^{st}\left(\frac{\cosh(\sqrt{s})\sinh(\sqrt{s}\xi)}{s\sinh(\sqrt{s})} + \frac{\cosh(\sqrt{s}\xi)}{s}\right)\right] = 1 - \lim_{s\to 0}\left[\frac{\sinh(\sqrt{s}\xi)}{\sinh(\sqrt{s})}\right] = 1-\xi \tag{6.123}$$

The solution to the second term in Eq. (6.122) is

$$\begin{aligned}
&\lim_{\substack{\sqrt{s}\to in\pi\\ s\to -n^2\pi^2}}\left[(s+n^2\pi^2)\exp[s\tau]\left(\frac{\cosh(\sqrt{s}\xi)}{s} - \frac{\cosh(\sqrt{s})\sinh(\sqrt{s}\xi)}{s\sinh(\sqrt{s})}\right)\right]\\
&= \exp[-n^2\pi^2\tau]\\
&\times \lim_{\substack{\sqrt{s}\to in\pi\\ s\to -n^2\pi^2}}\left[(s+n^2\pi^2)\frac{\cosh(in\pi\xi)}{n^2\pi^2} - (s+n^2\pi^2)\left(\frac{\cosh(\sqrt{s})\sinh(\sqrt{s}\xi)}{s\sinh(\sqrt{s})}\right)\right]
\end{aligned} \tag{6.124}$$

where the limit is determined using L'Hôpital's rule as follows:

$$\begin{aligned}
&\lim_{\substack{\sqrt{s}\to in\pi\\ s\to -n^2\pi^2}}\left[(s+n^2\pi^2)\frac{\cosh(in\pi\xi)}{n^2\pi^2} - (s+n^2\pi^2)\left(\frac{\cosh(\sqrt{s})\sinh(\sqrt{s}\xi)}{s\sinh(\sqrt{s})}\right)\right]\\
&= \lim_{\substack{\sqrt{s}\to in\pi\\ s\to -n^2\pi^2}}\left[-\frac{\cosh(\sqrt{s})\sinh(\sqrt{s}\xi)}{s}\,\frac{\frac{d}{ds}(s+n^2\pi^2)}{\frac{d}{ds}[\sinh(\sqrt{s})]}\right]\\
&= \lim_{\substack{\sqrt{s}\to in\pi\\ s\to -n^2\pi^2}}\left[-\frac{\cosh(\sqrt{s})\sinh(\sqrt{s}\xi)}{s}\,\frac{1}{\cosh(\sqrt{s})/2\sqrt{s}}\right]\\
&= \lim_{\substack{\sqrt{s}\to in\pi\\ s\to -n^2\pi^2}}\left[-\frac{2\sinh(\sqrt{s}\xi)}{\sqrt{s}}\right]\\
&= -\frac{2\sinh(in\pi\xi)}{in\pi} = \frac{2i\sin(n\pi\xi)}{in\pi} = -\frac{2\sin(n\pi\xi)}{n\pi}
\end{aligned} \tag{6.125}$$

Thus, the final equation representing the dimensionless solute concentration membrane 1 is

$$\widetilde{C}_1 = 1-\xi-\frac{2}{\pi}\sum_{n=1}^{\infty}\frac{\sin(n\pi\xi)}{n}\exp[-n^2\pi^2\tau] \tag{6.126}$$

The cumulative mass transfer from the vehicle to membrane 1 can then be determined from

$$\widetilde{M}_{V,1} = \int_0^{\tau} \widetilde{J}_1\Big|_{\xi=0} d\tau = -\int_0^{\tau} \frac{\partial \widetilde{C}_1}{\partial \xi}\bigg|_{\xi=0} d\tau \tag{6.127}$$

which gives the following result

$$\begin{aligned}\widetilde{M}_{V,1} &= \int_0^{\tau}\left(1 + 2\sum_{n=1}^{\infty}\cos(n\pi\xi)|_{\xi=0}\exp[-n^2\pi^2\tau]\right)d\tau \\ &= \tau + 2\sum_{n=1}^{\infty}\int_0^{\tau}\exp[-n^2\pi^2\tau]d\tau \\ &= \tau + 2\sum_{n=1}^{\infty}\frac{1}{n^2\pi^2} - 2\sum_{n=1}^{\infty}\frac{\exp[-n^2\pi^2\tau]}{n^2\pi^2}\end{aligned} \tag{6.128}$$

This expression can be simplified further

$$\widetilde{M}_{V,1} = \tau + \frac{1}{3} - 2\sum_{n=1}^{\infty}\frac{\exp[-n^2\pi^2\tau]}{n^2\pi^2} \tag{6.129}$$

by recognizing that the second term is a constant represented by the identity (see Section 6.3.3)

$$\sum_{n=0}^{\infty}\frac{1}{(n\pi)^2} = \frac{1}{6} \tag{6.130}$$

As illustrated above, derivation of the cumulative mass transfer across a membrane boundary for membrane concentrations involving infinite summations of $\exp[-\lambda_n^2\tau]$ produces an infinite summation at $\tau = 0$, which is a constant, designated τ_o, representing the lead (for the upstream boundary) or lag (for the downstream boundary) time. In models that are more complex than this example, a simplified representation of the summation at $\tau = 0$ may not be easily recognized. For these problems, it is often convenient to derive τ_o in the Laplace domain by performing the operation specified by Eq. (6.103) as described in Section 6.3.2.1. The expression for τ_o is then included in the general representation of the cumulative mass transfer across boundary *i–j*:

$$\widetilde{M}_{i,j} = \widetilde{J}^{ss}(\tau - \tau_o) + \widetilde{M}_{i,j}^{uss} \tag{6.131}$$

where $\widetilde{M}_{i,j}^{uss}$ is the unsteady-state part (i.e. the third term in Eq. (6.129)). For Model A1,

$$\mathcal{L}\{\widetilde{J}_1\} = -\frac{d}{d\xi}\mathcal{L}\{\widetilde{C}_1\} = \sqrt{s}\left(-\frac{\sinh(\sqrt{s}\xi)}{s} + \frac{\cosh(\sqrt{s})\cosh(\sqrt{s}\xi)}{s\sinh(\sqrt{s})}\right) \tag{6.132}$$

and

$$\widetilde{J}^{ss} = \lim_{s\to 0} s\,\mathcal{L}\{\widetilde{J}_1\} = \lim_{s\to 0}\left[s\sqrt{s}\left(-\frac{\sinh(\sqrt{s}\xi)}{s} + \frac{\cosh(\sqrt{s})\cosh(\sqrt{s}\xi)}{s\sinh(\sqrt{s})}\right)\right] = 1 \tag{6.133}$$

As a result, at the boundary between the vehicle and membrane 1, τ_o (the lead time) is

$$\tau_o = \lim_{s\to 0}\left[\frac{1}{s}\left(1 - s\,\mathcal{L}\left\{\widetilde{J}_1\Big|_{\xi=0}\right\}\Big/\widetilde{J}^{ss}\right)\right] = \lim_{s\to 0}\left[\frac{1}{s} - \frac{\cosh(\sqrt{s})}{\sqrt{s}\sinh(\sqrt{s})}\right] = -\frac{1}{3} \tag{6.134}$$

while at the boundary between membrane 1 and the receptor, τ_o (the lag time) is

$$\tau_o = \lim_{s\to 0}\left[\frac{1}{s}\left(1 - s\,\mathcal{L}\left\{\widetilde{J}_1\Big|_{\xi=1}\right\}\Big/\widetilde{J}^{ss}\right)\right] = \lim_{s\to 0}\left[\frac{1}{s} - \frac{\cosh^2(\sqrt{s})}{\sqrt{s}\sinh(\sqrt{s})} + \frac{\sinh(\sqrt{s})}{\sqrt{s}}\right] = \frac{1}{6} \tag{6.135}$$

6.3.2.4 Example B – Model A10

This example illustrates the Laplace transform solution to a problem described by two coupled differential equations along with inversion by method of residues. In this example, the dimensionless differential equations and the initial and boundary conditions describing the one-dimensional diffusion through one finite membrane, initially solute-free that is exposed at time zero to a well-stirred finite-volume vehicle with a known initial solute concentration and a receptor solution containing no solute are

$$\frac{\partial \widetilde{C}_1}{\partial \tau} = \frac{\partial^2 \widetilde{C}_1}{\partial \xi^2} \qquad \text{for } 0 \leq \xi \leq 1, \quad \tau > 0 \tag{6.136}$$

$$\frac{d\widetilde{C}_V}{d\tau} = \frac{1}{\Lambda_V} \left.\frac{\partial \widetilde{C}_1}{\partial \xi}\right|_{\xi=0} \qquad \text{for } \tau > 0 \tag{6.137}$$

$$\text{IC 1}: \widetilde{C}_1 = 0 \qquad \text{at } \tau = 0, \quad 0 \leq \xi \leq 1 \tag{6.138}$$

$$\text{IC 2}: \widetilde{C}_V = 1 \qquad \text{at } \tau = 0 \tag{6.139}$$

$$\text{BC 1}: \widetilde{C}_1 = \widetilde{C}_V \qquad \text{at } \xi = 0, \quad \tau > 0 \tag{6.140}$$

$$\text{BC 2}: \widetilde{C}_1 = \widetilde{C}_R = 0 \qquad \text{at } \xi = 1, \quad \tau > 0 \tag{6.141}$$

The Laplace transforms of Eqs. (6.136) and (6.137) are

$$s\mathcal{L}\{\widetilde{C}_1\} - \left.\widetilde{C}_1\right|_{\tau=0} = \frac{d^2\mathcal{L}\{\widetilde{C}_1\}}{d\xi^2} \tag{6.142}$$

$$s\mathcal{L}\{\widetilde{C}_V\} - \left.\widetilde{C}_V\right|_{\tau=0} = \frac{1}{\Lambda_V} \left.\frac{d\mathcal{L}\{\widetilde{C}_1\}}{d\xi}\right|_{\xi=0} \tag{6.143}$$

which after applying the initial conditions, $\left.\widetilde{C}_1\right|_{\tau=0} = 0$ and $\left.\widetilde{C}_V\right|_{\tau=0} = 1$, simplify to

$$\frac{d^2\mathcal{L}\{\widetilde{C}_1\}}{d\xi^2} - s\mathcal{L}\{\widetilde{C}_1\} = 0 \tag{6.144}$$

$$\mathcal{L}\{\widetilde{C}_V\} = \frac{1}{s}\left[1 + \frac{1}{\Lambda_V} \left.\frac{d\mathcal{L}\{\widetilde{C}_1\}}{d\xi}\right|_{\xi=0}\right] \tag{6.145}$$

The constants A and B in the solution to Eq. (6.144)

$$\mathcal{L}\{\widetilde{C}_1\} = A\sinh(\sqrt{s}\xi) + B\cosh(\sqrt{s}\xi) \tag{6.146}$$

are derived from the Laplace transformed boundary conditions for Eq. (6.146):

$$\text{BC 1}: \mathcal{L}\{\widetilde{C}_1\} = \mathcal{L}\{\widetilde{C}_V\} \qquad \text{at } \xi = 0 \tag{6.147}$$

$$\text{BC 2}: \mathcal{L}\{\widetilde{C}_1\} = 0 \qquad \text{at } \xi = 1 \tag{6.148}$$

Applying first the boundary condition at $\xi = 0$ gives the following expression for B:

$$\mathcal{L}\{\widetilde{C}_1\} = B = \mathcal{L}\{\widetilde{C}_V\} = \frac{1}{s}\left[1 + \frac{A\sqrt{s}}{\Lambda_V}\right] \tag{6.149}$$

which includes

$$\left.\frac{d\mathcal{L}\{\widetilde{C}_1\}}{d\xi}\right|_{\xi=0} = A\sqrt{s} \tag{6.150}$$

Substituting for B into Eq. (6.146) and then applying the boundary condition for $\xi = 1$ yields

$$\mathcal{L}\{\widetilde{C}_1\} = A\sinh(\sqrt{s}) + \frac{1}{s}\left[1 + \frac{A\sqrt{s}}{\Lambda_V}\right]\cosh(\sqrt{s}) = 0 \tag{6.151}$$

which combined with the expression for B in Eq. (6.149) gives

$$A = -\frac{\Lambda_V \cosh(\sqrt{s})}{\sqrt{s}\cosh(\sqrt{s}) + \Lambda_V\, s\, \sinh(\sqrt{s})} \tag{6.152}$$

$$B = \frac{\Lambda_V \sinh(\sqrt{s})}{\sqrt{s}\cosh(\sqrt{s}) + \Lambda_V\, s\, \sinh(\sqrt{s})} \tag{6.153}$$

Thus, the solution for $\widetilde{C}_1$ in the Laplace s-domain is

$$\mathcal{L}\{\widetilde{C}_1\} = \Lambda_V \frac{\sinh(\sqrt{s})\cosh(\sqrt{s}\xi) - \cosh(\sqrt{s})\sinh(\sqrt{s}\xi)}{\sqrt{s}\cosh(\sqrt{s}) + \Lambda_V\, s\, \sinh(\sqrt{s})} \tag{6.154}$$

which using trigonometric identities is simplified to

$$\mathcal{L}\{\widetilde{C}_1\} = \frac{\Lambda_V \sinh[\sqrt{s}(1-\xi)]/\sqrt{s}}{\cosh(\sqrt{s}) + \Lambda_V\sqrt{s}\sinh(\sqrt{s})} \tag{6.155}$$

The first derivative of $\mathcal{L}\{\widetilde{C}_1\}$ is then determined

$$\frac{d\mathcal{L}\{\widetilde{C}_1\}}{d\xi} = -\frac{\Lambda_V\sqrt{s}\cosh[\sqrt{s}(1-\xi)]}{\sqrt{s}\cosh(\sqrt{s}) + \Lambda_V\, s\, \sinh(\sqrt{s})} \tag{6.156}$$

and evaluated at $\xi = 0$

$$\left.\frac{d\mathcal{L}\{\widetilde{C}_1\}}{d\xi}\right|_{\xi=0} = -\frac{\Lambda_V\sqrt{s}\cosh(\sqrt{s})}{\sqrt{s}\cosh(\sqrt{s}) + \Lambda_V\, s\, \sinh(\sqrt{s})} \tag{6.157}$$

to obtain an expression for $\mathcal{L}\{\widetilde{C}_V\}$

$$\mathcal{L}\{\widetilde{C}_V\} = \frac{1}{s}\left[1 - \frac{\cosh(\sqrt{s})}{\cosh(\sqrt{s}) + \Lambda_V\sqrt{s}\sinh(\sqrt{s})}\right] \tag{6.158}$$

which simplifies to

$$\mathcal{L}\{\widetilde{C}_V\} = \frac{\Lambda_V \sinh(\sqrt{s})/\sqrt{s}}{\cosh(\sqrt{s}) + \Lambda_V\sqrt{s}\sinh(\sqrt{s})} \tag{6.159}$$

Note that both $\mathcal{L}\{\widetilde{C}_1\}$ and $\mathcal{L}\{\widetilde{C}_V\}$ are equal to zero when $s = 0$. Equations (6.155) and (6.159) for $\mathcal{L}\{\widetilde{C}_1\}$ and $\mathcal{L}\{\widetilde{C}_V\}$can be inverted using the method of residues because all values of s in these equations are single-valued except at the poles (i.e. nonzero values of s that are singular) as shown in Example A; i.e. see Eqs. (6.120) and (6.121). Poles for $\mathcal{L}\{\widetilde{C}_1\}$ and $\mathcal{L}\{\widetilde{C}_V\}$ are all first-order and occur at values of s satisfying Eq. (6.160):

$$\cosh(\sqrt{s}) + \Lambda_V\sqrt{s}\sinh(\sqrt{s}) = 0 \tag{6.160}$$

which, using the trigonometric identities

$$\sinh(i\sqrt{s}/i) = i\sin(\sqrt{s}/i) \quad \text{and} \quad \cosh(i\sqrt{s}/i) = \cos(\sqrt{s}/i) \tag{6.161}$$

where $i = \sqrt{-1}$, can be written as

$$\cos(\sqrt{s}/i) + \Lambda_V\sqrt{s}\,i\sin(\sqrt{s}/i) = 0 \tag{6.162}$$

Poles (singularities) in Eq. (6.162) occur at multiple values of s, expressed as $\sqrt{s} = i\lambda_n$ (or equivalently, $s = -\lambda_n^2$) for integer values of $n > 0$, which satisfy Eq. (6.163):

$$\lambda_n \tan\lambda_n = 1/\Lambda_V \tag{6.163}$$

The inversion of Eq. (6.155) by method of residues is

$$\widetilde{C}_1(\tau,\xi) = \sum_{n=1}^{\infty}\text{Res}(e^{s\tau}\mathcal{L}\{\widetilde{C}_1(\tau,\xi)\}; s_n = -\lambda_n^2) \tag{6.164}$$

where the residues at $s = -\lambda_n^2$ are determined from the expression:

$$\begin{aligned}
&\text{Res}(e^{s\tau}\mathcal{L}\{\widetilde{C}_1(\tau,\xi)\}; s_n = -\lambda_n^2)\\
&= \lim_{\substack{\sqrt{s}\to i\lambda_n\\ s\to-\lambda_n^2}}\left[(s-s_n)\exp[s\tau]\left(\frac{\Lambda_V\sinh[\sqrt{s}(1-\xi)]/\sqrt{s}}{\cosh(\sqrt{s})+\Lambda_V\sqrt{s}\sinh(\sqrt{s})}\right)\right]\\
&= \Lambda_V\exp[-\lambda_n^2\tau]\lim_{\substack{\sqrt{s}\to i\lambda_n\\ s\to-\lambda_n^2}}\left[(s+\lambda_n^2)\left(\frac{\sinh[\sqrt{s}(1-\xi)]/\sqrt{s}}{\cosh(\sqrt{s})+\Lambda_V\sqrt{s}\sinh(\sqrt{s})}\right)\right]
\end{aligned} \tag{6.165}$$

Given the following limits,

$$\lim_{\substack{\sqrt{s}\to i\lambda_n\\ s\to-\lambda_n^2}}[s+\lambda_n^2] = 0 \tag{6.166}$$

$$\lim_{\substack{\sqrt{s}\to i\lambda_n\\ s\to-\lambda_n^2}}[\cosh(\sqrt{s}) + \Lambda_V\sqrt{s}\sinh(\sqrt{s})] = 0 \tag{6.167}$$

$$\lim_{\substack{\sqrt{s}\to i\lambda_n\\ s\to-\lambda_n^2}}[\sinh[\sqrt{s}(1-\xi)]] = \sinh[i\lambda_n(1-\xi)] = i\sin[\lambda_n(1-\xi)] \neq 0 \tag{6.168}$$

we can simplify Eq. (6.165) as follows:

$$\begin{aligned}
\text{Res}(e^{s\tau}\mathcal{L}\{\widetilde{C}_1(\tau,\xi)\}; s_n = -\lambda_n^2) &= \Lambda_V\exp[-\lambda_n^2\tau]\left(\frac{i\sin[\lambda_n(1-\xi)]}{i\lambda_n}\right)\\
&\quad\times\lim_{\substack{\sqrt{s}\to i\lambda_n\\ s\to-\lambda_n^2}}\left[\frac{s+\lambda_n^2}{\cosh(\sqrt{s})+\Lambda_V\sqrt{s}\sinh(\sqrt{s})}\right]
\end{aligned} \tag{6.169}$$

The remaining limit is determined by applying L'Hôpital's rule and trigonometric identities to be:

$$\lim_{\substack{\sqrt{s}\to i\lambda_n\\ s\to-\lambda_n^2}}\left[\frac{s+\lambda_n^2}{\cosh(\sqrt{s})+\Lambda_V\sqrt{s}\sinh(\sqrt{s})}\right]$$

$$= \lim_{\substack{\sqrt{s}\to i\lambda_n \\ s\to -\lambda_n^2}} \left[\frac{2\sqrt{s}}{\Lambda_V\sqrt{s}\cosh(\sqrt{s}) + (1+\Lambda_V)\sinh(\sqrt{s})}\right]$$

$$= \frac{2i\lambda_n}{i\lambda_n\Lambda_V\cosh(i\lambda_n) + (1+\Lambda_V)\sinh(i\lambda_n)}$$

$$= \frac{2\lambda_n}{\lambda_n\Lambda_V\cos(\lambda_n) + (1+\Lambda_V)\sin(\lambda_n)} \tag{6.170}$$

Therefore, after simplification

$$\mathrm{Res}(e^{s\tau}\mathcal{L}\{\widetilde{C}_1(\tau,\xi)\}; s_n = -\lambda_n^2)$$
$$= \Lambda_V\exp[-\lambda_n^2\tau]\frac{2\sin[\lambda_n(1-\xi)]}{\lambda_n\Lambda_V\cos(\lambda_n) + (1+\Lambda_V)\sin(\lambda_n)} \tag{6.171}$$

A further simplification is made using the identity specified in Eq. (6.163), which defines the values for λ_n. Substituting for $\cos(\lambda_n) = \Lambda_V\lambda_n\sin(\lambda_n)$ in the denominator of Eq. (6.171) gives

$$\lambda_n\Lambda_V\cos(\lambda_n) + (1+\Lambda_V)\sin(\lambda_n) = [\Lambda_V^2\lambda_n^2 + 1 + \Lambda_V]\sin(\lambda_n)$$
$$= (1+\Lambda_V+\Lambda_V^2\lambda_n^2)\sin(\lambda_n) \tag{6.172}$$

and thus

$$\mathrm{Res}(e^{s\tau}\mathcal{L}\{\widetilde{C}_1(\tau,\xi)\}; s_n = -\lambda_n^2) = \Lambda_V\exp[-\lambda_n^2\tau]\frac{2\sin[\lambda_n(1-\xi)]}{(1+\Lambda_V+\Lambda_V^2\lambda_n^2)\sin(\lambda_n)} \tag{6.173}$$

Therefore, the dimensionless concentration of solute in membrane 1 is

$$\widetilde{C}_1(\tau,\xi) = 2\Lambda_V\sum_{n=1}^{\infty}\frac{\sin[\lambda_n(1-\xi)]}{(1+\Lambda_V+\Lambda_V^2\lambda_n^2)\sin(\lambda_n)}\exp[-\lambda_n^2\tau] \tag{6.174}$$

By a similar process, the method of residues can be applied to Eq. (6.159) to determine the dimensionless vehicle concentration.

$$\widetilde{C}_V(\tau) = \sum_{n=1}^{\infty}\mathrm{Res}(e^{s\tau}\mathcal{L}\{\widetilde{C}_V(\tau)\}; s_n = -\lambda_n^2) \tag{6.175}$$

The derivation of the equation for the residues at $s = -\lambda_n^2$ is shown below.

$$\mathrm{Res}\,(e^{s\tau}\mathcal{L}\{\widetilde{C}_V(\tau)\}\,; s_n = -\lambda_n^2)$$
$$= \exp[-\lambda_n^2\tau]\lim_{\substack{\sqrt{s}\to i\lambda_n \\ s\to -\lambda_n^2}}\left[(s+\lambda_n^2)\left(\frac{\Lambda_V\sinh(\sqrt{s})/\sqrt{s}}{\cosh(\sqrt{s}) + \Lambda_V\sqrt{s}\sinh(\sqrt{s})}\right)\right]$$
$$= \Lambda_V\exp[-\lambda_n^2\tau]$$
$$\times \lim_{\substack{\sqrt{s}\to i\lambda_n \\ s\to -\lambda_n^2}}\left[(s+\lambda_n^2)\left(\frac{\sinh(\sqrt{s})}{\sqrt{s}}\right)\left(\frac{1}{\cosh(\sqrt{s}) + \Lambda_V\sqrt{s}\sinh(\sqrt{s})}\right)\right]$$
$$= \Lambda_V\exp[-\lambda_n^2\tau]\left(\frac{\sinh(i\lambda_n)}{i\lambda_n}\right)\lim_{\substack{\sqrt{s}\to i\lambda_n \\ s\to -\lambda_n^2}}\left[\frac{s+\lambda_n^2}{\cosh(\sqrt{s}) + \Lambda_V\sqrt{s}\sinh(\sqrt{s})}\right]$$

$$= \Lambda_V \exp[-\lambda_n^2 \tau] \left(\frac{\sin(\lambda_n)}{\lambda_n} \right) \lim_{\substack{\sqrt{s} \to i\lambda_n \\ s \to -\lambda_n^2}} \left[\frac{s + \lambda_n^2}{\cosh(\sqrt{s}) + \Lambda_V \sqrt{s} \sinh(\sqrt{s})} \right] \tag{6.176}$$

where the limit is given by Eq. (6.170). Therefore,

$$\text{Res}(e^{s\tau} \mathcal{L}\{\widetilde{C}_V(\tau)\}; s_n = -\lambda_n^2) = \frac{2\Lambda_V \sin(\lambda_n) \exp[-\lambda_n^2 \tau]}{\lambda_n \Lambda_V \cos(\lambda_n) + (1 + \Lambda_V) \sin(\lambda_n)} \tag{6.177}$$

which can be simplified further using the identity in Eq. (6.163) to give

$$\begin{aligned} \text{Res}(e^{s\tau} \mathcal{L}\{\widetilde{C}_V(\tau)\}; s_n = -\lambda_n^2) &= \frac{2\Lambda_V \sin(\lambda_n) \exp[-\lambda_n^2 \tau]}{(1 + \Lambda_V + \Lambda_V^2 \lambda_n^2) \sin(\lambda_n)} \\ &= \frac{2\Lambda_V \exp[-\lambda_n^2 \tau]}{1 + \Lambda_V + \Lambda_V^2 \lambda_n^2} \end{aligned} \tag{6.178}$$

Thus, the dimensionless solute concentration in the vehicle is described by the expression

$$\widetilde{C}_V(\tau) = 2\Lambda_V \sum_{n=1}^{\infty} \frac{\exp[-\lambda_n^2 \tau]}{1 + \Lambda_V + \Lambda_V^2 \lambda_n^2} \tag{6.179}$$

which is equal to $\widetilde{C}_1$ at $\xi = 0$ as required by the boundary condition, Eq. (6.140).

6.3.3 Useful Identities

The following identities are useful for evaluating the $\tau = 0$ limit in solutions involving summations with exponentials of τ.

$$\sum_{n=1}^{\infty} \frac{1}{(n\pi)^2} = \frac{1}{6} \tag{6.180}$$

$$\sum_{n=1}^{\infty} \frac{(-1)^n}{(n\pi)^2} = -\frac{1}{12} \tag{6.181}$$

$$\sum_{n=0}^{\infty} (-1)^n \frac{2}{(2n+1)\pi} = \frac{1}{2} \tag{6.182}$$

$$\sum_{n=0}^{\infty} \left(\frac{2}{(2n+1)\pi} \right)^2 = \frac{1}{2} \tag{6.183}$$

$$\sum_{n=0}^{\infty} (-1)^n \left(\frac{2}{(2n+1)\pi} \right)^3 = \frac{1}{4} \tag{6.184}$$

References

1 Frasch, H.F. and Bunge, A.L. (2015). The transient dermal exposure II: post-exposure absorption and evaporation of volatile compounds. *J. Pharm. Sci.* 104: 1499–1507.

2 Carslaw, H.S. and Jaeger, J.C. (1959). *Conduction of Heat in Solids*. Oxford, Great Britain: Oxford University Press.

3 McCarley, K.D. and Bunge, A.L. (1998). Physiologically relevant one-compartment pharmacokinetic models for skin. 1. Development of models. *J. Pharm. Sci.* 87: 470–481.

4 McCarley, K.D. and Bunge, A.L. (1998). Physiologically relevant one-compartment pharmacokinetic models for skin. 1. Development of models, Correction. *J. Pharm. Sci.* 87: 1264.
5 Franz, T.J., Lehman, P.A., Franz, S.F. et al. (1993). Percutaneous penetration of n-nitrosodiethanolamine through human skin (in-vitro): comparison of finite and infinite dose applications from cosmetic vehicles. *Fundam. Appl. Toxicol.* 21: 213–221.
6 Crank, J. (1975). *The Mathematics of Diffusion*. Oxford: Oxford Science Publication.
7 Parks, J.M., Cleek, R.L., and Bunge, A.L. (1997). Chemical release from topical formulations across synthetic membranes: infinite dose. *J. Pharm. Sci.* 86: 187–192.
8 Parks JM. (1996). Release kinetics of soil contaminants: development of experimental methods and theoretical descriptions. M.S. Thesis, Colorado School of Mines, Golden, Colorado.
9 Cleek, R.L. and Bunge, A.L. (1993). A new method for estimating dermal absorption from chemical exposure. 1. General approach. *Pharm. Res.* 10: 497–506.
10 Cain, G. and Meyer, G.H. (2006). *Separation of Variables for Partial Differential Equations – An Eigenfunction Approach*. Boca Raton, FL: Chapman & Hall/CRC.
11 Cannon, J.R. (1984). *The One-Dimensional Heat Equation*. Menlo Park, CA: Addison-Wesley Publishing Company.
12 McCarley, K.D. and Bunge, A.L. (2000). Physiologically relevant two-compartment models for skin. *J. Pharm. Sci.* 89: 1212–1235.
13 Siegel, R.A. (1991). Algebraic, differential, and integral relations for membranes in series and other multilaminar media – permeabilities, solute consumption, lag times, and mean 1st passage times. *J. Phys. Chem.* 95: 2556–2565.
14 Siegel, R.A. (1986). A Laplace transform technique for calculating diffusion time lags. *J. Membr. Sci.* 26: 251–262.

7

Understanding Drug Delivery Outcomes: Progress in Microscopic Modeling of Skin Barrier Property, Permeation Pathway, Dermatopharmacokinetics, and Bioavailability

Guoping Lian[1,2], Tao Chen[1], Panayiotis Kattou[1], Senpei Yang[3], Lingyi Li[3], and Lujia Han[3]

[1]*Department of Chemical and Process Engineering, University of Surrey, Guildford, UK*
[2]*Unilever Research Colworth, Bedford, UK*
[3]*College of Engineering, China Agricultural University, Beijing, P.R. China*

7.1 Introduction

Understanding the heterogeneous barrier property and permeation pathway of the skin is important for skin health [1, 2], transdermal drug delivery [3–6], cosmetic care [7–9] as well as the protection of the skin from environmental exposure to air pollution and harmful chemicals [10–12]. The barrier property of the skin is related to the complex heterogeneous structures and compositions at both cellular and molecular levels. Early studies revealed the stratum corneum (SC), the outmost layer of the skin as the main barrier for percutaneous absorption and permeation. The heterogeneous structure of the SC barrier leads to multiple pathways for percutaneous permeation and heterogeneous absorption. For the SC, the cornified lipid envelopes surrounding the corneocytes were considered as the main pathway for transdermal permeation. A strong correlation between the SC permeability and the hydrophobicity of the permeant (measured by octanol–water partition coefficient) has been observed [13, 14]. However, experimental studies showed that solute uptake by the SC occurred in both the lipid and corneocyte domains [15, 16]. The existence of a transverse SC lipid membrane and corneocytes transport is required to account for the uptake of solute in the corneocyte domain [17, 18]. This is referred to as the transcellular pathway. In addition to the intercellular pathway and transcellular pathway, human skin appendages such as hair follicles and sweat ducts also provide a significant route for transdermal permeation of solutes [19, 20].

Much of the challenge is in quantifying the effect of the microscopic structure and the heterogeneity of the SC on its barrier property and the relative importance of various permeation pathways [21]. Here, we review recent progress in microscopic modeling of the heterogeneous barrier property and permeation pathway of the SC barrier and their effects on the dermatopharmacokinetics and bioavailability of dermal exposure. First, we discuss the general mathematical framework used for microscopic modeling of the heterogeneous barrier property and permeation pathway of the SC. Second, we discuss the progress in obtaining tissue specific properties of the SC barrier. Third, we discuss the progress in applying microscopic modeling for *in silico* prediction of the dermatopharmacokinetic and bioavailability. The aim is to highlight the advantages and challenges in the microscopic modeling approach.

Fundamentals of Drug Delivery, First Edition.
Edited by Heather A. E. Benson, Michael S. Roberts, Adrian C. Williams, and Xiaowen Liang.

7.2 Governing Equation

7.2.1 Homogenized Model

The SC is comprised of 10–20 layers of flattened corneocyte cells interspersed in the continuous lipid matrix [22, 23]. This microstructure is often represented by the so-called "bricks and mortar" model [22, 24–26] as shown in Figure 7.1.

With the idealized "bricks-and-mortar" microstructure of the SC barrier, three permeation pathways can be envisaged: the intercellular pathway through the continuous and tortuous lipid matrix, the transcellular pathway transversing multiple layers of lipid membranes and corneocytes and the shunt pathway of the hair follicles and/or sweat ducts. The overall permeability of the SC barrier can be expressed as follows:

$$P_{sc} = P_l + P_t + P_s \tag{7.1}$$

where P_l, P_t, and P_s denote the permeability components of the intercellular lipid pathway, the transcellular pathway, and the shunt pathway.

In one dimension (1D), the SC barrier has been frequently modeled as a homogenized membrane. The governing equation for solute transfer across the SC can be described by the following equation of Fick's second law of diffusion:

$$\frac{\partial C_{sc}}{\partial t} = \frac{\partial}{\partial y}\left(D_{sc}\frac{\partial C_{sc}}{\partial y}\right) \tag{7.2}$$

where C_{sc} and D_{sc} are homogenized concentration and effective diffusivity of solute in the SC membrane.

The homogenized concentration of solute in the SC can be defined as follows:

$$C_{sc} = f_{sc\text{-}w}C_{sc\text{-}w} + f_{sc\text{-}l}C_{sc\text{-}l} + f_{sc\text{-}p}C_{sc\text{-}p} + f_{sc\text{-}sh}C_{sc\text{-}sh} = K_{sc/w}C_{sc\text{-}w} \tag{7.3}$$

where $K_{sc/w} = (f_{sc\text{-}w} + f_{sc\text{-}l}K_{l/w} + f_{sc\text{-}p}K_{p/w} + f_{sc\text{-}sh}K_{sh/w})$ is the partition coefficient of solute between the SC and water, $K_{l/w}$, $K_{p/w}$, and $K_{sh/w}$ are the partition coefficients of solute between SC lipid and water, corneocyte keratin and water, and shunt media and water respectively, $f_{sc\text{-}w}$, $f_{sc\text{-}l}$, $f_{sc\text{-}p}$, and $f_{sc\text{-}sh}$ are the volume fractions of water, lipids, protein, and shunt media (e.g. sebum) of the SC barrier. Estimation of the mass fractions of the SC lipid, water, protein, and sebum can be found elsewhere [27].

The effective diffusivity of solute in the SC is related to the overall permeability as follows:

$$D_{sc} = \frac{P_{sc}h_{sc}}{K_{sc/w}} \tag{7.4}$$

where h_{sc} is the thickness of the SC.

At steady-state, the flux of solute across the SC membrane can be calculated by the following equation:

$$J_{ss} = P_{sc}(C_{w1} - C_{w2}) \tag{7.5}$$

where C_{w1} and C_{w2} are the concentrations of solute in the aqueous solutions of the donor fluid and receptor fluid, respectively, both in contact with the SC membrane.

By default, Eq. (7.5) is valid only when the donor and receptor fluids in contact with the SC membrane are aqueous fluids. If nonaqueous solvents are used, the following more general form of solute flux across the SC membrane can be used:

$$J_{ss} = \frac{P_{sc}}{K_{v/w}}(C_{v1} - C_{v2}) \tag{7.6}$$

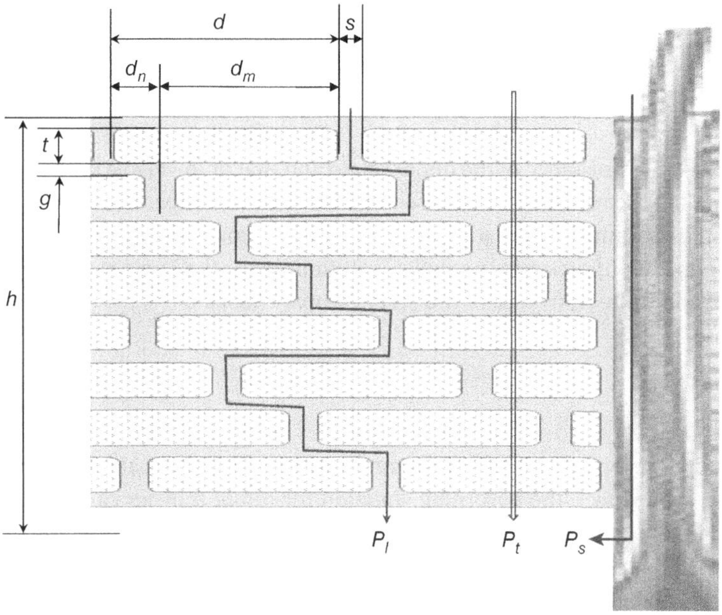

Figure 7.1 Bricks and mortar microstructure of human SC and permeation pathways. Source: Refs. [22, 24, 26].

where C_{v1} and C_{v2} are the concentrations of solute in the donor and receptor fluids in contact with the SC membrane, $K_{v/w}$ is the partition coefficient of solute between the vehicle (donor and receptor fluids) and water.

With Eq. (7.6), the maximum flux is reached when the concentration of solute in the donor fluid is saturated and the receptor fluid is an infinite sink. The underlying assumption of the above steady-state flux equation is that the resistance of both the donor phase and receptor phase is sufficiently small compared to that of the SC membrane.

At transient condition, analytical solution to Eq. (7.2) can be obtained under certain relatively simple boundary conditions. For instance, for the case of an infinite source of constant concentration C_d of the donor fluid and an infinite sink of zero concentration of the receptor fluid, the following Crank equation of diffusion across the SC membrane sheet applies [28]

$$C_{sc}(y) = \frac{K_{Sc/w}}{K_{v/w}} C_d \left[1 - \frac{y}{h_{sc}} + \frac{\pi}{2} \sum_{1}^{\infty} \frac{\cos(n\pi)}{n} \sin\left(\frac{n\pi y}{h_{sc}}\right) \exp\left(-D_{sc} n^2 \pi^2 \frac{t}{h_{sc}^2}\right) \right] \tag{7.7}$$

The transient flux of the solute diffusing out of the SC membrane is given by

$$J_t(y) = \frac{K_{Sc/w}}{K_{v/w}} C_d \frac{D_{sc}}{h_{sc}} \left\{ 1 + \frac{2}{\pi^2} \sum_{1}^{\infty} \cos(n\pi) n\pi^2 \exp\left(-D_{sc} n^2 \pi^2 \frac{t}{h_{sc}^2}\right) \right\} \tag{7.8}$$

and the accumulative amount is

$$Q_t(y) = \frac{K_{Sc/w}}{K_{v/w}} C_d \left[\frac{D_{sc}}{h_{sc}} t - \frac{h_{sc}}{6} - \frac{2h_{sc}}{\pi^2} \sum_{1}^{\infty} \frac{\cos(n\pi)}{n} \exp\left(-D_{sc} n^2 \pi^2 \frac{t}{h_{sc}^2}\right) \right] \tag{7.9}$$

At $t \to \infty$, the accumulative amount approaches a straight line given by

$$Q_t(y) = \frac{K_{Sc/w}}{K_{v/w}} C_d \left(\frac{D_{sc}}{h_{sc}} t - \frac{h_{sc}}{6} \right) \tag{7.10}$$

This line has an intercept on the time axis at

$$t_{lag} = \frac{h_{sc}^2}{6D_{sc}} \tag{7.11}$$

This is the so-called "lag time."

In the above transient analysis, the resistance of the donor phase and that of the viable dermis is considered to be negligibly small. If the resistance of those compartments are comparable to that of the SC barrier, or the donor phase is finite, a more sophisticated transient analysis is required. For instance, Anissimov and Roberts [29, 30] presented analytical solutions using the Laplace transformation method for finite dose and finite resistance of the other compartments.

7.2.2 Microscopic Model

7.2.2.1 Solute Diffusion in SC Lipid

At the microscopic level, solute transfer through the SC barrier has three-dimensional (3D) pathways. To model the 3D microscopic effect, the governing equation for solute transfer can be formulated separately for the SC lipid, corneocytes, and appendages. At molecular level, solute transfer in the cornified envelope of the lipid bilayer is heterogeneous and anisotropic. The following equations of solute diffusion (D) and partition (K) in the lipid bilayer apply [31]:

$$D(z) = \lim_{t\to\infty} \langle [z(t) - z(0)]^2 \rangle / 2t \tag{7.12}$$

$$K(z) = \exp\left(-\frac{\Delta G(z)}{k_B T}\right) \tag{7.13}$$

where z is the location in the lipid bilayer, $\Delta G(z)$ is the free energy difference, t is time, k_B is the Boltzmann constant, and T is the temperature. These two equations of solute diffusion and partition in lipid bilayers are based on the solubility-diffusion model of the Overton's rule.

Molecular dynamic simulations of solute permeation through a single lipid bilayer have been reported extensively [32–34]. Both experimental and computer simulation studies have shown that for undisrupted lipid bilayers, solute diffusion normal to the bilayer plane (transverse bilayer) is a few orders of magnitudes slower than that along the bilayer plane (parallel to bilayer).

The cornified SC lipid membrane is c. 6 lipid bilayers thick [35]. There is still a lack of detailed information of the possible disruptions and overlaps of the SC lipid bilayer envelopes. Two ideal situations can be envisaged: perfect lipid bilayer envelopes wrapping the interspersed corneocytes and imperfect lipid bilayer envelopes with high degrees of disruptions and overlaps. In the case that the cornified lipid envelopes are assembled as perfect lipid bilayers, solute diffusion in the SC lipid will be highly anisotropic. Permeation across the SC membrane via the lipid pathway will need to overcome ~100 transverse SC lipid bilayers and, hence, become rate limited by transverse lipid bilayer diffusion, which is several orders (up to 9, [14]) of magnitudes slower than lateral diffusion along the lipid bilayer plane. The microscopic model adopted by Wang et al. [36] assumed the cornified envelopes of the SC lipids are perfect lipid bilayers. They introduced a lipid mass transfer coefficient for solute diffusion transversing the SC lipid membrane to corneocytes, but still used the lateral diffusion for solute transfer through the tortuous lipid pathway. In the case of imperfect lipid bilayers of the cornified SC lipid envelopes containing high degrees of disruptions and overlaps, solute can leak from one lipid bilayer to another adjacent lipid bilayer. Diffusion through the SC lipid is becoming more isotropic and rate limited by the lateral diffusion along the lipid bilayer planes. This lateral diffusion model was first proposed by Johnson et al. [14] and adopted by many others, e.g. Mitragotri [37] and Chen et al. [27, 38].

When lateral bilayer diffusion is the primary mechanism of solute transport in the SC lipid, the following partial differential equation of diffusion applies

$$\frac{\partial C_l}{\partial t} = \frac{\partial}{\partial x}\left(D_{l\text{-}lat}\frac{\partial C_l}{\partial x}\right) + \frac{\partial}{\partial y}\left(D_{l\text{-}lat}\frac{\partial C_l}{\partial y}\right) + \frac{\partial}{\partial z}\left(D_{l\text{-}lat}\frac{\partial C_l}{\partial z}\right) \tag{7.14}$$

where C_l is the solute concentration in the SC lipid, $D_{l\text{-}lat}$ is the lateral diffusivity of solute in the SC lipid.

7.2.2.2 Solute Diffusion in SC Corneocytes

Corneocytes can be regarded as porous media of fibrous keratin. For solute transfer in the corneocyte, binding of solute to the keratin needs to be considered. The volume fraction of the porous fluid depends on the hydration level. The mass balance equation of both the Fick diffusion of solute in the porous fluid and binding to keratin can be derived as

$$\theta_w \frac{\partial C_w}{\partial t} = D_c\left(\frac{\partial^2 C_w}{\partial x^2} + \frac{\partial^2 C_w}{\partial y^2} + \frac{\partial^2 C_w}{\partial z^2}\right) - \theta_p k_a C_w + \theta_p k_d C_b \tag{7.15}$$

where C_w is the concentration of solute in the porous fluid of the SC corneocytes, C_b is the concentration of solute bound to keratin, D_c is the diffusivity of solute in the porous fluid of the SC corneocytes, k_a and k_d are rate constants for solute binding (adsorption) and dissociation (desorption) from the SC keratin; θ_w and θ_p are the volume fractions of water and keratin in the corneocytes, respectively.

Generally, solute binding and dissociation to the SC keratin is sufficiently faster than solute diffusion. It is reasonable to assume that $C_b = K_{p/w}C_w$. Equation (7.18) can therefore be rearranged as follows:

$$(\theta_w + \theta_p K_{p/w})\frac{\partial C_w}{\partial t} = D_c\left(\frac{\partial^2 C_w}{\partial x^2} + \frac{\partial^2 C_w}{\partial y^2} + \frac{\partial^2 C_w}{\partial z^2}\right) \tag{7.16}$$

where $C_c = (\theta_w + \theta_p K_{p/w})C_w$ is the overall concentration of solute in the SC corneocytes.

7.2.2.3 Solute Diffusion in Appendages

By analogous reasoning, the governing equation of solute transport in the shunt pathway can be derived. The media for solute transport in the axillary ducts of sweat glands and hair follicles can be regarded as homogenous. For solute transport in hair follicle sebum, the partial differential equation can be expressed as follows:

$$\frac{\partial C_s}{\partial t} = D_s\left(\frac{\partial^2 C_s}{\partial x^2} + \frac{\partial^2 C_s}{\partial y^2} + \frac{\partial^2 C_s}{\partial z^2}\right) \tag{7.17}$$

where C_s is the solute concentration and D_s is the diffusivity of solute in sebum.

7.2.3 Numerical Methods

To model the microscopic effect of the SC barrier, the governing equations of solute transport in the SC lipids, corneocytes, and axillary ducts are solved numerically. Typical numerical methods include finite element method, finite difference method, and finite volume method. All numerical methods require subdividing each of the heterogeneous domains of the SC barrier into sufficiently fine meshes. With the finite volume method, solute transfer from a meshed grid i to a neighboring grid j can be calculated by the following equation:

$$q_{ij} = \frac{A_{ij}}{\frac{\delta_i}{D_i} + \frac{K_{i/w}\delta_j}{K_{j/w}D_j}}\left(C_i - \frac{K_{i/w}}{K_{j/w}}C_j\right) \tag{7.18}$$

where q_{ij} is the flux of solute from grid i to grid j, $K_{i/w}$ and $K_{j/w}$ are the partition coefficients of solute between grid i and water and between grid j and water, C_i and C_j are the concentration of solute, δ_i and δ_j are the diffusion lengths of grid i and j, and D_i and D_j are the diffusivity of solute in grid i and j, respectively.

If the grid lies in the SC lipid or shunt pathway, the following mass balance equation can be obtained for the grid

$$\frac{dC_i}{dt} = -\frac{\sum_j q_{ij}}{V_i} \tag{7.19}$$

Otherwise, for a grid that lies in the corneocytes, the following mass balance equation applies

$$(\theta_w + \theta_p K_{p/w})\frac{dC_i}{dt} = -\frac{\sum_j q_{ij}}{V_i} \tag{7.20}$$

Together, there will be n simultaneous ordinary differential equations (ODEs), where n is the total number of the meshed grids. This set of n ODEs can be solved by using numerical methods, such as the Runge–Kutta methods widely available in commercial and open-source ODE solvers.

7.3 Input Parameters

7.3.1 SC Microstructure

The microscopic structure parameters of the human SC barrier depend on a number of factors such as body site, age, and sex. Typical microstructural parameters of the human SC barrier are listed in Table 7.1. This set of microstructural parameters of the human SC are adopted from Johnson et al. [35] and are used by others [21, 27, 39]. Wang et al. [36, 40] proposed a slightly different geometry of the "bricks and mortar" model, where the corneocytes are represented by parallelograms rather than rectangles. The use of parallelograms for the corneocytes makes the diffusion path in the tortuous lipid pathway slightly longer. Nevertheless, the two microscopic representations of the SC microstructure are very similar.

An early mathematical analysis of the effect of the microscopic heterogeneity of the SC on the barrier property was reported by Michaels et al. [25]. Two pathways were considered: the intercellular lipid pathway and the transcellular pathway. The permeability of the two pathways were derived as follows:

$$P_l = \frac{K_{l/w} D_{l\text{-}lat}}{h_{sc}\tau} \tag{7.21}$$

$$P_t = \frac{K_{l/w} D_{l\text{-}lat}}{h_{sc}} \frac{t+g}{\left(\frac{K_{l/w} D_{l\text{-}lat}}{D_c} t + g\right)} \tag{7.22}$$

with

$$\tau = \frac{d}{2g}\left[1 + \frac{d+g}{2(t+g)}\right] \tag{7.23}$$

The tortuosity factor (τ) of Eq (7.23) for the diffusion pathlength was derived by considering a two-dimensional geometry of the "bricks and mortar." Johnson et al. [35] evaluated the diffusion path length of the intercellular lipid for a 3D geometry of the "bricks and mortar" and proposed the following equation for the tortuosity factor of the SC barrier

$$\tau = 1 + \frac{2g}{h_{sc}} ln\left(\frac{d}{2s}\right) + \frac{Ndt}{h_{sc}s} + \left(\frac{d}{1+\omega}\right)^2 \left(\frac{\omega}{h_{sc}g}\right)(N-1) \tag{7.24}$$

where N is the number of the corneocyte layers, $\omega = d_m/d_n$ is the offset ratio.

Table 7.1 Typical "bricks-and-mortar" microstructure parameters of human SC.

Microstructure parameter	Value
Number of corneocyte layers (N)	16
Width of corneocytes (d)	40 μm
Height of corneocytes (t)	0.8 μm
Gap of intercellular lipids (g)	0.075 μm
Spacing between corneocytes (s)	0.075 μm
Offset ratio (w)	8

Source: Based on Johnson et al. [35].

For typical geometrical parameters of the SC barrier listed in Table 7.1, Johnson et al. [35] estimated the value of the tortuosity factor of human skin to be $\tau = 2740$, which corresponds to a diffusion path length of $\tau h = 3.6$ cm.

7.3.2 SC Lipid–Water Partition

The SC lipid is the main pathway for transdermal permeation. A number of studies have been reported on the partition and diffusion properties of solute in the SC lipid. Direct measurements of solute partition and diffusion in the SC lipid are rare. The partition coefficients of SC lipid–water have been mostly determined by comparing solute partition in delipidized SC and normal SC [15, 16]. It was observed that the solute partition between the SC lipid and water scales to the octanol–water partition coefficients. The following relationship is widely used:

$$K_{l/w} = K_{o/w}^{n} \tag{7.25}$$

where $K_{o/w}$ is octanol–water partition coefficient, n is power index, a fitted parameter.

The most quoted values of the power index are close to $n \approx 0.7$ [35, 39, 41]. Nitsche et al. [18] proposed a slightly different correlation with a prefactor of 0.48 and a slightly higher power index $n = 0.81$. Hansen et al. [41] proposed to use D_{ow}, the octanol–water distribution coefficient, to correlate with the experimental data of SC lipid–water partition. Figure 7.2 compares some of the widely used models of SC lipid–water partition coefficients against the available experimental data. The model of Johnson et al. [35] and Hansen et al. [41] are very similar. By using a pH-dependent octanol–water partition coefficient, Hansen et al. [41] was able to improve the prediction of ionizable compounds. The model of Nitsche et al. [18] appears to underestimate the SC lipid–water partition coefficient of hydrophilic solute by an order of magnitude.

7.3.3 Diffusivity in SC Lipids

Johnson et al. [14] reported a study on the diffusivity of chemicals in the SC lipids using imaging of fluorescence recovery after photobleaching (FRAP). Measuring the concentration intensity allowed the estimation of the effective diffusivity of the imaged chemicals in the SC lipids. They referred to this as the lateral bilayer diffusivity, implying the cornified lipid bilayer envelops surrounding the corneocytes are imperfect lipid bilayers with high level of disruptions [14, 35]. In most published studies, solute diffusivity in the SC lipids is estimated from the measured transdermal permeability

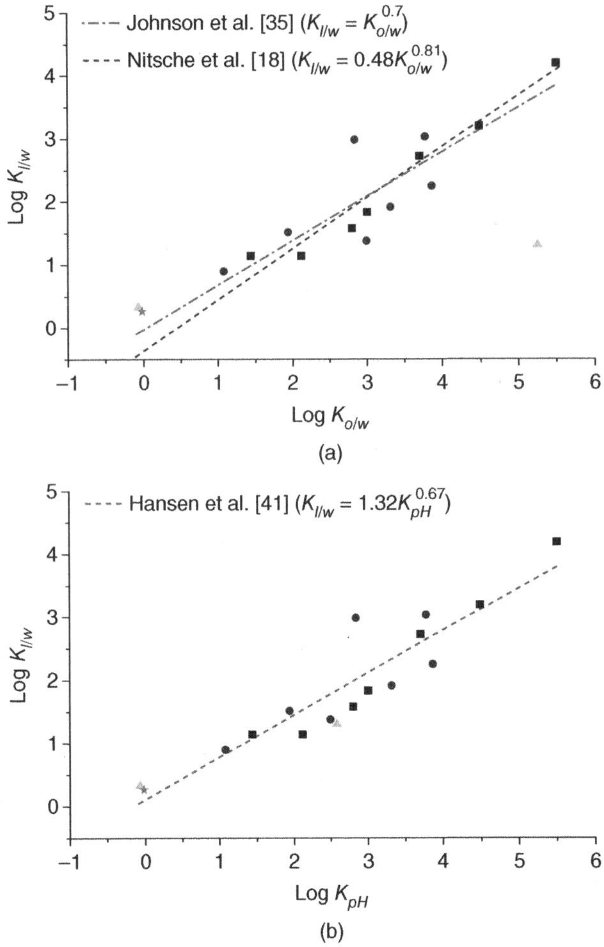

Figure 7.2 SC lipid–water partition coefficients: experimental data of Raykar et al. [15] (filled square), Johnson et al. [35] (filled circle), Hansen et al. [41] (filled triangle), Wang et al. [42] (filled star), and (a) correlation with octanol-water partition coefficients Log K_{ow} and (b) correlation with distribution coefficients Log K_{pH} at pH = 7.4.

using, for instance, the Franz diffusion cell system. The shunt pathway was considered to have negligible contribution. The transverse corneocytes pathway is also regarded as negligibly small, which is the case when the permeants are hydrophobic. The overall skin permeability is approximated by Eq. (7.26):

$$P_{sc} \approx P_l + P_t = \frac{K_{l/w}D_{l\text{-}lat}}{h_{sc}\tau}\left[1 + \frac{(t+g)\tau}{\left(\frac{K_{l/w}D_{l\text{-}lat}}{D_c}t + g\right)}\right] \approx \frac{K_{l/w}D_{l\text{-}lat}}{h_{sc}\tau} \tag{7.26}$$

Equation (7.26) is met if $\frac{K_{l/w}D_{l\text{-}lat}}{D_c} \gg 1$. The lateral diffusivity of the SC lipids can be determined from the measured skin permeability using the following approximation:

$$D_{l\text{-}lat} = \frac{P_{sc}h_{sc}\tau}{K_{l/w}}. \tag{7.27}$$

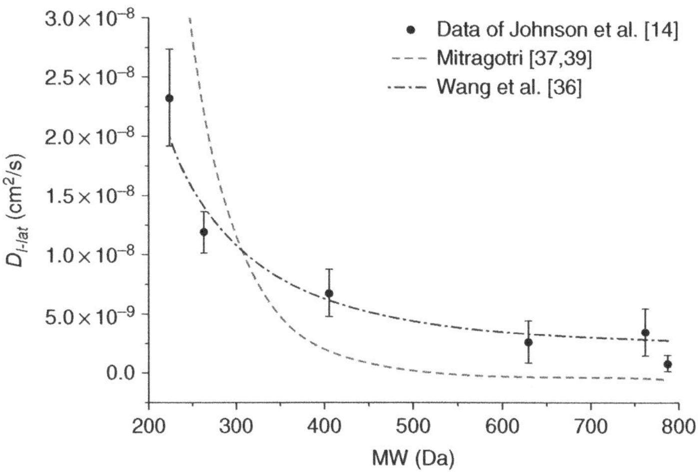

Figure 7.3 Lateral diffusivity of solute in SC lipid: comparison of different models with experimental data. Source: Based on Mitragotri [39].

Data of skin permeability have been reported for a large number of chemicals (e.g. Flynn [43]) and others. Estimation of solute diffusivity in the SC lipid from the measured SC permeability is prone to errors of both the tortuosity factor and the lipid–water partition coefficients.

Molecular weight has the most important effect on the diffusivity of chemicals in the SC lipids. Mitragotri [37, 39] applied the scaled particle theory to predict the diffusivity of solutes in the SC lipids. Based on the fundamental features of molecular diffusion in lipid bilayers, the following equation was proposed for the lateral diffusivity of solute in the SC lipids

$$D_{l\text{-}lat} = 2.0 \times 10^{-5} \exp(-0.46 r_s^2) \qquad (7.28)$$

where $r_s^3 = 0.217\text{MW}$ is the molecular radius of the solute in Å and the diffusivity is in cm^2/s.

Wang et al. [36] proposed a different equation as follows:

$$D_{l\text{-}lat} = 0.008\,98\text{MW}^{-2.43} + 2.34 \times 10^{-9} \qquad (7.29)$$

where the diffusivity is also in cm^2/s.

Figure 7.3 compares the two models against the published experimental data available. Here, only the directly measured data of the SC lipid is presented. The theoretical model of Mitragotri [39] overestimates the diffusivity of small molecules yet significantly underestimates that of large molecules. For this reason, Mitragotri suggested that for large molecules (MW > 380 Da) their lateral diffusivity in the SC lipid is size-independent. The model proposed by Wang et al. [36] appears to fit the experimental data well.

7.3.4 Binding to Keratin

Solute binding to SC keratin has been measured using delipidized SC for a number of chemicals [15]. Chemical binding to keratin also scales to octanol–water partition coefficients. The following relationship has been frequently used for correlating the binding coefficients of chemicals to keratin:

$$B_{p/w} = a K_{o/w}^b \qquad (7.30)$$

where a and b are the prefactor and power index. Their values are obtained by fitting Eq. (7.30) to measured data. Reported values are relatively close, e.g. $a = 5.6$, and $b = 0.27$ by Raykar et al. [15], $a = 4.2$ and $b = 0.31$ by Wang et al. [42] and $a = 4.47$, $b = 0.32$ by Hansen et al. [44]. In a separate review, Hansen et al. [41] used the pH-dependent octanol–water distribution coefficient D_{ow} and their reported fitting parameters were $a = 5.35$, $b = 0.32$. The use of pH-dependent octanol–water distribution coefficient assumes that charged ions do not bind to keratin. However, recent molecular dynamics simulation and isothermal titration calorimetry studies showed that electrostatic attractions between dissociated ions and charged amino acid groups of keratin can lead to chemical binding to keratin, although the electrostatic interaction is secondary to the hydrophobic interaction [45, 46].

Comparison of various models for predicting the experimental data of solute binding to SC keratin shows that these models are not significantly different (Figure 7.4).

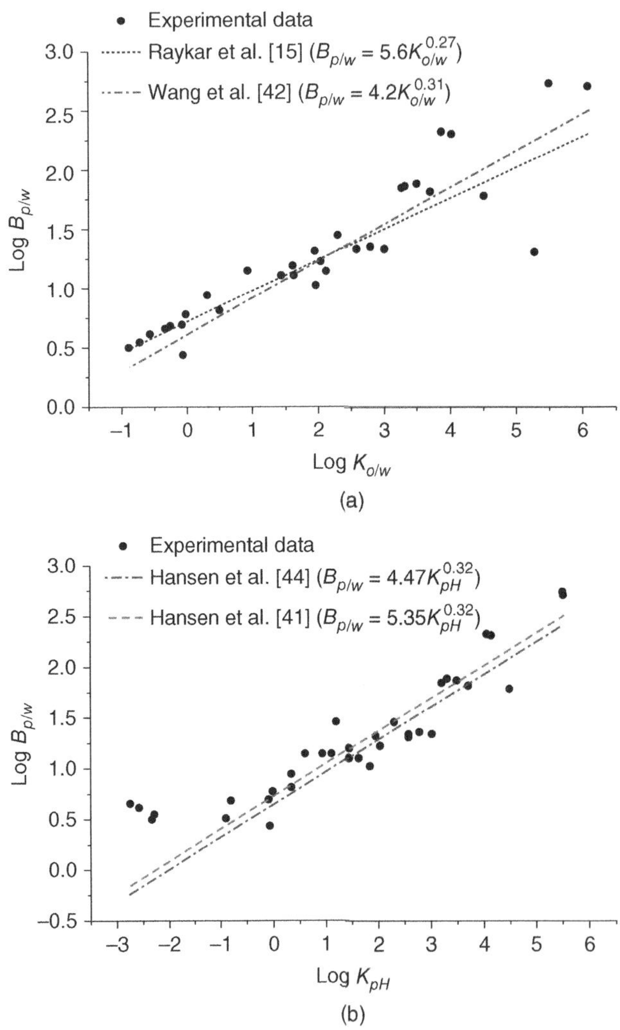

Figure 7.4 Solute binding to corneocyte keratin: comparison of various models and measured data.

7.3.5 Diffusivity in Corneocytes

Experimental data for solute diffusivity in the SC corneocytes are not available. In the absence of experimental data, attempts have been made to estimate the diffusivity of solutes in corneocytes theoretically. The model proposed by Wang et al. [36] was derived by considering the corneocytes as consisting of keratin fibrils embedded in a continuous aqueous fluid. Keratin fibrils are not accessible to solute diffusion and this is referred to as steric obstruction. Solute diffusivity in the aqueous phase of the corneocytes is also retarded in confined pores and this is referred as hydrodynamic hindrance. Combining both of the steric obstruction and hydrodynamic hindrance, Wang et al. [36] proposed the following equation (Eq. (7.31a)) for the diffusivity of solute in hydrated corneocytes ($\theta_p = 0.19$)

$$D_c = D_w(1 - \theta_p)(0.9999 - 1.2762\lambda + 0.0718\lambda^2 + 0.1195\lambda^3) \tag{7.31a}$$

For partially hydrated SC corneocytes ($\theta_p = 0.60$), a slightly different equation was proposed

$$D_c = D_w(1 - \theta_p)(1.0001 - 2.4497\lambda + 1.141\lambda^2 + 0.5432\lambda^3) \tag{7.31b}$$

where D_w is the diffusivity of solute in water, θ_p is the volume fraction of keratin protein, $\lambda = r_s/r_f$ with r_s the radius of the solute, and r_f the radius of keratin fiber.

Equation (7.31a) was obtained by fitting to the data of fibrous membranes reported by Phillips et al. [47]. Chen et al. [27, 38] proposed a slightly different model which also considered the steric obstruction of the keratin fibrils and the hydrodynamics hindrance of the confined water in fibrous pores. The equation was derived by Johnson et al. [48] for polymer gels and adopted for SC corneocytes as follows:

$$D_c = D_w \frac{\exp\left[-(1 - \theta_b)^{\nu}\left(1 + \frac{r_s}{r_f}\right)^{2\nu}\right]}{1 + \frac{r_s}{\sqrt{k}} + r_s^2/k} \tag{7.32}$$

where $k = \beta r_f^2(1 - \theta_b)^{\gamma}$, α, β, γ, ν are fitting parameters.

By applying the model to fit the measured data of skin permeability of eight representative chemicals, Chen et al. [27] obtained $\alpha = 9.47$, $\beta = 9.32 \times 10^{-8}$, $\gamma = -1.17$, $\nu = 1.09$.

Figure 7.5 compares the two theoretical models of Wang et al. [40] and Chen et al. [27, 38]. The predicted value of solute diffusivity in the corneocytes differ by several orders of magnitudes. The model of Wang et al. predicted the solute diffusivity in the corneocytes very close to that in water, and several orders of magnitudes higher than that of lateral diffusion in the SC lipids. The model of Chen et al., however, predicted solute diffusivity in the corneocytes 2 order of magnitudes slower than that in the SC lipids and 4–5 orders of magnitude slower than that in water. Clearly, due to the lack of direct measurement of solute diffusion in the SC corneocytes, uncertainty remains for the barrier property of the SC corneocytes.

7.3.6 Solute Diffusivity and Partition in Sebum

Although the importance of shunt pathway for transdermal permeation has been stressed from time to time [19, 24, 49, 50], studies are very limited on the transport properties of solute in shunt pathways. By assuming the hair follicle infundibulum is filled with sebum, Valiveti et al. [51] reported an experimental study on the sebum–water partition coefficient of a number of model drug chemicals using artificial sebum. Measured data of sebum–water partition coefficients were correlated to the octanol–water partition coefficients.

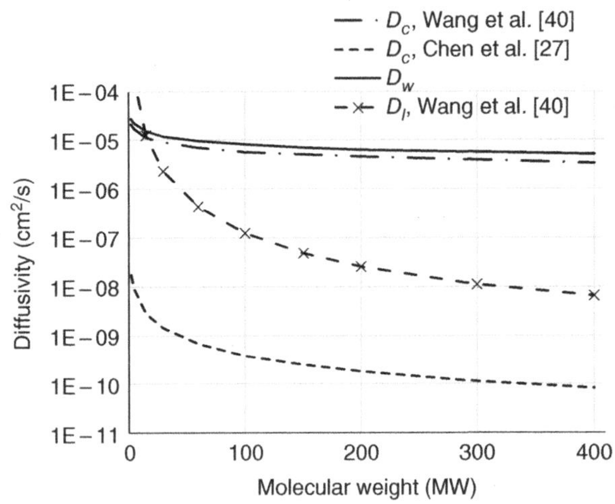

Figure 7.5 Solute diffusivity in SC corneocytes as a function of molecular weight of solute: comparison of theoretical models. Source: Refs. [27, 38, 40].

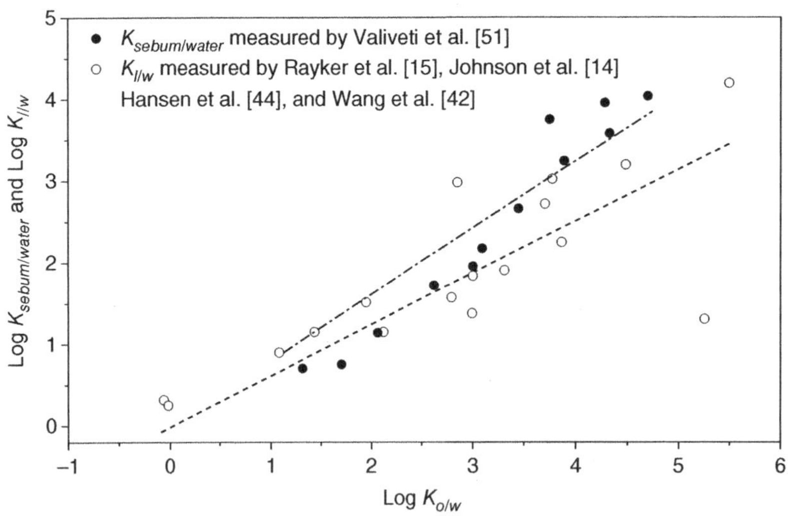

Figure 7.6 Sebum–water partition coefficients vs. SC lipid–water partition coefficients. Source: Based on Barbero and Frasch [21].

Valiveti and Lu [52] also reported an experimental study on the permeation of solute through artificial sebum film. Solute diffusivity in sebum was derived from the measured permeability and sebum–water partition coefficients.

Figure 7.6 compares the measured sebum–water partition coefficients with the measured SC lipid–water partition coefficients. They are very similar. For highly hydrophobic compounds, sebum–water partition coefficients are slightly higher than the SC lipid–water partition coefficient. In a modeling study, the property of sebum has been modeled as SC lipid [21].

7.4 Application

7.4.1 Steady-State

Early studies of microscopic modeling of transdermal permeation focused on the lipid pathway, using the homogenized 1D model of solute permeation across the SC membrane at steady state. Transdermal permeation reaches steady-state when the concentration of solute in the heterogeneous SC lipid, corneocytes, shunt pathway, and vehicle become independent of time, i.e.

$$\frac{\partial C_l}{\partial t} = \frac{\partial C_w}{\partial t} = \frac{\partial C_s}{\partial t} = 0 \tag{7.33}$$

At steady state, the flux of solute across a unit area of the SC barrier is related the permeability and given by

$$j = (P_l + P_t + P_s)(C_d - C_r) \tag{7.34}$$

where C_d and C_r are the solute concentrations in the donor and receptor fluids, respectively.

Understanding the relative importance of the various permeation pathways has been the focus of many studies. Two similar microscopic models were reported for predicting the relative contributions of the tortuous lipid pathway and the transcorneocytes pathway. The microscopic model developed by Wang et al. [36] and that of Chen et al. [27, 38] both used similar "bricks-and-mortar" microstructures and similar properties of solute partition and diffusion in SC lipid and corneocytes, as outlined in Section 7.3. The main difference is that, for the transcellular solute transfer across the lipid membrane sandwiched between corneocytes, the microscopic model of Wang et al. assumed an additional translipid bilayer resistance denoted by k_{tran}, while that of Chen et al. used, the same lateral diffusivity in the SC lipid, ignoring possible additional transverse lipid bilayer resistance. Using published skin permeability data, Wang et al. [36] derived the values of additional translipid bilayer transfer coefficients for a number of chemical compounds. A major contradiction of the model is that although transverse mass transfer is introduced for transverse corneocyte pathway, it is ignored for the tortuous lipid pathway. If SC lipid envelopes are perfect bilayers without disruptions and overlaps, permeation though the tortuous lipid pathway will also need to overcome ~100 lipid bilayers in addition to lateral diffusion [35, 39]. The model also predicted that the transverse corneocyte pathway dominates for most compounds, a surprising departure from the widely accepted view [13, 14, 39] that SC corneocytes are the main barrier for hindering transdermal permeation.

The study by Chen et al. [27] revealed that the relative contribution of the transverse corneocytes pathway depended on the hydrophobicity of the permeants. For highly lipophilic compounds of $\log K_{ow} > 2$, the continuous tortuous lipid pathway contributed more than 75% of transdermal permeation which is consistent with many other microscopic studies [13, 14, 39]. Only for hydrophilic compounds, does the transverse corneocytes pathway become important. For hydrophilic compounds, when not considering the transcellular permeation pathway, skin permeability of hydrophilic compounds was significantly underpredicted. Inclusion of the transverse corneocytes permeation pathway significantly improved the prediction of measured transdermal permeability. This is also the case for all reported microscopic models [39]. The microscopic model of Chen et al. [27, 38] used the same geometrical parameters of the SC "bricks and mortar" reported by Johnson et al. [35] which were also adopted by Mitragotri [39]. For hydrophobic compounds, the skin permeability predicted by Chen et al. [38] agreed well with that predicted by the analytical model of Mitragotri, suggesting the equation for the tortuosity factor derived by Johnson et al. [35] is sufficiently accurate. For hydrophobic compounds, the skin permeability predicted by the microscopic model of Chen et al. [27], the analytical model of Mitragotri [39] and the empirical correlation of Potts and Guy [13] all agreed very well with published experimental data.

7.4.2 Dermatopharmacokinetics

In most exposure applications, steady-state is rare. Most dermal exposure is dynamic and time-dependent. Experimental studies of transdermal permeation under dynamic exposure conditions have been reported, but there have been very limited studies of microscopic modeling of the time-dependent dermatopharmacokinetics. A few studies used compartment models to fit experimental data, but the approach did not consider the microscopic effect of the SC barrier and hence lacks predictive and extrapolation capabilities. Lian et al. [53] first applied the microscopic model of Chen et al. [27] to predict the *in vivo* tape striping data of 4-cyanophenol (4-CP) reported by Reddy et al. [54] and the *in vivo* Raman confocal data of retinol collected by the authors. The study showed that for moderately hydrophobic solute 4-CP, transcellular permeation still contributed significantly to the overall transdermal permeation. For the first time, the microscopic model was used to predict the disposition of 4-CP in both domains of the SC lipids and corneocytes. The predicted concentration of 4-CP in the corneocytes reached approximately half of that in the SC lipids. The predicted overall mass of 4-CP disposition in the SC coenocytes contributed approximately two-third of the total in the SC. When the disposition of 4-CP in the corneocytes was excluded, the predicted total mass of 4-CP in the SC was not able to account for the measured amount of 4-CP from tape stripping.

The microscopic modeling study of the Raman confocal data by Lian et al. [53] showed that although retinol is highly hydrophobic and its skin permeability is high, the transdermal delivery of retinal from a 30 : 70 (v/v) ethanol-propylene glycol vehicle to deeper dermis and viable dermis is relatively small. The predicted dermatopharmacokinetics showed that retinol penetrated rapidly into the SC but was retained in the SC region due to its high partition coefficient to the SC lipids and its high affinity of binding to keratin in the corneocytes. Penetration of retinol to deep dermis and dermis was predicted to be rather slow because of the aqueous continuous environment. It appears that for highly hydrophobic compounds, the SC layer can serve as a reservoir for sustained retention and slow penetration into dermis and viable dermis as well as systemic circulation.

7.4.3 Systemic Pharmacokinetics

Predicting systemic pharmacokinetics following dermal exposure is important for the purpose of efficacy and safety assessment. When the solute reaches the dermis, it is cleared from the skin through both diffusion into deeper tissues and the capillary blood flow. It is the latter pathway that makes the solute bioavailable systemically. The dermis clearance pathways have been considered in several microscopic, diffusion-based models [55–59], in which the focus was to improve the prediction accuracy within the skin but the systemic pharmacokinetics were not explicitly modeled. In contrast, systemic circulation was included in many compartment-based pharmacokinetic models [60, 61], but they have limited predictive capability due to the lack of sufficient consideration of microscopic structure and property of the skin.

More recently, the latest multiphase microscopic model of Chen et al. [62] was further extended to include the modeling of systemic pharmacokinetics [63]. Specifically, the diffusion and partition properties in the "bricks-and-mortar" structure of the SC, and those in the homogenized viable epidermis and dermis all followed the microscopic model developed in Liu et al. [60]. The transport of solutes into the capillary blood is modeled by using the assumption of solute equilibrium between dermis and blood capillaries, based on the method in Bookout et al. [59]. The "brick-and-mortar" of the SC as well as the dermis and viable dermis were meshed using the finite element method. Each computational mesh (resulting from discretizing the partial differential equation for diffusion) in the dermis domain is assumed to be in contact with blood capillaries, and the solute in the blood circulation has a uniform concentration. Based on these assumptions, within each dermis mesh, the solute flow from the capillary into and out of the mesh is given by $Q_{b,i}C_b$, where $Q_{b,i}$ is the blood flow into the dermis grid i and C_b is the solute concentration in the blood. When the capillary flow leaves the dermis, the capillary concentration is assumed to be in equilibrium with the concentration in the dermis grid, and thus the solute flux taken away by the capillary is given by $Q_{b,i}C_i/K_{db}$, where K_{db} is the partition coefficient from dermis to blood. Therefore, for a dermis mesh i, the mass balance equation is

$$V_i \frac{dC_i}{dt} = -\sum_j q_{ij} + Q_{b,i}\left(C_b - \frac{C_i}{K_{db}}\right) \quad (7.35)$$

where C_i is the concentration in mesh i, V_i is its volume, q_{ij} is the rate of mass transfer from mesh i to a neighboring mesh j due to diffusion given in Section 7.2.3.

The above equilibrium assumption can be viewed as a simplification of the capillary clearance models reported in [55, 56], in which transient equations were used to model the permeation of solute from dermis into capillary. The transient equations are more complex since the permeability between dermis and capillary needs to be considered, in addition to the partition coefficient.

The volumetric blood flow, $Q_{b,i}$, can be estimated according to physiology. It is known that the average resting cardiac output is c. 5.6 l/min for a human male and 4.9 l/min for a female [64], and the overall blood flow to skin is estimated to be 5% of cardiac output [59]. The systemic circulation and clearance is described by Eq. (7.36):

$$V_b \frac{dC_b}{dt} = N\sum_i Q_{b,i}\left(\frac{C_i}{K_{db}} - C_b\right) - kC_b \quad (7.36)$$

where V_b is the volume of whole-body blood vessel, and kC_b is the first-order clearance that may include transport into other tissues and metabolism. The summation is with respect to all dermis meshes, and N is the ratio of actual topical application area to the area being simulated. For computational efficiency, the lateral length being simulated usually covers a few corneocytes, and thus the simulated area is much smaller than the actual application area. The clearance rate, k, of many pharmaceutically or toxicologically important chemicals is available from clinical pharmacokinetic studies of oral and/or intravenous delivery routes.

The above model has been applied to predict the published clinical data of transdermal delivery of nicotine [65, 66] and caffeine [19]. Without fitting model parameters to the data being predicted, the prediction is in good agreement with the published data of systemic kinetics.

7.4.4 Shunt Pathway

The contribution of the shunt pathway to transdermal penetration has been confirmed and highlighted in several *in vitro* and *in vivo* experimental studies [19, 20, 49, 50, 67]. Despite this, there are very few microscopic modeling studies in the open literature that have considered this shunt pathway. Scheuplein [68] reported an early theoretical analysis of the shunt pathway where the solute diffusion coefficient in the hair appendage was estimated from the experimental measurement on wet hair. The theoretical analysis suggested that hair follicle pathway dominated early stages. Ho [69] reported a theoretical analysis on the three penetration routes, intercellular pathway, sweat duct pathway, and follicular pathway, by assuming aqueous-phase for sweat ducts and oil-phase hair follicular sebum. Keister and Kasting [70] used a transient model of two routes to fit the *in vitro* diffusion cell data of ibuprofen through excised human skin and concluded that hair follicular pathway accounted for 25% of the overall penetration. In those theoretical studies, experimental data included the shunt pathway and SC pathway, and no independent data of shunt pathway blocking were compared.

Experimental studies on shunt pathway were reported using *in vitro* microscopic imaging [49, 50], *in vitro* permeation study using sandwiched skin [71], and *in vivo* clinical with hair follicle blocking [19, 67]. While microscopic imaging and *in vitro* clinical studies showed increased contribution of hair follicular pathway with increasing hydrophobicity of solute, the *in vitro* permeation study showed an opposite trend, i.e. contribution of follicular pathway decreased for increasing hydrophobicity of the solutes. With *in vitro* permeation study, it is possible for water to penetrate into hair follicles and form an aqueous barrier between sandwiched skin samples.

Liu et al. [60] reported a study using a pharmacokinetic model to fit the *in vivo* data of Otberg et al. [19], where the caffeine plasma concentration–time profiles after topical application into the chest of volunteers with and without hair follicle blocking were collected. The fitted pharmacokinetic parameters showed that hair follicles contributed significantly to the overall penetration of caffeine across the skin barrier into the systemic circulation at early times after application.

Recently, a multiphase microscopic model similar to that of Wang et al. [36] was reported and the impact of hair follicle was modeled [21]. In this model, the transcellular permeation through multilayers (ML) of SC lipid membrane and corneocytes was considered as the dominant pathway and the hair follicle sebum was assumed to have the same partition property and lateral diffusivity of the SC lipid. Unfortunately, the formulated mathematical equation for the transcellular permeability (Eq. (7.11)) was incorrect (the lipid–water partition coefficient, the most important parameter of transdermal permeability was omitted and the corneocyte binding coefficient was incorrectly introduced). There was no validation of the model prediction.

A main challenge in modeling the shunt pathway is in obtaining the transport property. Hair follicular pathway is regarded as through sebum. The partition and diffusion properties of chemicals in artificial sebum were recently investigated by Valiveti et al. [51, 52] for a number of compounds of the 4-hydroxybenzoate series. Artificial sebum has been proven as a suitable mimic for real sebum physiochemistry for the purpose of such studies [72].

Using sebum partition and diffusion properties determined separately from artificial sebum, Kattou et al. [73] recently extended the multiphase microscopic model of Chen et al. [27, 38] to include the impact of the follicular pathway. The transcellular and intercellular pathways through the SC were modeled as in Chen et al. [27]. The follicular pathway was modeled by a sebum cylinder filling the gap between the skin and hair outer root sheath. The model was used to predict the *in vivo* clinical data of caffeine delivery reported by Otberg et al. [19]. The predicted bioavailability of caffeine in blood circulation following the topical application on the chest of volunteers with and without follicular blocking demonstrated good agreement with experimental data. The model simulations confirm that for caffeine, the follicular route is important; the maximum bioavailable concentration of caffeine in systemic circulation with open hair follicles was predicted to be 20% higher than that when hair follicles are blocked.

7.5 Perspective

Understanding transdermal barrier property, permeation pathway, dermatopharmacokinetics, and bioavailability is important for a number of applications in human health, hygiene, and safety, including the treatment of skin disease such as dermatitis, transdermal drug delivery, cosmetic care, and protection from environmental exposure. With the ban of animal testing for cosmetic products and ingredients and the restriction (reduce, replace, and refine) of animal testing for medical research, there is an increasing interest in developing *in silico* mathematical modeling capabilities. Much of the challenge in mathematical modeling is including the microstructure and property effect. Early studies demonstrated a strong correlation of skin permeability and the hydrophobicity of the solute for a large data set of chemicals. This led to the conjecture that the tortuous SC lipids were the dominant pathway for transdermal permeation and the corneocytes were the main barrier of physical obstruction. Further experimental study using FRAP by Johnson et al. [14] confirmed that the tortuous SC lipid was the primary pathway for transdermal permeation. A theoretical analysis of the FRAP data [35] showed that lateral diffusion parallel to the lipid bilayer appears to the dominant mechanism of solute permeation in the SC lipids. If the cornified lipid bilayer envelopes are perfect bilayers without overlaps and disruptions, solute permeation through the SC lipid will need to overcome the transverse resistance of approximately 100 lipid bilayers. Transverse lipid bilayer diffusion is several orders magnitudes slower than lateral diffusion. Including the transverse resistance will lead to significantly slower transdermal permeation than observed experimentally. Mitragotri [37, 39] presented a theoretical analysis of solute permeation through lipid bilayers using scaled particle theory and further supported that solute permeation in the SC lipid pathway is dominated by lateral diffusion.

Experimental studies have shown significant solute uptake by delipidized SC [15, 16]. These studies revealed the existence of the transverse corneocyte pathway. Microscopic modeling of transverse corneocyte permeation has been reported recently and controversy still exists regarding how the pathway should be modeled. For transverse corneocyte permeation pathway, solute is required to diffuse through approximately 16 layers of sandwiched SC lipids membranes and corneocytes. No direct experimental data of solute diffusivity for transverse corneocyte transfer are available. Instead, theoretical equations of solute diffusion in porous fibrous network or gels have been adopted for estimating the diffusivity of solute in the SC corneocytes. Two similar equations were proposed separately by Wang et al. [36] and Chen et al. [38], but the predicted values significantly different. The equation of Wang et al. predicted solute diffusivity in corneocytes very close to that in water and approximately 2 orders of magnitudes higher than that in SC lipids. On the other hand, the equation of Chen et al. predicted solute diffusivity in the SC corneocytes approximately 5 orders of magnitudes slower than that in water.

There is also a debate on how solute diffuses across the SC lipid to corneocytes for the transverse corneocyte pathway. Mitragotri [39] proposed the possible existence of aqueous pores in the SC lipids for connecting two adjacent corneocytes. Wang et al. [36, 40] regarded the cornified SC lipid membrane between adjacent corneocytes as perfect lipid bilayers and proposed an anisotropic translipid bilayer mass transfer barrier. Chen et al. [38] regarded the cornified SC lipid membrane as imperfect bilayers with high degrees of disruptions and overlaps and used the same lateral diffusivity for both the tortuous lipid pathway and transverse corneocyte pathway.

Although the two multiphase microscopic models are similar in representation to the "bricks and mortar" structure, the use of different transport properties for the transverse corneocyte pathway leads to significant controversy regarding the relative importance of the tortuous lipid pathway and transverse corneocyte pathway. Wang et al. [40] have used their microscopic model to fit the published skin permeability data and concluded that the transverse corneocytes pathway is the dominant permeation pathway for most of the chemicals they examined. This is in striking departure from the established view that skin permeation is dominated by the tortuous lipid pathway, especially for hydrophobic compounds. On the other hand, Chen et al. [27] showed that the relative contribution of the tortuous lipid pathway and transverse corneocyte pathway depended on the hydrophobicity of the compounds. For hydrophobic compounds of $\log K_{ow} > 2$, the tortuous lipid pathway dominates. For hydrophilic compounds log $K_{ow} < -1$, the transverse corneocyte pathway dominates. For chemicals in between, both tortuous lipid pathway and transverse corneocyte pathways are important. The molecular weight of the solute also has significant impact on the relative contribution. When the molecular size decreases, the relative contribution of the tortuous lipid pathway increases, suggesting smaller molecules prefer the lipid pathway. The microscopic study of Chen et al. is consistent with the widely accepted view. The experimental study of Johnson et al. [14] confirmed lateral diffusion in the tortuous SC lipids as the main mechanism of skin permeation. The fact that the skin permeability scales highly to the hydrophobicity indisputably suggests that the tortuous lipid pathway is the main permeation pathway for hydrophobic solutes.

Much of the challenge in resolving the debate and controversy is in determining the microscopic properties of the SC barrier. Only when direct measurement data of microscopic transport properties are available, many of the controversies regarding the relative importance of the tortuous lipid pathway and transverse corneocytes pathway can be resolved. This is also true for the shunt pathway. Recent progress in quantifying the partition and diffusion properties of sebum has allowed more accurate modeling of the hair follicular pathway [73]. Once the microscopic properties of the heterogeneous skin barrier are quantified, more realistic microscopic modeling can be made for the skin barrier property, permeation pathway, dermatopharmacokinetics, and bioavailability. This is evident with the initial applications of microscopic models for predicting transient *in vivo* exposure conditions [51, 64, 73].

References

1 Yosipovitch, G., Xiong, G.L., Haus, E. et al. (1998). Time-dependent variations of the skin barrier function in humans: transepidermal water loss, stratum corneum hydration, skin surface pH, and skin temperature. *J. Invest. Dermatol.* 110: 20–23.

2 van Logtestijn, M.D.A., Caspers, P.J., Kezic, S. et al. (2016). Water resistance profile as a marker of skin barrier damage in atopic dermatitis patients. *J. Dermatol. Sci.* 81: 126–128.

3 Kushner Iv, J., Kim, D., So, P.T.C. et al. (2007). Dual-channel two-photon microscopy study of transdermal transport in skin treated with low-frequency ultrasound and a chemical enhancer. *J. Invest. Dermatol.* 127: 2832–2846.

4 Azagury, A., Khoury, L., Enden, G., and Kost, J. (2014). Ultrasound mediated transdermal drug delivery. *Adv. Drug Delivery Rev.* 72: 127–143.

5 Zhai, Y. and Zhai, G. (2014). Advances in lipid-based colloid systems as drug carrier for topic delivery. *J. Controlled Release* 193: 90–99.

6 Sala, M., Elaissari, A., and Fessi, H. (2016). Advances in psoriasis physiopathology and treatments: up to date of mechanistic insights and perspectives of novel therapies based on innovative skin drug delivery systems (ISDDS). *J. Controlled Release* 239: 182–202.

7 Heinrich, K., Heinrich, U., and Tronnier, H. (2014). Influence of different cosmetic formulations on the human skin barrier. *Skin Pharmacol. Physiol.* 27: 141–147.

8 Rawlings, A.V. and Matts, P.J. (2005). Stratum corneum moisturization at the molecular level: an update in relation to the dry skin cycle. *J. Invest. Dermatol.* 124: 1099–1110.

9 Gautier, S., Xhauflaire-Uhoda, E., Gonry, P., and Piérard, G.E. (2008). Chitin-glucan, a natural cell scaffold for skin moisturization and rejuvenation. *Int. J. Cosmet. Sci.* 30: 459–469.

10 Goldsmith, L.A. (1996). Skin effects of air pollution. *Otolaryngol. Head Neck Surg.* 114: 217–219.

11 Mancebo, S.E. and Wang, S.Q. (2015). Recognizing the impact of ambient air pollution on skin health. *J. Eur. Acad. Dermatol. Venereol.* 29: 2326–2332.

12 Kim, K.E., Cho, D., and Park, H.J. (2016). Air pollution and skin diseases: adverse effects of airborne particulate matter on various skin diseases. *Life Sci.* 152: 126–134.

13 Potts, R.O. and Guy, R.H. (1992). Predicting skin permeability. *Pharm. Res.* 9: 663–669.

14 Johnson, M.E., Berk, D.A., Blankschtein, D. et al. (1996). Lateral diffusion of small compounds in human stratum corneum and model lipid bilayer systems. *Biophys. J.* 71: 2656–2668.

15 Raykar, P.V., Fung, M.C., and Anderson, B.D. (1988). The role of protein and lipid domains in the uptake of solutes by human stratum Corneum. *Pharm. Res.* 5: 140–150.

16 Anderson, B.D., Higuchi, W.I., and Raykar, P.V. (1988). Heterogeneity effects on permeability–partition coefficient relationships in human stratum corneum. *Pharm. Res.* 5: 566–573.

17 Kasting, G.B., Miller, M.M., and Talreja, P.S. (2005). Evaluation of stratum corneum heterogeneity. In: *Percutaneous Absorption* (eds. R.L. Bronaugh and H.I. Maibach), 193–212. New York: Taylor & Francis.

18 Nitsche, J.M., Wang, T.F., and Kasting, G.B. (2006). A two-phase analysis of solute partitioning into the stratum corneum. *J. Pharm. Sci.* 95: 649–666.

19 Otberg, N., Patzelt, A., Rasulev, U. et al. (2008). The role of hair follicles in the percutaneous absorption of caffeine. *Br. J. Clin. Pharmacol.* 65: 488–492.

20 Knorr, F., Lademann, J., Patzelt, A. et al. (2009). Follicular transport route – research progress and future perspectives. *Eur. J. Pharm. Biopharm.* 71: 173–180.

21 Barbero, A.M. and Frasch, H.F. (2017). Effect of stratum corneum heterogeneity, anisotropy, asymmetry and follicular pathway on transdermal penetration. *J. Controlled Release* 260: 234–246.

22 Scheuplein, R.J. and Blank, I.H. (1971). Permeability of the skin. *Physiol. Rev.* 51: 702–747.

23 Elias, P.M., Cooper, E.R., Korc, A., and Brown, B.E. (1981). Percutaneous transport in relation to stratum corneum structure and lipid composition. *J. Invest. Dermatol.* 76: 297–301.

24 Scheuplein, R.J. (1967). Mechanism of percutaneous absorption. II. Transient diffusion and the relative importance of various routes of skin penetration. *J. Invest. Dermatol.* 48: 79–88.

25 Michaels, A.S., Chandrasekaran, S.K., and Shaw, J.E. (1975). Drug permeation through human skin: theory and in vitro experimental measurement. *AlChE J.* 21: 985–996.

26 Heisig, M., Lieckfeldt, R., Wittum, G. et al. (1996). Non steady-state descriptions of drug permeation through stratum corneum. I. The biphasic brick-and-mortar model. *Pharm. Res.* 13: 421–426.

27 Chen, L., Lian, G., and Han, L. (2010). Modeling transdermal permeation. Part I. Predicting skin permeability of both hydrophobic and hydrophilic solutes. *AlChE J.* 56: 1136–1146.

28 Crank, J. (1975). *The Mathematics of Diffusion*. Oxford: Oxford University Press.

29 Anissimov, Y.G. and Roberts, M.S. (2001). Diffusion modeling of percutaneous absorption kinetics: 2. Finite vehicle volume and solvent deposited solids. *J. Pharm. Sci.* 90: 504–520.

30 Anissimov, Y.G. and Roberts, M.S. (1999). Diffusion modeling of percutaneous absorption kinetics. 1. Effects of flow rate, receptor sampling rate, and viable epidermal resistance for a constant donor concentration. *J. Pharm. Sci.* 88: 1201–1209.

31 Marrink, S.-J. and Berendsen, H.J.C. (1994). Simulation of water transport through a lipid membrane. *J. Phys. Chem.* 98: 4155–4168.

32 Bassolino-Klimas, D., Alper, H.E., and Stouch, T.R. (1993). Solute diffusion in lipid bilayer membranes: an atomic level study by molecular dynamics simulation. *Biochemistry* 32: 12624–12637.

33 Marrink, S.J. and Berendsen, H.J.C. (1996). Permeation process of small molecules across lipid membranes studied by molecular dynamics simulations. *J. Phys. Chem.* 100: 16729–16738.

34 Shinoda, W. (2016). Permeability across lipid membranes. *Biochim. Biophys. Acta, Biomembr.* 1858: 2254–2265.

35 Johnson, M.E., Blankschtein, D., and Langer, R. (1997). Evaluation of solute permeation through the stratum corneum: lateral bilayer diffusion as the primary transport mechanism. *J. Pharm. Sci.* 86: 1162–1172.

36 Wang, T.F., Kasting, G.B., and Nitsche, J.M. (2006). A multiphase microscopic diffusion model for stratum corneum permeability. I. Formulation, solution, and illustrative results for representative compounds. *J. Pharm. Sci.* 95: 620–648.

37 Mitragotri, S. (2002). A theoretical analysis of permeation of small hydrophobic solutes across the stratum corneum based on Scaled Particle Theory. *J. Pharm. Sci.* 91: 744–752.

38 Chen, L., Lian, G., and Han, L. (2008). Use of "bricks and mortar" model to predict transdermal permeation: model development and initial validation. *Ind. Eng. Chem. Res.* 47: 6465–6472.

39 Mitragotri, S. (2003). Modeling skin permeability to hydrophilic and hydrophobic solutes based on four permeation pathways. *J. Controlled Release* 86: 69–92.

40 Wang, T.F., Kasting, G.B., and Nitsche, J.M. (2007). A multiphase microscopic diffusion model for stratum corneum permeability. II. Estimation of physicochemical parameters, and application to a large permeability database. *J. Pharm. Sci.* 96: 3024–3051.

41 Hansen, S., Lehr, C.M., and Schaefer, U.F. (2013). Improved input parameters for diffusion models of skin absorption. *Adv. Drug Delivery Rev.* 65: 251–264.

42 Wang, L., Chen, L., Lian, G., and Han, L. (2010). Determination of partition and binding properties of solutes to stratum corneum. *Int. J. Pharm.* 398: 114–122.

43 Flynn, G.L. (1990). Physiochemical determinants of skin absorption. In: *Principles of Route-to-Route Extrapolation for Risk Assessment* (eds. T.R. Gerrity and C.J. Henry), 93–127. New York: Elsevier.

44 Hansen, S., Henning, A., Naegel, A. et al. (2008). In-silico model of skin penetration based on experimentally determined input parameters. Part I: experimental determination of partition and diffusion coefficients. *Eur. J. Pharm. Biopharm.* 68: 352–367.

45 Marzinek, J.K., Lian, G., Marzinek, J.K. et al. (2013). Molecular and thermodynamic basis for EGCG-Keration interactions – Part I: molecular dynamics simulations. *AIChE J.* 59: 4816–4823.

46 Zhao, Y., Chen, L., Han, L. et al. (2013). Molecular and thermodynamic basis for EGCG-keratin interactions – Part II: experimental investigation. *AIChE J.* 59: 4824–4827.

47 Phillips, R.J., Deen, W.M., and Brady, J.F. (1989). Hindered transport of spherical macromolecules in fibrous membranes and gels. *AIChE J.* 35: 1761–1769.

48 Johnson, E.M., Berk, D.A., Jain, R.K., and Deen, W.M. (1996). Hindered diffusion in agarose gels: test of effective medium model. *Biophys. J.* 70: 1017–1026.

49 Grams, Y. and Bouwstra, J.A. (2002). Penetration and distribution of three lipophilic probes in vitro in human skin focusing on the hair follicle. *J. Controlled Release* 83: 253–262.

50 Grams, Y., Alaruikka, S., Lashley, L. et al. (2003). Permeant lipophilicity and vehicle composition influence accumulation of dyes in hair follicles of human skin. *Eur. J. Pharm. Sci.* 18: 329–336. https://doi.org/10.1016/S0928-0987(03)00035-6.

51 Valiveti, S., Wesley, J., and Lu, G.W. (2008). Investigation of the drug partition property in artificial sebum. *Int. J. Pharm.* 346: 10–16.

52 Valiveti, S. and Lu, G.W. (2007). Diffusion properties of model compounds in artificial sebum. *Int. J. Pharm.* 345: 88–94.

53 Lian, G., Chen, L., Pudney, P.D.A. et al. (2010). Modeling transdermal permeation. Part II. Predicting the dermatopharmacokinetics of percutaneous solute. *AIChE J.* 56: 2551–2560.

54 Reddy, M.B., Stinchcomb, A.L., Guy, R.H., and Bunge, A.L. (2002). Determining dermal absorption parameters in vivo from tape strip data. *Pharm. Res.* 19: 292–298.

55 Anissimov, Y.G. and Roberts, M.S. (2011). Modelling dermal drug distribution after topical application in human. *Pharm. Res.* 28: 2119–2129.

56 Kretsos, K. and Kasting, G.B. (2007). A geometric model of dermal capillary clearance. *Math. Biosci.* 208: 430–453.

57 Kretsos, K., Miller, M.A., Zamora-Estrada, G., and Kasting, G.B. (2008). Partitioning, diffusivity and clearance of skin permeants in mammalian dermis. *Int. J. Pharm.* 346: 64–79.

58 Ibrahim, R., Nitsche, J.M., and Kasting, G.B. (2012). Dermal clearance model for epidermal bioavailability calculations. *J. Pharm. Sci.* 101: 2094–2108.

59 Bookout, R.J. Jr.,, McDaniel, C.R., Quinn, D.W., and McDougal, J.N. (1996). Multilayered dermal subcompartments for modelling chemical absorption. *SAR QSAR Environ. Res.* 5: 133–150.

60 Liu, X., Grice, J.E., Lademann, J. et al. (2012). Hair follicles contribute significantly to penetration through human skin only at times soon after application as a solvent deposited solid in man. *Br. J. Clin. Pharmacol.* 72: 768–774.

61 Polak, S., Ghobadi, C., Mishra, H. et al. (2012). Prediction of concentration-time profile and its inter-individual variability following the dermal drug absorption. *J. Pharm. Sci.* 101: 2584–2595.

62 Chen, L., Han, L., Saib, O., and Lian, G. (2015). In silico prediction of percutaneous absorption and disposition kinetics of chemicals. *Pharm. Res.* 32: 1779–1793.

63 Chen, T., Lian, G., and Kattou, P. (2016). In silico modelling of transdermal and systemic kinetics of topically applied solutes: model development and initial validation for transdermal nicotine. *Pharm. Res.* 33: 1602–1614.

64 Guyton, A.C. and Hall, J.E. (2006). *Textbook of Medical Physiology*, 11e, vol. 25. Philadelphia, PA: Elsevier.

65 Bannon, Y.B., Corish, J., Corrigan, O.I. et al. (1989). Transdermal delivery of nicotine in normal human volunteers: a single dose and multiple dose study. *Eur. J. Clin. Pharmacol.* 37: 285–290.

66 DeVeaugh-Geiss, A.M., Chen, L.H., Kotler, M.L. et al. (2010). Pharmacokinetic comparison of two nicotine transdermal systems, a 21-mg/24-hour patch and a 25-mg/16-hour patch: a randomized, open-label, single-dose, two-way crossover study in adult smokers. *Clin. Ther.* 32: 1140–1148.

67 Lademann, J., Otberg, N., Richter, H. et al. (2001). Investigation of follicular penetration of topically applied substances. *Skin Pharmacol. Appl. Skin Physiol.* 14 (Suppl 1): 17–22.

68 Scheuplein, R.J. (1965). Mechanism of percutaneous adsorption: I. Routes of penetration and the influence of solubility. *J. Invest. Dermatol.* 45: 334–346.

69 Ho, C.K. (2004). Probabilistic modeling of percutaneous absorption for risk-based exposure assessments and transdermal drug delivery. *Stat. Method.* 1: 47–69.

70 Keister, J.C. and Kasting, G.B. (1986). The use of transient diffusion to investigate transport pathways through skin. *J. Controlled Release* 4: 111–117.

71 Frum, Y., Bonner, M.C., Eccleston, G.M., and Meidan, V.M. (2007). The influence of drug partition coefficient on follicular penetration: in vitro human skin studies. *Eur. J. Pharm. Sci.* 30 (3–4): 280–287.

72 Lu, G.W., Valiveti, S., Spence, J. et al. (2009). Comparison of artificial sebum with human and hamster sebum samples. *J. Pharm. Sci.* 367: 37–43.

73 Kattou, P., Lian, G., Glavin, S. et al. (2017). Development of a two-dimensional model for predicting transdermal permeation with the follicular pathway: demonstration with a caffeine study. *Pharm. Res.* 34: 2036–2048.

8

Role of Membrane Transporters in Drug Disposition

Hong Yang and Yan Shu

Department of Pharmaceutical Sciences, School of Pharmacy, University of Maryland at Baltimore, Baltimore, MD, USA

8.1 Introduction

Movement of endogenous and exogenous chemicals across the biological membrane plays a central role to achieve the physiological function of these compounds and to dictate their fate of elimination. Due to the special structure of a lipid bilayer, the permeability of the biological membrane toward different chemicals is mainly dependent on their physical and chemical properties. Some small size, noncharged, hydrophobic compounds could easily move across membranes by simple diffusion; on the contrary, facilitated transportation is necessitated for large size, charged, and/or hydrophilic compounds to move across the membrane.

Transporters are specialized, membrane-bound proteins, which serve as carriers for the transportation of chemicals in and out of the cell, either actively or passively. The human genome encodes more than 400 membrane transporters, the majority of which can be categorized into two superfamilies: the adenosine triphosphate (ATP)-binding cassette (ABC) transporters and the solute carriers (SLCs) [1–3]. ABC transporters transport chemicals with a primary active mechanism, which means that they transport chemicals against their concentration and/or electrochemical gradient by using the energy directly from ATP hydrolysis. In contrast, the modes of transport by SLC transporters are more diverse. Some of them are named as facilitative SLC transporters since they transport chemicals down their concentration and/or electrochemical gradient across the membrane. Other SLC transporters are typically secondary and tertiary active transporters. Similar to the ABC transporters, they could uphill transport chemicals against a gradient; however, unlike the ABC transporters, the energy (driving force) comes from coupling to a co-transported solute or ion, either in the same (symport) or the opposite direction (antiport) [3]. In addition, transporters may also be grouped as electrogenic and nonelectrogenic, depending on whether a translocation of net charge occurs or not across the membrane.

The expression of these membrane transporters is found in tissues throughout the body, with particularly high abundance in the epithelial cells of major absorption and secretion organs such as intestine, liver, and kidney; and also in sanctuary organs such as placenta, testis, and brain [3]. While some transporters are exclusively expressed in specific tissues, others apparently have a broader distribution in a variety of tissues. Both endogenous and exogenous chemicals could be substrates of these membrane transporters. In particular, some are named as drug transporters due to not only their high expression in the major organs (intestine, liver, and kidney) for drug

Fundamentals of Drug Delivery, First Edition.
Edited by Heather A. E. Benson, Michael S. Roberts, Adrian C. Williams, and Xiaowen Liang.

absorption, metabolism, and elimination but also their capability to transport those exogenous drugs. From 2007 to 2018, after extensive discussion, the International Transporter Consortium (ITC), consisting of experts from the industry, academy, and regulatory agency, reached consensus on a panel of 10 human membrane transporters for which there is compelling evidence that they are clinically relevant in drug absorption, distribution, and excretion, and should be investigated prospectively during drug development [1, 4, 5]. This panel of transporters includes P-glycoprotein (P-gp), breast cancer resistance protein (BCRP), organic cation transporter 1 and 2 (OCT1/2), multidrug and toxin extrusion protein 1 and 2-K (MATE1/2-K), organic anion transporter 1/3 (OAT1/3), and organic anion transporting polypeptide 1B1/1B3 (OATP1B1/1B3) (Table 8.1). In addition, five other drug transporters, including bile-salt export pump (BSEP), multidrug resistance protein 2/3/4 (MRP2/3/4), and OATP2B1 were recommended for retrospective evaluation on the basis of preclinical and clinical observations. The ITC agrees that the inclusion of a transporter into such a panel will continue to evolve as new evidence becomes available from the dynamic field of transporter pharmacology. Of note, even though these transporters have been well established in drug disposition, we should recognize that they may also participate in transportation of endogenous chemicals, by which they possibly exert a fundamental role in physiology [6].

In this chapter, we will mainly focus on the role of drug transporters in drug disposition (absorption, distribution, metabolism, and elimination, ADME). The distribution of drug transporters in major organs will be described first, followed by an extensive discussion on the role of drug transporters in drug disposition. The main effort will be made toward these drug transporters proposed by the ITC.

8.2 Distribution of Major Drug Transporters in Human Tissues

The highest expression of drug transporters is usually found in the major absorption and excretion organs including intestine, liver, and kidney, and also the organs protected by the endothelial barriers separating them from the circulating blood, such as brain, testis, and placenta. The distribution of major drug transporters in the intestine, liver, kidney, brain, and other tissues in humans will be delineated separately in this section.

8.2.1 Major Drug Transporters in the Intestine

Due to the low cost and the better compliance by patients, administration of drugs through the gastrointestinal (GI) tract is the most commonly used route for drug treatment. The physiological features (pH value, expression of drug transporters, and CYP450 enzymes) and anatomical properties (length, surface area, food, and drug retention time) dictate the intestine as the premier site in drug absorption and metabolism throughout the GI tract [7, 8]. Although many drugs are deemed to make entry by passive diffusion, drug transporters have been proved to play a vital role in drug absorption, by either facilitating the absorption of drugs into enterocytes and then the circulation, or transporting the absorbed drugs back into the intestinal lumen.

8.2.1.1 Drug Transporters Expressed in the Apical (Luminal) Membrane

Both uptake and efflux drug transporters are expressed in the apical membrane of the enterocytes (Figure 8.1). In general, uptake drug transporters may increase oral bioavailability by transporting the chemicals from the intestinal lumen into the enterocytes, while the efflux transporters reduce the bioavailability by expelling absorbed chemicals back into the intestinal lumen.

Table 8.1 Drug transporters recommended by the International Transporter Consortium (ITC).

Transporter	Organ(s)	Reason for select	Recommendation	ITC workshop
P-gp	Intestine, kidney, liver and brain	Role in drug absorption, distribution and excretion	Prospective investigation	1st
BCRP	Intestine, liver, kidney, brain, placenta and mammary glands	Role in drug absorption, distribution and excretion	Prospective investigation	1st
OATP1B1	Liver	Role in drug distribution and excretion	Prospective investigation	1st
OATP1B3	Liver	Role in drug distribution and excretion	Prospective investigation	1st
OAT1	Kidney and placenta	Role in drug distribution and excretion	Prospective investigation	1st
OAT3	Kidney and brain	Role in drug distribution and excretion	Prospective investigation	1st
OCT2	Kidney and brain	Role in drug distribution and excretion	Prospective investigation	1st
OCT1	Liver and intestine	Role in drug distribution and excretion	Prospective investigation	3rd
MATE1	Kidney, liver and skeletal muscle	Role in drug distribution and excretion	Prospective investigation	2nd
MATE2-K	Kidney	Role in drug distribution and excretion	Prospective investigation	2nd
MRP2	Liver, kidney and intestine	Role in drug absorption, distribution and excretion	Retrospective investigation	2nd
MRP3	Liver and intestine	Role in drug distribution	Retrospective investigation	2nd
MRP4	Liver, kidney brain and platelets	Role in drug distribution and excretion	Retrospective investigation	2nd
BSEP	Liver	Role in drug excretion	Retrospective investigation	2nd
OATP2B1	Liver	Role in drug distribution and excretion	Retrospective investigation	3rd

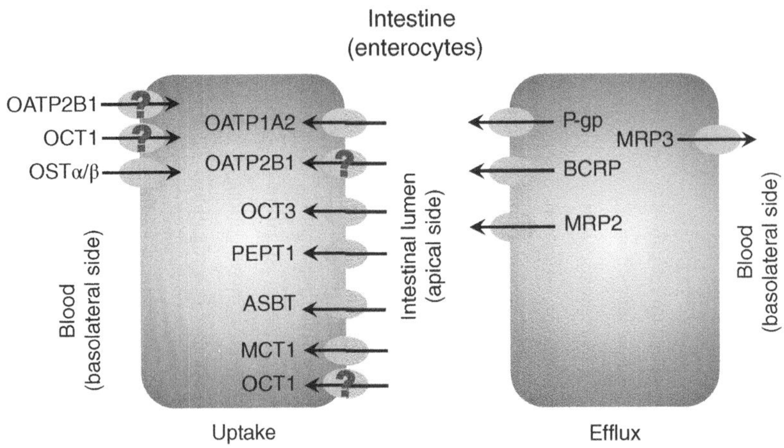

Figure 8.1 Subcellular localization of uptake and efflux transport proteins in human intestine (enterocytes). Sources: Giacomini et al. [1], Klaassen and Aleksunes [3].

8.2.1.1.1 Uptake Transporters

Several members of organic anion transporting polypeptide, including OATP1A2 and OATP2B1, have been detected in human intestine at both transcriptional level [9] and protein level [9, 10]. The organic cation transporter 3 (OCT3) has also been pinpointed in the apical membrane as an uptake transporter [11]. In addition, peptide transporter 1 (PEPT1) and PEPT2 have been found at the transcriptional level; however, the protein expression has been only confirmed for PEPT1, suggesting PEPT1 as the main PEPT isoform in the intestine [12–14]. Other uptake transporters, including the apical sodium/bile acid cotransporter (ASBT) [15] and the monocarboxylic transporter 1 (MCT1) [8], are also localized in the apical membrane, but with relatively limited importance in drug absorption. The localization of OCT1 and OATP2B1 in human intestine is rather controversial since evidences exist to support their localization in both apical and basolateral membrane. Some inclined to believe its localization on apical side since low but considerable oral bioavailability (10–20%) for metformin cannot be simply explained just by passive diffusion [16]. Likewise, some researchers can localize OATP2B1 in apical side of human enterocytes and Caco-2 cells [17, 18], while others provided compelling evidence to show its expression in basolateral membrane [19]. The localization of OATP2B1 at apical side is more acceptable by considering its role in substrate absorption from the preclinical and clinical studies [20–23].

8.2.1.1.2 Efflux Transporters

Both P-gp and BCRP are the well-characterized efflux drug transporters expressed in the apical membrane of human enterocytes, which play a major role in determining drug bioavailability [24, 25]. Another efflux drug transporter with confirmed expression in the intestine is MRP2 [24].

8.2.1.2 Drug Transporters Expressed in the Basolateral Membrane

The expression of uptake and efflux transporters has also been found in the basolateral membrane of enterocytes (Figure 8.1); interestingly, the classes of transporters identified are significantly less than the apical side [1, 3]. Different from those at the apical membrane, the efflux transporters at the basal membrane will help the absorbed chemicals to enter into the circulation system, while the uptake transporters either function to dispose drugs from the blood into the enterocytes, or serve as the first step to eliminate chemicals into the intestinal lumen.

8.2.1.2.1 Uptake Transporters

The studies using immunolabeling localization as well as the pharmacokinetics (PK) in mice have provided evidence to support that OCT1 was expressed on the basolateral membrane [11]. However, this has been challenged by another study which presented compelling data to argue that OCT1 was actually expressed in the apical membrane [26]. Further investigation is needed to solve the controversy. In addition, heteromeric organic solute transporters α/β (OSTα/β) are also located in the basolateral membrane, which are considered to be the main transporters for the absorption of bile acids and several drugs [27, 28].

8.2.1.2.2 Efflux Transporters

Solid data supports that MRP3 is located in the basolateral membrane, though its function in enterocytes is still unclear [29, 30].

8.2.1.3 Expression of Drug Transporters in Different Intestinal Regions

Human intestine functionally consists of two major segments: the small intestine and the large intestine (colon). Further, the small intestine can be divided into three distinct regions, the duodenum, jejunum, and ileum. As shown in Table 8.2, the transcriptional level of transporters is significantly lower in the colon as compared to the small intestine. Likewise, the protein expression can be detected for all of the listed transporters in the small intestine, while only half of the transporters (mainly the efflux transporters expressed in the apical membrane) have been confirmed in the colon. This distinct pattern of transporter expression in the intestine is consistent with the well-established function of the small intestine as the major absorptive region in the GI tract, while the expression of mere efflux transporters in the colon implicates its role as an elimination site for drugs and other chemicals.

8.2.2 Major Drug Transporters in the Liver

Liver has been recognized well as the crucial organ involving in all aspects of drug disposition, including absorption, metabolism, distribution, and elimination. Aside from the known abundant expression of drug-metabolizing enzymes (e.g. cytochromes P450 or CYP450), a variety of drug transporters are also located in the basolateral and apical membrane of hepatocytes. The drug transporters function either independently, or in tandem with the metabolizing enzymes in determining the fate of a drug. In addition, some of the drug transporters are also present in the epithelia of the bile duct.

8.2.2.1 Drug Transporters Expressed in the Apical Membrane of Hepatocytes

Intriguingly, only efflux transporters have been identified in the apical membrane of hepatocytes (Figure 8.2), including P-gp, BCRP, MRP2, BSEP, and MATE1 [33–36]. Nevertheless, the drug excretion into the bile should not be deemed as a terminal or single trip since the excreted drugs could reenter into the circulation by intestinal reabsorption (passive diffusion and/or transporter-mediated uptake), a process named enterohepatic recirculation. The absence of uptake transporters in the apical membrane might be hypothetically explained by the existence of an effective intestinal reabsorption system as well as by the possibility that the excreted chemicals serve as communicating messengers between the liver and the intestine.

Table 8.2 Expression of drug transporters in different intestinal regions.

	mRNA					Protein				
Transporters	**Duodenum**	**Jejunum**	**Ileum**	**Colon**	**References**	**Duodenum**	**Jejunum**	**Ileum**	**Colon**	**References**
OATP1A2	Positive/−	−	−	−	[9, 31]	+	\	\	\	[9]
OATP2B1	+	+	+	+	[31]	+	+	+	\	[18][a)]
PEPT1	+++	+++	+++	+	[31]	\	\	++	+	[14]
OCT1	−	+	\	+	[9, 12]	\	+	\	\	[11]
OCT3	\	+	\	−	[12]	\	+	\	\	[11]
ASBT	++	\	+++	−	[15, 31]	+	\	++	\	[15]
P-gp	Positive	++	++	+	[9, 12]	++	++	++	++	[9, 24]
BCRP	+++	+++	+++	+	[9, 12, 32]	++	++	++	++	[25][b)]
MRP2	Positive	+++	+++	−	[9, 12, 24]	\	++	+	+/−	[24]
MRP3	Positive	+	\	+	[9, 12]	+	\	+	+	[29]

"−, +, ++, +++," relative expression levels ranging from negative to high; "Positive," detected without relative expression data; "\," no data available.

a) Small intestine but no specific intestinal region was examined for OATP2B1 protein expression.

b) Small intestine but no specific intestinal region was examined for BCRP protein expression.

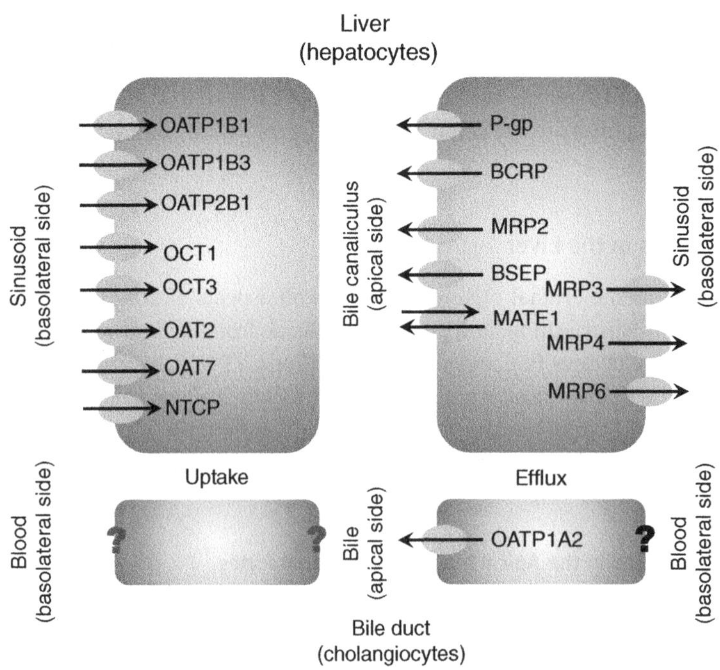

?, no data is available with respect to the expression of transporters except OATP1A2 in humans.

Figure 8.2 Subcellular localization of uptake and efflux transport proteins in human liver hepatocytes and cholangiocytes. Sources: Burt et al. [33], Vildhede et al. [34], Otsuka et al. [35].

8.2.2.2 Drug Transporters Expressed in the Basolateral (Sinusoidal) Membrane of Hepatocytes

Many uptake and efflux transporters have been confirmed to reside in the basolateral membrane of hepatocytes (Figure 8.2). The uptake transporters move the drugs absorbed from the GI tract and from the circulation into hepatocytes, either functioning as the first step of drug elimination into the bile or collaborating with the drug-metabolizing enzymes for drug biotransformation. Since there are no identified uptake transporters in the apical membrane, the main task of the efflux transporters located in the basolateral membrane is probably to transport the biotransformed chemicals back to the circulation, thereafter, to their sites of action and/or elimination.

8.2.2.2.1 Uptake Transporters

Accumulating data have indicated that OCT1, OCT3, OAT2, OAT7, OATP1B1, OATP1B3, OATP2B1, and sodium/taurocholate cotransporting peptide (NTCP) are located in the basolateral membrane [33]. Among them, proteomics data from human liver tissue and human hepatocytes have revealed that OATB1B1 and OATB1B3 have the highest abundance of expression, while OAT2 and OCT3 the lowest [34, 36].

8.2.2.2.2 Efflux Transporters

Only the members from MRP family, including MRP3, MRP4, and MRP6, have been identified as the efflux transporters in the basolateral membrane. MRP3 and MRP6 have a similar abundance as compared to OCT1, NTCP, and OAT7, while MRP4 has the lowest abundance among all the basolateral transporters [33, 34].

8.2.2.3 Drug Transporters Expressed in the Bile Duct Epithelia (Cholangiocytes)

Other than the hepatocytes, the expression of transporters has also been found in the epithelial cells (cholangiocytes) along the bile ducts in the liver (Figure 8.2). For instance, OATP1A2 is located in the apical side of cholangiocytes in human liver [37]. In addition, certain uptake transporters, such as Asbt and Pept1, and efflux transporters, like Mate1 and Mdr1a, are located in the apical membrane of cholangiocytes in rodents [3, 35, 38–41]. even though there is no such localization data available from human liver tissues. These drug transporters, along with other ion channels, may also function physiologically to modify the content of hepatocyte-derived bile in response to different signals.

8.2.3 Major Drug Transporters in the Kidney

Kidney is the major organ for elimination of hydrophilic drugs and metabolites. Clinically, more than 30% of the top 200 prescribed drugs in the United States are predominantly eliminated through the urine [2, 42]. In general, renal drug elimination is the net effect of three concurrent processes: glomerular filtration; tubular secretion; and tubular reabsorption. Drug transporters, primarily located in the proximal tubule cells in the renal cortex, are actively involved in the processes of tubular secretion and absorption. Due to the polarized distribution from the basolateral to the apical membrane, these drug transporters can function in tandem to vectorially transport drugs from the blood into the urine. For instance, OCT2 in the basolateral membrane works collaboratively with MATE1 and MATE2-K in the apical membrane for the elimination of many organic cationic drugs.

8.2.3.1 Drug Transporters Expressed in the Apical Membrane of Proximal Tubule Cells

The proximal tubule cells have two major physiological or pharmacological roles: one is to reabsorb crucial chemicals back into the circulation; while the other is to secrete the unwanted into the urine. Both uptake and efflux transporters have been localized in the apical membrane along the proximal tubule cells to fulfill the obligation (Figure 8.3).

8.2.3.1.1 Uptake Transporters

Various transporters, including OAT4, organic cation/ergothioneine transporter 1 (OCTN1), OCTN2, PEPT1, PEPT2, and urate transporter 1 (URAT1), have been confirmed with an expression in the apical membrane of proximal tubular cells [43–48]. The abundance of OAT4, OCTN1, and OCTN2 was significantly lower (around one-tenth) than that of the others located on the basolateral membrane, such as OAT1, OAT3, and OCT2 [49].

8.2.3.1.2 Efflux Transporters

The efflux transporters, including P-gp, MRP2, MRP4, and BCRP of the ABC family, are located in the apical membrane [50–53], all with a high abundance except BCRP [49] Furthermore, the MATE (MATE1 and MATE2-K) transporters of the SLC family have been localized in the apical membrane with an even higher abundance than the ABC transporters [35]. Although the MATEs are bi-directional transporters *per se*, the slightly acidified primary urine (pH = 6.3) in the physiological status renders them as efficient efflux transporters for certain organic cations.

8.2.3.2 Drug Transporters Expressed in the Basolateral Membrane of Proximal Tubule Cells

8.2.3.2.1 Uptake Transporters

Uptake drug transporters are extensively expressed in the basolateral membrane [43, 49, 54] for the transport of both organic anions and cations (Figure 8.3). Like OCT1 in hepatocytes, OCT2 is the single organic cation transporter identified in human kidney. The mRNA of OCT3 has been detected but with a much lower level, and its protein expression and localization remain unclear [49]. In contrast, three isoforms of OATs (OAT1/2/3) are well present for the uptake of organic anions. This raises an interesting question to be addressed as to why there are several isoforms of OATs but only one OCT present evolutionally in human kidney.

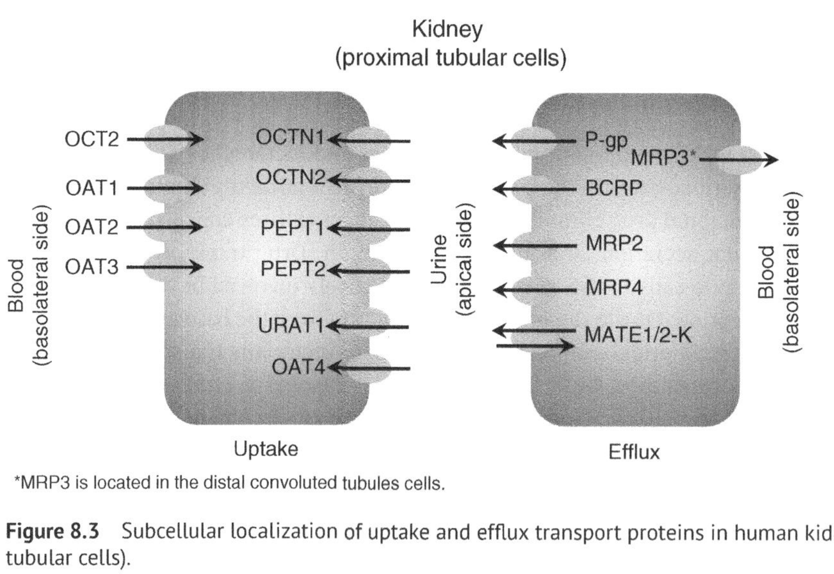

Figure 8.3 Subcellular localization of uptake and efflux transport proteins in human kidney (proximal tubular cells).

8.2.3.2.2 *Efflux Transporters*

Only MRP3 has been identified in human kidney; however, the immunohistological staining has showed that MRP3 is only present in the distal convoluted tubules but not in the proximal tubules [29]. The virtual absence of drug efflux transporters in the basolateral membrane agrees with the primary role of kidney as an organ for drug elimination.

8.2.4 Major Drug Transporters in the Central Nervous System (CNS)

The central nervous system (CNS) is separated and protected well from the systemic circulation by the impervious barriers formed by capillary endothelial cells (known as the blood–brain barrier, BBB) and by choroid plexus epithelial cells (known as the blood–cerebrospinal fluid barrier, BCSFB). While the impenetrable barriers ensure the CNS to function in a relatively stable environment, they inevitably post a great challenge for the delivery of drugs into the brain [55]. The barriers formed by the cells and tight junctions can easily block the entry of lipophobic drugs. While the lipophilic drugs may still make their entry by simple diffusion, their CNS accessibility will be critically determined by the drug transporters equipped at both barriers [56, 57].

8.2.4.1 Drug Transporters Expressed in the Capillary Endothelial Cells of BBB

The BBB is the main interface between the systemic circulation and the extracellular fluid surrounding neurons and other supporting cells (Figure 8.4). Drug transporters are almost solely located in the apical side (luminal side) of the capillary endothelial cells [1, 3]. The majority are

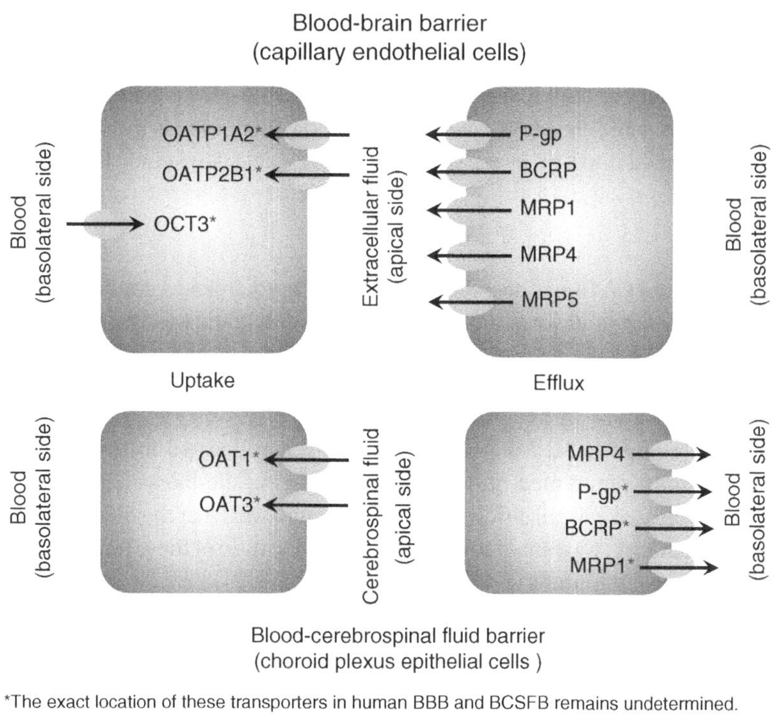

Figure 8.4 Subcellular localization of uptake and efflux transport proteins in human blood–brain barrier (BBB, capillary endothelial cells) and blood–cerebrospinal fluid barrier (BCSFB, choroid plexus epithelial cells).

efflux transporters, though both the uptake and efflux ones are identified. The expression pattern is consistent with the major role of BBB as a protective shield for the brain.

8.2.4.1.1 Efflux Transporters

A variety of ABC transporters, including P-gp, BCRP, MRP1, MRP4, and MRP5, are located in the apical side of BBB [53, 58–61]. The transcriptional expression of MRP2, MRP3, and MRP6 has also been reported, but no evidence of protein localization is available [56, 60]. Interestingly, both the mRNA and protein of MATE1 have been detected in the BBB, though the exact location is still unclear [62]. Overall, the presence of multiple efflux transporters (covering almost all types of organic chemicals) renders the BBB a super-efficient pump to drive unwelcome chemicals out of the brain.

8.2.4.1.2 Uptake Transporters

A limited number of uptake drug transporters are confirmed in the capillary endothelial cells of BBB, including OATP1A2, OATB2B1, and OCT3 [62–64]. There is no concrete evidence showing the exact location of these transporters in the BBB. However, based on their location in other tissues such as intestine and liver, OATP1A2 and OATB2B1 are likely located at the luminal side, while OCT3 is at the albuminal side.

8.2.4.2 Drug Transporters Expressed in the Choroid Plexus Epithelial Cells of BCSFB

The brain is surrounded by the cerebrospinal fluid (CSF) which provides both a mechanical protection (cushion effect) and a homeostatic environment for the brain. The choroid plexus, a blood vessel-enriched area in the ventricles of the brain, produces CSF at a constant rate. The drug concentration in CSF is a key determinant of the efficacy when the drug target is adjacent to the ventricles of the brain, for instance, using antibiotics to treat the meningitis caused by viral or bacterial infection. Like the capillary endothelia cells of BBB, the choroid plexus epithelial cells have multiple drug transporters as well (Figure 8.4), which could serve dual roles as the scavenger (to eliminate the metabolites excreted into CSF by nerve cells) and the shield (to exclude unwanted chemicals from the blood).

8.2.4.2.1 Uptake Transporters

The protein expression of OAT1 and OAT3 has been detected in the choroid plexus epithelial cells [65]. Despite lacking human evidence of localization, data from pig study indicated their location in the membrane facing the CSF [66].

8.2.4.2.2 Efflux Transporters

Multiple ABC transporters including MRP1, MRP4, BCRP, and P-gp are located in the BCSFB, among which only MRP4 has been pinpointed on the membrane facing the blood [25, 60, 67]. Data obtained from pig study indicated a localization of P-gp, BCRP, and MRP4 at the blood-facing membrane [66]. Further investigation is warranted to clarify the exact location of these transporters in the choroid plexus epithelial cells in human.

8.2.5 Major Drug Transporters in Other Tissues

In addition to the tissues described above, drug transporters have also been detected in a variety of other tissues, such as placenta, testis, lung, heart, mammary glands, and the immune and stem cells [3]. Among them, the expression of transporters in placenta, mammary gland, and testis are the most studied and best understood.

8.2.5.1 Drug Transporters Expressed in Placenta Villi Epithelial Cells (Syncytiotrophoblasts)

The placenta is responsible for material (nutrition and waste) exchange and serves as a protective shield between the fetus and the mother. Therefore, the structure of the placenta should allow the maximal exchange of nutrients and waste between them, while minimizing the exposure of toxic chemicals to the fetus. The drug transporters located in the epithelial cells of placenta villi, also called syncytiotrophoblasts, are vital modulators for these purposes (Figure 8.5). Multiple efflux transporters, including P-gp, BCRP, and MRP2/3/4, have been confirmed to be located in the apical side (facing the mother blood) of syncytiotrophoblast [68–71]. Similar to their function in other tissues, these efflux pumps effectively drive the exogenous chemicals (drugs and toxics) back into the maternal circulation. MRP1 is uniquely found in both the apical and basolateral side of syncytiotrophoblasts, which makes its true function remain elusive [72]. Intriguingly, the expression of P-gp decreases gradually initially and then dramatically while gestation is advancing, implicating a growing vulnerability of the fetus to toxic insults [69, 70]. In addition, the uptake transporter OCTN2, responsible for delivering carnitine to the fetus during gestation, is also located in the apical membrane of syncytiotrophoblasts [73]. Unlike liver and kidney, OCT3 is the primary isoform present in basolateral side of syncytiotrophoblasts, and was reported to facilitate fetal exposure of metformin during pregnancy [74].

8.2.5.2 Drug Transporters Expressed in Mammary Glands

It has been reported that human mammary glands can concentrate certain drugs such as cimetidine and rosuvastatin in the milk [75, 76], which unequivocally demonstrates an involvement of drug transporters-facilitate transportation. The most compelling evidence goes to the efflux transporter BCRP, which is extensively expressed and located in the apical membrane of mammary glands across different species [77], (Figure 8.6) while the most widely expressed P-gp is absent [3]. The function of BCRP in drug excretion into the milk has been extensively studied and well understood from Bcrp-null mice model [78, 79], and also from human studies [80]. This function is generally described as a double-edged sword: preventing accumulation of exogenous chemicals in the mother, while delivering those potentially toxics to the newborn simultaneously [3]. Several hypotheses have been proposed to explain the physiological role of BCRP-mediated excretion of chemicals into the milk; for instance, one conceptually attractive hypothesis is to prepare the

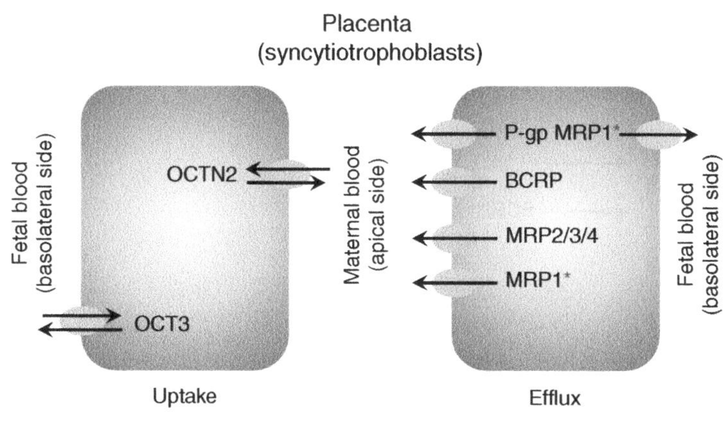

Figure 8.5 Subcellular localization of uptake and efflux transport proteins in human placenta (syncytiotrophoblasts).

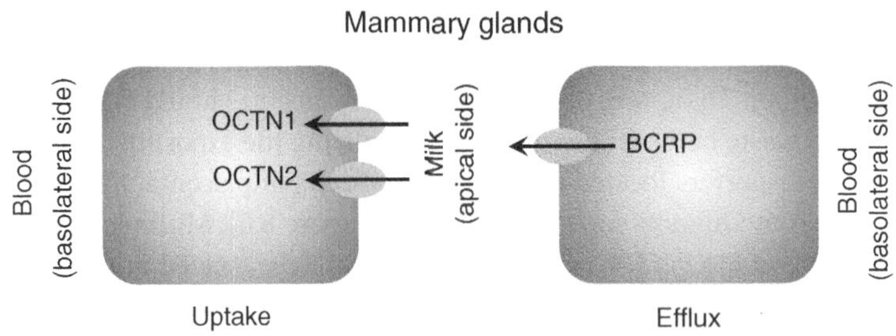

Figure 8.6 Subcellular localization of uptake and efflux transport proteins in human mammary glands.

suckling pups for the switch to solid food [81]. Besides the BCRP, the uptake transporters OCTN1 and OCTN2 are also reported to be expressed in mammary glands, which might be crucial for the delivery of carnitine into the milk [82]. The mRNA expression of a bunch of other uptake and efflux transporters have also been detected and compared in mammary epithelial cells (MECs) from lactating and non-lactating humans though neither the evidence of their protein expression is available from the literatures nor their function in drug milk excretion was confirmed *in vivo* in human [83, 84]. It has to be mentioned that significant mRNA fold increase of OCT1, PEPT2, and CNT1/3 was found in human lactating MEC, strongly indicating a role of these transporters in altering the composition of milk. It was reported that the culture model of human MEC cells exhibited a 45% higher secretory tetraethyl ammonium (TEA) transport, which was consistent with the substantial increase of OCT1 during lactation [85].

8.2.5.3 Drug Transporters Expressed in the Blood–Testis-Barrier (BTB)

Blood–testis-barrier (BTB) is mainly composed of the Sertoli-cell-barrier (SCB), which is anatomically formed by the Sertoli cells with tight junctions. In addition, the myoid cell layer and the capillary endothelia cells are also viewed as parts of BTB. All these structures function together to separate the lumen (apical side, in contact with the sperm cells) of seminiferous tubules from its basolateral side (in contact with the blood) [86]. Like the brain in the CNS, the testis is protected by the BTB, which could prevent the entry of many exogenous toxins to influence spermatogenesis. As such, efflux transporters are extensively expressed in the BTB [87] (Figure 8.7). Solid data shows that P-gp and BCRP are expressed in the myoid cells and the capillary endothelia cells

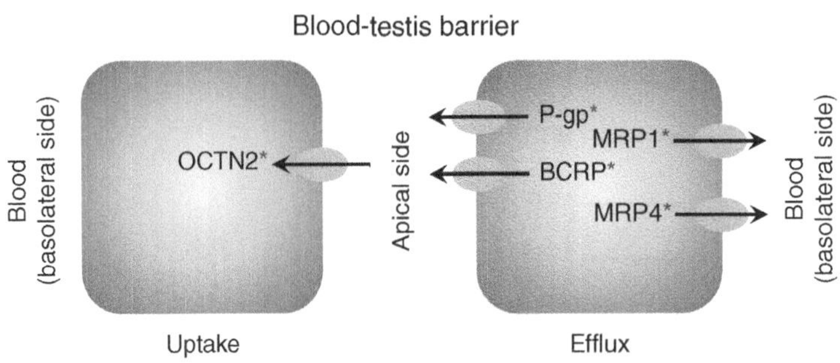

*P-gp and BCRP are expressed in the myoid cells and the capillary endothelia cells of the BTB;
MRP1 and MRP4 are located in the basolateral membrane of the Sertoli cells of the BTB;
OCTN2 is located in the apical membrane of the Sertoli cells of the BTB.

Figure 8.7 Subcellular localization of uptake and efflux transport proteins in human blood–testis barrier (BTB).

[88, 89], which probably serve as the first mechanism to expel exogenous chemicals back into the blood. MRP1 and MRP4, located in the basolateral membrane of the Sertoli cells, could also drive the unwelcome chemicals out of the seminiferous tubules [88, 90]. In contrast, the uptake drug transporters are almost absent in the BTB, except the carnitine transporter OCTN2 which is well expressed in the Sertoli cells [91].

8.3 Role of Drug Transporters in Drug Disposition

Over the past decades, it has become increasingly recognized that drug transporters play a crucial, and sometimes determinative, role in drug absorption, metabolism, distribution, and excretion. The importance of drug transporters is particularly reflected in decision-making during drug development. For instance, both additional preclinical and clinical studies may be warranted if a new candidate drug meets the criteria described in the transporter decision trees recommended by the U.S. Food and Drug Administration (FDA) [1, 4, 5]. This section will begin with a general introduction on how the drug transporters influence drug disposition, followed by a detailed description for each of the major drug transporters proposed by the ITC.

Absorption: The process of drug absorption determines the amount of drugs which could eventually enter the body. In particular, drugs administrated orally are largely subject to the first-pass elimination. As described in Section 8.2, intestinal epithelial cells express a variety of uptake and efflux drug transporters. These efflux transporters including P-gp, BCRP, and MRP2, may extensively drive their substrates back into intestinal lumen, resulting in a reduced absorption rate and bioavailability that may be substantially improved by concomitant administration of inhibitors of these efflux transporters. The uptake transporters, on the contrary, can facilitate the entry of their substrates and improve the bioavailability, which may also be blocked by the inhibitors concomitantly administrated.

Metabolism: Drug transporters do not metabolize drugs directly, however, some of the drug transporters expressed in hepatocytes play a key role in determining the fate of metabolism of many drugs. Hepatocytes are the main site for drug biotransformation due to its high abundance of various phase I and phase II metabolizing enzymes. The delivery of drugs into the hepatocytes serves as the first and sometimes the rate-limiting step for their metabolism. The uptake drug transporters residing in the basolateral membrane of hepatocytes could then function in tandem with the metabolizing enzymes for drug biotransformation. The inhibitors and malfunctional variants of these uptake transporters could dramatically reduce the extent of metabolism, thus increase the systemic exposure of these drugs. In addition, the efflux transporters could increase the exposure of drugs to the metabolizing enzymes in the enterocytes by repeatedly expelling them into the intestinal lumen for reabsorption, further reducing the oral bioavailability of a drug.

Distribution: Drug transporters are present throughout the body, not just in the tissues we mentioned above in Section 8.2 but also other tissues including heart, lungs, skin, muscles, and so on. The drug transporters can either facilitate or restrict the entry of drugs, making them the vital determinants of drug distribution or drug concentration in these tissues. Those drugs, which could inhibit or induce the activity of drug transporters, would thus significantly alter the distribution of victim drugs when they are administrated concomitantly, leading to the consequence of either therapeutic failure or toxicity. In addition, it is likely that a transporter-mediated change in drug distribution may not be readily reflected by alteration in PK parameters such as the area under curve (AUC), maximal plasma concentration (C_{max}), and clearance [1].

Excretion: Kidney and liver are the two major organs responsible for the elimination of drugs and their metabolites. The drug transporters located in the basolateral and apical membrane of renal tubular cells and hepatocytes could work in tandem to transport many drugs and their metabolites into the urine and the bile for elimination, respectively. Any factors, such as inhibitors, inducers, and genetic variants that could alter the activity of these transporters, will change the excretion rate of the drugs. For instance, the inhibitor of an efflux drug transporter in the apical membrane of proximal tubular cells could result in a significant increase of substrate drug concentration both in the tubular cells (distribution) and plasma (systemic exposure), and a significant decrease of renal clearance (elimination). A prolonged half-life might also be observed due to the slowdown of drug excretion.

8.3.1 Role of P-gp in Drug Disposition

As described in Section 8.2, P-gp is substantially expressed in the apical membrane of epithelial cells in the intestine (enterocytes), liver (hepatocytes), kidney (proximal tubular cells), placenta (syncytiotrophoblast), and brain (capillary endothelial cells). The strategic distribution could, on the one hand, restrict cellular uptake of exogenous chemicals from the intestinal tract into the enterocytes (drug absorption) and from the blood into the brain and the placenta (drug distribution); on the other hand, to facilitate the elimination of chemicals from the hepatocytes and the proximal tubular cells into the bile and the urine, respectively (drug elimination). In addition, even though P-gp does not metabolize drugs, it could increase the exposure of drugs to the metabolizing enzymes in the enterocytes by repeatedly expelling them into the intestinal lumen for reabsorption (drug metabolism) [92].

The generation of *Mdr1 (−/−)* knockout mice has provided a great tool to thoroughly examine the role of P-gp in drug disposition. Sparreboom et al. found that the AUC of paclitaxel increased by twofold after its intravenous administration in *Mdr1a (−/−)* mice as compared with *Mdr1a (+/+)* mice [93]. They further verified that the elimination of paclitaxel into the bile was significantly reduced in *Mdr1a (−/−)* mice, with only 2.5% of the dose found in the bile as compared to 11% in *Mdr1a (+/+)* mice. Meanwhile, a significantly higher increase of AUC (sixfold) was observed by oral administration of paclitaxel, which could be explained by the combination of a reduced elimination and an increased absorption from the intestinal lumen. The role of P-gp in drug distribution in the brain and the placenta was evidenced in *Mdr1 (−/−)* mice as well. An 87-fold higher accumulation of ivermectin [94] and a 55-fold higher accumulation of digoxin [95] in the brain were observed in *Mdr1a (−/−)* mice than in *Mdr1a (+/+)* mice, whereas the increase of these drugs in other tissues such as liver and kidney was rather limited, with less than fourfold change. Similarly, but to a less extent, the accumulation of drugs in mouse fetus was increased by 2.5-, 5-, and 16-fold for digoxin, saquinavir, and paclitaxel, respectively, in *Mdr1a/1b (−/−)* double knockout mice [96]. In addition, the pretreatment of P-gp inducer dexamethasone significantly increases the intestinal first-pass metabolism of indinavir [97], supporting the concept that P-gp expression has an impact on drug metabolism.

It is clinically unrealistic to evaluate the role of P-gp in drug disposition by knocking out *Mdr1* in humans. Though a nonfunctional mutant (T3587G) of *MDR1* has been identified in humans, no clinical data is available regarding its effect on drug disposition probably due to its low frequency (<1%) [98]. Therefore, the role of P-gp in human drug disposition has mainly been implied by the studies with P-gp chemical inhibitors. For instance, the oral bioavailability of docetaxel increased from 8% to 88%, and the AUC increased by sevenfold, when coadministrated with the P-gp inhibitor cyclosporin in human subjects [99]. A reduced renal clearance was also observed for digoxin when it was coadministrated with P-gp inhibitors such as quinidine, verapamil, and clarithromycin [100–102]. Positron emission tomography (PET) imaging with a radiolabeled P-gp substrate, such as ^{11}C-verapamil, provides a clinically feasible approach to study the role of P-gp in drug distribution [103, 104]. For instance, it was observed that the penetration of verapamil into the brain was dramatically enhanced while it was cotreated with the P-gp inhibitors tariquidar and quinidine [105, 106]. Since only selected data were presented in Table 8.3, one could refer to other more comprehensive reviews summarizing drugs whose disposition is influenced by P-gp activity [92, 170–173].

8.3.2 Role of BCRP in Drug Disposition

There is significant overlap of tissue distribution and localization between BCRP and P-gp, both of which could be found in the apical membrane of epithelia in the intestine (enterocytes), liver (hepatocytes), placenta (syncytiotrophoblast), and brain (capillary endothelial cells). Interestingly, unlike P-gp, BCRP is also located in the mammary glands in which its expression could be highly induced during lactation [77]. Implied from its cellular location in various tissues, BCRP may play a crucial role in limiting drug absorption from the intestine, restricting drug penetration (distribution) into the brain, fetus, and testes, and facilitating drug excretion into the bile and milk.

The role of BCRP in drug disposition has been well-established after the generation of *Bcrp (−/−)* mice. Kusuhara et al. found that the AUC of sulfasalazine increased by eightfold when cotreated with a strong BCRP inhibitor curcumin in *Bcrp (+/+)* mice, while the increase was not seen in *Bcrp (−/−)* mice [112]. This provided direct evidence that inhibition of BCRP in the enterocytes could substantially increase the absorption, thus oral bioavailability of drugs that are substrates of BCRP. Likewise, coadministration of BCRP inhibitor pantoprazole was reported to significantly reduce the clearance of methotrexate in *Bcrp (+/+)* mice, but not in *Bcrp (−/−)* mice [119]. In contrast, an unequivocal role of BCRP in drug penetration through the BBB has yet to be established. It was found that the brain-to-plasma AUC ratio of BCRP substrate topotecan was quite similar between *Bcrp (−/−)* mice and *Bcrp (+/+)* mice [113]. However, a significant increase of the ratio (1.6-fold) was seen in *Mdr1a/1b/Bcrp (−/−)* triple-knockout mice as compared to *Mdr1a/1b (−/−)* double-knockout mice. Topotecan is not only a good substrate of BCRP but also a weaker substrate of P-gp. The P-gp, which is a dominantly expressed efflux transporter in the BBB, is probably efficient enough to restrict the entry of topotecan into the brain when BCRP is absent. This may explain the result of limited drug entry when BCRP and P-gp were absent from the BBB. In addition, besides its role in the BBB, BCRP could also limit the entry of drugs (e.g. topotecan, phytoestrogens) into other sanctuary sites, such as placenta and testes, as reported previously [115–118].

Clinically, the role of BCRP in drug disposition is primarily confirmed from inhibitor studies as well. The oral availability of topotecan increased from 40% to 97% when it was coadministrated with GF120918 [120]. Even though GF120918 is an inhibitor for both P-gp and BCRP, the increase of oral availability might mainly be attributed to the inhibition of BCRP since topotecan is a rather weak substrate of P-gp. Besides topotecan, inhibition of BCRP could also increase the intestinal absorption and systemic exposure of other BCRP substrates such as rosuvastatin and sulfasalazine [109, 112]. The cotreatment of rosuvastatin with the BCRP inhibitors tipranavir and ritonavir not only significantly increased the AUC and C_{max} of rosuvastatin but also reduced its clearance and prolonged its elimination half-life, which clearly demonstrated the role of BCRP in drug elimination. A case report showed that rosuvastatin is present in human milk with a higher concentration than that in the serum [75, 76]. Similar to the mouse studies, the role of BCRP in human BBB is only appreciated after inhibition of P-gp. Bauer et al. showed that the brain

Table 8.3 Role of major drug transporters in drug disposition (selected data from literature).

Transporter (gene)	Pre-/clinical	Absorption	Metabolism	Distribution	Excretion
P-gp (*ABCB1*)	Preclinical	***Paclitaxel***: AUC was sixfold higher in *Mdr1a (−/−)* mice than *Mdr1a (+/+)* mice [93]	***Indinavir***: Intestinal first-pass metabolism increased from 6% to 34% when rats were pretreated with dexamethasone [97]	***Ivermectin and digoxin***: Brain concentration was 87- and 55-fold higher in *Mdr1a (−/−)* mice than *Mdr1a (+/+)* mice [94, 95] ***Digoxin, saquinavir, and paclitaxel***: Drug accumulation in fetus was 2.5-, 5-, and 16-fold higher in *Mdr1a/1b (−/−)* mice than *Mdr1a (+/+)* mice [96]	***Paclitaxel***: Bile excretion increased from 2.5% of the dose in *Mdr1a (−/−)* mice to 11% in *Mdr1a (+/+)* mice [93]
	Clinical	***Digoxin***: AUC, C_{max}, and absorption rate constant increased by 76%, 75%, and 30%, respectively, when cotreated with quinidine [107] ***Aliskiren***: AUC and C_{max} increased by 400% and 200%, respectively, when cotreated with cyclosporine [108] ***Atorvastatin***: AUC and C_{max} was around ninefold higher when co-administrated with cyclosporine, or tipranavir plus ritonavir [109, 110] ***Docetaxel***: AUC and oral bioavailability was 7- and 10-fold higher when co-administrated with cyclosporine [99]	No data available	***Verapamil***: The distribution in brain increased by 273% and 60% when cotreated with tariquidar and quinidine, respectively [105, 106] ***Daunorubicin***: Concentration in P-gp positive leukemic cells increased by 52% after cotreatment with P-gp inhibitor valspodar [111]	***Digoxin***: Clearance of digoxin decreased by 64% and 53% when cotreated with quinidine and cyclosporine, respectively [101, 102] ***Aliskiren***: Elimination half-life prolonged from 25–45 h when co-treated with cyclosporine [108] ***Everolimus***: Clearance of everolimus decreased by 70% when cotreated with verapamil [100]

BCRP (*ABCG2*)	Preclinical	***Sulfasalazine***: AUC increased by eightfold when cotreated with curcumin in *Bcrp (+/+)* mice, but not in *Bcrp (−/−)* mice [112]	No data available	***Topotecan and imatinib***: Brain-to-plasma AUC ratio increased by 3- to 10-fold in *Mdr1a/1b/Bcrp (−/−)*, while only twofold in *Mdr1a/1b (−/−)* mice [113, 114] ***Topotecan, glyburide, and genistein***: Fetal-to-maternal plasma AUC ratio increased by two- to fivefold in *Bcrp (−/−)* mice [115–117] ***PhIP, dantrolene, prazosin, and phytoestrogens***: Testes-to-plasma AUC ratio increased by 15-fold in *Bcrp (−/−)* mice [115, 118]	***Methotrexate***: Clearance decreased by twofold when pantoprazole was intravenously coadministrated in *Bcrp (+/+)* mice, while was not affected in *Bcrp (−/−)* mice [119] ***Nitrofurantoin***: Milk-to-plasma ratio was nearly 80-fold higher in *Bcrp (+/+)* lactating mice than in *Bcrp (−/−)* lactating mice [79]
	Clinical	***Topotecan***: Oral bioavailability increased from 40% to 97% when coadministrated with GF120918 [120] ***Rosuvastatin***: AUC and C_{max} increased by 1.5- to 5-fold when cotreated with lopinavir/ritonavir and tipranavir/ritonavir [109, 121] ***Sulfasalazine***: AUC and C_{max} increased by twofold when cotreated with curcumin [112]	No data available	***Elacridar***: Brain distribution increased when cotreated with tariquidar [122]	***Rosuvastatin***: Clearance decreased by 30%; elimination half-life increased by twofold when cotreated with tipranavir/ritonavir [109] ***Sulfasalazine***: Clearance decreased by 50–75% when cotreated with curcumin [112] ***Rosuvastatin***: Concentration in milk was higher than in serum [76]
MRP2 (*ABCC2*)	Preclinical	***Paclitaxel***: Mrp2 had no significant impact on intestinal absorption of paclitaxel [123]	No data available	No data available	***Methotrexate and irinotecan***: Biliary excretion dramatically reduced in Mrp2-deficiency rats (Eisai hyperbilirubinemic rats) compared to normal rats [124, 125]

(Continued)

Table 8.3 (Continued)

Transporter (gene)	Pre-/clinical	Absorption	Metabolism	Distribution	Excretion
					Doxorubicin: Biliary excretion decreased by 2-,10-, and 50-fold in *Mrp2* (−/−), *Mdr1a/1b* (−/−), and *Mdr1a/1b/Mrp2* (−/−) mice, respectively [30] ***Paclitaxel***: Bile excretion decreased by 80% in *Mrp2* (−/−) mice compared to wild-type mice [123]
	Clinical	No data available	No data available	***Doxorubicin and cisplatin***: The patients with MRP2-positive esophageal squamous cell carcinoma showed a much lower five-year survival rate (25.6% vs. 55.7%) than MRP2-negative patients [126]	No data available
MRP3 (*ABCC3*)	Preclinical	No data available	No data available	***Morphine and acetaminophen***: A nearly 3- and 20-fold increase of glucuronide conjugates was observed in the liver of *Mrp3* (−/−) mice compared to *Mrp3* (+/+) mice [127, 128]	***Morphine***: A major shift of elimination route from urine (in *Mrp3* (+/+) mice) to feces (*Mrp3* (−/−) mice) was observed [128] ***Acetaminophen***: Elimination of acetaminophen-glucuronide in bile was 10-fold higher in *Mrp3* (−/−) mice compared to *Mrp3* (+/+) mice [127]
	Clinical	No data available	No data available	No data available	No data available
MRP4 (*ABCC4*)	Preclinical	No data available	No data available	***Topotecan***: Brain and cerebrospinal fluid levels were elevated in *Mrp4* (−/−) mice compared to *Mrp4* (+/+) mice [30]	***Adefovir and tenofovir***: Renal luminal efflux reduced to 37 and 46% in *Mrp4* (−/−) mice, respectively, compared to *Mrp4* (+/+) mice [129]

	Clinical	No data available	No data available	No data available	No data available
OATP1B1/1B3 (*SLCO1B1/1B3*)	Preclinical	No data available	***Simvastatin and atorvastatin***: Clearance decreased by 50–80% in *Oatp1a/b* (*−/−*) mice than *Oatp1a/b* (*+/+*) mice [130]	***Rosuvastatin***: Liver distribution was reduced by 54% when cotreated with rifampin in rats [131]	***Rosuvastatin***: AUC increased by eight fold, and nonrenal clearance decreased by 50% in *Oatp1a/b* (*−/−*) mice than *Oatp1a/b* (*+/+*) mice [132]
	Clinical	No data available	***Atorvastatin***: Clearance decreased by 80% when cotreated with rifampin [133]	No data available	***Rosuvastatin***: AUC and C_{max} were twofold higher when cotreated with gemfibrozil [134] ***Pravastatin and simvastatin acid***: AUC was 2- to 3-fold higher in *OATP1B1 521CC* genotype than *521TT* [135, 136]
OATP2B1 (*SLCO2B1*)	Preclinical	***Fexofenadine:*** AUC and C_{max} were 41% and 70% lower in Oatp2b1(*−/−*) mice compared to wild type. Grape juice and apple juice reduce AUC by 35% and 70%, and reduce C_{max} by 80% and 88% in wild type mice [23]	No data available	No data available	***Fexofenadine and rosuvastatin:*** AUC and C_{max} were similar among wild type and Oatp2b1(*−/−*) mice [23]
	Clinical	***Fexofenadine, aliskiren, and celiprolol:*** AUC and C_{max} were decreased by 1.5- to 2-fold by grapefruit, apple and orange juice [20–22]. ***Rosuvastatin:*** AUC was decreased by 50% when with OATP2B1 inhibitor ronacaleret [137]	No data available	No data available	***Rosuvastatin:*** AUC was increased by 1.9-fold when with a potent inhibitor of OATP1B1 and 2B1 asunaprevir [138]
OAT1/3 (*SLC22A6/A8*)	Preclinical	No data available	No data available	***Cidofovir***: Concentration in the cortex of kidney decreased by 70% when cotreated with probenecid in rabbits [139, 140]	***Furosemide***: Renal clearance decreased by 70% in *Oat1* (*−/−*) mice compared to *Oat1* (*+/+*) mice [143]

(Continued)

Table 8.3 (Continued)

Transporter (gene)	Pre-/clinical	Absorption	Metabolism	Distribution	Excretion
				Cidofovir: Renal toxicity was significantly improved by coadministration of probenecid in monkeys [141] ***Tenofovir***: Renal toxicity was significantly lower in *Oat1 (−/−)* mice compared to *Oat1 (+/+)* mice [142]	***Benzylpenicillin***: Clearance was decreased by 50–70% in *Oat3 (−/−)* mice compared to *Oat1 (+/+)* mice [144]
	Clinical	No data available	No data available	***Cidofovir***: Renal toxicity was significantly reduced when cotreated with probenecid [145]	***Furosemide, fexofenadine, and cidofovir***: Renal clearance decreased by 66%, 70%, and 32%, respectively, when cotreated with probenecid [139, 146–148]
OCT1/2 (*SLC22A1/A2*)	Preclinical	No data available	No data available	***Metformin***: Hepatic metformin accumulation decreased by nearly 70% in *Oct1 (−/−)* mice compared to *Oct1 (+/+)* mice [149] ***Metformin***: Renal and hepatic metformin accumulation was unchanged in *Oct1/2 (−/−)* mice [150]	***Metformin and cisplatin***: Renal elimination decreased by 60% in *Oct1/2 (−/−)* mice compared to *Oct1/2 (+/+)* [150, 151]
	Clinical	No data available	***Fenoterol***: AUC increased by twofold in *OCT1* null allele carriers [152]. ***Ondansetron and tropisetron***: AUC increased by two to fourfold in *OCT1* null allele carriers [153]	***Metformin***: Volume of distribution decreased by 50% in *OCT1* variants [154] ***Cisplatin***: Variants in *OCT2* was associated with reduced cisplatin-induced nephrotoxicity [151, 155, 156]	***Metformin***: Plasma clearance decreased by 37% in *OCT1* variants [154] ***Metformin***: Renal clearance by secretion decreased by 35–60% in *OCT2* variants [157, 158]

MATE1/2-K (*SLC47A1/A2*)	Preclinical	No data available	No data available	***Metformin and cisplatin***: Kidney accumulation significantly increased in *Mate1 (−/−)* mice compared to *Mate1 (+/+)* mice, and in *Mate1 (−/−)* mice co-treated with ondansetron [159–161]	***Metformin***: Renal clearance by secretion decreased by nearly 80% in *Mate1 (−/−)* mice compared to *Mate1 (+/+)* mice [162]
	Clinical	No data available	No data available	***Metformin***: The glucose-lowing effect of metformin reduced significantly in *MATE1* variant (*rs2289669 G>A*) [163] ***Cisplatin***: *MATE1* variant (*rs2289669 G>A*) has no effect on cisplatin-mediated nephrotoxicity [155]	***Metformin and creatinine***: Renal clearance decreased by 35% and 20%, respectively, when cotreated with pyrimethamine and ondansetron [164, 165] ***Metformin, cephalexin, procainamide, and dofetilide***: Renal clearance decreased by 27–41% when cotreated with cimetidine or cetirizine [166–169]

Sources: Lin et al. [92], Lund et al. [170], Wessler et al. [171].

distribution of elacridar, a dual substrate of P-gp, and BCRP, was increased but to a lower extent than verapamil, a well-established P-gp substrate, when P-gp was inhibited by tariquidar [122]. A more comprehensive review about the role of BCRP in drug disposition is available from Mao et al. [174].

8.3.3 Role of BSEP in Drug-Induced Cholestatic Liver Injury

BSEP is exclusively expressed in the apical membrane of hepatocytes, functioning as the pump to excrete bile salts. The main substrate of BSEP is conjugated monovalent bile salts [175]. Due to its narrow substrate specificity, BSEP is not regarded as a major drug transporter participating in drug disposition [1, 4]. The main consideration of the ITC in their recommendation of BSEP for drug interaction is the crucial role of this transporter in drug-induced cholestatic liver injury [4]. The BESP-mediated bile salt transportation is well recognized as the rate-limiting step for bile salt excretion, though some potential but understudied backup systems like MRP2 in apical membrane and MRP3/4 and OSTα/β in basolateral side are thought to contribute [5]. Hence, certain drugs, that perturb the function of BSEP, or patients with deficient BSEP mutation, would potentially increase the buildup of hepatotoxic bile salts in hepatocytes, thus causing cholestatic liver injury [176]. For example, bosentan, cyclosporine, glibenclalmide, rifampin, and troglitazone have been recognized to cause drug-induced cholestatic liver injury, at least partially, by inhibiting BSEP in humans [177, 178]. It is therefore deemed as a necessity by the ITC to examine the BSEP-drug interaction *in vitro* if there is any clinical evidence of cholestatic liver injury by the drug of interest, even though prospective BSEP testing is not required at this moment.

8.3.4 Role of MRPs (MRP2, MRP3, and MRP4) in Drug Disposition

MRP2, also known as the canalicular multispecific organic anion transporter, resides in the apical membrane of hepatocytes, enterocytes of the small intestine, and renal proximal tubular cells in humans [179]. MRP2 has received the most attention among the nine members in the MRP family, due to its crucial role in the elimination of a variety of anionic drugs and their conjugates, including bilirubin glucuronides, leukotriene C4, methotrexate, and paclitaxel [4]. The role of MRP2 in drug disposition was first identified and characterized in a Mrp2-deficient rat (Eisai hyperbilirubinemic rat, a model for Dubin–Johnson syndrome) [30]. The biliary excretion of methotrexate and irinotecan were found to decrease substantially in Eisai hyperbilirubinemic rats as compared to the wild type controls [124, 125]. Similarly, a decreased biliary excretion of doxorubicin and paclitaxel was also observed in *Mrp2 (−/−)* mice as compared to *Mrp2 (+/+)* mice [123]. However, the MRP2 expressed in the small intestine seems not to play an important role in drug absorption. The oral bioavailability of paclitaxel, which was dramatically increased in *Mdr1a/1b (−/−)* mice, was nevertheless not enhanced in *Mrp2 (−/−)* mice when compared to wild type mice [123]. Likewise, there is no evidence supporting a major role of MRP2 in drug elimination from the kidney. Clinically, no typical drug–drug interaction (DDI) and drug-induced toxicity has been attributed to the perturbation of MRP2 so far. Therefore, a retrospective examination of MRP2 inhibition by the drug of interest will only be warranted, as recommended by the ITC, if there is evidence of drug-induced hyperbilirubinemia in preclinical and clinical studies.

The expression of MRP3 is also found in intestinal epithelial cells and hepatocytes; however, unlike MRP2, MRP3 is exclusively located in the basolateral membrane [4]. The glucuronic acid conjugates of drugs and endogenous chemicals, especially bilirubin mono- and bis-glucuronides are good substrates of MRP3. Due to their strategic location, MRP3 along with MRP2 are involved in the occurrence of conjugated hyperbilirubinemia [30]. While a reduced activity of MRP2 either due to genetic mutants or chemical perpetrators may lead to a decreased excretion of bilirubin conjugates into the bile, the expression of MRP3 could be dramatically induced by the bilirubin conjugates accumulated in hepatocytes. The upregulated expression of MRP3 may then transport the excessive bilirubin conjugates into the blood to reduce their hepatotoxicity, leading to conjugated hyperbilirubinemia. This has been supported by the observation from animal studies. The plasma level of bilirubin glucuronides was significantly lower in *Mrp3 (−/−)* mice than *Mrp3 (+/+)* mice during the cholestasis caused by bile-duct ligation [180, 181]. In addition, the role of MRP3 in drug distribution and excretion is well exemplified by the administration of morphine and acetaminophen in *Mrp3 (−/−)* mice [127, 128]. The liver accumulation of morphine and acetaminophen increased by 3- and 10-fold, respectively, in *Mrp3 (−/−)* mice as compared to *Mrp3 (+/+)* mice, accompanied by a dramatic shift of elimination route from the urine to the bile. However, there is still no concrete clinical evidence to justify the examination of potential DDI with respect to MRP3 activity.

MRP4 has a much broader substrate specificity as compared to other MRP members, partially due to its high affinity to glutathione (GSH) conjugated substrates [30]. MRP4 may be localized to either the apical or basolateral membrane in different cell types. For instance, the protein expression of MRP4 could be found in the apical membrane of BBB and proximal tubules, but in the basolateral membrane of hepatocytes. MRP4 may be regarded as a physiological backup to BSEP in bile salt transportation, although it lacks efficiency. A fourfold increase of bile acid level was observed in the liver of *Mrp4 (−/−)* mice, which suffered a more serious liver toxicity than *Mrp4 (+/+)* mice after bile duct ligation [182]. MRP4 is also an important determinant of drug distribution through the BBB and drug elimination in the kidney. For instance, the concentration of topotecan in the brain and CSF was significantly increased [30], while the renal excretion of adefovir and tenofovir was substantially reduced in *Mrp4 (−/−)* mice as compared to *Mrp4 (+/+)* mice [129]. Nevertheless, similar to MRP3, the clinical evidence is still absent in the regard of MRP4-medicated DDI.

8.3.5 Role of OATPs (OATP1B1, OATP1B3, and OATP2B1) in Drug Disposition

OATP1B1 and OATP1B3 are selectively and almost exclusively located in the basolateral membrane of the hepatocytes in human liver. Both of the OATPs can transport a variety of structurally unrelated amphiphilic chemicals, though with a preference toward anionic compounds. There is an overlap of substrates between the two OATPs (e.g. valsartan, bosentan, and olmesartan), however, specific substrates for each (e.g. rosuvastatin, pitavastatin, estrone-3-sulfate for OATP1B1, and telmisartan, docetaxel, cell counting kit-8 [CCK-8] for OATP1B3) have been reported [183]. The OATP-mediated transport is ATP- and sodium-independent, and is electroneutral, implicating an anion exchange. However, the exact driving force is still controversial and warranted for further investigation [184].

OATPs-mediated hepatic uptake has been regarded as the rate-limiting step for drug substrates that are predominantly targeted, metabolized, or eliminated in the liver. For instance, the metabolism and elimination of atorvastatin, which is extensively metabolized by CYP3A4 in the liver, is subject to the regulation of OATPs-mediated hepatic uptake of this drug. This has been substantiated in both mouse and human studies [130, 133]. The clearance of atorvastatin was reduced by nearly 80% in *Oatp1a/b (−/−)* mice than *Oatp1a/b (+/+)* mice, and in human subjects who were cotreated with the OATP inhibitor rifampin compared to controls. Rosuvastatin and pravastatin are also good substrates of OATPs, but have not undergone any significant metabolic changes before excretion into the bile. A significant reduction in the clearance of these two statins were seen in *Oatp1a/b (–/–)* mice and in humans coadministrated with an OATP inhibitor [132, 135]. Interestingly, a dramatic increase of systemic exposure of statins has been frequently observed when these statins are cotreated with OATP perpetrators. The change in liver distribution, however, is rather limited [132]. This might be explained by the PK of these statins. When taken orally, the statins are predominantly accumulated in the liver; in contrast less than 5% of the dose is found in the plasma [131, 132]. The reduction in liver accumulation due to the inhibition of OATPs could be compensated and masked by the relatively high-drug concentration in the portal vein. However, a small reduction in liver accumulation could then be translated into a dramatic increase (five to ninefold) in plasma exposure. The DDI in this regard is crucial to the occurrence of statin side effects. The most serious adverse effect of statins is rhabdomyolysis, which is closely related to their systemic (plasma) concentration. For instance, the patients with *OATP1B1 521CC* genotype, which could lead to two- to threefold increase of simvastatin systemic exposure than the *521TT* genotype [136], and would suffer an 18% incidence of myopathy as compared to 0.6% in the patients with the *521TT* genotype during a five-year treatment of simvastatin [185].

OATP2B1 has a broad substrate specificity and exhibits a broader tissue expression in liver, intestine, and blood–brain barrier. It is recognized as the main OATP isoform in the intestine, and has a similar abundance as to OATP1B3 in liver, suggesting its important role in drug disposition at both organs. Preclinical data from *Oatp2b1* knockout mice showed that AUC and C_{max} were 41% and 70% lower than wild type mice after oral administration of fexofenadine [23]. Likewise, the human studies also support the role of OATP2B1 in drug absorption. It was reported that fruit juice decreased substrates (fexofenadine, aliskiren, and celiprolol) exposure by 1.5- to 2-fold [20–23], a similar effect was also observed in mice study [23]. In contrast, the role of OATP2B1 in hepatic drug clearance is still elusive. No difference of systemic exposure was observed between wild type and *Oatp2b1* knockout mice when fexofenadine was intravenously administrated [23]. Likewise, there is no confirmative evidence from human studies either. A recommendation of retrospective evaluation is thus issued by the ITC to examine the role OATP2B1, based on the intestinal or liver DDI which cannot be attributed to more common mechanisms (e.g. intestinal P-gp/BCRP and hepatic OATP1B1/1B3) [5].

8.3.6 Role of OATs (OAT1 and OAT3) in Drug Disposition

In humans, OAT1 and OAT3 are substantially and majorly expressed on the basolateral side of the renal tubular cells, mediating the renal elimination of organic anions by using the outward dicarboxylate gradient as the driving force [2]. OATs could accept a variety of structurally unrelated organic anions as their substrates, including antivrials, antibiotics, H2-antagonists, diuretics [186]. Although a large overlap of substrates exists between OAT1 and OAT3, they do show specificity toward different drugs. For instance, the antivirals cidofovir and acyclovir are transported by OAT1 but not OAT3 [187], and benzylpenicillin on the contrary [144].

OAT1 and OAT3 are crucial determinants of renal distribution and elimination for their substrates, which is implicated by their selective localization in the kidney. For instance, the renal clearance of furosemide and benzylpenicillin were decreased by 70% and 50–70% in *Oat1 (–/–)* and *Oat3 (–/–)* mice, respectively, as compared to wild type mice [143, 144]. A similar reduction of renal clearance was also observed in humans when furosemide was coadministrated with the OAT inhibitor probenecid [146], which was first used in World War II as the penicillin-sparing agent to slow down the loss of the precious antibiotics from the urine. The other substrate of OATs that perhaps benefits the most from OAT inhibition is cidofovir, a drug used to treat cytomegalovirus infection. The use of cidofovir is dose-limiting by its severe nephrotoxicity due to its accumulation in the proximal tubular cells [139]. OAT inhibition by probenecid could significantly reduce the accumulation of cidofovir in the cortex of the kidney, reduce its renal clearance to the level of glomerular filtration, and dramatically ameliorate its renal toxicity in both pre- and clinical studies [140, 141, 145]. Therefore, coadministration of probenecid is required by US FDA as an approach to avoid cidofovir-induced nephrotoxicity.

8.3.7 Role of OCTs (OCT1 and OCT2)/MATEs (MATE1 and MATE2-K) in Drug Disposition

OCTs and MATEs are pairs of transporters expressed abundantly in both liver and kidney for vectorial transport of organic cations across the epithelia. In humans, OCT1 and OCT2 are almost exclusively expressed in the basolateral side of hepatocytes and renal tubular cells respectively, while MATE1 is found in the apical membrane of both cells [1]. In addition, a functional MATE2 splice variant MATE2-K resides exclusively in the apical membrane of renal tubular cells, with a relatively lower abundance as compared to MATE1. The distribution of the orthologs is slightly different in rodents. Mouse Oct1 and Mate1 are highly expressed in the liver, while Oct1, Oct2, and Mate1 are all significantly resided in the kidney [2, 3]. The driving force of transportation differs between OCTs and MATEs. The OCTs-mediated transport (uptake) from the blood into the cell is driven by the internal negative membrane potential (~−70 mV), while the MATEs-mediated transport (efflux) from the cell into the bile and the urine is driven by the outward proton gradient (H^+/organic cation antiport).

As implicated from their substantial expression in both the liver and the kidney, OCTs and MATEs can be critical determinants in the distribution and elimination of organic cationic drugs. Shu et al. found that the hepatic accumulation of metformin decreased by almost 70% in *Oct1 (−/−)* mice as compared to *Oct1 (+/+)* mice [149]. Subjects with impaired OCT1 mutation showed a reduced hepatic metformin accumulation while with no obvious alteration in systemic exposure [188]. Likewise, a nearly 50% decrease in oral volume of distribution of metformin was seen in healthy volunteers who carried an *OCT1* allele of reduced function, implicating an impaired liver partition of metformin in those subjects [154]. In addition, a reduced glucose-lowering effect of metformin was also observed in the healthy carriers of *OCT1* variants [149], albeit the same *OCT1* variants were not found to attenuate the reductive effect of metformin on plasma hemoglobin A1C in a larger cohort study involving patients with type 2 diabetes [189]. The controversial results suggest complex anti-diabetic effects of metformin, which probably could not be fully attributed to its action in hepatocytes. Since metformin is exclusively eliminated from kidney, and is not a sensitive pharmacokinetic probe for OCT1 inhibition, efforts were made to identify more sensitive probes clinically. For example, carriers with OCT1 null alleles exhibited a twofold increase of systemic exposure of fenoterol, an OCT1 substrate [152]. Likewise, a two to fourfold increase of systemic exposure of ondansetron and tropisetron was observed in subjects who have impaired OCT1 mutation [153]. The accumulating clinical evidences persuaded the ITC to include OCT1 into the list of transporters. A prospective evaluation is required during drug development. A more sophisticated review can be found in the ITC issued white paper in 2018 [5].

The role of OCTs in drug distribution and elimination in the kidney has been established as well by using *Oct1/2 (−/−)* double knockout mice and in human subjects carrying the *OCT2* alleles of reduced function. The renal elimination of metformin was significantly reduced in *Oct1/2 (−/−)* mice as compared to the controls [150], and also in the humans with *OCT2* variants [183]. Interestingly, metformin accumulation in the liver and kidney of *Oct1/2 (−/−)* mice was unchanged. The decrease in tissue accumulation was probably canceled out by the threefold increase of systemic metformin exposure. In addition, the renal elimination and distribution of cisplatin, a substrate of OCTs and MATEs, decreased substantially in *Oct1/2 (−/−)* mice [151]. Cisplatin-mediated nephrotoxicity, which was dose-limiting, was significantly attenuated in the patients who carried *OCT2* variants, implicating a reduced accumulation of cisplatin in the kidney by a decreased activity of OCT2 [155].

Compelling evidence also supports a critical role of MATEs in drug disposition. For instance, the accumulation of metformin in the kidney of *Mate1 (−/−)* mice was increased substantially [159], while the renal clearance of metformin by active secretion decreased by approximate 80% as compared to *Mate1 (+/+)* mice [162]. In addition, the selective MATEs inhibitors pyrimethamine and ondansetron could increase renal accumulation, while reduce renal elimination of metformin in both mice and humans [160, 164, 165]. DDI studies have been performed with the OCTs/MATEs perpetrator cimetidine in humans, observing a 27–41% decrease of renal clearance for metformin, cephalexin, procainamide, and dofetilide [166–169]. The DDIs were previously attributed to the inhibition of OCT2; however, it has been widely accepted as to the perpetration of MATEs due to a more potent inhibitory effect of cimetidine on MATE1/2-K than OCT2. In addition, MATE activities might also determine cisplatin-induced nephrotoxicity. The renal accumulation of cisplatin was significantly enhanced in *Mate1 (−/−)* mice and in the wild-type mice cotreated with ondansetron [160]. Nevertheless, the patients carrying *SLC47A1* variant (*rs2289669 G>A*) showed no improvement on cisplatin-mediated nephrotoxicity [155], though the same variant was associated with a reduced anti-diabetic effect of metformin in another clinical trial [163].

For decades, creatinine clearance has been clinically used to reflect renal function during estimating glomerular filtration rate (GFR) [190]. The recognition of OCTs and MATEs as the transporters of creatinine (account for 20% of creatinine clearance) [191, 192] suggests that the perpetrators of OCTs and MATEs could significantly reduce creatinine clearance without any real renal damage. For instance, pyrimethamine could decrease creatinine clearance by 20% in humans [165]. Consistently, a systemic review with meta-analysis also indicated that endogenous creatinine clearance is inaccurate for the measurement of GFR [193]. Hence, the usefulness of creatinine clearance as the index of renal function has been compromised in many clinical scenarios of drug therapy.

8.4 Closing Remarks

Membrane transporters are present in all tissues in the body with varying abundance. It would be a prerequisite to clarify the precise localization and abundance of transporters in different tissues, before elucidating the role of transporters in drug disposition. In this chapter, we begun with a description of the distribution and localization of major drug transporters in various tissues, followed by a discussion of their role in drug disposition. We tried to deliver unequivocal information. For instance, the presence of transporters in specific tissues should be evidenced by protein expression; and likewise, the precise localization of transporters should be confirmed by immunohistochemistry with specific antibodies. In addition, the role of each transporter in drug disposition is validated by the evidence from both pre- and clinical studies. However, our current review is limited by the discussion solely on the major drug transporters proposed in the ITC panel. In addition, considering the length of the chapter, we only presented selected data from the literature (Table 8.3), inevitably compromising the comprehensiveness of this chapter.

Classically, alteration in drug disposition is reflected by changes in systemic drug exposure (e.g. AUC, C_{max}) and clearance (plasma clearance, renal clearance, half-life), which rely heavily on the measurement of plasma drug concentrations. The plasma concentrations measured, nevertheless, may fail to accurately reflect the true drug levels in the tissues, which is regulated by the uptake and efflux transporters and the key determinant of drug efficacy and adverse effects in most cases. It remains a challenge to clinically measure drug concentrations (distribution) in tissues. Identification of more suitable substrate tracers of these transporters for imaging studies like PET-computed tomography (CT) may provide useful information in this regard. Moreover, the lack of selective probe substrates and inhibitors for specific drug transporters introduced enormous complexity to interpretation of *in vivo* data. Significant progress in this regard is needed to deepen our understanding of the role of transporters in drug disposition.

References

1 Giacomini, K.M., Huang, S.M., Tweedie, D.J. et al. (2010). Membrane transporters in drug development. *Nat. Rev. Drug Discovery* 9 (3): 215–236.

2 Yin, J. and Wang, J. (2016). Renal drug transporters and their significance in drug–drug interactions. *Acta Pharm. Sin. B* 6 (5): 363–373.

3 Klaassen, C.D. and Aleksunes, L.M. (2010). Xenobiotic, bile acid, and cholesterol transporters: function and regulation. *Pharmacol. Rev.* 62 (1): 1–96.

4 Hillgren, K.M., Keppler, D., Zur, A.A. et al. (2013). Emerging transporters of clinical importance: an update from the International Transporter Consortium. *Clin. Pharmacol. Ther.* 94 (1): 52–63.

5 Zamek-Gliszczynski, M.J., Taub, M.E., Chothe, P.P. et al. (2018). Transporters in drug development: 2018 ITC recommendations for transporters of emerging clinical importance. *Clin. Pharmacol. Ther.* 104 (5): 890–899.

6 Nigam, S.K. (2018). The SLC22 transporter family: a paradigm for the impact of drug transporters on metabolic pathways, signaling, and disease. *Annu. Rev. Pharmacol. Toxicol.* 58: 663–687.

7 Kimura, T. and Higaki, K. (2002). Gastrointestinal transit and drug absorption. *Biol. Pharm. Bull.* 25 (2): 149–164.

8 Thummel, K.E. (2007). Gut instincts: CYP3A4 and intestinal drug metabolism. *J. Clin. Invest.* 117 (11): 3173–3176.

9 Glaeser, H., Bailey, D.G., Dresser, G.K. et al. (2007). Intestinal drug transporter expression and the impact of grapefruit juice in humans. *Clin. Pharmacol. Ther.* 81 (3): 362–370.

10 Drozdzik, M., Gröer, C., Penski, J. et al. (2014). Protein abundance of clinically relevant multidrug transporters along the entire length of the human intestine. *Mol. Pharm.* 11 (10): 3547–3555.

11 Muller, J., Lips, K.S., Metzner, L. et al. (2005). Drug specificity and intestinal membrane localization of human organic cation transporters (OCT). *Biochem. Pharmacol.* 70 (12): 1851–1860.

12 Hilgendorf, C., Ahlin, G., Seithel, A. et al. (2007). Expression of thirty-six drug transporter genes in human intestine, liver, kidney, and organotypic cell lines. *Drug Metab. Dispos.* 35 (8): 1333–1340.

13 Daniel, H. and Kottra, G. (2004). The proton oligopeptide cotransporter family SLC15 in physiology and pharmacology. *Pflugers Arch.* 447 (5): 610–618.

14 Ziegler, T.R., Fernandez-Estivariz, C., Gu, L.H. et al. (2002). Distribution of the H+/peptide transporter PepT1 in human intestine: up-regulated expression in the colonic mucosa of patients with short-bowel syndrome. *Am. J. Clin. Nutr.* 75 (5): 922–930.

15 Hruz, P., Zimmermann, C., Gutmann, H. et al. (2006). Adaptive regulation of the ileal apical sodium dependent bile acid transporter (ASBT) in patients with obstructive cholestasis. *Gut* 55 (3): 395–402.

16 Müller, J., Keiser, M., Drozdzik, M. et al. (2017). Expression, regulation and function of intestinal drug transporters: an update. *Biol. Chem.* 398 (2): 175–192.

17 Sai, Y., Kaneko, Y., Ito, S. et al. (2006). Predominant contribution of organic anion transporting polypeptide OATP-B (OATP2B1) to apical uptake of estrone-3-sulfate by human intestinal Caco-2 cells. *Drug Metab. Dispos.* 34 (8): 1423–1431.

18 Kobayashi, D., Nozawa, T., Imai, K. et al. (2003). Involvement of human organic anion transporting polypeptide OATP-B (SLC21A9) in pH-dependent transport across intestinal apical membrane. *J. Pharmacol. Exp. Ther.* 306 (2): 703–708.

19 Mooij, M., E A de Koning, B., Lindenbergh-Kortleve, D.J. et al. (2016). Human intestinal PEPT1 transporter expression and localization in preterm and term infants. *Drug Metab. Dispos.* 44 (7): 1014–1019.

20 Ieiri, I., Doi, Y., Maeda, K. et al. (2012). Microdosing clinical study: pharmacokinetic, pharmacogenomic (SLCO2B1), and interaction (grapefruit juice) profiles of celiprolol following the oral microdose and therapeutic dose. *J. Clin. Pharmacol.* 52 (7): 1078–1089.

21 Tapaninen, T., Neuvonen, P.J., and Niemi, M. (2011). Orange and apple juice greatly reduce the plasma concentrations of the OATP2B1 substrate aliskiren. *Br. J. Clin. Pharmacol.* 71 (5): 718–726.

22 Imanaga, J., Kotegawa, T., Imai, H. et al. (2011). The effects of the SLCO2B1 c.1457C > T polymorphism and apple juice on the pharmacokinetics of fexofenadine and midazolam in humans. *Pharmacogenet. Genomics* 21 (2): 84–93.

23 Medwid, S., Li, M.M.J., Knauer, M.J. et al. (2019). Fexofenadine and rosuvastatin pharmacokinetics in mice with targeted disruption of organic anion transporting polypeptide 2B1. *Drug Metab. Dispos.* 47 (8): 832–842.

24 Berggren, S., Gall, C., Wollnitz, N. et al. (2007). Gene and protein expression of P-glycoprotein, MRP1, MRP2, and CYP3A4 in the small and large human intestine. *Mol. Pharmaceutics* 4 (2): 252–257.

25 Maliepaard, M., Scheffer, G.L., Faneyte, I.F. et al. (2001). Subcellular localization and distribution of the breast cancer resistance protein transporter in normal human tissues. *Cancer Res.* 61 (8): 3458–3464.

26 Han, T.K., Everett, R.S., Proctor, W.R. et al. (2013). Organic cation transporter 1 (OCT1/mOct1) is localized in the apical membrane of Caco-2 cell monolayers and enterocytes. *Mol. Pharmacol.* 84 (2): 182–189.

27 Ballatori, N., Christian, W.V., Lee, J.Y. et al. (2005). OSTalpha-OSTbeta: a major basolateral bile acid and steroid transporter in human intestinal, renal, and biliary epithelia. *Hepatology* 42 (6): 1270–1279.

28 Rao, A., Haywood, J., Craddock, A.L. et al. (2008). The organic solute transporter alpha-beta, Ostalpha-Ostbeta, is essential for intestinal bile acid transport and homeostasis. *Proc. Natl. Acad. Sci. U.S.A.* 105 (10): 3891–3896.

29 Scheffer, G.L., Kool, M., de Haas, M. et al. (2002). Tissue distribution and induction of human multidrug resistant protein 3. *Lab. Invest.* 82 (2): 193–201.

30 Kruh, G.D., Belinsky, M.G., Gallo, J.M., and Lee, K. (2007). Physiological and pharmacological functions of Mrp2, Mrp3 and Mrp4 as determined from recent studies on gene-disrupted mice. *Cancer Metastasis Rev.* 26 (1): 5–14.

31 Meier, Y., Eloranta, J.J., Darimont, J. et al. (2007). Regional distribution of solute carrier mRNA expression along the human intestinal tract. *Drug Metab. Dispos.* 35 (4): 590–594.

32 Gutmann, H., Hruz, P., Zimmermann, C. et al. (2005). Distribution of breast cancer resistance protein (BCRP/ABCG2) mRNA expression along the human GI tract. *Biochem. Pharmacol.* 70 (5): 695–699.

33 Burt, H.J., Riedmaier, A.E., Harwood, M.D. et al. (2016). Abundance of hepatic transporters in Caucasians: a meta-analysis. *Drug Metab. Dispos.* 44 (10): 1550–1561.

34 Vildhede, A., Wisniewski, J.R., Noren, A. et al. (2015). Comparative proteomic analysis of human liver tissue and isolated hepatocytes with a focus on proteins determining drug exposure. *J. Proteome Res.* 14 (8): 3305–3314.

35 Otsuka, M., Matsumoto, T., Morimoto, R. et al. (2005). A human transporter protein that mediates the final excretion step for toxic organic cations. *Proc. Natl. Acad. Sci. U.S.A.* 102 (50): 17923–17928.

36 Drozdzik, M., Busch, D., Lapczuk, J. et al. (2019). Protein abundance of clinically relevant drug transporters in the human liver and intestine: a comparative analysis in paired tissue specimens. *Clin. Pharmacol. Ther.* 105 (5): 1204–1212.

37 Lee, W., Glaeser, H., Smith, L.H. et al. (2005). Polymorphisms in human organic anion-transporting polypeptide 1A2 (OATP1A2): implications for altered drug disposition and central nervous system drug entry. *J. Biolumin. Chemilumin.* 280 (10): 9610–9617.

38 Gigliozzi, A., Fraioli, F., Sundaram, P. et al. (2000). Molecular identification and functional characterization of Mdr1a in rat cholangiocytes. *Gastroenterology* 119 (4): 1113–1122.

39 Knutter, I., Rubio-Aliaga, I., Boll, M. et al. (2002). H+-peptide cotransport in the human bile duct epithelium cell line SK-ChA-1. *Am. J. Physiol. Gastrointest. Liver Physiol.* 283 (1): G222–G229.

40 Lazaridis, K.N., Pham, L., Tietz, P. et al. (1997). Rat cholangiocytes absorb bile acids at their apical domain via the ileal sodium-dependent bile acid transporter. *J. Clin. Invest.* 100 (11): 2714–2721.

41 Luo, Z.L., Cheng, L., Wang, T. et al. (2019). Bile acid transporters are expressed and heterogeneously distributed in rat bile ducts. *Gut Liver* 13 (5): 569–575.

42 Morrissey, K.M., Stocker, S.L., Wittwer, M.B. et al. (2013). Renal transporters in drug development. *Annu. Rev. Pharmacol. Toxicol.* 53: 503–529.

43 Ekaratanawong, S., Anzai, N., Jutabha, P. et al. (2004). Human organic anion transporter 4 is a renal apical organic anion/dicarboxylate exchanger in the proximal tubules. *J. Pharm. Sci.* 94 (3): 297–304.

44 Noshiro, R., Anzai, N., Sakata, T. et al. (2006). The PDZ domain protein PDZK1 interacts with human peptide transporter PEPT2 and enhances its transport activity. *Kidney Int.* 70 (2): 275–282.

45 Tamai, I., Nakanishi, T., Kobayashi, D. et al. (2004). Involvement of OCTN1 (SLC22A4) in pH-dependent transport of organic cations. *Mol. Pharmaceutics* 1 (1): 57–66.

46 Tamai, I., Ohashi, R., Nezu, J. et al. (1998). Molecular and functional identification of sodium ion-dependent, high affinity human carnitine transporter OCTN2. *J. Biolumin. Chemilumin.* 273 (32): 20378–20382.

47 Tamai, I., Yabuuchi, H., Nezu, J. et al. (1997). Cloning and characterization of a novel human pH-dependent organic cation transporter, OCTN1. *FEBS Lett.* 419 (1): 107–111.

48 Liu, W., Liang, R., Ramamoorthy, S. et al. (1995). Molecular cloning of PEPT 2, a new member of the H+/peptide cotransporter family, from human kidney. *Biochim. Biophys. Acta* 1235 (2): 461–466.

49 Prasad, B., Johnson, K., Billington, S. et al. (2016). Abundance of drug transporters in the human kidney cortex as quantified by quantitative targeted proteomics. *Drug Metab. Dispos.* 44 (12): 1920–1924.

50 Huls, M., Brown, C.D., Windass, A.S. et al. (2008). The breast cancer resistance protein transporter ABCG2 is expressed in the human kidney proximal tubule apical membrane. *Kidney Int.* 73 (2): 220–225.

51 van Aubel, R.A., Smeets, P.H., Peters, J.G. et al. (2002). The MRP4/ABCC4 gene encodes a novel apical organic anion transporter in human kidney proximal tubules: putative efflux pump for urinary cAMP and cGMP. *J. Am. Soc. Nephrol.* 13 (3): 595–603.

52 Schaub, T.P., Kartenbeck, J., Konig, J. et al. (1999). Expression of the MRP2 gene-encoded conjugate export pump in human kidney proximal tubules and in renal cell carcinoma. *J. Am. Soc. Nephrol.* 10 (6): 1159–1169.

53 Thiebaut, F., Tsuruo, T., Hamada, H. et al. (1987). Cellular localization of the multidrug-resistance gene product P-glycoprotein in normal human tissues. *Proc. Natl. Acad. Sci. U.S.A.* 84 (21): 7735–7738.

54 Motohashi, H., Sakurai, Y., Saito, H. et al. (2002). Gene expression levels and immunolocalization of organic ion transporters in the human kidney. *J. Am. Soc. Nephrol.* 13 (4): 866–874.

55 Banks, W.A. (2016). From blood–brain barrier to blood–brain interface: new opportunities for CNS drug delivery. *Nat. Rev. Drug Discovery* 15 (4): 275–292.

56 Stieger, B. and Gao, B. (2015). Drug transporters in the central nervous system. *Clin. Pharmacokinet.* 54 (3): 225–242.

57 Lee, G., Dallas, S., Hong, M., and Bendayan, R. (2001). Drug transporters in the central nervous system: brain barriers and brain parenchyma considerations. *Pharmacol. Rev.* 53 (4): 569–596.

58 Nwaozuzu, O.M., Sellers, L.A., and Barrand, M.A. (2003). Signalling pathways influencing basal and H_2O_2-induced P-glycoprotein expression in endothelial cells derived from the blood–brain barrier. *J. Neurochem.* 87 (4): 1043–1051.

59 Cooray, H.C., Blackmore, C.G., Maskell, L., and Barrand, M.A. (2002). Localisation of breast cancer resistance protein in microvessel endothelium of human brain. *Neuroreport* 13 (16): 2059–2063.

60 Nies, A.T., Jedlitschky, G., Konig, J. et al. (2004). Expression and immunolocalization of the multidrug resistance proteins, MRP1–MRP6 (ABCC1–ABCC6), in human brain. *Neuroscience* 129 (2): 349–360.

61 Virgintino, D., Robertson, D., Errede, M. et al. (2002). Expression of P-glycoprotein in human cerebral cortex microvessels. *J. Histochem. Cytochem.* 50 (12): 1671–1676.

62 Geier, E.G., Chen, E.C., Webb, A. et al. (2013). Profiling solute carrier transporters in the human blood–brain barrier. *Clin. Pharmacol. Ther.* 94 (6): 636–639.

63 Gao, B., Hagenbuch, B., Kullak-Ublick, G.A. et al. (2000). Organic anion-transporting polypeptides mediate transport of opioid peptides across blood–brain barrier. *J. Pharmacol. Exp. Ther.* 294 (1): 73–79.

64 Gao, B., Vavricka, S.R., Meier, P.J., and Stieger, B. (2015). Differential cellular expression of organic anion transporting peptides OATP1A2 and OATP2B1 in the human retina and brain: implications for carrier-mediated transport of neuropeptides and neurosteriods in the CNS. *Pflugers Arch.* 467 (7): 1481–1493.

65 Alebouyeh, M., Takeda, M., Onozato, M.L. et al. (2003). Expression of human organic anion transporters in the choroid plexus and their interactions with neurotransmitter metabolites. *J. Pharm. Sci.* 93 (4): 430–436.
66 Uchida, Y., Goto, R., Takeuchi, H. et al. (2020). Abundant expression of OCT2, MATE1, OAT1, OAT3, PEPT2, BCRP, MDR1, and xCT transporters in blood-arachnoid barrier of pig and polarized localizations at CSF- and blood-facing plasma membranes. *Drug Metab. Dispos.* 48 (2): 135–145.
67 Leggas, M., Adachi, M., Scheffer, G.L. et al. (2004). Mrp4 confers resistance to topotecan and protects the brain from chemotherapy. *Mol. Cell. Biol.* 24 (17): 7612–7621.
68 Yeboah, D., Sun, M., Kingdom, J. et al. (2006). Expression of breast cancer resistance protein (BCRP/ABCG2) in human placenta throughout gestation and at term before and after labor. *Can. J. Physiol. Pharmacol.* 84 (12): 1251–1258.
69 St-Pierre, M.V., Serrano, M.A., Macias, R.I. et al. (2000). Expression of members of the multidrug resistance protein family in human term placenta. *Am. J. Physiol. Regul. Integr. Comp. Physiol.* 279 (4): R1495–R1503.
70 Sun, M., Kingdom, J., Baczyk, D. et al. (2006). Expression of the multidrug resistance P-glycoprotein, (ABCB1 glycoprotein) in the human placenta decreases with advancing gestation. *Placenta* 27 (6–7): 602–609.
71 Azzaroli, F., Mennone, A., Feletti, V. et al. (2007). Clinical trial: modulation of human placental multidrug resistance proteins in cholestasis of pregnancy by ursodeoxycholic acid. *Aliment Pharmacol. Ther.* 26 (8): 1139–1146.
72 Nagashige, M., Ushigome, F., Koyabu, N. et al. (2003). Basal membrane localization of MRP1 in human placental trophoblast. *Placenta* 24 (10): 951–958.
73 Grube, M., Meyer Zu Schwabedissen, H., Draber, K. et al. (2005). Expression, localization, and function of the carnitine transporter octn2 (slc22a5) in human placenta. *Drug Metab. Dispos.* 33 (1): 31–37.
74 Lee, N., Hebert, M.F., Wagner, D.J. et al. (2018). Organic cation transporter 3 facilitates fetal exposure to metformin during pregnancy. *Mol. Pharmacol.* 94 (4): 1125–1131.
75 Somogyi, A. and Gugler, R. (1979). Cimetidine excretion into breast milk. *Br. J. Clin. Pharmacol.* 7 (6): 627–629.
76 Schutte, A.E., Symington, E.A., and du Preez, J.L. (2013). Rosuvastatin is transferred into human breast milk: a case report. *Am. J. Med.* 126 (9): e7–e8.
77 Jonker, J.W., Merino, G., Musters, S. et al. (2005). The breast cancer resistance protein BCRP (ABCG2) concentrates drugs and carcinogenic xenotoxins into milk. *Nat. Med.* 11 (2): 127–129.
78 Merino, G., Alvarez, A.I., Pulido, M.M. et al. (2006). Breast cancer resistance protein (BCRP/ABCG2) transports fluoroquinolone antibiotics and affects their oral availability, pharmacokinetics, and milk secretion. *Drug Metab. Dispos.* 34 (4): 690–695.
79 Merino, G., Jonker, J.W., Wagenaar, E. et al. (2005). The breast cancer resistance protein (BCRP/ABCG2) affects pharmacokinetics, hepatobiliary excretion, and milk secretion of the antibiotic nitrofurantoin. *Mol. Pharmacol.* 67 (5): 1758–1764.
80 Malfará, B.N., Benzi, J.R.L., de Oliveira Filgueira, G.C. et al. (2019). ABCG2 c.421C>A polymorphism alters nifedipine transport to breast milk in hypertensive breastfeeding women. *Reprod. Toxicol.* 85: 1–5.
81 Vlaming, M.L., Lagas, J.S., and Schinkel, A.H. (2009). Physiological and pharmacological roles of ABCG2 (BCRP): recent findings in Abcg2 knockout mice. *Adv. Drug Delivery Rev.* 61 (1): 14–25.

82 Kwok, B., Yamauchi, A., Rajesan, R. et al. (2006). Carnitine/xenobiotics transporters in the human mammary gland epithelia, MCF12A. *Am. J. Physiol. Regul. Integr. Comp. Physiol.* 290 (3): R793–R802.

83 Alcorn, J., Lu, X., Moscow, J.A., and McNamara, P.J. (2002). Transporter gene expression in lactating and nonlactating human mammary epithelial cells using real-time reverse transcription-polymerase chain reaction. *J. Pharmacol. Exp. Ther.* 303 (2): 487–496.

84 Garcia-Lino, A.M. et al. (2019). Transporters in the mammary gland-contribution to presence of nutrients and drugs into milk. *Nutrients* 11 (10).

85 Kimura, S., Morimoto, K., Okamoto, H. et al. (2006). Development of a human mammary epithelial cell culture model for evaluation of drug transfer into milk. *Arch. Pharm. Res.* 29 (5): 424–429.

86 Miller, S.R. and Cherrington, N.J. (2018). Transepithelial transport across the blood-testis barrier. *Reproduction* 156 (6): R187–R194.

87 Mruk, D.D., Su, L., and Cheng, C.Y. (2011). Emerging role for drug transporters at the blood–testis barrier. *Trends Pharmacol. Sci.* 32 (2): 99–106.

88 Bart, J., Hollema, H., Groen, H.J. et al. (2004). The distribution of drug-efflux pumps, P-gp, BCRP, MRP1 and MRP2, in the normal blood–testis barrier and in primary testicular tumours. *Eur. J. Cancer* 40 (14): 2064–2070.

89 Melaine, N., Lienard, M.O., Dorval, I. et al. (2002). Multidrug resistance genes and P-glycoprotein in the testis of the rat, mouse, guinea pig, and human. *Biol. Reprod.* 67 (6): 1699–1707.

90 Klein, D.M., Wright, S.H., and Cherrington, N.J. (2014). Localization of multidrug resistance-associated proteins along the blood–testis barrier in rat, macaque, and human testis. *Drug Metab. Dispos.* 42 (1): 89–93.

91 Koepsell, H., Lips, K., and Volk, C. (2007). Polyspecific organic cation transporters: structure, function, physiological roles, and biopharmaceutical implications. *Pharm. Res.* 24 (7): 1227–1251.

92 Lin, J.H. and Yamazaki, M. (2003). Role of P-glycoprotein in pharmacokinetics: clinical implications. *Clin. Pharmacokinet.* 42 (1): 59–98.

93 Sparreboom, A., van Asperen, J., Mayer, U. et al. (1997). Limited oral bioavailability and active epithelial excretion of paclitaxel (Taxol) caused by P-glycoprotein in the intestine. *Proc. Natl. Acad. Sci. U.S.A.* 94 (5): 2031–2035.

94 Schinkel, A.H., Smit, J.J., van Tellingen, O. et al. (1994). Disruption of the mouse mdr1a P-glycoprotein gene leads to a deficiency in the blood–brain barrier and to increased sensitivity to drugs. *Cell* 77 (4): 491–502.

95 Schinkel, A.H., Wagenaar, E., van Deemter, L. et al. (1995). Absence of the mdr1a P-glycoprotein in mice affects tissue distribution and pharmacokinetics of dexamethasone, digoxin, and cyclosporin A. *J. Clin. Invest.* 96 (4): 1698–1705.

96 Smit, J.W., Huisman, M.T., van Tellingen, O. et al. (1999). Absence or pharmacological blocking of placental P-glycoprotein profoundly increases fetal drug exposure. *J. Clin. Invest.* 104 (10): 1441–1447.

97 Lin, J.H., Chiba, M., Chen, I.W. et al. (1999). Effect of dexamethasone on the intestinal first-pass metabolism of indinavir in rats: evidence of cytochrome P-450 3A [correction of P-450 A] and p-glycoprotein induction. *Drug Metab. Dispos.* 27 (10): 1187–1193.

98 Mutoh, K., Mitsuhashi, J., Kimura, Y. et al. (2006). A T3587G germ-line mutation of the MDR1 gene encodes a nonfunctional P-glycoprotein. *Mol. Cancer Ther.* 5 (4): 877–884.

99 Malingre, M.M., Richel, D.J., Beijnen, J.H. et al. (2001). Coadministration of cyclosporine strongly enhances the oral bioavailability of docetaxel. *J. Clin. Oncol.* 19 (4): 1160–1166.

100 U.S. Food and Drug Administration. Clinical Pharmacology and Biopharmaceutics Review(s). NDA 22-334 Afinitor.

101 Dorian, P., Strauss, M., Cardella, C. et al. (1988). Digoxin-cyclosporine interaction: severe digitalis toxicity after cyclosporine treatment. *Clin. Invest. Med.* 11 (2): 108–112.

102 Ochs, H.R., Bodem, G., and Greenblatt, D.J. (1981). Impairment of digoxin clearance by coadministration of quinidine. *J. Clin. Pharmacol.* 21 (10): 396–400.

103 Bauer, M., Zeitlinger, M., Karch, R. et al. (2012). Pgp-mediated interaction between (*R*)-[^{11}C]verapamil and tariquidar at the human blood–brain barrier: a comparison with rat data. *Clin. Pharmacol. Ther.* 91 (2): 227–233.

104 Sasongko, L., Link, J.M., Muzi, M. et al. (2005). Imaging P-glycoprotein transport activity at the human blood–brain barrier with positron emission tomography. *Clin. Pharmacol. Ther.* 77 (6): 503–514.

105 Bauer, M., Karch, R., Zeitlinger, M. et al. (2015). Approaching complete inhibition of P-glycoprotein at the human blood–brain barrier: an (*R*)-[^{11}C]verapamil PET study. *J. Cereb. Blood Flow Metab.* 35 (5): 743–746.

106 Liu, L., Collier, A.C., Link, J.M. et al. (2015). Modulation of P-glycoprotein at the human blood–brain barrier by quinidine or rifampin treatment: a positron emission tomography imaging study. *Drug Metab. Dispos.* 43 (11): 1795–1804.

107 Pedersen, K.E., Christiansen, B.D., Klitgaard, N.A., and Nielsen-Kudsk, F. (1983). Effect of quinidine on digoxin bioavailability. *Eur. J. Clin. Pharmacol.* 24 (1): 41–47.

108 Rebello, S., Compain, S., Feng, A. et al. (2011). Effect of cyclosporine on the pharmacokinetics of aliskiren in healthy subjects. *J. Clin. Pharmacol.* 51 (11): 1549–1560.

109 Pham, P.A., la Porte, C.J., Lee, L.S. et al. (2009). Differential effects of tipranavir plus ritonavir on atorvastatin or rosuvastatin pharmacokinetics in healthy volunteers. *Antimicrob. Agents Chemother.* 53 (10): 4385–4392.

110 Hermann, M., Asberg, A., Christensen, H. et al. (2004). Substantially elevated levels of atorvastatin and metabolites in cyclosporine-treated renal transplant recipients. *Clin. Pharmacol. Ther.* 76 (4): 388–391.

111 Tidefelt, U., Liliemark, J., Gruber, A. et al. (2000). P-Glycoprotein inhibitor valspodar (PSC 833) increases the intracellular concentrations of daunorubicin in vivo in patients with P-glycoprotein-positive acute myeloid leukemia. *J. Clin. Oncol.* 18 (9): 1837–1844.

112 Kusuhara, H., Furuie, H., Inano, A. et al. (2012). Pharmacokinetic interaction study of sulphasalazine in healthy subjects and the impact of curcumin as an in vivo inhibitor of BCRP. *Br. J. Pharmacol.* 166 (6): 1793–1803.

113 de Vries, N.A., Zhao, J., Kroon, E. et al. (2007). P-Glycoprotein and breast cancer resistance protein: two dominant transporters working together in limiting the brain penetration of topotecan. *Clin. Cancer Res.* 13 (21): 6440–6449.

114 Oostendorp, R.L., Buckle, T., Beijnen, J.H. et al. (2009). The effect of P-gp (Mdr1a/1b), BCRP (Bcrp1) and P-gp/BCRP inhibitors on the in vivo absorption, distribution, metabolism and excretion of imatinib. *Invest. New Drugs* 27 (1): 31–40.

115 Enokizono, J., Kusuhara, H., and Sugiyama, Y. (2007). Effect of breast cancer resistance protein (Bcrp/Abcg2) on the disposition of phytoestrogens. *Mol. Pharmacol.* 72 (4): 967–975.

116 Jonker, J.W., Buitelaar, M., Wagenaar, E. et al. (2002). The breast cancer resistance protein protects against a major chlorophyll-derived dietary phototoxin and protoporphyria. *Proc. Natl. Acad. Sci. U.S.A.* 99 (24): 15649–15654.

117 Zhou, L., Naraharisetti, S.B., Wang, H. et al. (2008). The breast cancer resistance protein (Bcrp1/Abcg2) limits fetal distribution of glyburide in the pregnant mouse: an Obstetric-Fetal Pharmacology Research Unit Network and University of Washington Specialized Center of Research Study. *Mol. Pharmacol.* 73 (3): 949–959.

118 Enokizono, J., Kusuhara, H., Ose, A. et al. (2008). Quantitative investigation of the role of breast cancer resistance protein (Bcrp/Abcg2) in limiting brain and testis penetration of xenobiotic compounds. *Drug Metab. Dispos.* 36 (6): 995–1002.

119 Breedveld, P., Zelcer, N., Pluim, D. et al. (2004). Mechanism of the pharmacokinetic interaction between methotrexate and benzimidazoles: potential role for breast cancer resistance protein in clinical drug–drug interactions. *Cancer Res.* 64 (16): 5804–5811.

120 Kruijtzer, C.M., Beijnen, J.H., Rosing, H. et al. (2002). Increased oral bioavailability of topotecan in combination with the breast cancer resistance protein and P-glycoprotein inhibitor GF120918. *J. Clin. Oncol.* 20 (13): 2943–2950.

121 Kiser, J.J., Gerber, J.G., Predhomme, J.A. et al. (2008). Drug/drug interaction between lopinavir/ritonavir and rosuvastatin in healthy volunteers. *J. Acquir. Immune Defic. Syndr.* 47 (5): 570–578.

122 Bauer, M., Romermann, K., Karch, R. et al. (2016). Pilot PET study to assess the functional interplay between ABCB1 and ABCG2 at the human blood–brain barrier. *Clin. Pharmacol. Ther.* 100 (2): 131–141.

123 Lagas, J.S., Vlaming, M.L., van Tellingen, O. et al. (2006). Multidrug resistance protein 2 is an important determinant of paclitaxel pharmacokinetics. *Clin. Cancer Res.* 12 (20 Pt 1): 6125–6132.

124 Chu, X.Y., Kato, Y., Niinuma, K. et al. (1997). Multispecific organic anion transporter is responsible for the biliary excretion of the camptothecin derivative irinotecan and its metabolites in rats. *J. Pharmacol. Exp. Ther.* 281 (1): 304–314.

125 Masuda, M., I'Izuka, Y., Yamazaki, M. et al. (1997). Methotrexate is excreted into the bile by canalicular multispecific organic anion transporter in rats. *Cancer Res.* 57 (16): 3506–3510.

126 Yamasaki, M., Makino, T., Masuzawa, T. et al. (2011). Role of multidrug resistance protein 2 (MRP2) in chemoresistance and clinical outcome in oesophageal squamous cell carcinoma. *Br. J. Cancer* 104 (4): 707–713.

127 Manautou, J.E., de Waart, D.R., Kunne, C. et al. (2005). Altered disposition of acetaminophen in mice with a disruption of the *Mrp3* gene. *Hepatology* 42 (5): 1091–1098.

128 Zelcer, N., van de Wetering, K., Hillebrand, M. et al. (2005). Mice lacking multidrug resistance protein 3 show altered morphine pharmacokinetics and morphine-6-glucuronide antinociception. *Proc. Natl. Acad. Sci. U.S.A.* 102 (20): 7274–7279.

129 Imaoka, T., Kusuhara, H., Adachi, M. et al. (2007). Functional involvement of multidrug resistance-associated protein 4 (MRP4/ABCC4) in the renal elimination of the antiviral drugs adefovir and tenofovir. *Mol. Pharmacol.* 71 (2): 619–627.

130 Higgins, J.W., Bao, J.Q., Ke, A.B. et al. (2014). Utility of Oatp1a/1b-knockout and OATP1B1/3-humanized mice in the study of OATP-mediated pharmacokinetics and tissue distribution: case studies with pravastatin, atorvastatin, simvastatin, and carboxydichlorofluorescein. *Drug Metab. Dispos.* 42 (1): 182–192.

131 He, J., Yu, Y., Prasad, B. et al. (2014). PET imaging of Oatp-mediated hepatobiliary transport of [^{11}C] rosuvastatin in the rat. *Mol. Pharmaceutics* 11 (8): 2745–2754.

132 Iusuf, D., van Esch, A., Hobbs, M. et al. (2013). Murine Oatp1a/1b uptake transporters control rosuvastatin systemic exposure without affecting its apparent liver exposure. *Mol. Pharmacol.* 83 (5): 919–929.

133 Lau, Y.Y., Huang, Y., Frassetto, L., and Benet, L.Z. (2007). Effect of OATP1B transporter inhibition on the pharmacokinetics of atorvastatin in healthy volunteers. *Clin. Pharmacol. Ther.* 81 (2): 194–204.

134 Schneck, D.W., Birmingham, B.K., Zalikowski, J.A. et al. (2004). The effect of gemfibrozil on the pharmacokinetics of rosuvastatin. *Clin. Pharmacol. Ther.* 75 (5): 455–463.

135 Niemi, M., Pasanen, M.K., and Neuvonen, P.J. (2006). SLCO1B1 polymorphism and sex affect the pharmacokinetics of pravastatin but not fluvastatin. *Clin. Pharmacol. Ther.* 80 (4): 356–366.

136 Pasanen, M.K., Neuvonen, M., Neuvonen, P.J., and Niemi, M. (2006). SLCO1B1 polymorphism markedly affects the pharmacokinetics of simvastatin acid. *Pharmacogenet. Genomics* 16 (12): 873–879.

137 Johnson, M., Patel, D., Matheny, C. et al. (2017). Inhibition of intestinal OATP2B1 by the calcium receptor antagonist ronacaleret results in a significant drug-drug interaction by causing a 2-fold decrease in exposure of rosuvastatin. *Drug Metab. Dispos.* 45 (1): 27–34.

138 Eley, T., Han, Y.H., Huang, S.P. et al. (2015). Organic anion transporting polypeptide-mediated transport of, and inhibition by, asunaprevir, an inhibitor of hepatitis C virus NS3 protease. *Clin. Pharmacol. Ther.* 97 (2): 159–166.

139 Cundy, K.C., Petty, B.G., Flaherty, J. et al. (1995). Clinical pharmacokinetics of cidofovir in human immunodeficiency virus-infected patients. *Antimicrob. Agents Chemother.* 39 (6): 1247–1252.

140 Cundy, K.C., Li, Z.H., and Lee, W.A. (1996). Effect of probenecid on the distribution, metabolism, and excretion of cidofovir in rabbits. *Drug Metab. Dispos.* 24 (3): 315–321.

141 Lacy, S.A., Hitchcock, M.J., Lee, W.A. et al. (1998). Effect of oral probenecid coadministration on the chronic toxicity and pharmacokinetics of intravenous cidofovir in cynomolgus monkeys. *Toxicol. Sci.* 44 (2): 97–106.

142 Kohler, J.J., Hosseini, S.H., Green, E. et al. (2011). Tenofovir renal proximal tubular toxicity is regulated by OAT1 and MRP4 transporters. *Lab. Invest.* 91 (6): 852–858.

143 Eraly, S.A., Vallon, V., Vaughn, D.A. et al. (2006). Decreased renal organic anion secretion and plasma accumulation of endogenous organic anions in OAT1 knock-out mice. *J. Biolumin. Chemilumin.* 281 (8): 5072–5083.

144 Vanwert, A.L., Bailey, R.M., and Sweet, D.H. (2007). Organic anion transporter 3 (Oat3/Slc22a8) knockout mice exhibit altered clearance and distribution of penicillin G. *Am. J. Physiol. Renal Physiol.* 293 (4): F1332–F1341.

145 Lalezari, J.P., Drew, W.L., Glutzer, E. et al. (1995). (*S*)-1-[3-hydroxy-2-(phosphonylmethoxy) propyl]cytosine (cidofovir): results of a phase I/II study of a novel antiviral nucleotide analogue. *J. Infect. Dis.* 171 (4): 788–796.

146 Li, M., Anderson, G.D., and Wang, J. (2006). Drug–drug interactions involving membrane transporters in the human kidney. *Expert Opin. Drug Metab. Toxicol.* 2 (4): 505–532.

147 Liu, S., Beringer, P.M., Hidayat, L. et al. (2008). Probenecid, but not cystic fibrosis, alters the total and renal clearance of fexofenadine. *J. Clin. Pharmacol.* 48 (8): 957–965.

148 Yasui-Furukori, N., Uno, T., Sugawara, K., and Tateishi, T. (2005). Different effects of three transporting inhibitors, verapamil, cimetidine, and probenecid, on fexofenadine pharmacokinetics. *Clin. Pharmacol. Ther.* 77 (1): 17–23.

149 Shu, Y., Sheardown, S.A., Brown, C. et al. (2007). Effect of genetic variation in the organic cation transporter 1 (OCT1) on metformin action. *J. Clin. Invest.* 117 (5): 1422–1431.

150 Higgins, J.W., Bedwell, D.W., and Zamek-Gliszczynski, M.J. (2012). Ablation of both organic cation transporter (OCT)1 and OCT2 alters metformin pharmacokinetics but has no effect on tissue drug exposure and pharmacodynamics. *Drug Metab. Dispos.* 40 (6): 1170–1177.

151 Filipski, K.K., Mathijssen, R.H., Mikkelsen, T.S. et al. (2009). Contribution of organic cation transporter 2 (OCT2) to cisplatin-induced nephrotoxicity. *Clin. Pharmacol. Ther.* 86 (4): 396–402.

152 Tzvetkov, M.V., Matthaei, J., Pojar, S. et al. (2018). Increased systemic exposure and stronger cardiovascular and metabolic adverse reactions to fenoterol in individuals with heritable OCT1 deficiency. *Clin. Pharmacol. Ther.* 103 (5): 868–878.

153 Tzvetkov, M.V., Saadatmand, A.R., Bokelmann, K. et al. (2012). Effects of OCT1 polymorphisms on the cellular uptake, plasma concentrations and efficacy of the 5-HT(3) antagonists tropisetron and ondansetron. *Pharmacogenom. J.* 12 (1): 22–29.

154 Shu, Y., Brown, C., Castro, R.A. et al. (2008). Effect of genetic variation in the organic cation transporter 1, OCT1, on metformin pharmacokinetics. *Clin. Pharmacol. Ther.* 83 (2): 273–280.

155 Iwata, K., Aizawa, K., Kamitsu, S. et al. (2012). Effects of genetic variants in SLC22A2 organic cation transporter 2 and SLC47A1 multidrug and toxin extrusion 1 transporter on cisplatin-induced adverse events. *Clin. Exp. Nephrol.* 16 (6): 843–851.

156 Zhang, J. and Zhou, W. (2012). Ameliorative effects of SLC22A2 gene polymorphism 808 G/T and cimetidine on cisplatin-induced nephrotoxicity in Chinese cancer patients. *Food Chem. Toxicol.* 50 (7): 2289–2293.

157 Chen, Y., Li, S., Brown, C. et al. (2009). Effect of genetic variation in the organic cation transporter 2 on the renal elimination of metformin. *Pharmacogenet. Genomics* 19 (7): 497–504.

158 Song, I.S., Shin, H.J., Shim, E.J. et al. (2008). Genetic variants of the organic cation transporter 2 influence the disposition of metformin. *Clin. Pharmacol. Ther.* 84 (5): 559–562.

159 Toyama, K., Yonezawa, A., Masuda, S. et al. (2012). Loss of multidrug and toxin extrusion 1 (MATE1) is associated with metformin-induced lactic acidosis. *Br. J. Pharmacol.* 166 (3): 1183–1191.

160 Li, Q., Guo, D., Dong, Z. et al. (2013). Ondansetron can enhance cisplatin-induced nephrotoxicity via inhibition of multiple toxin and extrusion proteins (MATEs). *Toxicol. Appl. Pharmacol.* 273 (1): 100–109.

161 Nakamura, T., Yonezawa, A., Hashimoto, S. et al. (2010). Disruption of multidrug and toxin extrusion MATE1 potentiates cisplatin-induced nephrotoxicity. *Biochem. Pharmacol.* 80 (11): 1762–1767.

162 Tsuda, M., Terada, T., Mizuno, T. et al. (2009). Targeted disruption of the multidrug and toxin extrusion 1 (mate1) gene in mice reduces renal secretion of metformin. *Mol. Pharmacol.* 75 (6): 1280–1286.

163 Becker, M.L., Visser, L.E., van Schaik, R.H. et al. (2009). Genetic variation in the multidrug and toxin extrusion 1 transporter protein influences the glucose-lowering effect of metformin in patients with diabetes: a preliminary study. *Diabetes* 58 (3): 745–749.

164 Li, Q., Yang, H., Guo, D. et al. (2016). Effect of ondansetron on metformin pharmacokinetics and response in healthy subjects. *Drug Metab. Dispos.* 44 (4): 489–494.

165 Kusuhara, H., Ito, S., Kumagai, Y. et al. (2011). Effects of a MATE protein inhibitor, pyrimethamine, on the renal elimination of metformin at oral microdose and at therapeutic dose in healthy subjects. *Clin. Pharmacol. Ther.* 89 (6): 837–844.

166 Somogyi, A., McLean, A., and Heinzow, B. (1983). Cimetidine-procainamide pharmacokinetic interaction in man: evidence of competition for tubular secretion of basic drugs. *Eur. J. Clin. Pharmacol.* 25 (3): 339–345.

167 van Crugten, J., Bochner, F., Keal, J., and Somogyi, A. (1986). Selectivity of the cimetidine-induced alterations in the renal handling of organic substrates in humans. Studies with anionic, cationic and zwitterionic drugs. *J. Pharmacol. Exp. Ther.* 236 (2): 481–487.

168 Abel, S., Nichols, D.J., Brearley, C.J., and Eve, M.D. (2000). Effect of cimetidine and ranitidine on pharmacokinetics and pharmacodynamics of a single dose of dofetilide. *Br. J. Clin. Pharmacol.* 49 (1): 64–71.

169 Somogyi, A., Stockley, C., Keal, J. et al. (1987). Reduction of metformin renal tubular secretion by cimetidine in man. *Br. J. Clin. Pharmacol.* 23 (5): 545–551.

170 Lund, M., Petersen, T.S., and Dalhoff, K.P. (2017). Clinical implications of P-glycoprotein modulation in drug-drug interactions. *Drugs* 77 (8): 859–883.

171 Wessler, J.D., Grip, L.T., Mendell, J., and Giugliano, R.P. (2013). The P-glycoprotein transport system and cardiovascular drugs. *J. Am. Coll. Cardiol.* 61 (25): 2495–2502.

172 Fromm, M.F. (2003). Importance of P-glycoprotein for drug disposition in humans. *Eur. J. Clin. Invest.* 33 (Suppl 2): 6–9.

173 Elmeliegy, M., Vourvahis, M., Guo, C. et al. (2020). Effect of P-glycoprotein (P-gp) inducers on exposure of P-gp substrates: review of clinical drug-drug interaction studies. *Clin. Pharmacokinet.* 59 (6): 699–714.

174 Mao, Q. and Unadkat, J.D. (2015). Role of the breast cancer resistance protein (BCRP/ABCG2) in drug transport-an update. *AAPS J.* 17 (1): 65–82.

175 Alrefai, W.A. and Gill, R.K. (2007). Bile acid transporters: structure, function, regulation and pathophysiological implications. *Pharm. Res.* 24 (10): 1803–1823.

176 Jansen, P.L. et al. (1999). Hepatocanalicular bile salt export pump deficiency in patients with progressive familial intrahepatic cholestasis. *Gastroenterology* 117 (6): 1370–1379.

177 Konig, J., Muller, F., and Fromm, M.F. (2013). Transporters and drug–drug interactions: important determinants of drug disposition and effects. *Pharmacol. Rev.* 65 (3): 944–966.

178 Morgan, R.E., Trauner, M., van Staden, C.J. et al. (2010). Interference with bile salt export pump function is a susceptibility factor for human liver injury in drug development. *Toxicol. Sci.* 118 (2): 485–500.

179 Nies, A.T. and Keppler, D. (2007). The apical conjugate efflux pump ABCC2 (MRP2). *Pflugers Arch.* 453 (5): 643–659.

180 Belinsky, M.G., Dawson, P.A., Shchaveleva, I. et al. (2005). Analysis of the in vivo functions of Mrp3. *Mol. Pharmacol.* 68 (1): 160–168.

181 Zelcer, N., van de Wetering, K., de Waart, R. et al. (2006). Mice lacking Mrp3 (Abcc3) have normal bile salt transport, but altered hepatic transport of endogenous glucuronides. *J. Hepatol.* 44 (4): 768–775.

182 Mennone, A., Soroka, C.J., Cai, S.Y. et al. (2006). Mrp4−/− mice have an impaired cytoprotective response in obstructive cholestasis. *Hepatology* 43 (5): 1013–1021.

183 Maeda, K. (2015). Organic anion transporting polypeptide (OATP)1B1 and OATP1B3 as important regulators of the pharmacokinetics of substrate drugs. *Biol. Pharm. Bull.* 38 (2): 155–168.

184 Roth, M., Obaidat, A., and Hagenbuch, B. (2012). OATPs, OATs and OCTs: the organic anion and cation transporters of the SLCO and SLC22A gene superfamilies. *Br. J. Pharmacol.* 165 (5): 1260–1287.

185 Link, E., Parish, S., Armitage, J. et al. (2008). SLCO1B1 variants and statin-induced myopathy--a genomewide study. *N. Engl. J. Med.* 359 (8): 789–799.

186 Robertson, E.E. and Rankin, G.O. (2006). Human renal organic anion transporters: characteristics and contributions to drug and drug metabolite excretion. *Pharmacol. Ther.* 109 (3): 399–412.

187 Cihlar, T., Lin, D.C., Pritchard, J.B. et al. (1999). The antiviral nucleotide analogs cidofovir and adefovir are novel substrates for human and rat renal organic anion transporter 1. *Mol. Pharmacol.* 56 (3): 570–580.

188 Sundelin, E., Gormsen, L.C., Jensen, J.B. et al. (2017). Genetic polymorphisms in organic cation transporter 1 attenuates hepatic metformin exposure in humans. *Clin. Pharmacol. Ther.* 102 (5): 841–848.

189 Zhou, K., Donnelly, L.A., Kimber, C.H. et al. (2009). Reduced-function SLC22A1 polymorphisms encoding organic cation transporter 1 and glycemic response to metformin: a GoDARTS study. *Diabetes* 58 (6): 1434–1439.

190 Levey, A.S., Coresh, J., Balk, E. et al. (2003). National Kidney Foundation practice guidelines for chronic kidney disease: evaluation, classification, and stratification. *Ann. Intern. Med.* 139 (2): 137–147.

191 Urakami, Y., Kimura, N., Okuda, M., and Inui, K. (2004). Creatinine transport by basolateral organic cation transporter hOCT2 in the human kidney. *Pharm. Res.* 21 (6): 976–981.

192 Imamura, Y., Murayama, N., Okudaira, N. et al. (2011). Prediction of fluoroquinolone-induced elevation in serum creatinine levels: a case of drug-endogenous substance interaction involving the inhibition of renal secretion. *Clin. Pharmacol. Ther.* 89 (1): 81–88.

193 Soveri, I., Berg, U.B., Bjork, J. et al. (2014). Measuring GFR: a systematic review. *Am. J. Kidney Dis.* 64 (3): 411–424.

Part II

Challenges in Controlled Drug Delivery and Advanced Delivery Technologies

9

Advanced Drug Delivery Systems for Biologics

May Wenche Jøraholmen, Selenia Ternullo, Ann Mari Holsæter, Gøril Eide Flaten, and Nataša Škalko-Basnet

Drug Transport and Delivery Research Group, Department of Pharmacy, Faculty of Health Sciences, University of Tromsø The Arctic University of Norway, Tromsø, Norway

9.1 Introduction

Biologics are biologically originated drugs often also referred to as biologicals, biopharmaceuticals, and biological products. They can be defined as any drug produced using biotechnology, comprising peptides, proteins, antigens, antibodies, nucleic acids, and cell therapies [1]. Although antibodies and their derivatives, particularly human monoclonal antibodies (mAbs) are one of the most rapidly growing categories of targeted therapeutic agents [2], small interfering RNA (siRNA), cytokines, enzymes, and a variety of peptide drugs are gaining increased attention. Biologics are often referred to as a new field; however, the origins of biologics date back to the early insulin therapy produced from animal sources [3, 4]. Their main advantages over conventional drugs lie in their high selectivity, potent therapeutic efficacy, and limited side effects. In addition, they exhibit more predictable behavior in *in vivo* conditions [5].

Protein therapeutics offer a highly specific and rather complex set of functions not always achievable by a biomimetic approach. They exhibit a limited interference with the normal biological processes, low immunogenicity, offer an alternative to gene therapy, and are, from an industrial point of view, linked to faster clinical development and approval time as well as better patent protection [6, 7]. Biologics are prone to aggregation, deamidation, isomerization, hydrolysis, oxidation, and denaturation due to their chemical structure. To overcome these stability issues, several stabilizers have been proposed such as trehalose, dextrans, and polysorbates. The inherent physicochemical and biological instability needs to be overcome by the advanced drug delivery systems [8]. Moreover, their high molecular mass affects their permeability across biological barriers such as mucosal cell membranes, as well as skin, resulting in very limited intracellular targeting and a need for parenteral formulations [9]. Most biologics are designed and formulated for parenteral routes of administration. Intravenous, intramuscular, and subcutaneous injections are most commonly used delivery routes. Consequently, many biologics suffer from short plasma half-lives, resulting in the need for frequent administration and ultimately poor patient compliance [10]. For pharmaceutical and biotechnological companies, biologics are attractive formulation targets as an increasing number of protein-based biologics are coming off patent in the next few years opening opportunities for patented drug products and devices [11].

New technologies have been proposed to enable improved delivery of biologics including needle-free devices, nanoparticles, and smart nanomaterials [12, 13]. The introduction of the

Fundamentals of Drug Delivery, First Edition.
Edited by Heather A. E. Benson, Michael S. Roberts, Adrian C. Williams, and Xiaowen Liang.

concept of personalized medicine, has led to a faster market growth of biologics [14]. Improving pharmacokinetic profiles, especially assuring sustained/controlled drug delivery have been a focus of research. [7, 15, 16].

One of the most challenging lines in biologics product development is the targeted delivery systems for siRNA. siRNA has confirmed superiority in numerous *in vitro* studies but failed to achieve targeted, as well as intracellulary efficient, delivery in various *in vivo* studies [17]. Advanced delivery strategies could provide an improvement in the targeted delivery of RNA drugs, enabling maximized drug potency while minimizing off target toxicity and immunogenicity [18].

The formulation development strategies applied to small molecular weight (SMW) drugs cannot be readily transferred into formulating biologics, especially when considering controlled and targeted release. In addition, specific challenges in nonclinical safety testing of biologics need to be addressed and optimized [12, 19].

9.2 Considerations in Biologics Product Development

9.2.1 Challenges Specific to the Route of Administration

Most marketed biologics have been designed for the parenteral route of administration. Due to their specificity and affinity to numerous biological targets, mAbs are the fastest-growing class of new medicines. Although available on the market for over 20 years, they are still administered in an acute care setting through intravenous infusion. Recently, mAbs are gaining attention due to proven potential to serve as targeting moieties. Antibody-based bioconjugates, partnering of mAbs with biologic (protein, peptide, oligonucleotide, or another mAbs), enable more precise drug therapy. For more details on the potential of antibodies in targeted drug therapy, refer to recent review by Kennedy et al. [20]. Considering the limitations of a needle-based delivery, there is an urgent need for alternative routes of biologics administration.

9.2.2 Challenges Related to Parenteral Administration

One of the main concerns in product development is the exponential relationship between the concentrations of biologics and viscosity of the formulation; highly viscous formulations that are often required to assure a desired concentration, cannot be readily injected. In addition, the pH and osmolality of the formulation need to be considered [21]. The rapid development of nanotechnology and nanomedicine, and numerous delivery systems such as micro- and nano-particles, opened the possibilities to benefit from specially designed lipid- or polymer-based protein carriers [9]. It is worth noting that some of the polymeric particles used as protein carriers are designed to release the incorporated molecules by the diffusion-controlled mechanism, which often requires particles in the micron rather than the nano size range [22].

9.2.3 Optimization of Dosage Regimens

The pharmacokinetic limitations mentioned earlier require frequent administrations and rather larger doses for injections of biologics. Controlled, or at least sustained release formulations could provide greater therapeutic efficacy, and consequently, safety of biologics, as well as reduced frequency of administration and improved patient compliance. One approach to circumvent limitations of biologics is to chemically modify and redesign the molecule. However, this would lead to de facto formation of a new chemical entity and thus require de novo registration of a new entity. The pharmaceutical industry therefore prefers modern encapsulation technologies in product development. However, those technologies require increased manufacturing costs.

9.3 Administration Routes for Biologics Delivery

9.3.1 Parenteral Route

Sustained release biologics formulations for the parenteral routes avoid any chemical or fusion modification of the protein, while relying on subcutaneous (*sc*) or intramuscular depots. It is important to consider that the development of injectable depot formulations for small molecules and peptides cannot be directly translated into products for protein-based biologicals [16].

Some of the more promising approaches to achieve sustained/controlled delivery of biologics via parenteral routes include (i) liquid gelling or self-assembling systems, (ii) biodegradable micro- and nanoparticles based on polymers such as poly(lactic) acid (PLA) or poly(lactic-*co*-glycolic) acid (PLGA), (iii) lipid-based systems, (iv) protein microcrystals, (v) micro- and nanoparticles made of hydrophilic or amphiphilic polymers, (vi) biodegradable or nonbiodegradable but removable solid implants [5]. Formulation comprising a water-soluble triblock copolymer (PLGA–PEG–PLGA), exhibiting low viscosity at room temperature, and high viscosity when injected, provided the sustained release of protein [16]. Several other systems based on Pluronic® triblock polymers complexed with short PLA chains, or photopolymerisable hydrogels, as well as alginate- or dextran-based polysaccharides also appeared promising in the preclinical studies [23, 24]. Multiblock Pluronic copolymers linked by D-lactide and L-lactide oligomers of different spacer lengths form physically cross-linked Pluronic hydrogels with superior critical gelation concentrations and temperatures. This system allowed the prolonged release of human growth hormone (hGH) for up to 13 days in *in vitro* conditions. The release was based on the diffusion/erosion coupled mechanism and followed zero-order pharmacokinetics (PK) [23]. PLGA microspheres-based hGH formulation was the first product approved in 1999, and later removed in 2004, probably due to high costs of manufacturing [16]. The technology was later modified and revived [25]. PLGA copolymers exhibit limited compatibility with proteins [16]. A poly(ether–ester) multiblock copolymer composed of poly(ethylene glycol) terephthalate and poly(butylene terephthalate) (PEGT/PBT) has been introduced for the production of protein-containing microspheres [26]. This system exhibited superior controlled release properties, however, its biocompatibility was found to be dependent on the size and shape of the implant, as evaluated in rabbit studies. Furthermore, the microspheres were suitable for the sterilization via gamma irradiation. Insulin-containing lipid microparticles assured insulin release for more than 14 days in an *in vitro* study [22]. The glycerol triplamitate microparticles manufactured by spray-congealing, incorporated a high amount of insulin and assured its release for over 28 days in *in vitro* conditions [27]. Several more systems, based on lipids and prepared in the absence of organic solvents during the manufacturing, seem to be suitable for protein delivery [14]. However, clinical evaluation needs to be performed. Water-based particulate systems utilize poly-electrolyte-coated protein microcrystals dispersed in aqueous vehicle for injection, or hydrogels [16]. The nanoparticulate Medusa® formulation, based on the poly-L-glutamate backbone grafted with hydrophobic α-tocopherol molecules, is a colloidal dispersion of nanoparticles (10–50 nm) in water [28]. The particle dispersion forms a *sc* gel-like implant after *sc* injection, providing a depot for protein release. Some of these formulations are already in Phase II clinical trials [16]. Sodium hyaluronate microparticles produced by spray-drying technology and containing LB03002, administered once-weekly achieved sustained-release of growth hormone, passed a Phase II clinical trial conducted globally on pediatric patients [29]. Interesting technology involves solid implants designed and developed to release proteins over longer period. The Duros® delivery technology comprises sterile, nonbiodegradable single-use solid implants. This system provides controlled protein release over 3–12 months and can be removed at any time during therapy, quickly and safely. Phase I, II, and III clinical studies have proven the safety and efficacy of this novel technology [16].

9.3.2 Oral Route

For SMW drugs, oral administration is the preferable route, however, oral delivery of proteins and peptides remains one of the greatest challenges in pharmaceutical development. Despite employing the newest technologies to deliver peptides orally, only limited bioavailability could be achieved [30]. Several limiting factors need to be addressed to overcome the limited success of oral therapy, such as

(1) Harsh pH conditions.
(2) Extensive enzymatic degradation (proteolytic enzymes).
(3) Poor absorption due to the large molecular size of some of biologics.
(4) First-pass effect.

Transport across the intestinal epithelium remains the most critical barrier for successful oral drug delivery [31]. Moreover, gastrointestinal tract (GI) emptying and transit time, especially colon emptying time, can negatively influence the bioavailability of biologics, as well as inter- and intra-patient variations. In addition, peptides, unless modified, do not cross cell membranes, and exhibit very limited permeability [32]. Recently, the role of bacterial microbiota has gained deserved attention when designing and developing peptides with improved oral bioavailability [33]. The challenges of industrial manufacturing of biologics and the comprehensive overview of the current status of oral peptide delivery, including patented formulations, have been summarized by Maher et al. [34]. Although controllable drug delivery remains the ultimate goal, many of the formulations discussed in this and the following chapters are sustained delivery systems at the stage of pharmaceutical development.

A recent review [32] refers to more than 140 peptides in the clinical stage of pharmaceutical development, with several dozen already approved by the Food and Drug Administration (FDA). Only eight currently marketed oral biologics are peptides [30]. The first oral cyclosporine was formulated as a soft gelatin capsule and later improved into a self-nanoemulsifying drug delivery system (SNEDDS). SNEDDS comprise small oil droplets (less than 150 nm) which facilitate improved solubilization while the emulsifier composition increases intestinal permeability [35]. Among the oral peptides that act locally in the intestine but are not absorbed, linaclotide has been packaged in microcrystalline cellulose spheres with an outer enteric coating, formulated into a hard gelatin capsule [36]. The properties of commercially available peptides with respect to their formulation challenges have recently been summarized [30].

Although peptidase inhibition and PEGylation seem to be common strategies to enhance oral bioavailability of peptides, nanoparticles as peptide carriers recently gained increased attention [31, 37]. Nanoparticles composed of chitosan and gamma poly(glutamic acid) have been proposed for insulin delivery [38, 39]. Improvements have also been made using mucoadhesive intestinal patches [12, 40].

Solid lipid nanoparticles (SLNs) can be used to enhance oral drug delivery. TrabiOral® comprises encapsulation and conjugation with biologically active ligands to enhance their uptake [41]. Targeted delivery can also be achieved by utilizing poly(lactic acid)-*b*-poly(ethylene glycol) (PLA–PEG) block copolymer-based nanoparticles with Fc-thiol surface ligand groups for targeted oral insulin delivery [42].

"Robotic-pill" a capsule that consists of two chemical compartments filled with citric acid and sodium bicarbonate, respectively, is one of the novel technologies for improved oral delivery. When the capsule faces a pH close to neutral, the barriers between the two substances erode and allow a chemical reaction that pushes drug coated sugar-based micro-needle through the outer capsule layer to penetrate the epithelium [43]. This is a similar technology to the "needle pill" which is based on stainless steel that was gradually exposed as the pH in the GI tract wore off the out layer of the capsule [44].

Various novel oral drug delivery technologies are currently being investigated or have completed Phase I clinical studies, ranging from the capsule technology comprising bioadhesive calcium phosphate nanoparticles enteric coated with cellulose acetate phthalate for insulin delivery [45] to silica nanoparticles with an adsorbed polysaccharide and peptide or protein, suspended in oil and incorporated into enteric capsules [34]. An interesting novel technology comprises a solid dose platform based on medium chain fatty acid salts and their derivatives targeting absorption from the small intestine. It consists of three different enteric-coated formats [30, 46]. More details on novel technologies for oral delivery of biologics are available from the review by Aguirre et al. [30]. The authors provide an extensive overview of different technologies in the different stages of drug development. Interestingly, most of these technologies were developed for oral insulin delivery.

To enhance the absorption through the intestinal epithelium, various approaches have been proposed such as absorption enhancers, use of polyamines, use of chitosan to open intracellular tight junction, and aid paracellular transport of proteins and cell-penetrating peptides (CPPs) [5]. Liposomes, one of the most-studied nanosystems, have also been proposed as absorption enhancers for biologics. This includes bilosomes for oral delivery of cholera toxin B subunit for targeted oral immunization against hepatitis B [47], liposomes containing bile salts for insulin delivery [48], lectin-modified insulin-containing liposomes [49], and pectin-liposome nanocomplexes [50].

How successful will these technologies be in clinical evaluations remains to be seen. So far, the oral bioavailability for most of the proposed technologies remains limited.

9.3.3 Buccal Route

Buccal epithelium is a stratified squamous epithelium which provides direct access to the lingual, facial, and retromandibular veins, which lead into the jugular vein, thus avoiding the first-pass effect [51]. The stratified epithelium is the major barrier for successful drug absorption which has to be overcome to permit the penetration of biologics, which are often macromolecules. Among the studied dosage forms and delivery systems for buccal drug delivery, novel delivery systems such as films, wafers, gels, and nanoparticle-based systems seems to be the most promising [3]. Nanostructure-based delivery systems can be tailored to enhance the targeting and biodistribution of biologics [51]. Recent approaches in enhancing the buccal mucosal transport of peptides and proteins are summarized by Caon et al. [52].

Insulin remains the most attractive biologic for buccal delivery. The Oral-lyn®, a buccal insulin for the treatment of both type I and II diabetes, seems to be promising as an alternative insulin administration as it is already approved in selected markets [53]. The product is a micellar dispersion comprising a blend of surfactants that act both as stabilizers and permeation enhancers [53].

The dispersion is dispensed by the help of a spray actuator. Morales et al. [3] provide an extensive overview of dosage forms for buccal biologics delivery including preclinical and clinical data. The studied dosage forms and delivery systems vary from films to microneedles. From a physicochemical point of view, dissolving biologics directly in a film or hydrogel matrix can lead to their instability; therefore, using a carrier system such as nanoparticles is advisable [54–56]. Combined delivery systems, such as liposomes and microneedles, appear to be useful for vaccination [57]. Similar prospects are expected from the inkjet printing and hot-melt extrusion for buccal delivery of biologics [58, 59]. Inkjet printing could also be applied to manufacture novel delivery systems for buccal drug delivery [60]. Recent developments promise wider use of this novel technology in biologics development [61].

9.3.4 Sublingual Route

Sublingually absorbed drugs achieve faster onset of action as compared to the oral delivery; the route offers avoidance of first-pass metabolism, lower exposure to metabolizing enzymes and increased bioavailability. What is particularly promising is the possibility to control/optimize the microenvironment, such as local pH, to enhance the absorption [62]. Considering the often-large molecular size of biologics, membrane permeability within the sublingual region remains rather low, and many biologics exhibit highly hydrophilic properties, additional hindrance to absorption. However, attempts have been made to administer insulin in alginic acid nanoparticles with nicotinamide as an absorption enhancer [63].

Sublingual immunotherapy (SLIT) [64] and exenatide (GLP-1) [65] therapy seem to benefit from sublingual absorption.

9.3.5 Pulmonary Route

Significant advances in safe and efficacious aerosol formulations in the past several decades has led to increased focus on pulmonary drug delivery, particularly in respect to macromolecules, including siRNA and DNA [66]. Using the respiratory tract for targeted drug delivery offers a noninvasive alternative for both local and systemic conditions. This route provides rather fast onset and favorable PK, avoiding harsh conditions in GI tract and first-pass metabolism [67]. Pulmonary drug delivery enables enhanced local action of inhaled drugs and reduction in the required dosage, providing a decrease in side effects. The surface area available for possible drug absorption can be as large as 100 m^2 and the thinness of alveolar epithelium of as little as 0.5 μm enables efficient drug absorption. Moreover, lungs exhibit limited metabolic/enzymatic activity, as compared to the GI tract [3]. Depreter et al. [68] reviewed the challenges and potentials of inhaled proteins for either topical or systemic applications. The success of pulmonary delivery of proteins and peptides will be dependent on physiological factors (including breathing pattern and alveolar macrophages) and specific properties of biologicals (including molecular weight and limited lipophilicity) [5, 69]. The protective mucus layer covering the airway epithelium is an additional barrier to drug absorption. Determination of the optimal particle properties of aerosol-delivered nanoparticles such as the size, size distribution, surface characteristics, and drug load, is a prerequisite for successful protein-based therapy. Although the failure of Exubera® (short-acting insulin loaded within a dry powder inhaler [DPI] system) has slowed the development of delivery systems within this field, several promising pipelines might overcome the current limitations of this route (reviewed by [70]). It is important to recognize that delivery of biologics via this route bears different challenges as compared to conventional drugs for asthma treatment; most of the biologics have narrower therapeutic indexes and are rather expensive [3]. Biologics also require the inclusion of stabilizers in their formulations which need to be optimized during preformulation stage of product development [71]. The size of the carriers should be optimized to assure that the particles/drug can reach the deep lungs [72]. Moreover, the particles need to avoid lung clearance to assure the required residence time. How fast the particle is cleared from the lungs is directly determined by its size and site of deposition, and also by the breathing patterns, the specific anatomy of the individual's airways and the disease state [72]. Thus, it is crucial to optimize each individual formulation destined for pulmonary delivery.

When considering systemic absorption, macromolecules, due to their large size, face additional challenges other than common mucociliary clearance including macrophages attack and enzymatic degradation [3]. It is interesting to note that optimal pulmonary delivery has been proposed to be by nonspherical particles (worm-like) which can inhibit the phagocytosis of the particles by minimizing the size-normalized curvature of particles, at least in cell experiments [73]. Pulmonary delivery can be improved by various means; for example by the inclusion of protease inhibitors (including bacitracin and Na-glycocholate) in the formulations of biologics [74] or approaches such as enhanced transport-mediated transcytosis [66]. Morales et al. [75] summarized the most commonly used surfactants in pulmonary formulations which are acting as absorption enhancers by mechanisms which are not fully understood up to now.

Several inhaled biologics have reached clinical studies and the market and are extensively summarized by Morales et al. [3].

Pulmozyme®, a recombinant human deoxyribonuclease I (rhDNase) used in cystic fibrosis treatment, was approved early in 1993; alpha-1-antitrypsin for the treatment of emphysema and cystic fibrosis, and hGH, and interferon gamma for idiopathic pulmonary fibrosis and mycobacterial disease are among the most promising pipelines [3, 76].

9.3.5.1 Additional Concerns in Pulmonary Delivery

Dosage forms and delivery systems destined for pulmonary delivery require optimal delivery devices to assure enhanced bioavailability. The devices are generally categorized as pressurized meter dose inhalers (pMDIs), nebulizers, and DPIs. Currently, no commercially available biologics are delivered via pMDI. One product, Pulmozyme is commercially available to be administered via nebulizer and nebulized alpha-1-antitrypsin is in phase II/III clinical studies ([77]; https://clinicaltrials.gov/ct2/show/NCT01217671). Two DPI insulin products have passed FDA requirements so far, Afrezza® and Exubera [3].

The respiratory system remains a challenging target for effective gene therapy due to issues of nanoparticles stability in physiological fluids and their ability to overcome biological barriers found in the lungs. In addition the safety profile needs to be confirmed. For example, the delivery of DNA via poly(ß-amino esters) (PBAE) NPs decorated with PEG (mucus-penetrating) for intratracheal delivery [78] and viral-mediated pulmonary targeted gene delivery [79] have been reported.

9.3.6 Intranasal Route

The nasal route offers several advantages over the parenteral route with respect to the delivery of protein and peptide-based biopharmaceuticals. Although this route remains very promising for SMW drugs, for rather large molecules such as proteins, several limitations need to be overcome in order to fully utilize this route. The mucosal and enzymatic barriers, rapid clearance and low residence time are just some of the obstacles remaining to be overcome [80]. The absorption promoters such as surfactants, enzymatic inhibitors, or multifunctional polymers have been proposed to improve bioavailability. The use of mucoadhesive delivery systems such as bioadhesive microspheres, chitosan-based delivery systems and similar, have not, up to now, fulfilled expectations. Nasal route is an attractive means to deliver biologics to brain [81].

A lipid-based colloidal delivery system, Pheroid technology, is a new strategy for nasal delivery of biologics. Pheroid technology consists of a unique submicron emulsion based on the essential fatty acids and thus has potential to avoid safety issues related to polymer use [70]. Pheroid vesicles and microsponges have been studied for nasal delivery of human growth factor (hGF) [82] and calcitonin [83]. Pheroid vesicles and microsponges have been studied for nasal delivery of hGF and showed promising absorption enhancement in a rat model, although not yet superior to the chitosan-based absorption enhancer [82]. Similar observations were made when delivering calcitonin both via nasal and oral routes. Although improvements were noticeable, chitosan-based enhancers were found to be superior, particularly in the nasal route [83].

Moreover, nasal delivery is an attractive route for brain targeting, overcoming the blood–brain barrier [84]. Intranasal delivery of large molecular weight biologics (proteins, gene vectors to name some) is an attractive means to treat a variety of CNS disorders such as multiple sclerosis, Parkinson's disease, epilepsy, and Alzheimer's disease [85].

9.3.7 Trans(dermal) Delivery

Successful transdermal drug delivery can assure continuous drug delivery over a prolonged period of time (even days) providing controllable drug levels. This is of a great importance for potent drugs, drugs with short half-life, and drugs that might not survive the GI tract and first-pass metabolism [3]. The transdermal route is also a noninvasive route compared to the parenteral routes. So far, the FDA has approved two dozen drugs as transdermal patches [3], however, the full potential of the transdermal route for biologics remains to be exploited.

The transdermal route of drug delivery relies on passive diffusion of molecules through the lipid-rich *stratum corneum* and is, therefore, directly dependent on the drug's properties such as its size, physicochemical characteristics, and susceptibility to metabolism by skin enzymes [5]. *Stratum corneum*, often described as a "brick-and-mortar" structure (described in Chapter 6), represents a tight barrier for drug penetration. The drugs that can penetrate the "mortar" part, extracellular lipids, have higher chances to be formulated as transdermal delivery systems, however, peptides and proteins have a limited potential to penetrate due to their physicochemical properties [3]. The transdermal delivery of peptide and protein drugs is summarized in an extensive review by Schuetz et al. [86] and more recently [3, 87, 88].

Some of the promising strategies proposed to overcome the limitations of transdermal delivery of peptides and proteins were summarized by Morales et al. [3]. Penetration of a drug can be optimized by acting on its molecular structure, its interaction with the skin microstructures or the penetration pathways. Alternatively, the *stratum corneum* could be punctured or partially removed to permit drug penetration [3]. Utilizing membrane-penetrating peptides can also lead to improved skin penetration of drugs [89–92].

To increase the skin permeability, several chemical and physical methods acting on the disruption of the *stratum corneum* structure (barrier) have been proposed. These chemical enhancers need to be designed to be a compromise between potential skin irritancy and increased skin permeability [93] and are, up to now, not successfully enhancing the delivery of biologics. The use of short, high-voltage pulses to the skin causes skin electroporation, which leads to increased skin permeability [94]. Peptides can penetrate via this rearranged *stratum corneum* structure as shown by Sen et al. [95]. Moreover, the electric pulse creates an electrophoretic driving force allowing the charged molecules to penetrate [96]. Similarly, ultrasound can be applied to generate and oscillate the cavitation bubbles that, upon collapsing, disrupt the *stratum corneum* structure [97]. Insulin [98], heparin [99], tetanus toxoid [100], and siRNA [101] are some of the biologics successfully tested for ultrasound-assisted skin delivery.

Enhancement in skin penetration can be achieved by utilizing chemical enhancers (trypsin) or nanoparticles (gold nanoparticles); however, those approaches are at very early stages of development [102, 103]. Iontophoresis (continuous supply of low voltage) provides electrically driven transport and has been tested for leuprolide [104] and insulin [105] skin delivery.

A rather challenging approach to increase the skin permeability is the creation of micron-scale pores in the *stratum corneum*, assuring the delivery of larger proteins with minimum skin damage [106]. One method to achieve this is to use thermal ablation that causes micron scale "craters" in the *stratum corneum*. The heat is generated in a controllable manner by laser or electrical resistive heating. This approach has been evaluated for hGH [107] and interferon alpha-2b (*IFN*α-2b) delivery [108]. Mechanical microscraping has been proposed for vaccine delivery [109, 110]. Microdermabrasion, a sandblasting of the skin surface, was tested as mean for insulin delivery [111].

Recently, much attention has been focused on the use of microneedles to deliver macromolecules via the skin [112]. Microneedles are painless, bloodless, and cause minimum or no skin irritation and can be self-administered [113–115]. Skin targeting enables rapid uptake of molecules into the circulation and increased immunogenicity of vaccines [116–118]. Currently, the clinical trials using microneedle patches for delivery of parathyroid hormone [119], insulin [120], and influenza vaccine [121] seem to be promising.

Although several delivery systems have been proposed to serve as optimal carriers for percutaneous delivery of growth factors [122], their efficacy in *in vivo* conditions has yet to be confirmed.

A lyophilized hydrogel patch system has been developed for microneedle-mediated insulin delivery. This system demonstrated provision of an appropriate reservoir for sustained release of insulin via microneedle-mediated transdermal delivery in animal studies in diabetic rats [123].

9.3.7.1 Gene Delivery

Carbon nanotubes functionalized noncovalently with succinated polyethyleimine have been employed for transdermal delivery of siRNA. *In vivo* delivery studies in a murine melanoma model have confirmed successful delivery of siRNA through the skin and into the tumor. Moreover, siRNA to BRAF, an important gene for regulating cell growth and proliferation, incorporated in the carbon nanotubes has been tested in melanoma-bearing mice. A significant attenuation of tumor growth was observed 25 days after the treatment [124].

Chen et al. [125] also reported enhancement in siRNA skin penetration. siRNA was incorporated in ethosomes composed of skin permeating and cell entering-peptide (SPACE-peptide) in combination with 1,2-dioleoyl-3-trimethylammonium-propane (DOTAP). Skin penetration studies were performed both *in vitro* and *in vivo* in full thickness pigskin and in female BALB/C mice, respectively. The formulation enhanced the siRNA penetration in the epidermis to a higher extent compared to siRNA aqueous solution. Higher uptake of siRNA in human epidermal keratinocytes was also observed for the siRNA-loaded SPACE-peptide-decorated cationic ethosomes. To achieve a synergistic effect in treating skin inflammation, siRNA-based therapy has also been combined with other drugs. Desai et al. [126] designed a novel nanocarrier system for the co-delivery of siRNA against tumor necrosis factor-α (*TNF*α) and capsaicin. The nanocarrier consisted of biodegradable lipid-polymer hybrid NPs (size around 150 nm), containing cationic amphiphiles with cyclic pyrrolidinium head group and PLGA. The novel nanocarrier enhanced skin penetration of both drugs up to the dermis layer. Moreover, the system was noncytotoxic to human cells. The efficacy of the formulation was tested in a psoriatic plaque-like mouse model where the formylation assured a significant reduction in the expression of various inflammatory cytokines.

Gold nanoparticles densely conjugated with siRNA are able to penetrate both mouse skin and human epidermis and assured a 100% uptake in human keratinocytes within two hours [127]. Moreover, the nanosystem exhibited high efficacy in abolishing epidermal growth factor receptor (EGFR) protein expression in both mouse and human skin, with low immunogenicity and toxicity.

A combination of nanotechnology and iontophoresis has been used to deliver antisense oligonucleotides (ASOs) [128]. The generation 4 polyamidoamine dendrimer was selected as a nanocarrier for delivery of B-cell lymphoma/leukemia 2 (*BCL2*) ASO for treatment of skin cancer. *In vitro* skin penetration studies on pigskin showed that anodal iontophoresis in combination with dendrimers successfully delivered *BCL2* ASO in viable epidermis. Moreover, the dendrimer enhanced the uptake of ASO in human epidermoid carcinoma cells and caused gene silencing *in vivo*.

SECosomes have also been employed for transdermal delivery of siRNA. Geusens et al. [129] have developed an ultraflexible nanocarrier consisting of a cationic lipid (DOTAP), an edge activator (sodium cholate), a stabilizer (cholesterol), and 30% ethanol. SECosomes assured siRNA penetration into the epidermis in *in vitro* skin penetration studies using human skin. Moreover, the SECosomes efficiently delivered siRNA in cultured primary melanocytes and keratinocytes. More recently, Bracke et al. [130] incorporated siRNA in SECosomes to target human beta defensin to treat psoriasis. A significant improvement in the psoriatic phenotype and normalization of the skin architecture were observed in a bioengineered skin-humanized mouse model. Cationic elastic liposomes have also been used to improve the transdermal delivery of antisense oligodeoxynucleotides for interleukin-13 mRNA for the treatment of atopic dermatitis into the mouse [131]. A significant reduction in the skin thickness and inflammatory cells was observed. Similarly, cationic elastic liposomes were used as nanocarriers for interleukin-4 (IL-4) encoding plasmid DNA for the treatment of psoriasis [132]. The nanoformulation assured an enhanced expression of IL-4 reducing the hyperplastic and inflamed skin vessels. siRNA has also been incorporated in edge-activated DOTAP liposomes for melanoma treatment [133]. The nanocarrier deposited siRNA into the upper dermis of human skin which could internalize it within the melanoma cells knocking down target protein expression and inhibiting melanoma cell proliferation. Cationic lipid nanoparticles also show potential to deliver plasmid DNA to the skin, mainly due to the charge-mediated interaction between the plasmid DNA-loaded nanoparticles and the skin [134].

Desmet et al. [135] applied elastic liposomes, SECosomes, in delivery of RNA interference (RNAi) into the skin. To improve the penetration and transfection ability, the combination of DOTAP, dioleoylphosphatidylethanolamine (DOPE), cholesterol, and ethanol was applied in formulating SECosomes. SECosomes deliver RNAi into the epidermis, avoiding the dermis or circulatory system. The transfection efficiency was confirmed in a psoriatic tissue model, where downregulation of the psoriasis marker human beta-defensin 2 was mediated by the liposomal formulation.

Recently, the so-called "gemini surfactants" (*N*,*N*-bis(dimethylalkyl)-α,ω-alkanediammonium surfactants) have gained interest for therapeutic gene delivery. These surfactants can condense plasmid DNA and self-assemble into small particles with a net positive charge. Lipid nanoparticles consisting of DOPE and cationic gemini surfactant and incorporating tdTomato plasmid DNA exhibited high transfection efficiency in a mouse epidermal keratinocyte cell line, PAM-212 [136].

9.3.7.2 Vaccine Delivery

Dermal vaccination is of particular interest as the skin contains a network of immune cells (Langerhans and dendritic cells) and is highly responsive to antigens. Nanodelivery systems have shown to be efficient for vaccine delivery through the skin [137].

Rosada et al. [138] proposed elastic phosphatidylcholine-based liposomes as a gene-based vaccine for tuberculosis, whereas Chopra and Cevc [139] suggested the same type of vesicles for tetanus vaccine. More recently, transcutaneous immune-boosts with the subunit antigen Ag85a (immune-dominant antigens of *Mycobacterium tuberculosis*), delivered by deformable vesicles, activated leucocytes, and promoted T cell proliferation in *Bacillus* Calmette–Guerin-primed mice [140]. Chen et al. [141] confirmed the adjuvant effect of porcine interleukin 2 (pIL-2) when encapsulated in SLNs. SLNs, consisting of hydrogenated castor oil and polylactide-*co*-glycolide, exhibited a sustained release of pIL-2 and significantly enhanced the foot and mouth disease virus-specific antibody level as compared to recombinant pIL-2 alone in an *in vivo* study in mice [141].

Nanoemulsions and microemulsions have also shown promising results in animal studies. Nanoemulsion incorporating a fluorescein isothiocyanate-bovine serum albumin conjugate (FITC-BSA) as vaccine model exhibited higher mouse skin permeation compared to both FITC-BSA in solution and FITC-BSA in surfactants excipients [142]. Stein et al. [143] have developed an imiquimod-containing emulsion gel (IMI-Gel) in which the crystalline imiquimod was nano-dispersed in a hydrophilic polyacrylate gel containing a lipophilic dispersed phase. The IMI-Gel was superior to subcutaneous and oral route administration and as effective as the commercial product, Aldara®, in a therapeutic tumor vaccination model, indicating the induction of high-quality memory T cell responses.

Nanocarriers have also been combined with microneedles to transport vaccine through the skin [144]. Poly(lactide-*co*-glycolide) microneedle arrays were coated with multilayer films via layer-by-layer assembly of a biodegradable cationic PBAE and negatively charged interbilayer-cross-linked multilamellar lipid vesicles. Protein antigen was incorporated in the system and the formulation was deposited in the skin mice via microneedle application. Confocal images revealed that the nanocapsules were phagocytosed by the skin immune cells. Moreover, the microneedle-nanocapsules combination promoted higher antigen-specific humoral immune responses compared to the orally administered antigen or antigen-loaded nanocapsules via intradermal injection.

9.3.7.3 Protein Delivery

Nanosystems have also been employed for transdermal delivery of proteins, which, due to their size and hydrophilicity, do not easily penetrate the skin due to the lipophilic *stratum corneum* barrier. Choi et al. [145] encapsulated proteins, such as insulin, bovin serum albumin (BSA), and β-galactosidase, in nanogels (70 nm) composed of Pluronic F127 crosslinking conjugated with chitosan. Full thickness human skin penetration studies revealed that the loaded proteins penetrated the skin to a higher extent than the physical mixture of protein and chitosan. Moreover, the proteins remained biologically active as confirmed by enzymatic activity of β-galactosidase.

Transdermal delivery of insulin using gold nanorods in combination with near-infrared light irradiation was confirmed in an *in vivo* mice model [146]. The nanorods enhanced the permeation of insulin through the mouse skin and decreased blood glucose levels in the diabetic mice. Recently, Yu et al. [147] developed a microneedle-array containing glucose-responsive vesicles as a "smart" patch for transdermal delivery of insulin. The hypoxia-sensitive hyaluronic acid vesicles incorporating insulin and glucose oxidase enzyme loaded in a microneedle array patch effectively regulated blood glucose levels in a diabetic mouse model.

9.3.8 Dermal Delivery of Growth Hormones

Dermal delivery of proteins is an attractive mode to deliver growth hormones into the wound and enhance wound healing. Table 9.1 provides an overview of representative nanosystems employed for delivery of growth factors to treat local skin diseases, particularly wounds.

PLGA-alginate microspheres with recombinant epidermal growth factor (rhEGF) to treat chronic wounds were developed by Gainza et al. [148]. The polymeric-based microspheres

Table 9.1 Overview of drug delivery systems (DDSs) incorporating growth factors (GFs) for local skin therapy.

DDSs	GFs	Outcomes	References
Polymeric micro- and nano-particles (NPs)			
PLGA-alginate microspheres	rhEGF	Improved wound healing in rats: reduction of wound area, re-epithelization, and resolution of inflammatory processes.	[148]
PLGA microspheres	PDGF-BB and chlorexidine	Decreased levels of infection and increased wound healing in rats.	[149]
PLGA NPs	VEGF	Induced re-epithelization, angiogenesis, and formation of granulation tissue in mouse full thickness excision model.	[150]
Chitosan NPs in fibrin gel	rhEGF	Sustained *in vitro* release of rhEGF; rhEGF bioactivity maintained for up to 7 days.	[151]
Aggrecan-mimetic NPs	FGF-2	Stable FGF2 activity for more than three weeks *in vitro*.	[152]
Fibrin NPs in chitosan-hyaluronic acid sponge	VEGF	Sustained release of VEGF for up to seven days *in vitro*. The formulation induced angiogenesis *in vitro*.	[153]
PLA NPs in hydrogel	EGF and curcumin	Enhanced wound closure with increased granulation tissue formation, collagen deposition, and angiogenesis in rat model.	[154]
Xanthan-galactomannan/ DODAB liposomes	EGF	Fivefold higher sustained release of EGF than one-layer liposomes.	[155]
Lipid-based nanoparticles			
SLN (precirol) and NLC (precirol and mygliol)	rhEGF	Higher bioactivity than free rhEGF. Improved wound closure, re-epithelisation grade and suppression of the inflammatory process in mice full thickness wound model. rhEGF-NLC: improvement of wound closure, percentage of healed wounds and wound healing quality in porcine full-thickness excisional wound model.	[156]; [157]
SLN and NLC in semi-solid hydrogels (Noveon AA-1 or Pluronic F127) or fibrin-based solid scaffolds	rhEGF	Both nanocarriers remained in upper skin layers for at least 48 h postapplication onto pigskin *in vitro*; no penetration to deeper layers. rhEGF release from the fibrin-based scaffolds was less than that from the reconstituted NP suspensions.	[158]
Conventional liposomes (soybean lecithin and cholesterol)	bFGF	Fast tissue collagen generation and dermal cell proliferation expression in *in vivo* studies in rats with second-degree burns.	[159]
Conventional liposomes (hydrogenated lecithin)	rhEGF	Enhanced permeation of rhEGF in rat skin.	[122]
Cationic elastic liposomes (hydrogenated phosphatidylcholine-cholesterol-DOTAP)	EGF, IGF, and PDGF-A	Accelerated and synergistic wound closure in diabetic mouse model.	[160]

Table 9.1 (Continued)

DDSs	GFs	Outcomes	References
Nanofibers			
PELA nanofibers	bFGF	Sustained release of bFGF. Higher wound recovery rate in diabetic rats.	[161]
PLLCL and gelatin nanofibers	Multiple EIF: EGF, insulin, hydrocortisone, retinoic acid	More sustained release of the drugs. Formulation was bioactive *in vitro*.	[162]
Chitosan and PEO nanofibers	VEGF and PDGF-BB-loaded PLGA nanoparticles	Biphasic release of both growth factors; fast release of VEGF over 24 h, sustained release of PDGF-BB. Accelerated wound healing in *in vivo* rat studies.	[163]
Chitosan and PEO nanofibers	PRP	Enhanced proliferation of human keratinocytes and dermal fibroblasts.	[164]
PLGA nanofibers	rhEGF and *Aloe vera*	Improvement in wound closure and re-epithelisation in mice.	[165]
PLGA nanofibers	VEGF and TGF-β3	Enhanced bioactivity of GFs. Reduction of necrosis and enhanced blood perfusion in skin flap areas in *in vivo* mouse skin flap model.	[166]

Abbreviations: bFGF, basic fibroblast growth factor; DDSs, drug delivery systems; DODAB, dioctadecyldimethylammonium bromide; DOTAP, 1,2-dioleoyl-3-trimethylammonium-propane; EGF, epidermal growth factor; EIF, epidermal induction factors; FGF-2, fibroblast growth factor-2; GFs, growth factors; IGF, insulin-like growth factor; NLC, nanostructured lipid carriers; NPs, nanoparticles; PDGF-A, platelet-derived growth factor-A; PDGF-BB, platelet-derived growth factor-BB; PELA, poly(ethylene glycol)-poly(DL-lactide); PEO, poly(ethylene oxide); PLA, polylactic acid; PLGA, poly(lactic-*co*-glycolic acid); PLLCL, poly(L-lactic acid)-*co*-poly-(ε-caprolactone); PRP, platelet-rich plasma; rhEGF, recombinant epidermal growth factor; SLN, solid lipid nanoparticles; TGF-β3, transforming growth factor beta 3; VEGF, vascular endothelial growth factor.

retained the biological activity of rhEGF and assured its sustained release on the wound surface. The *in vivo* testing in a full thickness wound rat model revealed decreased wound area and a complete re-epithelization and resolution of the inflammatory process within 11 days. PLGA microspheres were also used for dual delivery of platelet-derived growth factor (PDGF)-BB and chlorhexidine. Wounds treated with the dual delivery system showed increased healing and decreased levels of infection in a rat model [149]. Chereddy et al. [150] developed PLGA-based nanoparticles for the delivery of vascular endothelial growth factor (VEGF). *In vivo* wound healing in a mouse model showed the nanoparticle-based formulation assured significant granulation tissue formation with higher collagen content and angiogenesis compared with untreated and empty nanoparticles.

The rhEGF-loaded chitosan nanoparticles incorporated in a fibrin gel matrix were prepared as a sustained drug delivery system [151]. Its bioactivity was maintained for at least seven days.

Place et al. [152] developed glycosaminoglycan nanoparticles with fibroblast growth factor (FGF-2). This polymer was used to mimic aggrecan, which regulates growth factor–receptor interactions at the cell surface. The FGF-2 bioactivity could be preserved for more than three weeks.

VEGF in fibrin nanoparticles was incorporated in a chitosan-hyaluronic acid sponge as a wound dressing for diabetic wounds [153]. *In vitro* release studies revealed an initial burst release of VEGF followed by a sustained release up to seven days. The formulation overcame the problem of short half-life of the growth factor, which usually requires high doses to achieve therapeutic effects and frequent administrations which reduced patient compliance. The formulation was also found to be effective in inducing angiogenesis in human dermal fibroblasts and in human umbilical vein endothelial cells.

Co-encapsulation of epidermal growth factor (EGF) and curcumin in PLA NPs and loaded in hydrogel was recently suggested by Li et al. [154] as a supportive matrix for the regenerative tissue and a sustained drug depot for both EGF and curcumin. The formulation enhanced wound closure by increasing granulation tissue formation, collagen deposition, and angiogenesis in rat model. Sustained release of EGF was also obtained from liposomes covered by a layer-by-layer structure composed by xanthan and galactomannan polymers overcoming the low stability of the drug [155].

The rhEGF has been encapsulated in SLNs and nanostructured lipid carriers (NLCs) by Gainza et al. [156]. Proliferation assays on Balb/c 3T3 fibroblasts, human skin fibroblasts, and human keratinocytes confirmed increased mitogenic capability of all tested cell lines. *In vivo* studies in the full thickness wound model in db/db mice revealed improved wound closure, reduction of the inflammatory response and good re-epithelization. The effectiveness of rhEGF-loaded lipid NPs applied topically (10 and 20 μg of rhEGF) was comparable to a single intralesional administration of 75 μg of rhEGF-loaded PLGA and alginate microspheres [158], where both NPs were formulated as wound dressings comprising either hydrogels or fibrin-based solid scaffolds. *In vitro* pigskin penetration studies confirmed a sustained release of rhEGF and the ability of NPs to act as drug-depot [157].

Basic fibroblast growth factor (bFGF) encapsulated in lecithin/cholesterol liposomes evaluated in rats with second-degree burns exhibited fast tissue collagen generation and dermal cell proliferation [159].

Cationic elastic liposomes have also been proposed as delivery system for growth factors (GFs), namely EGF, insulin-like growth factor (IGF), and PDGF, genetically conjugated with a low-molecular-weight protamine to increase skin permeability. The proposed formulations accelerated wound closure in a diabetic mouse model, assuring both rapid and prolonged effects [160].

Nanofibers have also shown promise in delivery of GFs for wounds healing. Yang et al. [161] adopted emulsion electrospinning to imbed bFGF into poly(ethylene glycol)-poly(DL-lactide) (PELA) nanofibers. The nanofibers assured an initial burst release of bFGF followed by a gradual release over four weeks. *In vivo* evaluation of skin regeneration in diabetic rats revealed a significantly higher wound recovery compared to the free bFGF.

Jin et al. [162] encapsulated multiple epidermal induction factors (EGF, insulin, hydrocortisone, and retinoic acid) in core-shell nanofibers composed of poly(L-lactic acid)-*co*-poly-(ε-caprolactone) (PLLCL) and gelatin. The nanofibers allowed a more sustained drug release, which could promote the adipose-derived stem cell differentiation. Xie et al. [163] incorporated two GFs, namely VEGF and PDGF-BB, in chitosan/poly(ethylene oxide) nanofibers. PDGF-BB was first incorporated in PLGA NPs to ensure a slower release of the drug compared to VEGF. *In vitro* studies revealed a burst release for VEGF, while a sustained release was observed for PDGF-BB. VEGF helped angiogenesis in an early stage of wound healing, whereas PDGF-BB improved re-epithelization, collagen deposition, and tissue remodeling. Bertoncelj et al. [164] used the same nanofiber composition to incorporate platelet-rich plasma (PRP), a natural source of growth factors. The formulation significantly induced proliferation of human keratinocytes and dermal fibroblasts. Recently, PLGA nanofibers containing rhEGF and *Aloe vera*, have been developed by Garcia-Orue et al. [165]. The combination of *A. vera*, which possesses antibacterial properties, and nanofibrous dressings, which provide a physical barrier toward the entry of microorganisms, prevented bacterial growth. Moreover, *in vivo* studies in a full thickness wound mice model revealed an improvement in wound healing.

PLGA nanofibers containing VEGF and transforming growth factor beta3 (TGF-β3) improved neovascularization and enhanced blood perfusion in a mouse skin flap [166].

Other biologicals have been incorporated into various nanosystems for treatment of skin diseases. PLGA nanoparticles encapsulating LL37, an endogeneous human host defense peptide that modulates wound healing and angiogenesis, improved wound healing by assuring higher re-epithelialization, granulation tissue formation and immunomodulation [167]. Recently, LL37 was incorporated in NLC [168]. Its encapsulation in NLC did not affect the bioactivity. Moreover, the nanoformulation was found effective in improving wound healing in mice.

Platelet lysate containing porous silicon microparticles have been used to enhance cell proliferation in wound healing [169]. Ex vivo studies on human skin revealed an enhancement in the proliferation and regeneration process of wound healing.

Hormone melatonin has been loaded into lecithin/chitosan nanoparticles to treat wounds [170]. Lecithin/chitosan nanoparticles prolonged the release of melatonin and a *in vitro* scratch wound healing assay on human keratinocytes showed that the formulation promoted re-epithelialization.

9.3.9 Vaginal Route

The vaginal route of drug administration is mainly considered as an alternative to oral therapy in the treatment of local vaginal infections and reproductive conditions. However, the rich blood supply and large surface area of the vagina could be utilized for systemic drug delivery. Progress in the development of vaginal drug delivery includes both local and systemic therapy with a wide range of compounds, including peptides and proteins [171]. The physiology of the vagina must be considered in the development and optimization of vaginal delivery systems. The specificities such as anatomy, presence of mucus and the changes in vaginal fluid, pH, and the thickness of epithelium influence the efficacy of the applied drugs and delivery systems [172, 173]. Increased vaginal pH due to menopause or the presence of semen, menstrual blood, or disease may affect the solubility and stability of the drug. The composition, volume, and viscosity of vaginal fluids are influenced by the age, menstrual cycle, and sexual arousal, thus affecting the drug release pattern. The distribution, penetration, and residence time of vaginal delivery systems are greatly affected by the mucus. The surface properties of delivery systems, especially nanoparticles, and their ability to involve in either mucoadhesion or mucuspenetration can classify nanoparticles as drug delivery systems in three categories [174], namely;

(i) conventional (nonmucoadhesive),
(ii) mucoadhesive, and
(iii) mucuspenetrating (PEGylated) delivery systems.

To achieve an efficient delivery to mucosal tissue and deeper epithelium, the drug needs to overcome the mucosal barrier. The surface-modified nanoparticles have been proposed for this aim [175–178].

Most drugs applied vaginally are low molecular weight drugs. However, more recently this particular site is attaining growing attention as a route for the delivery of biologics [179]. Primarily, lipid-based nanoparticles, such as liposomes, and PLGA NPs have been studied as vaginal drug delivery systems for biologics (Table 9.2).

The ability of liposomes to protect drug from degradation in the presence of vaginal fluids makes them suitable for vaginal application of biologics. Moreover, liposomes offer a sustained and controlled release of the entrapped drug at the vaginal site [174]. Foldvari and Moreland introduced liposomes as a vaginal drug delivery system 20 years ago, by clinically evaluating the

Table 9.2 Advanced drug delivery systems investigated for vaginal administration of biologics.

Delivery system	Drug	Target/outcome	References
Lipid-based nanoparticles			
Conventional liposomes	IFN α-2b	Preliminary clinical testing, vaginal treatment of HPV in female patient. Cervical lesions completely resolved, viral load in vaginal tissue decreased.	[180]
PEGylated liposomes	IFN α-2b	*In vitro* mucoadhesive studies, ex vivo penetration studies on sheep vaginal tissue. Confirmation of mucoresistant properties of PEGylated liposomes, enhanced IFN α-2b tissue penetration.	[181]
PEGylated lipoplex-entrapped alginate scaffold system	siRNA	*In vitro* and *in vivo* release and cell uptake. Controlled siRNA release *in vitro* and *in vivo*. Increased uptake in vaginal tissue, enhanced knockdown of targeted genes in vagina of mice.	[182]
Lyophilized liposomal gel-rods	CN54gp140	Liposomal gel-based formulation development. Mucoadhesive, mechanical, and rheological properties of reformed gels have potential for intravaginal delivery of proteins.	[183]
Polymeric nanoparticles			
PLGA nanoparticles	siRNA	*In vitro* cell uptake, *in vivo* gene silencing. Sustained release, efficient cell uptake *in vitro*. Gene knockdown throughout vaginal tract of mice for at least 2 wk. Nanoparticles penetrated deep into the epithelial tissue.	[184]
PLGA nanoparticles	siRNA	*In vitro* release, *in vivo* efficacy in mice. Confirmed controlled release of siRNA. Prevention of lethal intravaginal HSV-2 infection in mice and increased survival for up to 4 wk.	[185]
PLGA nanoparticles	PSC-RANTES	*In vitro* release, *ex vivo* penetration using human vaginal tissue. Sustained release and a fivefold increase in tissue uptake. Enhanced tissue permeation for the PSC-RANTES PLGA nanoparticles.	[186]

Abbreviations: HPV, human papillomavirus; HSV, herpes simplex virus; IFN α-2b, interferon alpha 2b; PEG, polyethylene glycol; PLGA, poly lactic-co-glycolic acid; PSC-RANTES, engineered amino terminus-modified analog of the chemokine RANTES (Regulated upon Activation, Normal T Expressed and Secreted); siRNA, small interfering RNA.

applicability of IFN α-2b-containing liposomes in the topical therapy of genital papillomavirus infections [180]. Their preliminary clinical testing showed that vaginal administration of liposomal IFN α-2b resulted in complete resolution of cervical lesions in a female patient at the end of treatment. Moreover, a reduced viral load in vaginal tissue was confirmed after the therapy. In a later study, Foldvari et al. investigated biphasic vesicles for topical application of IFN α for the treatment of external genital warts; the obtained results were equally successful [187].

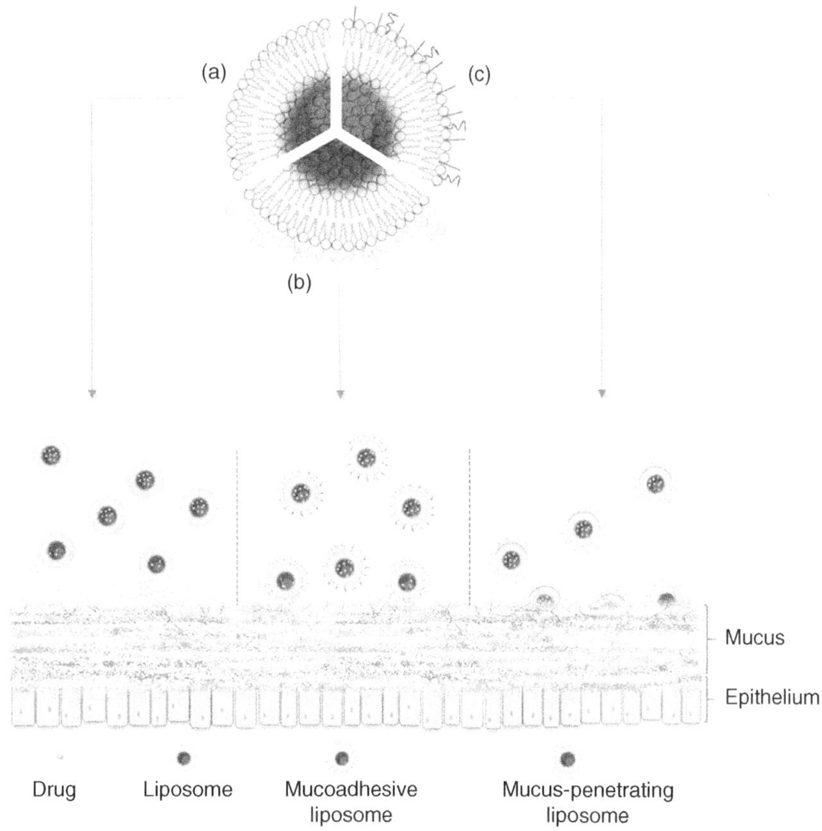

Figure 9.1 The effect of liposomal surface modification on drug delivery to mucosal tissue. (a) Plain, nonmodified liposomes, deliver drugs to the mucosa; (b) Mucoadhesive polymer-coated liposomes (e.g. chitosan-coated liposomes) bind to mucosa and assure prolonged residence time; (c) Mucus-penetrating liposomes (e.g. PEGylated liposomes) enable mucus-penetration and are a promising strategy to assure drug delivery in the close proximity to epithelial layer.

Surface modifications of liposomes or their incorporation in suitable hydrogel systems have been prominent approaches in improving drug delivery at mucosal sites; it was agreed that mucoadhesive polymers prolong the residence time at vaginal site [174]. Recent findings, however, indicates that new drug delivery systems that are able to penetrate mucus assure closer contact with the epithelium and improved drug delivery [188]. PEG, an uncharged hydrophilic polymer widely applied in pharmaceutical formulations, has been proposed for topical vaginal therapy. The surface density and the molecular weight of PEG affect the diffusion rate through mucus and findings indicate that nanoparticles densely coated with low molecular weight PEG, avoid the adhesive interactions with mucus, and are able to reach deeper layers of epithelium [175, 189, 190]. This enables a closer contact with the mucosal surface, improving the drug effectiveness. The surface modifications of liposomes, as an example of surface-modified nanosystems, are summarized in Figure 9.1.

The ability of IFN α-2b entrapped in PEGylated liposomes to penetrate through the vaginal tissue has been demonstrated [181]. *Ex vivo* penetration studies performed on vaginal tissue obtained from pregnant sheep, confirmed enhanced IFN α-2b penetration from PEGylated liposomes as compared to IFN α-2b in solution. Additionally, *in vitro* mucoadhesive studies confirmed the mucoresistant properties of PEGylated liposomes, enabling mucus-penetration. This indicates that the

PEGylated liposomal formulation represents a promising strategy to assure protein delivery in the close proximity to the vaginal epithelium. In another study, a novel alginate scaffold system containing mucoresistant PEGylated lipoplexes was developed to provide a sustained vaginal delivery of siRNA [182]. The enhanced delivery of nucleic acids into the vaginal tissues from the PEGylated lipoplexes was confirmed as compared to the conventional liposomes both *in vitro* and *in vivo*. Further, a siRNA-mediated knockdown of the target gene in the vagina of mice was observed; the delivery system presents an interesting approach in vaginal administration of biologics. However, there is a possibility of induced cell toxicity and tissue inflammation by lipoplexes at higher concentrations [191], and the cytotoxicity of the system should be assessed, especially *in vivo*.

Liposomes also bear potential as a delivery system for mucosal vaccine development, and the formulation for the intravaginal immunization has been proposed by Gupta and colleagues; it comprised liposomal gel formulation containing HIV-1 envelope glycoprotein, CN54gp140 [183]. Liposomally entrapped CN54gp140 was incorporated into hydroxyethyl cellulose aqueous gel and further lyophilized to yield a rod-shaped solid dosage form for vaginal application. The liposomal gel exhibited suitable mucoadhesive properties that were sustained, and even improved, in the presence of simulated vaginal fluid. Even though the effectiveness of the freeze-dried liposomal gel remains to be confirmed, the dosage form presents an innovative approach for mucosal immunization.

Most of the polymeric nanoparticles destined for topical vaginal therapy were designed for improving microbicide and, recently, siRNA delivery [192]. Particularly, PLGA nanoparticles have been investigated for the vaginal administration of biologics. The group of Saltzman developed biodegradable polymer nanoparticles densely loaded with siRNA to obtain efficient and sustained gene silencing [184]. A single application of siRNA-loaded PLGA nanoparticles into the reproductive tract of mice resulted in gene knockdown throughout the vaginal tract for at least two weeks. Further, the nanoparticles were able to penetrate deeply in the vaginal mucosa. In another study by the same group, the administration of siRNA-loaded PLGA nanoparticles prevented lethal intravaginal herpes simplex virus (HSV-2) infection in mice and increased their survival for up to four weeks [185]. The toxicity of the PLGA nanoparticles has been examined in both cell culture and vaginal tissue. The formulation was found nontoxic to HeLa cells and no prominent inflammation in vaginal tissue samples was observed.

PLGA nanoparticles have also been investigated for the vaginal delivery of PSC RANTES, an analog of the chemokine Regulated upon Activation, Normal T Expressed and Secreted (RANTES), to exploit its antiviral activity in HIV prevention [186]. *In vitro* release and ex vivo penetration studies on human vaginal tissue proved controlled release as well as increased tissue distribution and penetration from PSC RANTES PLGA nanoparticles compared to control solution. The anti-HIV activity of the analogue was confirmed, however, the PLGA encapsulation did not increase this activity. Further, no toxicity was observed in HeLa cells in the presence of the delivery system, nor did the PLGA cause any prominent morphological changes in the excised human ectocervical tissue.

The abovementioned preliminary studies indicate that the vagina as a site of drug administration deserves to be further exploited. It not only bears specific challenges but also offers opportunities for innovative formulation development. Many of the genital disorders would benefit from efficient local drug therapy.

We have described the main challenges in delivery of biologics with a focus on the noninvasive routes of administration and related specificities. Each of the route requires a careful tailoring in design and development of novel biologics formulations. Noninvasive routes hold a promise which, hopefully, will be fully utilized in the near future.

9.4 Conclusion

Biologics are a drug category with increasing importance in pharmaceutical development. Biologics represent a challenge in formulation development due to the issues discussed in this chapter. However, they also have great potential for development of innovative formulations and devices, to fully utilize the potential of various routes of biologics delivery. Currently, the focus in this field is on extracellular targets due to limited success in intracellular delivery so far. Perspectives are within the patient-friendly and effective delivery of biologics via noninvasive routes. Although the pulmonary and skin routes seem to be very promising, vaginal administration deserves more attention.

References

1 Kling, J. (2011). Fresh from the biotech pipeline. *Nat. Biotechnol.* 30: 128–131.

2 Nelson, A.L., Dhimolea, E., and Reichert, J.M. (2010). Development trends for human monoclonal antibody therapeutics. *Nat. Rev. Drug Discovery* 9: 767–774.

3 Morales, J.O., Fathe, K.R., Brunaugh, A. et al. (2017). Challenges and future prospects for the delivery of biologics: oral mucosal, pulmonary and transdermal routes. *AAPS J.* 19: 652–668.

4 Quianzon, C.C. and Cheikh, I. (2012). History of insulin. *J. Community Hosp. Intern. Med. Perspect.*: 2. https://doi.org/10.3402/jchimp.v2i2.18701.

5 Škalko-Basnet, N. (2014). Biologics: the role of delivery systems in improved therapy. *Biol. Targets Ther.* 8: 107–114.

6 Leader, B., Baca, Q.J., and Golan, D.E. (2008). Protein therapeutics: a summary and pharmacological classification. *Nat. Rev. Drug Discovery* 7: 21–39.

7 Wahlich, J., Desai, A., Greco, F. et al. (2019). Nanomedicines for the delivery of biologics. *Pharmaceutics* 11: 210.

8 Matougui, N., Boge, L., Groo, A.-C. et al. (2016). Lipid-based nanoformulations for peptide delivery. *Int. J. Pharm.* 502: 80–97.

9 Mitragotri, S., Burke, P.A., and Langer, R. (2014). Overcoming the challenge in administering biopharmaceuticals: formulation and delivery strategies. *Nat. Rev. Drug Discovery* 13: 655–672.

10 Ezan, E. (2013). Pharmacokinetic studies of protein drugs: past, present and future. *Adv. Drug Delivery Rev.* 65: 1065–1073.

11 Urquhart, L. (2018). Market watch: top drugs and companies sales in 2017. *Nat. Rev. Drug Discovery* 17: 232.

12 Vllasaliu, D., Thanou, M., Stolnik, S., and Fowler, R. (2018). Recent advances in oral delivery of biologics: nanomedicine and physical modes of delivery. *Expert Opin. Drug Delivery* 15: 759–770.

13 Xu, X. and Vugmeyster, Y. (2012). Challenges and opportunities in absorption, distribution, metabolism and excretion studies of therapeutic biologicals. *AAPS J.* 14: 781–791.

14 Martin-Moe, S., Lim, F.J., Wong, R.L. et al. (2011). A new roadmap for biopharmaceutical drug product development: integrating development, validation and quality by design. *J. Pharm. Sci.* 100: 3031–3043.

15 Duncan, R. and Gaspar, R. (2011). Nanomedicine(s) under the microscope. *Mol. Pharmaceutics* 8: 2101–2141.

16 Richard, J. (2012). Parenteral biologics delivery: recent progresses, key challenges and perspectives. *Eur. J. Parent. Pharm. Sci.* 17: 94–109.

17 Burnett, J.C. and Rossi, J.J. (2012). RNA-based therapeutics: current progress and future prospects. *Chem. Biol.* 19: 60–71.
18 Kronenberg, S., Baumann, A., de Haan, L. et al. (2013). Current challenges and opportunities in nonclinical safety testing of biologics. *Drug Discovery* 18: 1138–1143.
19 Burgess, B.E. (2012). Optimizing drug delivery for modern biological. *Pharm. Technol.* 25: 30–32.
20 Kennedy, P.J., Oliveira, C., Granja, P.L., and Sarmento, B. (2017). Antibodies and associates: partners in targeted drug delivery. *Pharmacol. Ther.* 177: 129–145.
21 Inoue, N., Takai, E., Arakawa, T., and Shiraki, K. (2014). Arginine and lysine reduce the high viscosity of serum albumin solutions for pharmaceutical injection. *J. Biosci. Bioeng.* 117: 539–543.
22 Reithmeier, H., Hermann, J., and Gőpferich, A. (2001). Lipid microparticles as a parentral controlled release device for peptides. *J. Controlled Release* 73: 339–350.
23 Chung, H.J., Lee, Y., and Park, T.G. (2008). Thermo-sensitive and biodegradable hydrogels based on stereocomplexed Pluronic multi-block copolymers for controlled protein delivery. *J. Controlled Release* 127: 22–30.
24 Van Tomme, S.R., Strom, G., and Hennink, W.E. (2008). In situ gelling hydrogels for pharmaceutical and biomedical applications. *Int. J. Pharm.* 355: 1–18.
25 Jordan, F., Naylor, A., Kelly, C.A. et al. (2010). Sustained release hGH microsphere formulation produced by a novel supercritical fluid technology: in vivo studies. *J. Controlled Release* 141: 153–160.
26 van Dijkhuizen-Radersma, R., Hesseling, S.C., Kaim, P.E. et al. (2002). Biocompatibility and degradation of poly(ether-ester) microspheres: in vitro and in vivo evaluation. *Biomaterials* 23: 4719–4729.
27 Maschke, A., Becker, C., Eyrich, D. et al. (2007). Development of a spray congealing process for the preparation of insulin-loaded lipid microparticles and characterization thereof. *Eur. J. Pharm. Biopharm.* 65: 175–187.
28 Chan, Y.-P., Meyruiex, R., Kravtzoff, R. et al. (2007). Review on Medusa®: a polymer-based sustained release technology for protein and peptide drugs. *Expert Opin. Drug Delivery* 4: 441–451.
29 Péter, F., Bidlingmaier, M., Savoy, C. et al. (2012). Three-year efficacy and safety of LB03002, a once-weakly sustained-release growth hormone (GH) preparation, in prepubertal children with GH deficiency (GHD). *J. Clin. Endocrinol. Metab.* 97: 400–407.
30 Aguirre, T.A.S., Teijeiro-Osorio, D., Rosa, M. et al. (2016). Current status of selected oral peptide technologies in advanced preclinical development and in clinical trials. *Adv. Drug Delivery Rev.* 106: 223–241.
31 Pridgen, E.M., Alexis, F., and Farokhzad, O.C. (2014). Polymeric nanoparticle technologies for oral drug delivery. *Clin. Gastroenterol. Hepatol.* 12: 1605–1610.
32 Fosgerau, K. and Hoffmann, T. (2015). Peptide therapeutics: current status and future directions. *Drug Discovery Today* 20: 122–128.
33 Brayden, D.J. (2014). Principles underlying peptide, protein and macromolecule interactions with non-injected mucosae: from oral to buccal and nasal delivery. *Bull. Tech. Gattfosse* 107: 11–28.
34 Maher, S., Ryan, B., Duffy, A., and Brayden, D.J. (2014). Formulation strategies to improve oral peptide delivery. *Pharm. Pat. Anal.* 3: 313–336.
35 Ritschel, W.A. (1996). Microemulsion technology in the reformulation of cyclosporine: the reason behind the pharmacokinetic properties of Neoral. *Clin. Transplant.* 10: 364–373.

36 Frentzen, A., Witowski, S., Grossi, A. et al. (2014). Stable solid formulation of GC-C receptor agonist polypeptide suitable for oral administration. US Patent 8802628 B2, filed 25 February 2010.

37 Frank, L.A., Contri, R.V., Pohlmann, A.R., and Guterres, S.S. (2015). Improved drug biological effects by encapsulation into polymeric nanocapsules. *Wiley Interdiscip. Rev. Nanomed. Nanobiotechnol.* 7: 623–639.

38 Sonaje, K., Lin, Y.-H., Juang, J.-H. et al. (2009). In vivo evaluation of safety and efficacy of self-assembled nanoparticles for oral insulin delivery. *Biomaterials* 30: 2329–2339.

39 Sonaje, K., Chen, Y.J., Chen, H.-L. et al. (2010). Enteric-coated capsules filled with freeze-dried chitosan/poly(γ-glutamic acid) nanoparticles for oral insulin delivery. *Biomaterials* 31: 3384–3394.

40 Whitehead, K., Shen, Z., and Mitragotri, S. (2004). Oral delivery of macromolecules using intestinal patches: applications for insulin delivery. *J. Controlled Release* 98: 37–45.

41 Rao, K.K. (2007). Polymerized solid lipid nanoparticles for oral or mucosal delivery of therapeutic proteins and peptides. US Patent WO2007113665, filed 11 October 2007.

42 Pridgen, E.M., Alexis, F., Kuo, T.T. et al. (2013). Transepithelial transport of Fc-targeted nanoparticles by the neonatal Fc receptor for oral delivery. *Sci. Transl. Med.* 5, 213ra167.

43 Hay, T. (2014). Can "robotic" pills replace injections? http://www.ranitherapeutics.com/assets/WSJ_article_Feb_2014 (accessed 01 September 2017).

44 Traverso, G., Schoellhammer, C.M., Schroeder, A. et al. (2015). Microneedles for drug delivery via gastrointestinal tract. *J. Pharm. Sci.* 104: 362–367.

45 Bell, S. and Morco, T. (2001). Therapeutic calcium phosphate particles and methods of manufacture and use. US Patent 20010048925 A1, filed 06 December 2001.

46 Walsh, E.G., Adamczyk, B.E., Chalasani, K.B. et al. (2011). Oral delivery of macromolecules: rationale underpinning Gastrointestinal Permeation Enhancement Technology (GIPET). *Ther. Delivery* 2: 1595–1610.

47 Shukla, A., Katare, O.P., Singh, B., and Vyas, S.P. (2010). M-cell targeted delivery of recombinant hepatitis B surface antigen using cholera toxin B subunit conjugated bilosomes. *Int. J. Pharm.* 385: 47–52.

48 Niu, M., Lu, Y., Hovgaard, L., and Wu, W. (2011). Liposomes containing glycocholate as potential oral insulin delivery systems: preparation, in vitro characterization, and improved protection against enzymatic degradation. *Int. J. Nanomed.* 6: 1155–1166.

49 Zhang, N., Ping, Q.N., Huang, G.H., and Xu, W.F. (2005). Investigation of lectin-modified insulin liposomes as carriers for oral administration. *Int. J. Pharm.* 294: 247–259.

50 Thirawong, N., Thongborisute, J., Takeuchi, H., and Sriamornsak, P. (2008). Improved intestinal absorption of calcitonin by mucoadhesive delivery of novel pectin-liposome nanocomplexes. *J. Controlled Release* 125: 236–245.

51 Morales, J.O. and McConville, T. (2014). Novel strategies for the buccal delivery of macromolecules. *Drug Dev. Ind. Pharm.* 40: 579–590.

52 Caon, T., Jin, L., Simoes, C.M.O. et al. (2015). Enhancing the buccal mucosal delivery of peptide and protein therapeutics. *Pharm. Res.* 32: 1–21.

53 Bernstein, G. (2008). Delivery of insulin to the buccal mucosa utilizing the RapidMist™ system. *Expert Opin. Drug Delivery* 5: 1047–1055.

54 Giovino, C., Ayensu, I., Tetteh, J., and Boateng, J.S. (2012). Development and characterization of chitosan films impregnated with insulin loaded PEG-*b*-PLA nanoparticles (NPs): a potential approach for buccal delivery of macromolecules. *Int. J. Pharm.* 428: 143–151.

55 Giovino, C., Ayensu, I., Tetteh, J., and Boateng, J.S. (2013). An integrated buccal delivery system combining chitosan films impregnated with peptide loaded PEG-*b*-PLA nanoparticles. *Colloids Surf., B* 112: 9–15.

56 Morales, J.O., Huang, S., Williams, R.O. III, and McConville, J.T. (2014). Films loaded with insulin-coated nanoparticles (ICNP) as potential platforms for peptide buccal delivery. *Colloids Surf., B* 122: 38–45.

57 Zhen, Y., Wang, N., Gao, Z. et al. (2015). Multifunctional liposomes constituting microneedles induced robust systemic and mucosal immunoresponses against the loaded antigens via oral mucosal vaccination. *Vaccine* 33: 4330–4340.

58 Cossé, A., König, C., Lamprecht, A., and Wagner, K.G. (2017). Hot melt extrusion for sustained protein release: matrix erosion and in vitro release of PLGA-based implants. *AAPS PharmSciTech.* 18: 15–26.

59 Genina, N., Janßen, E.M., Breitenbach, A. et al. (2013). Evaluation of different substrates for inkjet printing of rasagiline mesylate. *Eur. J. Pharm. Biopharm.* 85: 1075–1083.

60 Buanz, A.B.M., Saunders, M.H., Basit, A.W., and Gaisford, S. (2011). Preparation of personalized-dose salbutamol sulphate oral films with thermal ink-jet printing. *Pharm. Res.* 28: 2386–2392.

61 Montenegro-Nicolini, M., Miranda, V., and Morales, J.O. (2017). Inkjet printing of proteins: an experimental approach. *AAPS J.* 19: 234–243.

62 Durfee, S., Messina, J., and Khankari, J. (2006). Fentanyl effervescent buccal tablets. *Am. J. Drug Delivery* 4: 1.5.

63 Taylor, D.R. (2013). Single-dose fentanyl sublingual spray for breakthrough cancer pain. *Clin. Pharmacol.* 5: 131–141.M.

64 Durham, S.R., Yang, W.H., Pedersen, M.R. et al. (2006). Sublingual immunotherapy with once-daily grass allergen tablets: a randomized controlled trail in seasonal allergic rhinoconjunctivities. *J. Alergy Clin. Immunol.* 117: 802–809.

65 Bowen, T., Greenbaum, J., Charbonneau, Y. et al. (2004). Canadian trial of sublingual swallow immunotherapy for ragweed rhinoconjunctivities. *Ann. Allergy, Asthma Immunol.* 93: 425–430.

66 Weers, J.G., Bell, J., Chan, H.-K. et al. (2010). Pulmonary formulations: what remains to be done? *J. Aerosol Med. Pulm. Drug Delivery* 23, S-5-23.

67 Agu, R.U., Ugwoke, M.I., Armand, M. et al. (2001). The lung as a route for systemic delivery of therapeutic proteins and peptides. *Respir. Res.* 2: 198–209.

68 Depreter, F., Pilcer, G., and Amighi, K. (2013). Inhaled proteins: challenges and perspectives. *Int. J. Pharm.* 447: 251–280.

69 Graf, M., Ziegler, C.E., Gregoritza, M., and Goepferich, A.M. (2019). Hydrogel microspheres evading alveolar macrophages for sustained pulmonary protein delivery. *Int. J. Pharm.* 566: 652–661.

70 Chung, S.W., Hil-lal, T.A., and Byun, Y. (2012). Strategies for non-invasive delivery of biologics. *J. Drug Targeting* 20: 481–501.

71 Bosquillon, C., Lombry, C., Préat, V., and Vanbever, R. (2001). Influence of formulation excepinets and physical characteristics of inhalation dry powders on their aerosolization performance. *J. Controlled Release* 70: 329–339.

72 Patton, J.S. and Byron, P.R. (2007). Inhaling medicines: delivering drugs to the body through the lungs. *Nat. Rev. Drug Discovery* 6: 67–74.

73 Champion, J.A. and Mitragotri, S. (2009). Shape induced inhibition of phagocytosis of polymer particles. *Pharm. Res.* 26: 244–249.

74 Morita, T., Yamammoto, A., Takakura, Y. et al. (1994). Improvement of the pulmonary absorption of ($Asu^{1,7}$)-Eel calcitonin by various protease inhibitors in rats. *Pharm. Res.* 11: 909–913.
75 Morales, J.O., Peters, J.I., and Williams R.0. Jr. (2011). Surfactants: their critical role in enhancing drug delivery to the lungs. *Ther. Delivery* 2: 623–641.
76 Hickey, A.J. (2013). Back to the future: inhaled drug products. *J. Pharm. Sci.* 102: 1165–1172.
77 Hertel, S.P., Winter, G., and Friess, W. (2015). Protein stability in pulmonary drug delivery via nebulization. *Adv. Drug Delivery Rev.* 93: 79–94.
78 Mastorakos, P., da Silva, A., Chisholm, J. et al. (2015). Highly compacted biodegradable DNA nanoparticles capable of overcoming the mucus barrier for inhaled lung gene therapy. *Proc. Natl. Acad. Sci. U.S.A.* 112: 8720–8725.
79 Feng, F., Harper, R., and Reynolds, P.N. (2016). BMPR2 gene delivery reduces mutation-related PAH and counteracts TGF-ß-mediated pulmonary cell signaling. *Respirology* 21: 526–532.
80 Illum, L. (2003). Nasal drug delivery-possibilities, problems and solutions. *J. Controlled Release* 87: 187–198.
81 Schwarz, B. and Merkel, O.M. (2019). Nose-to-brain delivery of biologics. *Ther. Delivery* 10: 207–210.
82 Steyn, D., du Plessis, L., and Kotzé, A. (2010). Nasal delivery of recombinant human growth hormone: in vivo evaluation with Pheroid technology and *N*-trimethyl chitosan chloride. *J. Pharm. Pharm. Sci.* 13: 263–273.
83 du Plessis, L.H., Lubbe, J., Strauss, T., and Kotzé, A.F. (2010). Enhancement of nasal and intestinal calcitonin delivery by the novel Pheroid™ fatty acid based delivery system, and by *N*-trimethyl chitosan chloride. *Int. J. Pharm.* 385: 181–186.
84 Dhuria, S.V., Hanson, L.R., and Frey, W.H. (2010). Intranasal delivery to the central nervous system: mechanism and experimental considerations. *J. Pharm. Sci.* 93: 2851–2862.
85 Lochhead, J.J. and Thorne, R.G. (2012). Intranasal delivery of biologics to the central nervous system. *Adv. Drug Delivery Rev.* 64: 614–628.
86 Schuetz, Y.B., Naik, A., Guy, R.H., and Kalia, Y.N. (2005). Emerging strategies for the transdermal delivery of peptide and protein drugs. *Expert Opin. Drug Delivery* 2: 533–548.
87 Schoellhammer, C.M., Blankschtein, D., and Langer, R. (2014). Skin permeabilization for transdermal drug delivery: recent advances and future prospects. *Expert Opin. Drug Delivery* 11: 393–407.
88 Singh, N., Kalluri, H., Herwadkar, A. et al. (2012). Transcending the skin barrier to deliver peptides and proteins using active technologies. *Crit. Rev. Ther. Drug Carrier Syst.* 29: 265–298.
89 Chen, Y., Shen, Y., Guo, X. et al. (2006). Transdermal protein delivery by a coadministred peptide identified via phage display. *Nat. Biotechnol.* 24: 455–460.
90 Hsu, T. and Mitragotri, S. (2011). Delivery of siRNA and other macromolecules into skin and cells using a peptide enhancer. *Proc. Natl. Acad. Sci. U.S.A.* 108: 15816–15821.
91 Kim, Y.-C., Ludovice, P.J., and Prausnitz, M.R. (2007). Transdermal delivery enhanced by magainin pore-forming peptide. *J. Controlled Release* 122: 375–383.
92 Rothbard, J., Garlington, S., Lin, Q. et al. (2000). Conjugation of arginine oligomers to cyclosporine A facilitates topical delivery and inhibition of inflammation. *Nat. Med.* 6: 1253–1257.
93 Karande, P. and Mitragotri, S. (2009). Enhancement of transdermal drug delivery via synergistic action of chemicals. *Biochim. Biophys. Acta* 1788: 2362–2373.
94 Ita, K. (2016). Perspectives on transdermal electroporation. *Pharmaceutics* 8: 9.
95 Sen, A., Daly, M.E., and Hui, S.W. (2002). Transdermal insulin delivery using lipid enhanced electroporation. *Biochim. Biophys. Acta* 1564: 5–8.

96 Prausnitz, M.R., Edelman, E.R., Gimm, J.A. et al. (1995). Transdermal delivery of heparin by skin electroporation. *Biotechnology* 13: 1205–1209.

97 Azagury, A., Khoury, L., Enden, G., and Kost, J. (2014). Ultrasound mediated transdermal drug delivery. *Adv. Drug Delivery Rev.* 72: 127–143.

98 Mitragotri, S., Blankschtein, D., and Langer, R. (1995). Ultrasound-mediated transdermal protein delivery. *Science* 269: 850–853.

99 Mitragotri, S. and Kost, J. (2001). Transdermal delivery of heparin and low-molecular weight heparin using low-frequency ultrasound. *Pharm. Res.* 18: 1151–1156.

100 Dahlan, A., Alpar, H.O., Stickings, P. et al. (2009). Transcutaneous immunization assisted by low-frequency ultrasound. *Int. J. Pharm.* 368: 123–128.

101 Tran, M.A., Gowda, R., Park, E.-J. et al. (2009). Noninvasive drug delivery using ultrasound: targeting melanoma using siRNA against mutant (V600E) B-Raf. *AIP Conf. Proc.* 1113: 423–427.

102 Huang, Y., Yu, F., Park, Y.-S. et al. (2010). Co-administration of protein drugs with gold nanoparticles to enable percutaneous delivery. *Biomaterials* 31: 9086–9091.

103 Li, Y.-Z., Quan, Y.-S., Jin, M.-N. et al. (2009). Trypsin as a novel potential absorption enhancer for improving the transdermal delivery of macromolecules. *J. Pharm. Pharmacol.* 61: 1005–1012.

104 Kochhar, C. and Imanidis, G. (2004). In vitro transdermal iontophoretic delivery of leuprolide under constant current application. *J. Controlled Release* 98: 25–35.

105 Langkjær, L., Brange, J., Grodsky, G.M., and Guy, R.H. (1998). Iontophoresis of monomeric insulin analogues in vitro: effects of insulin charge and skin pretreatment. *J. Controlled Release* 51: 47–56.

106 Arora, A., Prausnitz, M.R., and Mitragotri, S. (2008). Micro-scale devices for transdermal drug delivery. *Int. J. Pharm.* 364: 227–236.

107 Levin, G., Gershonowitz, A., Sacks, H. et al. (2005). Transdermal delivery of human growth hormone through RF-microchannels. *Pharm. Res.* 22: 550–555.

108 Badkar, A.V., Smith, A.M., Eppstein, J.A., and Banga, A.K. (2007). Transdermal delivery of interferon alpha-2b using microporation and iontophoresis in hairless rats. *Pharm. Res.* 24: 1389–1395.

109 Glenn, G.M., Flyer, D.C., Ellingsworth, L.R. et al. (2007). Transcutaneous immunization with heat-labile enterotoxin: development of a needle-free vaccine patch. *Expert Rev. Vaccines* 6: 809–819.

110 Mikszta, J., Alarcon, J.B., Brittingham, J.M. et al. (2002). *Nat. Med.* 8: 415–419.

111 Andrews, S., Lee, J.W., Choi, S.-O., and Prausnitz, M.R. (2011). Transdermal insulin delivery using microneedles. *Pharm. Res.* 28: 2110–2118.

112 Caffarel-Salvador, E. and Donnelly, R.F. (2016). Transdermal drug delivery mediated by microneedle arrays: innovations and barriers to success. *Curr. Pharm. Des.* 22: 1105–1117.

113 Arya, J. and Prausnitz, M.R. (2016). Microneedle patches for vaccination in developing countries. *J. Controlled Release* 240: 135–141.

114 Gill, H.S., Denson, D.D., Burris, B.A., and Prausnitz, M.R. (2008). Effect of microneedle design on pain in human subjects. *Clin. J. Pain* 24: 585–594.

115 Norman, J.J., Arya, J.M., McClain, M.A. et al. (2014). Microneedle patches: usability and acceptability for self-vaccination against influenza. *Vaccine* 32: 1856–1862.

116 Norman, J.J., Brown, M.R., Raviele, N.A. et al. (2013). Faster pharmacokinetics and increased patient acceptance of intradermal insulin delivery using a single hollow microneedle in children and adolescents with type 1 diabetes. *Pediatr. Diabetes* 14: 459–465.

117 Rini, C.J., McVey, E., Sutter, D. et al. (2015). Intradermal insulin infusion achieves faster insulin action than subcutaneous infusion for 3-day wear. *Drug Deliv. Transl. Res.* 5: 332–345.
118 Skountzou, I. and Compans, R.W. (2015). Skin immunization with influenza vaccines. *Curr. Top. Microbiol. Immunol.* 386: 343–369.
119 Cosman, F., Lane, N.E., Bolognese, M.A. et al. (2010). Effect of transdermal teriparatide administration on bone mineral density in postmenopausal women. *J. Clin. Endocrinol. Metab.* 95: 151–158.
120 Pettis, R.J. and Harvey, A.J. (2012). Microneedle delivery: clinical studies and emerging medical applications. *Ther. Delivery* 3: 357–371.
121 Hirobe, S., Azukizawa, H., Hanafusa, T. et al. (2015). Clinical study and stability assessment of a novel transcutaneous influenza vaccination using a dissolving microneedle patch. *Biomaterials* 57: 50–58.
122 Jeon, S.-O., Hwang, H.-J., Oh, D.-H. et al. (2012). Enhanced percutaneous delivery of recombinant human epidermal growth factor employing nano-liposome system. *J. Microencapsulation* 29: 234–241.
123 Qiu, Y., Qin, G., Zhang, S. et al. (2012). Novel lyophilized hydrogel patches for convenient and effective administration of microneedle-mediated insulin delivery. *Int. J. Pharm.* 437: 51–56.
124 Siu, K.S., Chen, D., Zheng, X. et al. (2014). Non-covalently functionalized single-walled carbon nanotube for topical siRNA delivery into melanoma. *Biomaterials* 35: 3435–3442.
125 Chen, M., Zakrewsky, M., Gupta, V. et al. (2014). Topical delivery of siRNA into skin using SPACE-peptide carriers. *J. Controlled Release* 179: 33–41.
126 Desai, P.R., Marepally, S., Patel, A.R. et al. (2013). Topical delivery of anti-TNFα siRNA and capsaicin via novel lipid-polymer hybrid nanoparticles efficiently inhibits skin inflammation in vivo. *J. Controlled Release* 170: 51–63.
127 Zheng, D., Giljohann, D.A., Chen, D.L. et al. (2012). Topical delivery of siRNA-based spherical nucleic acid nanoparticle conjugates for gene regulation. *Proc. Natl. Acad. Sci. U.S.A.* 109: 11975–11980.
128 Venuganti, V.V., Saraswathy, M., Dwivedi, C. et al. (2015). Topical gene silencing by iontophoretic delivery of an antisense oligonucleotide-dendrimer nanocomplex: the proof of concept in a skin cancer mouse model. *Nanoscale* 7: 3903–3914.
129 Geusens, B., Van Gele, M., Braat, S. et al. (2010). Flexible nanosomes (SECosomes) enable efficient siRNA delivery in cultured primary skin cells and in the viable epidermis of ex vivo human skin. *Adv. Funct. Mater.* 20: 4077–4090.
130 Bracke, S., Carretero, M., Guerrero-Aspizua, S. et al. (2014). Targeted silencing of DEFB4 in a bioengineered skin-humanized mouse model for psoriasis: development of siRNA SECosome-based novel therapies. *Exp. Dermatol.* 23: 199–201.
131 Kim, S.T., Lee, K.-M., Park, H.-J. et al. (2009). Topical delivery of interleukin-13 antisense oligonucleotides with cationic elastic liposome for the treatment of atopic dermatitis. *J. Gene Med.* 11: 26–37.
132 Li, J., Li, X., Zhang, Y. et al. (2010). Gene therapy for psoriasis in the K14-VEGF transgenic mouse model by topical transdermal delivery of interleukin-4 using ultradeformable cationic liposome. *J. Gene Med.* 12: 481–490.
133 Dorrani, M., Garbuzenko, O.B., Minko, T., and Michniak-Kohn, B. (2016). Development of edge-activated liposomes for siRNA delivery to human basal epidermis for melanoma therapy. *J. Controlled Release* 228: 150–158.
134 Jin, S.-E. and Kim, C.-K. (2014). Charge-mediated topical delivery of plasmid DNA with cationic lipid nanoparticles to the skin. *Colloids Surf., B* 116: 582–590.

135 Desmet, E., Bracke, S., Forier, K. et al. (2016). An elastic liposomal formulation for RNAi-based topical treatment of skin disorders: proof-of-concept in the treatment of psoriasis. *Int. J. Pharm.* 500: 268–274.

136 Gharagozloo, M., Rafiee, A., Chen, D.W., and Foldvari, M. (2015). A flow cytometric approach to study the mechanism of gene delivery to cells by gemini-lipid nanoparticles: an implication for cell membrane nanoporation. *J. Nanobiotechnol.* 13: 62.

137 Jain, S., Patel, N., Shah, M.K. et al. (2017). Recent advances in lipid-based vesicles and particulate carriers for topical and transdermal application. *J. Pharm. Sci.* 106: 423–445.

138 Rosada, R.S., Silva, C.L., Andrade Santana, M.H. et al. (2012). Effectiveness, against tuberculosis, of pseudo-ternary complexes: peptide-DNA-cationic liposome. *J. Colloid Interface Sci.* 373: 102–109.

139 Chopra, A. and Cevc, G. (2014). Non-invasive, epicutaneous immunisation with toxoid in deformable vesicles protects mice against tetanus, chiefly owing to a Th2 response. *Eur. J. Pharm. Sci.* 56: 55–64.

140 Kumar, G., Sharma, N., Gupta, P. et al. (2016). Improved protection against tuberculosis after boosting the BCG-primed mice with subunit Ag 85a delivered through intact skin with deformable vesicles. *Eur. J. Pharm. Sci.* 82: 11–20.

141 Chen, G., Zeng, S., Jia, H. et al. (2014). Adjuvant effect enhancement of porcine interleukin-2 packaged into solid lipid nanoparticles. *Res. Vet. Sci.* 96: 62–68.

142 Ledet, G., Pamujula, S., Walker, V. et al. (2014). Development and in vitro evaluation of a nanoemulsion for transcutaneous delivery. *Drug Dev. Ind. Pharm.* 40: 370–379.

143 Stein, P., Gogoll, K., Tenzer, S. et al. (2014). Efficacy of imiquimod-based transcutaneous immunization using a nano-dispersed emulsion gel formulation. *PLoS One* 9, e102664.

144 DeMuth, P.C., Moon, J.J., Suh, H. et al. (2012). Releasable layer-by-layer assembly of stabilized lipid nanocapsules on microneedles for enhanced transcutaneous vaccine delivery. *ACS Nano* 6: 8041–8051.

145 Choi, W.I., Lee, J.H., Kim, J.-Y. et al. (2012). Efficient skin permeation of soluble proteins via flexible and functional nano-carrier. *J. Controlled Release* 157: 272–278.

146 Nose, K., Pissuwan, D., Goto, M. et al. (2012). Gold nanorods in an oil-base formulation for transdermal treatment of type 1 diabetes in mice. *Nanoscale* 4: 3776–3780.

147 Yu, J., Zhang, Y., Ye, Y. et al. (2015). Microneedle-array patches loaded with hypoxia-sensitive vesicles provide fast glucose-responsive insulin delivery. *Proc. Natl. Acad. Sci. U.S.A.* 112: 8260–8265.

148 Gainza, G., Aguirre, J.J., Pedraz, J.L. et al. (2013). rhEGF-loaded PLGA-Alginate microspheres enhance the healing of full-thickness excisional wounds in diabetised Wistar rats. *Eur. J. Pharm. Sci.* 50: 243–252.

149 Jiang, B., Zhang, G., and Brey, E.M. (2013). Dual delivery of chlorexidine and platelet-derived growth factor-BB for enhanced wound healing and infection control. *Acta Biomater.* 9: 4976–4984.

150 Chereddy, K.K., Lopes, A., Koussoroplis, S. et al. (2015). Combined effects of PLGA and vascular endothelial growth factor promote the healing of non-diabetic and diabetic wounds. *Nanomedicine* 11: 1975–1984.

151 Zhou, W., Zhao, M., Zhao, Y., and Mou, Y. (2011). A fibrin gel loaded with chitosan nanoparticles for local delivery of rhEGF: preparation and in vitro release studies. *J. Mater. Sci. - Mater. Med.* 22: 1221–1230.

152 Place, L.W., Sekyi, M., and Kipper, M.J. (2014). Aggrecan-mimetic, glycosaminoglycan-containing nanoparticles for growth factor stabilization and delivery. *Biomacromolecules* 15: 680–689.

153 Mohandas, A., Anisha, B.S., Chennazhi, K.P., and Jayakumar, R. (2015). Chitosan–hyaluronic acid/VEGF loaded fibrin nanoparticles composite sponges for enhancing angiogenesis in wounds. *Colloids Surf., B* 127: 105–113.

154 Li, X., Ye, X., Qi, J. et al. (2016). EGF and curcumin co-encapsulated nanoparticle/hydrogel system as potent skin regeneration agent. *Int. J. Nanomed.* 11: 3993–4009.

155 Kaminski, G.A.T., Sierakowski, M.R., Pontarolo, R. et al. (2016). Layer-by-layer polysaccharide-coated liposomes for sustained delivery of epidermal growth factor. *Carbohydr. Polym.* 140: 129–135.

156 Gainza, G., Pastor, M., Aguirre, J.J. et al. (2014). A novel strategy for the treatment of chronic wounds based on the topical administration of rhEGF-loaded lipid nanoparticles: In vitro bioactivity and in vivo effectiveness in healing-impaired db/db mice. *J. Controlled Release* 185: 51–61.

157 Gainza, G., Bonafonte, D.C., Moreno, B. et al. (2015a). The topical administration of rhEGF-loaded nanostructured lipid carriers (rhEGF-NLC) improves healing in a porcine full-thickness excisional wound model. *J. Controlled Release* 197: 41–47.

158 Gainza, G., Chu, W.S., Guy, R.H. et al. (2015b). Development and in vitro evaluation of lipid nanoparticle-based dressings for topical treatment of chronic wounds. *Int. J. Pharm.* 490: 404–411.

159 Xiang, Q., Xiao, J., Zhang, H. et al. (2011). Preparation and characterization of bFGF-encapsulated liposomes and evaluation of wound-healing activities in the rat. *Burns* 37: 886–895.

160 Choi, J.U., Lee, S.W., Pangeni, R. et al. (2017). Preparation and in vivo evaluation of cationic elastic liposomes comprising highly skin-permeable growth factors combined with hyaluronic acid for enhanced diabetic wound-healing therapy. *Acta Biomater.* 57: 197–215.

161 Yang, Y., Xia, T., Zhi, W. et al. (2011). Promotion of skin regeneration in diabetic rats by electrospun core-sheath fibers loaded with basic fibroblast growth factor. *Biomaterials* 32: 4243–4254.

162 Jin, G., Prabhakaran, M.P., Kai, D., and Ramakrishna, S. (2013). Controlled release of multiple epidermal induction factors through core–shell nanofibers for skin regeneration. *Eur. J. Pharm. Biopharm.* 85: 689–698.

163 Xie, Z., Paras, C.B., Weng, H. et al. (2013). Dual growth factor releasing multi-functional nanofibers for wound healing. *Acta Biomater.* 9: 9351–9359.

164 Bertoncelj, V., Pelipenko, J., Kristl, J. et al. (2014). Development and bioevaluation of nanofibers with blood-derived growth factors for dermal wound healing. *Eur. J. Pharm. Biopharm.* 88: 64–74.

165 Garcia-Orue, I., Gainza, G., Gutierrez, F.B. et al. (2017). Novel nanofibrous dressings containing rhEGF and *Aloe vera* for wound healing applications. *Int. J. Pharm.* 523: 556–566.

166 Lee, M.S., Ahmad, T., Lee, J. et al. (2017). Dual delivery of growth factors with coacervate-coated poly(lactic-co-glycolic acid) nanofiber improves neovascularization in a mouse skin flap model. *Biomaterials* 124: 65–77.

167 Chereddy, K.K., Her, C.-H., Comune, M. et al. (2014). PLGA nanoparticles loaded with host defense peptide LL37 promote wound healing. *J. Controlled Release* 194: 138–147.

168 Garcia-Orue, I., Gainza, G., Girbau, C. et al. (2016). LL37 loaded nanostructured lipid carriers (NLC): a new strategy for the topical treatment of chronic wounds. *Eur. J. Pharm. Biopharm.* 108: 310–316.

169 Fontana, F., Mori, M., Riva, F. et al. (2016). Platelet lysate-modified porous silicon microparticles for enhanced cell proliferation in wound healing applications. *ACS Appl. Mater. Interfaces* 8: 988–996.

170 Blažević, F., Milekić, T., Duvnjak Romić, M. et al. (2016). Nanoparticle-mediated interplay of chitosan and melatonin for improved wound epithelialisation. *Carbohydr. Polym.* 146: 445–454.

171 Vanić, Ž. and Škalko-Basnet, N. (2013). Nanopharmaceuticals for improved topical vaginal therapydelivery: can they deliver? *Eur. J. Pharm. Sci.* 50: 29–41.

172 das Neves, J., Amiji, M., and Sarmento, B. (2011). Mucoadhesive nanosystems for vaginal microbicide development: friend or foe? *WIREs Nanomed. Nanobiotechnol.* 3: 389–399.

173 Machado, R.M., Palmeira-de-Oliveira, A., Gaspar, C. et al. (2015). Studies and methodologies on vaginal drug permeation. *Adv. Drug Delivery Rev.* 92: 14–26.

174 Vanić, Ž. and Škalko-Basnet, N. (2014). Mucosal nanosystems for improved topical drug delivery: vaginal route of administration. *J. Drug Delivery Sci. Technol.* 24: 435–444.

175 Ensign, L.M., Tang, B.C., Wang, Y.-Y. et al. (2012). Mucus-penetrating nanoparticles for vaginal drug delivery protect against herpes simplex virus. *Sci. Transl. Med.* 4: 138–179.

176 Ensign, L.M., Cone, R., and Hanes, J. (2014). Nanoparticle-based drug delivery to the vagina: a review. *J. Controlled Release* 190: 500–514.

177 Gupta, S., Gabrani, R., Ali, J., and Dang, S. (2011). Exploring novel approaches to vaginal drug delivery. *Recent Pat. Drug Deliv. Formul.* 5: 82–94.

178 Mallipeddi, R. and Rohan, L.C. (2010). Progress in antiretroviral drug delivery using nanotechnology. *Int. J. Nanomed.* 5: 533.

179 das Neves, J. (2014). Vaginal delivery of biopharmaceuticals. In: *Mucosal Delivery of Biopharmaceuticals* (eds. J. das Neves and B. Sarmento), 261–280. New York, NY: Springer.

180 Foldvari, M. and Moreland, A. (1997). Clinical observations with topical liposome-encapsulated interferon alpha for the treatment of genital papillomavirus infections. *J. Liposome Res.* 7: 115–126.

181 Jøraholmen, M.W., Basnet, P., Acharya, G., and Škalko-Basnet, N. (2017). PEGylated liposomes for topical vaginal therapy improve delivery of interferon alpha. *Eur. J. Pharm. Biopharm.* 113: 132–139.

182 Wu, S.Y., Chang, H.-I., Burgess, M., and McMillan, N.A.J. (2011). Vaginal delivery of siRNA using a novel PEGylated lipoplex-entrapped alginate scaffold system. *J. Controlled Release* 155: 418–426.

183 Gupta, P.N., Pattani, A., Curran, R.M. et al. (2012). Development of liposome gel based formulations for intravaginal delivery of the recombinant HIV-1 envelope protein CN54gp140. *Eur. J. Pharm. Sci.* 46: 315–322.

184 Woodrow, K.A., Cu, Y., Booth, C.J. et al. (2009). Intravaginal gene silencing using biodegradable polymer nanoparticles densely loaded with small-interfering RNA. *Nat. Mater.* 8: 526.

185 Steinbach, J.M., Weller, C.E., Booth, C.J., and Saltzman, W.M. (2012). Polymer nanoparticles encapsulating siRNA for treatment of HSV-2 genital infection. *J. Controlled Release* 162: 102–110.

186 Ham, A.S., Cost, M.R., Sassi, A.B. et al. (2009). Targeted delivery of PSC-RANTES for HIV-1 prevention using biodegradable nanoparticles. *Pharm. Res.* 26: 502–511.

187 Foldvari, M., Badea, I., Kumar, P. et al. (2011). Biphasic vesicles for topical delivery of interferon alpha in human volunteers and treatment of patients with human papillomavirus infections. *Curr. Drug Delivery* 8: 307–319.

188 Sims, L.B., Miller, H.A., Halwes, M.E. et al. (2019). Modeling of nanoparticle transport through the female reproductive tract for the treatment of infectious diseases. *Eur. J. Pharm. Biopharm.* 138: 37–47.

189 Lai, S.K., O'Hanlon, D.E., Harrold, S. et al. (2007). Rapid transport of large polymeric nanoparticles in fresh undiluted human mucus. *Proc. Natl. Acad. Sci. U.S.A.* 104: 1482–1487.

190 Mert, O., Lai, S.K., Ensign, L. et al. (2012). A poly (ethylene glycol)-based surfactant for formulation of drug-loaded mucus penetrating particles. *J. Controlled Release* 157: 455–460.

191 Khazanov, E., Simberg, D., and Barenholz, Y. (2006). Lipoplexes prepared from cationic liposomes and mammalian DNA induce CpG-independent, direct cytotoxic effects in cell cultures and in mice. *J. Gene Med.* 8: 998–1007.

192 Rodríguez-Gascón, A., del Pozo-Rodríguez, A., Isla, A., and Solinís, M.A. (2015). Vaginal gene therapy. *Adv. Drug Delivery Rev.* 92: 71–83.

10

Recent Advances in Cell-Mediated Drug Delivery Systems for Nanomedicine and Imaging

Li Li and Zhi Qi

Australian Institute for Bioengineering and Nanotechnology (AIBN), The University of Queensland, Brisbane, QLD, Australia

10.1 Introduction

Drug delivery systems have been used to carry various therapeutic agents to improve their efficacy and reduce their side effects. With advances in nanotechnology, a variety of drug delivery systems have been developed including liposomes, nanoemulsions, micelles, polymers, dendrimers, viral-liked nanoparticles (NPs), and inorganic nanoparticles (gold, silver, silica, iron oxides [IOs], etc.) [1, 2]. These drug delivery systems have improved drug accumulation in the target sites, protected the drugs from enzyme degradation, and prolonged their circulation time. However, their clinical application is very limited due to the off-target problem and the safety of these nanoparticle delivery systems. A recent report has clearly stated that only 0.5–1.4% of drugs can reach the tumor site whether they have been delivered passively or actively by nanoparticles [3]. Moreover, there has been a great debate over the years on nanoparticle application due to their unsure toxicity and immunogenicity in animals and humans. Thus, development of novel delivery strategies to improve target efficiency and reduce side effects is highly critical.

In the 1970s, Ihler et al. were the first to develop the erythrocyte (red blood cell [RBC])-based therapeutic strategy for Gaucher's disease. They successfully loaded enzymes (β-glucosidase and β-galactosidase) into RBCs and demonstrated that the enzymes remained active after the cells were lysed [4]. Various cells such as erythrocytes, leukocytes, and stem cells have been isolated and used to carry various therapeutic agents to the targeted organs [5]. Compared to traditional drug delivery systems, cell-based delivery systems have offered great advantages such as specifically delivering drugs to their site of action, enhanced drug efficiency, prolonged drug circulation, limited immunogenicity, and cytotoxicity. In addition, cell-based delivery systems are intrinsically biodegradable and biocompatible. Thus, they have been utilized in a variety of disease treatments including inflammatory diseases, stroke, cancer, neurodegenerative and cardiovascular diseases, and osteoarthritis. So far, cells used for therapeutic delivery include erythrocytes, platelets, leukocytes, stem cells, fibroblasts, islets, hepatocytes, and cancer cells (Figure 10.1).

Inspired by the cell-mediated delivery systems, cell membranes (CM) have been extracted from cells and used to deliver drugs to mimic the cell function. The CM from the cells have shown similar efficacy compared to the cells but offered high drug loading efficiency and improved immunogenicity properties [6–8]. In this chapter, we have summarized the recent advances in cell-mediated drug delivery including cell types, cell modification, and encapsulation approaches. Then, the application of cell-mediated drug delivery systems in disease treatment and imaging are reviewed and the

Fundamentals of Drug Delivery, First Edition.
Edited by Heather A. E. Benson, Michael S. Roberts, Adrian C. Williams, and Xiaowen Liang.

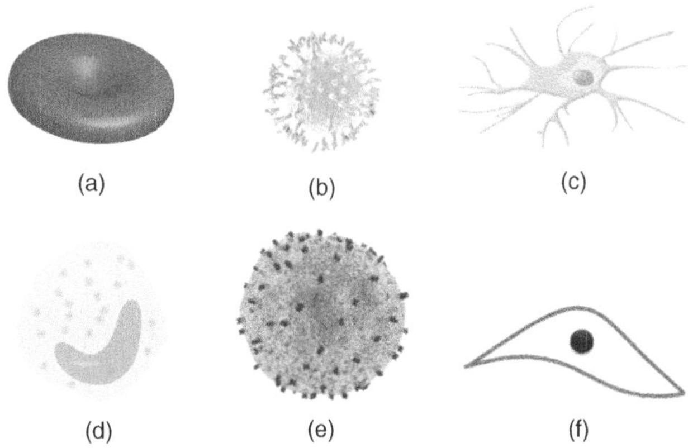

Figure 10.1 The common cells used for drug delivery. (a) Erythrocytes (red blood cells [RBCs]), (b) leukocytes (white blood cells [WBCs]), (c) dendritic cells (DCs), (d) monocytes, (e) macrophages, and (f) mesenchymal stem cell (MSCs).

most recent mechanisms of cell-mediated drug delivery systems. Finally, the current drug administration methods for cell-mediated delivery systems and their preclinical and clinical applications are discussed.

10.2 Cell Types and Modification for Therapeutic Agent Delivery

Various cells have been used as carriers to deliver drugs and biomolecules efficiently to targeted sites. These cells include erythrocytes, platelets, leukocytes, stem cells, fibroblasts, islets, hepatocytes, and cancer cells. The functions of the common cells and their applications for drug delivery are summarized in Table 10.1.

10.2.1 Cell Types

10.2.1.1 Blood Cells

Blood cells including erythrocytes, leukocytes, and platelets have been used as cell vehicles for cancer therapy and inflammation due to their unique properties such as migration to the inflammatory sites, the capability to penetrate several cell barriers, controlled release of the therapeutic agents, and excellent biocompatibility and biodegradation [6].

Erythrocytes, also named RBCs, are the most abundant blood cells with a biconcave discoid shape and no organelles. RBCs can be isolated from blood cells through conventional centrifugation and the diameter of RBCs is approximately 6–8 μm. RBCs have a long lifespan of around 100–120 days. Senescent RBCs can be cleared by macrophages of the reticuloendothelial system (RES) [7, 8].

RBC-based delivery systems have been frequently used as carriers of therapeutic agents and imaging probes owing to their long circulation half-life, excellent immune biocompatibility and biodegradation, and a simple preparation process. They can be loaded with molecular drugs, proteins, and genes by adsorption, surface conjugation, reversible osmotic lysis method, chemical perturbations, and mechanical stress. RBCs can encapsulate tumor-associated antigens and deliver them to antigen-presenting cells (APCs) to induce antigen-specific immune responses against the tumor [10]. Interferon alpha (IFN-α) and ribavirin (RIBA) were encapsulated into RBCs to activate monocytes to target liver macrophages for hepatitis C virus (HCV) infection [11]. Recently,

Table 10.1 Summary of cell types and their application for drug delivery.

Cell types	Size (μm)	Lifespan (d)	Shapes	Target organs	Advantages	Disadvantages
RBCs	6–8	100–120	Oval, biconcave disks, excellent flexibility	The reticuloendothelial system such as liver, spleen, and bone morrow; the vascular system	Biocompatibility, complete biodegradability, longer blood circulation, target delivery, high cargo-loading capacity, enhanced cargo stability, and reduced toxicity to the health cells	Drug leakage during the circulation, low efficiency for infect-ion, inflammation, or cancer treatment due to the low capability of RBCs to overcome the endothelial barrier
WBCs	7–15	Up to 20	Irregular shapes subject to change	Inflammatory tissues, bones, brains, lung, tumors, etc.	Quick migration and transportation of the drugs, target delivery resolving inflammation, and repairing tissues damage	Short lifespan time; intracellular degradation
Platelets	2–4	8–10	Discoid shapes that can be changed to round octopus, and extend long filaments when stimulating	Tumors, cardiovascular	Numerous receptors on the cell membranes and other biochemical entities, orchestrating tissue repair, and regenerative mechanisms	Relatively short shelf-life *in vitro* and circulation time *in vivo*
MSCs	15–50	Long-time up to years	Spindle	Brains, various tumors, and inflammatory sites	Migrating to the sites of inflammation; crossing over BBB; relatively easy to harvest and culture *in vitro*	Low drug-loading efficiency; drug release at the target site; migration efficiency toward the tumor target
NSCs	10–16	Long-time up to years	Various shapes	Tumors	Migrating to specific tissues, particularly tumors and CNS tissues	

(*continued*)

Table 10.1 (Continued)

Cell types	Size (μm)	Lifespan (d)	Shapes	Target organs	Advantages	Disadvantages
DCs	6–9; 10–15	Only a few days	Irregular nuclei and short cytoplasmic protrusion	Inflammatory tissues, infection organs, and tumors	The ability to cross over several biological barriers such as BBB	Complex process to get T or B cells, vulnerable, difficult to harvest and handles, relative short lifespan time
T/B cells	7	An relative short time	Sparkling sphere	Tumors, brain, spleen, tonsils, and lymph nodes	Localized drug targeting; complementation and amplification of cytotoxic T cell effects by antitumor drugs, detection and attack of metastases not yet diagnosed, autonomously working system, high specificity, and generation of long-term memory T cells [9]	
Macrophages	20–80	A long life span, ranging from months to years	Irregular sphere	Various tissues, particularly the central nervous systems; tumors	The ability to cross over several biological barriers such as BBB	

several nanoparticle (NP) delivery systems have encapsulated drugs which then are internalized by the carrier cells [12–14]. Encapsulation of NPs into RBCs has enhanced drug-targeting properties, prolonged the circulation time of NPs, and protected them from systematic clearance by the RES.

Platelets are tiny blood cells (2–4 μm) generated in the bone marrow along with RBCs and white blood cells (WBCs). Platelets can be isolated from autologous or allogeneic blood and help the body form clots to stop bleeding. Their lifespan in the blood circulation is short at around 8–10 days. The "old" platelets can be degraded by the spleen or liver macrophages. Platelets can regulate the hemostasis–thrombosis balance, promote tumor growth and metastasis, and in addition, they are of interest for application in inflammation treatment, cancer therapies, and coagulation or thrombotic disorders [15–17]. They can encapsulate therapeutic agents through *ex vivo* processing and be reinjected in the circulation. Xu et al. encapsulated doxorubicin (DOX) into platelets to treat lymphoma Raji cells and demonstrated that DOX-platelet improved the therapeutic efficacy and reduced the cardiotoxicity of DOX. DOX-platelet has shown a longer retention time than synthetic drug delivery systems [18]. Platelets can also encapsulate NPs with thrombolytic agents for the targeted delivery of clot-augmenting bio-therapeutics [19, 20].

Leukocytes, also called WBCs, play an important role in the immune system. They protect the body against infection, inflammation and tumor growth, and clean cellular debris. The average size of WBCs is 7–15 μm and their lifespan is up to 20 days, shorter than that of RBCs. WBCs have specific functions including immune response, cellular interaction, cell adhesion, and migration to inflammatory tissues, which make them an attractive delivery system. They can be classified into five subtypes including neutrophils, eosinophils, basophils, monocyte, and lymphocytes. WBCs have the capability to interact with tumor cells both in the bloodstream and at the site of solid tumors. These interactions can be utilized to effectively deliver drugs to tumor cells, which can reduce both the amount of drug required and various nonspecific cytotoxic effects within the body [21]. Monocytes and macrophages have been utilized as "Trojan Horses" to cross the brain–blood barrier (BBB) and deliver therapeutics to metastatic deposits in the brain [22]. In addition, WBCs can also deliver imaging contrast agents to targeted tissues, such as the brain and tumors for imaging [23–25].

10.2.1.2 Stem Cells

Stem cells have the capability to self-renew and differentiate into a wide variety of functional cells and migrate to the sites of injury, inflammation, and tumors. They have been attractive as cell vehicles to carry genes, nanoparticles and drugs for immunotherapy, tissue regeneration/repair, and tumor treatments. The stem cells used as cell vehicles include mesenchymal stem cells (MSCs), neural stem cells (NSCs), and endothelial progenitor cells (EPCs). MSCs and NSCs are able to migrate to specific tissues and track various cytokines. As reported in the literature, these stem cells are able to overcome the BBB to deliver therapeutic agents efficiently [26].

MSCs can be easily isolated from bone marrow, adipose tissue, and umbilical cord, but the isolation process from the original tissues involves extremely invasive procedures [27]. They have been thoroughly investigated for immunotherapy in cancer models due to their innate homing ability [28–31]. MSC-based therapies have shown excellent immunosuppressive effect *in vitro* and *in vivo* due to their modulation of the immune system at various levels [32, 33]. They induce the formation of T-regulatory cells but also interfere with the generation of cytotoxic T cells, helper T, B, and dendritic cells (DCs). Recently, Wang et al. have developed MSC-based gene therapies for malignant glioma treatment and demonstrated that using engineered MSCs as a gene carrier significantly reduced the tumor growth *in vitro* and *in vivo* by increasing apoptosis [34].

NSCs can be derived from bone marrow and have shown an excellent homing ability to inflammatory sites and great capacity to repair damaged tissues of the central and peripheral nervous systems. Thus, NSCs have been used as carriers in particular for the treatment of neurological-associated diseases such as gliomas and neuroblastoma [35]. Recent studies have shown that NSCs isolated from mammalian central nervous system (CNS) including human, can be propagated *in vitro* and then implanted into the brain of animal models of human neurological disorders [36]. For murine models of lysosomal storage diseases, stroke, Parkinson disease, and Huntington disease, F3 human NSCs implanted into the brain were found to migrate throughout the lesion sites, differentiate into neurons and glial cells, and restore functional deficits found in these neurological disorders [36]. Docetaxel loaded nanoparticles have been encapsulated into NSCs (docetaxel-NP@NSCs) and the efficiency was evaluated by monitoring tumor microvasculature, tumor cell proliferation, and apoptosis over a period of seven days. They found that docetaxel-NP@NSCs decreased tumor vessel density and altered cell proliferation compared to the free drug [37].

10.2.1.3 Antigen Presenting Cells (APCs)

APCs are a heterogeneous group of immune cells including DCs, B cells, and macrophages. They can be identified in every tissue, however, they have a short circulation time in the body and their isolation and culturing process is very tedious.

APCs play an important role in the immune response and have been attractive as carriers for immunotherapy. DCs and T cells have been used in clinical trials to induce immune responses directed against specific antigens [23, 38]. DCs have the capability to capture, process, and present foreign antigens to stimulate innate and adaptive antigen-specific immune responses. They can be transduced to express and release different proteins [38, 39]. Various strategies such as DC-tumor cell fusion have been developed to deliver TAAs to DCs as cancer vaccines [40–42]. T-cells have been studied extensively for cell therapies for cancer treatment [43]. In early clinical trials, infusing modified T cells showed promising results for the treatment of cancer and chronic inflammation.

10.2.1.4 Cell Membranes

CMs derived from various cells have attracted much interest as delivery vehicles, as they have retained several functions of living cells such as target specificity, prolonged circulation times, and good biocompatibility [44–47]. CMs are phospholipid bilayer, spherical-shaped vesicles consisting of lipid, proteins, and specific receptors with a diameter size ranging from 40 to 200 nm [48]. The nanosized CMs can enhance cellular uptake and evade the immune system clearance due to their smaller size compared to cells [47]. Most research focuses on development of CM-cloaking nanoparticles for drugs, protein, and genetic material delivery by leveraging the advantages of both natural cells and nanoparticles [44, 49]. They can also load contrast agents for *in vivo* cell tracking [50, 51]. For example, platelet-mimicking nanoparticles are invisible to the immune system and selectively target the damaged vasculature and cancer tissue, thereby effectively alleviating the immunogenicity of exogenous nanoparticles, leading to an enhanced therapeutic effect [52–54]. Neutrophil membrane-coated poly(lactic-*co*-glycolic acid) (PLGA) nanoparticles have inhibited synovial inflammation and alleviated joint damage in inflammatory arthritis [55].

Recently, due to specific homotypic binding to the source cells, cancer CMs have also been used as delivery systems. Cancer CMs have the capability to specifically target primary tumors as well as metastatic sites to enhance drug efficiency for cancer treatment [56–59]. In addition, they have been used as effective therapeutic vaccine platforms to efficiently deliver tumor-specific antigens to stimulate a robust anti-cancer immune response [60]. As reported, cancer CM-coated PLGA NPs were used to deliver TAAs directly to DCs, then after taking up, processing, and presenting the antigen, induced a specific anti-tumor immune response [7, 57]. This system can also be tailored to codeliver both the antigens and adjuvants together, as simultaneous delivery is a preferred enhancement strategy for vaccination [60].

10.2.2 Cargo Loading Methods

Several approaches have been developed to load therapeutic agents into cell-mediated delivery systems. These approaches can be classified into noncovalent and covalent methods. Therapeutic agents such as drugs, siRNA, proteins, and nanoparticles have been loaded into cells, either on the surface or inside cells, via adsorption on the cell surface, internalization into the cells and conjugation on the cells (Figure 10.2). The details of the cargo loading approaches and cellular modifications have been reported by Timin and Joshi [9, 26].

Noncovalent methods include adsorption, hypotonic and isotonic method, and electroporation. The drugs or nanoparticles are adsorbed or attached on the surface of cells (Figure 10.2a) through coculture of the cells with the drugs or nanoparticles. These cargoes are attached on cells via electrostatic interaction and hydrophobic insertion. For example, the hydrophobic groups of the drugs can insert into the hydrophobic interior of the lipid bilayer of CMs, locking them on the cell surface. It is a very simple process but the loading efficiency and stability are low. As shown in Figure 10.2b, small drugs can also be encapsulated into the cells through hypotonic and isotonic methods, where nonphagocytic cells like erythrocytes are reversibly opened under hypotonic/isotonic conditions to encapsulate exogenous substances upon resealing of membrane pores [61–63]. Large molecules such as plasmid DNA, proteins, and nanoparticles are incorporated into the cells through electroporation [64, 65]. The electroporation technique has been used to introduce gadolinium chelates and superparamagnetic iron oxide nanoparticles (SPIONs) into the cells to effectively label the cells for cell tracking and imaging [66, 67]. However, these methods may cause a slight change of the structure and composition of the cells during the loading process, resulting in rapid clearance by the RES organs. In addition, many nanoparticles like polymers, liposomes, and inorganic nanoparticles can be internalized inside the cells via endocytosis or phagocytosis by coculture with the cells. The internalization efficiency of these nanoparticles into the cells mainly depends on cell types and physiochemical properties of nanoparticles including composition, size, charge, surface properties, hydrophobicity/hydrophilicity, and shape. In general, positively charged nanoparticles are

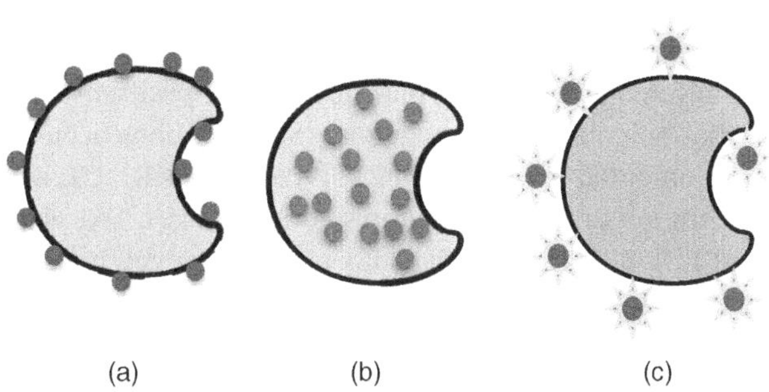

Figure 10.2 The cargo loading into the cells by adsorption (a), internalization and encapsulation (b), and conjugation on the cell surface (c).

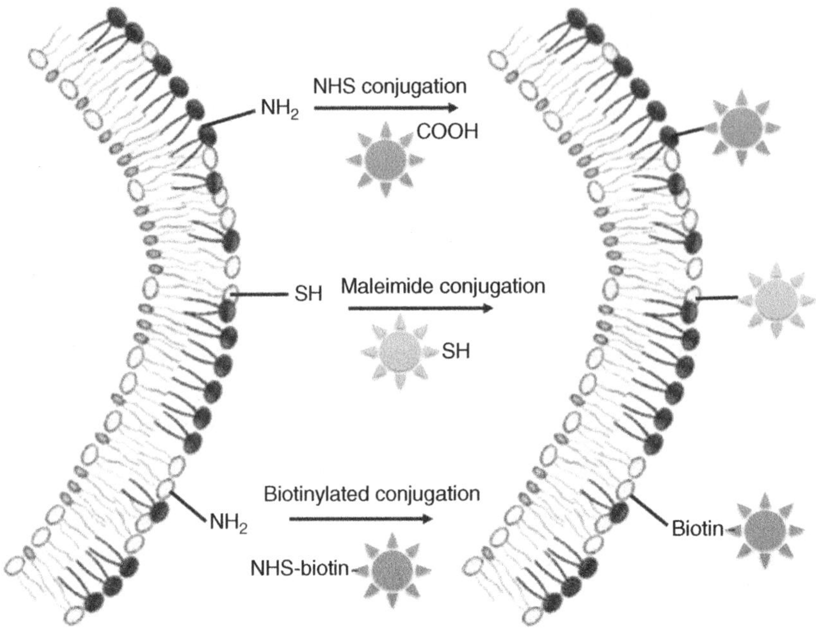

Figure 10.3 Covalent conjugation of drugs/nanoparticles on cells.

internalized into cells faster than negatively charged nanoparticles. Macrophages prefer to uptake nanoparticles in the range of a few nm to μm.

To improve the stability of nanoparticle loading with drugs, the covalent method has been developed to conjugate the active cargos on cells (Figure 10.2c). Most cells possess functional groups on the surface such as thiol, amine, and biotin. Common conjugation methods include (i) carbodiimide conjugation for the primary amine group of membrane protein, (ii) maleimide–thiol conjugation for the thiol groups of cysteine containing membrane proteins, (iii) biotinylated conjugation with streptavidin or avidin on the CM (Figure 10.3).

10.3 Imaging and Tracking of Cell-Based Delivery Systems

Recently, several imaging techniques have been developed and applied to tracking the migration and homing of cell-mediated drug delivery systems *in vivo* such as magnetic resonance imaging (MRI), fluorescent flow cytometry, near-infrared fluorescence imaging (NIRF), positron emission tomography (PET)/single-photon emission computed tomography (SPECT), photoacoustic imaging (PAI), ultrasonography (US), and computed tomography (CT) [9, 65, 68]. MRI, CT, and PET/SPECT can provide deep tissue penetration and obtain high-resolution images. PAI gives the best spatial resolution. NIRF, PAI, and PET/SPECT are sufficiently sensitive for cell imaging at the concentrations of contrast agents ranging from picomolar to nanomolar. In addition, the delivery efficiency of cell-mediated therapies has been evaluated via radiolabeling [20, 21], bioluminescence [22–25], fluorescent protein expression [26–30], and exogenous fluorescence labels [18, 19, 31, 32]. For example, radionuclides are the clinical standard for cellular imaging approved by US Food and Drug Administration (FDA). They have been used to label DCs for scintigraphic imaging.

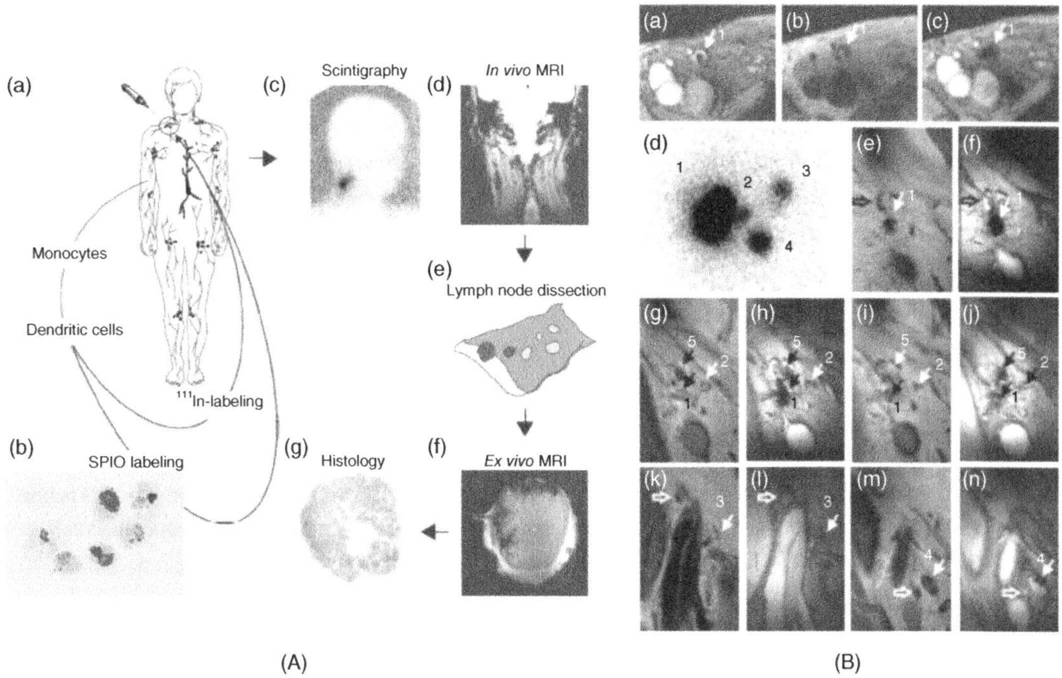

Figure 10.4 (A) Cell-tracking study protocol of SPIO and ^{111}In-labeled monocytes. (a–f) Monocytes from cytopheresis in stage-III melanoma patients (a) were labeled with SPIO particles and ^{111}In (b) and then injected intranodally into a (either cervical, inguinal, or axillary) lymph node basin by scintigraphy (c) and MRI at 3 T (d). (e–g) The lymph node basis is resected and separate lymph nodes are visualized with high-resolution MRI at 7 T (f) and histology (g); (B) *In vivo* scintigraphic and MRI. (a–c) Monitoring of the delivery of dendritic cells labeled with SPIO and ^{111}In by MRI before and after intranodal injection in patient 1 (a–c) and patient 3 (d–n). Source: From de Vries et al. [68]. © 2005, Springer Nature.

10.3.1 MRI

SPIONs and gadolinium (Gd) chelates, have been commonly used as MRI agents for cell tracking [13, 69–73]. For example, SPIONs and Gd chelates have been loaded into RBCs for MRI and X-ray imaging. These MRI agents have long circulation time and high stability in the blood circulatory system. Gadopentetic acid (Gd-DTPA) has been loaded into RBC for prolonged intravascular tissue enhancement via MRI [8]. Zhang et al. encapsulated gadolinium oxide nanoparticles (Gd_2O_3) or gadolinium hybrid (GH) nanoparticles within RBCs and found that the Gd_2O_3 and GH RBCs exhibited high relaxation rates and revealed high sensitivity for T_1 MRI [13]. Choi et al. developed a macrophage-mediated delivery system containing iron oxides (IOs) and doxorubicin encapsulated liposomes (LP-DOX) for cancer treatment. They confirmed that macrophage-IO could migrate and infiltrate into tumors effectively by MRI [23]. Autologous DCs have been labeled with a clinical superparamagnetic iron oxide (SPIO) formulation or ^{111}In-oxine via co-injection under ultrasound guidance in a lymph node for melanoma patients (Figure 10.4) [68]. MRI results have shown the accuracy of DC delivery and inter- and intranodal cell migration patterns. MRI cell tracking using iron oxides appears clinically safe and well suited to monitor cellular therapy in humans. Human MSCs were labeled *in vitro* with SPION with no effect on differentiation potential, proliferation, survival, or migration of the cells. Iron-labeled MSCs were intravenously injected into the mice and tracked *in vivo* to multiple lung metastases using MRI [73].

10.3.2 PET

PET can provide precise quantification imaging and excellent tissue penetration with excellent sensitivity. Huang et al. constructed ^{64}Cu labeled hyaluronic acid on mesoporous silica NPs (HA-MSN-^{64}Cu) and then incorporated HA-MSN-^{64}Cu into the MSC. The mesenchymal stem cell nanoparticle (MSC-NP) system overcame the blood–brain barrier and actively targeted orthotropic glioma. PET imaging demonstrated that there was a 5.2 ± 1.3-fold higher tumor accumulation in the glioma by MSC-HA-MSN-^{64}Cu within 24 hours after injection than the free HA-MSN-^{64}Cu [33].

10.3.3 X-Ray Imaging

Time-resolved dynamic imaging of biofluids can provide valuable information for clinical diagnosis and treatment of circulatory disorders [70]. Ahn et al. encapsulated modified Au NPs (20 nm) into RBCs to make a contrast-enhanced flow tracer for biomedical imaging of blood flows [70]. They found that Au@RBCs generated enhanced dynamic X-ray imaging and effectively captured the dynamic motions of blood flows. They demonstrated that the neutral surface on Au NPs is beneficial for NPs to penetrate into RBCs regardless of hydrophilic or hydrophobic properties of NPs.

10.3.4 Multimodal Imaging Techniques

Recently, multimodal imaging techniques have been developed to track the cells with different imaging agents, to obtain a more precise status of administrated cells. Giesel et al. have used gadoflurine M and Cy3.5 agents to label MSCs and tracked them via MRI and optical imaging techniques [74]. Gadoflurine M–Cy3.5 were internalized into the cytosol through coculture with MSCs and provided clear visualization of cells via MRI and fluorescent microscopy in the rat brain.

10.4 Cell-Mediated Drug Delivery Systems for Disease Treatment

10.4.1 Cancer Therapy

Extracted CMs have attracted much interest as carriers to encapsulate therapeutic agents and coat nanoparticles for various cancer therapies and imaging. Sun et al. reported 4T1 CM-coated DOX-gold nanocages (CDAuNs) to significantly inhibit tumor growth and metastasis of breast cancer as shown in Figure 10.5 [58]. An *in vitro* study has shown that CDAuNs can selectively target homotypic tumor cells, trigger DOX release due to the hyperthermia under near-infrared imaging (NIR) irradiation, and improve drug efficiency via chemo/photothermal combinational therapy. Furthermore, these CM-coated combinational therapy inhibited the tumor growth with 98.9% and 98.5% inhibition rates of tumor volume and metastatic nodules.

Recently, Choi et al. developed macrophage bearing liposomal-DOX (macrophage-LP-DOX) for targeted chemotherapy to improve cancer therapeutic efficacy through deeper penetration into the tumor, even into the hypoxic region behind tumor blood vessels [23].

10.4.2 Immunotherapy

Immunotherapy uses the patient's own immune system to fight cancer including chimeric antigen receptor (CAR) T-cell therapy, monoclonal antibody therapy, immune checkpoint inhibitor therapy, and cancer vaccines. DCs have been commonly used as adjuvants to deliver antigens

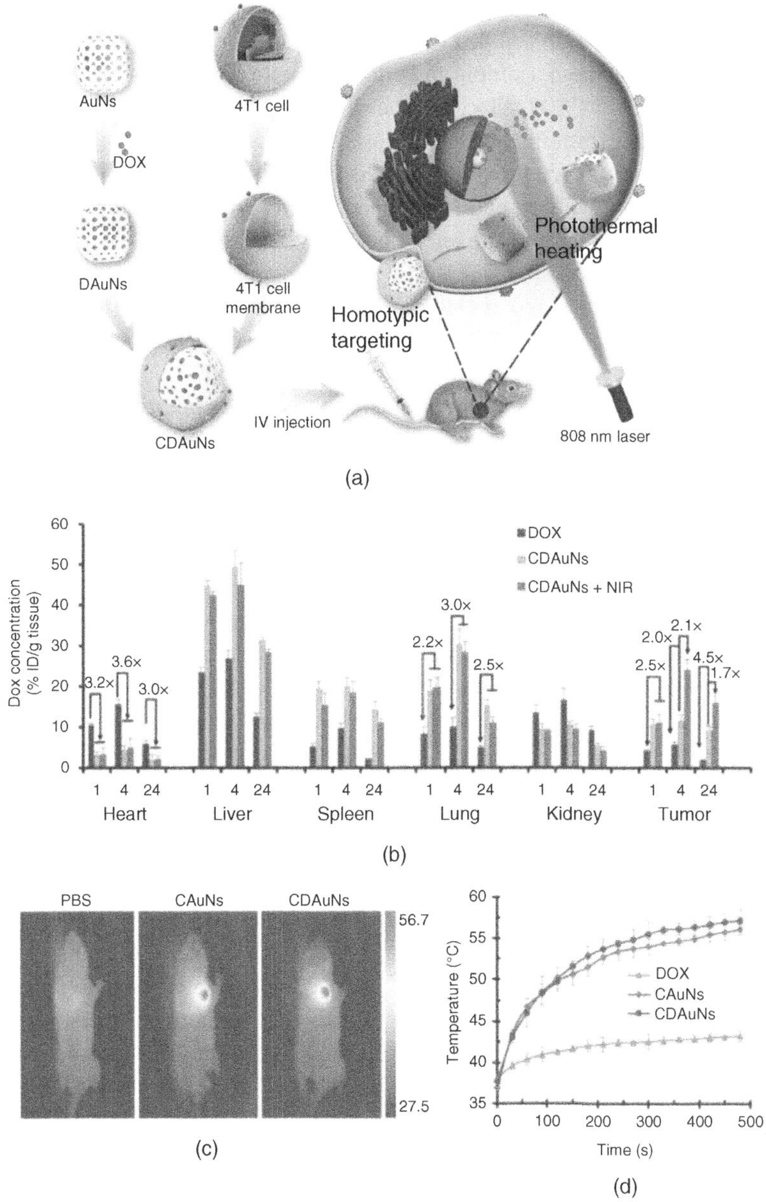

Figure 10.5 (a) Schematic illustration of the 4T1 CM-coated gold nanocage (CDAuNs) for homotypic-targeted combinational therapy of breast tumor growth and metastasis. (b) Quantitative *in vivo* biodistribution analysis of DOX in different formulations at 1, 4, and 24 hours after intravenous injection in nude mice bearing metastatic 4T1 breast tumor. (c) Infrared thermographic images of 4T1 tumor-bearing nude mice after NIR irradiation. (d) Temperature increase behaviors of the tumor tissues in the mice receiving intravenous injection with DOX, CAuNs, and CDAuNs. Source: From Sun et al. [58]. © 2016, John Wiley & Sons.

against cancer. *In vivo* micemodel studies have shown that tumor antigen-loaded DCs can induce therapeutic and protective antitumor immunity as DCs can regulate T-cell immunity. Immature DCs and mature DCs have been used to carry proteins and peptides to the tumor via fusion of DCs with tumor cells [75]. Cole et al. have developed retroviral vector-loaded antigen-specific T cells to control tumor growth [43]. They used a simple and effective method to adhere MFG-LacZ retroviral particles on the surface of $CD8^+$ OT-1 T cells. These viruses were delivered to tumor sites after *in vivo* T-cell transfer in fully immunocompetent hosts and subsequently released from the T-cell surface to enhance cytotoxic efficacy, providing high levels of selectivity for viral transfer to metastatic tumors *in vivo* [43]. Recently, Banz et al. developed a novel antigen-delivery system using RBCs to improve cancer immunotherapy. The antigens were loaded in the RBCs to target DCs and induce cytotoxic T-cell responses [10].

Yang et al. have used human-induced pluripotent stem cell (iPSC)-derived NSCs as a cellular delivery carrier for gene therapy. iPSC-derived NSCs were transduced with a baculoviral vector containing the herpes simplex virus thymidine kinase suicide gene and injected via the tail vein to 4T1 breast tumor-bearing mice, showing inhibition of the growth of the orthotopic 4T1 breast tumor and the metastatic spread of the cancer cells in the presence of ganciclovir [76].

10.4.3 Brain-Related Diseases

Stem cells can actively cross the BBB and enter into the inflammatory sites in the brain. NSCs have been used to carry and deliver therapeutic agents (drugs, enzymes, genes, growth factors, etc.) for neurological diseases such as Alzheimer's disease, Parkinson's disease, Huntington's disease, and brain cancers. Tyle et al. used NSCs as carriers to specifically deliver a conditionally replicating adenovirus to U87MG glioma and demonstrated that NSCs delivery system enhanced intratumoral distribution of an oncolytic vector [77]. Huang et al. designed and prepared a MSC-based multifunctional platform to target orthotopic glioblastoma by integrating the tumor-targeted delivery of MSC and multimodal imaging agents-loaded mesoporous silica nanoparticles (MSNs) [33]. The MSN-MSC platform has shown rapid cellular uptake, long retention time, and stability of particles. *In vivo* multimodal imaging (NIR, MRI, and PET) showed the feasibility of tumor tropism delivery of multifunctional MSC platform with improved tumor delivery of NPs compared with free particles.

Klyachko et al. have developed an antioxidant- or catalase-loaded backpack (BP) macrophage platform for the treatment of brain inflammation [78]. BPs loaded with a potent antioxidant or catalase were attached to autologous macrophages and transported across the BBB in a mouse model of lipopolysaccharide (LPS)-induced encephalitis, enhancing the therapeutic effects. These results suggested that immune cells such as monocyte and macrophages as drug carriers may be new strategy for treating a range of neuroinflammatory and neurodegenerative diseases [78].

10.4.4 Inflammatory Diseases

Ko et al. loaded palmitate protein G (PPG) into MSCs with antibodies to target the inflamed colon and draining mesenteric lymph nodes for inflammatory bowel diseases (IBDs). Antibody-fused MSCs enhanced PPG accumulation in the inflamed sites and reduced inflammation [79].

Anselmo et al. designed cellular BPs attached to monocytes for inflammatory disease treatment [21]. They revealed BP-loaded monocytes retained monocyte functions and transported BP efficiently to the target inflamed tissues, enhancing the treatment effectiveness [21].

In addition, APCs such as macrophages and DCs play an important role in the immune response, and they can migrate to the acute inflammatory tissues. Zhang et al. studied acute cytokine response to systemic adenoviral vectors (Ad) *in vivo* mediated by DC and macrophages via intravenous injection in mice model [40]. They found that Ad-modified DC and macrophages accumulated in the splenic marginal zone one hour after intravenous (IV) injection. Inflammatory cytokines including interleukin-6 (IL-6), IL-12, and tumor necrosis factor-alpha (TNF-α) were detected *in vivo*. These early activated splenic DCs spontaneously produced high levels of IL-6 and IL-12 but activated splenic macrophages spontaneously secreted only IL-6, suggesting DC and macrophages play different roles in Ad-triggered acute inflammatory response in terms of their abilities to produce distinct patterns of inflammatory cytokines [40].

10.4.5 Theranostic Application

Baharak et al. have developed indocyanine green (ICG)-loaded RBC membrane for NIR and therapeutic applications. They have demonstrated ICG-loaded RBC membrane significantly inhibited tumor growth via photothermal destruction and displayed clear NIR imaging, which may have potential application in personal nanomedicine [14].

Image-guided chemo-photodynamic combination therapy has also been developed by loading chlorine e6-iron oxide nanoparticles and DOX into RBCs for cancer treatment, achieving excellent synergistic therapeutic effects in an animal tumor model [62].

10.4.6 Others

CM coating improves nanoparticles and therapeutic agents with perfect biocompatibility and highly specific targeting ability to homologous tumors, enhancing the efficiency of cancer therapy. Single CM has commonly been used to coat the therapeutic agents and nanoparticles. Nevertheless, Diana et al. have recently developed dual-membrane-coated nanoparticles using RBC membrane and platelet membrane ([RBC-P]NPs), as shown in Figure 10.6 [80]. RBC-P was prepared via fusion of RBC membrane with a platelet membrane at a ratio of nearly 1 : 1, retaining the cell function well. PLGA cores were coated with RBC-P via sonication method. [RBC-P]NPs retained functionality of each individual cell type well. [RBC-P]NPs were highly bound to metastatic MDA-MB231 human breast cancer cells within the spectrum between red blood cell nanoparticle (RBCNPs) and PNPs, which exhibited low and high amounts of binding, respectively. The fusion of different CMs combined with nanoparticles offer great potential to marry functions across disparate fields, including targeted drug delivery, immune modulation, and detoxification [80].

10.5 The Mechanism of Cell-Mediated Delivery Systems for the Cell Therapies

Cell–cell communication plays a critical role in cell-mediated delivery systems. Lu et al. investigated the cell interaction between bone marrow mesenchymal stem cells (BM-MSCs) and nucleus pulposus cells (NPCs) used for stem cell therapies for intervertebral disk degeneration [81]. They found that when coculturing NPC exosomes with BM-MSCs, NPC exosomes promoted BM-MSC migration, and induced BM-MSC differentiation to a nucleus pulposus-like phenotype; when coculturing BM-MSC exosomes with NPCs, BM-MSC exosomes promoted NPC proliferation, and

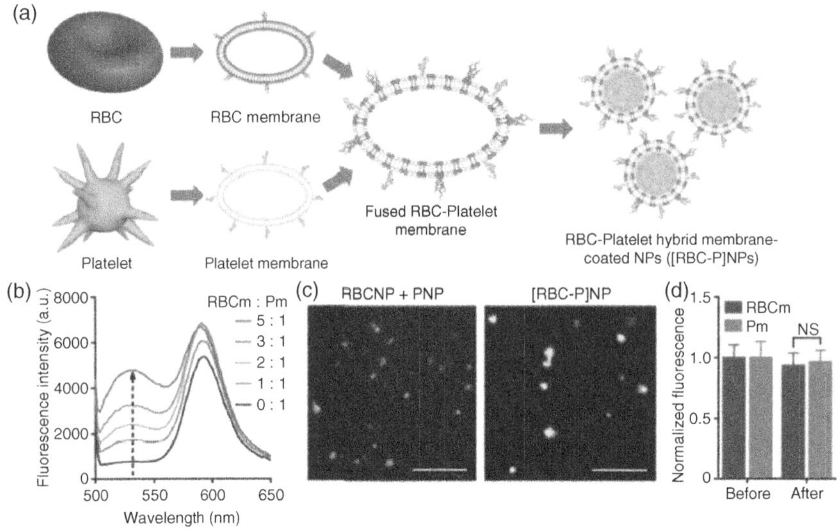

Figure 10.6 (a) Schematic of membrane fusion and coating. Membrane material is derived from both RBCs and platelets and then fused together. The resulting fused membrane is used to coat poly(lactic-*co*-glycolic acid) (PLGA) polymeric cores to produce [RBC-P]NPs. (b) Platelet membrane was doped with a Förster resonance energy transfer (FRET) pair of fluorescent probes and mixed with increasing amounts of RBC membrane. The recovery of fluorescence emission from the donor at the lower emission peak (534 nm) was monitored (RBC_m:P_m = RBC membrane to platelet membrane protein ratio). (c) Confocal fluorescent microscopy images of either a mixture of RBCNPs and platelet nanoparticles (PNPs) or of the [RBC-P]NPs (red = RBC membrane, green = platelet membrane; scale bar = 10 μm). (d) Fluorescent quantification of dye-labeled RBCm and Pm inputted to [RBC-P]NPs at a 1 : 1 protein ratio before the coating process and after purification of the coated nanoparticles ($n = 3$; mean ± SD). Source: From Dehaini et al. [80]. © 2017, John Wiley & Sons.

healthier extracellular matrix production in the degenerate NPCs [81]. Their findings suggested that cell–cell communication should play an important part in the delivery process and treatment. As summarized from previous research, the major advantage of cell-mediated delivery systems includes detoxification, adhesion, and homing of cells to target sites.

10.5.1 Detoxification

As reported, neutralization of toxins, especially α-toxin, may increase proinflammatory cytokine production and enhance bacterial clearance [44]. Recent research on cell-mediated delivery systems has demonstrated that RBC-mediated nanoparticle delivery systems can bind bacterial pore-forming toxins (PFTs) and pathological autoantibodies to prevent their attack of healthy RBCs and macrophage-coated NPs can bind and neutralize endotoxins to reduce the immune activation [6, 44, 45, 53, 55, 82, 83].

Zhang et al. developed neutrophil membrane-coated PLGA NPs to inhibit synovial inflammation and alleviate joint damage in inflammatory arthritis. They confirmed neutrophil-coated PLGA efficiently targeted activated human chondrocytes or human umbilical vein endothelial cells (HUVECs) due to specific interactions between lymphocyte function-associated antigen 1 (LFA-1) on the neutrophil membrane and the intercellular activation molecule 1 (ICAM-1) overexpressed on activated chondrocytes and HUVECs. The binding and neutralizing of neutrophil-coated PLGA to ICAM-1 resulted in the improvement of cartilage protection efficacy and anti-arthritis

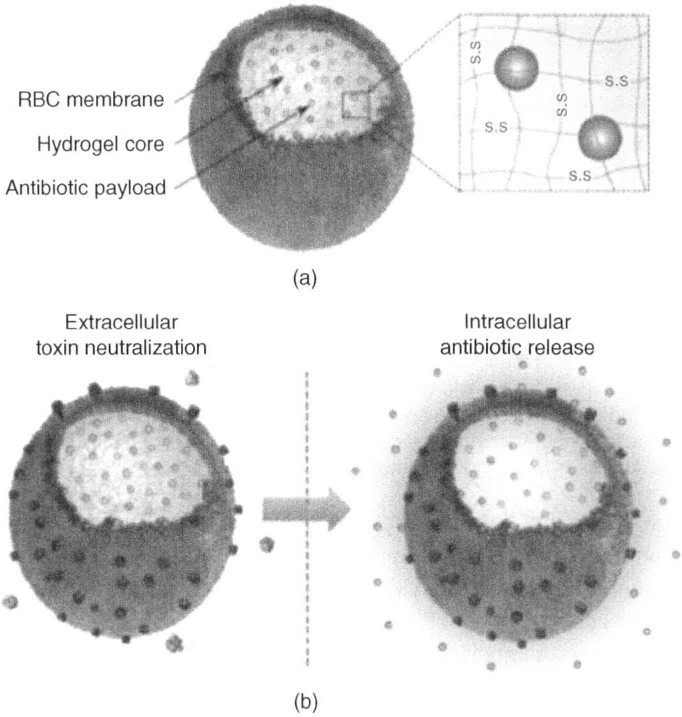

Figure 10.7 Schematic illustrations of a red blood cell (RBC) membrane-coated nanogel (RBC-nanogel) system for combinatorial extracellular toxin neutralization and intracellular redox-responsive antibiotic release. (a) The RBC-nanogel is constructed with a natural RBC membrane shell and a hydrogel core containing antibiotic payload. The hydrogel core is cross-linked with disulfide bonds, cleavable in reducing environment. (b) In extracellular environment, the RBC-nanogel absorbs and neutralizes pore-forming toxins (PFTs) secreted by bacteria. Such toxin sequestration facilitates the intracellular uptake of the toxin-secreting bacteria and the RBC-nanogel. Once inside the cell, the RBC-nanogel rapidly releases its encapsulated antibiotic payload triggered by a redox reaction. Source: From Zhang et al. [44]. © 2017, Elsevier.

effectiveness *in vivo* [55]. A redox-responsive RBC membrane coated-nanogel (RBC-nanogel) platform has been designed for the combinatorial antivirulence and responsive antimicrobial delivery against methicillin-resistant *Staphylococcus aureus* (MRSA) infection [44]. As shown in Figure 10.7, RBC shells of RBC-nanogels have absorbed and neutralized PFTs secreted by bacteria to increase proinflammatory cytokine production and enhance bacteria clearance by immune cells.

10.5.2 Adhesive Mechanism

Several researchers have demonstrated that cell migration and accumulation to the disease sites may be attributed to strong adhesion of circulating cells. Thomas studied the neutrophil interactions with activated surface-adherent platelets in hydrodynamic shear flow and in chemotaxis [84]. Platelets bound to thrombogenic surfaces supported activation-dependent firm adhesion of neutrophils in hydrodynamic shear flow or chemotaxis. They found that neutrophils from patients without β2-integrin receptors can not develop firm adhesion to activated platelets though P-selectin-mediated rolling-interactions; however, healthy neutrophils tether and roll on adherent, activated platelets in shear flow arrest after several seconds and develop firm

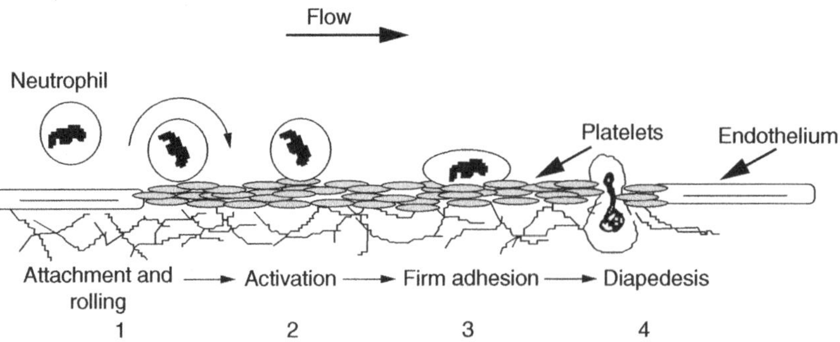

Figure 10.8 Proposed sequential adhesion model of neutrophil attachment to and transmigration across activated, surface-adherent platelets. P-selectin mediates the initial attachment and rolling in flow (1); surface receptor(s) or surface-bound cytokine(s) induce neutrophil activation (2); the β2-integrin Mac-1 promotes firm adhesion (3), and also mediates chemoattractant-induced diapedesis (4). Source: From Diacovo et al. [84]. © 1996, The American Society of Hematolog.

adhesion. The firm adhesion interaction between neutrophils and activated platelets is ascribed to β2-integrins Mac-1 that promotes neutrophil accumulation on and migration across platelets deposited at vascular injury sites. They have proposed and demonstrated an adhesive mechanism of the neutrophil–platelet interaction (Figure 10.8). As shown in Figure 10.8, P-selectin induces neutrophil rolling adhesions, playing an important role in neutrophil recruitment to site of platelet deposition (step 1). The subsequent arrest of rolling neutrophils on surface-adherent platelets is induced by surface receptors or cytokines such as the β2-integrins (step 2 in Figure 10.8). These molecules promote firm adhesion on endothelium (step 3 in Figure 10.8) and mediate chemoattractant-induced diapedesis (step 4).

10.5.3 Homing Mechanism

Stem cells have shown great potential in delivering therapeutic agents to the tumor sites for cancer therapy as they can actively immigrate to the inflammatory sites. Moreover, APCs have displayed the homing ability to hypoxic, ischemic, and necrotic environments that are associated with diseases [26]. Loebinger et al. reported to deliver tumor necrosis receptor-related apoptosis-inducing ligand (TRAIL) protein delivered by MSCs for metastatic breast cancer treatment [73]. They further used iron oxide NPs to track the delivery and migration of transduced MSCs to the tumor via MRI. They proposed that the mechanism of MSCs to the tumor may involve chemokine ligands and receptors with the recruitment of leukocytes to area of inflammation [73].

In addition, Timin et al. demonstrated that leukocytes could be used to access the microenvironment deep within tumors as monocytes, tumor-associated macrophages, and tumor-infiltrating lymphocytes are frequently found within tumor microenvironments, including hypoxic and necrotic regions. The leukocyte-mediated delivery system has the ability to target the tumor sites due to tumor-homing mechanisms of leukocytes, which would offer new approaches for cell-mediated drug delivery [9].

10.6 The Administration Approach of Cell-Assist Drug Delivery System

Cell-mediated delivery systems can be administrated by various routes including intravenous (IV), intraperitoneal (IP), and subcutaneous (SC) routes for disease treatment. For example, DCs have been used in clinical trials via ID, IV, intranodal injection, and SC injection [41]. Recently, intradermal (ID) injection has been developed and used for cell therapies. As reported, ID and intranodal administration of antigen-loaded DCs were more effective to induce an immune response [85, 86]. Furthermore, SC or ID injection of DCs also indicated strong immune response [41].

The inhalation route has also been used for cell-mediated delivery system. Gupta et al. administered fasudil-encapsulated nanoerythrosomes (NERs) by the inhalation route [87]. Though the maximum concentration of fasudil in NERs administrated intratracheally (IT) was lower than that administrated intravenously (IV), fasudil in NERs via IT administration was available in the circulation even after 24 hours, but plain fasudil was not detectable after 6 hours. Moreover, the formulation safety studies indicated that NERs had no significant toxicity and can be used as cell-derived carriers for inhalational delivery of fasudil [87]. They further modified NERs with cell permeable peptide (CAR-NERs) and evaluated the CAR-NERs as inhalable carriers of for fasudil the treatment of pulmonary arterial hypertension (PAH) [88]. CAR-NERs-Fasudil were effective in pulmonary selective vasodilation with little effect on mean systemic arterial pressure (MSAP). *In vivo* drug absorption studies demonstrated that NERs extended the half-life of fasudil by approximately six- to eightfold compared with plain fasudil administered via either IV or IT route.

10.7 Clinical Application of Cell-Based Delivery Systems

DCs and RBCs were the first cell-based delivery systems to be used in preclinical and clinical trials. DCs can actively induce antigen specific T-cell immunity *in vivo* to control tumor growth, even for the heavily pretreated cancer patients. Furthermore, some autologous DC-mediated immunotherapy for cancer treatment has been approved by the FDA. In 2001, Foley et al. reviewed the genetically modified DCs in preclinical and clinical application for malignancies, including melanoma and prostate cancer [89]. Tumor antigen-loaded DC vaccines have been introduced in the clinic and have proven feasible and nontoxic, and both immunological and clinical responses have been observed. The gene-modified DCs encapsulated multiantigens or peptides were intravenously injected to the patients without significant toxicity. The injected modified DCs have induced antitumor immune responses in patients with cancer. RBCs have been used to encapsulate therapeutic agents (enzymes and drugs) for leukemia and neurodegenerative diseases [90].

Portnow et al. performed a first-in-human study on NSC-mediated anticancer gene therapy via intracranial administration in high-grade glioma patients and found cytosine deaminase-neural stem cells (CD-NSCs) migrated to distant tumor sites and mediated 5-fluorouracil (FU) in the brain [91]. They demonstrated initial safety and the ability of NSCs to target brain tumors *in vivo*.

Recently, Leoni et al. have developed an automated ID injection device for fibroblast-based cell therapy and performed several preclinical tests to evaluate their safety and efficiency for drug delivery [92]. They confirmed that the automated ID injection consistently delivered repeated doses to the ID layer in an *ex vivo* skin model. In addition, platelet-mediated delivery systems have also been developed and evaluated in preclinical trials, particularly for the targeting of tumor cells.

10.8 Conclusion and Outlook

Cell-mediated delivery systems have shown great potential to become implemented in the clinic for the treatment of several diseases. As discussed in Section 10.2, blood cells, stem cells, APC cells, and cancer cells are able to load therapeutic agents efficiently and transport these agents to their desired sites. Cell-mediated therapy has significantly enhanced the effectiveness of disease treatments and reduced the side effects of the therapeutic agents. Thus, cell-mediated delivery systems have been developed and used for various disease treatments and imaging. They have sparked intense interest in their preclinical/clinical application. Moreover, CM-cloaking nanoparticle delivery systems have been developed and become a promising therapeutic platform. CM coating has improved the delivery of nanoparticles and therapeutic agents with perfect biocompatibility and highly specific targeting ability to homologous tumors. Recent research has demonstrated that cell–cell interaction and communication play an important role on cell migration to targeted cells. However, the efficacy of these cellular therapies is ultimately contingent upon their ability to appropriately control the fate and function of the therapeutic cells [93]. The main challenges for cell-mediated delivery system are to load therapeutic agents into selected cells without changes of cell properties, transport them to the target sites, and retain them. Modification of the cell surface with various ligands helps to transport cells to the target sites and control the release of the therapeutic agents. Though there have been advances in modification of cell surface to enhance the target efficiency, the modification processes are complicated and cell properties have been altered during the process, resulting in the induction of immunogenicity to clear the delivery system. Development of simple approaches to modify the cells without any functional change is highly important. CM-coating nanoparticles with therapeutic agents and imaging probe encapsulation are able to transport and track the therapeutic agents to the target sites and monitor the release of the therapeutic agents. This cell-mimicking strategy offers great opportunity to develop a novel cell-based therapy to improve their effectiveness and reduce the side effects in clinical application.

References

1 Siegel, R. (1997). *Controlled Drug Delivery Challenges and Strategies* (ed. K. Park). Washington, DC: American Chemical Society.

2 Mitragotri, S. and Lahann, J. (2009). Physical approaches to biomaterial design. *Nat. Mater.* 8: 15.

3 Wilhelm, S., Tavares, A.J., Dai, Q. et al. (2016). Analysis of nanoparticle delivery to tumours. *Nat. Rev. Mater.* 1: 16014.

4 Ihler, G.M., Glew, R.H., and Schnure, F.W. (1973). Enzyme loading of erythrocytes. *Proc. Natl. Acad. Sci. U.S.A.* 70: 2663.

5 Millán, C.G., Gandarillas, C.I.C., Marinero, M.L.S., and Lanao, J.M. (2012). Cell-based drug-delivery platforms. *Ther. Deliv.* 3: 25.

6 Sun, Y., Su, J., Liu, G. et al. (2017). Advances of blood cell-based drug delivery systems. *Eur. J. Pharm. Sci.* 96: 115.

7 Tan, S., Wu, T., Zhang, D., and Zhang, Z. (2015). Cell or cell membrane-based drug delivery systems. *Theranostics* 5: 863.

8 Burnouf, T., Burnouf, P.A., Wu, Y.W. et al. (2018). Circulatory-cell-mediated nanotherapeutic approaches in disease targeting. *Drug Discovery Today* 23: 934.

9 Timin, A.S., Litvak, M.M., Gorin, D.A. et al. (2018). Cell-based drug delivery and use of nano- and microcarriers for cell functionalization. *Adv. Healthcare Mater.* 7: 1700818.

10 Banz, A., Cremel, M., Rembert, A., and Godfrin, Y. (2010). In situ targeting of dendritic cells by antigen-loaded red blood cells: a novel approach to cancer immunotherapy. *Vaccine* 28: 2965–2972.

11 Sulkowski, M.S. (2003). Anemia in the treatment of hepatitis C virus infection. *Clin. Infect. Dis.* 37: S315.

12 Sahoo, K., Karumuri, S., Hikkaduwa Koralege, R.S. et al. (2017). Molecular and biocompatibility characterization of red blood cell membrane targeted and cell-penetrating-peptide-modified polymeric nanoparticles. *Mol. Pharmaceutics* 14: 2224.

13 Zhang, K., Cao, Y., Kuang, Y. et al. (2016). Gd_2O_3 and GH combined with red blood cells to improve the sensitivity of contrast agents for cancer targeting MR imaging. *Biomater. Sci.* 5: 46.

14 Bahmani, B., Bacon, D., and Anvari, B. (2013). Erythrocyte-derived photo-theranostic agents: hybrid nano-vesicles containing indocyanine green for near infrared imaging and therapeutic applications. *Sci. Rep.* 3: 2180.

15 Howard, M., Zern, B.J., Anselmo, A.C. et al. (2014). Vascular targeting of nanocarriers: perplexing aspects of the seemingly straightforward paradigm. *ACS Nano* 8: 4100.

16 Al Ghumlas, A. and Abdel Gader, A.G. (2013). The blood platelet: an intriguing cell. *J. Appl. Hematol.* 4: 1–12.

17 Felding-Habermann, B. (2003). Targeting tumor cell-platelet interaction in breast cancer metastasis. *Pathophysiol. Haemos. Thromb.* 33 (Suppl. 1): 56.

18 Xu, P., Zuo, H., Chen, B. et al. (2017). Doxorubicin-loaded platelets as a smart drug delivery system: an improved therapy for lymphoma. *Sci. Rep.* 7: 42632.

19 Hansen, C.E., Myers, D.R., Baldwin, W.H. et al. (2017). Platelet-microcapsule hybrids leverage contractile force for targeted delivery of hemostatic agents. *ACS Nano* 11: 5579.

20 Pastore, C. (2017). Drug delivery hitching a platelet ride. *Nat. Nanotechnol.* 12: 613.

21 Anselmo, A.C., Gilbert, J.B., Kumar, S. et al. (2015). Monocyte-mediated delivery of polymeric backpacks to inflamed tissues: a generalized strategy to deliver drugs to treat inflammation. *J. Controlled Release* 199: 29.

22 Mitchell, M.J. and King, M.R. (2015). Leukocytes as carriers for targeted cancer drug delivery. *Expert Opin. Drug Deliv.* 12: 375.

23 Choi, J., Kim, H.Y., Ju, E.J. et al. (2012). Use of macrophages to deliver therapeutic and imaging contrast agents to tumors. *Biomaterials* 33: 4195.

24 Eisenblatter, M., Ehrchen, J., Varga, G. et al. (2009). In vivo optical imaging of cellular inflammatory response in granuloma formation using fluorescence-labeled macrophages. *J. Nucl. Med.* 50: 1676.

25 Lisy, M.R., Schuler, E., Lehmann, F. et al. (2006). Diagnosis of peritonitis using near-infrared optical imaging of in vivo labeled monocytes-macrophages. *J. Biomed. Optics* 11: 9.

26 Joshi, B.P., Hardie, J., and Farkas, M.E. (2018). Harnessing biology to deliver therapeutic and imaging entities via cell-based methods. *Chem. Eur. J.* 24: 8717.

27 Pittenger, M.F., Mackay, A.M., Beck, S.C. et al. (1999). Multilineage potential of adult human mesenchymal stem cells. *Science* 284: 143.

28 Lee, W.Y.W., Zhang, T., Lau, C.P.Y. et al. (2013). Immortalized human fetal bone marrow-derived mesenchymal stromal cell expressing suicide gene for anti-tumor therapy in vitro and in vivo. *Cytotherapy* 15: 1484.

29 Jing, W., Chen, Y., Lu, L. et al. (2014). Human umbilical cord blood-derived mesenchymal stem cells producing IL15 eradicate established pancreatic tumor in syngeneic mice. *Mol. Cancer Ther.* 13: 2127.

30 Andreas, K. (2008). Mesenchymal stem cell delivery routes and fate. *Int. J. Stem Cells* 1: 1.

31 Meng, X., Ichim, T.E., Zhong, J. et al. (2007). Endometrial regenerative cells: a novel stem cell population. *J. Transl. Med.* 5: 57.

32 Young, J.S., Morshed, R.A., Kim, J.W. et al. (2014). Advances in stem cells, induced pluripotent stem cells, and engineered cells: delivery vehicles for anti-glioma therapy. *Expert Opin. Drug Deliv.* 11: 1733.

33 Huang, X., Zhang, F., Wang, H. et al. (2013). Mesenchymal stem cell-based cell engineering with multifunctional mesoporous silica nanoparticles for tumor delivery. *Biomaterials* 34: 1772.

34 Wang, X.J., Xiang, B.Y., Ding, Y.H. et al. (2017). Human menstrual blood-derived mesenchymal stem cells as a cellular vehicle for malignant glioma gene therapy. *Oncotarget* 8: 58309.

35 Fliervoet, L.A.L. and Mastrobattista, E. (2016). Drug delivery with living cells. *Adv. Drug Delivery Rev.* 106: 63.

36 Kim, S.U. (2007). Genetically engineered human neural stem cells for brain repair in neurological diseases. *Brain Dev.* 29: 193.

37 Ayer, M. and Klok, H.-A. (2017). Cell-mediated delivery of synthetic nano- and microparticles. *J. Controlled Release* 259: 92.

38 Koido, S. (2016). Dendritic-tumor fusion cell-based cancer vaccines. *Int. J. Mol. Sci.* 17: 828.

39 Agrahari, V., Agrahari, V., and Mitra, A.K. (2017). Next generation drug delivery: circulatory cells-mediated nanotherapeutic approaches. *Expert Opin. Drug Deliv.* 14: 285.

40 Zhang, Y., Chirmule, N., Gao, G.P. et al. (2001). Acute cytokine response to systemic adenoviral vectors in mice is mediated by dendritic cells and macrophages. *Mol. Ther.* 3: 697.

41 Nencioni, A., Grünebach, F., Schmidt, S.M. et al. (2008). The use of dendritic cells in cancer immunotherapy. *Crit. Rev. Oncol. Hematol.* 65: 191.

42 Banchereau, J. and Palucka, A.K. (2005). Dendritic cells as therapeutic vaccines against cancer. *Nat. Rev. Immunol.* 5: 296.

43 Cole, C., Qiao, J., Kottke, T. et al. (2005). Tumor-targeted, systemic delivery of therapeutic viral vectors using hitchhiking on antigen-specific T cells. *Nat. Med.* 11: 1073.

44 Zhang, Y., Zhang, J., Chen, W. et al. (2017). Erythrocyte membrane-coated nanogel for combinatorial antivirulence and responsive antimicrobial delivery against *Staphylococcus aureus* infection. *J. Controlled Release* 263: 185.

45 Hu, C.M.J., Fang, R.H., Copp, J. et al. (2013). A biomimetic nanosponge that absorbs pore-forming toxins. *Nat. Nanotechnol.* 8: 336.

46 Mohandas, N. and Gallagher, P.G. (2008). Red cell membrane: past, present, and future. *Blood* 112: 3939.

47 Rao, L., Bu, L.L., Xu, J.H. et al. (2015). Red blood cell membrane as a biomimetic nanocoating for prolonged circulation time and reduced accelerated blood clearance. *Small* 11: 6225.

48 Pang, L., Zhang, C., Qin, J. et al. (2017). A novel strategy to achieve effective drug delivery: exploit cells as carrier combined with nanoparticles. *Drug Deliv.* 24: 83.

49 Yuan, D., Zhao, Y., Banks, W.A. et al. (2017). Macrophage exosomes as natural nanocarriers for protein delivery to inflamed brain. *Biomaterials* 142: 1.

50 Lim, K.S., Lee, D.Y., Valencia, G.M. et al. (2017). Cell surface-engineering to embed targeting ligands or tracking agents on the cell membrane. *Biochem. Biophys. Res. Commun.* 482: 1042.

51 Youniss, F.M., Sundaresan, G., Graham, L.J. et al. (2014). Near-infrared imaging of adoptive immune cell therapy in breast cancer model using cell membrane labeling. *PLoS One* 9: e109162.

52 Hu, C.M., Fang, R.H., Wang, K.C. et al. (2015). Nanoparticle biointerfacing by platelet membrane cloaking. *Nature* 526: 118.

53 Hu, C.M., Fang, R.H., Luk, B.T., and Zhang, L. (2013). Nanoparticle-detained toxins for safe and effective vaccination. *Nat. Nanotechnol.* 8: 933.

54 Jia, Y., Sheng, Z., Hu, D. et al. (2018). Highly penetrative liposome nanomedicine generated by a biomimetic strategy for enhanced cancer chemotherapy. *Biomater. Sci.* 6: 1546.

55 Zhang, Q., Dehaini, D., Zhang, Y. et al. (2018). Neutrophil membrane-coated nanoparticles inhibit synovial inflammation and alleviate joint damage in inflammatory arthritis. *Nat. Nanotechnol.* 13: 1182.

56 Lv, Y., Liu, M., Zhang, Y. et al. (2018). Cancer cell membrane-biomimetic nanoprobes with two-photon excitation and near-infrared emission for intravital tumor fluorescence imaging. *ACS Nano* 12: 1350.

57 Fang, R.H., Hu, C.M., Luk, B.T. et al. (2014). Cancer cell membrane-coated nanoparticles for anticancer vaccination and drug delivery. *Nano Lett.* 14: 2181.

58 Sun, H.P., Su, J.H., Meng, Q.S. et al. (2017). Cancer cell membrane-coated gold nanocages with hyperthermia-triggered drug release and homotypic target inhibit growth and metastasis of breast cancer. *Adv. Funct. Mater.* 27: 1604300.

59 Sun, H., Su, J., Meng, Q. et al. (2016). Cancer-cell-biomimetic nanoparticles for targeted therapy of homotypic tumors. *Adv. Mater.* 28: 9581.

60 Banskota, S., Yousefpour, P., and Chilkoti, A. (2017). Cell-based biohybrid drug delivery systems: the best of the synthetic and natural worlds. *Macromol. Biosci.* 17: 1600361.

61 Hughes, R.M., Marvin, C.M., Rodgers, Z.L. et al. (2016). Phototriggered secretion of membrane compartmentalized bioactive agents. *Angew. Chem. Int. Ed.* 55: 16080.

62 Wang, C., Sun, X., Cheng, L. et al. (2014). Multifunctional theranostic red blood cells for magnetic-field-enhanced in vivo combination therapy of cancer. *Adv. Mater.* 26: 4794.

63 Favretto, M.E., Cluitmans, J.C.A., Bosman, G.J.C.G.M., and Brock, R. (2013). Human erythrocytes as drug carriers: loading efficiency and side effects of hypotonic dialysis, chlorpromazine treatment and fusion with liposomes. *J. Controlled Release* 170: 343.

64 Baron, S., Poast, J., Rizzo, D. et al. (2000). Electroporation of antibodies, DNA, and other macromolecules into cells: a highly efficient method. *J. Immunol. Methods* 242: 115.

65 Arbab, A.S., Janic, B., Haller, J. et al. (2009). In vivo cellular imaging for translational medical research. *Curr. Med. Imaging Rev.* 5: 19.

66 Arbab, A.S., Liu, W., and Frank, J.A. (2006). Cellular magnetic resonance imaging: current status and future prospects. *Expert Rev. Med. Dev.* 3: 427.

67 Walczak, P., Kedziorek, D.A., Gilad, A.A. et al. (2005). Instant MR labeling of stem cells using magnetoelectroporation. *Magn. Reson. Med.* 54: 769.

68 de Vries, I.J., Lesterhuis, W.J., Barentsz, J.O. et al. (2005). Magnetic resonance tracking of dendritic cells in melanoma patients for monitoring of cellular therapy. *Nat. Biotechnol.* 23: 1407.

69 Brähler, M., Georgieva, R., Buske, N. et al. (2006). Magnetite-loaded carrier erythrocytes as contrast agents for magnetic resonance imaging. *Nano Lett.* 6: 2505.

70 Ahn, S., Jung, S.Y., Seo, E., and Lee, S.J. (2011). Gold nanoparticle-incorporated human red blood cells (RBCs) for X-ray dynamic imaging. *Biomaterials* 32: 7191.

71 Foroozesh, M., Hamidi, M., Zarrin, A. et al. (2011). Preparation and in vitro characterization of tramadol-loaded carrier erythrocytes for long-term intravenous delivery. *J. Pharm. Pharmacol.* 63: 322.

72 Ferrauto, G., Delli Castelli, D., Di Gregorio, E. et al. (2014). Lanthanide-loaded erythrocytes as highly sensitive chemical exchange saturation transfer MRI contrast agents. *J. Am. Chem. Soc.* 136: 638.

73 Loebinger, M.R., Kyrtatos, P.G., Turmaine, M. et al. (2009). Magnetic resonance imaging of mesenchymal stem cells homing to pulmonary metastases using biocompatible magnetic nanoparticles. *Cancer Res.* 69: 8862.

74 Giesel, F.L., Stroick, M., Griebe, M. et al. (2006). Gadofluorine m uptake in stem cells as a new magnetic resonance imaging tracking method: an in vitro and in vivo study. *Invest. Radiol.* 41: 868.

75 Tacken, P.J., de Vries, I.J., Torensma, R., and Figdor, C.G. (2007). Dendritic-cell immunotherapy: from ex vivo loading to in vivo targeting. *Nat. Rev. Immunol.* 7: 790.

76 Yang, J., Lam, D.H., Goh, S.S. et al. (2012). Tumor tropism of intravenously injected human-induced pluripotent stem cell-derived neural stem cells and their gene therapy application in a metastatic breast cancer model. *Stem Cells* 30: 1021.

77 Tyler, M.A., Ulasov, I.V., Sonabend, A.M. et al. (2008). Neural stem cells target intracranial glioma to deliver an oncolytic adenovirus in vivo. *Gene Ther.* 16: 262.

78 Klyachko, N.L., Polak, R., Haney, M.J. et al. (2017). Macrophages with cellular backpacks for targeted drug delivery to the brain. *Biomaterials* 140: 79.

79 Ko, I.K., Kim, B.G., Awadallah, A. et al. (2010). Targeting improves MSC treatment of inflammatory bowel disease. *Mol. Ther.* 18: 1365.

80 Dehaini, D., Wei, X., Fang, R.H. et al. (2017). Erythrocyte–platelet hybrid membrane coating for enhanced nanoparticle functionalization. *Adv. Mater.* 29: 1606209.

81 Lu, K., Li, H.Y., Yang, K. et al. (2017). Exosomes as potential alternatives to stem cell therapy for intervertebral disc degeneration: in vitro study on exosomes in interaction of nucleus pulposus cells and bone marrow mesenchymal stem cells. *Stem Cell Res. Ther.* 8: 108.

82 Xu, P., Wang, R., Wang, X., and Ouyang, J. (2016). Recent advancements in erythrocytes, platelets, and albumin as delivery systems. *OncoTargets Ther.* 9: 2873.

83 Gutiérrez Millán, C., Bravo, D.G., and Lanao, J.M. (2017). New erythrocyte-related delivery systems for biomedical applications. *J. Drug Delivery Sci. Technol.* 42: 38.

84 Diacovo, T.G., Roth, S.J., Buccola, J.M. et al. (1996). Neutrophil rolling, arrest, and transmigration across activated, surface-adherent platelets via sequential action of P-selectin and the beta 2-integrin CD11b/CD18. *Blood* 88: 146.

85 Fong, L., Brockstedt, D., Benike, C. et al. (2001). Dendritic cells injected via different routes induce immunity in cancer patients. *J. Immunol.* 166: 4254.

86 Mick, R., Bedrosian, I., Xu, S. et al. (2003). Intranodal administration of peptide-pulsed mature dendritic cell vaccines results in superior CD8+ T-cell function in melanoma patients. *J. Clin. Oncol.* 21: 3826.

87 Gupta, N., Patel, B., and Ahsan, F. (2014). Nano-engineered erythrocyte ghosts as inhalational carriers for delivery of fasudil: preparation and characterization. *Pharm. Res.* 31: 1553.

88 Gupta, N., Patel, B., Nahar, K., and Ahsan, F. (2014). Cell permeable peptide conjugated nanoerythrosomes of fasudil prolong pulmonary arterial vasodilation in PAH rats. *Eur. J. Pharm. Biopharm.* 88: 1046.

89 Foley, R., Tozer, R., and Wan, Y. (2001). Genetically modified dendritic cells in cancer therapy: implications for transfusion medicine. *Transfus. Med. Rev.* 15: 292.

90 Chessa, L., Leuzzi, V., Plebani, A. et al. (2014). Intra-erythrocyte infusion of dexamethasone reduces neurological symptoms in ataxia teleangiectasia patients: results of a phase 2 trial. *Orphanet J. Rare Dis.* 9: 5.

91 Portnow, J., Synold, T.W., Badie, B. et al. (2017). Neural stem cell-based anticancer gene therapy: a first-in-human study in recurrent high-grade glioma patients. *Clin. Cancer Res.* 23: 2951.

92 Leoni, G., Lyness, A., Ginty, P. et al. (2017). Preclinical development of an automated injection device for intradermal delivery of a cell-based therapy. *Drug Delivery Transl. Res.* 7: 695.

93 Csizmar, C.M., Petersburg, J.R., and Wagner, C.R. (2018). Programming cell–cell interactions through non-genetic membrane engineering. *Cell Chem. Biol.* 25: 931.

11

Overcoming the Translational Gap – Nanotechnology in Dermal Drug Delivery

Christian Zoschke and Monika Schäfer-Korting

Institute of Pharmacy (Pharmacology and Toxicology), Freie Universität Berlin, Berlin, Germany

11.1 Nanotechnology – Failure or Future in Drug Delivery?

The improved understanding of disease pathways enables the identification of relevant pharmacological targets and the development of drugs addressing these targets. Yet, many drugs still lack the needed selectivity to address off-targets adverse effects or reach only limited concentrations at the target site. This holds true for both systemic and topical treatments, with the latter being the focus of this chapter. In the treatment of skin diseases, while cutaneous application allows for a reduction of systemic adverse drug reactions, the *stratum corneum* limits access to the disease site. Moreover, spill-over into dermal blood vessels can still result in severe systemic toxicity. Drug loading to nanoparticles may offer an approach to overcome these problems. The technological advances in the production and characterization of nano-sized particles has created a plethora of expectations in controlled drug delivery. Along with this vast number of expectations, the requirements for nanoparticles have also increased. Nanotechnological drug delivery should be smart and therefore overcome physiological barriers of the skin, mucosa, or tumor environment; nanoparticles must carry high drug payloads, release effectively upon a trigger, and should be easily eliminated with minimal toxicity.

There is no shortage of innovative concepts in nanotechnology, but the translation from bench to bedside is scarce, even today. Despite 40 000 studies on active tumor targeting by nanotechnology and despite the major improvements in the understanding of the biochemical interactions of particles with cells and tissues, none of the actively targeted nanoparticles has passed clinical trials [1]. Successful nanosystems that have obtained marketing authorization are PEGylated liposomal doxorubicin, liposomal daunorubicin, and nanoparticle albumin-bound paclitaxel [2], but these do not involve active targeting.

Although a great deal of the focus for active targeted nanotechnology has been on anti-cancer drugs, the need for active targeting nanotechnology and the same translational gap applies to many other diseases. Nanotechnology for topical and cutaneous delivery faces similar challenges with a vast array of technological improvements, but few marketed products [3]. The use of nanotechnology in drug delivery might profit from understanding the reasons for this failure and will contribute to making the latest nanotechnology, like nanorobots [4], more rapidly available for the patient.

Four major challenges stand out in the explanation of the low translational rate of advanced nanotechnology into clinical application. First, nanotechnology fails due to limited data on stability, and poorly characterized by-products [5]. Second, nanotechnology fails due to unknown toxicity

Fundamentals of Drug Delivery, First Edition.
Edited by Heather A. E. Benson, Michael S. Roberts, Adrian C. Williams, and Xiaowen Liang.

including the individual effects of a certain type of nanoparticle under assay conditions, as well as the unknown applicability of existing risk assessment for nanotechnology and missing benchmark materials [6].

Third, nanotechnology fails because certain concepts are not helpful to enter cells. In fact, nanoparticles need to be less than 50 kDa (~10 nm) to enter the nucleus via diffusion [7]. Physiological barriers such as cutaneous, alveolar, or gastrointestinal epithelia prevent nanoparticles reaching their pharmacological target, and even intravenous administration does not guarantee sufficient amounts of the nanoparticle at the target site. In particular, skin with its low blood supply is unlikely to benefit from intravenous nanoparticle administration. Taking cancer as an example, the endothelial hyperpermeability and the lymphatic dysfunction of the tumor tissue was described as an opportunity to accumulate nanoparticles [8]. The endothelial permeability of the hepatic and splenic reticuloendothelial system, however, removes nanoparticles from systemic circulation most efficiently and thus contributes to low nanoparticle concentrations within the tumor. In addition, the lymphatic dysfunction of the tumor tissue limits the nanoparticle uptake. Finally, the tumor microenvironment, i.e. the tumor cell surrounding stroma and parenchymal cells, contributes not only to the failure of nanotechnology but also to the failure of anti-cancer drugs in general (Figure 11.1). Drugs gain access to the perivascular regions of the tumor mass, while hypoxia and acidic pH limit drug efficacy in central regions of the tumor mass [9].

Fourth, nanotechnology and drug candidates fail because human diseases are not appropriately represented in preclinical models. The overall probability of success ranges between 3% and 34% for investigational new drugs in clinical trials, questioning current approaches to preclinical drug candidate evaluation [10]. Thus, the translational success remains poor in drug development in general and not only for nanotechnology. Drug candidates also fail in humans due to the low predictive capacity of preclinical models. In particular, the predictive capacity of animal models has been questioned [11–13]. Recent meta-analyses unraveled poor preclinical study design, since randomization, blinding, and consideration of drop-outs are the exception (Figure 11.2) [14].

Animal models and human cell-based models have rarely been assessed for their suitability in pharmacological research. Moreover, preclinical models frequently lack the patient heterogeneity, and thus exclude factors such as age, ethnic background, sex, and the individual disease burden.

Herein, we discuss the latest concepts in nanotechnology, the latest developments in human-cell based preclinical models, as well as cutting edge methods to track nanoparticles and their payload within cells and tissues. We selected the cutaneous use of nanotechnology as an example, since numerous concepts have been developed and skin models are validated for the evaluation of local substance effects [15, 16]. The leading position of human based testing on effects induced to the skin results from the demands of modern society to avoid animal experiments in the risk assessment of chemicals, and in particular in the development of cosmetic products.

11.2 Identification of the Clinical Need

Important diseases of the skin include the inflammatory diseases acne, psoriasis, and dermatitis. Acne is the number one reason for dermatologic consultations worldwide. Fragility of the skin affects humans around the world. In addition, infections and tumors are common [17], and poorly healing ulcers on feet in diabetes are a major medical need. The pathophysiology and treatment of some common skin conditions is described below.

Atopic dermatitis is a chronically relapsing skin disorder characterized by inflammation, itching, psychological stress, and sleep disturbance. This is caused by impaired mechanical and immunological barrier integrity. The impaired keratinization and order of stratum corneum

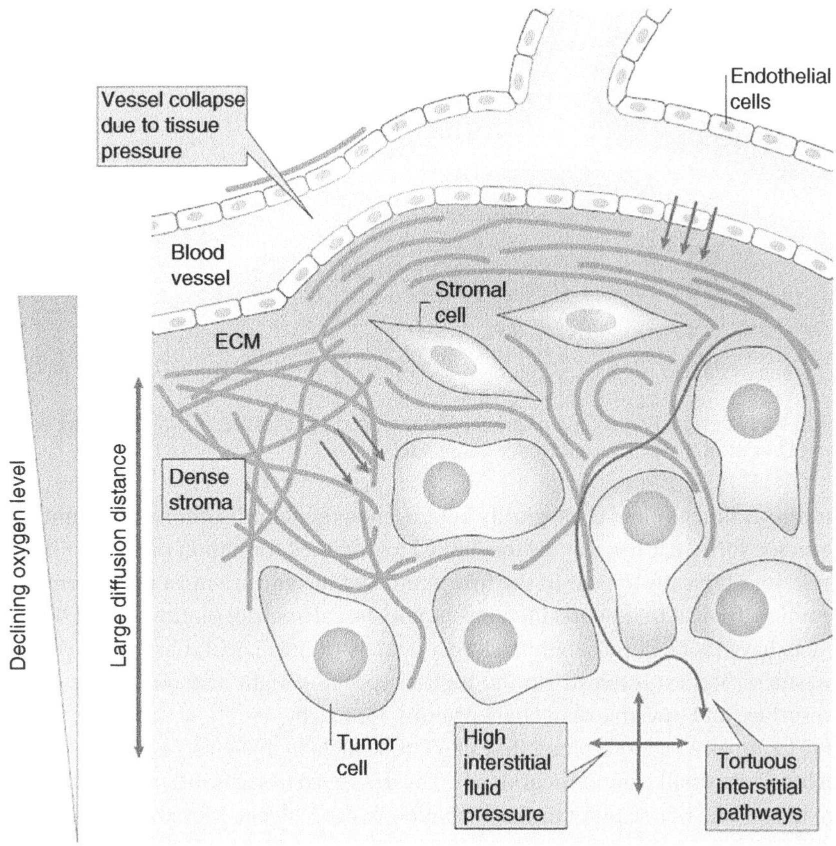

Figure 11.1 Limitations of drug transport in extravascular tissue. A drug must pass several potential barriers in order to reach tumor cells. High tissue pressures may cause vessel collapse, restricting blood flow. The endothelial cells lining microvessels restrict extravasation of drug. Dense stroma, consisting of extracellular matrix (ECM) and cells such as fibroblasts, can be a physical barrier, particularly for large molecules and nanoparticles, and is a binding site for some drugs. The pathway for transport within tumors may be tortuous due to stroma and parenchymal cells. Large transport distances may result in incomplete drug distribution. High interstitial fluid pressure (IFP) acting in all directions (denoted by the crossed arrows) may reduce fluid flow and restrict drug delivery by convection. Light gray background shading represents the oxygen level decreasing with distance from the blood vessel. Source: From Dewhirst and Secomb [9]. © 2017, Springer.

lipids allows access of environmental agents to viable skin layers. The immunological defense declines, for example by lower amounts of the antimicrobial peptide human β-defensin, with respective alterations of the dermal microbiome. Preventive interventions include specific skin care including avoiding irritants and allergens. Acute exacerbations need topical drug treatment with agents such as glucocorticoids and/or tacrolimus or pimecrolimus.

Psoriasis is a chronic inflammatory disease affecting mainly the skin, a severe form is psoriasis arthritis with additional involvement of the joints. Cure is not possible, but drugs adjusted to the severity of the symptoms allow for control. The discovery of the involvement of various cytokines has opened new horizons in the systemic drug therapy by specific antibodies and soluble receptor proteins. This approach is gaining increasing relevance in severe psoriasis because of the risk of tumor induction by PUVA (psoralen plus ultraviolet A radiation) therapy.

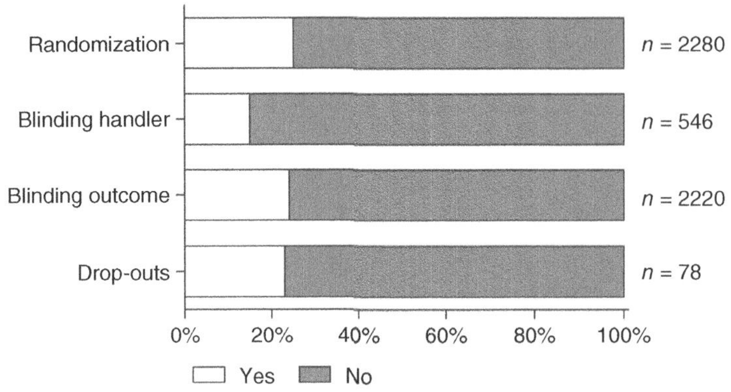

Figure 11.2 Preclinical study design. Percentage of primary animal studies assessed per validity item; yes or no. Source: From van Luijk et al. [14]. Licensed Under CC By 4.0.

Ichthyoses are a group of clinically and etiologically heterogenous genetic disorders of keratinization. In the most common form, ichthyosis vulgaris, linked to impaired formation of the cornified envelope, scaly skin lesions typically appear in the first months of life and are most prominent on the trunk and extremities. Topical treatments include emollients, calcipotriol ointment, and liarozole cream that seem to have therapeutic benefit and a good safety profile. Local substitution of the missing protein appears of highest interest, yet the high-molecular-weight and protein-cleaving enzymes at the skin surface and in viable skin challenge this approach.

Wounds, either due to impaired venous blood flow or a late sequela in diabetes, result in a major decline in quality of life. Even small wounds heal slowly. The destroyed tissue is inflamed and there is a high risk of persistent infection leading to the formation of deep ulcers. Moreover, ulcers are often very painful.

Infections of human skin are predominantly induced by *Streptococci* and *Staphylococci* (impetigo). Biofilm formation is of particular concern, as the altered bacterial metabolism often results in drug resistance. Bacteria are of relevance with respect to acne development and in atopic dermatitis due to alterations of the skin microbiome. The use of disinfectants is often preferred in skin infections, as these agents are less prone to induce the development of resistance and allergic contact dermatitis.

The cure of bacterial infections of the skin is limited by sheer forces eliminating a topically applied drug. Nanomaterials are developed in order to improve drug adherence and the delivery to the microorganisms.

UV damage and skin cancer: The most important skin tumors comprise melanoma because of the high mortality, and squamous and basal cell carcinoma due to the very high incidence. Tumor development is related to the extended UV exposure of the skin. Surgical removal of basal and squamous carcinoma efficiently clears single lesions, but is problematic for recurrent disease and fails to cure field cancerization in squamous cell carcinoma. Topically applied drugs against squamous and basal cell carcinoma include 5-amino levulinic acid for photodynamic therapy, imiquimod for toll-like receptor 7 activation, and 5-fluorouracil for blocking the action of thymidylate synthase. All drugs induce severe local adverse reactions. Melanoma treatment relies on surgery and systemic application of anti-cancer drugs.

Hair disorders: Alopecia areata affects men and women with disfiguring hair loss. In order to overcome systemic as well as local adverse effects, an optimized balance between interfollicular permeation and follicular uptake is aimed for by hair follicle targeting.

11.3 Nanoparticle Design and Physicochemical Characterization

Nanoparticles are natural, incidental, or manufactured particles with a size of less than 100 nm in at least one dimension and in more than 50% of the particles. Nanoparticles for pharmaceutical use are composed of a wide range of materials such as lipids and polymers [18] (Figure 11.3).

Lipid nanoparticles: Lipid particles can increase dermal penetration by improved delivery, for example by mixing with skin surface lipids or fluidizing skin lipids. Liposomal amphotericin B (Ambisome®) improved the pharmacotherapy of systemic fungal infections by reducing amphotericin's nephrotoxicity. Topically applied econazole liposomes (Pevaryl 1% Lipogel®) enhance the skin penetration of the drug, but fail to improve the clinical outcome, as azole antimycotics are already highly efficient and well tolerated.

Due to superior stability compared to liposomes, the nano- and in particular microemulsions and the particulate systems appear preferable. Solid lipid nanoparticles are well tolerated by the skin and eye as determined from the Episkin-SM™ skin irritation test and HET-CAM assay [19]; yet offer only a limited loading efficiency for drugs in imperfect regions of the particle. Nanostructured lipid carriers, a mixture of fluid and solid lipids, improve drug loading [20]. Yet, the penetration enhancement of a lipophilic dye into excised human skin compared to a cream following nanostructured lipid carriers and nanoemulsions was less as compared to solid lipid nanoparticles [21]. Of note, loading to nanostructured lipid carriers failed to enhance clobetasol propionate activity in a rat paw edema test compared to a commercial gel [22].

Various lipid nanoparticles were investigated for glucocorticoid delivery. Studying human skin *ex vivo*, epidermal targeting was observed with drugs or dyes attached to the particle surface but not when embedded into the lipid matrix. The mode of association of the guest molecule and the particle can be studied by electron paramagnetic resonance spectroscopy [23, 24] and fluorescence spectroscopy [21] proving glucocorticoid association to the surface of solid lipid nanoparticles and nanostructured lipid carriers.

For an insight into the mechanism of penetration enhancement by solid lipid nanoparticles, three glucocorticoids varying in structure and lipophilicity (prednisolone, diester prednicarbate, and monoester betamethasone 17-valerate) were loaded onto solid lipid nanoparticles and glucocorticoid-particle interaction, drug release and skin penetration were investigated. An oil-in-water cream served for reference. Both with solid lipid nanoparticles and cream, drug release ranked prednisolone > prednicarbate > betamethasone valerate. With the cream, the rank order did not change when studying skin penetration. Consequently, glucocorticoid release appears to be of highest relevance for drug delivery to the skin. However, the differences in steroid penetration seen with the cream were almost lost with the solid lipid nanoparticles indicating that the solid lipid nanoparticles influence skin penetration by an intrinsic mechanism linked to a specific interaction of the drug-carrier complex and the skin surface. This is due to the lipid nature and nanosize of the carrier and is not derived by testing drug release. Interestingly, loading to solid lipid nanoparticles even resulted in epidermal targeting of prednisolone and prednicarbate [25, 26] which may be of interest when aiming to influence epidermal dysfunction in particular.

The enhanced dexamethasone penetration into porcine skin *ex vivo* did not change when embedding lipid nanoparticles into hydroxyethyl cellulose gel. Drug levels in the stratum corneum and viable epidermis increased for 4 hours when applying the reference cream but doubled after 24 hours exposure compared to 4 hours exposure following the nanoparticles with spin-labeled dexamethasone associated to the carrier's surface. The nanoparticles provided drug reservoirs both within the *stratum corneum* and in the hair follicle [27]. Because of the higher glucocorticoid

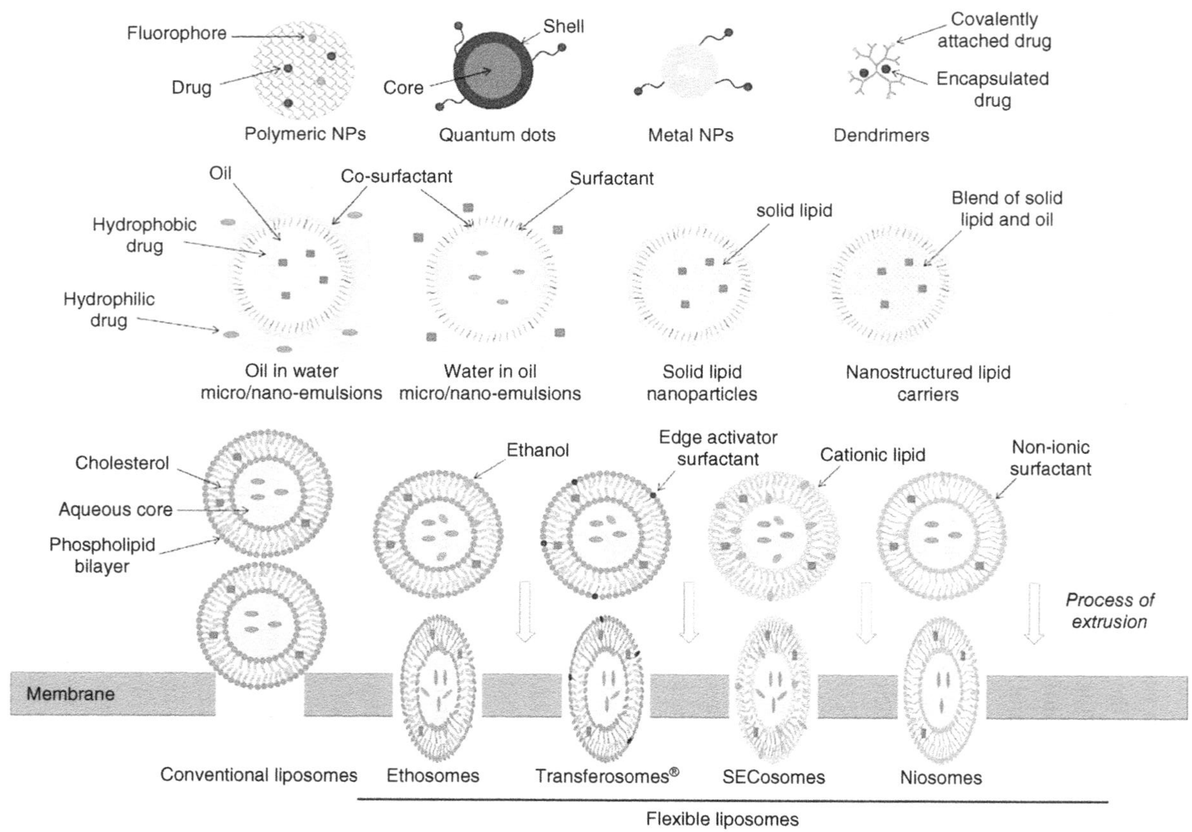

Figure 11.3 Nanodelivery systems investigated for dermal and transdermal drug delivery. Source: From Roberts et al. [3]. © 2017, Elsevier.

levels, an improved anti-inflammatory effect appears possible, yet, efficacy and adverse drug reactions induced by nanoparticle-based formulations in human awaits to be shown.

Taken together, lipid carriers are of high interest for drug delivery in skin diseases, and respective studies were performed in various disease models [19, 28].

Polymeric nanoparticles/nanocapsules: These are container systems, carrying drug payloads in a core space, surrounded by a polymer shell [29]. The modification of the shell lipids creates a large versatility for how nanocapsules interact with their environment and thus functionalizes them in terms of trigger.

A point rarely investigated is the penetration of the nanotransporter itself into the skin, independent of the drug penetration profile. De Brum et al. investigated the correlation between the flexibility of biodegradable nanoparticles and the depth of their skin penetration. The impact of the chemical composition, the surface chemistry, or the shape and size distribution on the results was minimized using almost the same materials and their concentrations. The anionic polymeric nanocapsules and the lipid-core nanocapsules were close in size (about 180 nm) and zeta potential. Studying pig skin *ex vivo*, the rather flexible nanocapsules reached the dermis, whereas the more rigid lipid-core nanocapsules were retained in the epidermis [30]. This puts the general statement on the restricted skin penetration of nanoparticles >10 nm in size [31] into a new perspective.

Polymeric nanoparticles can form a drug reservoir retaining the drug locally at the site of skin disease. The biodegradable and well-tolerated poly(glycolic acid) (PGA), and poly(lactic acid) (PLA), as well as the respective copolymer (poly(lactic-*co*-glycolic) acid [PLGA]) are most often studied. Recently, core–multishell (CMS) nanoparticles have received major interest because they mimic the structure of liposomes while being stable. Using pH-sensitive or temperature controlled nanogels, drug targeting to inflamed skin may become an option in the near future. Examples are presented here.

Polymeric nanoparticles have potential to control the release and penetration of glucocorticoids and thus to maximize the treatment efficacy. Dexamethasone release, skin penetration, and safety were studied. Eudragit® RS nanoparticles that were smaller and positively charged had lower dexamethasone loading capacity (0.3–0.7%) than ethyl cellulose nanoparticles (1.4–2.2%). Drug release and penetration significantly declined with decreasing drug to polymer ratio and increased when Eudragit RS was blended with ethyl cellulose. Drug release and penetration from the nanoparticles were slower than from a conventional cream. Dexamethasone penetration should be enhanced by the interaction of the cationic surface of the particles with *stratum corneum* lipids, yet this effect was not observed [32].

However, Eudragit RS appears to induce glucocorticoid targeting to viable epidermis which was aimed for and observed in a previous study applying dexamethasone loaded Eudragit RS nanocapsules embedded in hydrogels. The formulation promoted controlled dexamethasone release and increased the drug delivered into viable epidermis, the main target tissue to topical glucocorticoid action in dermatitis [33]. The nanoparticles bear no cytotoxicity potential except for reactive oxygen species (ROS) generation potential at high ethyl cellulose concentration [32].

CMS nanotransporters built by a hydrophilic poly(ethylene imine) or polyglycerol amine core (i.e. a dendrimer), and covalently linked to long-chain dicarbonic acids (lipophilic inner shell) which form ester groups with polyethylene glycol (PEG) (outer shell) mimic the structure of (stealth) liposomes (Figure 11.4) [34].

CMS nanotransporters are tolerable by skin and eye [19]. The delivery of the loaded lipophilic dye Nile red to pig skin *ex vivo* by CMS nanoparticles (aggregates 20–30 nm in size) clearly exceeded

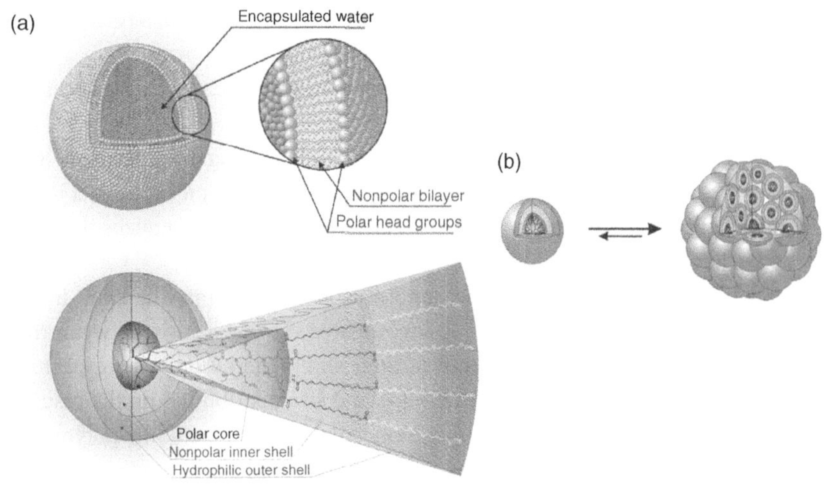

Figure 11.4 Core–multishell (CMS) nanotransporters. (a) Dendritic core–multishell architecture in the bottom panel with poly(ethylene imine) core, inner hydrophobic segment, and terminal mPEG chain. On the top panel, the typical liposome structure to reveal the structural resemblance to CMS nanotransporters. (b) Dynamics between CMS nanotransporter unimers and their aggregates those are responsible for encapsulation. Source: From Quadir et al. [34]. © 2008, Elsevier.

the delivery by solid lipid nanoparticles, both particles being internalized by primary human keratinocytes [35]. Hydrophilic substances can be loaded to CMS nanotransporters, too, and penetrate the skin as shown by spin probe [36].

Dye-tagged and Nile red loaded CMS nanotransporters, respectively, were tested for penetration in normal and stripped human skin *ex vivo*, normal reconstructed human skin (RHS), and *in vitro* skin disease models. Nile red was delivered rapidly into the viable epidermis and dermis of normal skin, whereas the CMS nanotransporters remained solely in the stratum corneum after 6 hours but penetrated into deeper skin layers after 24 hours exposure. Fluorescence lifetime imaging microscopy proved a stable dye-tag and revealed striking nanotransporter–skin interactions. The viable layers of stripped skin were penetrated more efficiently by dye-tagged CMS nanotransporters and the cargo compared to normal skin. Normal RHS reflected the penetration of Nile red and CMS nanotransporters in human skin. Importantly, the non-hyperkeratotic non-melanoma skin cancer and hyperkeratotic peeling skin disease models come along with altered absorption in the skin diseases [37].

As with the nanoparticles and nanogels (see below), the interaction of CMS nanotransporters with immune cells was investigated to exclude immunotoxicity. The nanoparticle did not induce cytotoxicity, genotoxicity nor cause ROS or an immunological response. As with the nanogels, CMS nanotransporters were efficiently taken up by Langerhans cells via divergent endocytic pathways and localized within the lysosomes. In addition, an energy-dependent exocytosis of the native CMS nanotransporters into the extracellular environment was observed, too [38]. Animal studies support the good tolerability. Applied to normal SKH-1 mice and to mice with oxazolone-induced atopic dermatitis, CMS nanotransporters accumulated in the *stratum corneum* without penetration into viable epidermis indicating that barrier alterations in atopic dermatitis had no influence on the penetration of the nanoparticles. Following subcutaneous injection, the CMS nanotransporter was deposited in the regional lymph nodes as well as in liver, spleen, lung, and kidney. *In vitro* toxicity tests, clinical data, and morphometry-assisted histopathological analyses yielded no evidence of any toxic or otherwise adverse local or systemic effects [39].

Since nanotransporters can be degraded in the skin, the conventional amide-based CMS nanotransporters may release toxic polyglycerol amines. To circumvent this issue, tailored ester-based CMS nanotransporters of varying inner alkyl chain length were synthesized and characterized for biodegradation and dexamethasone delivery to the skin. *In vitro* and *in vivo* toxicity studies indicated best biocompatibility of the ester-based CMS nanotransporters when building the inner shell by C15-dicarbonic acid. Ester-based CMS nanotransporters also delivered the drug efficiently into human skin and the pro-inflammatory cytokine interleukin (IL)-8 markedly declined when testing in a model of tumor necrosis factor (TNF)-induced skin inflammation. Efficacy of penetration and anti-inflammatory response exceeded those of a commercial cream. This systematic approach of a rational nanotransporter design and screening for biocompatibility, dermal delivery, and therapeutic efficacy may be useful when aiming for other nanotransporter systems for topical applications [40].

Loaded with peptides/proteins, larger nanoparticles may open up transdermal vaccination strategies. CMS nanotransporters can access viable skin layers and be taken up by Langerhans cells [38, 41].

Nanogels are colloidal hydrogel nanoparticles formed through physical or chemical crosslinking. Nanogels swelling by contact with water can act as penetration enhancers.

Critical for the use of nanogels for drug delivery are the loading capacity and encapsulation, in particular for hydrophobic drugs. Nanogel-peptide conjugates were evaluated for their ability to encapsulate temoporfin (log P 9.17), the photosensitizer that is approved for systemic application in head and neck cancer. The peptide was selected based on combinatorial means and conjugated to polyglycerol as the nanogel precursor. Peptide conjugation enhanced the loading capacity of nanogels 16 times and penetration of viable skin increased as well [42]. This study indicates that nanogel-peptide conjugates could be used as unique nanotransporters with high loading capacity for hydrophobic compounds, which provides the basis for the design of advanced topical drug delivery systems.

Dendritic polyglycerol crosslinked to thermoresponsive polymers (poly(*N*-isopropylacrylamide), p(di(ethylene glycol) methyl ether methacrylate-*co*-oligo ethylene glycol methacrylate, poly(glycidyl methyl ether-co-ethyl glycidyl ether); size around 150 nm) penetrates into the *stratum corneum* and disturbs the organization of barrier proteins and lipids, which was attributed to hydration effects. The ability of nanogels to enhance the penetration of loaded fluorescein into human skin *ex vivo* correlated with the induced *stratum corneum* hydration. Nanogels having additional or more exposed polyether structure in their chemical composition proved most effective [43].

Deeply penetrating the stratum corneum, nanogels may interact with viable keratinocytes and dendritic cells, especially in diseased skin. Considering those particles for drug delivery in inflammatory skin disease, any potential toxicological risks have to be excluded. Nanogels based on dendritic polyglycerol, poly(*N*-isopropylacrylamide), and poly(glycidyl methyl ether-*co*-ethyl glycidyl ether) cores entered into the cells and localized predominantly within lysosomal compartments of keratinocytes. The nanogels displayed neither irritant nor cytotoxic effects in the cells of interest, but induced ROS. The specific uptake mechanisms of the nanogels by dendritic cell lines depend on cloud point temperature which determines the phase transition of the nanoparticle. The major process, caveolae-mediated endocytosis was most active at 37 °C, which is above the cloud point temperature. At cloud point temperature (29 °C), which is characterized by two different states of the thermoresponsive nanogels, macropinocytosis were also involved [44, 45].

Facing the tolerability and the specific interaction with immune cells, indodicarbocyanine-tagged and fluorescein-loaded thermoresponsive nanogels were investigated for penetration and release properties upon topical application on *ex vivo* human skin with intact or disrupted barrier, as well as for temperature-triggered effects and particle internalization by skin cells. Topically applied nanogel penetrated into the stratum corneum of both intact and disrupted skin explants. Yet, significant amounts of released dye and particle penetrated in the epidermis and dermis two hours after application only in barrier-disrupted skin. A thermal trigger (infrared radiation) significantly enhanced penetration. Penetrated particles were internalized by skin cells in the dermis, in particular by antigen presenting cells. Thus, thermoresponsive nanogels represent promising carrier systems for the treatment of inflammatory skin diseases [46] and alopecia areata as described below.

Polymers for pH triggered drug release: Normal skin surface pH being 5.24 ± 0.40 increases to 6.13 ± 0.52 on lesional skin and to 5.80 ± 0.41 on perilesional skin of patients suffering from atopic dermatitis [47]. In parallel to the increased skin surface pH, barrier function declines. Consequently, nanoparticles for pH triggered drug release adapted to clinical needs are of high interest. pH-sensitive, dexamethasone-loaded Eudragit L nanoparticles and cellulose phthalate nanoparticles were characterized in detail for pH-dependent swelling, erosion, dissolution, and drug release kinetics as well as for tolerability. The particles proved devoid of toxicity by the ROS and MTT assays [48]. Spin-labeled dexamethasone loaded to Eudragit L 100 nanoparticles was rapidly released above pH 5.9 [49] (Figure 11.5).

The nanoparticles considerably improved the penetration into the viable skin layers of intact pig ear skin *ex vivo*. Even more enhancement was seen with barrier-disrupted (stripped) skin, targeted drug delivery to lesional skin appears possible, avoiding side effects [49]. Due to the impaired barrier function of atopic skin, this result in stripped pig skin should translate to human pathologies. Drug release can also be triggered by temperature, as described for thermoresponsive nanogels [46, 50].

11.4 Biomedical Studies

Qualified methods are needed in order to study the potential of various delivery systems to human skin, both normal and diseased. This holds true both with respect to the analytical tools and the tissues used for testing, being neglected quite often. Aiming for standardized hazard investigation due to the percutaneous absorption of chemical compounds making contact with the skin, OECD Guideline 428 describes the essentials of *ex vivo* test procedures. Human skin and pig skin appear appropriate, rat skin is over-predictive [51]. Murine skin *ex vivo* might also be suitable [52]. Recently, the predictive capacity of SKH-1 mice skin as well as of RHS has been demonstrated [53]. The anti-inflammatory potency of topical glucocorticoids [54, 55] and the risk of skin thinning in RHS [55] were in accordance with human *in vivo*. Moreover, the effects of *Lactococcus lactis* lysate in reconstructed human epidermis proved predictive for human *in vivo* as shown by a consecutive clinical study [56].

RHS is generated from isolated juvenile human keratinocytes and fibroblasts, expanded and then cultured under defined conditions to generate a stratified skin tissue representing the essential epidermal layers including the *stratum corneum* (for review see [57]). The approach was extended to various disease models for atopic dermatitis, psoriasis, ichthyosis, and skin cancer.

11.4.1 Atopic Dermatitis

The use of polymeric nanoparticles, hydrogel nanoparticles, liposomes, ethosomes, solid lipid nanoparticles, and nanoemulsion for delivery in atopic skin and evidence for their therapeutic superiority has been reviewed before [58].

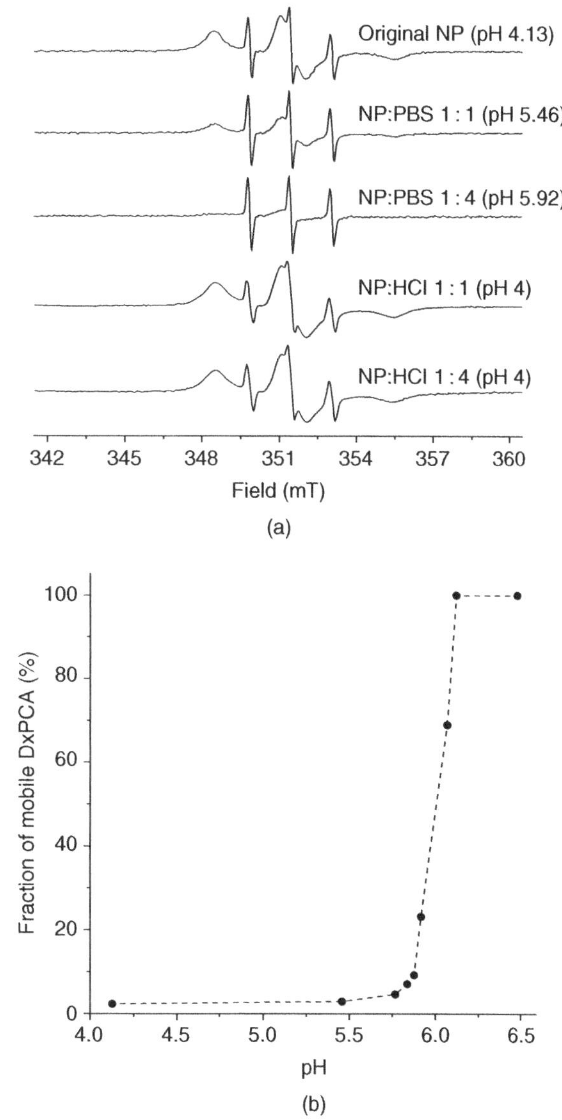

Figure 11.5 Triggered drug release. (a) Change of EPR spectra of DxPCA loaded in pH-sensitive nanoparticles as a function of pH. (b) The fraction of mobile DxPCA in the nanoparticles dispersion as a function of pH calculated from simulated EPR spectra. Each spectrum is normalized by the height of the central peak. Source: From Dong et al. [49]. © 2018, Elsevier.

More recently, a cyclodextrin-coated thermoresponsive nanogel has been shown to enhance the penetration of dexamethasone into a RHS-based skin model of atopic dermatitis as compared to a cream with the identical drug concentration. The improved penetration enhanced nuclear translocation of the glucocorticoid receptor and decreased the proinflammatory thymic stromal lipoprotein, demonstrating the superior efficacy induced by the cyclodextrin-coated nanogel [50].

11.4.2 Psoriasis

For the treatment of inflammatory skin diseases, psoriasis and atopic dermatitis in particular, Boisgard et al. developed five topical semi-solid formulations incorporating PLA nanoparticles close to 200 nm in size which were labeled by encapsulation of the lipophilic fluorescent dye BODIPY. Embedding the nanoparticles into well-tolerated gel vehicles reduced BODIPY penetration and the permeation into and through normal and inflamed BALB/c mouse skin (imiquimod-induced psoriasis model). Whereas no differences in the dye amounts within normal and inflamed skin were observed, the permeation appeared least with formulations based on Viscarin for gel formation. Moreover, the permeation of inflamed mouse skin clearly exceeded the permeation of normal skin [59]. If this interesting result holds true with respect to inflamed human skin and in particular if the efficacy of drug-loaded preparations in psoriasis is predictable by the mouse model, a potential effect of new technology is to be awaited.

In order to reduce the risk of sunburn and tumor induction by PUVA treatment, psoralen was loaded in cationic liposomes and anionic liposomes with particle size of nearly 100 nm. Both enhanced the penetration and permeation in BALB/c mice skin as compared to the solution used for reference. Symptoms of psoriasis and levels of key psoriatic cytokines (TNF-α, IL-17, and IL-22) declined in the imiquimod-induced psoriatic model [60].

Loading methoxsalen to nanoemulsions embedded in chitosan gel showed strong dependency on chitosan molecular weight [61]. The coupling of heparin to chitosan results in chitosan-heparin polyplexes and improved gene release into cells, going along with higher transfection and more gene silencing efficiency *in vitro* [62] Applying a nanoemulsion based on clove essential oil, retention in viable pig skin was almost twofold higher than with a compounded cream used for reference [63]. Efficacy and safety of the liposomal formulations have to be compared to the currently preferred PUVA bath, not only to the oral therapy. Considering the high risk of tumor induction by PUVA therapy, however, the new biologicals targeting psoriasis-specific cytokines appear most promising, although high costs currently limit the use of these innovative agents and the high molecular weight currently excludes their topical use.

Facing the potential of nanogels, protein delivery to the skin has been aimed for. The soluble TNFα receptor fusion protein etanercept (molecular mass around 150 kDa) approved for subcutaneous injection in severe arthritis and recalcitrant psoriasis was loaded to nanogels having polyglycerol or poly(*N*-isopropylacrylamide) cores, respectively. Etanercept induced an anti-inflammatory effect in filaggrin deficient RHS [64] with less translocation of the glucocorticoid receptor into the nucleus and reduced formation of TNFα, thymic stromal lymphopoietin (TSLP), and intercellular activation molecule 1 (ICAM-1) [65]. A respective study in psoriatic patients is needed to prove translation of the effect seen in the disease model. This given, a new horizon will open for carrier-supported drug delivery. The design of this disease model was closely related to an ichthyosis model developed before.

11.4.3 Ichthyosis

An RHS-based model of congenital ichthyosis, representing severe epidermal barrier dysfunction was generated by knock-down of disease-related genes, transglutaminase-1 in particular in primary human keratinocytes as well as by using keratinocytes and fibroblasts from patients with congenital ichthyosis. The disturbed barrier function became obvious by increased permeation of testosterone and caffeine, particularly in transglutaminase knock-down models compared to control models. Enhanced penetration of the model dye Nile red incorporated into solid lipid nanoparticles and CMS nanotransporters, respectively, was evident in the disease models. The findings indicate that nanocarriers may be useful in new topical therapeutic approaches for the currently very limited treatment of congenital ichthyosis [66]. Of note, the thermoresponsive poly(*N*-isopropylacrylamide)-polyglycerol-based nanogel provided transglutaminase-1 delivery to the ichthyosis model and thereby restored the deficient skin barrier. The thermal trigger point of 35 °C of the nanogel favors drug skin penetration: at ≥35 °C, the particle size (~200 nm) instantly declined by 20% and 93% of the protein was released. The nanogels themselves do not penetrate viable skin [67].

11.4.4 Wound Healing

For efficient pain reduction in severe skin wounds, topically applied opioids may be an option. By stimulating keratinocyte migration, opioids may also accelerate wound healing. As conventional formulations fail to consistently provide sufficient pain control, morphine delivery by solid lipid was considered as an option. Besides enhanced healing, such a formulation may allow for long intervals of painful wound dressing changes. Morphine-loaded solid lipid nanoparticles and CMS nanotransporters enhanced the closure of experimentally induced wounds in human keratinocyte cultures (scratch test), with efficacy corresponding to transforming growth factor-β used for control. Loading to solid lipid nanoparticles improved the closure of standardized wounds induced by CO_2-laser irradiation of RHS. Keratinocytes almost completely covered the dermis equivalent after four days, which was not the case when applying the vehicle. Yet, unloaded solid lipid nanoparticles also accelerated reepithelization, which indicates an intrinsic activity of the lipids [19, 28].

Due to the induction of gingiva hyperplasia, an adverse reaction after oral administration, the antiepileptic drug phenytoin is also of interest for enhanced wound healing. Yet, because of several severe adverse reactions, percutaneous absorption should be minimal. Phenytoin loaded nanocapsules and nanoemulsions embedded into chitosan gels were investigated for the healing process of cutaneous wounds in rats. The hydrogels were of the physiological skin surface pH value of 4.9–5.6. Penetration and permeation of phenytoin was compared with hydrogels containing non-encapsulated drug. Enabling adhesion to wounded skin and controlled phenytoin release, drug loading to the nanomaterials enhanced the concentrations within the tissue. Due to a declined permeation of the skin *ex vivo*, nanoencapsulation may also reduce systemic availability [68].

The described results indicate a future use of nanoparticles in regenerative medicine. Other than nanoparticle-based drug formulations, wound management by biomaterials covering the lesion and forming an environment stimulating healing are already introduced into the market (for review see [69, 70]).

The endogenous host defense peptide LL37 possessing antimicrobial activity is also involved in the modulation of wound healing. A combination with the elastase inhibitor Serpin A1, a protein having wound-healing properties should induce a synergistic effect. Topical application by a formulation for controlled sustained delivery of both agents at the synergistic ratio should promote wound closure, reduce bacterial contamination and enhance anti-inflammatory activity. Loading to a solid lipid nanoparticles formulation proved synergistic in scratch test on fibroblasts and keratinocytes. Antibacterial activity against *Staphylococcus aureus* and *Escherichia coli* was also enhanced by the combination to LL37 or A1 [71]. The effects in 3D wound models and in patients are encouraging.

11.4.5 Infections

Inspired by delivery systems mimicking marine mussels for adhesion, Zhang et al. developed a bioadhesive nanoparticles embedding hydrogel to enhance localized antimicrobial drug delivery. Ciprofloxacin was loaded to PLGA nanoparticles. The particles were embedded into a 3D polyethylene dopamine methacrylamide hydrogel, the catechol groups being the bioadhesive moiety. The process allowed tailoring the properties of nanoparticles and hydrogel network independently. A bioadhesive nanoparticle-gel showed superior adhesion and antibiotic retention under high shear stress on a bacterial film, a mammalian cell monolayer and mouse skin. The nanoparticle gel inhibited the formation of an *E. coli* bacterial film [72]. In addition, ciprofloxacin was used for a gel-type material based on a biocompatible dendrimer rich in carboxylic groups which formed ciprofloxacin complexes by an acid–base reaction with the amine groups of the drug. Embedding the ciprofloxacin-loaded dendrimers into a hydrogel, the formulation proved suitable for application on mucosal membranes and skin due to high biocompatibility and a lack of skin irritation.

Sustained ciprofloxacin release was attributed to ion exchange phenomenon and to the drug reservoir effect. *In vitro* growth inhibition of *S. aureus* and *Pseudomonas aeruginosa* corresponded to 38% and 58% of the respective solution at similar ciprofloxacin concentrations. Drug delivery was appropriate, both in terms of magnitude and velocity to allow for a bactericidal effect [73].

Recently, there has been increasing interest in antimicrobial peptides for topical drug therapy. Resistance development of these microbicide agents is most unlikely because of the rapid metabolic inactivation. Disruption of the bacterial membrane causes the bactericidal effect, destroying human cell membranes the pronounced cytotoxicity. To overcome the low stability of antimicrobial peptides, high toxicity and inefficient delivery to the target site, peptide covalent conjugation to gold nanoparticles was investigated. A poly(ethylene glycol) linker delayed proteolytic digestion and increased by ~15-fold the activity of the free esculentin-1a against *P. aeruginosa* without being toxic to human keratinocytes. Moreover esculentin-1a-coated gold nanoparticles may improve wound healing as seen in a monolayer culture of the keratinocyte cell line HaCaT applying the scratch test [74].

Eliminating the pathogen and healing the open wounds seen in ecthyma is an important clinical need. The causative *Streptococci* and *Staphylococci* generally enter the skin through areas of minor trauma (atopic dermatitis, arthropod bites). Currently, ecthyma cure is based on beta-lactam antibiotics applied orally or parenterally as well as topical use of antiseptics.

Antifungal drugs loaded to solid lipid nanoparticles already described in the literature include terbinafine and several azole antimycotics among others. Despite increased skin levels [75], the therapeutic benefit compared to the highly efficacious and well-tolerated conventional formulations remains an open question.

11.4.6 Skin Cancer

Imiquimod was loaded to poly ε-caprolactone nanocapsules, co-encapsulating antiproliferative copaiba oil. Moreover, using copaiba oil and Brazilian cupuaçu seed butter, imiquimod loaded nanostructured lipid carriers were prepared by high-pressure homogenization, the solid lipid being sorbitan monooleate. Controlled imiquimod release was obtained, the release being slower from the particles based on Brazilian lipids. Both formulations were well tolerated by monolayer cultures of the HaCaT cell line and tested for drug penetration into abdominal pig skin *ex vivo*. The nanocapsules improved the retention of imiquimod within the skin and reduced skin permeation. Because of higher imiquimod concentrations in viable skin, the nanocapsules were considered of interest for the use in skin carcinoma [76]. Next, imiquimod-loaded nanocapsules should be compared to the commercial product.

Considering the generally multifactorial tumor development, a combination of synergistic drugs appears promising. The hypothesis was tested by dye-tagged paclitaxel loaded microemulsions and nanoemulsions that were co-loaded with dye-tagged C6 ceramide. In addition, the formulation contained tributyrin for increased cytotoxicity. The formulations were compared for their ability to co-localize the drugs in pig ear skin *ex vivo*. At four and eight hours post-application, the nanoemulsion enhanced paclitaxel delivery into viable skin about twofold compared to the microemulsion. C6 ceramide was enhanced 1.9-fold at eight hours. The drugs were co-localized mainly in the epidermis. Based on this result, the nanoemulsion was selected for evaluation of histological changes in a monolayer culture and a 3D melanoma model. In a monolayer culture of two melanoma cell lines, drug encapsulation decreased EC_{50} by about 4-fold (paclitaxel) and 13-fold (ceramide), demonstrating an improved drug delivery compared to the respective solutions in propylene glycol. Co-encapsulation of paclitaxel and ceramide further decreased EC_{50} by 2.5 to 4.5-fold, indicating a synergistic effect. Incorporation of a melanoma cell line into the epidermal layer of RHS served for the generation of the 3D melanoma model. Topical administration of the nanoemulsion for 48 hours promoted marked epidermis destruction. The destruction of both melanoma cells and keratinocytes suggests nonselective cytotoxicity [77]. If, however, topical melanoma therapy may become a realistic goal remains open.

The heptapeptide DEETGEF can prevent DNA damage by interfering with oxidative processes. The peptide was loaded on solid lipid nanoparticles made from shea butter. On human skin *ex vivo*, the formulation induced antioxidative enzymes (e.g. NAD(P)H quinone oxidoreductase and heme oxygenase-1) and protected the skin against UV induced stress. The number of sun burn cells declined and less 8-hydroxy-2′-deoxyguanosine was formed compared to placebo [78].

11.4.7 Alopecia Areata

Larger nanoparticles are known to penetrate into the hair follicle and can deliver their loaded drug close to an area rich in immunocompetent cells. The motion of a single rigid nanoparticle inside a hair follicle is induced by oscillating radial hair motion, with the velocity being dependent both on frequency and particle size [79]. Specifically, a dendritic polyglycerol and *N*-isopropylamide based thermoresponsive nanogel and hydroxyethyl cellulose nanogels sized 300–500 nm applied to porcine ear skin efficiently released the model dye coumarin 6 within the follicle [80].

Loading clobetasol propionate to nanospheres, nanocapsules, and lipid-core nanocapsules has been shown to provide sustained drug release. Follicular uptake was highest for the application of nanocapsules in dispersion and after application of massage. Both the glucocorticoid and the nanocarrier (Rhodamine B labeled polymer) were recovered in the hair follicle. In contrast, negligible drug amounts were detected in the hair follicle following a solution or hydrogel, regardless of any massage. Moreover, loading to nanoparticles reduced glucocorticoid permeation of human epidermis. Thus, these formulations might offer an advantage over conventional formulations for inflammatory-based skin/scalp diseases with respect to efficacy and safety [81].

11.5 Approaches to Fill the Translational Gaps in Nanotechnology

To unleash the full clinical potential of nanotechnology in drug delivery, we suggest the need to first compile all available data on nanotechnology in drug delivery. This database should exceed existing reviews and should be not restricted to a specific subset [82–87]. A first approach was published as cancer nanomedicine repository [88]. The database recommended here would include all publicly available data on nanotechnology applications in drug delivery, irrespective of chemical architecture, substance to be delivered, administration site or clinical purpose. Next, this database should be used as the scientific basis for machine learning purposes. Machine learning is being integrated into a growing number of drug development challenges (Figure 11.6), as fast parallel processing becomes broadly available [89].

In particular, artificial intelligence could identify the nanotechnological concept with the highest payload for a given drug or the best drug release trigger for a given substance. Moreover, the structure and the loadability of nanoparticles can be modeled *in silico* [90]. The better defined the biochemical problem is, the more likely machine learning approaches could overcome the particular challenge. Taken together, machine learning offers the opportunity to analyze the existing, vast data set from decades of research and thus will help to prioritize future research. However, the successful application of machine learning techniques depends on the training data set as well as on

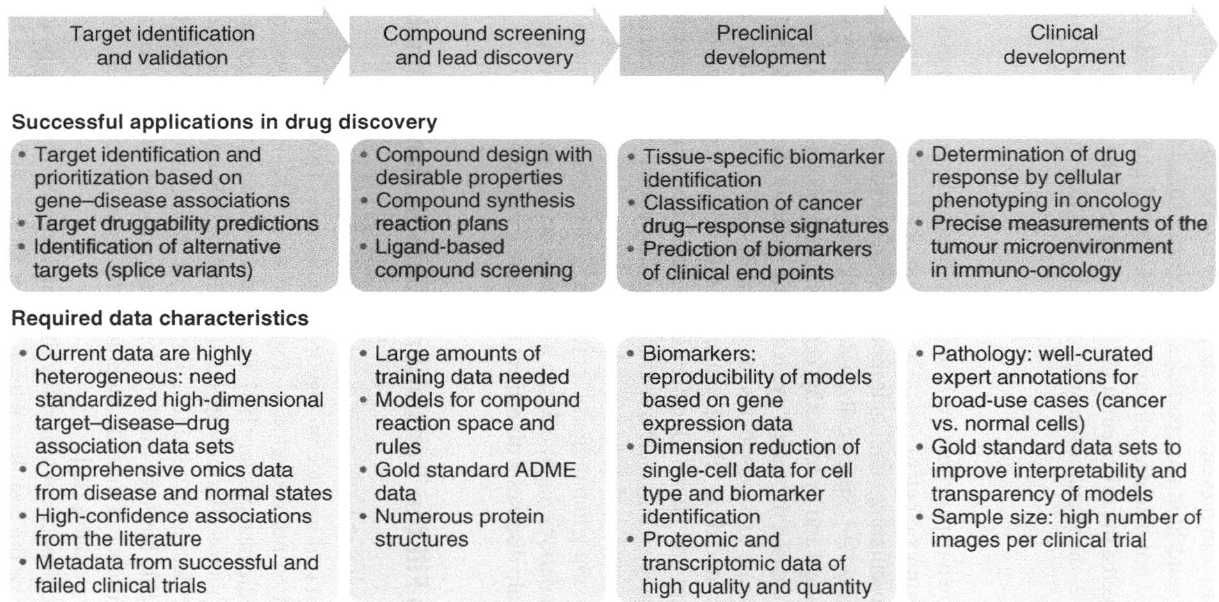

Figure 11.6 Several successful applications of machine learning in various stages of the drug development pipeline in pharmaceutical companies have been published. However, within each data domain, there are still challenges related to the standard of data quality and data quantity needed to capitalize on the full potential of these methods for discovery. ADME, absorption, distribution, metabolism, and excretion. Source: From Vamathevan et al. [89]. © 2019, Springer Nature.

the weights of the parameters we as a research community define. Thus, machine learning does not replace data generation and curation. To obtain meaningful data, nanotechnology must be tested on relevant and reliable models prior to their use in clinical trials. The authors already proposed an integrated test strategy for nanoparticles with milestones after the physico-chemical characterization, after the local safety and efficacy evaluation, as well after systemic safety assessment [91]. Issues like purity, stability, payload capacity, and basic toxicity should be evaluated before the efficacy assessment of nanoparticles to ensure reproducibility and to terminate uncontrollable nanotechnology at the earliest time point.

The selection of a relevant and reliable model depicts one of the most critical steps for the translational success of investigational new drugs or nanoparticles. The models used should be as simple as possible, answer a well-defined research question, and should have proven their predictive capacity before. The concept of performance standards proved useful to evaluate the applicability of preclinical models in modern toxicology including skin irritation, corrosion, and sensitization [92].

Although monolayer cell culture has been the standard for *ex vivo* models for decades, accumulating evidence suggests the use of 3D cell culture systems [93]. Anti-cancer drugs like doxorubicin or paclitaxel showed different potency in 3D compared with 2D culture of some tumor cell lines of the NCI60 panel [94, 95]. In addition to these gradual changes in drug potency, we observed a major decline in efficacy for the 15-lipoxygenase metabolite 13-HpOTrE when comparing 3D vs. 2D cell culture effects [96]. Advanced 3D cell culture systems emulate also the tumor microenvironment including fibroblasts and immune cells [97]. First approaches to mimic squamous cell carcinoma in 3D skin disease models proved the effects of cancer-associated fibroblasts on tumor cells [98] as well as the effects of tumor cells on the skin barrier function [99]. However, skin disease models using juvenile skin and normal collagen might not represent the patient situation of aged and multimorbid persons. Thus, we investigated the epidermal effects of advanced glycated end-products which accumulate in the skin over the lifespan and especially in diabetes patients. Dermal advanced glycated end products markedly changed the epidermal morphology, underlining the importance of matrix–cell interactions [100]. UVB irradiation damaged keratinocytes in another skin model result in increased epidermal penetration of CMS nanotransporters [101].

Intrinsically aged fibroblasts in culture develop a secretory phenotype that is different from the classical senescence-associated secretory phenotype [102]. In fact, only 5% of fibroblasts are senescent in intrinsically aged skin [103]. In RHS, intrinsically aged fibroblasts affect the expression of proteins involved in wound healing [104]. Next, the integration of immune cells will help to better understand many diseases, not limited to cancer. Meanwhile Langerhans-like cells and T-cells can be grown in RHS [105, 106]. A deeper insight into tumor heterogeneity becomes feasible by patient-derived xenografts (grown in mice) or patient derived organoids (grown *ex vivo*) and retain the hallmarks of the individual cancer [107].

Finally, more information from these models has to be retrieved. A most detailed insight into skin penetration of drugs and CMS nanotransporters becomes possible by stimulated X-ray microscopy, clearly demonstrating that nanoparticles remain in the *stratum corneum,* whereas the penetration of dexamethasone into viable epidermis is enhanced compared to standard formulations (Figure 11.7) [108]. The relevance of the method was recently proven by a head-to-head comparison of X-ray microscopy to liquid chromatography tandem mass spectrometry [53].

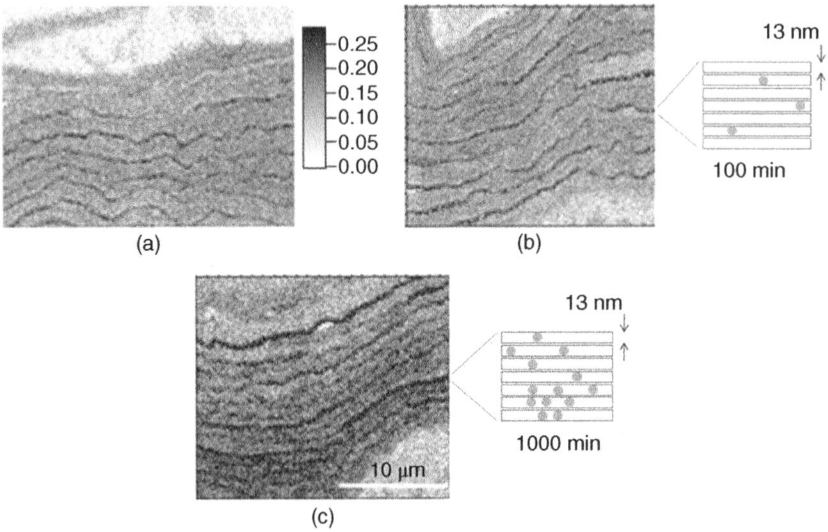

Figure 11.7 Detection of CMS nanotransporters in skin. High resolution X-ray micrographs of the *stratum corneum* of human skin samples derived from differential absorption recorded at 531.4 and 535.9 eV, respectively. Differential absorption is represented by a gray scale, where a gain in absorption corresponds to black color, which turns above a threshold value of 0.28 into dark gray: application of dexamethasone-loaded CMS nanotransporters for (a) reference sample containing only hydroxyethylcellulose (HEC) gel; (b) 100 minutes; and (c) 1000 minutes. The viable epidermis is located at the bottom of each micrograph. Schematic diagrams are shown in (b) and (c) symbolizing the location of the CMS nanotransporters within the 13 nm lipid lamellae. Source: From Yamamoto et al. [108]. © 2016, Elsevier.

Data from X-ray microscopy enable the prediction of very early drug penetration as well as of long-time permeation. Time-dependent drug penetration was modeled from drug concentration profiles by the 1D general diffusion equation that accounts for spatial variations in the diffusivity and free energy [109]. Despite the unprecedented spatial resolution with this label-free detection method, the experimental setup does not yet allow high-throughput sample preparation. Fluorescence lifetime imaging microscopy requires a fluorescent probe or a fluorescence label of the probe and a special experimental setting, yet the sample preparation can be adapted to higher throughput of samples [110]. In addition, the spatial correlation of fluorescence lifetimes with tissue structures allows tracking drugs and nanoparticles within tissues [111, 112]. The range of detection methods for drugs and nanotransporters has expanded greatly and goes beyond the scope of this chapter. Recent reviews summarize developments from matrix-assisted laser desorption/ionization time-of-flight imaging mass spectrometry to atomic-force microscopy-based techniques with a spatial resolution below 10 nm [113–115].

In conclusion, the translational success of nanotechnology will profit from innovative concepts and from the use of fit-for-purpose preclinical models in combination with high-content readouts (Figure 11.8). Machine learning for the identification of relevant interactions and the rational qualification of preclinical disease models will enhance value of nanotechnology for clinical application. Despite of the long and stony pathway for decades, nanoparticles will have a future in drug delivery. According to the clinical needs, the focus in dermatology lies on drugs for inflammatory skin diseases and skin cancer.

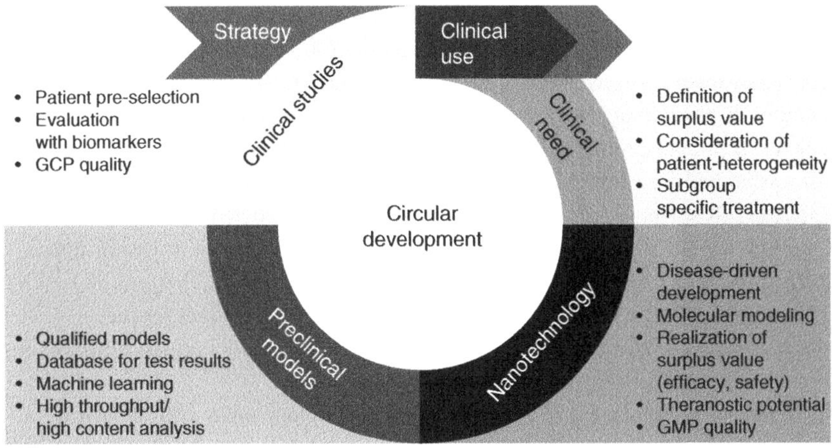

Figure 11.8 Circular approach of nanotechnology development. Transition from linear nanotechnology development from bench to bin into a circular development considering all disciplines from bench to bedside. Following the identification of a clinical need (light gray) nanotechnology should be developed to create a surplus value to current therapeutic standards (black). If the preclinical evaluation serves as bottle-neck for nanotechnology development (dark gray), clinical studies will be more successful (white).

References

1 Rosenblum, D., Joshi, N., Tao, W. et al. (2018). Progress and challenges towards targeted delivery of cancer therapeutics. *Nat. Commun.* 9: 1410.

2 Jain, R.K. and Stylianopoulos, T. (2010). Delivering nanomedicine to solid tumors. *Nat. Rev. Clin. Oncol.* 7: 653–664.

3 Roberts, M.S., Mohammed, Y., Pastore, M.N. et al. (2017). Topical and cutaneous delivery using nanosystems. *J. Controlled Release* 247: 86–105.

4 Li, J., Esteban-Fernández de Ávila, B., Gao, W. et al. (2017). Micro/nanorobots for biomedicine: delivery, surgery, sensing, and detoxification. *Sci. Rob.* 2: eaam6431.

5 Laux, P., Tentschert, J., Riebeling, C. et al. (2018). Nanomaterials: certain aspects of application, risk assessment and risk communication. *Arch. Toxicol.* 92: 121–141.

6 Haase, A. and Luch, A. (2016). Genotoxicity of nanomaterials in vitro: treasure or trash? *Arch. Toxicol.* 90: 2827–2830.

7 Barua, S. and Mitragotri, S. (2014). Challenges associated with penetration of nanoparticles across cell and tissue barriers: a review of current status and future prospects. *Nano Today* 9: 223–243.

8 Chauhan, V.P. and Jain, R.K. (2013). Strategies for advancing cancer nanomedicine. *Nat. Mater.* 12: 958.

9 Dewhirst, M.W. and Secomb, T.W. (2017). Transport of drugs from blood vessels to tumour tissue. *Nat. Rev. Cancer* 17: 738.

10 Wong, C.H., Siah, K.W., and Lo, A.W. (2018). Estimation of clinical trial success rates and related parameters. *Biostatistics* 20: 273–286.

11 Bailey, J., Thew, M., and Balls, M. (2014). An analysis of the use of animal models in predicting human toxicology and drug safety. *Altern. Lab. Anim.* 42: 181–199.

12 Seok, J., Warren, H.S., Cuenca, A.G. et al. (2013). Genomic responses in mouse models poorly mimic human inflammatory diseases. *Proc. Natl. Acad. Sci. U.S.A.* 110: 3507–3512.
13 Takao, K. and Miyakawa, T. (2015). Genomic responses in mouse models greatly mimic human inflammatory diseases. *Proc. Natl. Acad. Sci. U.S.A.* 112: 1167–1172.
14 van Luijk, J., Bakker, B., Rovers, M.M. et al. (2014). Systematic reviews of animal studies; missing link in translational research? *PLoS One* 9: e89981.
15 OECD 431 2014 (2014). Test No. 431: In vitro skin corrosion: reconstructed human epidermis (Rhe) test method. *OECD Guidelines for the Testing of Chemicals, Section 4: Health Effects [cited 2019/07/12].* https://doi.org/10.1787/9789264224193-en.
16 OECD 439 2013 (2013). Test No. 439: In vitro skin irritation. *OECD Guidelines for the Testing of Chemicals, Section 4: Health Effects [cited 2019/07/12].* https://doi.org/10.1787/9789264090958-en.
17 Blume-Peytavi, U., Bagot, M., Tennstedt, D. et al. (2019). Dermatology today and tomorrow: from symptom control to targeted therapy. *J. Eur. Acad. Dermatol. Venereol.* 33 (Suppl 1): 3–36.
18 Sala, M., Diab, R., Elaissari, A., and Fessi, H. (2018). Lipid nanocarriers as skin drug delivery systems: properties, mechanisms of skin interactions and medical applications. *Int. J. Pharm.* 535: 1–17.
19 Wolf, N.B., Küchler, S., Radowski, M.R. et al. (2009). Influences of opioids and nanoparticles on in vitro wound healing models. *Eur. J. Pharm. Biopharm.* 73: 34–42.
20 Müller, R.H., Radtke, M., and Wissing, S.A. (2002). Solid lipid nanoparticles (SLN) and nanostructured lipid carriers (NLC) in cosmetic and dermatological preparations. *Adv. Drug Delivery Rev.* 54 (Suppl 1): S131–S155.
21 Lombardi Borgia, S., Regehly, M., Sivaramakrishnan, R. et al. (2005). Lipid nanoparticles for skin penetration enhancement-correlation to drug localization within the particle matrix as determined by fluorescence and parelectric spectroscopy. *J. Controlled Release* 110: 151–163.
22 Nagaich, U. and Gulati, N. (2016). Nanostructured lipid carriers (NLC) based controlled release topical gel of clobetasol propionate: design and in vivo characterization. *Drug Delivery Transl. Res.* 6: 289–298.
23 Saeidpour, S., Lohan, S.B., Solik, A. et al. (2017). Drug distribution in nanostructured lipid particles. *Eur. J. Pharm. Biopharm.* 110: 19–23.
24 Sivaramakrishnan, R., Nakamura, C., Mehnert, W. et al. (2004). Glucocorticoid entrapment into lipid carriers – characterisation by parelectric spectroscopy and influence on dermal uptake. *J. Controlled Release* 97: 493–502.
25 Santos Maia, C., Mehnert, W., Schaller, M. et al. (2002). Drug targeting by solid lipid nanoparticles for dermal use. *J. Drug Targeting* 10: 489–495.
26 Schlupp, P., Blaschke, T., Kramer, K.D. et al. (2011). Drug release and skin penetration from solid lipid nanoparticles and a base cream: a systematic approach from a comparison of three glucocorticoids. *Skin Pharmacol. Physiol.* 24: 199–209.
27 Lohan, S.B., Saeidpour, S., Solik, A. et al. (2017). Investigation of the cutaneous penetration behavior of dexamethasone loaded to nano-sized lipid particles by EPR spectroscopy, and confocal Raman and laser scanning microscopy. *Eur. J. Pharm. Biopharm.* 116: 102–110.
28 Küchler, S., Wolf, N.B., Heilmann, S. et al. (2010). 3D-wound healing model: influence of morphine and solid lipid nanoparticles. *J. Biotechnol.* 148: 24–30.
29 Iyisan, B. and Landfester, K. (2019). Modular approach for the design of smart polymeric nanocapsules. *Macromol. Rapid Commun.* 40: e1800577.
30 De Brum, T.L., Fiel, L.A., Contri, R.V. et al. (2015). Polymeric nanocapsules and lipid-core nanocapsules have diverse skin penetration. *J. Nanosci. Nanotechnol.* 15: 773–780.

31 Scientific Committee on Cosmetic Products (2007). The Scientific Committee on Cosmetic Products (SCCP) opinion on safety of nanomaterials in vosmetic products [cited 2019/07/12]. http://ec.europa.eu/health/ph_risk/committees/04_sccp/docs/sccp_o_123.pdf.

32 Balzus, B., Sahle, F.F., Hönzke, S. et al. (2017). Formulation and ex vivo evaluation of polymeric nanoparticles for controlled delivery of corticosteroids to the skin and the corneal epithelium. *Eur. J. Pharm. Biopharm.* 115: 122–130.

33 Beber, T.C., de Andrade, D.F., dos Santos Chaves, P. et al. (2016). Cationic polymeric nanocapsules as a strategy to target dexamethasone to viable epidermis: skin penetration and permeation studies. *J. Nanosci. Nanotechnol.* 16: 1331–1338.

34 Quadir, M.A., Radowski, M.R., Kratz, F. et al. (2008). Dendritic multishell architectures for drug and dye transport. *J. Controlled Release* 132: 289–294.

35 Küchler, S., Radowski, M.R., Blaschke, T. et al. (2009). Nanoparticles for skin penetration enhancement – a comparison of a dendritic core–multishell-nanotransporter and solid lipid nanoparticles. *Eur. J. Pharm. Biopharm.* 71: 243–250.

36 Haag, S.F., Fleige, E., Chen, M. et al. (2011). Skin penetration enhancement of core–multishell nanotransporters and invasomes measured by electron paramagnetic resonance spectroscopy. *Int. J. Pharm.* 416: 223–228.

37 Alnasif, N., Zoschke, C., Fleige, E. et al. (2014). Penetration of normal, damaged and diseased skin – an in vitro study on dendritic core–multishell nanotransporters. *J. Controlled Release* 185: 45–50.

38 Edlich, A., Volz, P., Brodwolf, R. et al. (2018). Crosstalk between core–multishell nanocarriers for cutaneous drug delivery and antigen-presenting cells of the skin. *Biomaterials* 162: 60–70.

39 Radbruch, M., Pischon, H., Ostrowski, A. et al. (2017). Dendritic core–multishell nanocarriers in murine models of healthy and atopic skin. *Nanoscale Res. Lett.* 12: 64.

40 Hönzke, S., Gerecke, C., Elpelt, A. et al. (2016). Tailored dendritic core–multishell nanocarriers for efficient dermal drug delivery: a systematic top-down approach from synthesis to preclinical testing. *J. Controlled Release* 242: 50–63.

41 Frombach, J., Unbehauen, M., Kurniasih, I.N. et al. (2019). Core–multishell nanocarriers enhance drug penetration and reach keratinocytes and antigen-presenting cells in intact human skin. *J. Controlled Release* 299: 138–148.

42 Zabihi, F., Wieczorek, S., Dimde, M. et al. (2016). Intradermal drug delivery by nanogel-peptide conjugates; specific and efficient transport of temoporfin. *J. Controlled Release* 242: 35–41.

43 Giulbudagian, M., Rancan, F., Klossek, A. et al. (2016). Correlation between the chemical composition of thermoresponsive nanogels and their interaction with the skin barrier. *J. Controlled Release* 243: 323–332.

44 Edlich, A., Gerecke, C., Giulbudagian, M. et al. (2017). Specific uptake mechanisms of well-tolerated thermoresponsive polyglycerol-based nanogels in antigen-presenting cells of the skin. *Eur. J. Pharm. Biopharm.* 116: 155–163.

45 Gerecke, C., Edlich, A., Giulbudagian, M. et al. (2017). Biocompatibility and characterization of polyglycerol-based thermoresponsive nanogels designed as novel drug-delivery systems and their intracellular localization in keratinocytes. *Nanotoxicology* 11: 267–277.

46 Rancan, F., Giulbudagian, M., Jurisch, J. et al. (2017). Drug delivery across intact and disrupted skin barrier: identification of cell populations interacting with penetrated thermoresponsive nanogels. *Eur. J. Pharm. Biopharm.* 116: 4–11.

47 Knor, T., Meholjic-Fetahovic, A., and Mehmedagic, A. (2011). Stratum corneum hydration and skin surface pH in patients with atopic dermatitis. *Acta Dermatovenerol. Croat.* 19: 242–247.

48 Sahle, F.F., Gerecke, C., Kleuser, B., and Bodmeier, R. (2017). Formulation and comparative in vitro evaluation of various dexamethasone-loaded pH-sensitive polymeric nanoparticles intended for dermal applications. *Int. J. Pharm.* 516: 21–31.

49 Dong, P., Sahle, F.F., Lohan, S.B. et al. (2019). pH-sensitive Eudragit® L 100 nanoparticles promote cutaneous penetration and drug release on the skin. *J. Controlled Release* 295: 214–222.

50 Giulbudagian, M., Hönzke, S., Bergueiro, J. et al. (2017). Enhanced topical delivery of dexamethasone by beta-cyclodextrin decorated thermoresponsive nanogels. *Nanoscale* 10: 469–479.

51 OECD 428 2004 (2004). Test No. 428: Skin absorption: in vitro method. *OECD Guidelines for the Testing of Chemicals, Section 4: Health Effects [cited 2019/07/12]*. https://doi.org/10.1787/9789264071087-en.

52 Guth, K., Schäfer-Korting, M., Fabian, E. et al. (2015). Suitability of skin integrity tests for dermal absorption studies in vitro. *Toxicol. in Vitro* 29: 113–123.

53 Wanjiku, B., Yamamoto, K., Klossek, A. et al. (2019). Qualifying X-ray and stimulated raman spectromicroscopy for mapping cutaneous drug penetration. *Anal. Chem.* 91: 7208–7214.

54 Weindl, G., Castello, F., and Schäfer-Korting, M. (2011). Evaluation of anti-inflammatory and atrophogenic effects of glucocorticoids on reconstructed human skin. *Altern. Lab. Anim.* 39: 173–187.

55 Zöller, N.N., Kippenberger, S., Thaci, D. et al. (2008). Evaluation of beneficial and adverse effects of glucocorticoids on a newly developed full-thickness skin model. *Toxicol. in Vitro* 22: 747–759.

56 Hausmann, C., Hertz-Kleptow, D., Zoschke, C. et al. (2019). Reconstructed human epidermis predicts barrier-improving effects of *Lactococcus lactis* emulsion in humans. *Skin Pharmacol. Physiol.* 32: 72–80.

57 Niehues, H., Bouwstra, J.A., El Ghalbzouri, A. et al. (2018). 3D skin models for 3R research: the potential of 3D reconstructed skin models to study skin barrier function. *Exp. Dermatol.* 27: 501–511.

58 Shao, M., Hussain, Z., Thu, H.E. et al. (2016). Drug nanocarrier, the future of atopic diseases: advanced drug delivery systems and smart management of disease. *Colloids Surf., B* 147: 475–491.

59 Boisgard, A.S., Lamrayah, M., Dzikowski, M. et al. (2017). Innovative drug vehicle for local treatment of inflammatory skin diseases: ex vivo and in vivo screening of five topical formulations containing poly(lactic acid) (PLA) nanoparticles. *Eur. J. Pharm. Biopharm.* 116: 51–60.

60 Doppalapudi, S., Jain, A., Chopra, D.K., and Khan, W. (2017). Psoralen loaded liposomal nanocarriers for improved skin penetration and efficacy of topical PUVA in psoriasis. *Eur. J. Pharm. Sci.* 96: 515–529.

61 Barradas, T.N., Senna, J.P., Cardoso, S.A. et al. (2018). Formulation characterization and in vitro drug release of hydrogel-thickened nanoemulsions for topical delivery of 8-methoxypsoralen. *Mater. Sci. Eng., C* 92: 245–253.

62 Pilipenko, I., Korzhikov-Vlakh, V., Sharoyko, V. et al. (2019). pH-Sensitive chitosan-heparin nanoparticles for effective delivery of genetic drugs into epithelial cells. *Pharmaceutics* 11: 317. https://doi.org/10.3390/pharmaceutics11070317.

63 Oliveira, C.A., Gouvea, M.M., Antunes, G.R. et al. (2018). Nanoemulsion containing 8-methoxypsoralen for topical treatment of dermatoses: development, characterization and ex vivo permeation in porcine skin. *Int. J. Pharm.* 547: 1–9.

64 Hönzke, S., Wallmeyer, L., Ostrowski, A. et al. (2016). Influence of Th2 cytokines on the cornified envelope, tight junction proteins, and ss-defensins in filaggrin-deficient skin equivalents. *J. Invest. Dermatol.* 136: 631–639.

65 Giulbudagian, M., Yealland, G., Hönzke, S. et al. (2018). Breaking the barrier – potent anti-inflammatory activity following efficient topical delivery of etanercept using thermoresponsive nanogels. *Theranostics* 8: 450–463.

66 Eckl, K.M., Weindl, G., Ackermann, K. et al. (2014). Increased cutaneous absorption reflects impaired barrier function of reconstructed skin models mimicking keratinisation disorders. *Exp. Dermatol.* 23: 286–288.

67 Witting, M., Molina, M., Obst, K. et al. (2015). Thermosensitive dendritic polyglycerol-based nanogels for cutaneous delivery of biomacromolecules. *Nanomedicine* 11: 1179–1187.

68 Cardoso, A.M., de Oliveira, E.G., Coradini, K. et al. (2019). Chitosan hydrogels containing nanoencapsulated phenytoin for cutaneous use: skin permeation/penetration and efficacy in wound healing. *Mater. Sci. Eng., C* 96: 205–217.

69 Das, S. and Baker, A.B. (2016). Biomaterials and nanotherapeutics for enhancing skin wound healing. *Front. Bioeng. Biotechnol.* 4: 82.

70 Xue, M., Zhao, R., Lin, H., and Jackson, C. (2018). Delivery systems of current biologicals for the treatment of chronic cutaneous wounds and severe burns. *Adv. Drug Delivery Rev.* 129: 219–241.

71 Fumakia, M. and Ho, E.A. (2016). Nanoparticles encapsulated with LL37 and serpin A1 promotes wound healing and synergistically enhances antibacterial activity. *Mol. Pharmaceutics* 13: 2318–2331.

72 Zhang, Y., Zhang, J., Chen, M. et al. (2016). A bioadhesive nanoparticle-hydrogel hybrid system for localized antimicrobial drug delivery. *ACS Appl. Mater. Interfaces* 8: 18367–18374.

73 Garcia, M.C., Cuggino, J.C., Rosset, C.I. et al. (2016). A novel gel based on an ionic complex from a dendronized polymer and ciprofloxacin: evaluation of its use for controlled topical drug release. *Mater. Sci. Eng., C* 69: 236–246.

74 Casciaro, B., Moros, M., Rivera-Fernandez, S. et al. (2017). Gold-nanoparticles coated with the antimicrobial peptide esculentin-1a(1-21)NH2 as a reliable strategy for antipseudomonal drugs. *Acta Biomater.* 47: 170–181.

75 Santos, R.S., Loureiro, K., Rezende, P. et al. (2018). Innovative nanocompounds for cutaneous administration of classical antifungal drugs: a systematic review. *J. Dermatolog. Treat.*: 1–37.

76 Venturini, C.G., Bruinsmann, F.A., Contri, R.V. et al. (2015). Co-encapsulation of imiquimod and copaiba oil in novel nanostructured systems: promising formulations against skin carcinoma. *Eur. J. Pharm. Sci.* 79: 36–43.

77 Carvalho, V.F.M., Migotto, A., Giacone, D.V. et al. (2017). Co-encapsulation of paclitaxel and C6 ceramide in tributyrin-containing nanocarriers improve co-localization in the skin and potentiate cytotoxic effects in 2D and 3D models. *Eur. J. Pharm. Sci.* 109: 131–143.

78 Suter, F., Schmid, D., Wandrey, F., and Zulli, F. (2016). Heptapeptide-loaded solid lipid nanoparticles for cosmetic anti-aging applications. *Eur. J. Pharm. Biopharm.* 108: 304–309.

79 Radtke, M., Patzelt, A., Knorr, F. et al. (2017). Ratchet effect for nanoparticle transport in hair follicles. *Eur. J. Pharm. Biopharm.* 116: 125–130.

80 Sahle, F.F., Giulbudagian, M., Bergueiro, J. et al. (2017). Dendritic polyglycerol and *N*-isopropylacrylamide based thermoresponsive nanogels as smart carriers for controlled delivery of drugs through the hair follicle. *Nanoscale* 9: 172–182.

81 Mathes, C., Melero, A., Conrad, P. et al. (2016). Nanocarriers for optimizing the balance between interfollicular permeation and follicular uptake of topically applied clobetasol to minimize adverse effects. *J. Controlled Release* 223: 207–214.

82 Bernardos, A., Piacenza, E., Sancenon, F. et al. (2019). Mesoporous silica-based materials with bactericidal properties. *Small* 15: e1900669.

83 Hillman, Y., Lustiger, D., and Wine, Y. (2019). Antibody-based nanotechnology. *Nanotechnology* 30: 282001.

84 Matos, A.I., Carreira, B., Peres, C. et al. (2019). Nanotechnology is an important strategy for combinational innovative chemo-immunotherapies against colorectal cancer. *J. Controlled Release* 307: 108–138.

85 Nayak, R., Meerovich, I., and Dash, A.K. (2019). Translational multi-disciplinary approach for the drug and gene delivery systems for cancer treatment. *AAPS PharmSciTech* 20: 160.

86 Prajnamitra, R.P., Chen, H.C., Lin, C.J. et al. (2019). Nanotechnology approaches in tackling cardiovascular diseases. *Molecules* 24 (10): 2017. https://doi.org/10.3390/molecules24102017.

87 Tosi, G., Pederzoli, F., Belletti, D. et al. (2019). Nanomedicine in Alzheimer's disease: amyloid beta targeting strategy. *Prog. Brain Res.* 245: 57–88.

88 Wilhelm, S., Tavares, A.J., Dai, Q. et al. (2016). Analysis of nanoparticle delivery to tumours. *Nat. Rev. Mater.* 1: 16014.

89 Vamathevan, J., Clark, D., Czodrowski, P. et al. (2019). Applications of machine learning in drug discovery and development. *Nat. Rev. Drug Discovery* 18: 463–477.

90 Weber, M., Zoschke, C., Sedighi, A. et al. (2014). Free energy simulations of cargo-carrier interactions for core–multishell nanotransporters. *J. Nanomed. Nanotechnol.* 5: 234.

91 Zoschke, C., Schilrreff, P., Romero, E.L. et al. (2015). Dendritic nanoparticles for cutaneous drug delivery – testing in human skin and reconstructed human skin. *Curr. Pharm. Des.* 21: 2784–2800.

92 OECD (2019). Performance standards. http://www.oecd.org/chemicalsafety/testing/performance-standards.htm (accessed 12 July 2019).

93 Fitzgerald, K.A., Malhotra, M., Curtin, C.M. et al. (2015). Life in 3D is never flat: 3D models to optimise drug delivery. *J. Controlled Release* 215: 39–54.

94 Nirmalanandhan, V.S., Duren, A., Hendricks, P. et al. (2010). Activity of anticancer agents in a three-dimensional cell culture model. *Assay Drug Dev. Technol.* 8: 581–590.

95 Selby, M., Delosh, R., Laudeman, J. et al. (2017). 3D models of the NCI60 cell lines for screening oncology compounds. *SLAS Discov.* 22: 473–483.

96 Wolff, C., Zoschke, C., Kumar Kalangi, S. et al. (2019). Tumor microenvironment determines drug efficacy in vitro – apoptotic and anti-inflammatory effects of 15-lipoxygenase metabolite, 13-HpOTrE. *Eur. J. Pharm. Biopharm.* 142: 1–7.

97 Gu, L. and Mooney, D.J. (2016). Biomaterials and emerging anticancer therapeutics: engineering the microenvironment. *Nat. Rev. Cancer* 16: 56–66.

98 Commandeur, S., Ho, S.H., de Gruijl, F.R. et al. (2011). Functional characterization of cancer-associated fibroblasts of human cutaneous squamous cell carcinoma. *Exp. Dermatol.* 20: 737–742.

99 Zoschke, C., Ulrich, M., Sochorová, M. et al. (2016). The barrier function of organotypic non-melanoma skin cancer models. *J. Controlled Release* 233: 10–18.

100 Balansin Rigon, R., Kaessmeyer, S., Wolff, C. et al. (2018). Ultrastructural and molecular analysis of ribose-induced glycated reconstructed human skin. *Int. J. Mol. Sci.* 19 (11): 3521. https://doi.org/10.3390/ijms19113521.

101 Löwenau, L.J., Zoschke, C., Brodwolf, R. et al. (2017). Increased permeability of reconstructed human epidermis from UVB-irradiated keratinocytes. *Eur. J. Pharm. Biopharm.* 116: 149–154.
102 Waldera Lupa, D.M., Kalfalah, F., Safferling, K. et al. (2015). Characterization of skin aging-associated secreted proteins (SAASP) produced by dermal fibroblasts isolated from intrinsically aged human skin. *J. Invest. Dermatol.* 135: 1954–1968.
103 Biran, A., Zada, L., Abou Karam, P. et al. (2017). Quantitative identification of senescent cells in aging and disease. *Aging Cell* 16: 661–671.
104 Hausmann, C., Zoschke, C., Wolff, C. et al. (2019). Fibroblast origin shapes tissue homeostasis, epidermal differentiation, and drug uptake. *Sci. Rep.* 9: 2913.
105 Bock, S., Said, A., Müller, G. et al. (2018). Characterization of reconstructed human skin containing Langerhans cells to monitor molecular events in skin sensitization. *Toxicol. in Vitro* 46: 77–85.
106 van den Bogaard, E.H., Tjabringa, G.S., Joosten, I. et al. (2014). Crosstalk between keratinocytes and T cells in a 3D microenvironment: a model to study inflammatory skin diseases. *J. Invest. Dermatol.* 134: 719–727.
107 Schütte, M., Risch, T., Abdavi-Azar, N. et al. (2017). Molecular dissection of colorectal cancer in pre-clinical models identifies biomarkers predicting sensitivity to EGFR inhibitors. *Nat. Commun.* 8: 14262.
108 Yamamoto, K., Klossek, A., Flesch, R. et al. (2016). Core–multishell nanocarriers: transport and release of dexamethasone probed by soft X-ray spectromicroscopy. *J. Controlled Release* 242: 64–70.
109 Schulz, R., Yamamoto, K., Klossek, A. et al. (2017). Data-based modeling of drug penetration relates human skin barrier function to the interplay of diffusivity and free-energy profiles. *Proc. Natl. Acad. Sci. U.S.A.* 114: 3631–3636.
110 Volz, P., Brodwolf, R., Zoschke, C. et al. (2018). White-light supercontinuum laser-based multiple wavelength excitation for TCSPC-FLIM of cutaneous nanocarrier uptake. *Z. Phys. Chem.* 232: 671–688.
111 Alexiev, U., Volz, P., Boreham, A., and Brodwolf, R. (2017). Time-resolved fluorescence microscopy (FLIM) as an analytical tool in skin nanomedicine. *Eur. J. Pharm. Biopharm.* 116: 111–124.
112 Volz, P., Schilrreff, P., Brodwolf, R. et al. (2017). Pitfalls in using fluorescence tagging of nanomaterials: tecto-dendrimers in skin tissue as investigated by Cluster-FLIM. *Ann. N.Y. Acad. Sci.* 1405: 202–214.
113 Graf, C. and Rühl, E. (2019). Imaging techniques for probing nanoparticles in cells and skin. In: *Biological Responses to Nanoscale Particles: Molecular and Cellular Aspects and Methodological Approaches* (eds. P. Gehr and R. Zellner), 213–239. Cham: Springer International Publishing.
114 Solon, E.G., Schweitzer, A., Stoeckli, M., and Prideaux, B. (2010). Autoradiography, MALDI-MS, and SIMS-MS imaging in pharmaceutical discovery and development. *Aaps J.* 12: 11–26.
115 Ucal, Y., Durer, Z.A., Atak, H. et al. (2017). Clinical applications of MALDI imaging technologies in cancer and neurodegenerative diseases. *Biochim. Biophys. Acta, Proteins Proteomics* 1865: 795–816.

12

Theranostic Nanoparticles for Imaging and Targeted Drug Delivery to the Liver

Haolu Wang, Haotian Yang, Qi Ruan, Michael S. Roberts, and Xiaowen Liang

Therapeutics Research Group, The University of Queensland Diamantina Institute, The University of Queensland, Translational Research Institute, Brisbane, QLD, Australia

12.1 Introduction

The liver plays a major role in maintaining body homeostasis by the regulation of digestion and detoxification. Before entering the liver, blood flows from the gastrointestinal tract and heart via the portal vein and hepatic arteries, and into the liver through the sinusoids (specialized vascular vessels between hepatocyte cords). Blood then accumulates in the central veins and leaves the liver through the hepatic veins. The key functional cells during these processes are hepatocytes, which constitute 70–85% of liver by cell population [1]. They contain abundant mitochondria, endoplasmic reticulum, and enzymes, enabling metabolic, endocrine and excretory functions. Other non-parenchymal cells in the liver include Kupffer cells, specialized macrophages sited on the wall of sinusoid and stellate cells [1]. Bile is secreted by hepatocytes, which is drained out of liver through bile ducts lined with biliary epithelial cells [2].

Liver diseases are frequently encountered in the clinic, with high morbidity and mortality worldwide, such as hepatocellular carcinoma (HCC), viral hepatitis, and cirrhosis [3, 4]. Most drugs are taken up and cleared from the circulation without targeting to the liver due to first pass-effect of the liver [5], and some of them are eliminated by the large population of macrophages in the liver. Although many diagnostic and therapeutic drugs have been developed, most drugs cannot target specific cells resulting in adverse effects in other liver cells or organs [1]. For example, free doxorubicin used to treat HCC has side effects including cardiac toxicity, oral mucositis, and hair loss [6]. Hence, it is imperative to explore new efficient targeting pharmacological interventions for diagnosis and treatment of liver diseases.

Nanoparticles (NPs) with a diameter of less than 100 nm have been used as drug carriers in liver targeting delivery [7]. NPs have shown great potential in targeting drugs to the liver either passively or actively due to their unique size and surface characteristics [1]. These carriers significantly improve conventional chemotherapeutic agents with enhanced solubility, less side effects, and longer half-life [8]. Theranostics is a new strategy with clinical potential that combines diagnostics and therapeutics [9], which will provide a promising therapeutic strategy for liver disease. NPs can serve as a novel platform for drug delivery and diagnostic imaging for liver diseases, enabling a more personalized method and providing more accurate diagnosis, better outcomes, as well as improved safety [9]. This chapter systematically summarizes the NP-based approaches for theranostic applications in liver diseases. The schematic diagram of various types of theranostic NPs is shown in Figure 12.1.

Fundamentals of Drug Delivery, First Edition.
Edited by Heather A. E. Benson, Michael S. Roberts, Adrian C. Williams, and Xiaowen Liang.

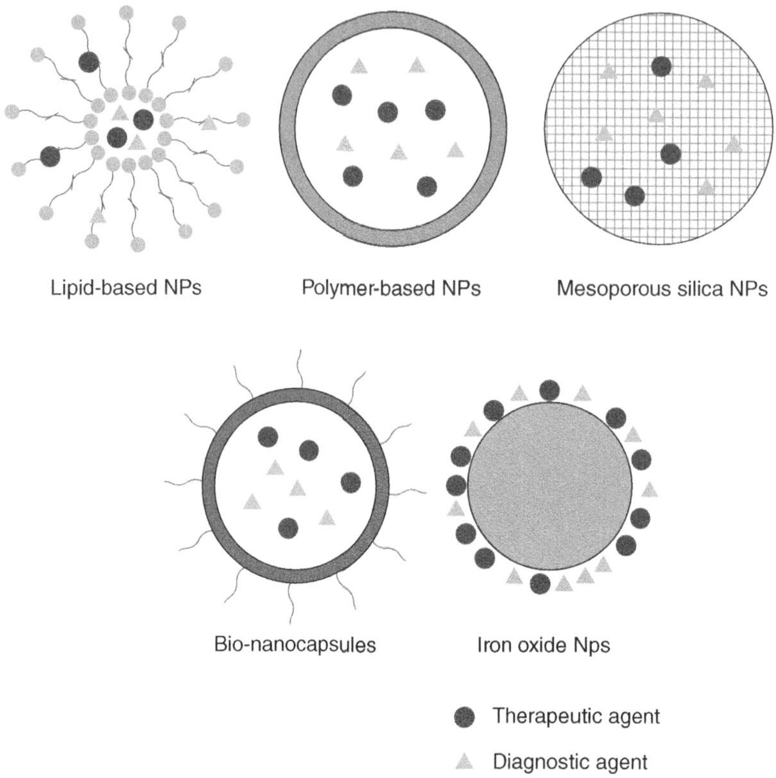

Figure 12.1 Schematic diagram of theranostic NPs include lipid- and polymer-based NPs, mesoporous silica NPs, bio-nanocapsules, and iron oxide NPs.

12.2 The Types of Theranostic NPs

12.2.1 Lipid- and Polymer-Based NPs

Liposome and polymeric NPs are two important platforms for nanotheranostics in cancer [10]. Liposomes are spherical vesicles composed of a lipid bilayer, which is produced with phospholipids and have properties of low toxicity and high productivity [11]. Liposome-based NPs have been widely used as a drug and imaging agents carrier and already been commercialized in cancer treatments (e.g. Doxil®, Myocet®) due to their ability of optimizing size, surface charges, and internal environment, as well as their capability of incorporating hydrophilic agents inside and hydrophobic agents between the lipid bilayers [11]. Polymer-based NPs are commonly applied in clinics, which use variable biodegradable and biocompatible materials for surface coating and conjugation with drugs. Thus, polymer-based NPs are a class of NPs for many drug delivery applications and include the polymers poly(DL-lactide-*co*-glycolide) (PLGA), poly(ethyleneglycol) (PEG), Poly-L-lysine (PLL), polyethyleneimine (PEI), and chitosan [12]. PLGA carriers were approved by the FDA to load dyes such as indocyanine green (ICG) for theranostic application.

12.2.2 Mesoporous Silica NPs

The first family of mesoporous silica NPs (MSNs) was developed in the 1990s. They have attracted increasing interest due to their larger surface area and adjustable pore size compared to other

NPs. Moreover, MSN surface can be easily modified, enabling to control and target drug delivery [13]. MSNs have been reported to effectively prevent drug resistance via endosomal delivery or co-delivery of genes and drugs in cancer treatment [13, 14].

12.2.3 Bio-nanocapsules

Bio-nanocapsules are approximately 50 nm hollow particles consisting of hepatitis B virus (HBV) surface antigen, showing a human hepatocyte-recognizing molecule [15]. These NPs have been used as hepatitis B vaccine for a long time. Bio-nanocapsules can incorporate various payloads such as genes and drugs by electroporation and deliver them to human hepatocytes specifically [15]. The recombinant envelop large (L) proteins of HBV produced from yeast cells were used as hollow particles for genes and drugs delivery. With an average diameter of 80 nm and free of viral genomes, it showed a high specificity and efficiency of delivering genes and drugs to hepatic tumor cells *in vitro* and in mouse xenograft models. The L particle was later modified by substituting the Gln-292 and Gly-308 in the L protein with Arg to reduce binding of endogenous immunoglobulins, thus allowing repetitive administration of BNCs [16]. Furthermore, bio-nanocapsules can also target other cells and tissues by modifying the N-terminal-half of the pre-S region on L proteins to antibody binding site and conjugate with antibodies. For example, they were able to effectively accumulate in epidermal growth factor receptor (EGFR) expressing glioma by modifying to express IgG Fc-binding site and conjugating with anti-EGFR antibodies [17].

12.2.4 Iron Oxide NPs

The most commonly used iron oxide nanoparticles (IONs) in biomedical applications are magnetite (Fe_3O_4) and maghemite (γ-Fe_2O_3) [18]. IONs are well known for their applications as contrast agents in magnetic resonance imaging (MRI). In HCC, the tumor contrast could be significantly increased by intravenous injection of IONs with average diameter of 150 nm, due to rapid sequestering of NPs by Kupffer cells in normal tissue of the liver while tumor has no Kupffer cells [19]. As a drug carrier, small size of IONs are easily produced and can be coated with various polymers for targeting. Additionally, it can actively target specific site using external magnetic field [20].

12.3 Mechanisms of NPs Targeting the Liver

In general, NP delivery strategies can be divided into passive and active targeting. The passive targeting strategy employs the physicochemical properties of NPs include size and surface charge and routes of administration of NPs. In contrast, active targeting of NP employs specific ligands or coating to recognize and bind to certain types of cells, minimizing nonspecific effects on other types of cells in the liver [1] (Figure 12.2).

12.3.1 Passive Targeting to the Liver

Nonspecific NPs are mostly accumulated in liver and spleen, with a small fraction distributed in other organs [21]. Physiological and anatomical characteristics of the liver enable the NP accumulation with specific size and surface properties. NPs tend to interact with serum proteins in the blood stream after intravenous injection, a process called "protein corona," leading to rapid uptake by Kupffer cells [22]. NPs smaller than 3 nm in diameter extravasate different tissues nonspecifically

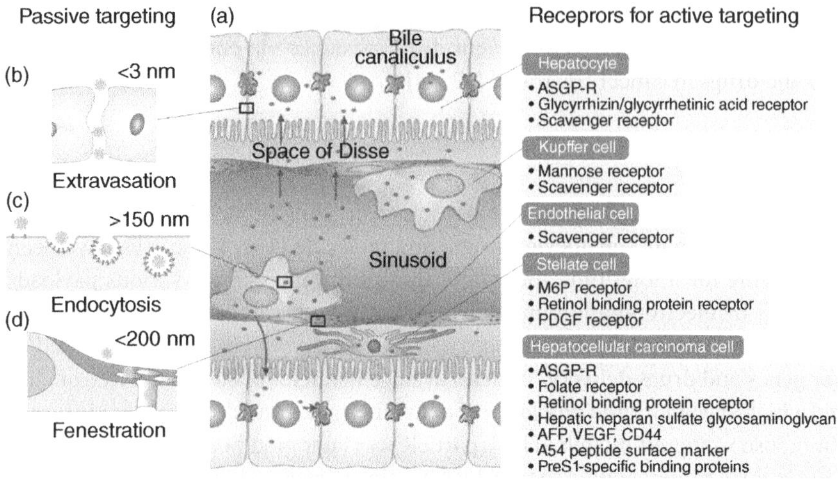

Figure 12.2 Mechanisms of NPs passively and actively targeting the liver. (a) Structure of liver sinusoid in hepatic lobules. (b) NPs smaller than 3 nm in diameter could extravasate tissues nonspecifically. (c) NPs with negative surface charge or large than 150 nm could be captured by Kupffer cells. (d) NPs smaller than 200 nm could penetrate through sinusoidal fenestration. Receptors or cellular targets on different types of liver cells were also summarized for actively targeting of NPs. Source: From Wang et al. [1]. © 2015,Royal Society of Chemistry.

(Figure 12.2b), while NP uptake by Kupffer cells increases with the increase of NP size [21]. Thus, larger NPs are preferentially recognized and taken up by Kupffer cells in the liver with shorter half-life (Figure 12.2c). It has been suggested that NPs smaller than 200 nm could avoid capture by Kupffer cells, diffuse out of sinusoids and potentially reach hepatocytes [1] (Figure 12.2d).

Endothelial cells lining the liver sinusoids have fenestrae of 50–200 nm in diameter [23] and NPs smaller than 200 nm could pass through sinusoidal fenestrations. Under certain circumstance such as HCC, the fenestrae of endothelial cells can increase to 400–600 nm, resulting in enhanced permeability and extravasation of NPs [23]. It has been reported that the ideal size of NPs for cancer targeting and treatment is between 70 and 200 nm [24].

In addition to the size of NPs, there are also other factors can affect the distribution of NPs. The negatively charged NPs are more likely to be taken up by Kupffer cells and endothelial cells, while hepatocytes tend to take up positive ones [21]. Moreover, hydrophobic NPs can rapidly be removed by Kupffer cells compared to hydrophilic ones [25]. PEGylation has been reported to reduce protein corona, minimizing particle uptake by Kupffer cells. Overall, passive targeting of NPs to the liver mainly relies on the physiochemical properties of NPs and has limited applications.

12.3.2 Active Targeting to the Liver

There are several different receptors on each type of liver cells, which have been identified for possible targeting. NPs modified with specific ligands that can be recognized by their receptors or cellular targets in liver cells have been used for targeted imaging and drug delivery. These developed NPs are able to actively target different types of liver cells include hepatocytes, endothelial cells, Kupffer cells, hepatic stellate cells (HSCs), and cancer cells (Figure 12.2).

Asialoglycoprotein (ASGP) receptor is expressed in well differentiated forms of HCC cells and the membrane of hepatocytes, which have been widely used for specific delivery to liver and HCC cells. This receptor is binding to a broad range of molecules containing galactose and *N*-acetyl-galactosamine residues such as lactose, galactosamine, galactoside, and lactobionic acid, which can be conjugated to the NP surface for active targeting of liver cells [8]. In addition, glycyrrhizin or glycyrrhetinic acid receptor is also employed in the active targeting strategies for liver cells. NPs modified by glycyrrhizin or glycyrrhetinic can be taken up by hepatocytes through receptor-mediated endocytosis [26]. Since HCC cells have more glycyrrhetinic acid receptors than hepatocytes, it can be also used in cancer targeting imaging and therapy [27].

There are three groups of receptors overexpressed on HSCs, including mannose-6-phosphate (M6P)/insulin-like growth factor-II receptor, retinol binding protein receptor and platelet-derived growth factor (PDGF) receptor [8, 19]. Human serum albumin modified with M6P conjugated to NPs which can specifically target to HSCs. In hepatic fibrogenesis, activated HSCs can upregulate PDGF receptor-β, which has been applied as the HSC target.

Kupffer cell and endothelial cell have the similar characteristics for targeting and both of them recognized oxidized low-density lipoprotein and human serum albumin by scavenger receptor [28]. Mannose receptor is a transmembrane protein expressed on the surface of Kupffer cells and endothelial cells, which has been applied to design a cell-specific mannosylated NPs for active targeting of Kupffer cells [29].

12.3.3 Strategies for Combining Passive and Active Targeting

The passive and active targeting can be combined together for liver specific imaging and delivery. Guan et al. developed a lactosylnorcantharidin *N*-trimethyl chitosan (Lac-NCTD-TMC) NP-based novel liver targeting delivery system, which can control the particle size at about 120 nm to penetrate the sinusoidal fenestration and active targeting the tumor site of the liver. NCTD, a small molecule originally secreted by blister beetle, modified with a lactobiose-bearing galactose group can significantly enhance the tumor cell uptake through ASGP receptor recognition [30].

12.4 NPs in Liver Target Imaging

Imaging modalities play an important role in liver cancer screening, diagnosis, treatment planning, and monitoring of treatment response. NPs have been used as contrast agents in various noninvasive liver imaging techniques, including magnetic resonance imaging (MRI), computed tomography (CT), and fluorescence imaging.

12.4.1 NP-Based Contrast Agents in Liver MRI

Magnetic NPs have been applied as contrast agents for MRI by reducing the relaxation time of protons in absorbing tissues to generate contrast effects under external magnetic fields [1, 31]. There are two types of NP-based contrast agents for liver MRI with the improved sensitivity and reliability: superparamagnetic NPs used as T_2 contrast agents and can reduce the T_2 relaxation time and paramagnetic NPs that work through shortening the T_1 relaxation time [32]. In general, clinically used MRI contrast agents are gadolinium-based T_1 contrast agents. NP-based T_1 contrast agents such as MnO and Gd_2O_3 showed great advantage over conventional T_1 contrast agents, including better biocompatibility, higher plasma stability, signal enhancement, specific liver targeting, and better contrast imaging [33, 34].

Superparamagnetic IONs is a typical T_2 contrast agent and is the only NP-based T_2 contrast agent that has been clinically approved. This NP is made by an agglomerate of numerous iron oxide crystals and can be coated with dextran, carboxydextran, or PEG [35]. It has been reported that these NPs accumulated in normal liver tissue due to the uptake by Kupffer cells. Liver cancers including HCC, cholangiocarcinoma, and colorectal cancer liver metastasis have less Kupffer cells and less accumulation of superparamagnetic IONs. There are two NPs have been approved for clinical application: ferumoxides with a particle size of 120–180 nm and ferucarbotran of about 60 nm. Furthermore, these NPs modified with ASGP receptor has been employed for active targeting to hepatocytes to increase the sensitivity of diagnosis via ASGP receptor mediated approach [36].

Some NP-based MRI contrast agents can provide effects in both T_1 imaging with high tissue resolution and T_2 imaging with high feasibility for detecting hepatic lesions [1]. A single hybrid nanocrystal has been successfully synthesized to integrate T_1 contrast agent MnO NPs and T_2 contrast agent Fe_3O_4, which showed the more accuracy than that of single contrast agent [37]. The NP-based dual-modal contrast agents provide an innovative and promising approach for clinical application in diagnosis of liver diseases.

12.4.2 NP-Based Contrast Agents in Liver CT Imaging

NP-based contrast agents for CT imaging have better imaging efficacy, longer circulating half-life and lower toxicity compared to conventional iodine-based CT contrast agents [1]. Alkaline earth metal-based NPs (ExiTron nano 6000 and 12000®) have been assessed as liver specific contrast agents for CT imaging in animals [38]. These NPs with an average diameter of 110 nm and coated with polymer, which are mainly taken up by Kupffer cells and accumulate in normal liver tissues. Therefore, they can be used to clearly detect the region of tumors since tumor tissues do not take up these contrast agents [38]. Additionally, these NPs with long half-life allow long-term imaging after a single injection. For example, ExiTron nano 12000 can provide contrast signals for the liver up to several weeks after the injection [39]. However, no NP-based CT imaging contrast has been applied to active target the liver.

12.4.3 NPs for Near-Infrared Fluorescence Imaging in Liver

Near-infrared fluorescence (NIRF) imaging has numerous advantages for noninvasive imaging of the liver such as *in vivo* deep tissue penetration, multi-detection capability, high resolution, and sensitivity [40]. Since NIRF signals can be generated from the use of an near-infrared (NIR) laser source of 700–900 nm, exogenous contrast agents such as NP incorporated ICG can be used. Most human NIRF imaging employ ICG as imaging contrast within blood and vasculature. ICG is typically embedded into NPs composed of polymer and inorganic matrices through noncovalent or covalent combination within a core–shell structure [9]. Compared to conventional contrast agents in NIRF imaging, these NPs show enhanced photostability, biocompatibility, active targeting, low self-aggregation, and high fluorescence signals [9]. MSN loaded ICG NPs have been applied for detecting accurate tumor border delineation and tumor residuals during liver cancer surgery using NIRF imaging. High ICG-loading system results in high contrast between tumor tissues and normal liver tissues under NIRF imaging. This imaging system allows deep tissue imaging and simultaneous visualization at anatomy resolution within the region of interest [41].

12.5 NPs for Therapeutic and Drug Delivery in Liver Disease

12.5.1 NP Delivery System in HCC

NPs have attracted significant interest not only for diagnosis but also as therapeutic tools. Liver cancer is the second most lethal tumor worldwide [42]. Liver cancers can arise as a primary tumor, including HCC and intrahepatic cholangiocarcinoma, or as secondary metastasis from cancers including colorectal cancer liver metastases (CLM). Despite significant advances in the treatment of liver cancer, response rates of current chemotherapy are still low [43, 44] due to multidrug resistance of cancer cells and insufficient accumulation of drugs at the tumor. Therefore, NP-based therapeutic and drug delivery systems will increase the therapeutics targeting to and concentration at the site, and prolong exposure of cancer cells to therapeutics.

Anti-cancer drug-loaded polymer NPs have been reported to have more tumor inhibition efficacy and higher concentration in cancer cells. There are several studies have been employed polymer NPs to load anti-cancer drug such as 5-fluorouracil, cisplatin, doxorubicin, and mitoxantrone [45–47] for target delivery to liver cancer. These NPs have a size range of 55–200 nm without specific surface modification and ligand coating. There are two anti-cancer drug-loaded polymer NPs under clinical testing: doxorubicin-loaded (Livatag®) [47] and mitoxantrone-loaded NPs [48]. In a Phase II trial, the 18-month survival rate of Livatag (88.9%) is significantly higher than standard care group (54.5%). It has potential to treat HCC after a treatment failure or intolerance to Sorafenib. Also, in the Phase II clinical trial, NP-loaded mitoxantrone group has a higher median survival period (5.46 months) than mitoxantrone-free group (3.23 months) in patients with unresected HCC [48]. Polymer NPs can also be a gene delivery system to load short hairpin RNA (shRNA) or small interfering RNAs (siRNAs) to deliver to HCC [49, 50]. NP-Loaded shRNA can effectively inhibit vascular endothelial growth factor (VEGF) expression in cancer cells and suppress tumor growth in HCC models [50]. SiRNA-loaded polymer target delivery system is developed for HCC to improve the therapeutic approaches [49].

Liposomes are the most popular non-polymer nanocarrier for therapeutics delivery of liver cancer. As mentioned above, liposomes are spherically enclosed membrane vesicles and formed by lipids [11]. Lyso-thermosensitive liposomal doxorubicin (LTLD) is the only non-polymer NP that has been under clinical trial [1]. LTLD can rapidly accumulated in the liver and selectively release doxorubicin during radiofrequency ablation [51]. In Phase III Heat study, it was reported that adding LTLD to radiofrequency ablation was safe but did not increase progression-free survival or overall survival in the overall study population. However, subgroup analysis suggested that LTLD improved radiofrequency efficacy when it dwell time for a solitary lesion longer than 45 minutes [52]. In addition, safety and feasibility of ultrasound targeted doxorubicin delivery of LTLD in liver tumors has been tested in Phase I trial. The results showed the treatment has resulted in an average increase of 3.7 times doxorubicin concentrations in intratumoral biopsy after focused ultrasound. They suggest the combined treatment of LTLD and noninvasive focused ultrasound hyperthermia seems to be clinically feasible, safe, and able to enhance intratumoral drug delivery, which provide alternative treatment in liver cancer that is refractory to standard chemotherapy [53].

Apart from passive targeting to HCC, more studies have conjugated active targeting ligands to the surface of nanocarriers. Galactose-coating or galactosylated NPs have recently emerged as viable and versatile tools to targeted deliver therapeutics to liver cancer [54]. For example, 5-fluorouracil-loaded galactosylated NPs have been found to effectively inhibit tumor growth through the activation of p53 pathway in orthotopic HCC mouse model [55]. Galactosamine conjugated paclitaxel-loaded NPs have also shown the significant efficacy in the reduction of tumor size due to their accumulation in tumor sites and selective paclitaxel release in the tumor via a ligand-receptor recognition [56]. Some targeting strategies utilized the other proteins frequently overexpressed on the surface of liver cancer cells such as VEGF, folate receptor, and hepatitis B surface antigen [57, 58]. Although most of nanocarriers are organic NPs, some inorganic NPs have also been used to treat HCC such as mesoporous silica NPs and superparamagnetic IONs [59, 60]. Both of them have significant anti-tumor activity and low toxicity compared to free drugs.

12.5.2 NP Delivery System in Non-tumoral Liver Disease

Recently, several innovative NP-based therapeutic delivery systems have been explored for non-tumoral liver diseases such as viral hepatitis, ischemia-reperfusion injury, liver fibrosis, and failure [8]. HBV is a highly contagious DNA virus that can be transmitted through parental or mucosal exposure to infected blood. Currently, some approved strategies for HBV infections include interferon-α, nucleoside, and nucleotide, but all of them remain inadequate at achieving long-term resolution of the viral loads [8]. Solid lipid NPs have been applied to load adefovir dipivoxil and delivery to HepG2.2.15 cells, resulting in significant downregulation of hepatitis B surface antigen and HBV DNA [8]. Several notable cationic liposomal loaded anti-HBV siRNA delivery systems have also shoed potential for the treatment of HBV infections [61, 62]. For example, Kim et al. designed liposomes with apo A-I linked onto the surface to deliver siRNA to the liver to reduce viral protein expression [61]. Liposomes were mainly taken up by the liver through scavenger receptor-mediated endocytosis. Apo A-I is a protein component of high density lipoprotein which is an endogenous product and do not have the immunological side effects [61].

Liver fibrosis is excessive accumulation of extracellular matrix proteins such as collagen that occurs in most types of chronic liver diseases [63]. Advanced liver fibrosis could result in cirrhosis, and may lead to HCC. The activation of HSCs play an important role in fibrogenesis, which are the major target for treatment of liver fibrosis. However, conventional treatments for liver fibrosis are of little clinical success and until now no standard treatment is available [1]. Liposomes have been applied to treatment of liver fibrosis by actively targeting HSC expressed receptors including retinol binding protein receptors, mannose-6-phosphate receptor, and PDGF receptor-β. Retinol binding protein receptors can recognize vitamin A and certain cyclic peptide. Sato et al. applied vitamin A-coupled liposomes to deliver siRNA against gp46 (the rat homolog of human heat shock protein 47) of HSCs in rat models of acute and chronic liver fibrosis [64]. Their approach explores the key roles of HSCs in both fibrogenesis and uptake and storage of vitamin A and provides therapeutic potential for liver fibrosis and cirrhosis [64]. Another study employed reducible polyethyleneimine grafted hyaluronic acid (HA) to deliver siRNA therapeutics to HSCs and these NPs showed a feasible therapeutic effect on liver cirrhosis with a significantly reduced nodule formation, collagen content, and the number of activated HSCs [65]. Sorafenib is a tyrosine kinase inhibitor has been shown to be a potential anti-fibrotic agent. PLGA NPs have been designed to load sorafenib to treat fibrotic mouse models. Sorafenib-loaded PLGA NPs significantly shrank the abnormal blood vessels and decreased microvascular density, resulting in vessel normalization in fibrotic livers [66]. Apart from therapeutic drug and gene delivery, polymeric NPs are designed to transport and deliver nitric oxide into HSCs for potential treatment of liver fibrosis and portal hypertension. Vitamin A coated NPs can release nitric oxide in the liver to inhibit the expression of collagen I and α-smooth muscle actin, and fibrogenic genes in activated HSCs without any obvious cytotoxic effects [67].

12.6 Theranostic NPs in Liver Diseases

In order to have better diagnosis and therapy outcomes, new therapeutic NPs with multifunction such as diagnostic and therapeutic capabilities have emerged as promising protocols for the treatment of liver disease (Table 12.1). Photothermal therapy is an attractive tool for cancer therapy due to its minimal invasiveness compared to other cancer interventions such as chemotherapy, radiation, and surgery. Nowadays, NPs have been widely explored as photothermal agents including carbon NPs, gold NPs, and magnetite NPs, owing to their high optical extinction coefficients in the NIR

Table 12.1 Theranostic NP applications in liver diseases.

NP type	Surface coating	Size (nm)	Imaging technique	Therapy type	Liver diseases	Outcome	References
Magnetite	Poly(dopamine)	50–200	MRI	Photothermal	Liver cancer	More accumulation in tumor; more efficient for tumor ablation	Wu et al. [68]
Gold nanorods	Gadolinium oxide	<100	MR/X-ray	Photothermal	CLM	Efficient for site-directed photothermal therapy	Parchur et al. [69]
SPION	—	15	MRI	Chemotherapy and nanoblation	Liver cancer of VX2 rabbit bearing orthotopic model	Increased intra-tumoral doxorubicin uptake; limited off-targeted delivery	Mouli et al. [59]
Polymer	Anti-VEGF	70	MRI	Chemotherapy	HCC	High anti-tumor activities; selective accumulation in tumors; high spin-spin (T2) relaxivity	Huang et al. [57]
Polymer	Folate	84.7	MRI	Chemotherapy	HCC models of rats and rabbits	High bioavailability and efficacy of chemotherapeutics; high MRI sensitivity	Maeng et al. [6]
Liposomes	Cyclic peptide C*SRNLIDC*	83.5–86.9	Fluorescence imaging	Anti-fibrotic effects of IFN-γ	Fibrotic rats	Inhibited activated HSC proliferation; targeted anti-fibrotic treatment	Li et al. [70, 71]
Nanohybrid	Acetalated dextran	473 ± 13	CT	Porous silicon delivery of therapeutics	Acute liver failure	Enhanced CT imaging; increased *in situ* drug delivery	Liu et al. [72]
Apoferritin	—	11.9–18.8	MRI	Therapeutic anti-inflammatory	Acute hepatitis	Simultaneously monitoring apoferritin take up by cells and anti-inflammatory efficacy	Cutrin [73]

wavelength range [68]. Wu et al. developed a magnetite nanocluster@poly(dopamine)-PEG-loaded ICG nanobead for tumor MRI and photothermal therapy using a magnetite nanocluster core coated with poly(dopamine), that was then conjugated with PEG and ICG was absorbed on the surface of nanocluster [68]. These developed nanocluster had more accumulation in the tumor shown by MRI and photothermal ablation of liver tumors was more efficient at a relatively low dose. Another photothermal therapy NPs are composed of gold nanorods @ Gd_2O_3 with X-ray contrast and MRI contrast [69]. These NPs were infused into the rat liver by accessing the portal vein for site-specific delivery. MRI confirmed that NPs delivery via the hepatic portal vein can double the colorectal liver metastasis tumor-to-liver contrast. Photothermal ablation was largely damaged tumor region with minimal damage to the adjacent liver tissue [69]. These results suggest that newly developed theranostic NPs can be effectively employed for the site-specific photothermal therapy of colorectal liver metastasis by leveraging interventional radiology methods.

Superparamagnetic iron oxide nanoparticles (SPIONs) are usually used for liver diagnostic imaging by MRI as mentioned above [74]. Mouli et al. designed an innovative image-guided local delivery strategy using doxorubicin-loaded SPIONs, significantly increasing intra-tumoral drug uptake with limited off-target [59]. They employed nanoblation to be image-guided placement of electrodes into the tumor to electroporate tumor cells. This strategy could lead to the formation of temporary channels in the membrane of tumor cells, resulting rapidly entering tumor cells by NPs. In this study, nanoembolization is the image-guided carrier to deliver embolic agents and NPs onto the blood supply of tumors, which could minimize the off-target delivery and optimize the tumor cell uptake of drugs [59] (Figure 12.3). In addition, polymer NPs have also been widely applied to MRI-visible smart polymeric drug delivery system. As shown in Figure 12.4, Huang et al. designed a new pH-sensitive and MRI-visible polymer NPs which were further loaded adriamycin. Anti-cancer drugs are stable in the polymeric micelles at neutral pH, resembling the physiological environment, where they can release rapidly in acidic endosomal of tumor cells for cancer targeting therapy [57]. SPIONs were encapsulated inside the core of polymeric micelles for the purpose of MRI-guided therapy. These NPs were further modified to couple with anti-VEGF for specific identifying tumors more efficiently. The results demonstrated that these NPs accumulated in liver tumors and had a high spin-spin relaxivity, suggesting they are potential applied as theranostic NPs for liver cancer [57].Maeng et al. also designed a polymer composed of poly(ethylene oxide)-trimellitic anhydride chloride-folate carrying both doxorubicin and SPIONs using folate as the targeting moiety [6]. The efficacy of NPs was evaluated in liver cancer models of rats and rabbits, which showed increased bioavailability and efficacy of doxorubicin compared to free drug. Furthermore, they showed higher MRI sensitivity than a conventional MRI contrast agent (Resovist®) with lower iron content. All these results indicate these newly developed theranostic NPs are a promising candidate for liver cancer treatment and monitoring the treatment response by MRI [6].

Apart from liver cancer, theranostic liposomes have been developed for potential application in clinical therapy of hepatic fibrosis [70, 71]. As discussed above, one of major reasons for no standard clinical applied anti-fibrotic therapies is unable to specifically target the activated HSCs which is key cells responsible for fibrosis. Li et al. constructed targeted liposomes modified by a cyclic peptide with affinity for the PDGF receptor-β to deliver interferon-γ (IFN-γ) to HSCs [70, 71]. *In vitro* study showed that these NPs were more effective in suppressing activated HSC proliferation and inducing apoptosis of activated HSCs. *In vivo* study demonstrated the levels of collagen type IV-C in fibrotic rats were less in NP-treated group compared to those treated with IFN-γ and control group. All these data suggest that liposomes might be a more effective and anti-fibrotic agent and a new opportunity for clinical therapy for fibrosis. Recently, a novel nanohybrid based on porous silicon, gold NPs, and acetalated dextran (DPSi/DAu@AcDEX) is reported to encapsulate

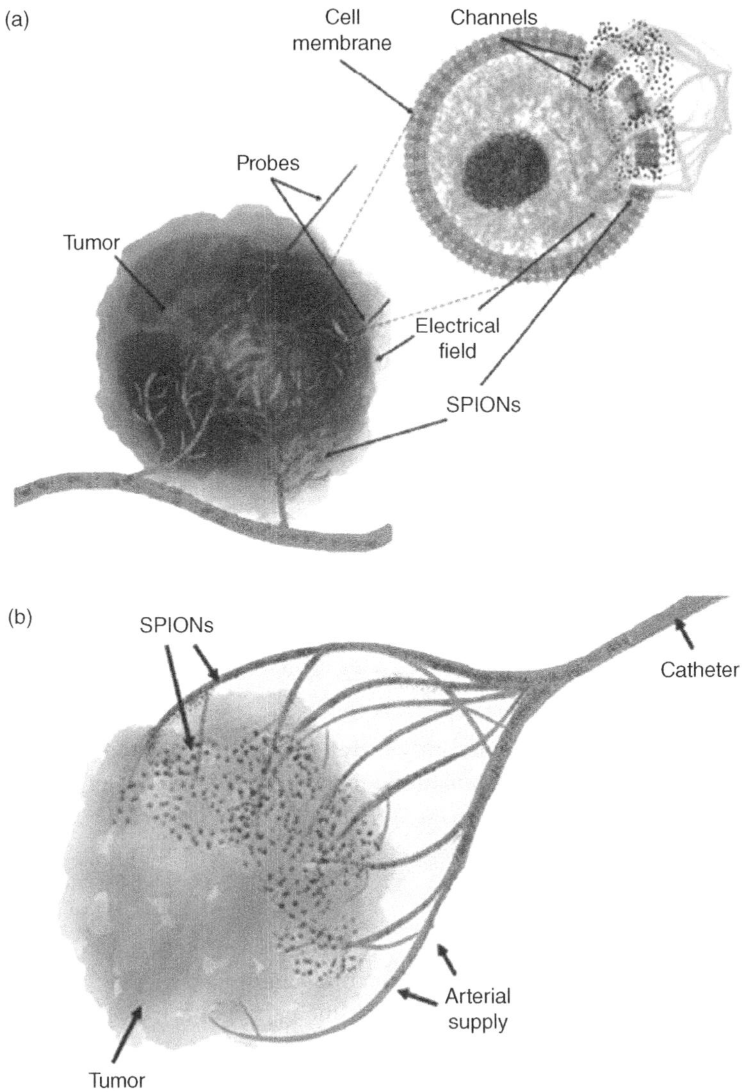

Figure 12.3 Schematic structure of nanoablation and nanoembolization. (a) In nanoablation, ablation probes are inserted into tumor tissue, which generates an electrical field around the tumor. (b) In nanoembolization. NPs and embolic agents are delivered to arterial supply of tumors by a microcatheter. Source: Reproduced with permission of Ref. Mouli et al. [59], © 2013, American Chemical Society.

and deliver drugs and increase CT signals for acute liver failure theranostics [72]. This system can identify pathological change in tissues and selectively deliver drugs to lesion area, increasing the local drug concentration. Moreover, the rest of gold NPs within the matrix further enhances CT imaging quality. Overall, this nanohybrid is a potential theranostic platform for acute liver failure [72]. Apoferritin has been explored to deliver simultaneously therapeutic and imaging agents to hepatocytes as this protein is efficiently taken up by hepatocyte scavenger receptor class A type 5 via the ferritin transporting route [73]. One study has performed to load MRI contrast agent GdHPDO3A and curcumin to ferritin for treatment of acute hepatitis [73]. The encapsulation

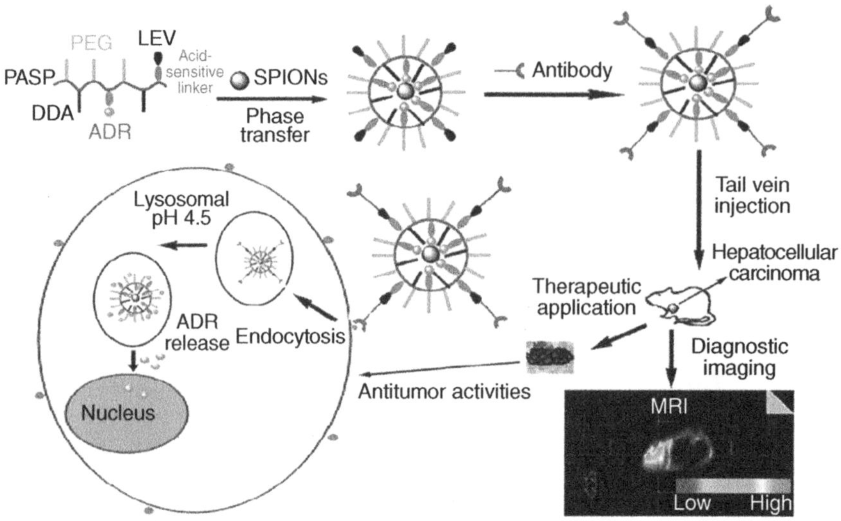

Figure 12.4 Schematic illustration of the formation of novel theranostic NPs and their dual-functional process. Adriamycin is stable in polymer at neutral pH of the physiological environment and release rapidly in acidic endosomal/lysosomal compartments of tumor cells after the breaking of acid-sensitive linker. The hydrophobic SPIONs are encapsulated inside the core of the polymer for MRI. PASP, polyaspartate; PEG, polyethylene glycols; LEV, levulinic acid; DDA, dodecylamine; ADR, adriamycin. Source: Reproduced with permission of ref. Huang et al. [57], © 2013, John Wiley & Sons.

of curcumin inside the apoferritin cavity can significantly increase its biostability and bioavailability while maintaining its therapeutic anti-inflammatory properties, which can be monitored by MRI.

12.7 Conclusions

Theranostics is a novel concept that refers to the combination of diagnosis and therapy, and the use of NPs for this purpose is emerging as a promising therapeutic paradigm in cancer. NPs, as a promising therapeutic agent, have been effective in liver disease treatment. They are a novel "Trojan horse" strategy to simultaneously deliver imaging contrast and therapeutic drugs or genes to target the liver that overcome the challenge of accessing and treating the specific cells in the liver and limitations of free drugs and genes such as poor biostability, nonspecific target and short half-life. By the physicochemical properties of NPs and specific ligands for active targeting, NPs can target and accumulate in the specific organ or cells which can minimize the first-pass effect and improve the therapeutic efficiency. As we described above, the targeting efficiency of theranostic NPs can be monitored by NIR, MRI, or CT imaging. Furthermore, they can be employed for various imaging-guided therapies for liver diseases include photothermal, chemotherapy, anti-fibrotic, and anti-inflammatory. Currently, liposomes and polymer NPs are widely used as theranostic NPs for liver diseases. In the near future, more efforts might focus on inorganic NPs such as calcium phosphate and organic-inorganic hybrid NPs, which have unique material- and size-dependent physicochemical properties and not assessed for liver disease theranostics yet.

It is still unclear why many NPs actively targeting the liver that showed positive results in animals do not reach the clinic. Although a few theranostic NPs are currently in the clinic trials such as Feridex®, Endorem®, Resovist, Livatag, and ThermoDox®, most of them have major side effects such as severe hypotension, severe fatigue, and so on [1]. One possible reason is that specific ligands for active targeting are commonly exogenous compounds, which may trigger immunological side effects for human applications. Hence, endogenous products such as apoferritin and vitamin A will be promising target ligands for clinic use. Another possible reason is that targeting efficiency of NPs in patients may not be as good as in animal models. Therefore, targeting efficiency has to be examined in more than one animal model and a proper animal model has to be chosen for this study.

There is still a long way for theranostic NPs toward clinic use, but it is our opinion that the forthcoming research efforts would focus on the development and synthesis of biosafety NPs with efficient active targeting to the liver. Currently, most studies focus on liver cancer, but it is worth applying theranostic NPs to other liver diseases. Also, these multifunctional NPs have stimulated the trend toward "personalized treatment," where the treatment response of an individual patient could be monitored and assessed by these NPs. Considering the inter-individual variability in therapeutic response, only the patients who demonstrate a high targeting efficiency and less side effects of theranostic NPs would continue to be treated. Finally, the novel techniques such as nanoablation and nanoembolization will provide opportunities for further effective liver therapy in the near future. By integrating imaging functionalities into therapy, these NPs could be revolutionary theranostic systems for therapy selection, therapeutic regimens, and monitoring of treatment response for liver diseases.

References

1 Wang, H., Thorling, C.A., Liang, X. et al. (2015). Diagnostic imaging and therapeutic application of nanoparticles targeting the liver. *J. Mater. Chem. B* 3: 939–958. https://doi.org/10.1039/C4TB01611D.

2 Stanger, B.Z. (2015). Cellular homeostasis and repair in the mammalian liver. *Annu. Rev. Physiol.* 77: 179–200. https://doi.org/10.1146/annurev-physiol-021113-170255.

3 Blachier, M., Leleu, H., Peck-Radosavljevic, M. et al. (2013). The burden of liver disease in Europe: a review of available epidemiological data. *J. Hepatol.* 58: 593–608. https://doi.org/10.1016/j.jhep.2012.12.005.

4 Sia, D., Villanueva, A., Friedman, S.L., and Llovet, J.M. (2017). Liver cancer cell of origin, molecular class, and effects on patient prognosis. *Gastroenterology* 152: 745–761. https://doi.org/10.1053/j.gastro.2016.11.048.

5 Poelstra, K., Prakash, J., and Beljaars, L. (2012). Drug targeting to the diseased liver. *J. Controlled Release* 161: 188–197. https://doi.org/10.1016/j.jconrel.2012.02.011.

6 Maeng, J.H., Lee, D.H., Jung, K.H. et al. (2010). Multifunctional doxorubicin loaded superparamagnetic iron oxide nanoparticles for chemotherapy and magnetic resonance imaging in liver cancer. *Biomaterials* 31: 4995–5006. https://doi.org/10.1016/j.biomaterials.2010.02.068.

7 Chen, Z.P., Xiao, L., Liu, W. et al. (2012). Novel materials which possess the ability to target liver cells. *Exp. Opin. Drug Deliv.* 9: 649–656. https://doi.org/10.1517/17425247.2012.679261.

8 Li, L.J., Wang, H.Y., Ong, Z.Y. et al. (2010). Polymer- and lipid-based nanoparticle therapeutics for the treatment of liver diseases. *Nano Today* 5: 296–312. https://doi.org/10.1016/j.nantod.2010.06.007.

9 Wang, H., Li, X., Tse, B.W. et al. (2018). Indocyanine green-incorporating nanoparticles for cancer theranostics. *Theranostics* 8: 1227–1242. https://doi.org/10.7150/thno.22872.

10 Luk, B.T., Fang, R.H., and Zhang, L. (2012). Lipid- and polymer-based nanostructures for cancer theranostics. *Theranostics* 2: 1117–1126. https://doi.org/10.7150/thno.4381.

11 Kasuya, T. and Kuroda, S. (2009). Nanoparticles for human liver-specific drug and gene delivery systems: in vitro and in vivo advances. *Exp. Opin. Drug Deliv.* 6: 39–52. https://doi.org/10.1517/17425240802622096.

12 Pathak, A., Vyas, S.P., and Gupta, K.C. (2008). Nano-vectors for efficient liver specific gene transfer. *Int. J. Nanomed.* 3: 31–49.

13 Wang, Y., Zhao, Q., Han, N. et al. (2015). Mesoporous silica nanoparticles in drug delivery and biomedical applications. *Nanomed. Nanotechnol. Biol. Med.* 11: 313–327. https://doi.org/10.1016/j.nano.2014.09.014.

14 Poonia, N., Lather, V., and Pandita, D. (2018). Mesoporous silica nanoparticles: a smart nanosystem for management of breast cancer. *Drug Discovery Today* 23: 315–332. https://doi.org/10.1016/j.drudis.2017.10.022.

15 Jung, J., Matsuzaki, T., Tatematsu, K. et al. (2008). Bio-nanocapsule conjugated with liposomes for in vivo pinpoint delivery of various materials. *J. Controlled Release* 126: 255–264. https://doi.org/10.1016/j.jconret.2007.12.002.

16 Jung, J., Somiya, M., Jeong, S.Y. et al. (2018). Low immunogenic bio-nanocapsule based on hepatitis B virus escape mutants. *Nanomed. Nanotechnol. Biol. Med.* 14: 595–600. https://doi.org/10.1016/j.nano.2017.11.017.

17 Tsutsui, Y., Tomizawa, K., Nagita, M. et al. (2007). Development of bionanocapsules targeting brain tumors. *J. Controlled Release* 122: 159–164. https://doi.org/10.1016/j.jconrel.2007.06.019.

18 Vangijzegem, T., Stanicki, D., and Laurent, S. (2019). Magnetic iron oxide nanoparticles for drug delivery: applications and characteristics. *Exp. Opin. Drug Deliv.* 16: 69–78. https://doi.org/10.1080/17425247.2019.1554647.

19 Reddy, L.H. and Couvreur, P. (2011). Nanotechnology for therapy and imaging of liver diseases. *J. Hepatol.* 55: 1461–1466. https://doi.org/10.1016/j.jhep.2011.05.039.

20 Gupta, A.K. and Gupta, M. (2005). Synthesis and surface engineering of iron oxide nanoparticles for biomedical applications. *Biomaterials* 26: 3995–4021. https://doi.org/10.1016/j.biomaterials.2004.10.012.

21 Liang, X.W., Grice, J.E., Zhu, Y. et al. (2015). Intravital multiphoton imaging of the selective uptake of water-dispersible quantum dots into sinusoidal liver cells. *Small* 11: 1711–1720. https://doi.org/10.1002/smll.201402698.

22 Walkey, C.D. and Chan, W.C.W. (2012). Understanding and controlling the interaction of nanomaterials with proteins in a physiological environment. *Chem. Soc. Rev.* 41: 2780–2799. https://doi.org/10.1039/c1cs15233e.

23 Monkemoller, V., Schuttpelz, M., McCourt, P. et al. (2014). Imaging fenestrations in liver sinusoidal endothelial cells by optical localization microscopy. *Phys. Chem. Chem. Phys.* 16: 12576–12581. https://doi.org/10.1039/c4cp01574f.

24 Gaumet, M., Vargas, A., Gurny, R., and Delie, F. (2008). Nanoparticles for drug delivery: the need for precision in reporting particle size parameters. *Eur. J. Pharm. Biopharm.* 69: 1–9. https://doi.org/10.1016/j.ejpb.2007.08.001.

25 Cheng, S.H., Li, F.C., Souris, J.S. et al. (2012). Visualizing dynamics of sub-hepatic distribution of nanoparticles using intravital multiphoton fluorescence microscopy. *ACS Nano* 6: 4122–4131. https://doi.org/10.1021/nn300558p.

26 Negishi, M., Irie, A., Nagata, N., and Ichikawa, A. (1991). Specific binding of glycyrrhetinic acid to the rat-liver membrane. *Biochim. Biophys. Acta* 1066: 77–82. https://doi.org/10.1016/0005-2736(91)90253-5.

27 He, Z.Y., Zheng, X., Wu, X.H. et al. (2010). Development of glycyrrhetinic acid-modified stealth cationic liposomes for gene delivery. *Int. J. Pharm.* 397: 147–154. https://doi.org/10.1016/j.ijpharm.2010.06.029.

28 Kamps, J.A.A.M., Morselt, H.W.M., Swart, P.J. et al. (1997). Massive targeting of liposomes, surface-modified with anionized albumins, to hepatic endothelial cells. *Proc. Natl. Acad. Sci. U.S.A.* 94: 11681–11685. https://doi.org/10.1073/pnas.94.21.11681.

29 Opanasopit, P., Sakai, M., Nishikawa, M. et al. (2002). Inhibition of liver metastasis by targeting of immunomodulators using mannosylated liposome carriers. *J. Controlled Release* 80: 283–294. https://doi.org/10.1016/S0168-3659(02)00006-8.

30 Guan, M., Zhou, Y., Zhu, Q.L. et al. (2012). *N*-Trimethyl chitosan nanoparticle-encapsulated lactosyl-norcantharidin for liver cancer therapy with high targeting efficacy. *Nanomed. Nanotechnol. Biol. Med.* 8: 1172–1181. https://doi.org/10.1016/j.nano.2012.01.009.

31 Klenk, C., Gawande, R., Uslu, L. et al. (2014). Ionising radiation-free whole-body MRI versus ^{18}F-fluorodeoxyglucose PET/CT scans for children and young adults with cancer: a prospective, non-randomised, single-centre study. *Lancet Oncol.* 15: 275–285. https://doi.org/10.1016/S1470-2045(14)70021-X.

32 Zhou, Z.J., Wang, L.R., Chi, X.Q. et al. (2013). Engineered iron-oxide-based nanoparticles as enhanced T-1 contrast agents for efficient tumor imaging. *ACS Nano* 7: 3287–3296. https://doi.org/10.1021/nn305991e.

33 Chen, Z.J., Yu, D.X., Wang, S.J. et al. (2009). Biocompatible nanocomplexes for molecular targeted MRI contrast agent. *Nanoscale Res. Lett.* 4: 618–626. https://doi.org/10.1007/s11671-009-9286-x.

34 Tian, X.M., Shao, Y.Z., He, H.Q. et al. (2013). Nanoamplifiers synthesized from gadolinium and gold nanocomposites for magnetic resonance imaging. *Nanoscale* 5: 3322–3329. https://doi.org/10.1039/c3nr00170a.

35 Tanimoto, A. and Kuribayashi, S. (2006). Application of superparamagnetic iron oxide to imaging of hepatocellular carcinoma. *Eur. J. Radiol.* 58: 200–216. https://doi.org/10.1016/j.ejrad.2005.11.040.

36 Selim, K.M.K., Ha, Y.S., Kim, S.J. et al. (2007). Surface modification of magnetite nanoparticles using lactobionic acid and their interaction with hepatocytes. *Biomaterials* 28: 710–716. https://doi.org/10.1016/j.biomaterials.2006.09.014.

37 Im, G.H., Kim, S.M., Lee, D.G. et al. (2013). Fe_3O_4/MnO hybrid nanocrystals as a dual contrast agent for both T-1- and T-2-weighted liver MRI. *Biomaterials* 34: 2069–2076. https://doi.org/10.1016/j.biomaterials.2012.11.054.

38 Boll, H., Nittka, S., Doyon, F. et al. (2011). Micro-CT based experimental liver imaging using a nanoparticulate contrast agent: a longitudinal study in mice. *PLoS One* 6, ARTN e25692, https://doi.org/10.1371/journal.pone.0025692.

39 Boll, H., Figueiredo, G., Fiebig, T. et al. (2013). Comparison of Fenestra LC, ExiTron nano 6000, and ExiTron nano 12000 for micro-CT imaging of liver and spleen in mice. *Acad. Radiol.* 20: 1137–1143. https://doi.org/10.1016/j.acra.2013.06.002.

40 Yi, X.M., Wang, F.L., Qin, W.J. et al. (2014). Near-infrared fluorescent probes in cancer imaging and therapy: an emerging field. *Int. J. Nanomed.* 9: 1347–1365. https://doi.org/10.2147/Ijn.S60206.

41 Zeng, C.T. et al. (2016). Intraoperative identification of liver cancer microfoci using a targeted near-infrared fluorescent probe for imaging-guided surgery. *Sci. Rep.* 6, ARTN 21959, https://doi.org/10.1038/srep21959.

42 Villanueva, A. (2019). Hepatocellular carcinoma. *New Engl. J. Med.* 380: 1450–1462. https://doi.org/10.1056/NEJMra1713263.
43 Thomas, M.B. and Zhu, A.X. (2005). Hepatocellular carcinoma: the need for progress. *J. Clin. Oncol.* 23: 2892–2899. https://doi.org/10.1200/Jco.2005.03.196.
44 Wang, C. and Li, J. (2012). An update on chemotherapy of colorectal liver metastases. *World J. Gastroenterol.* 18: 25–33. https://doi.org/10.3748/wjg.v18.i1.25.
45 Lai, L.F. and Guo, H.X. (2011). Preparation of new 5-fluorouracil-loaded zein nanoparticles for liver targeting. *Int. J. Pharm.* 404: 317–323. https://doi.org/10.1016/j.ijpharm.2010.11.025.
46 Li, X.L., Lie, R.T., Qian, X.P. et al. (2008). Superior antitumor efficiency of cisplatin-loaded nanoparticles by intratumoral delivery with decreased tumor metabolism rate. *Eur. J. Pharm. Biopharm.* 70: 726–734. https://doi.org/10.1016/j.ejpb.2008.06.016.
47 Merle, P., Ahmed, S., Habersetzer, F. et al. (2006). Phase 1 study of intra-arterial hepatic (IAH) delivery of Doxorubicin-Transdrug (R) (DT) for patients with advanced hepatocellular carcinoma (HCC). *Liver Int.* 26: 61–62.
48 Zhou, Q.H., Sun, X., Zeng, L.Y. et al. (2009). A randomized multicenter phase II clinical trial of mitoxantrone-loaded nanoparticles in the treatment of 108 patients with unresected hepatocellular carcinoma. *Nanomed. Nanotechnol.* 5: 419–423. https://doi.org/10.1016/j.nano.2009.01.009.
49 Farra, R., Musiani, F., Perrone, F. et al. (2018). Polymer-mediated delivery of siRNAs to hepatocellular carcinoma: variables affecting specificity and effectiveness. *Molecules* 23, ARTN 777, https://doi.org/10.3390/molecules23040777.
50 Huang, Z., Dong, L., Chen, J.J. et al. (2012). Low-molecular weight chitosan/vascular endothelial growth factor short hairpin RNA for the treatment of hepatocellular carcinoma. *Life Sci.* 91: 1207–1215. https://doi.org/10.1016/j.lfs.2012.09.015.
51 Poon, R.T.P. and Borys, N. (2009). Lyso-thermosensitive liposomal doxorubicin: a novel approach to enhance efficacy of thermal ablation of liver cancer. *Expert Opin. Pharmacother.* 10: 333–343. https://doi.org/10.1517/14656560802677874.
52 Tak, W.Y., Lin, S.M., Wang, Y.J. et al. (2018). Phase III HEAT study adding lyso-thermosensitive liposomal doxorubicin to radiofrequency ablation in patients with unresectable hepatocellular carcinoma lesions. *Clin. Cancer Res.* 24: 73–83. https://doi.org/10.1158/1078-0432.Ccr-16-2433.
53 Lyon, P.C., Gray, M.D., Mannaris, C. et al. (2018). Safety and feasibility of ultrasound-triggered targeted drug delivery of doxorubicin from thermosensitive liposomes in liver tumours (TARDOX): a single-centre, open-label, phase 1 trial. *Lancet Oncol.* 19: 1027–1039. https://doi.org/10.1016/S1470-2045(18)30332-2.
54 Jain, A., Jain, A., Parajuli, P. et al. (2018). Recent advances in galactose-engineered nanocarriers for the site-specific delivery of siRNA and anticancer drugs. *Drug Discovery Today* 23: 960–973. https://doi.org/10.1016/j.drudis.2017.11.003.
55 Cheng, M.R., He, B., Wan, T. et al. (2012). 5-Fluorouracil nanoparticles inhibit hepatocellular carcinoma via activation of the p53 pathway in the orthotopic transplant mouse model. *PLoS One* 7, ARTN e47115, https://doi.org/10.1371/journal.pone.0047115.
56 Liang, H.F., Chen, C.T., Chen, S.C. et al. (2006). Paclitaxel-loaded poly(gamma-glutamic acid)-poly(lactide) nanoparticles as a targeted drug delivery system for the treatment of liver cancer. *Biomaterials* 27: 2051–2059. https://doi.org/10.1016/j.biomaterials.2005.10.027.

57 Huang, H., Li, Y., Li, C. et al. (2013). A novel anti-VEGF targeting and MRI-visible smart drug delivery system for specific diagnosis and therapy of liver cancer. *Macromol. Biosci.* 13: 1358–1368. https://doi.org/10.1002/mabi.201300137.

58 Kraft, J.C., Freeling, J.P., Wang, Z., and Ho, R.J. (2014). Emerging research and clinical development trends of liposome and lipid nanoparticle drug delivery systems. *J. Pharm. Sci.* 103: 29–52. https://doi.org/10.1002/jps.23773.

59 Mouli, S.K., Tyler, P., McDevitt, J.L. et al. (2013). Image-guided local delivery strategies enhance therapeutic nanoparticle uptake in solid tumors. *ACS Nano* 7: 7724–7733. https://doi.org/10.1021/nn4023119.

60 Li, L.L., Tang, F.Q., Liu, H.Y. et al. (2010). In vivo delivery of silica nanorattle encapsulated docetaxel for liver cancer therapy with low toxicity and high efficacy. *ACS Nano* 4: 6874–6882. https://doi.org/10.1021/nn100918a.

61 Kim, S.I., Shin, D., Choi, T.H. et al. (2007). Systemic and specific delivery of small interfering RNAs to the liver mediated by apolipoprotein A-I. *Mol. Ther.* 15: 1145–1152. https://doi.org/10.1038/sj.mt.6300168.

62 Mevel, M., Kamaly, N., Carmona, S. et al. (2010). DODAG; a versatile new cationic lipid that mediates efficient delivery of pDNA and siRNA. *J. Controlled Release* 143: 222–232. https://doi.org/10.1016/j.jconrel.2009.12.001.

63 Bataller, R. and Brenner, D.A. (2005). Liver fibrosis. *J. Clin. Invest.* 115: 209–218. https://doi.org/10.1172/JCI24282.

64 Sato, Y., Murase, K., Kato, J. et al. (2008). Resolution of liver cirrhosis using vitamin A-coupled liposomes to deliver siRNA against a collagen-specific chaperone. *Nat. Biotechnol.* 26: 431–442. https://doi.org/10.1038/nbt1396.

65 Park, K., Hong, S.W., Hur, W. et al. (2011). Target specific systemic delivery of TGF-beta siRNA/(PEI-SS)-g-HA complex for the treatment of liver cirrhosis. *Biomaterials* 32: 4951–4958. https://doi.org/10.1016/j.biomaterials.2011.03.044.

66 Lin, T.T., Gao, D.Y., Liu, Y.C. et al. (2016). Development and characterization of sorafenib-loaded PLGA nanoparticles for the systemic treatment of liver fibrosis. *J. Controlled Release* 221: 62–70. https://doi.org/10.1016/j.jconrel.2015.11.003.

67 Duong, H.T.T. and Dong, Z.X. (2015). The use of nanoparticles to deliver nitric oxide to hepatic stellate cells for treating liver fibrosis and portal hypertension. *Small* 11: 2291–2304. https://doi.org/10.1002/smll.201402870.

68 Wu, M., Wang, Q.T., Zhang, D. et al. (2016). Magnetite nanocluster@poly(dopamine)-PEG@ indocyanine green nanobead with magnetic field-targeting enhanced MR imaging and photothermal therapy in vivo. *Colloids Surf., B* 141: 467–475. https://doi.org/10.1016/j.colsurfb.2016.02.022.

69 Parchur, A.K., Sharma, G., Jagtap, J.M. et al. (2018). Vascular interventional radiology-guided photothermal therapy of colorectal cancer liver metastasis with theranostic gold nanorods. *ACS Nano* 12: 6597–6611. https://doi.org/10.1021/acsnano.8b01424.

70 Li, F., Li, Q.H., Wang, J.Y. et al. (2012). Effects of interferon-gamma liposomes targeted to platelet-derived growth factor receptor-beta on hepatic fibrosis in rats. *J. Controlled Release* 159: 261–270. https://doi.org/10.1016/j.jconrel.2011.12.023.

71 Li, Q.H., Yan, Z.Q., Li, F. et al. (2012). The improving effects on hepatic fibrosis of interferon-gamma liposomes targeted to hepatic stellate cells. *Nanotechnology* 23, ARTN 265101, https://doi.org/10.1088/0957-4484/23/26/265101.

72 Liu, Z.H., Li, Y.Z., Li, W. et al. (2018). Multifunctional nanohybrid based on porous silicon nanoparticles, gold nanoparticles, and acetalated dextran for liver regeneration and acute liver failure theranostics. *Adv. Mater.* 30, doi:, ARTN 1703393, https://doi.org/10.1002/adma.201703393.

73 Cutrin, J.C., Crich, S.G., Burghelea, D. et al. (2013). Curcumin/Gd loaded apoferritin: a novel "theranostic" agent to prevent hepatocellular damage in toxic induced acute hepatitis. *Mol. Pharm.* 10: 2079–2085. https://doi.org/10.1021/mp3006177.

74 Gao, L., Xie, L.S., Long, X.J. et al. (2013). Efficacy of MRI visible iron oxide nanoparticles in delivering minicircle DNA into liver via intrabiliary infusion. *Biomaterials* 34: 3688–3696. https://doi.org/10.1016/j.biomaterials.2013.01.094.

13

Toxicology and Safety of Nanoparticles in Drug Delivery System

Klintean Wunnapuk

Toxicology Division, Department of Forensic Medicine, Faculty of Medicine, Chiang Mai University, Chiang Mai, Thailand

13.1 Introduction

Nanoparticles have been rapidly developed in recent years for pharmaceutical and medicinal purposes as drug delivery systems [1–7]. Nanoparticles applied for drug delivery systems range in size from 3 to 200 nm in diameter, and are classified into several different compound categories including biological origin such as phospholipids, lipid-based or biodegradable polymeric, and metal and inorganic carriers [8–10]. The advantages of nanoparticle drug delivery system include improvement of drug bioavailability and delivery efficiency mainly due to their unique characteristics [8, 10] such as their large functional surface area per unit volume which can be modified to alter biodistribution or to achieve target-specific drug delivery [8, 10]. In addition, they have the ability to penetrate smaller capillaries, cell membranes, and other biological barriers due to their small size that facilitates drug distribution and drug accumulation at the target site [10]. Moreover, the use of degradable nanoparticles allows sustained drug release within the target site leading to sufficient therapeutic dosage [10, 11]. Significant numbers of research studies have demonstrated the enormous benefits of nanoparticles in drug delivery systems; however, the use of these carriers may impose adverse side effects in clinical practice leading to unexpected or delayed toxicity in patients. The mechanisms of some nanoparticles underlying nanotoxicity in humans are still not well studied. In this chapter, the following topics are addressed. Firstly, biotransformation, biodistribution, and clearance of nanoparticle drug delivery systems are reviewed, as these factors may be implicated in toxicity. Secondly, the toxicities of different types of biological origin or chemical nanoparticles will be discussed. In addition, recent findings regarding the molecular mechanism of nanoparticle toxicities are illustrated, with recommendations for future studies and applications.

13.2 Lipid-Based Nanocarrier: Liposomes

Liposomes are phospholipid and cholesterol vesicles consisting of one or more concentric lipid bilayers encapsulating an aqueous core with size ranging from less than 100 nm (small unilamellar vesicles, SUVs) to larger than 500 nm (multilamellar vesicles, MLVs) [12–14] (Figure 13.1).

Fundamentals of Drug Delivery, First Edition.
Edited by Heather A. E. Benson, Michael S. Roberts, Adrian C. Williams, and Xiaowen Liang.

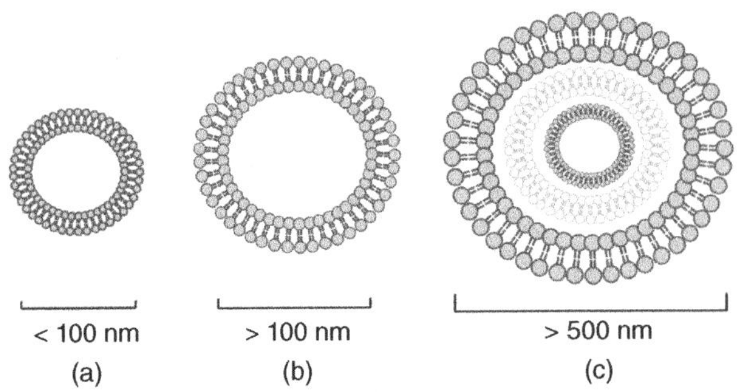

Figure 13.1 (a) Small unilamellar vesicles (SUV), (b) large unilamellar vesicles (LUV), and (c) multilamellar vesicles (MLVs).

The unique property of liposome is the ability to transport both lipophilic and hydrophilic molecules due to the hydrophilic heads and hydrophobic long tails of the phospholipid bilayers similar to lipid bilayers of cell membrane [12, 14]. This natural-like character of liposomes leads to minimal toxicity and pharmacologically inactive. Liposome could be classified based on following factors: [1] number of bilayers, [2] size, and [3] functionalization and formulation [14, 15].

There are four well-known types of liposomal delivery systems: [1] conventional or empty liposomes, [2] stealth or sterically stabilized liposomes, [3] ligand-targeted liposomes, and [4] theranostic and nanoparticle hybrid liposomes (Figure 13.2) [12, 15, 16]. First, liposome formulations compose of charged or neutral phospholipids and cholesterol are defined as conventional liposomes (Figure 13.2a) [12, 15]. Second, stealth liposomes or PEGylated liposomes (Figure 13.2b) compose of phospholipids, cholesterol, and polyethylene glycol (PEG). Third, ligand-targeted liposomes conjugated with antibodies, peptides/proteins, and carbohydrates on the lipid bilayer of the conventional or PEGylated liposomes (Figure 13.2c). These conjugated molecules are specific and selective to drugs and the site of diseases. Lastly, theranostic (Figure 13.2d) or nanoparticle hybrid liposomes are the system consist of specific targeted therapy based on specific targeted diagnostic tests such as liposome-quantum dot hybrid vesicles, which are promising nanoconstructs for cell imaging, loaded with doxorubicin (Figure 13.2e–g) [12, 15, 16].

13.3 Cellular Uptake Mechanism of Liposomes

There are several intercellular internalization pathways of liposomes including (a) clathrin-mediated endocytosis, (b) caveolin-mediated endocytosis, (c) macro pinocytosis, and (d) pH-dependent membrane fusion (Figure 13.3) [17–19]. The surface charges and sizes of liposomes affect the mechanism of cellular internalization. Cationic and anionic liposomes can be taken up through macro pinocytosis; on the other hand, neutral liposomes enter into the cell via caveolin-mediated endocytosis pathway. All liposomes can be taken up by cells via clathrin-mediated endocytosis [17, 18, 20]. Cellular uptake of the larger liposomes (~100–60 nm in diameter) is mainly through clathrin-mediated pathway, while the smaller liposomes (~40 nm in diameter) are mediated by a dynamin-dependent pathway. Medium-size liposomes (~70 nm in diameter) can translocate through clathrin-, caveolin-, dynamin-dependent endocytosis and macro pinocytosis [21, 22].

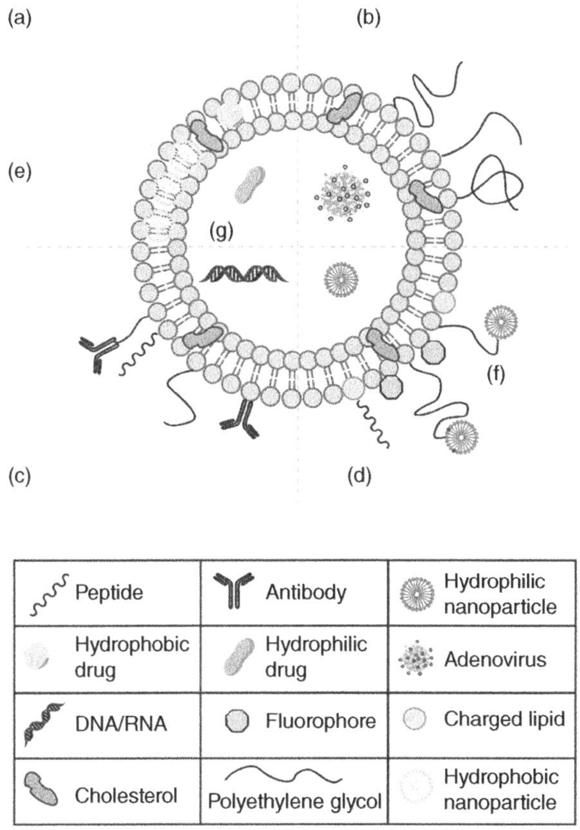

Figure 13.2 Schematic representation of different types of liposomal formulation in drug delivery system. (a) Conventional liposome, (b) PEGylated liposome, (c) ligand-targeted liposome, (d) theranostic liposome, (e) embedded NP, (f) conjugated/absorbed NP, and (g) encapsulated NP. Source: Sercombe et al. [12]; Monteiro et al. [15]; Al-Jamal and Kostarelos [16].

13.4 Biodistribution, Clearance and Toxicity of Liposomes

13.4.1 Effect of Lipid Compositions on Liposome Distribution and Blood Circulation

The pharmacokinetics and pharmacodynamics of liposomes depend on their size, surface charge, formulations, dose, route of administration, and membrane packing density [23–28]. Generally, conventional liposomes have very short half-life and are rapidly eliminated by mononuclear phagocytes of the reticular endothelia system (RES) resulting in limited and less effects in terms of therapeutic efficacy as well as accumulation and toxicity [12, 14, 28, 29]. Allen and Hansen [30] demonstrated that the circulation half-lives of liposomes size ranged from 92 to 123 nm in diameter were prolonged as dose increased. For example, tissue distribution of liposomes in mice were studied using a 100-fold concentration range of liposomes (0.1–10 μmol phospholipid in mouse) at several time points over a 48-hours time period. The result indicated that the $t_{1/2\alpha}$ of the studied liposomes were 0.7–3.2 hours and $t_{1/2\beta}$ were 1.8–13 hours depending on liposome

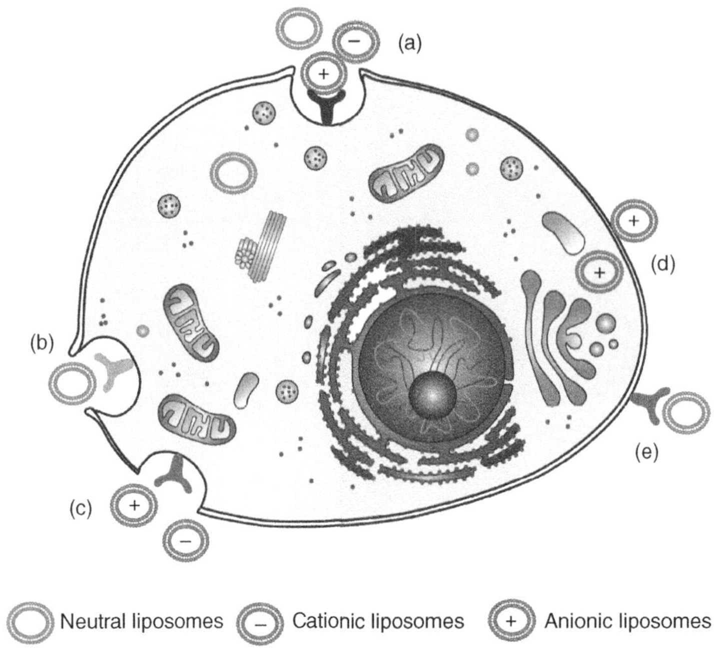

Figure 13.3 Cellular uptake mechanism of liposomes: (a) clathrin-mediated endocytosis (~ 100–160 nm), (b) caveolin-mediated endocytosis (~70 nm), (c) macro pinocytosis (~70 nm), (d) pH-dependent membrane fusion, and (e) dynamin-mediated pathway. Source: Kamaly and Miller [17]; Kang et al. [18]; Raz et al. [19].

concentrations. Moreover, the elimination rate constant and total body clearance of the empty liposomes were smaller when the dose of the liposomes was higher, indicating dose-dependent clearance [30]. Non-formulated liposomes are more likely to accumulate to liver, spleen, kidney, lungs, bone marrow, and lymph nodes, which are the main site of non-formulated liposome accumulation after systemic administration due to the fenestration of their blood vessel [12]. The pore size of these capillaries (100–800 nm) in the RES system is large enough for empty liposomes or drug-loaded liposomes (50–1000 nm in size) filtration. Moreover, serum components and opsonins, including complement protein, immunoglobulin, albumin, and fibrinogen, play an important role in liposome distribution, accumulation, and elimination [31, 32]. During the blood circulation non-surface modified liposomes bind to serum opsonins subsequently phagocytosis by resident macrophage [32]. Lipid composition of liposomes such as phospholipids, cholesterol, and sphingomyelin also involve in liposome circulation [33]. Encapsulated liposomes, stealth liposomes, theranostic liposomes, and other modified liposomes can increase circulation time and prolong accumulation of the drug at the site of action, leading to higher therapeutic benefits and lower toxicity of encapsulated drugs. Modification liposomes with gangliosides and sialic acid derivatives, such as monosialoganglioside (GM1) or PEG can extend their circulation time in blood [25, 28, 30]. However, PEG has been widely used to incorporate with the empty liposomes to obtain more therapy effectiveness [34]. PEG is biocompatible and eliminable [35], which can be dissolved in both lipid and aqueous without toxicity [34]. Modification of liposomes by PEGylated prolong blood circulation up to 45 hours after administration as surface-PEG-coated molecules generate steric hindrance effect that inhibits the electrostatic and hydrophobic reactions with serum proteins/complements to avoid RES uptake of liposomes [12, 34, 36, 37]. Another explanation is that the iC3b, fragment of the third component of complement system, cannot deposit on the modified-surface-liposomes. As a result, there is not enough specific ligand for complement receptor type 3 (CR3) on the Kupffer cell surface [31, 34]. Likewise, PEG-containing liposomes that bound with serum proteins create a nonspecific surface-shielding property that reduces the interactions between liposomes and macrophage cells [38] or bind to some unknown serum components (known as dysopsonins) that inhibit the phagocytosis [31].

However, repeated injections of empty PEGlated-liposomes can induce the accelerated blood clearance (ABC) phenomenon that cause dramatic decrease of liposome levels in circulation but accumulate more in the liver in rat [12, 31, 39]. The mechanism underlying this behavior may relate to the opsonins, immune responses, lipid dose, PEG surface density, and the time-interval between the first and following injections [12, 31]. This phenomenon has not been observed in patients receiving PEGylated-liposomal doxorubicin (Doxil®) probably due to the high-administrated doses of the phospholipid in Doxil (15 μmol phospholipid/kg) used in Clinical trails [40, 41]. In accordance with the report from Ishida et al. [24], ABC reduction was observed to some extent after increasing the initial dose of PEGlated-liposomes greater than 1 μmol phospholipid/kg. Encapsulated liposomal doxorubicin and amphotericin allow better therapeutic effect with less toxicity than the standard care since their pharmacokinectics and biodistribution are improved to enhance drug delivery efficiency to diseased tissues compared to free drugs [12, 42–45]

13.4.2 Effect of Surface Charge on Liposome Distribution and Blood Circulation

Phagocytic cells are able to uptake liposomes with or without opsonization. Serum proteins increase the uptake of anionic and neutral liposomes by macrophages but not with the cationic liposomes. Negative charged phospholipid phosphatidylserine and phosphatidylglycerol can cross the hepatic sinusoidal endothelial lining and these anionic particles are recognized by resident macrophages resulting in high elimination [25, 32]. Positive charged liposomes are commonly encapsulated with DNA for gene therapy [46–49]. Unfortunately, this type of liposomes complex induces reactive oxygen species (ROS), interferes energy and lipid amino acid metabolisms, and disrupts several signally pathway subsequently cell death. In HepG2 cells, cationic liposomes interrupted arachidonic acid metabolism, D-glutamine metabolism, sphingolipids metabolism, pantothenate and coenzyme A (CoA) biosynthesis, glutathione (GSH) metabolism, the citric acid (TCA) cycle, phenylalanine metabolism, glycerophospholipid metabolism, and purine metabolism [50]. Moreover, cationic liposomes loading with stearylamine induced apoptotic cell death and mitochondrial disruption resulting in releasing cytochrome c, caspase-3 and -8, and protein kinase C [32, 51]. Therefore, conventional/modified liposomes with neutral and/or negatively charged lipids plus cholesterol are preferable for clinical application such as DaunoXome® and AmBisome® [32, 34].

13.4.3 Effect of Size on Liposome Distribution and Blood Circulation

Not only the surface charge impacts pharmacokinetics and pharmacodynamics of the liposomes but also the size effects on their distribution and blood circulation. Larger liposomes are cleared from the circulation faster than smaller liposomes [31]. For example, stealth neutral liposomes size range of about 140–220 nm prolonged blood circulation ($t_{1/2} = \sim 24$ hours), whereas the larger size of about 300 nm were rapidly accumulated and eliminated by the RES system ($t_{1/2} = \sim 9$ hours) [52]. The large liposomes (>200 nm) are more likely to be opsonized better than the smaller ones [31]. Small size of liposomal doxorubicin (≤100 nm) has longer blood circulation and lower doxorubicin-induced toxicity [33, 53]. This phenomenon can be explained by the fact that liposomes with the diameter ≤100 nm can pass through hepatic sinusoidal capillaries and deeply penetrate into hepatic parenchymal cells [33, 54–56]. Gabizon reported that hydrogenated phosphatidylinositol doxorubicin encapsulated liposomes with the mean diameters of 80 and 13 nm could reduce toxic effects of doxorubicin. However, larger doxorubicin encapsulated liposomes with the diameter of 210 nm increased the deaths in animals [57]. Similar to the work done by Zou et al. using annamycin loaded liposomes, less animals died when the size of encapsulated liposomes decreased from 1560 to 30 nm [58]. These results indicate that the toxicity of liposomal antitumor drugs depends on the size and the formulations. Commercial products of doxorubicin (PEGylated-liposomal doxorubicin) such as DaunoXome (diameter 60–80 nm) and Doxil (diameter ~100 nm) showed less myelosuppression [33].

However, Palmar Plantar Erythrodysthesia (PPE) or "foot and hand syndrome" is a significant side effect of Doxil, which have not seen in conventional doxorubicin. The symptoms of PPE syndrome include redness, tenderness, and peeling of the skin. Infusion of Doxil can induce flushing and shortness of breath in patients. The severity of PPE increases with Doxil dose, C_{max} and the dose intervals. For example, reduction in the dose from 60 to 45 mg/m^2 demonstrated less impact on the severity of skin toxicity. Moreover, the symptoms normally are pronounced after two courses of Doxil or three-week intervals. These adverse effects are actually the result of a complement activation-related pseudo-allergy or acute hypersensitivity [42, 59].

13.5 Application of Liposomes in Drug Delivery

A number of commercial liposomal loading drugs have been approved by Food and Drug Administration (FDA) and are available on the market [12, 60–62] as shown in Table 13.1.

Doxil/Caelyx® and AmBisome have been successfully applied in clinical use [12, 28, 60]. Both liposomal formulation modify the effectiveness and the toxicity of doxorubicin and amphotericin B by altering the pharmacokinetic profiles and extend the circulation times [28]. For example, Doxil has a slow plasma clearance and a small volume of distribution and can accumulate in Kaposi's sarcoma and head and neck cancers longer than free doxorubicin [61] as a result of reduction of the RES affinity. Apart from the mucosal and cutaneous toxicities, PEGylated liposomal doxorubicin cause mild myelosuppression, low cardiotoxicity, and minimal alopecia [61]. Traditional doxorubicin has greater cardiotoxicity than Doxil as part of doxorubicin-induced cardiotoxicity appears to be related to a high peak plasma concentration of free doxorubicin. Various mechanisms of doxorubicin cardiotoxicity have been proposed including cardiomyopathy due to doxorubicin-induced apoptosis and necrosis [67] The minimal toxicities of Doxil allow patients to increase treatment duration for a longer accumulation of the drug in cancer cells [42]. Moreover, these liposomal drugs with the mean size range of 80–100 nm [45, 61, 62] are small enough to penetrate into malignant cells via high permeable blood vessels in tumors [68]. The longer retention period of drugs in target sites is known as the enhanced permeability and retention (EPR) effect in which the vascular has high permeability that supplies to pathological tissues such as cancer and inflammation cells [12, 28]. The discontinuous endothelial junctions in pathological tissue may be ranged from 300 to 4700 nm in size allowing extravasation of macromolecules and prolonged retention of these molecules within the pathological site [69]. The small size (<100 nm) of AmBisome allows drug distribution to different organs due to an increase circulation time in blood. Furthermore, the lipid compositions of liposomal amphotericin B products are designed to specifically bind to fungal cell walls, which can minimize adverse effects of amphotericin B by allowing drug-fungal direct interaction [62].

Table 13.1 Approved liposomal and lipid-based products.

Product	**Drug**	**Disease**	**Type of liposomal-based delivery system**	**References**
AmBisome®	Amphotericin B	Fungal infections Leishmaniasis	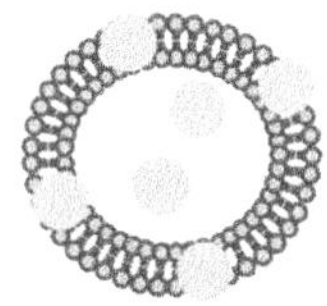 Unilamellar liposome	[60, 62]
Doxil®/Caelyx®	Doxorubicin	Kaposi's sarcoma Ovarian cancer Breast cancer Lung cancer	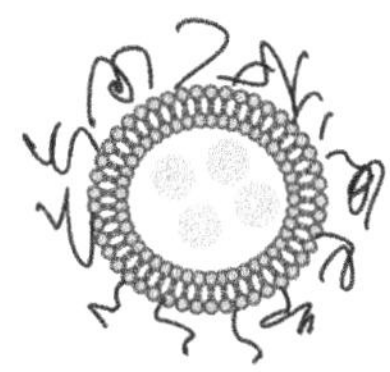 PEGylated liposome	[12, 60]
DaunoXome®	Daunorubicin	Kaposi's sarcoma	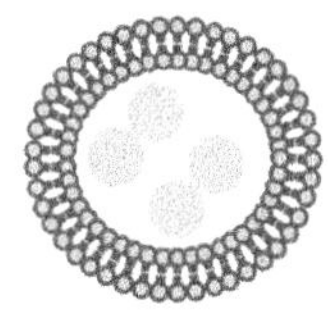 Entrapped liposome	[12, 60]
Amphotec®	Amphotericin B	Invasive aspergillosis	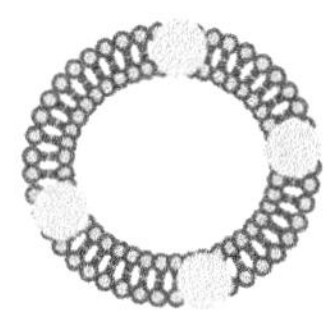 Colloidal dispersion	[63]
Abelcet®	Amphotericin B	Aspergillosis	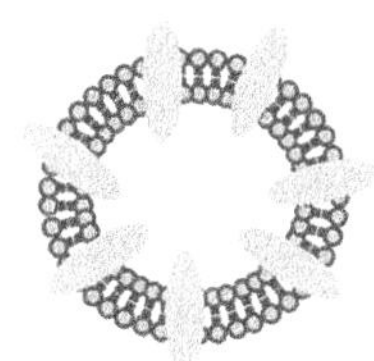 Ribbon-like lipid complex	[64]

(Continued)

Table 13.1 (Continued)

Product	Drug	Disease	Type of liposomal-based delivery system	References
Depodur®	Morphine sulfate	Pain relief	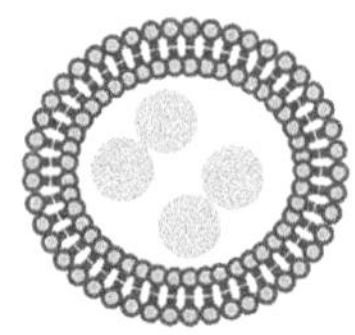 Entrapped liposome	[65]
DepoCyt®	Cytosine	Lymphomatous	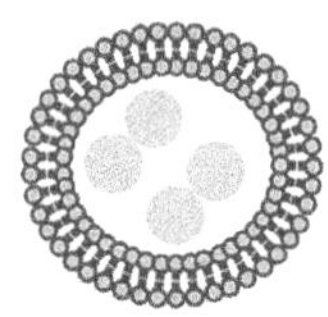 Entrapped liposome	[66]
Myocet®	Doxorubicin cyclophos-phamide	Breast cancer	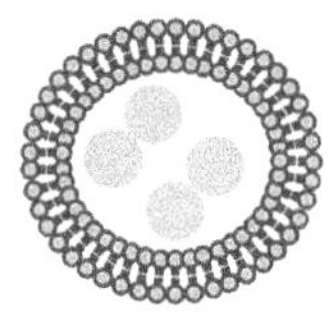 Entrapped liposome	[60]

13.6 Inorganic Nanocarrier: Carbon Nanotubes

Carbon nanotubes (CNTs) are carbon-based tubes composed of benzene rings (Figure 13.4a–c) which have high aspect ratios corresponding to 100–300 nm in length, anisotropic character and adaptable size-dependent properties. CNTs have been applied in medicine as sensors for detecting DNA and protein, as diagnostic devices for the discrimination of various proteins and carriers of vaccines or drugs in drug delivery system [70, 71]. There are three main types of CNTs, including single-walled carbon nanotubes (SWCNTs) with a single cylindrical carbon wall and diameters of 0.4–3.0 nm (Figure 13.4a), double-walled carbon nanotubes (DWCNTs) (Figure 13.4b) with a double cylindrical carbon wall and diameters of 1–2 nm. Lastly, multi-walled carbon nanotubes (MWCNTs) which have cylinders nested within other cylinders with diameters of 2–100 nm (Figure 13.4c) [72, 73]. Generally, CNTs demonstrate low chemical reactivity and are not soluble in water [74]. However, the introduction of a functional group at the ends and side-walls of the graphite surface or bound with carboxylic acids can increase their water solubility and nonspecific protein-resistant properties [8, 70, 74]. These functionalizations enable their adsorption or attachment of various molecules such as peptides, proteins, plasmid DNA, small interfering RNA, and drugs [9, 74].

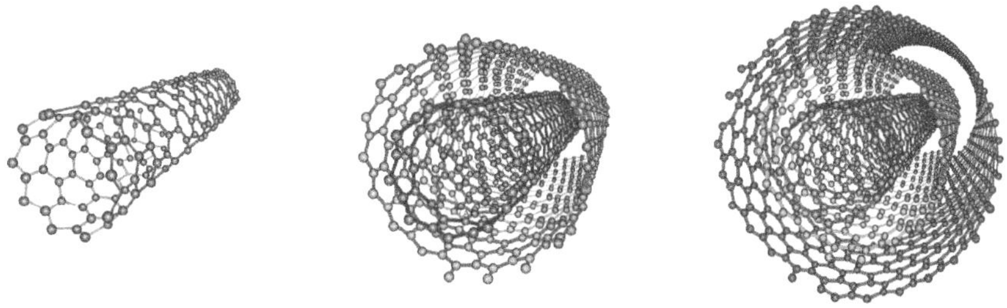

(a) (b) (c)

Figure 13.4 Types of carbon nanotubes: (a) Single-walled carbon nanotubes, (b) double-walled carbon nanotubes, and (c) multi-walled carbon nanotubes.

13.7 Cellular Uptake Mechanism of Carbon Nanotubes

Intracellular uptake, distribution and elimination of CNTs, and their potential impact on metabolism depend on the nature of functionalization, their shape, size, coating, metallic impurities, hydrophilicity, and surface charge [75–79]. Three mechanisms of cellular uptake of CNTs have been proposed: energy-independent nanopenetration, endocytosis, and phagocytosis [71, 73]. First, functionalized nanotubes directly transport across cell membrane as nano-needle by insertion, membrane fusion, direct pore transport, and passive diffusion through lipid bilayers without causing cell damage or death [73, 77, 80, 81]. Internalization of CNTs via the pathway consists of a two-step process in which the hydrophilic functionalized nanotubes are spontaneously adsorbed onto, and merge into, the membrane surface with their long axis parallel to the lipid bilayers. Subsequently, the tubes align and fuse to become a transmembrane pore as shown in Figure 13.5 [82].

The second mechanism involves non-receptor mediated endocytosis, an energy- and temperature-dependent process (Figure 13.6) [73, 77, 83]. This mechanism involves the engulfing of CNTs within a segment of the cell membrane to form a vesicle, subsequently, CNT internalization into the cytoplasm. The final mechanism also involves endocytosis with receptor-mediated process such as clathrin-, caveolae-, stabilin-mediated endocytosis and macropinocytosis [84–86]

Modification and functionalization of both ends and side-walls of SWCNTs and MWCNTs facilitates their penetration into cells [74, 77]. The specialized functional groups on CNTs surface provide recognized signals to cells, resulting in interactions with the cells. For example, PEG-dispersed nanotubes trigger macrophage opsonophagocytosis [72]. Generally, non-modified-SWCNT do not carry recognizable signals thus are poorly taken-up by cells, resulting in aggregation and toxicity.

13.8 Biodistribution, Clearance, and Toxicity of Carbon Nanotubes

Several animal studies demonstrate that functionalized CNTs can be rapidly taken up and distributed to the whole body within 0.5–6 hours after intravenous administration. Soon after administration, these functionalized CNTs undergo renal elimination via glomerular filtration [87–94]. Functionalized CNTs are mainly excreted via the urinary system by direct penetration through glomerular pores, passing into Bowman's space and afterward to the bladder and finally into the urine [85, 89, 92]. It is interesting how cylindrical nanoparticles can be excreted intact through

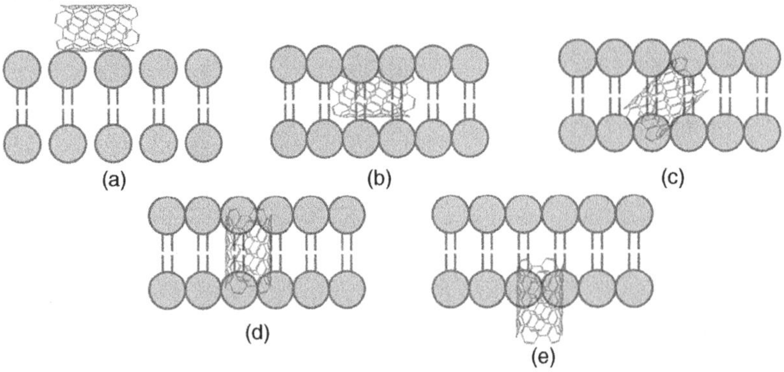

Figure 13.5 An energy independent translocated mechanism: (a) hydrophilic CNT adsorbed and accommodated onto the lipid bilayers of the cell membrane, (b) membrane fusion, (c) reoriented toward the core of the membrane, (d) completely relocated between the membrane layers, and (e) translocated into the cell cytoplasm. Source: Adapted from Lopez et al. [82].

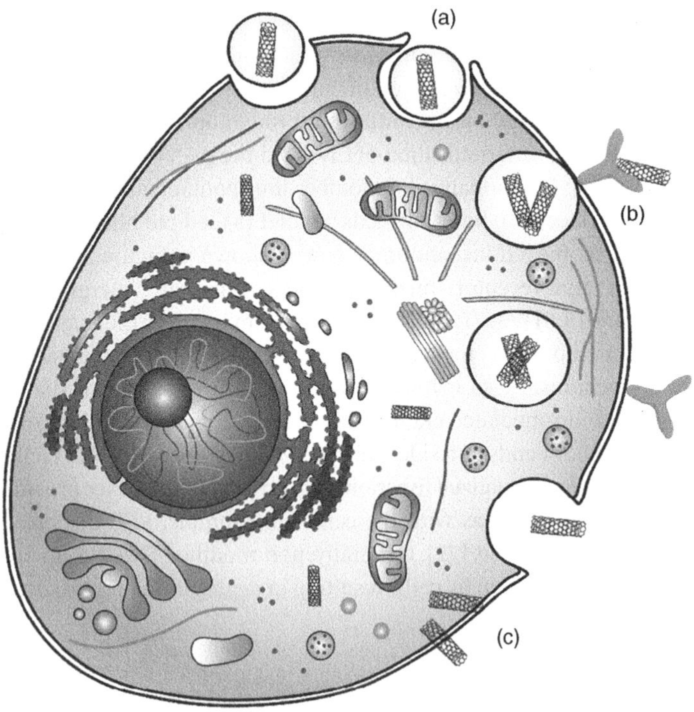

Figure 13.6 Uptake mechanism: (a) non-receptor mediated endocytosis, (b) receptor-mediated endocytosis, and (c) passive diffusion. Source: Rastogi et al. [73]; Klumpp et al. [77]; Kam and Dai [83].

the renal route. The glomerular capillary wall is composed of three layers, including fenestrated endothelium with the minimum diameter of is 30–60 nm, the thickness of the glomerular basement membrane in rodents and humans is 200–400 nm, and the width of the epithelial podocyte filtration slits is 25–60 nm and podocyte slit membrane pore is 5–20 nm [88, 95]. Whereas, the average diameter and length of SWCNTs are approximately 1 and 300–1000 nm, respectively, while MWCNTs are approximately 20–30 nm in diameter and 500–2000 nm in length. The SWCNTs or

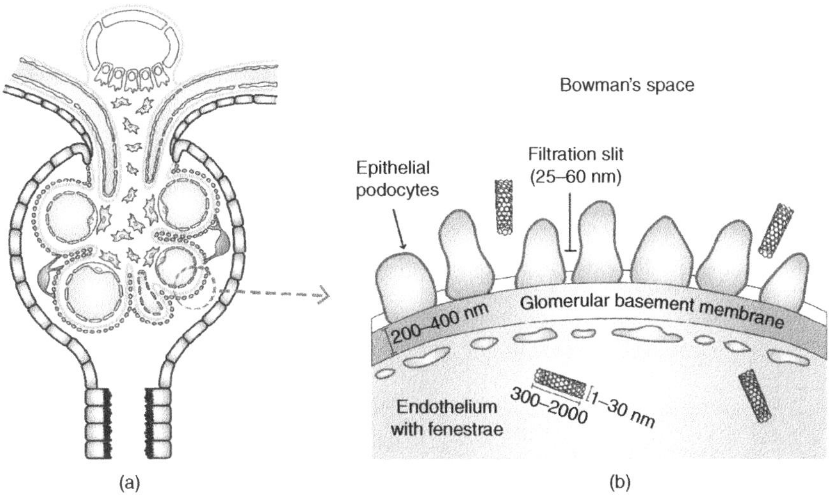

Figure 13.7 The mechanism of renal elimination of the nanotubes. (a) Basic structure of the glomerular and the glomerular filtration barrier. Glomerulus is a cluster of intertwined capillaries. The Bowman's capsule or glomerular capsule is a thin-walled structure that surrounds the glomerulus and is composed of visceral and parietal layers. On the urinary side of the glomerular basement membrane are podocytes. (b) At the glomerular basement membrane, only short CNTs (300–2000 nm in length and 1–30 nm in diameter) with favorable charge characteristics and a proper alignment can filtrate through the membrane.

MWCNTs seem to be able to longitudinally transport through the podocyte-slit membrane. This hypothesis was confirmed by several studies [88, 96], which indicated the modified SWCNTs or MWCNTs were rapidly eliminated intact by glomerular filtration, accumulated in the proximal tubule brush border and glomerulus, and progressively translocated to cytoplasm and nucleus in tubular cells. In addition, the nanotubes were up taken higher from the apical membrane than from the basolateral membrane of proximal tubular cells. Mathematical modeling revealed that a high degree of alignment with the longitudinal axis allows the nanotubes to vertically orient to the glomerular pores [96]. However, the dimension, shape, and degree of functional group of the individual nanoparticles are extremely important to urinary elimination [88]. For example, the higher degree of chemical functionalization (over 0.99 mmol g/l) provide better renal clearance and less accumulation in critical organs [91, 97]. Similarly, carboxyl-MWCNTs with a density greater than 3 μmol/mg and length less than 500 nm are rapidly cleared via the kidney, resulting in low toxicity in an animal model [98]. The mechanism of CNT elimination through the glomerular filtration system is illustrated in Figure 13.7.

However, some hydrophilic SWCNTs such as polyethylene-glycol-coated nanotubes excreted via both urinary system and biliary systems [85, 91, 92]. Normally, molecules larger than 200 nm cannot cross the fenestrations of the hepatic sinusoids, but modified nanotubes can translocate by passing through fenestrae along their vertical axis as in the kidney [72]. The CNTs with negative surface charge are likely to be taken up by Kupffer cells and endothelial cells; on the other hand, the CNTs with positive surface charge trend to be up taken by hepatocytes [99]. The mechanism of biliary elimination of nanotubes can be described as follows: functionalized CNTs translocate from sinusoid to the fenestrated sinusoidal epithelial cells via Stabilin receptors, the scavenger receptors expressed by liver sinusoidal endothelium. Moreover, some portion of CNTs are passively vertically diffused from the Disse space into the biliary canaliculi and afterward excreted into the bile as illustrated in Figure 13.8 [85].

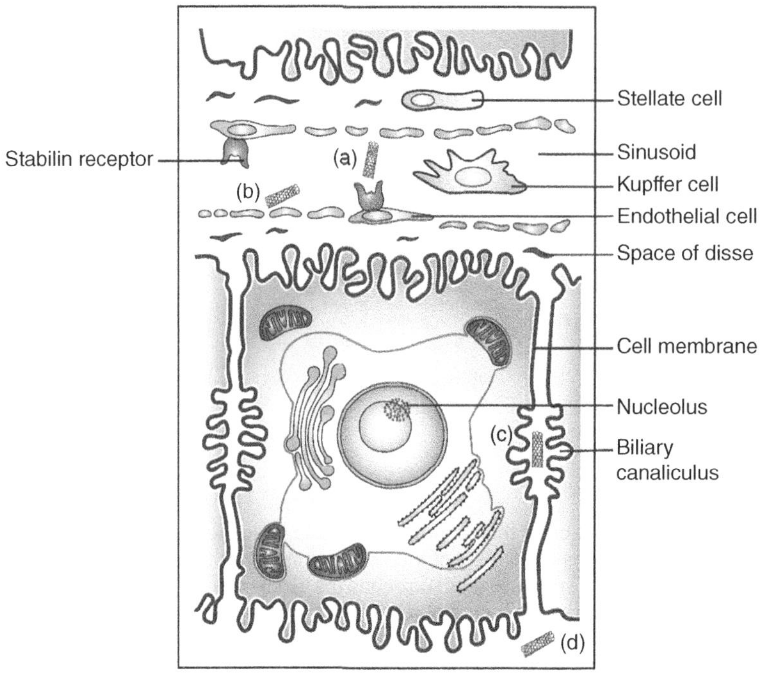

Figure 13.8 The mechanism of CNT elimination viabiliary system. (a, b) CNTs translocate from sinusoid to the fenestrated sinusoidal epithelial cells via Stabilin receptors. (c) CNTs are passive transported from the Disse space into the hepatocyte through the biliary canaliculus. (d) Excretion of the CNTs into bile. Source: Based on Alidori et al. [85].

Nonetheless, when orally or intraperitoneal administered CNTs with lengths exceeding 10 μm, granulomas can be formed in the reticuloendothelial system, resulting in inflammation and eventually acute toxicity [100]. Intermediate lengths of CNTs, less than 10 μm, can be recognized and taken up by macrophages and subsequent opsonophagocytosis and accumulation in the reticuloendothelial system. Short-length CNTs (<300 nm) can be rapidly eliminated via kidney and bile duct without any side effects [92, 100]. Pristine CNTs have been found to be cytotoxic to various mammalian cell lines *in vitro*. They can also cause acute toxicity dominantly in the skin, brain, and lungs *in vivo* [37, 71, 78, 101–105]. An important adverse effect of unmodified CNTs is the generation of ROS [104]. Overproduction of ROS can induce oxidative stress, resulting in lipid peroxidation and eventually cell death as well as cancer initiation [78, 104, 106]. Pristine nanotubes tend to aggregate due to their insolubility; therefore, such adverse effects are observed. Over the last decade, numerous work has been done in order to improve the biocompatibility of these nanotubes *in vivo* [71, 78, 93, 100, 107]. A remarkable reduction of their skin irritation and pulmonary inflammation and genotoxicities can be demonstrated using hydroxyl-, carboxy-, or amino-surface modification [77, 78, 101]. However, their toxicities are also dependent on the route of exposure. Several studies have reported no toxicity in animal models after intravenous or intraperitoneal administration but adverse effects were reported *in vivo* after pulmonary exposure [108–112]. Biodistribution, toxicities, and elimination of some functionalized CNTs are listed in Table 13.2.

Table 13.2 Pharmacokinetic and pharmacodynamic profiles of functionalized carbon nanotubes.

Functionalized nanotube	Type	Diameter, Length (nm)	Route	Model	Accumulation	Elimination	$t_{1/2}$ (h)	Toxicity	References
DTPA	SWCNT	1, 300–1000	i.v.	Mice	No	Kidney	3–3.5	No	[87]
DTPA	MWCNT	20–30, 500–2000	i.v.	Mice	No	Kidney	—	No	[87]
Glucosamine	MWCNT	20–40	i.p.	Mice	No	Kidney and liver	5.5	No	[107]
PEG	SWCNT	1–5, 50–200	i.v.	Mice	Liver and spleen	—	—	No	[90]
PEGylated Phospholipids	SWCNT	1–2, 100	i.v.	Mice	Liver and spleen	Kidney and liver	—	No	[92]
PMHC18-mPEG	SWCNT	17–56, 200	i.v.	Mice	No	Reticuloendothelial system	22.1 h	No	[79]
DOTA	SWCNT	—	i.p.	Mice	No	Kinney	—	No	[94]

DTPA, Diethylentria*minepentaacetic dianhydride*; *PEG*, Polyethylene-glycol; PMHC18-mPEG, Poly (maleicanhydride-alt- 1-octadecene)-PEG methyl ethers; DOTA, Lysine-1,4,7,10-tetraazacyclododecane-1,4,7,10-tetraacetic acid.

13.9 Application of Carbon Nanotubes in Drug Delivery

Recent research has highlighted the impact of functionalized CNTs on drug delivery based on their unique properties, including high drug-carrying capacities, prolonged circulation time, specific target accumulation, and remarkably fast elimination. SWCNTs conjugated with doxorubicin and hyaluronic acid with gadolinium loading (Gd/SWCNTs-HA-ss-DOX) presented tumor targeting properties, redox-responsive releasing function, and antitumor efficacy [113]. Likewise, phospholipid-branched PEG and tetrahydropyranyl-adriamycin coated SWNCT walls (SWCNT-THP) showed the minimal side effects without loss of therapeutic efficacy in human bladder cancer *in vitro* [114]. Furthermore, when the cannabinoid, 2-arachidonoylglycerol (2-AG) was conjugated with MWCNTs in order to improve the pharmacological profile of 2-AG for colitis treatment, MWCNT loaded 2-AG provided a substantial concentration and longer-lasting therapeutic effects than the drug alone [115]. The functionalized SWCNTs and MWCNTs were also applied in neurology, for instance, cationic polymer polyethyleneimine (PEI) conjugated with SWCNTs showed the quality to promote neurite outgrowth and branching, although to a smaller extent when compared with PEI substrate alone [116]. In addition, CNTs have been studied as skin cancer therapeutic carriers and mediators as well as transdermal drug delivery devices such as in DWCNTs membrane-based transdermal nicotine delivery [117–119].

13.10 Conclusion

Nanoparticles have become a new material and approach in drug delivery system. There are still controversies over the use of nanoparticle platform in clinical application as several pros and cons of the nanoparticle platform have been reported in animal model and some clinical trials. Some nanotherapeutics have been approved by FDA and are currently used in the clinic such as liposomal-loaded-doxorubicin (Doxil). These liposomal drugs reduce the toxicity of the encapsulated drugs resulting in less side effects on patients. These lipid-based products allow the maximum therapeutic effect on the disease targets. However, several inorganic nanoparticles can induce molecular toxicity caused by the components coated on the nanoparticle surface such as metallic impurities on CNT wall. These impurities induce oxidative stress in molecular levels leading to cell injury and cell death. Further investigations on nanoparticle toxicity and their side effects in animal models are needed before clinical use.

References

1 Shen, S., Wu, Y., Liu, Y., and Wu, D. (2017). High drug-loading nanomedicines: progress, current status, and prospects. *Int. J. Nanomed.* 12: 4085–4109.

2 Fojtu, M., Gumulec, J., Stracina, T. et al. (2017). Reduction of doxorubicin-induced cardiotoxicity using nanocarriers: a review. *Curr. Drug Metab.* 18 (3): 237–263.

3 Foerster, F., Bamberger, D., Schupp, J. et al. (2016). Dextran-based therapeutic nanoparticles for hepatic drug delivery. *Nanomedicine* 11 (20): 2663–2677.

4 Alam, T., Khan, S., Gaba, B. et al. (2017). Nanocarriers as treatment modalities for hypertension. *Drug Delivery* 24 (1): 358–369.

5 Alam, F., Naim, M., Aziz, M., and Yadav, N. (2015). Unique roles of nanotechnology in medicine and cancer-II. *Ind. J. Cancer* 52 (1): 1–9.

6 Albisa, A., Espanol, L., Prieto, M., and Sebastian, V. (2017). Polymeric nanomaterials as nanomembrane entities for biomolecule and drug delivery. *Curr. Pharm. Des.* 23 (2): 263–280.

7 Alphandery, E., Grand-Dewyse, P., Lefevre, R. et al. (2015). Cancer therapy using nanoformulated substances: scientific, regulatory and financial aspects. *Expert Rev. Anticancer Ther.* 15 (10): 1233–1255.

8 De Jong, W.H. and Borm, P.J. (2008). Drug delivery and nanoparticles: applications and hazards. *Int. J. Nanomed.* 3 (2): 133–149.

9 Cho, K., Wang, X., Nie, S. et al. (2008). Therapeutic nanoparticles for drug delivery in cancer. *Clin. Cancer Res.* 14 (5): 1310–1316.

10 Sahoo, S.K. and Labhasetwar, V. (2006). Nanoparticle interface: an important determinant in nanoparticle-mediated drug/gene delivery. In: *Nanoparticle Technology for Drug Delivery*, vol. 159 (eds. R. Gupta and U.B. Kompella), 141–148. New York, NY: Taylor & Francis Group.

11 Panyam, J. and Labhasetwar, V. (2004). Targeting intracellular targets. *Curr. Drug Delivery* 1 (3): 235–247.

12 Sercombe, L., Veerati, T., Moheimani, F. et al. (2015). Advances and challenges of liposome assisted drug delivery. *Front. Pharmacol.* 6: 286.

13 Al-Jamal, W.T., Al-Jamal, K.T., Tian, B. et al. (2009). Tumor targeting of functionalized quantum dot-liposome hybrids by intravenous administration. *Mol. Pharm.* 6 (2): 520–530.

14 Conniot, J., Silva, J.M., Fernandes, J.G. et al. (2014). Cancer immunotherapy: nanodelivery approaches for immune cell targeting and tracking. *Front. Chem.* 2: 105.

15 Monteiro, N., Martins, A., Reis, R.L., and Neves, N.M. (2014). Liposomes in tissue engineering and regenerative medicine. *J. R. Soc. Interface* 11 (101): 20140459.

16 Al-Jamal, W.T. and Kostarelos, K. (2011). Liposomes: from a clinically established drug delivery system to a nanoparticle platform for theranostic nanomedicine. *Acc. Chem. Res.* 44 (10): 1094–1104.

17 Kamaly, N. and Miller, A.D. (2010). Paramagnetic liposome nanoparticles for cellular and tumour imaging. *Int. J. Mol. Sci.* 11 (4): 1759–1776.

18 Kang, J.H., Jang, W.Y., and Ko, Y.T. (2017). The effect of surface charges on the cellular uptake of liposomes investigated by live cell imaging. *Pharm. Res.* 34 (4): 704–717.

19 Raz, A., Bucana, C., Fogler, W.E. et al. (1981). Biochemical, morphological, and ultrastructural studies on the uptake of liposomes by murine macrophages. *Cancer Res.* 41 (2): 487–494.

20 Wrobel, I. and Collins, D. (1995). Fusion of cationic liposomes with mammalian cells occurs after endocytosis. *Biochim. Biophys. Acta* 1235 (2): 296–304.

21 Connor, J., Yatvin, M.B., and Huang, L. (1984). pH-sensitive liposomes: acid-induced liposome fusion. *Proc. Natl. Acad. Sci. U. S. A.* 81 (6): 1715–1718.

22 Andar, A.U., Hood, R.R., Vreeland, W.N. et al. (2014). Microfluidic preparation of liposomes to determine particle size influence on cellular uptake mechanisms. *Pharm. Res.* 31 (2): 401–413.

23 Wei, Y., Guo, J., Zheng, X. et al. (2014). Preparation, pharmacokinetics and biodistribution of baicalin-loaded liposomes. *Int. J. Nanomed.* 9: 3623–3630.

24 Ishida, T., Harada, M., Wang, X.Y. et al. (2005). Accelerated blood clearance of PEGylated liposomes following preceding liposome injection: effects of lipid dose and PEG surface-density and chain length of the first-dose liposomes. *J. Controlled Release* 105 (3): 305–317.

25 Kamps, J.A. and Scherphof, G.L. (2004). Biodistribution and uptake of liposomes in vivo. *Methods Enzymol.* 387: 257–266.

26 Charrois, G.J. and Allen, T.M. (2003). Rate of biodistribution of STEALTH liposomes to tumor and skin: influence of liposome diameter and implications for toxicity and therapeutic activity. *Biochim. Biophys. Acta* 1609 (1): 102–108.

27 Ellens, H., Morselt, H.W., Dontje, B.H. et al. (1983). Effects of liposome dose and the presence of lymphosarcoma cells on blood clearance and tissue distribution of large unilamellar liposomes in mice. *Cancer Res.* 43 (6): 2927–2934.

28 Zamboni, W.C. (2005). Liposomal, nanoparticle, and conjugated formulations of anticancer agents. *Clin. Cancer Res.* 11 (23): 8230–8234.

29 van Etten, E.W., ten Kate, M.T., Snijders, S.V., and Bakker-Woudenberg, I.A. (1998). Administration of liposomal agents and blood clearance capacity of the mononuclear phagocyte system. *Antimicrob. Agents Chemother.* 42 (7): 1677–1681.

30 Allen, T.M. and Hansen, C. (1991). Pharmacokinetics of stealth versus conventional liposomes: effect of dose. *Biochim. Biophys. Acta* 1068 (2): 133–141.

31 Ishida, T., Harashima, H., and Kiwada, H. (2002). Liposome clearance. *Biosci. Rep.* 22 (2): 197–224.

32 Kelly, C., Jefferies, C., and Cryan, S.A. (2011). Targeted liposomal drug delivery to monocytes and macrophages. *J. Drug Delivery* 2011: 727241.

33 Nagayasu, A., Uchiyama, K., and Kiwada, H. (1999). The size of liposomes: a factor which affects their targeting efficiency to tumors and therapeutic activity of liposomal antitumor drugs. *Adv. Drug Delivery Rev.* 40 (1–2): 75–87.

34 Immordino, M.L., Dosio, F., and Cattel, L. (2006). Stealth liposomes: review of the basic science, rationale, and clinical applications, existing and potential. *Int. J. Nanomed.* 1 (3): 297–315.

35 Powell, J.S., Nugent, D.J., Harrison, J.A. et al. (2008). Safety and pharmacokinetics of a recombinant factor VIII with pegylated liposomes in severe hemophilia A. *J. Thromb. Haemost.* 6 (2): 277–283.

36 Vert, M. and Domurado, D. (2000). Poly(ethylene glycol): protein-repulsive or albumin-compatible? *J. Biomater. Sci. Polym. Ed.* 11 (12): 1307–1317.

37 Ciofani, G., Raffa, V., Vittorio, O. et al. (2010). In vitro and in vivo biocompatibility testing of functionalized carbon nanotubes. *Methods Mol. Biol.* 625: 67–83.

38 Johnstone, S.A., Masin, D., Mayer, L., and Bally, M.B. (2001). Surface-associated serum proteins inhibit the uptake of phosphatidylserine and poly(ethylene glycol) liposomes by mouse macrophages. *Biochim. Biophys. Acta* 1513 (1): 25–37.

39 Dams, E.T., Laverman, P., Oyen, W.J. et al. (2000). Accelerated blood clearance and altered biodistribution of repeated injections of sterically stabilized liposomes. *J. Pharmacol. Exp. Ther.* 292 (3): 1071–1079.

40 Lyass, O., Uziely, B., Ben-Yosef, R. et al. (2000). Correlation of toxicity with pharmacokinetics of pegylated liposomal doxorubicin (Doxil) in metastatic breast carcinoma. *Cancer* 89 (5): 1037–1047.

41 Lyass, S., Perry, Y., Venturero, M. et al. (2000). Laparoscopic cholecystectomy: what does affect the outcome? A retrospective multifactorial regression analysis. *Surg. Endosc.* 14 (7): 661–665.

42 Barenholz, Y. (2012). Doxil® – the first FDA-approved nano-drug: lessons learned. *J. Controlled Release* 160 (2): 117–134.

43 Porche, D.J. (1996). Liposomal doxorubicin (Doxil). *J. Assoc. Nurses AIDS Care* 7 (2): 55–59.

44 Liu, M., Chen, M., and Yang, Z. (2017). Design of amphotericin B oral formulation for antifungal therapy. *Drug Delivery* 24 (1): 1–9.

45 Aversa, F., Busca, A., Candoni, A. et al. (2017). Liposomal amphotericin B (AmBisome®) at beginning of its third decade of clinical use. *J. Chemother.* 29 (3): 131–143.

46 Farhood, H., Gao, X., Son, K. et al. (1994). Cationic liposomes for direct gene transfer in therapy of cancer and other diseases. *Ann. N.Y. Acad. Sci.* 716: 23–34; discussion -5.

47 Yang, K., Clifton, G.L., and Hayes, R.L. (1997). Gene therapy for central nervous system injury: the use of cationic liposomes: an invited review. *J. Neurotrauma* 14 (5): 281–297.
48 Serikawa, T., Kikuchi, A., Sugaya, S. et al. (2006). In vitro and in vivo evaluation of novel cationic liposomes utilized for cancer gene therapy. *J. Controlled Release* 113 (3): 255–260.
49 Miller, A.D. (1998). Cationic liposome systems in gene therapy. *IDrugs* 1 (5): 574–583.
50 Yu, J., Zhang, H., Li, Y. et al. (2017). Metabolomics revealed the toxicity of cationic liposomes in HepG2 cells using UHPLC-Q-TOF/MS and multivariate data analysis. *Biomed. Chromatogr.* 31 (12):10.1002/bmc.4036. doi:https://doi.org/10.1002/bmc.4036.
51 Iwaoka, S., Nakamura, T., Takano, S. et al. (2006). Cationic liposomes induce apoptosis through p38 MAP kinase-caspase-8-Bid pathway in macrophage-like RAW264.7 cells. *J. Leukocyte Biol.* 79 (1): 184–191.
52 Awasthi, V.D., Garcia, D., Goins, B.A., and Phillips, W.T. (2003). Circulation and biodistribution profiles of long-circulating PEG-liposomes of various sizes in rabbits. *Int. J. Pharm.* 253 (1–2): 121–132.
53 Mayer, L.D., Tai, L.C., Ko, D.S. et al. (1989). Influence of vesicle size, lipid composition, and drug-to-lipid ratio on the biological activity of liposomal doxorubicin in mice. *Cancer Res.* 49 (21): 5922–5930.
54 Allen, T.M., Hansen, C., Martin, F. et al. (1991). Liposomes containing synthetic lipid derivatives of poly(ethylene glycol) show prolonged circulation half-lives in vivo. *Biochim. Biophys. Acta* 1066 (1): 29–36.
55 Liu, D., Mori, A., and Huang, L. (1992). Role of liposome size and RES blockade in controlling biodistribution and tumor uptake of GM1-containing liposomes. *Biochim. Biophys. Acta* 1104 (1): 95–101.
56 Oku, N. and Namba, Y. (1994). Long-circulating liposomes. *Crit. Rev. Ther. Drug Carrier Syst.* 11 (4): 231–270.
57 Gabizon, A.A. (1992). Selective tumor localization and improved therapeutic index of anthracyclines encapsulated in long-circulating liposomes. *Cancer Res.* 52 (4): 891–896.
58 Zou, Y., Ling, Y.H., Reddy, S. et al. (1995). Effect of vesicle size and lipid composition on the in vivo tumor selectivity and toxicity of the non-cross-resistant anthracycline annamycin incorporated in liposomes. *Int. J. Cancer* 61 (5): 666–671.
59 Rose, P.G. (2005). Pegylated liposomal doxorubicin: optimizing the dosing schedule in ovarian cancer. *Oncologist* 10 (3): 205–214.
60 Allen, T.M. and Cullis, P.R. (2013). Liposomal drug delivery systems: from concept to clinical applications. *Adv. Drug Delivery Rev.* 65 (1): 36–48.
61 Gabizon, A., Shmeeda, H., and Barenholz, Y. (2003). Pharmacokinetics of pegylated liposomal doxorubicin: review of animal and human studies. *Clin. Pharmacokinet.* 42 (5): 419–436.
62 Adler-Moore, J.P., Gangneux, J.P., and Pappas, P.G. (2016). Comparison between liposomal formulations of amphotericin B. *Med. Mycol.* 54 (3): 223–231.
63 Bowden, R., Chandrasekar, P., White, M.H. et al. (2002). A double-blind, randomized, controlled trial of amphotericin B colloidal dispersion versus amphotericin B for treatment of invasive aspergillosis in immunocompromised patients. *Clin. Infect. Dis.* 35 (4): 359–366.
64 Walsh, T.J., Hiemenz, J.W., Seibel, N.L. et al. (1998). Amphotericin B lipid complex for invasive fungal infections: analysis of safety and efficacy in 556 cases. *Clin. Infect. Dis.* 26 (6): 1383–1396.
65 Gambling, D., Hughes, T., Martin, G. et al. (2005). A comparison of Depodur, a novel, single-dose extended-release epidural morphine, with standard epidural morphine for pain relief after lower abdominal surgery. *Anesth. Analg.* 100 (4): 1065–1074.

66 Glantz, M.J., Jaeckle, K.A., Chamberlain, M.C. et al. (1999). A randomized controlled trial comparing intrathecal sustained-release cytarabine (DepoCyt) to intrathecal methotrexate in patients with neoplastic meningitis from solid tumors. *Clin. Cancer Res.* 5 (11): 3394–3402.

67 Tacar, O., Sriamornsak, P., and Dass, C.R. (2013). Doxorubicin: an update on anticancer molecular action, toxicity and novel drug delivery systems. *J. Pharm. Pharmacol.* 65 (2): 157–170.

68 Hobbs, S.K., Monsky, W.L., Yuan, F. et al. (1998). Regulation of transport pathways in tumor vessels: role of tumor type and microenvironment. *Proc. Natl. Acad. Sci. U. S. A.* 95 (8): 4607–4612.

69 Nehoff, H., Parayath, N.N., Domanovitch, L. et al. (2014). Nanomedicine for drug targeting: strategies beyond the enhanced permeability and retention effect. *Int. J. Nanomed.* 9: 2539–2555.

70 Bianco, A., Kostarelos, K., Partidos, C.D., and Prato, M. (2005). Biomedical applications of functionalised carbon nanotubes. *Chem. Commun.* 5: 571–577.

71 Firme, C.P. 3rd and Bandaru, P.R. (2010). Toxicity issues in the application of carbon nanotubes to biological systems. *Nanomedicine* 6 (2): 245–256.

72 Kotagiri, N. and Kim, J.W. (2014). Stealth nanotubes: strategies of shielding carbon nanotubes to evade opsonization and improve biodistribution. *Int. J. Nanomed.* 9 (Suppl 1): 85–105.

73 Rastogi, V., Yadav, P., Bhattacharya, S.S. et al. (2014). Carbon nanotubes: an emerging drug carrier for targeting cancer cells. *J. Drug Delivery* 2014: 670815.

74 Melanko, J., Pearce, M.E., and Salem, A. (2009). Nanotubes, nanorods, nanofibers, and fullerenes for nanoscale drug delivery. In: *Nanotechnology in Drug Delivery*, vol. 10 (eds. V. MMD, P. Aramwit and G.S. Kwon), 105–278. New York, NY: Springer.

75 Al Faraj, A., Fauvelle, F., Luciani, N. et al. (2011). In vivo biodistribution and biological impact of injected carbon nanotubes using magnetic resonance techniques. *Int. J. Nanomed.* 6: 351–361.

76 Baughman, R.H., Zakhidov, A.A., and de Heer, W.A. (2002). Carbon nanotubes--the route toward applications. *Science* 297 (5582): 787–792.

77 Klumpp, C., Kostarelos, K., Prato, M., and Bianco, A. (2006). Functionalized carbon nanotubes as emerging nanovectors for the delivery of therapeutics. *Biochim. Biophys. Acta* 1758 (3): 404–412.

78 Kostarelos, K. (2008). The long and short of carbon nanotube toxicity. *Nat. Biotechnol.* 26 (7): 774–776.

79 Prencipe, G., Tabakman, S.M., Welsher, K. et al. (2009). PEG branched polymer for functionalization of nanomaterials with ultralong blood circulation. *J. Am. Chem. Soc.* 131 (13): 4783–4787.

80 Cai, Y.Q., Cai, Y.E., Mou, S.F., and Lu, Y.Q. (2005). Multi-walled carbon nanotubes as a solid-phase extraction adsorbent for the determination of chlorophenols in environmental water samples. *J. Chromatogr. A* 1081 (2): 245–247.

81 Tran, P.A., Zhang, L., and Webster, T.J. (2009). Carbon nanofibers and carbon nanotubes in regenerative medicine. *Adv. Drug Delivery Rev.* 61 (12): 1097–1114.

82 Lopez, C.F., Nielsen, S.O., Moore, P.B., and Klein, M.L. (2004). Understanding nature's design for a nanosyringe. *Proc. Natl. Acad. Sci. U. S. A.* 101 (13): 4431–4434.

83 Kam, N.W. and Dai, H. (2005). Carbon nanotubes as intracellular protein transporters: generality and biological functionality. *J. Am. Chem. Soc.* 127 (16): 6021–6026.

84 Sun, W., Lu, Y., and Gu, Z. (2014). Advances in anticancer protein delivery using micro–/nanoparticles. *Part. Part. Syst. Char.* 31 (12): 1204–1222.

85 Alidori, S., Bowman, R.L., Yarilin, D. et al. (2016). Deconvoluting hepatic processing of carbon nanotubes. *Nat. Commun.* 7: 12343.

86 Maruyama, K., Haniu, H., Saito, N. et al. (2015). Endocytosis of multiwalled carbon nanotubes in bronchial epithelial and mesothelial cells. *Biomed. Res. Int.* 2015: 793186.

87 Singh, R., Pantarotto, D., Lacerda, L. et al. (2006). Tissue biodistribution and blood clearance rates of intravenously administered carbon nanotube radiotracers. *Proc. Natl. Acad. Sci. U. S. A.* 103 (9): 3357–3362.

88 Lacerda, L., Herrero, M.A., Venner, K. et al. (2008). Carbon-nanotube shape and individualization critical for renal excretion. *Small* 4 (8): 1130–1132.

89 Lacerda, L., Soundararajan, A., Singh, R. et al. (2008). Dynamic imaging of functionalized multi-walled carbon nanotube systemic circulation and urinary excretion. *Adv. Mater.* 20 (2): 225-+.

90 Schipper, M.L., Nakayama-Ratchford, N., Davis, C.R. et al. (2008). A pilot toxicology study of single-walled carbon nanotubes in a small sample of mice. *Nat. Nanotechnol.* 3 (4): 216–221.

91 Liu, Z., Cai, W., He, L. et al. (2007). In vivo biodistribution and highly efficient tumour targeting of carbon nanotubes in mice. *Nat. Nanotechnol.* 2 (1): 47–52.

92 Liu, Z., Davis, C., Cai, W. et al. (2008). Circulation and long-term fate of functionalized, biocompatible single-walled carbon nanotubes in mice probed by Raman spectroscopy. *Proc. Natl. Acad. Sci. U. S. A.* 105 (5): 1410–1415.

93 Wang, H., Wang, J., Deng, X. et al. (2004). Biodistribution of carbon single-wall carbon nanotubes in mice. *J. Nanosci. Nanotechnol.* 4 (8): 1019–1024.

94 Mulvey, J.J., Feinberg, E.N., Alidori, S. et al. (2014). Synthesis, pharmacokinetics, and biological use of lysine-modified single-walled carbon nanotubes. *Int. J. Nanomed.* 9: 4245–4255.

95 Kostarelos, K. (2010). Carbon nanotubes: fibrillar pharmacology. *Nat. Mater.* 9 (10): 793–795.

96 Ruggiero, A., Villa, C.H., Bander, E. et al. (2010). Paradoxical glomerular filtration of carbon nanotubes. *Proc. Natl. Acad. Sci. U. S. A.* 107 (27): 12369–12374.

97 Al-Jamal, K.T., Nunes, A., Methven, L. et al. (2012). Degree of chemical functionalization of carbon nanotubes determines tissue distribution and excretion profile. *Angew. Chem. Int. Ed. Engl.* 51 (26): 6389–6393.

98 Jain, S., Thakare, V.S., Das, M. et al. (2011). Toxicity of multiwalled carbon nanotubes with end defects critically depends on their functionalization density. *Chem. Res. Toxicol.* 24 (11): 2028–2039.

99 Wang, H.L., Thorling, C.A., Liang, X.W. et al. (2015). Diagnostic imaging and therapeutic application of nanoparticles targeting the liver. *J. Mater. Chem. B* 3 (6): 939–958.

100 Kolosnjaj-Tabi, J., Hartman, K.B., Boudjemaa, S. et al. (2010). In vivo behavior of large doses of ultrashort and full-length single-walled carbon nanotubes after oral and intraperitoneal administration to Swiss mice. *ACS Nano* 4 (3): 1481–1492.

101 Poulsen, S.S., Jackson, P., Kling, K. et al. (2016). Multi-walled carbon nanotube physicochemical properties predict pulmonary inflammation and genotoxicity. *Nanotoxicology* 10 (9): 1263–1275.

102 Smart, S.K., Cassady, A.I., Lu, G.Q., and Martin, D.J. (2006). The biocompatibility of carbon nanotubes. *Carbon* 44 (6): 1034–1047.

103 Shvedova, A.A., Castranova, V., Kisin, E.R. et al. (2003). Exposure to carbon nanotube material: assessment of nanotube cytotoxicity using human keratinocyte cells. *J. Toxicol. Environ. Health A* 66 (20): 1909–1926.

104 Shvedova, A.A., Pietroiusti, A., Fadeel, B., and Kagan, V.E. (2012). Mechanisms of carbon nanotube-induced toxicity: focus on oxidative stress. *Toxicol. Appl. Pharmacol.* 261 (2): 121–133.

105 Wang, C., Oh, S., Lee, H.A. et al. (2017). In vivo feasibility test using transparent carbon nanotube-coated polydimethylsiloxane sheet at brain tissue and sciatic nerve. *J. Biomed. Mater. Res. A* 105 (6): 1736–1745.

106 Fu, P.P., Xia, Q., Hwang, H.M. et al. (2014). Mechanisms of nanotoxicity: generation of reactive oxygen species. *J. Food Drug Anal.* 22 (1): 64–75.

107 Guo, J., Zhang, X., Li, Q., and Li, W. (2007). Biodistribution of functionalized multiwall carbon nanotubes in mice. *Nucl. Med. Biol.* 34 (5): 579–583.

108 Landsiedel, R., Sauer, U.G., Ma-Hock, L. et al. (2014). Pulmonary toxicity of nanomaterials: a critical comparison of published in vitro assays and in vivo inhalation or instillation studies. *Nanomedicine* 9 (16): 2557–2585.

109 Mittal, S., Sharma, P.K., Tiwari, R. et al. (2017). Impaired lysosomal activity mediated autophagic flux disruption by graphite carbon nanofibers induce apoptosis in human lung epithelial cells through oxidative stress and energetic impairment. *Part. Fibre Toxicol.* 14 (1): 15.

110 Wang, X., Mansukhani, N.D., Guiney, L.M. et al. (2016). Toxicological profiling of highly purified metallic and semiconducting single-walled carbon nanotubes in the rodent lung and *E. coli*. *ACS Nano* 10 (6): 6008–6019.

111 Vietti, G., Lison, D., and van den Brule, S. (2016). Mechanisms of lung fibrosis induced by carbon nanotubes: towards an adverse outcome pathway (AOP). *Part. Fibre Toxicol.* 13: 11.

112 Kasai, T., Umeda, Y., Ohnishi, M. et al. (2016). Lung carcinogenicity of inhaled multi-walled carbon nanotube in rats. *Part. Fibre Toxicol.* 13 (1): 53.

113 Hou, L., Yang, X.M., Ren, J.X. et al. (2016). A novel redox-sensitive system based on single-walled carbon nanotubes for chemo-photothermal therapy and magnetic resonance imaging. *Int. J. Nanomed.* 11: 607–624.

114 Chen, G., He, Y., Wu, X. et al. (2012). In vitro and in vivo studies of pirarubicin-loaded SWNT for the treatment of bladder cancer. *Braz. J. Med. Biol. Res.* 45 (8): 771–776.

115 Hassanzadeh, P., Arbabi, E., Atyabi, F., and Dinarvand, R. (2017). Application of carbon nanotubes as the carriers of the cannabinoid, 2-arachidonoylglycerol: towards a novel treatment strategy in colitis. *Life Sci.* 179: 66–72.

116 Nunes, A., Al-Jamal, K., Nakajima, T. et al. (2012). Application of carbon nanotubes in neurology: clinical perspectives and toxicological risks. *Arch. Toxicol.* 86 (7): 1009–1020.

117 Gulati, G.K., Chen, T., and Hinds, B.J. (2017). Programmable carbon nanotube membrane-based transdermal nicotine delivery with microdialysis validation assay. *Nanomedicine* 13 (1): 1–9.

118 Ding, L., Stilwell, J., Zhang, T. et al. (2005). Molecular characterization of the cytotoxic mechanism of multiwall carbon nanotubes and nano-onions on human skin fibroblast. *Nano Lett.* 5 (12): 2448–2464.

119 Ganesan, P. and Choi, D.K. (2016). Current application of phytocompound-based nanocosmeceuticals for beauty and skin therapy. *Int. J. Nanomed.* 11: 1987–2007.

Part III

Administrative Routes for Controlled Drug Delivery

14

Controlled Drug Delivery via the Ocular Route

Peter W.J. Morrison and Vitaliy V. Khutoryanskiy

Reading School of Pharmacy, University of Reading, Reading, UK

14.1 Introduction

The eyes are our body's sight sensing organs, their purpose to capture light, which in turn is converted into electrochemical impulses transmitted along the optic nerve to the primary visual cortex at the lower posterior region of the brain, where these signals are processed, giving a visual representation of our environment. Vision allows rapid and simultaneous sensation of our surroundings, sensing motion, distance, perspective, and color, allowing for interaction with the environment around us. Vision places more demand for resources by the brains processing power than all of our other senses combined [1].

The eye has evolved to be a delicate and intricate organ that is in many ways relatively isolated from the rest of the body by physiological barriers, effectively keeping the visual pathway optically clear and free from foreign matter. Being an accessible organ for introducing medicine it could be easily assumed that conditions of the eye are easily treated by simply instilling medication in the form of eye drops, thereby placing drugs just where they are needed; however, it is not quite that simple.

Superficial conditions such as conjunctivitis and blepharitis are more easily treated than disorders affecting the deeper structures of the organ, e.g. Fuchs endothelial dystrophy, macular degeneration, and retinitis pigmentosa [2, 3].

Ocular drug delivery is hampered by blink action, dilution and washout by tears, mechanisms of nasolacrimal drainage, and poor permeability of the cornea [4, 5]. Once conventional eye drops are instilled, there is only a short time where the medication is in contact with the eye, during this time drugs targeting intraocular tissue have to penetrate the cornea or sclera [6]. A small proportion of topically applied drugs are absorbed into the cornea, up to 5% but often much less [3, 7]. Most of the instilled dose is systemically absorbed by the nasolacrimal duct and conjunctiva from where it is degraded and eliminated by the body's metabolic processes [3, 8]. Developing ocular drug formulations with sufficient drug concentration that can remain on the eye long enough for efficacious drug delivery is very challenging to ocular medication scientists [9]. The three main routes for delivering medication to the eyes are systemic, topical, and intraocular; each method has advantages and disadvantages. Systemic drug delivery is inefficient due to "blood–eye barriers"; relatively high doses

Fundamentals of Drug Delivery, First Edition.
Edited by Heather A. E. Benson, Michael S. Roberts, Adrian C. Williams, and Xiaowen Liang.

are necessary in order to allow therapeutic levels of drug available to ocular tissue, this has an added drawback whereas all of the body is subject to the effects of the given medication and these effects can have undesirable consequences such as pulmonary and cardiovascular stimulation. Intraocular drug delivery by injection or implant is invasive and an undesired option for most patients, drug delivery by this method also carries the risk complications such as hemorrhage and retinal detachment; however, it offers a very effective means to place drugs just where they are needed while virtually eliminating systemic effects. For reasons of patient compliance and convenience, the topical route is preferred, and this method offers simplicity with the least inconvenience; most patients are capable of self-administering eye drops or ointment [2].

Recent advances in ocular drug delivery have established methods to improve ophthalmic drug retention and enhance corneal permeation as well as the development of controlled and sustained release systems [3].

Within this chapter, ocular physiology, some of the more common ocular conditions and the current state of the art of treating ocular disorders using controlled drug delivery are considered, together with an insight for future developments.

14.2 Physiology of the Eye

The eyes are paired organs embedded within the orbital cavities situated at the upper front of the skull, these bony cavities offer a high degree of physical protection. Each eye is free to move under the control of six extraocular muscles. Figure 14.1 shows the individual structures of the eye, the main ones will be described in more detail.

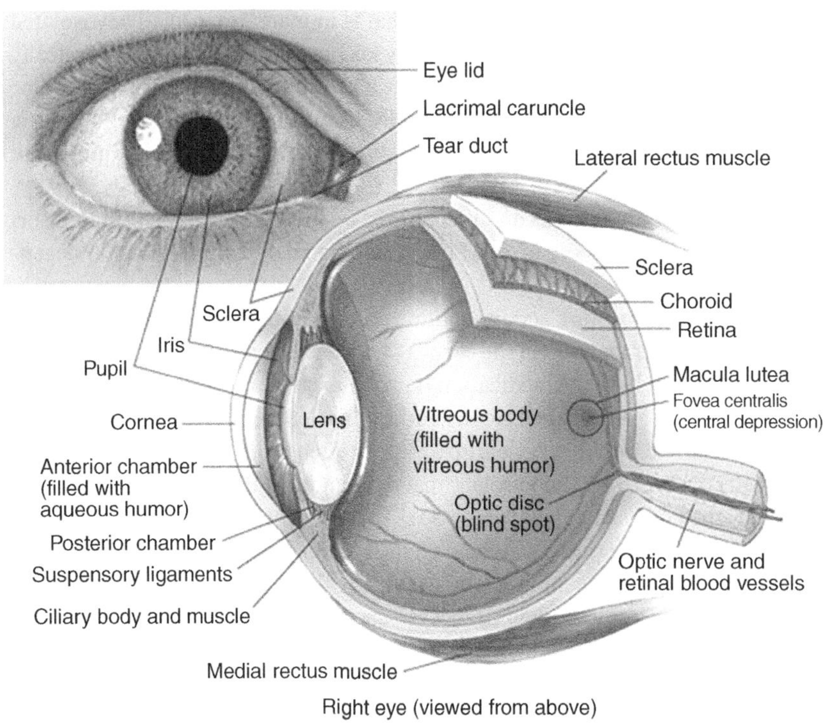

Figure 14.1 Anatomical detail of the human eye. Source: From BioGraphix [10]. © 2006, BioGraphix.

14.2.1 Ocular Membranes; Conjunctiva, Cornea, and Sclera

The contents of the eye are enclosed within tough membranes known as the "tunica fibrosa oculi," normal internal pressure in the range of 13–19 mm/Hg is necessary to maintain the correct shape of the organ. The cornea is an avascular and transparent membrane extending slightly forward from the main body and accounts for ~17% of the surface area. The remaining 83% surface area consists of a tough, opaque, and white membrane, known as sclera. A thin vascularized transparent and loosely attached mucosal membrane known as the conjunctiva covers the anterior part of the sclera known as the bulbar. The conjunctiva continues to line the inner surface of the eyelids forming the inferior and superior fornixes; this tissue offers protection and lubrication allowing the eye to freely move within the eye socket [11]. The limbus is a ring of tissue at the boundary where the cornea thickens and forms the transition to sclera. Limbal tissue is populated with stem cells which regenerate corneal epithelium, necessary due to a high epithelial cell turnover rate [12]. An important function of the epithelium is the protection it offers against ultraviolet radiation. Epithelia have high concentrations of tryptophan residues and ascorbate, which are considered important in absorbing ultraviolet radiation, thereby protecting deeper structures within the eye such as the crystalline lens and retina, tissues which are sensitive to potential damage from ultraviolet radiation [13]. The cornea allows light to enter the eye, and it provides ~60% of the eye's refractive power, accommodation is given by the crystalline lens which adjusts to maintain sharp focus on and near the fovea centralis, all other light falling on the retina provide for peripheral and vision in low light conditions [2, 11]. The cornea is a multi-layered membrane with lipophilic and hydrophilic properties, historically five layers were known; epithelium, Bowman's membrane, stroma, Descemet's membrane, and endothelium. An additional membrane some 15 μm thick between the stroma and Descemet's membrane named "Dua's layer" has recently been discovered in human corneal tissue, and this layer has potential to protect the endothelium during cornea transplant procedures by leaving the innermost cornea layers and endothelium intact [14]. Corneal epithelium is a lipophilic membrane offering ~90% resistance to hydrophilic drug transport and ~10% resistance to hydrophobic drugs; the stroma is a hydrophilic membrane in which hydrophilic drugs can easily diffuse, whereas highly lipophilic drug transport is resisted [15, 16]. Figure 14.2 shows a cross-section micrograph of the stratified arrangement of corneal tissue.

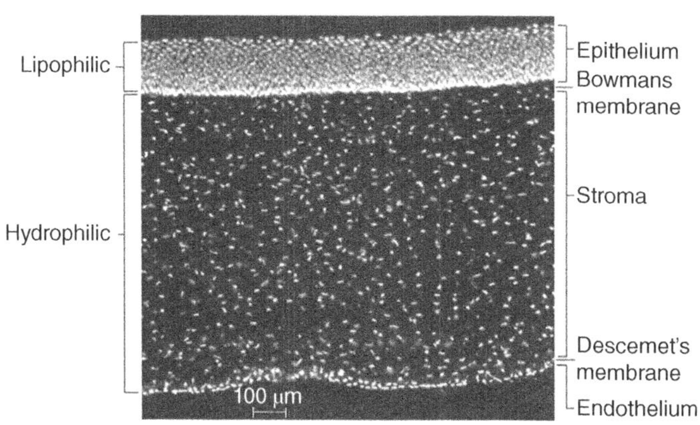

Figure 14.2 Cross section micrograph of bovine cornea showing the multi-layered structure. Source: Reprinted from Morrison et al. [15] with permission from © American Chemical Society (2013).

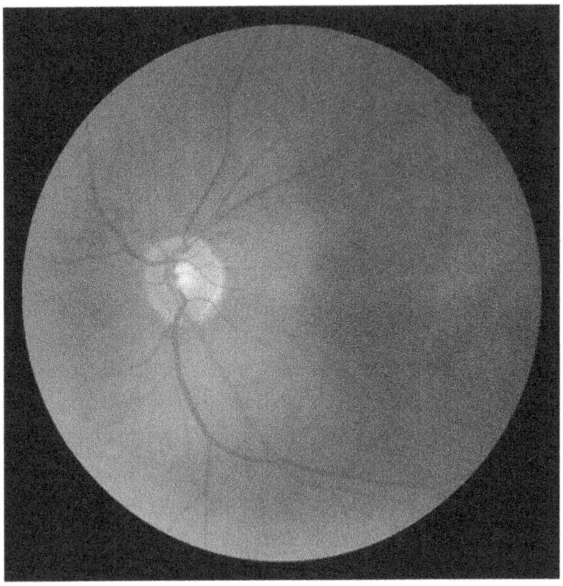

Figure 14.3 Vascular network within the uveal tract of a healthy individual, these blood vessels provide maintenance of intraocular metabolic function. Source: Ref. [18].

The cornea is nourished by the aqueous humor from inside the eye and is cleansed, lubricated, and oxygenated by mucus and the tear film at the outermost surface; further metabolic support is provided from limbal capillary blood vessels, although these vessels do not normally populate the cornea unless it becomes deprived of oxygen for an extended length of time, e.g. because of contact lens presence [11, 12].

14.2.2 Internal Ocular Structures

Known as the "*tunica vasculosa oculi*," the uveal tract forms a vascular network around the back of the eye that includes the iris, ciliary body, and choroid, covering the inside surface between the sclera and retina. The main function is to provide metabolic support to the retina, ciliary body, and iris [17]. Membranes within the vascular network are special in that they offer a "blood–eye barrier" to prevent ingress of substances that can compromise the clarity of the light transmitting pathways [6]. Figure 14.3 shows the network of blood vessels contained within the uveal tract in a healthy individual, these blood vessels provide maintenance of intraocular metabolic processes.

Adjacent to the choroid and kept in place by pressure from the vitreous body lies the retina, a membrane of highly specialized nerve tissue continuous to the optic nerve. It forms the light sensitive region of the eye and via the optic nerve links to the visual cortex, a specialized region of the brain dedicated to visual processing. The purpose of the retina is to convert light falling upon it into visual signals interpreted by the brain as images of the outside world [19]. The retina is a complex multi-layer structure ~0.5 mm thick lining the posterior inner surface of the eye, extending toward, but stopping just short of the ciliary body, the extent of coverage is shown in Figure 14.1. There is a short area some 4 mm long surrounding the front region of the retina extending between the front most edge of the retina up to the ciliary body, this area is known as the "pars plana," it is the thinnest part of the eyes sheath and the most likely region where medication can penetrate into the organ [20]. Specialized photoreceptor cells are embedded throughout the retina consisting of cones and rods. Cone cells offer high visual acuity and are associated with color perception, while rod cells detect light monochromatically and are sensitive even at low light levels; these are of particular importance in peripheral vision. There are three types of cone cell type acting as photoreceptive transducers employing the photosensitive molecule, rhodopsin, which responds to the primary colors, red, blue, and green. Color perception is mostly concentrated at or near the macula, while monochromatically sensitive rod cells are dispersed throughout the retina. Retinal nourishment and metabolic maintenance is regulated by the choroid and retinal pigmented epithelium (RPE) which itself is part of the blood–retina barrier (BRB), the RPE and BRB serve to regulate solute movement to the retina and into the body of the eye [11, 21].

14.2.3 Anterior Chamber, Lens, and Vitreous Body

The anterior chamber is the fluid filled void behind the cornea; this space holds aqueous humor, a clear aqueous fluid similar to blood plasma. The aqueous humor is secreted into the posterior chamber by the ciliary body and is continuously replenished, leaving the chamber via the trabecular meshwork and is drained through the canal of Schlemm. Slight resistance offered by the drainage mechanisms ensures normal intraocular pressure is maintained in a healthy individual. Situated between the anterior chamber and the posterior chamber is the iris which acts as a control for intensity of light entering the eye, the iris also responds to our emotional state and vigilance. The vitreous humor is an avascular, gel-like, and optically clear structure consisting of collagen, hyaluronic acid, polysaccharide, and vitrosin, the largest component being water at ~98%, it occupies ~80% of the internal volume of the eye at around 4 ml. Between the aqueous chamber and vitreous body is the crystalline lens which allows for accommodation between near and far vision. The optical clarity of the lens is due to a precise alignment of a single-cell type consisting of soluble protein fibers. It is contained within the capsular bag which itself is attached to the ciliary body which offers accommodation control when adjusting between near and far vision [6, 11, 17, 21].

14.3 Ocular Disorders

The eyes can be subject to many disorders, and this poses a major problem in an age where we are faced with a trend in which people are living longer and therefore more prone to age-related conditions. Many superficial disorders are relatively easy to treat, other more serious disorders affecting the deeper structures pose difficulties for treatment or when trying to prevent the symptoms getting worse. Listed in this section are some common conditions and their symptoms; note, many conditions often share similar indicators. Ocular disorders are either periocular (surface of the eye and surrounding tissue) or intraocular (within the eye). Periocular tissues include the surface area of the cornea and sclera, conjunctiva, eyelids, and nasolacrimal mucosa. Intraocular structures can be subdivided into those of the anterior chamber, and the posterior chamber encasing the vitreous humor and retina. Periocular conditions are easier to treat than intraocular disorders, which pose major challenges for treatment [2]. There are 285 million people worldwide living with serious vision disorders, 39 million being blind, further, they report that 75% of severe disorders and blindness could potentially be treated [22, 23].

14.3.1 Periocular Disorders

- *Blepharitis*: A common and often recurring inflammatory condition causing infection of the eyelids, *Staphylococcus aurus* being the most common pathogen. Associated conditions can manifest as dry eye, rosacea, and seborrhoea arising from infection of the lipid secreting glands, styes, or chalazion can develop [8].

- *Conjunctivitis*: An infection of the conjunctiva. Cause can be an allergenic response or bacterial infection. Symptoms include irritation, swelling, redness ("pink eye"), photophobia, and discharge. Allergenic conjunctivitis includes reaction to pollen, animal dander, or dust mites. Infectious conjunctivitis is the result of exposure to pathogens [24].
- *Dry eye*: Many factors can result in this condition, i.e. aqueous tears deficiency, Meibomian gland dysfunction, and keratoconjunctivitis sicca. Inadequate production or composition of the tears leads to irritation of the corneal surface. Foreign body sensation, dry feeling, pain, burning, and photophobia are typical symptoms. Dry, dusty, or smoky environments and activities which tend to reduce normal blink rates such as using video display units (PC's, laptops etc.) tend to make the condition worse [25].
- *Trachoma*: A contagious condition caused by infection of *Chlamydia trachomatis*. The condition is common in third world countries and often becomes chronic, the eyelid turns inward, subsequent blinking damages the cornea. If the condition remains untreated it often leads to blindness [26].
- *Keratitis*: An inflammatory corneal disorder. Cause could be due to injury, fungal, pathogenic or viral infection, and vitamin A deficiency can be a contributory factor. Symptoms include excessive tearing, pain, photophobia, and inflammation [27].

14.3.2 Intraocular Disorders

- *Cataract*: Changes in the crystalline lens causing reduced transparency. Mostly due to aging, other causes are congenital, exposure to some forms of radiation, trauma, or secondary to disease such as diabetes [28, 29].
- *Fuchs endothelial dystrophy*: Slowly progressing edema of the cornea due to inherited or mutagenic changes and die off of endothelial cells. Endothelium cells help maintain the delicate hydration balance within the cornea. Consequently, when affected, the stroma suffers from excessive hydration causing the stroma to become cloudy. Disease progression is in stages and continues for decades [30].
- *Keratoconus*: A progressive noninflammatory ectasia, a bilateral and asymmetric degeneration where corneal changes reduce corneal integrity, corneal thinning leads to a cone-shaped projection causing myopia and irregular astigmatism [31].
- *Iridocorneal endothelial syndrome*: A progressive unilateral disease involving changes to the endothelium, iris, swelling of the cornea, and associated development of glaucoma. The cause is unknown however herpes simplex virus has been thought to be implicated. Endothelial cells become more epithelia like, some break away and migrate to the iris and trabecular meshwork. Ocular hypertension often follows due to plugging of the drain mechanisms by migrated endothelial cells [32].
- *Retinitis pigmentosa*: An hereditary disease where degradation of photoreceptor cells occurs, initially night and peripheral vision are affected, degeneration progressively affects the central vision. Over 100 genes have been implicated with the disease [33].
- *Macular degeneration*: A leading cause of blindness in an aging population. Typically late-onset, implicating factors include hereditary, medical risk, and lifestyle. Neovascularization or ischemia affects the retina and RPE leading to severe vision loss [34, 35].
- *Diabetic macular edema*: A serious vitreoretinal disease where the retina becomes inflamed and swollen, the disorder can lead to retinal detachment. Age and diabetes are implicating factors and the condition can be hereditary [36–38].

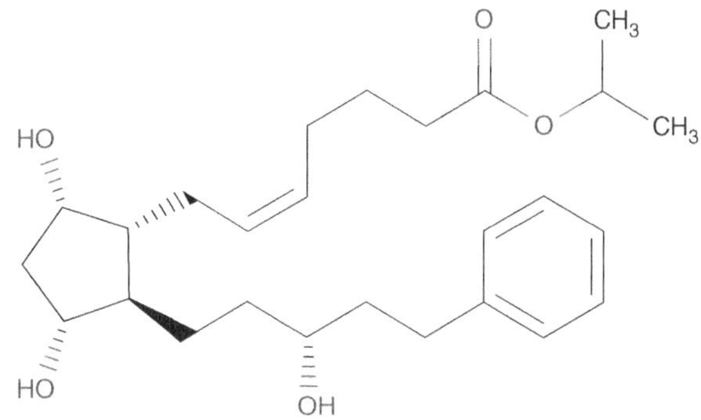

Figure 14.4 Chemical structure of the prodrug; latanoprost. Source: Latanoprost [40].

14.4 Controlled Drug Delivery Systems

Strategic approaches are employed to allow effective and controlled ocular drug delivery; these include formulation strategies to improve drug solubility/increased loading and improved bioavailability. Retention of instilled eye drops can be improved by inclusion of viscosity enhancing or mucoadhesive agents and drug transport into the eye can be improved by inclusion of penetration enhancing compounds [2, 3]. Prodrugs offer a means to treat a number of superficial and intraocular disorders, they are bio-reversible in that they are relatively inactive until metabolized within living tissue, once bio-activated they can achieve their desired therapeutic action at the target site [39]. One of the successful applications of prodrugs in ocular therapy is latanoprost, Figure 14.4, a prostaglandin, once taken up by the cornea, the prodrug is hydrolyzed by esterase's, transforming the prodrug into latanoprost acid, the active form of the drug [7, 41].

Nanotechnology techniques employ microparticles and nanoparticles as drug delivery vehicles capable of attaching to, or penetrating the cornea/sclera surface. They can be in the form of soluble, erodible/non-erodible solid microscopic units, microemulsions, liposomes, micelles, and niosomes [42, 43]. Very precise drug delivery can be achieved using intravitreal injection, drug depot implants, and degradable and injectable microparticles. Placing drug formulations or devices inside the eye virtually eliminates systemic complications and in some cases long-term sustained release can be achieved; however, it is the least desired option from the patient's perspective and there are risks of undesired complications such as hemorrhage and retinal detachment [2, 44–46]. Polymer films and contact lenses have potential to afford a means to deliver sustained release therapeutic agents in a way that can minimize systemic absorption by offering low-dose sustained drug delivery directly to the tear film on the eye surface, thereby reducing side effects often seen when using simple eye drop formulations [47]. Promising strategies currently under research and development include ultrasound (phonophoresis), iontophoresis, and microneedle, each of these techniques generate temporary disruption to penetration resistant membranes. Phonophoresis-enhanced drug delivery relies on high-frequency sound energy to drive active pharmaceutical agents across physical barriers such as the cornea and sclera [48], similarly, iontophoresis employs low-tension electrical energy to drive ionized medication across membranes, whereas microneedle technology renders membranes more permeable to medication by forming micro perforations in the membrane of interest, allowing opportunity for drugs to penetrate the tissue before natural healing allows the tissue to return to its normal state [49].

14.4.1 Formulation Strategies

Conventional aqueous eye drops are an inefficient means to treat ocular conditions due to limited drug solubility, dilution and washout by tears, and inadequate residence time. The application of eye drops tends to induce reflex tearing which quickly eliminates the dose. Eye drops typically administer a dose of ~30–75 μl with each drop and this quickly becomes diluted and drained leaving a normal resident volume of ~7 μl in the tear film, this remaining fluid is completely turned over in less than two minutes. The cul-de-sac between the eyelids and the ocular globe has capacity to hold up to 30 μl; therefore, designing eye drop formulations with therapeutically effective drug concentration delivered in small drop size 8–15 μl can be an effective strategy for this form of topical medication. Placement of the droplet in the lower fornix can minimize the problem from reflex tearing, thereby offering opportunity for the dose to remain in the tear film and superior/inferior fornixes long enough to allow better penetration into the cornea/sclera [9, 49, 50]. Pre-ocular retention can be enhanced by inclusion of various natural and synthetic hydrophilic polymers to increase formulation viscosity, thereby resisting premature elimination, examples include, cellulose derivatives, poly(vinyl alcohol), hyaluronic acid, dextran, gellan gum [9, 50]. Studies using rabbits have shown that formulating emulsions with positive charge is considered beneficial in the dose spreadability over the negatively charged epithelium, improving residence time, bioavailability, and penetration into ocular tissue [51]. Recent decades have seen developments in a diverse range topical medication in the form of water in oil (W/O) or oil in water (O/W) emulsions and microemulsions, these can allow lipophilic and hydrophilic drugs to be formulated, this strategy brings a dose form that is more resistant to washout and dilution [52, 53]. Ointments are a dose form offering more resistance to nonproductive losses. These are formulated using soft paraffin mixtures, with a melting point close to that of the ocular surface. This form of medication improves drug bioavailability and offers sustained release; however, there can be difficulties with patient compliance due to an initial blurring of vision until the dose form becomes uniformly spread on the cornea/sclera surface [54]. More recent formulation innovations have brought the development of atomized drug delivery systems where the dose form is sprayed onto closed eyes in the form of a fine mist, the medication then migrates into the tear film during blinking. This means of dosing is an effective and preferred method by some patients compared with traditional eye drops; it is a more appropriate dose form for superficial conditions such as dry eye and allergic reaction than deep tissue drug delivery [55].

14.4.2 Mucoadhesive Systems

More efficient formulation strategies designed to improve ocular medication retention are better able to resist dilution, washout, and nonproductive losses, this can bring more efficacious drug delivery aimed at the anterior and posterior segments which have remained challenging. Mucoadhesion could be described as interfacial interaction forces of polymeric material with mucosal tissue, several interacting theories have been described which combinations of these theories come into play [56, 57], namely;

- *Adsorption theory*: Relies of van der Waals' interactions and hydrogen bonding.
- *Diffusion theory*: Interpenetration of molecular chains of mucosal tissue, mucus, and the mucoadhesive dose form.
- *Electronic theory*: Charge dependant adhesion at the interface of mucosal tissue and some other substance.
- *Wetting theory*: Interaction between low viscosity mucoadhesive, mucus, and musosal tissue where the material has structural similarity to mucosa.

Mucoadhesion phenomenon can be employed to aid ocular drug retention and bioavailability at the eye surface or its surrounding tissue [56–58]. Traditional eye drop residence time is dependent on tear turnover; however, for mucoadhesive systems, residence time becomes under the control of mucus turnover, and this allows for increased drug retention and potentially improved bioavailability [2, 24]. For conditions such as "dry eye syndrome," medication may require frequent application, and for the patient, this can be inconvenient. Recent developments employing mucoadhesive nanoparticles in eye drop formulations have been shown to significantly reduce the need for frequent dosing with lower drug concentrations being necessary. Providing lower drug concentration in a dose form that resists clearance minimizes potential side effects without compromising therapeutic efficacy and helps improve patient compliance [59]. Research toward engineering mucoadhesive microparticles for ocular drug delivery have been shown to have the formulation retained in the ocular cul-de-sac for as long as 30 minutes, this is considerably longer than would be achieved using conventional eye drop dose forms [42]. Further research has investigated maleimide bearing nanogels as novel tuneable functionalized mucoadhesive materials, functionalized materials have potential to form covalent bond forming groups which offer effective sustained drug delivery properties to ocular mucosal tissue [60]. Irmukhametova et al. produced thiolated organosilica sub-100 nm nanoparticles in dilute aqueous dispersions offering excellent colloidal stability. Their work has shown them to have excellent mucoadhesive properties using the bovine cornea model, believed to be due to disulphide bonds forming between thiol groups of the nanoparticles and cysteine-rich domains of mucin. They conclude the nanoparticle dispersions could show promise for ocular drug delivery [61].

14.4.3 Solution to Gel *In Situ* Gelling Systems

Formulation strategies that allow precise liquid drop instillation which quickly undergoes a transition to retention enhanced gel are proving a beneficial innovation in ocular drug delivery. A solution phase can be described as a free flowing fluid, conversely, the gel phase is non-flowing. The solution to gel (sol–gel) system is designed to maintain its integrity for a duration whereby controlled drug delivery is achieved [62]. Sol–gel phase transition systems can be designed to respond to the following:

- *Temperature*: Dosage transforms from liquid at room temperature to a mucoadhesive gel at the temperature of the eye, ~34 °C.
- *pH*: Sol–gel transformation takes place when the dosage responds to the pH environment of the tear film, mean pH = 7.4.
- *Ion response*: Sol–gel transformation responds to ion concentration of the tear film [63, 64].

Research carried out by Al Khateb et al. investigated temperature responsive sol–gel systems for ocular drug delivery, their research was based on Pluronic F127 and Pluronic F68 and mixtures thereof, they found a 20% aqueous Pluronic F127 solution formed gel transformation at physiological temperature, the system was shown to be nonirritant with improved retention on bovine cornea. Figure 14.5 shows *in vivo* performance of fluorescein sodium loaded 20% pluronic formulations on rabbits. In Figure 14.5A, formulation retention is shown for 20% pluronic F127 (a), and 20% pluronic F68 (b). Figure 14.5B shows fluorescein sodium content of tear fluid as a function of time following ocular administration in vivo.

Gratieri et al. investigated chitosan/polaxomer drug delivery systems for ocular drug delivery of the antifungal drug, fluconazole using the porcine model. Chitosan is a biodegradable polymer with excellent ocular compatibility, presenting positive charged amine groups allowing for

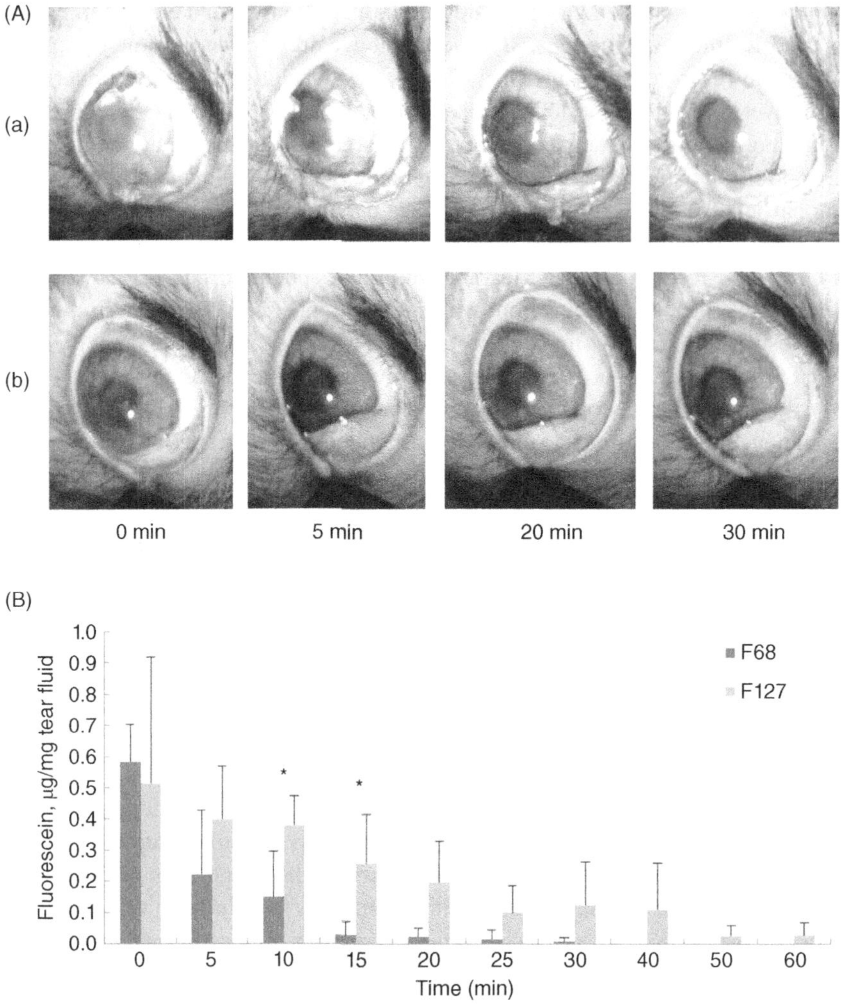

Figure 14.5 *In vivo* retention of pluronic formulations containing 1 mg/ml Fluorescein sodium solution. (A) exemplary images of rabbit eyes with 20 wt% F127 (a) and F68 (b) taken at different time intervals. Residual formulation containing fluorescein sodium is evident in Figure 14.5A (a) $t = 0-30$ minutes and (b) $t = 0$ minutes. (B) Content of fluorescein sodium in tear fluid as a function of time following ocular administration *in vivo*. (mean ± SD, $n = 4$, * represents $p < 0.05$ between groups). Source: Reproduced from Al Khateb et al. [63] with permission from International Journal of Pharmaceutics.

interaction with the negatively charged epithelium, therefore displaying good mucoadhesive properties [65, 66]. Polaxomers are thermo-responsive polymers translating to gel at physiological temperature. Their research has shown that chitosan/polaxomer drug delivery systems exhibited sustained delivery of fluconazole for treatment of fungal keratitis. Cao et al. investigated a thermosensitive sol-gel system using poly(*N*-isopropylacrylamide) with chitosan for ocular delivery of timolol maleate using *in vivo* rabbit model and they found the formulation was able to deliver a two-fold concentration of timolol maleate in the aqueous humor compared to conventional 0.5% eyedrops [67]. Mayol et al. investigated the synergistic effect of including hyaluronic acid in poloxamer formulations and they found that their platform was able to prolong and control release of acyclovir for more than six hours *in vitro*; they conclude that secondary forces such as hydrogen bonds induce pseudo plastic properties unlike Newtonian properties of the separate solutions compared to those of the poloxamer/hyaluronic acid formulation [68].

14.4.4 Penetration Enhancers

Inclusion of compounds that in some way alters corneal epithelia, thus rendering it more permeable to drug formulations is a line of attack to improve drug delivery into ocular tissues. Various preservative compounds have been included in ophthalmic medication to inhibit microbial contamination and to aid formulation stability. Ocular drug formulations are generally preserved by two categories of compounds; oxidizing agents and surface-active detergents (surfactants). Although preservatives are included to stabilize the formulation, they can have an adverse effect on ocular tissue by either causing damage to epithelial cell membranes or disrupting the tear film or protective mucin, especially when used to treat chronic conditions. Ironically, these disruptive effects can also aid drug penetration by removing protective barriers or damaging the integrity of the epithelia. Oxidizing preservatives penetrate cell lipid membranes and disrupt DNA, protein, and lipid components of bacterial cells; they also afford their action to ocular epithelial cells, although they offer less toxicity compared to surfactant preservatives. Regardless, oxidizing preservatives could still cause injury unless they are included in low concentrations as in current ocular drug formulations [69]. Benzalkonium chloride (BAC) is an example of topical formulation preservative based on its surfactant properties, a quaternary ammonium compound, and it is included in many ocular drug formulations, albeit in low concentrations in the range of 0.004–0.02%, BAC has been shown to adversely affect the cornea, conjunctiva and precorneal mucin [70]. A study by Baudouin and Lunardo investigated the effect of carteolol with and without BAC on healthy volunteers, they found the formulations where well tolerated. The preservative free group showed better tear film stability without loss of the drugs effect on intraocular pressure, tear film break up time was significantly reduced in the group tested using the formulation incorporating BAC although well tolerated in the healthy volunteers. However, they conclude that with long-term treatment in older patients effects could be more significant due to their prevalence to ocular surface disorders and tear irregularities [71]. A study by Wilson et al. investigating the effect of BAC on tear film dynamics in rabbit and man further support the negative impact on premature tear film break [72]. Kaur and Smitha report in a review that nonionic surfactants, amphipathic bile acids and salts, amphiphilic lysophosphatidilo-lipids, and glycosides are all substances with surfactant properties that are able to modify epithelial cell membranes and therefore enhance ocular drug penetration to some degree. They further emphasize that inclusion of penetration enhancers only bring limited improvement to drug bioavailability and this is improved by formulating topical ocular medication with viscosity enhancing bioadhesive polymers together with penetration enhancers, this synergistic approach promising better performance than one system alone [73]. Cyclodextrins (CDs) are water soluble cyclic oligosaccharides comprising of 6, 7, or 8 α-(1,4)-linked glucopyranose subunits. CDs are well researched for their drug solubilizing properties and penetration enhancement. They are container molecules that are able to form guest–host complexes with hydrophobic drugs [15]. Loftsson and co workers considered CDs in ocular drug formulations and report they are useful additives to enhance aqueous solubility, formulation stability, and bioavailability of ophthalmic drugs, while at the same time reduce ocular irritation. CDs are also efficient at allowing lipophilic drugs to be solubilized in aqueous media. These researchers recognized their effectiveness at enhancing drug permeation into ocular tissue, they report that CDs are known to be able to extract components such as cholesterol and phospholipids from biological membranes and state there is a positive correlation between the hemolytic activity of CDs and their ability to form complexes with cholesterol

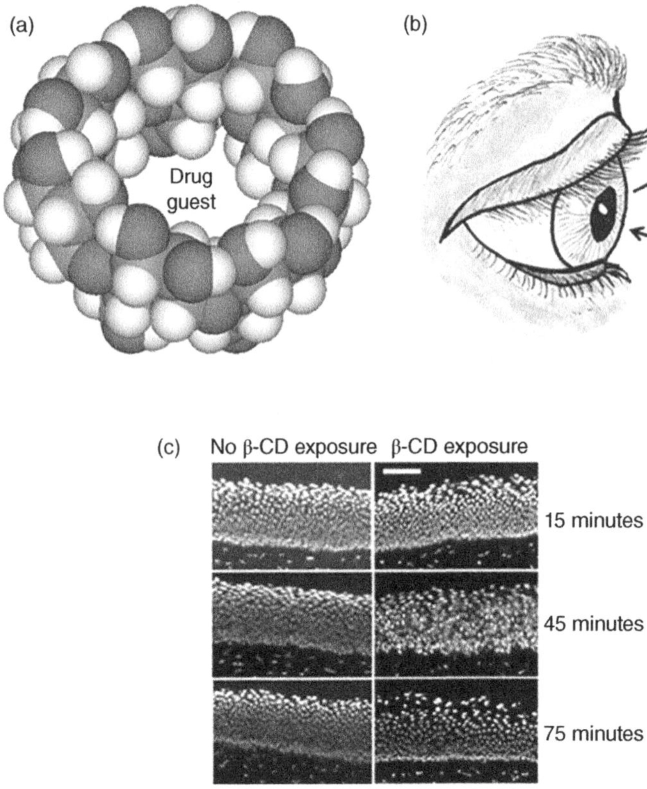

Figure 14.6 (a) Guest–host complex of β-CD/drug. (b) Preferential take up of cholesterol by β-CD. (c) β-CD mediated disruption of bovine cornea epithelium with increased time of exposure due to cholesterol extraction, scale bar = 100 μm. Source: (a, b) Based on Morrison et al. [15]. (c) Reprinted from Morrison et al. [15] with permission from American Chemical Society.

[74, 75]. However, research by Morrison et al. showed compelling evidence that β-CDs mode of action for enhancing ocular drug delivery was possible by the "triple action" brought about by enhanced drug solubility, preferential take up cholesterol, while simultaneously releasing its complexed drug payload which was then able to enter the epithelium through voids left by cholesterol extraction, Figure 14.6 [15].

β-CD has been investigated by Xu et al. as an added component to poly(2-hydroxyethyl methacrylate) (HEMA) polymers for sustained and controlled delivery of ophthalmic drugs from polymeric films and contact lenses. These materials were shown to have high loading potential of the drug puerarin due to the backbone affinity of the drug and the formation of inclusion complexes of puerarin with the β-CD component of the hydrogel. This controlled drug delivery system has been shown to deliver the drug more effectively than the same drug in eye drop formulation and the researchers conclude that this pHEMA/β-CD platform offers potential for hydrogel drug loading and sustained delivery via the ocular route [76]. Superficial corneal epithelial cells provide a very efficient barrier function due to the existence of cell binding sites to surrounding epithelial cells known as zonula occludens, below these tight junctions are further contact regions known as zonula adherens, and toward the cell base there are spot like contact zones known as desmosomes. The combination of these adhesive regions gives ocular epithelium superior barrier properties that prevent ingress of alien matter via the paracellular route. For integrity of the corneal epithelium barrier, an undetermined concentration of calcium ions are required in order to maintain tight junctions between epithelial cells. It follows that extracting calcium ions from the corneal epithelia environment is a strategy that could render the membrane more permeable to drug penetration. Calcium chelating agents are often added to ophthalmic formulations for stability purposes. Research has shown that incorporation of calcium chelating compounds in ocular drug formulations can help render corneal epithelia more permeable to applied drugs due to extraction of calcium ions which are necessary for tight junction functionality, and their inclusion in controlled drug delivery systems is a strategy to improve drug penetration into ocular tissues [16, 77]. Kaur and Smitha suggest that combining penetration enhancers such as calcium chelators with ocular bioadhesives offer a new avenue for controlled ocular drug delivery [73]. A recent study by Rodriguez et al. investigated the enhancement and inhibition effects on corneal permeability of timolol maleate using combinations of bioadhesive polymers with permeability enhancers; cyclodextrin and Ca^{2+} chelating agents. The researchers concluded that addition of permeability enhancers, i.e. ethylene glycol-bis(2-aminoethylether)-*N*,*N*,*N*′*N*′-tetraacetic acid (EGTA) and hydroxypropyl-β-cyclodextrin (HP-β-CD) enhanced the permeation of timolol maleate in the cornea using the bovine model *in vitro*, in agreement with previous reports. They also report that formulations of timolol maleate incorporating bioadhesive polymers, e.g. hyaluronic acid, chitosan, and alginate with or without inclusion of permeability enhancers reduced permeation of this drug. Further, microscopy analysis of the epithelia surface showed no evidence of epithelial disruption when exposing the cornea to combined formulations incorporating the investigated penetration enhancers and bioadhesive polymers, concluding that these polymer-enhancer interactions reduce drug solubilization and counteract any disruptive effects of permeability enhancers. Further analysis of epithelial interactions with mucoadhesive polymers showed that chitosan and hyaluronic acid formed stable bioadhesive interactions, whereas alginate gave poor mucoadhesive performance by comparison. It was concluded that chitosan–mucin interactions are due to electrostatic interactions and hydrogen bonding, while hyaluronan–mucin interactions in a stable state rely on polymer chain-interdiffusion and other non-covalent bonding interactions. Observations have shown that formulations having higher concentrations of mucoadhesive polymers offer reduced drug permeability compared to formulations without these polymers [78].

Crown ethers are cyclic molecules made up of a number of ether groups, these molecules have electron rich hydrophilic cavities and hydrophobic outer ring structures, and they are a class of container molecules that have an ability to sequester metal ions and are able to form guest-host complexes with neutral organic molecules. They have a unique property whereby the molecules can flex, allowing them to adapt to changing environments, showing solubility in aqueous and lipophilic solvents, and they have rapid and reversible ion-binding characteristics. Their ionophoric properties allow for transport across biological membranes with a range of hydrophilic and lipophilic barriers [79]. Complexes of drugs and crown ethers have been used therapeutically for tumor treatment and have been researched for drug delivery employing vesicular formulations [80–82]; however, until a recently published study, they had not been investigated for ocular drug delivery. Morrison et al. investigated crown ethers for their ability to enhance solubility and corneal permeability of ocular drugs, together with their ability to sequester Ca^{2+} ions. The researcher's *in vitro* studies showed that crown ethers were able to enhance the solubility and corneal permeability of riboflavin and had improved ability to sequester Ca^{2+} compared to a phosphate buffered saline (PBS) control solution. However, *in vivo* performance for riboflavin penetration employing the murine model proved inconclusive in enhancing corneal drug penetration; any permeation enhancement is likely to be due to crown ethers ionophoric properties as opposed to their Ca^{2+} sequestering ability. They conclude that further research is required in using crown ethers for ocular drug delivery [79].

14.4.5 Contact Lenses and Ocular Inserts

Contact lenses are optically clear polymer devices that fit directly on the cornea. Mostly, they are used to correct refractive irregularities. Generally, there are two classes of material used for making contact lenses; hard gas permeable, which are very durable and can last for many years, and more commonly, soft hydrogel, which tends to be more comfortable to wear. Hydrogel materials are soft polymers due to swelling upon taking up water, materials are available that can consist of up to 80% water while still having sufficient integrity to keep their form. This makes them ideal materials for taking in aqueous drug solutions and delivering this in a controlled manner just where the medication is needed, directly to the eye, providing a sustained low dose steadily diffusing from the polymer matrix to the tear film without the intrinsic overdose given by eye drop formulations. Any drug released in the pre-lens tear film will quickly be lost much like traditional eye drops. However, drug released into the post-lens tear film becomes entrapped between the lens and cornea, therefore in contact with the cornea for an extended time, allowing more bioavailability of the drug (Figure 14.7) [2]

Wichterle and Lim proposed using hydrogel contact lenses for drug delivery in their 1965 patent; they suggested hydrating the lenses in drug solutions prior to wear to offer extended drug availability compared to conventional delivery by eye drops [83]. Simply hydrating hydrogel lenses in aqueous drug solution is not a very efficient way to deliver drugs, there tends to be an initial burst release followed by a diminishing release over a short time, this is not ideal, but it is still an improvement over eye drop delivery. Further improvements from contact lens drug delivery were shown in a method reported by Gulsen and Chauhan; their technique was a pilot study which involved incorporating drug loaded silica shell nanoparticles into a p(HEMA) hydrogel during its formulation. This material is suitable for making contact lenses. They were able to show sustained drug release for several days from the hydrogel film, they anticipate a similar release profile from contact

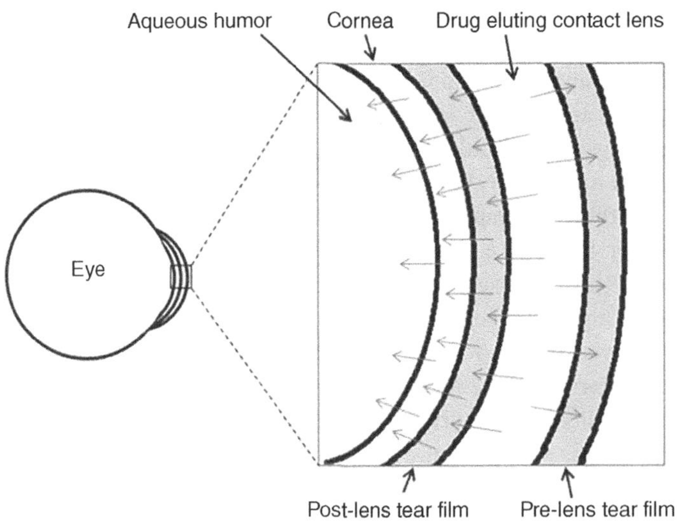

Figure 14.7 Scheme for ocular drug delivery via drug eluting contact lenses. Source: Adapted from Morrison et al. [2].

lenses made from the same hydrogel [84]. A study by Peng et al. investigated contact lens drug delivery of timolol from commercially available daily wear lenses (Acuvue® TrueEye™) *in vitro* and *in vivo*. They found that by loading vitamin E into the lenses prior to drug loading by soaking in a timolol maleate solution they were able to achieve the same level of reduction in intraocular pressure as from standard eye drops but with only 20% of drug loading. This would significantly reduce systemic side effect seen when using timolol eye drops [85]. Further advances have been made with molecular imprinting technology, this technique gives the polymer matrix a high affinity for specific drugs. In use, drug loaded molecular imprinted contact lenses allow for sustained release of their drug payload. Hiratani and Alvarez-Lorenzo developed a molecular imprinted system based on methacrylic acid and *N*,*N*,-diethylacrylamide for sustained delivery of timolol and they achieved sustained delivery over 48 hours *in vitro* [86]. Other researchers are following the same route using molecular imprinting techniques with similar success, it is proving to be a suitable approach to improve controlled drug delivery using hydrogel contact lenses [87–90].

Ocular inserts deliver drugs in a similar way to drug eluting contact lenses but instead of being placed on the cornea, they are positioned in the *cul-de-sac* on the conjunctiva between the eyelids and the sclera. Their purpose is to provide a drug reservoir from where they deliver drugs into the tear film, much of which will become available to the conjunctiva, sclera, and cornea. Drug release can be from a reservoir entrapped between semipermeable membranes, soluble or erodible matrixes. Insoluble devices are designed to be put in place for a specific amount of time, after which they are removed and discarded [2]. Seattone and Salminen summarized essential requirements for successful use of ocular insert devices [91], and these have been further evaluated here:

- Ophthalmologist and patient acceptance.
- Ease of use, insertion, and comfort.
- Once inserted they must stay in place for the intended duration.
- Must be nontoxic and nonirritating.
- Must not interfere with vision or oxygen availability to the eye.
- Reliable drug release kinetics.
- Suitable for a range of ocular drugs.
- Sterile and stable.
- Ease of manufacture and cost effective.
- Effective over sufficient time to offer desired treatment.

Ocular inserts are devices designed to provide prolonged and sustained drug delivery at a controlled rate. In the preferred scenario, these devices could be put in place and then release their drug payload by dissolution or erosion with no need for further intervention; however, many types of ocular insert devices are non-soluble drug delivery vehicles designed to offer precise controlled delivery over a specific period of time, the aim to extend drug bioavailability, and then they are removed once this is achieved [92]. Ocusert® was the first product to achieve acceptance for sustained and controlled pilocarpine delivery in treatment for ocular hypertension from such devices. Ocusert devices have their pilocarpine-alginate reservoir entrapped between semipermeable ethylene-vinyl acetate films designed to deliver pilocarpine at 20 or 40 μg/h [91, 93]. Ocufit SR® are rod-shaped silicone elastomeric sustained release devices that are placed in the conjunctival fornix of the lower eyelid. These devices are generally tolerated well by patients and have proved successful in controlled drug delivery. Antibiotic drugs, for example gentamicin and sulfisoxazole have been successfully delivered using the Minidisc system produced by Bausch & Lomb. These devices are drug loaded polymeric disc shaped devices ~4–5 mm diameter, and they can be erodible or non-erodible devices which are able to provide sustained drug delivery between

3 and 14 days dependant on which system is used, they are placed on the sclera in the upper or lower fornix and are well tolerated by the patient. New ophthalmic delivery system (NODS®) has been patented by Smith and Nephew Pharmaceutical Ltd to offer precise pilocarpine delivery from poly(vinyl alcohol) flags for treatment of glaucoma. The devices are designed to attach to ocular mucosa of the lower conjunctival *cul-de-sac*, fluid take up from the tear fluid allows sustained release into lacrimal fluid as the flag dissolves [91]. Hornof et al. report the benefits of an ocular insert drug delivery system based on a mucoadhesive thiolated poly(acrylic acid) designed for overnight delivery from the device placed in the lower conjunctival *cul-de-sac* whereupon the device swells by moisture uptake from the tear film. From this study *in vivo* it was shown that sustained delivery of fluorescein as a model drug was achieved for over eight hours compared with instillation of the same via eye drops which was significantly less efficacious. The researchers also report that the devices were well tolerated by volunteers employed for this study [94].

To date, ocular inserts have been poorly accepted by patients and practitioners in comparison to traditional liquid/sol–gel dose forms. However, the potential to offer sustained and controlled drug delivery by ocular inserts encourages much research in this area [95].

14.4.6 Intraocular Systems (Implants, Injectables, and Degradable Microparticles)

Due to effective barriers challenging drug penetration to the deeper structures of the eye, one of the most effective methods to get drugs into the eye involves intravitreal injection, albeit undesired by the patient and not without risk of hemorrhage or complications. The intravitreal route offers a very effective means to put drugs just where they are required, within the posterior segment [96]. Various conditions predominant from choroidal or retinal neovascularization, often leading to vascular leakage into the eye such as diabetic retinopathy and age-related macular degeneration are major diseases leading to severe vision loss and blindness in an aging population. Vascular endothelial growth factor (VEGF) is implicated in abnormal blood vessel formation, together with complications arising from vascular permeability (leaky blood vessels). The above mentioned disorders are potentially treated by regular intravitreal injections of "vascular endothelial growth factor" (VEGF) antagonisms at four to eight weeks intervals, examples of these drugs are ranibizumab (Lucentis), sodium pegaptanib (Macugen), aflibercept (Eylea), and bevacizumab (Avastin), these drugs are effective at inhibiting angiogenesis. This class of drugs decrease abnormal blood vessel formation associated with neovascular diseases. Although intravitreal injections can be an unpleasant experience, patients benefit from improved quality of life by such procedures [34, 97, 98]. In a safety evaluation of ocular drug delivery formulations, Short highlights sites and methods for intraocular drug delivery, and these include [99]:

- *Scleral plug*: A device that penetrates the sclera at the *Pars Plana* where it releases its payload in a controlled manner over an extended period of time.
- *Subconjunctival implant*: Inserted between the conjunctiva and sclera at sites away from the *Pars Plana*, offers controlled and sustained drug delivery.
- *Supra choroidal implant*: A device placed between the sclera and choroid, capable of sustained drug delivery.
- *Suprascleral injection*: An injection under the sclera, suitable for medium term therapy.
- *Intravitreal implant*: A sustained drug delivery device intruding directly into the vitreous humor for long term controlled drug delivery.
- *Intravitreal injection*: Drug injection directly into the vitreous humor. Repeat injections are often required.

Localized controlled and sustained drug delivery to the posterior segment is the subject of much research and drug delivery systems development employing implantable devices. Systems include reservoir implants, examples include Vitrasert, first approved in 1966 as a treatment of AIDS-related cytomegalovirus retinitis, implanted through a 5.5 mm *pars plana* incision, the device proved successful in providing sustained delivery of ganciclovir for up to eight months. The next generation reservoir implant system improving on the Vitrasert device came the Retisert, this device is implanted through a 3.5 mm *pars plana* incision and delivers fluocinolone acetonide for up to 30 months [100]. Implant technology discussed so far related to nondegradable reservoir devices. Biodegradable implantable devices offer an advantage over the former, they stay in place and slowly become absorbed, metabolized or expelled from the site of implantation as they deliver their drug payload, eliminating the need to have the device remain in the eye or require surgical removal once depleted. Biodegradable implantable devices employ biocompatible polymers, for example polyglycolic acid and polylactic acid that degrade into nontoxic by products within the body, that are then eliminated by natural metabolic processes. Drug delivery systems produced from these polymers allow for controlled and sustained drug delivery which can be tuned by adjusted the co-polymer matrix employed. Sustained drug release from biodegradable polymeric systems is a multifactorial process which follows a pattern of distinct phases; hydration – burst release – diffusion and chain scission – degradation and mass loss – fragmentation [101]. Rodrigues et al. further reports the benefits of using biodegradable drug delivery implantable devices over nondegradable reservoir devices include the fact that no surgical intervention is required following implantation, and this positively impacts on patient compliance, they further consider that issues of long term sustained drug delivery require further refinement [46].

14.4.7 Phonophoresis and Ionophoresis

Therapeutic phonophoresis (ultrasound mediated drug delivery) applied for transdermal application at 1–3 MHz and up to 2 W/cm^2, enhances drug delivery almost 10-fold. Mitragotri et al. have shown that adjusting phonophoresis power to 20 kHz, 125 mW/cm^2 in 100 ms pulses enhanced transdermal drug delivery of a range of drugs up to 1000-fold compared to therapeutic phonophoresis [102]. From this knowledge, Zderic et al. investigated the employment of phonophoresis for trans-corneal drug delivery. Their preliminary findings realized a fourfold enhancement for a range of β-blocker drugs using the rabbit model *in vitro*; however, they found this induced significant disruption to the epithelium and speculated this would be slow to heal (~5 days), with potential for complications. Further studies by the same team have confirmed that low power ultrasound enhanced drug delivery using the rabbit corneal model *in vitro* enhances drug penetration fourfold and it is speculated that this is due to reversible disruption of the superficial epithelial cell layers where tight junctions of non-disrupted cells maintain barrier function [48, 103].

Electromotive drug administration (iontophoresis) is a means to drive ionized (electrically charged) medication across biological membranes using an electric current as the driving force. This method has been used as early as the 1700s; however, it was not until the early 1900s before research in this field really started making progress [104]. More recently, iontophoresis has been the subject of research for ocular drug delivery via transcorneal and transscleral routes with relative success [30]. Hughes and Maurice reported a 100-fold increase in penetration of fluorescein and gentamicin into the aqueous chamber of rabbits using 1 mA. This method offered more controlled drug delivery compared to regular instillation. They conclude that a 30 seconds application at this power is sufficient to provide a therapeutic antibiotic dose in the anterior chamber for a 12 hours period [105]. Das and Suresh report a renewed interest in developing technology and hardware associated with ophthalmological iontophoresis, and this could potentially lead to further advances in this field [106]. Iontophoresis has a proven history for enhanced delivery of a range of drugs and these include fluorescein, antibiotics, antivirals, steroids, and macromolecular compounds, especially via the transscleral route. Generally regarded as a noninvasive procedure relying on a small amount of electrical energy as the driving force to push medication across otherwise relatively impervious membranes. Trans-scleral drug delivery to the choroid and retina has been shown to be significantly enhanced using iontophoresis with few side effects [107, 108]. However, drug delivery into the vitreous humor using the rabbit model has proved to be less promising with drug delivery only localized near the site of penetration [109]. Thrimawithana et al. considered potential side effects from iontophoresis, namely, epithelial edema, reduction in endothelial barrier function, inflammation, and burns. Higher current density potentially leads to choroidal and retinal damage. They further report that research has led to the development of improved iontophoresis devices which are easier to use and better tolerated by the patient [110]. For the current state of the art, it can be concluded that therapeutic phonophoresis and iontophoresis for ocular drug delivery is in the developmental stage of research where much refinement is required; however, this technique offers future potential.

14.4.8 Topical Prodrugs

Prodrugs can be described as pharmaceutically inactive derivatives of the active molecule. Prodrugs are often lipophilic, which can assist their delivery into epithelial tissue. Once they are delivered to the target, they are transformed into the pharmaceutically active form within the body, usually by action of enzymes or hydrolysis [7, 111]. Prodrugs provide a suitable approach in improving drug delivery and have proved useful in the delivery of ocular drugs, where they are converted into the active form by hydrolysis within the eye [112]. Prodrugs have been researched for ophthalmological use since the early 1970s. Desirable attributes for ocular prodrugs include good stability, solubility/lipophilic balance sufficient to allow aqueous solubility while retaining good corneal penetration, they must not induce irritation and they should release the active drug form at a therapeutic concentration [7]. Ali and Lehmussaari state further benefits when using prodrugs include enhanced bioavailability and potency, lower dose required, prolonged duration of action, and a reduction in side effects [113]. Prostaglandins are lipids with hormone like characteristics, their prodrug analogues; bimatoprost, latanoprost, and travoprost are effective, well-tolerated medications for topical drug delivery for controlling ocular hypertension and glaucoma [114]. Tirucherai and Mitra investigated the effect of HP-β-CD complexation on factors such as aqueous solubility, stability, and permeation through the cornea of Ganciclovir, a group of acyl ester prodrugs with antiviral activity. They determined that cyclodextrin complexation delivered increased solubility and stability of ganciclovir, although the corneal transport benefits are only realized for variants which are very lipophilic [115].

14.4.9 Microneedle Systems

Microneedle technology has been researched during the last 15 years for transdermal drug delivery and they can be in the form of; dissolving polymer, solid and hollow arrays and their mode of action dissolution of soluble polymer devices, dissolution of drug coating from solid microneedles and diffusion or injected drug delivery from hollow microneedles [116–118]. More recent research has followed from this technology in the use of microneedles for ocular intrascleral and intracorneal drug delivery. Although proof of concept has been successful, to date there are no devices commercially available employing drug coated and hollow microneedles. Shortcomings have been realized due to difficulties in fabricating devices and inconsistent drug loading/delivery [119]. Research by Singh et al. investigated the development of rapid dissolving drug loaded microneedles for ocular drug delivery to the anterior and posterior segment employing polyvinylpyrrolidone, a Food and Drug Administration (FDA) approved biocompatible polymer. They determined the mechanical properties, ability to penetrate the cornea and sclera, kinetics of dissolution and their ability to deliver a range of model drugs *in vitro*. Their study led to the conclusion that it was feasible to fabricate drug loaded microneedle arrays which rapidly dissolved, leaving deposits of model drug within scleral and corneal tissues from where the drugs were able to diffuse into the deeper ocular structures. The researcher's further report that the drug delivery platform has been shown to offer a suitable ocular drug delivery technique and realize follow on investigations are required to further develop from this their study [120].

14.5 Conclusions

The eye is easily accessible for instilling medication and one could consider that ocular disorders can be effectively treated by simply applying eyedrops, after all ~90% of topically applied medications are applied in this form. However, it is not quite that simple. Membrane barrier function resists penetration of alien material, and clearance mechanisms rapidly wash out material that challenges these barriers. With traditional eye drops, an intrinsic overdose is given, with up to 90% of the instilled dose lost within the first minute or so.

Superficial disorders for example conjunctivitis and blepharitis are more easily treated than conditions requiring medication to penetrate into the eye.

Decades of research and development have allowed many of the barriers to ocular drug delivery to be minimized. Strategies have been developed to overcome these challenges by improving ophthalmic drug retention and enhancing corneal and scleral permeation of drugs together with the development of controlled and sustained release systems. Bioavailability can be improved by enhancing drug solubility, allowing a higher drug loading in a formulation. Medication retention can be gained by inclusion of viscosity enhancing and mucoadhesive agents. Inclusion of penetration enhancers in a formulation, for example CDs, metal ion chelators, and surfactants can aid drug transport across ocular membranes. Nanotechnology employs microparticle and nanoparticle drug delivery vehicles allowing drugs to attach or penetrate the eye, developed in the form of soluble, erodible/non-erodible solid microscopic units, microemulsions, liposomes, micelles, and niosomes. Prodrugs offer a means to treat a number of superficial and intraocular disorders, they are bio-reversible analogues to therapeutic agents only becoming active when metabolized within the target tissue. Polymer films, inserts, and contact lenses offer improved bioavailability by providing a low dose, sustained delivery on the surface of the eye from where penetration into the cornea, sclera, and conjunctiva is possible.

Precision drug delivery can be achieved using intravitreal injection, drug depot implants, degradable, and injectable microparticles. This strategy places medication or drug depot devices just where they are needed, near the target tissue, virtually eliminating systemic complications. In some cases, controlled and sustained release can be achieved for several years using slow release implantable drug depots, a more desirable option than intraocular injections repeated every few months.

Future developments are likely to be achieved from current research into strategies aimed at getting drugs into the eye using minimally invasive techniques, for example phonopheresis, electrophoresis, and microneedles. These technologies rely on a temporary disruption to penetration resistant membranes. They are still in the development stages; however, drawbacks from these strategies are being overcome, and it is likely that these methods will be commercialized in the near future.

References

1 Hagen, S. The minds eye. *Rochester Review*. http://www.rochester.edu/pr/Review/V74N4/0402_brainscience.html (accessed 22 November 2016).

2 Morrison, P.W.J. and Khutoryanskiy, V.V. (2014). Advances in ophthalmic drug delivery. *Ther. Delivery* 5 (12): 1297–1315.

3 Morrison, P.W.J. and Khutoryanskiy, V.V. (2014). Anatomy of the eye and the role of ocular mucosa in drug delivery. In: *Mucoadhesive Materials and Drug Delivery Systems*, 1e (ed. V.V. Khutoryanskiy), 39–60. Wiley.

4 Gilhotra, R.M., Nagpal, K., and Mishra, D.N. (2011). Azithromycin novel drug delivery system for ocular application. *Int. J. Pharm. Invest.* 1: 22–28.

5 Kumaran, K.S.G.A., Karthika, K., and Padmapreetha, J. (2010). Comparative review on conventional and advanced ocular drug delivery formulations. *Int. J. Pharm. Pharm. Sci.* 2 (4): 1–5.

6 Washington, N., Washington, C., and Wilson, C.G. (2001). Ocular drug delivery. In: *Physiological Pharmaceutics: Barriers to Drug Absorption*, 2e, 249–269. New York, NY: Taylor and Francis, Inc.

7 Jarvinen, T. and Jarvinen, K. (1996). Prodrugs for improved ocular drug delivery. *Adv. Drug Delivery Rev.* 19 (2): 203–224.

8 Wilson, C.G., Zhu, Y.P., Kumala, P. et al. (2001). Ophthalmic drug delivery. In: *Drug Delivery and Targeting for Pharmacists and Pharmaceutical Scientists* (eds. A.M. Hillery, A.W. Lloyd and J. Swarbrick), 329–353. Taylor Francis, London.

9 Davies, N.M. (2000). Biopharmaceutical considerations in topical ocular drug delivery. *Clin. Exp. Pharmacol. Physiol.* 27 (7): 558–562.

10 BioGraphix. (2006). http://www.biographixmedia.com/human/eye-anatomy.html (accessed 17 December 2013).

11 Wilson, C.G., Semenova, E.M., Hughes, P.M., and Olenik, O. (2007). Eye structure and physiological functions. In: *Enhancement in Drug Delivery* (eds. E. Touitou and B.W. Barry), 473–487. CRC Press, Florida.

12 Kenchegowda, S. and Bazan, H.E.P. (2010). Significance of lipid mediators in corneal injury and repair. *J. Lipid Res.* 51 (5): 879–891.

13 Wang, L., Li, T., and Lu, L. (2003). UV-induced corneal epithelial cell death by activation of potassium channels. *Invest. Ophthalmol. Visual Sci.* 44 (12): 5095–5101.

14 Dua, H.S., Faraj, L.A., Said, D.G. et al. (2013). Human corneal anatomy redefined: a novel pre-descemet's layer (Dua's layer). *Ophthalmolology* 120 (9): 1778–1785.

15 Morrison, P.W.J., Connon, C.J., and Khutoryanskiy, V.V. (2013). Cyclodextrin-mediated enhancement of riboflavin solubility and corneal permeability. *Mol. Pharm.* 10 (2): 756–762.

16 Morrison, P.W.J. and Khutoryanskiy, V.V. (2014). Enhancement in corneal permeability of riboflavin using calcium sequestering compounds. *Int. J. Pharm.* 472 (1–2): 56–64.

17 Llobet, A., Gasull, X., and Gual, A. (2003). Understanding trabecular meshwork physiology: a key to the control of intraocular pressure. *News Physiol. Sci.* 18 (5): 205–209.

18 Specsavers Opticians, 2–4 Bridge Street, Saint Helens, Merseyside, UK.

19 Ress, D. and Heeger, D.J. (2003). Neuronal correlates of perception in early visual cortex. *Nat. Neurosci.* 6 (4): 414–420.

20 Atwal, E.S. (2015). Pars Plana approach can be used for Surgical Injection Technique. Ocular Surgery News US edition. http://www.healio.com/ophthalmology/cataract-surgery/news/print/ocular-surgery-news/%7Bc1d500b1-71a2-44e7-ac97-b7f9b3d52bb4%7D/pars-plana-approach-can-be-used-for-surgical-injection-technique (accessed 4 February 2017).

21 Abdulrazik, M., Beher-Cohen, F., and Benita, S. (2007). Drug delivery systems for enhanced ocular absorption. In: *Enhancement in Drug Delivery* (eds. E. Touitou and B.W. Barry). 489–526. CRC Press. Florida.

22 Mun, E., Morrison, P.W.J., and Khutoryanskiy, V.V. (2014). On the barrier functions of the cornea: a microscopic study of the penetration of fluorescently-labelled nanoparticles, polymers and sodium fluorescein. *Mol. Pharm.* 11 (10): 3556–3564.

23 Prevention of Blindness and Visual Impairment. (2017). World Health Organisation. http://www.who.int/blindness/history/en (accessed 16 January 2017).

24 Kaur, I.P. and Batra, A. (2007). Ocular penetration enhancers. In: *Enhancement in Drug Delivery* (eds. E. Touitou and B.W. Barry), 527–547. CRC Press, Florida.

25 Wright, P. and Mackie, I.A. (1977). Mucus in the healthy and diseased eye. *Trans. Ophthalmol. Soc. U.K.* 91 (1): 1–7.

26 Kreuter, J. (1990, 1990). Nanoparticles as bioadhesive ocular drug delivery systems. In: *Bioadhesive Drug Delivery Systems* (eds. V. Lenaerts and R. Gurny). 203–211. CRC Press, Florida.

27 Marriott, C. and Gregory, N.P. (1990). Mucus physiology and pathology. In: *Bioadhesive Drug Delivery Systems* (eds. V. Lenaerts and R. Gurny). 1–24. CRC Press, Florida.

28 Spector, A. (1984). The search for solution to senile cataracts, proctor lecture. *Invest. Ophthalmol. Visual Sci.* 25 (2): 130–146.

29 Sultana, Y., Jain, R., Aqil, M., and Ali, A. (2006). Review of ocular drug delivery. *Curr. Drug Delivery* 3 (2): 207–217.

30 Guadana, R., Krishna, H., Parenky, A., and Mitra, A.K. (2010). Ocular drug delivery. *AAPS J.* 12 (3): 348–360.

31 Wollensak, G., Spoerl, E., and Seiler, T. (2003). Riboflavin/ultraviolet-A-induced collagen crosslinking for the treatment of keratoconus. *Am. J. Ophthalmol.* 135 (5): 620–627.

32 Bourlais, C.L., Acer, L., Zia, H. et al. (1998). Ophthalmic drug delivery systems – recent advances. *Prog. Retinal Eye Res.* 17 (1): 33–58.

33 Hartong, D.T., Berson, E., and Dryja, T.P. (2006). Retinitis pigmentosa. *Lancet* 368: 1795–1809.

34 Ferrara, N., Damico, L., Shams, N. et al. (2006). Development of ranibizumab, an anti-vascular endothelial growth factor antigen binding fragment, as therapy for neovascular age-related macular degeneration. *Retina* 26 (8): 850–870.

35 Edwards, A.O., Ritter, R., Abel, K.J. et al. (2005). Complement factor H polymorphism and age-related macular degeneration. *Science* 308: 421–424.

36 Sivaprasad, S., McCluskey, P., and Lightman, S. (2006). Intravitreal steroids in the management of macula oedema. *Acta Ophthalmol. Scand.* 84: 722–733.

37 Haller, J.A., Kuppermann, B.D., Blumenkranz, M.S. et al. (2010). Randomized controlled trial of an intravitreous dexamethasone drug delivery system in patients with diabetic macular edema. *Arch. Ophthalmol.* 128 (3): 289–296.

38 Vadlapudi, A.D. and Mitra, A.K. (2013). Nanomicelles: an emerging platform for drug delivery to the eye. *Ther. Delivery* 4 (1): 1–3.

39 Routio, J., Kumpulainen, H., Heimbach, T. et al. (2008). Prodrugs: design and clinical applications. *Nat. Rev. Drug Discovery* 7: 255–270.

40 Latanoprost. https://www.drugbank.ca/structures/DB00654/image.svg (accessed 2 November 2017).

41 Russo, A., Riva, I., Pizzolante, T. et al. (2008). Latanoprost ophthalmic solution in a treatment of open angle glaucoma or raised intraocular pressure: a review. *Clin. Ophthalmol.* 2 (4): 897–905.

42 Choy, Y.B., Park, J.H., and Prausnitz, M.R. (2008). Mucoadhesive microparticles engineered for ophthalmic drug delivery. *J. Phys. Chem. Solids* 69 (5–6): 1533–1536.

43 Xu, Q., Kambhampati, S.P., and Kannan, R.M. (2013). Nanotechnology approaches for ocular drug delivery. *Middle East Afr. J. Ophthalmol.* 20 (1): 26–37.

44 Yasukawa, T., Tabata, Y., Kimura, H., and Ogura, Y. (2011). Recent advances in intraocular drug delivery systems. *Recent Pat. Drug Delivery Formulation* 5 (1): 1–10.

45 Shelke, N.B., Kadam, R., Tyagi, P. et al. (2011). Intravitreal poly(L-lactide) microparticles sustain retinal and choroidal delivery of TG-0054, a hydrophilic drug intended for neovascular diseases. *Drug Delivery Transl. Res.* 1 (1): 76–90.

46 Rodrigues da Silva, G., Fialho, S.L., Siqueira, R.C. et al. (2010). Implants as drug delivery devices for the treatment of eye diseases. *Braz. J. Pharm. Sci.* 46 (3): 585–595.

47 Khutoryanskaya, O.V., Morrison, P.W.J., Seilkhanov, S.K. et al. (2014). Hydrogen-bonded complexes and blends of poly(acrylic acid) and methyl cellulose: nanoparticles and mucoadhesive films for ocular delivery of riboflavin. *Macromol. Biosci.* 14 (2): 225–234.

48 Zderic, V., Clark, J.I., Martin, R.W., and Vaezy, S. (2004). Ultrasound-enhanced transcorneal drug delivery. *Cornea* 23 (8): 804–811.

49 Gaudana, R., Ananthula, H.K., Parenky, A., and Mitra, A.K. (2010). Ocular drug delivery. *Am. Assoc. Pharm. Sci.* 12 (3): 348–360.

50 Kumar, A., Malviya, R., and Sharma, P.K. (2011). Recent trends in ocular drug delivery: a short review. *Eur. J. Appl. Sci.* 3 (3): 86–92.

51 Klang, S., Abdulrazik, M., and Benita, S. (2000). Influence of emulsion droplet surface charge on indomethacin ocular tissue distribution. *Pharm. Dev. Technol.* 5 (4): 521–532.

52 Gupta, S. and Moulik, S.P. (2008). Biocompatible microemulsions and their prospective uses in drug delivery. *J. Pharm. Sci.* 97 (1): 22–45.

53 Czajkowska-Kosnik, A. and Sznitowska, M. (2009). Solubility of ocular therapeutic agents in self-emulsifying oils I. Self-emulsifying oils for ocular drug delivery: solubility of indomethacin, aciclovir and hydrocortisone. *Acta Poloniae Pharm.* 66 (6): 709–713.

54 Sasaki, H., Yamamura, K., Mukai, T. et al. (1999). Enhancement in ocular drug penetration. *Crit. Rev. Ther. Drug Carrier Syst.* 16 (1): 85–146.

55 Portes, A.J.F., de Silva, B.D.D., Vieira, L.B.R. et al. (2013). Perception of topical ocular drug delivery: comparison between eyedrop instillation in open eyes and vaporisation in closed eyes. *Rev. Bras. Oftalmol.* 73 (2): 98–102.

56 Shaikh, R., Singh, T.R.R., Garland, M.J. et al. (2011). Mucoadhesive drug delivery systems. *J. Pharm. BioAllied Sci.* 3 (1): 89–100.

57 Nazar, H., Roldo, M., Fatouros, D.G. et al. (2012). Hydrogela in mucosal delivery. *Ther. Delivery* 3 (4): 535–555.

58 Khutoryanskiy, V.V. (2007). Hydrogen-bonded interpolymer complexes as materials for pharmaceutical applications. *Int. J. Pharm.* 334 (1–2): 15–26.

59 Liu, S., Dozois, M.D., Chang, C.N. et al. (2016). Prolonged ocular retention of mucoadhesive nanoparticle eye drop formulation enables treatment of eye diseases using significantly reduced dosage. *Mol. Pharm.* 13 (9): 2897–2905.

60 Tonglairoum, P., Brannigan, R.P., Opanasopit, P., and Khutoryanskiy, V.V. (2016). Maleimide-bearing nanogels as novel materials for drug delivery. *J. Mater. Chem. B* 4: 6581–6587.

61 Irmukhametova, G.S., Mun, G.A., and Khutoryanskiy, V.V. (2011). Thiolated mucoadhesive and PEGylated nonmucoadhesive organosilica nanoparticles from 3-mercaptopropyltrimethoxysilane. *Langmuir* 27 (15): 9551–9556.

62 Jeong, B., Kim, S.W., and Bae, Y.H. (2012). Thermosensitive sol-gel reversible hydrogels. *Adv. Drug Delivery Rev.* 64: 154–162.

63 Al Khateb, K.A., Ozmukhametova, E.K., Mussin, M.N. et al. (2016). In situ gelling systems based on Pluronic F127/Pluronic F68 formulations for ocular drug delivery. *Int. J. Pharm.* 502 (1–2): 70–79.

64 Rupenthal, I.D., Alany, R.G., and Green, C.R. (2011). Ion-activated *in situ* gelling systems for antisense oligodeoxynucleotide delivery to the ocular surface. *Mol. Pharm.* 8 (6): 2282–2290.

65 Gratieri, T., Gelfuso, G.M., Rocha, E.M. et al. (2010). A poloxamer / chitosan *in situ* forming gel with prolonged retention time for ocular delivery. *Eur. J. Pharm. Biopharm.* 75 (2): 186–193.

66 Gratieri, T., Gelfuso, G.M., de Freitas, O. et al. (2011). Enhancing and sustaining the topical ocular delivery of fluconazole using chitosan solution and poloxamer/chitosan *in situ* forming gel. *Eur. J. Pharm. Biopharm.* 79 (2): 320–327.

67 Cao, Y., Zhang, C., Shen, W. et al. (2007). Poly(*N*-isopropylacrylamide)-chitosan as thermosensitive *in situ* gel-forming system for ocular drug delivery. *J. Controlled Release* 120 (3): 186–194.

68 Mayol, L., Quaglia, F., Borzacchiello, A. et al. (2008). A novel poloxamer/hyaluronic acid *in situ* forming hydrogel for drug delivery: rheological, mucoadhesive and *in vitro* release properties. *Eur. J. Pharm. Biopharm.* 70 (1): 199–206.

69 Freeman, P.D. and Kahook, M.Y. (2009). Preservatives in topical ophthalmic medications: historical and clinical perspectives. *Expert Rev. Ophthalmol.* 4 (1): 59–64.

70 Chung, S.H., Lee, S.K., Cristol, S.M. et al. (2006). Impact of short-term exposure of comercial eyedrops preserved with benzalonium chloride on precorneal mucin. *Mol. Vision* 12: 415–421.

71 Baudouin, C. and de Lunardo, C. (2013). Short term Comparitive study of topical 2% carteolol with and without benzalkonium chloride in healthy volunteers. *Br. J. Ophthalmol.* 82 (1): 39–42.

72 Wilson, W.S., Duncan, A.J., and Jay, J.L. (2013). Effect of benzalkonium chloride on the stability of the precorneal tear film in rabbit and man. *Br. J. Ophthalmol.* 59 (11): 667–669.

73 Kaur, I.P. and Smitha, R. (2002). Penetration enhancers and ocular bioadhesives: two new avenues for ophthalmic drug delivery. *Drug Dev. Ind. Pharm.* 28 (4): 353–369.

74 Loftsson, T. and Jarvinen, T. (1999). Cyclodextrins in ophthalmic drug delivery. *Adv. Drug Delivery Rev.* 36 (1): 59–79.

75 Loftsson, T. and Stefansson, E. (2002). Cyclodextrins in eye drop formulations: enhanced topical delivery of corticosteroids to the eye. *Acta Ophthalmol. Scand.* 80 (2): 144–150.

76 Xu, J., Li, X., and Sun, F. (2010). Cyclodextrin-containing hydrogels for contact lenses as a platform for drug incorporation and release. *Acta Biomater.* 6 (2): 486–493.

77 Grass, G.M., Wood, R.W., and Robinson, J.R. (1985). Effects of calcium chelating agents on corneal permeability. *Invest. Ophthalmol. Visual Sci.* 26 (1): 110–113.

78 Rodriguez, I., Vazquez, J.A., Pastrana, L., and Khutoryanskiy, V.V. (2017). Enhancement ad inhibition effects on the corneal permeability of timolol maleate: polymers, cyclodextrins and chelating agents. *Int. J. Pharm.* 529 (1–2): 168–177.

79 Morrison, P.W.J., Porfiryeva, N.N., Chahal, S. et al. (2017). Crown ethers: novel permeability enhancers for ocular drug delivery. *Mol. Pharm.* 14 (10): 3528–3538. https://doi.org/10.1021/acs.molpharmceut.7b00556.

80 Marjanović, M., Kralj, M., Supek, F. et al. (2017). Antitumor potential of crown ethers: structure-activity relationships, cell cycle disturbances, and cell death studies of a series of ionophores. *J. Med. Chem.* 50 (5): 1007–1018.

81 Muzzalupo, R., Nicoletta, F.P., Trombino, S. et al. (2007). A new crown ether as vesicular carrier for 5-fluoruacil: synthesis, characterization and drug delivery evaluation. *Colloids Surf., B* 58 (2): 197–202.

82 Darwish, I.A. and Uchegbu, I.F. (1997). The evaluation of crown ether based niosomes as cation containing and cation sensitive drug delivery systems. *Int. J. Pharm.* 159 (2): 207–213.

83 Wichterle, O. and Lim, D. (1965). Cross-linked hydrophilic polymers and articles made therefrom. Patent: 986 US 3220960.

84 Gulsen, D. and Chauhan, A. (2004). Ophthalmic drug delivery through contact lenses. *Invest. Ophthalmol. Visual Sci.* 45 (7): 2342–2347.

85 Peng, C.C., Burke, M.T., Carbia, B.E. et al. (2012). Extended drug delivery by contact lenses for glaucoma therapy. *J. Controlled Release* 162 (1): 152–158.

86 Hiratani, H. and Alvarez-Lorenzo, C. (2006). Timolol uptake and release by imprinted soft contact lenses made of *N,N*-diethylacrylamide and methacrylic acid. *J. Controlled Release* 83 (2): 223–230.

87 White, C.J. and Byrne, M.E. (2010). Molecularly imprinted therapeutic contact lenses. *Expert Opin. Drug Delivery Rev.* 7 (6): 765–780.

88 Salian, V.D., Vaughan, A.D., and Byrne, M.E. (2012). The role of living/controlled radical polymerization in the formation of improved imprinted polymers. *J. Mol. Recognit.* 25 (6): 361–369.

89 Byrne, M.E., Park, K., and Peppas, N.A. (2002). Molecular imprinting within hydrogels. *Adv. Drug Delivery Rev.* 54 (1): 149–161.

90 Alvarez-Lorenzo, C., Yanez, F., Barreiro-Iglesias, R., and Concheiro, A. (2006). Imprinted soft contact lenses as norfloxacin delivery systems. *J. Controlled Release* 113 (3): 236–244.

91 Seattone, M.F. and Salminen, L. (1995). Ocular inserts for topical delivery. *Adv. Drug Delivery Rev.* 16 (1): 95–106.

92 Rathore, K.S. and Nema, R.K. (2009). Review on ocular inserts. *Int. J. PharmTech Res.* 1 (2): 164–169.

93 Kuno, N. and Fujii, S. (2011). Recent advances in ocular drug delivery systems. *Polymers* 3 (1): 193–221.

94 Hornof, M., Weyenberg, W., Ludwig, A., and Bernkop-Schnurch, A. (2003). Mucoadhesive ocular inserts based on thiolated poly(acrylic acid): development and *in vivo* evaluation in humans. *J. Controlled Release* 89 (3): 419–428.

95 Karthikeyan, D., Bhowmick, M., Pandey, V.P. et al. (2013). The concept of ocular inserts as drug delivery systems: an overview. *Asian J. Pharm.* 2 (4): 192–200.

96 Ding, S. (1998). Recent developments in ophthalmic drug delivery. *Pharm. Sci. Technol. Today* 1 (8): 328–335.

97 Dong, A. (2013). Current and potential therapies for ocular neovascular diseases. *Clin. Exp. Pharmacol.* 3 (4).

98 Avery, R.L., Pieramici, D.J., Rabena, M.D. et al. (2006). Intravitreal bevacizumab (Avastin) for neovascular age-related macular degeneration. *Ophthalmology* 113 (3): 363–372.

99 Short, B.G. (2008). Safety evaluation of ocular drug delivery formulations: techniques and practical considerations. *Toxicol. Pathol.* 36: 49–62.

100 Stuart, A. The promise of implantable drug delivery systems. https://www.aao.org/eyenet/article/promise-of-implantable-drug-delivery-systems (accessed 20 July 2017).

101 Lee, S.S., Hughes, P., Ross, A.D., and Robinson, M.R. (2010). Biodegradable implants for sustained drug release in the eye. *Pharm. Res.* 27 (10): 2043–2053.

102 Mitragotris, S., Blankschtein, D., and Langer, R. (1996). Transdermal drug delivery using low frequency sonophoresis. *Pharm. Res.* 13 (3): 411–420.

103 Zderic, V., Vaesy, S., Martin, R.W., and Clark, J.I. (2002). Ocular drug delivery using 20-kHz ultrasound. *Ultrasound Med. Biol.* 28 (6): 823–829.

104 Watson, T. Iontophoresis. http://www.electrotherapy.org/modality/iontophoresis. (accessed 8 August 2017).

105 Hughes, L. and Maurice, D.M. (1984). A fresh look at iontophoresis. *Arch. Ophthalmol.* 102 (12): 1825–1829.

106 Das, S. and Suresh, P.K. (2010). Drug delivery to the eye: special reference to nanoparticles. *Int. J. Drug Delivery* 2 (1): 12–21.

107 Ghate, D. and Edelhauser, H.F. (2006). Ocular drug delivery. *Expert Opin. Drug Delivery* 3 (2): 275–287.

108 Sarraf, D. and Lee, D.A. (1984). The role of iontophresis in ocular drug delivery. *J. Ocul. Pharmacol.* 10 (1): 69–81.

109 Molokchia, S.A., Jeong, E.K., Higuchi, W.I., and Li, S.K. (2009). Transscleral iontophoretic and intravitreal delivery of a macromolecule: study of ocular distribution in vivo and postmortem with MRI. *Exp. Eye Res.* 88 (3): 418–425.

110 Thrimawithana, T.R., Young, S., Bunt, C.R. et al. (2011). Drug delivery to the posterior segment of the eye. *Drug Discovery Today* 16 (5–6): 270–277.

111 Stella, V.J. and Himmelstein, K.J. (1980). Prodrugs and site-specific drug delivery. *J. Med. Chem.* 23 (12): 1275–1282.

112 Thakur, R.R. and Kashiv, M. (2011). Modern delivery systems for ocular drug formulations: a comparative overview W.R.T. conventional dosage form. *Int. J. Res. Pharm. Biomed. Sci.* 2 (1): 8–18.

113 Ali, Y. and Lehmussaari, K. (2006). Industrial perspective in ocular drug delivery. *Adv. Drug Delivery Rev.* 58 (11): 1258–1268.

114 Alexander, C.L., Miller, S.J., and Abel, S.R. (2002). Prostaglandin analog treatment of glaucoma and ocular hypertension. *Ann. Pharmacother.* 36: 504–511.

115 Tirucherai, G.S. and Mitra, A.K. (2003). Effect of hydroxypropyl beta cyclodextrin complexation on aqueous solubility, stability, and corneal permeation of acyl ester prodrugs of ganciclovir. *AAPS PharmSciTech* 4 (3): 124–135.

116 Garland, M.J., Migalska, K., Mazlelaa, T. et al. (2011). Microneedle arrays as medical devices for enhanced transdermal drug delivery. *Expert Rev. Med. Devices* 8 (4): 459–482.

117 Gomaa, Y.A., Garland, M.J., McInnes, F. et al. (2012). Laser-engineered dissolving microneedles for active transdermal delivery of nadroparin calcium. *Eur. J. Pharm. Biopharm.* 82 (4): 299–307.

118 Martanto, W., Moore, J.S., Kashlan, O. et al. (2006). Microinfusion using hollow microneedles. *Pharm. Res.* 23 (1): 104–113.

119 Singh, T.R.R., Tekko, I., McAvoy, K. et al. (2017). Minimally-invasive microneedles for ocular drug delivery. *Expert Opin. Drug Delivery* 14 (4): 525–537.

120 Singh, T.R.R., Tekko, I., Al-Shammari, F. et al. (2016). Rapidly dissolving polymeric microneedles for minimally invasive intraocular drug delivery. *Drug Delivery Transl. Res.* 6 (6): 800–815.

15

Controlled Drug Delivery via the Otic Route

Jinsong Hao[1] *and S. Kevin Li*[2]

[1] *Department of Pharmaceutical Sciences, College of Pharmacy & Health Sciences, Campbell University, Buies Creek, NC, USA*
[2] *Division of Pharmaceutical Sciences, James L Winkle College of Pharmacy, University of Cincinnati, Cincinnati, OH, USA*

15.1 Introduction

Treatment of auditory and vestibular dysfunction has become increasingly dependent on inner ear drug delivery. In the past two decades, new drug delivery technologies have been developed for effective inner ear delivery. This chapter provides an overview of inner ear drug delivery via the otic route, describes the recent development in intratympanic and intracochlear strategies, and compares their advantages and disadvantages. The focus is on the controlled drug delivery systems for inner ear diseases. Drug delivery for treatments of disorders in the external canal and middle ear is not reviewed. Commercially available techniques for inner ear therapies as well as new applications of existing products in inner ear delivery are highlighted.

15.2 Anatomy and Physiology of the Otic Route

15.2.1 Anatomy of the Otic Route

The ear is divided into three parts: external ear, middle ear, and inner ear (Figure 15.1). External ear is composed of auricle or pinna and external auditory meatus (ear canal). The external ear is lined with skin; diseases of the external ear are mainly associated with skin disorders. The skin in the canal is supplied with ceruminous glands which produce cerumen (earwax). Cerumen produced in the ear canal can protect the tympanic membrane.

Middle ear is an air-filled cavity (tympanic cavity) separated from the external ear by the tympanic membrane (eardrum) which protects the middle and inner ear. The tympanic membrane vibrates to transmit sound signals from the external ear to the middle ear. The middle ear consists of an ossicular chain of three small connecting bones: malleus, incus, and stapes. The malleus is attached to the tympanic membrane and the footplate of the stapes is attached to the oval window of the inner ear. The middle ear is connected to the nasopharynx through the Eustachian tube, which is shorter and at a more horizontal angle in children than in adults. The tympanic membrane may be perforated due to diseases or trauma. Perforation of the tympanic membrane and otitis media is considered as disorders related to the middle ear.

Fundamentals of Drug Delivery, First Edition.
Edited by Heather A. E. Benson, Michael S. Roberts, Adrian C. Williams, and Xiaowen Liang.

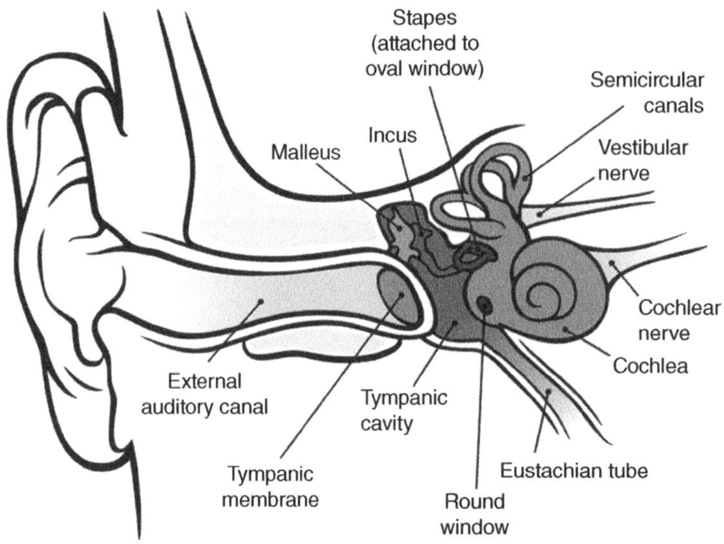

Figure 15.1 Anatomy of the otic route. Source: CNX OpenStax, Biology, Rice University. Available at https://commons.wikimedia.org/wiki/File:Figure_36_04_02.jpg. Licensed under CC-BY-4.0.

Inner ear (labyrinth) consists of two parts: the cochlea and the vestibular organ (Figure 15.2). The cochlea is a small complex spiral organ where the organ of Corti (containing the hair cells) lies in and is responsible for hearing. The cochlea can be considered as a long coiled tube with varying diameters along its length from the apex to the base. The cochlear sensory and neural cells do not regenerate. The degeneration of these cells caused by excessive noise, ototoxic drugs (e.g., gentamicin, neomycin sulfate, and tobramycin), or advancing age can lead to hearing loss. The vestibular organ is composed of three semicircular canals and is responsible for postural balance. The semicircular canals are three sets of balance receptors that detect horizontal and vertical head movements. Vertigo or dizziness occurs when inner ear malfunctions create an imbalance between the semicircular canals.

The inner ear is filled with different fluids in the bony labyrinth and the (inner) membranous labyrinth. The function of the inner ear is sensitive to small changes in the volume of the fluids. Endolymph is a clear and potassium-rich fluid that fills the membranous labyrinth. It has a composition similar to the intracellular environment. Perilymph, a fluid similar to the cerebrospinal fluid, surrounds the membranous labyrinth and fills the space between the membranous tunnel and the bony walls that surround it.

Oval window and round window are the two openings to the inner ear. The oval window is a membrane-covered opening which leads from the middle ear to the vestibule of the inner ear. The oval window is in contact with the stapes. The round window is a membrane-covered opening which leads from the middle ear to the cochlea of the inner ear. Both round window and oval window are considered as not only barriers but also passages for drug delivery to the inner ear.

15.2.2 Barriers Relevant to Inner Ear Drug Delivery

15.2.2.1 Blood Labyrinth Barrier

The blood labyrinth barrier (BLB) is a major barrier separating the inner ear from systemic circulation. It is made of capillary endothelial cells that line blood vessels in the stria vascularis [1, 2].

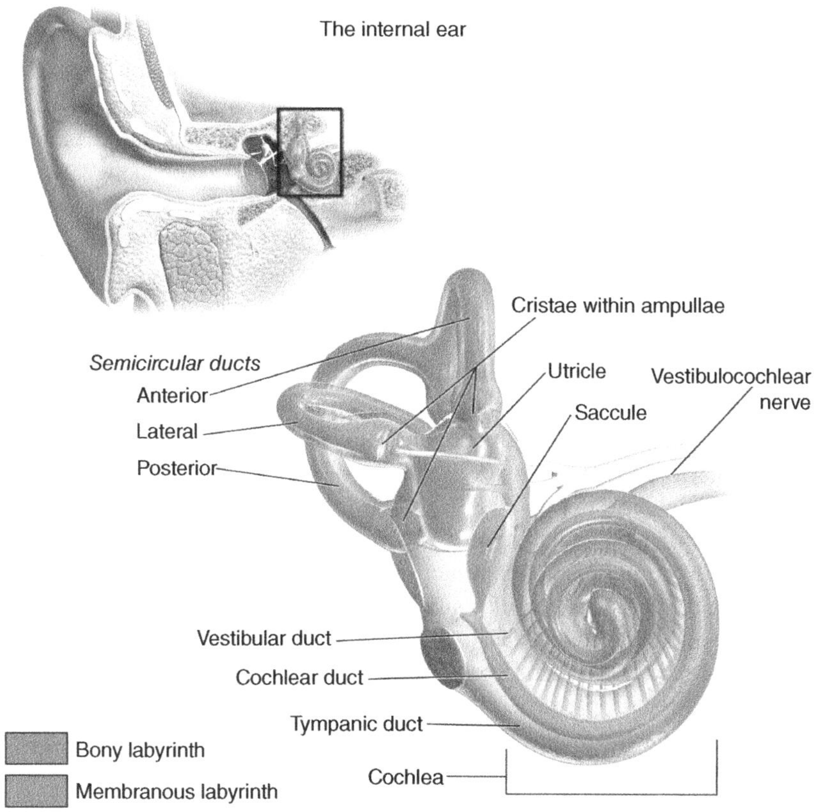

Figure 15.2 Anatomy of the inner ear (http://anatomypicture.blogspot.com). Source: ©1997 Encyclopaedia Britannica, Inc.

It also consists of tight junctions and generally permits only the permeation of small lipid-soluble molecules [3]. It plays an important role in maintaining the microhomeostasis of the inner ear fluid and protecting the inner ear integrity similar to the function of the blood brain barrier to the brain [4]. The BLB functions not only as a physical barrier but also as a biochemical barrier with efflux pump systems (P-glycoprotein [P-gp] and multidrug resistance protein-1 [MRP-1]) that further protect the inner ear [5]. Thus, the BLB is often considered as the rate-limiting barrier in the permeation of therapeutic agents from systemic circulation to the inner ear and prevents effective drug delivery into the inner ear following systemic administration.

A number of factors were found to cause tight junction opening of the BLB, including the presence of ototoxic drugs [6], noises [7], and inflammation [8]. It has been reported that aminoglycosides given systemically in combination with diuretics such as furosemide or ethacrynic acid result in hearing loss and inner ear damages faster than that of aminoglycosides alone [9, 10]. This suggests that the diuretics can enhance drug penetration across the BLB. Particularly, osmotic agents, such as glycerol, were suggested to induce endothelial cell "shrinkage" and thereby lead to the opening of BLB tight junction [11]. In addition to chemicals, hearing disorders such as autoimmune inner ear diseases, Meniere's disease, meningitis-associated labyrinthitis, and genetic diseases can also affect the BLB.

15.2.2.2 Round Window Membrane

The round window membrane (RWM) is a soft tissue barrier separating the middle ear from the inner ear. It has been considered as the main passage for drug delivery from the middle ear cavity to the inner ear. The RWM is made of three layers: an outer epithelial layer facing the middle ear cavity, a middle connection layer, and an inner cellular layer facing the scala tympani (ST) perilymph [12–14]. The RWM has an average thickness of 70 μm and a surface area of about 2.2 mm^2 in humans. However, the RWM shows various degrees of obstruction in about 1/5 to 1/3 of people. Standard or endoscopic tympanoscopy has been suggested for middle ear exploration and assessing the necessity of removing RWM obstructions to visualize the RWM in intratympanic injections for inner ear drug delivery. Raising the tympanomeatal flap in the standard middle ear exploration may cause bleeding. The variabilities in the width and shape of the ear canal as well as the limitations of endoscopes sometimes do not allow sufficient visualization of the RWM for inner ear drug delivery via intratympanic injections. The variable RWM thickness and conditions of the membrane across patient population are believed to be a factor leading to patient-to-patient variability in intratympanic inner ear drug delivery. The pseudomembrane of the RWM (plugs of connective or adipose tissue) also accounts for this variability [15].

The RWM acts like a semipermeable membrane. It is permeable primarily to low molecular weight (MW) molecules such as aminoglycoside antibiotics and corticosteroids. Large molecules such as horseradish peroxidase (MW 45000) can also diffuse through RWM under normal physiological conditions [16]. While larger molecules such as albumin (MW 70000) cannot diffuse across the RWM easily under normal conditions, such large molecules have been shown to penetrate the RWM into the inner ear during the early phase of inflammation [17]. Besides the molecular weight of the drug and disease state of the ear, other factors including the integrity of the RWM and drug physicochemical properties such as lipophilicity and charge also affect drug diffusion across the RWM [13]. In addition, the permeability of RWM is affected by local treatment with anesthetics, endotoxins and exotoxins [18], histamine [19], osmotic disturbances, and benzyl alcohol (a commonly used preservative) [20].

15.2.2.3 Oval Window

The oval window contains the footplate of the stapes and the annular ligament and is located toward the scala vestibule (SV). The oval window was found to be more permeable than the round window [21] and may be a promising alternative route to deliver drugs to the inner ear, particularly for the treatment of vestibular disorders. However, it has been widely accepted that drugs delivered to the middle ear space during local delivery primarily enter the inner ear via permeation across the RWM into ST, with subsequent diffusion into SV and the vestibule through the interstitial spaces of the spiral ligament.

15.2.2.4 Eustachian Tube

The Eustachian tube in adult humans is approximately 35 mm long and 3 mm in diameter. The two functions of Eustachian tube are pressure equalization and mucus drainage. Eustachian tube is closed normally and opens with swallowing, yawning, or chewing to equalize the pressure between the middle ear and the atmosphere. Pressure differences cause temporary conductive hearing loss due to the decreased motion of the tympanic membrane and ossicles of the ear. Eustachian tube drains mucus from the middle ear. Medications administered into the middle ear through the tympanic membrane can be cleared from the Eustachian tube into the nasopharynx. Clearance from Eustachian tube is thus considered as a barrier for effective drug delivery to the inner ear.

15.2.2.5 Tympanic Membrane

Tympanic membrane is a thin, cone-shaped membrane that separates the external ear from the middle ear. It has a surface area of about 80 mm^2 in adult and thickness of 100–150 μm depending on the location in the membrane. The membrane can be divided into two regions: the pars flaccida in the upper region (3–6% of the total area) and the pars tensa. The thick and tough pars tensa constitutes the larger part of the tympanic membrane and consists of three layers: an outer epidermal layer, a middle fibrous layer, and an inner mucosal layer. The pars flaccida lacks a fibrous layer and is relatively fragile. The outer epidermal layer of the tympanic membrane is continuous with the epidermis of the external canal. It is a keratinized skin with an outer stratum corneum and underlying stratum granulosum, stratum spinosum, and stratum basale layers resting on the basal lamina [22, 23]. The tympanic membrane is considered as a physical barrier for inner ear drug delivery. It can be expected that drugs administered in the external canal have limited permeability to the tympanic membrane and thus cannot effectively permeate through the tympanic membrane to the middle ear. To deliver drugs to the middle ear cavity for inner ear drug delivery, the integrity of the tympanic membrane has to be compromised during intratympanic administration. The perforation of the tympanic membrane may cause partial hearing loss and potential middle ear infections. Moreover, the shape of the external canal and the differences in the cone angle and depth of the tympanic membrane in humans [24] hinder the direct and precise administration of drugs to the RWM via the tympanic membrane. Intratympanic injection is usually performed by a trained otolaryngology specialist and precise delivery of drugs to the RWM is highly dependent of the procedure operation technique. Intratympanic injections may be a blind procedure if the RWM cannot be visualized.

15.3 Controlled Drug Delivery Systems

15.3.1 Intratympanic Administration

Intratympanic delivery to the inner ear was performed via the injection or perfusion of drug to the middle ear with the aim of drug diffusion through the RWM into the inner ear. This route of drug delivery was introduced more than half a century ago for the treatment of Meniere's disease with local anesthetics [25] and antibiotics [26, 27]. It has been widely used in clinics since 1990s. This approach possesses several advantages over systemic drug delivery as this local drug delivery method can bypass the BLB, and therefore result in higher concentrations in the inner ear fluids and avoid undesired systemic exposure [3, 28, 29]. A number of clinical studies have been published on intratympanic injections of corticosteroids for Meniere's diseases [30–32] and sudden sensorineural hearing loss (SSNHL) [33, 34]. Obstacles of intratympanic drug delivery include the anatomic absorption barriers such as the RWM, loss of drug through the Eustachian tube, and highly variable or unknown inner ear pharmacokinetic profiles of medications currently delivered via this route [15, 28, 35]. The percentage of drugs entering the inner ear following intratympanic injections can be relatively low. One example showed that gentamicin reaching the basal turn was in the order of 2.5% of the applied drug after intratympanic injection [36]. Drug clearance in the cochlea is also a major factor that leads to a base-apex concentration gradient. Due to this concentration gradient, it may be difficult to treat hearing disorders in the middle and lower frequency ranges (disorders in the apex area) by intratympanic drug delivery. In addition, it is difficult to predict the amounts of drugs delivered to the cells in different turns of the cochlea. Computational simulations have been used to study the base-apex concentration gradient to understand this process and assist in the prediction [37–39].

Effective drug delivery to the inner ear via the intratympanic route also relies on the contact time of the drug solution (or drug delivery system) with the RWM. Unfortunately, large portions of the administrated drugs are usually eliminated by the Eustachian tube following intratympanic drug delivery. There have been efforts to overcome this limitation through the development of devices and sustained-release drug delivery systems. Several cannula-based delivery systems are available commercially for sustained delivery of drugs to the middle ear. These devices include Silverstein MicroWick™ and Round Window Microcatheter. Potential problems and adverse effects of these devices include the persistent perforation of the tympanic membrane, risk of infection in the middle ear or external ear, and tissue growth in the middle ear either in the form of fibrosis or epithelial ingrowth leading to cholesteatoma [40, 41]. Sustained-release drug delivery systems can increase the residence time of a drug in the middle ear and provide controlled drug delivery to the inner ear. A number of systems have been studied for this application, including Gelfoam® and Seprapack™. These systems generally sustain drug delivery by the mechanisms of slow degradation of the materials, slow drug diffusion in the materials, or a combination of both.

15.3.1.1 Silverstein MicroWick

The MicroWick (Micromedics, Inc., Eagan, MN) is a single-wick device with a length of 9 mm and a diameter of 1 mm and made of absorbent polyvinyl acetate polymer. It is placed in the round window niche through a tympanostomy opening [42]. The procedure includes placing the MicroWick through a tiny ventilation tube in the tympanic membrane and then resting it on the RWM (Figure 15.3). The distal end of the wick is in the external auditory canal, where the patients can instill medications several times a day for several weeks. The reliability and consistency of gentamicin and methylprednisolone delivery by this device have been demonstrated [43]. Long-term follow-up of 69 Meniere's disease patients treated with gentamicin three times per day using this system has demonstrated vertigo control in 53 patients (76.8%) [44]. The long-term outcome of gentamicin inner-ear perfusion using the Silverstein MicroWick in the treatment of Meniere's disease was assessed [45]. Retrospective questionnaire survey of patients and review of medical records suggest that transtympanic gentamicin therapy using the Silverstein MicroWick can result in good vertigo control, although patients should be made fully aware of the risk of hearing deterioration. Overall, the self-treatment approach using the Silverstein MicroWick allows

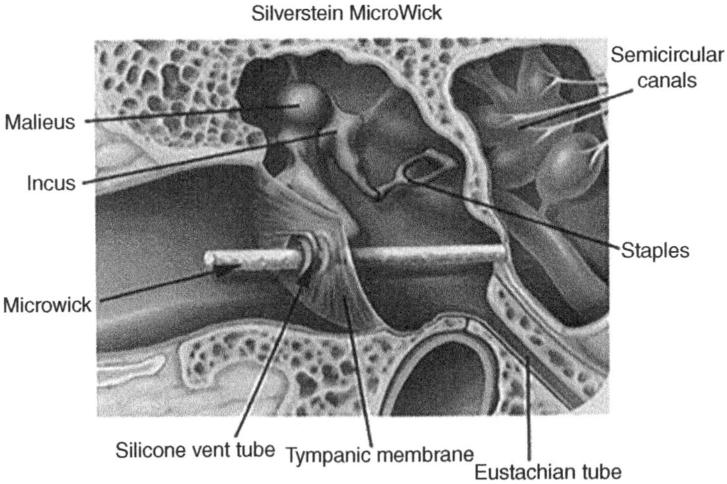

Figure 15.3 Illustration of Silverstein MicroWick procedure (http://ent-specialists.org).

the otologist to treat inner ear diseases in a similar way that the ophthalmologist treats eye diseases using eye drops [46].

15.3.1.2 Round Window Microcatheter (μCath)

The Round Window Microcatheter (IntraEar, Inc., Denver, CO, later acquired by Durect Corporation, Cupertino, CA) is a double lumen microcatheter designed to deliver drugs to the inner ear via the round window niche [47]. One of the two lumens of the catheter is for drug infusion and the other for fluid withdrawal. The catheter is placed in the round window niche after a tympanomeatal flap is raised. The other end of the catheter exits from the outer ear canal and is connected to a pump for drug infusion. The pump can be a micropump or an osmotic pump. This system has the advantage of delivering drugs to the middle ear continuously for up to several weeks. In the study by DeCicco et al., a total of 18 patients with Meniere's disease received microdose of gentamicin using the Round Window Microcatheter. Among them, tinnitus in 15 patients was improved significantly throughout the follow-up periods ranging from 6 to 18 months, vertigo was eliminated in all patients, and pressure was relieved in 17 patients [47]. The results in clinical studies by Hoffer et al. [48, 49] showed no significant reduction in cochlear or vestibular function observed in most patients. Long-term vertigo control was achieved using low dose gentamicin delivered by the catheter, while preserving hearing and vestibular function in the majority of patients [50]. This procedure appears to stabilize the vestibular function, enhancing the chance of effective vestibular rehabilitation. The use of the RWM catheter to deliver microdose gentamicin improves the therapy of Meniere's diseases over the normal cure rates reported with conventional intratympanic administration of gentamicin. The microcatheter has also been used for intratympanic infusion of steroids in patients with sensorineural hearing loss who have failed in the treatment with intravenous steroids and vasodilators [51].

15.3.1.3 Gelfoam

Gelfoam is a biodegradable gelatin polymer that was first introduced in middle ear surgery and recently used as a drug delivery system for the inner ear. A recent study has demonstrated improved outcome of Meniere's disease treatment by the placement of Gelfoam presoaked in gentamicin on the RWM that eliminated vertigo and tinnitus in 75% and 48% of the patients, respectively [52]. In another study, Havenith et al. applied Gelfoam infiltrated with brain-derived neurotrophic factor (BDNF) onto the RWM of deafened guinea pigs and evaluated the effect of this treatment by structural and functional measurements [53]. In the two weeks of Gelfoam BDNF treatment, survival of spiral ganglion neurons (SGNs) in the inner ear was observed in the low turn of the cochlea, but no significant improvement was observed in the apical turn of the cochlea, probably due to the small amount of drug reaching the apical turn. This suggests that local delivery of BDNF using Gelfoam can protect the SGNs in parts of the cochlea.

15.3.1.4 Seprapack

Hyaluronic acid (HA) is an anionic, nonsulfated glycosaminoglycan polysaccharide that is present in human body. It has been used as a material in otology due to its well-known safety profile [54]. A commercial hydrogel Seprapack (hyaluronic acid–carboxymethyl cellulose polymer, Genzyme Corporation) was primarily developed for nasal surgery and has been examined for the delivery of dexamethasone to the inner ear. The application of dexamethasone-loaded Seprapack on the RWM was found to reduce both low and high frequency hearing loss associated with trauma of cochlear implantation [55]. Similarly, *n*-acetyl cysteine with Seprapack was shown to be effective to protect the residual hearing after implant surgery in the high frequency hearing region near the point of implantation [56]. James et al. investigated the concentration of dexamethasone in the cochlea after intratympanic delivery with Seprapack and found significantly higher and sustained concentration of the drug in the cochlea than that by the drug alone without Seprapack [57]. These results were similar to those reported by Borden et al. who utilized the thiol-modified HA gelatin hydrogel as a drug delivery system [58]. Saber et al. suggested that HA gel had no toxic effect on the hair cells despite temporarily increasing the thickness of RWM [54].

15.3.1.5 Ozurdex as a RWM Implant

There is increasing evidence that intratympanic glucocorticoids are effective as salvage therapy for idiopathic sudden sensorineural hearing loss (ISSHL). Nevertheless, no drug has been approved for local therapy of the inner ear disorders. Ozurdex is a Food and Drug Administration (FDA)-approved dexamethasone implant for intravitreal use. Plontke et al. tested Ozurdex for inner ear delivery in patients with ISSHL and unsatisfactory recovery after systemic high dose glucocorticoid therapy [59]. In this clinical study, the implant was cut along the length into three to five pieces of approximately 0.8–1.5 mm each, according to the size of the human RWM. After raising a tympanomeatal flap, the implants were placed in the round window niche with drops of fibrin glue. The surgical placements of the implants generally presented no major difficulties but the sharp edges at the breaking points of the rigid implants were a concern and required special care. The mean hearing threshold of the patients significantly improved at follow-up (217 ± 62 days after implantation) by 31 ± 31 dB HL. Two of the five patients recovered completely. One patient showed partial hearing recovery with serviceable hearing. The promising results in patients with Ozurdex treatment demonstrate that the development of controlled release devices for RWM implantation would be beneficial, and appropriate device designs can ensure easy application and patient safety.

15.3.1.6 Propel Steroid-Eluting Stent

The propel mometasone-eluting stent (Intersect ENT) is an FDA-approved self-deploying device for delivering steroids into the ethmoid cavity following endoscopic sinus surgery in chronic sinusitis treatment. The implant is made of a biodegradable polylactide-*co*-glycolide polymer in a lattice pattern that expands in a spring-like fashion to conform to the walls of a dissected ethmoid cavity and contains a total of 370 μg of mometasone furoate (MF) designed for drug release over 30 days. The stent has been proven to be self-absorbable and tolerable in the nasal cavities of humans. This stent was recently explored for its use in inner ear drug delivery in guinea pigs after being cut into smaller miniature sticks of 5 mm in length to accommodate the small tympanic cavities of the animals. The MF-eluting miniature sticks were then implanted through a myringotomy incision into the ventral tympanic bulla in the animals to allow for prolonged drug release [60]. These treatments were able to provide prolonged local steroid delivery to the inner ear over two weeks, and no complications were observed in the middle ear. It was concluded that middle ear-sustained steroid delivery in humans might be possible using this method.

15.3.2 Trans-Oval Window Administration

RWM has been considered as the primary passage of drug entry to the inner ear. However, it has been reported that drugs can be also delivered into the inner ear via the oval window. One study [21] with gentamicin locally applied to either RWM or staples footplate in guinea pigs shows that gentamicin passed more easily through the oval window than through the round window, causing markedly higher levels of hearing loss between 8 and 32 kHz, a greater reduction of outer hair cells in the basal turn of the cochlea and fewer normal type I cells in the utricle in the vestibule. The results suggest the possibility of drug entry into the inner ear via the stapes footplate, and this route is more effective than the round window. Trans-oval window delivery may present a new approach of direct and targeted drug delivery to the inner ear.

The trans-oval window administration route may be of special interest for the delivery of aminoglycoside antibiotics to treat vestibular disorders such as Meniere's disease. Aminoglycoside antibiotics are toxic to the sensory cells in both the vestibule and cochlea, which cause the favorable suppression of vertigo symptoms by ablating the vestibule system but undesired hearing loss by damaging the cochlea, respectively. While intratympanic administration of aminoglycoside antibiotics has been an effective therapy for Meniere's disease, hearing loss is a major concern for this treatment. It is known that the risk of hearing loss increases with gentamicin concentration in the cochlea. If the vestibule can be treated directly by applying drugs onto the stapes footplate, the drug concentration in the cochlea will be lowered than that of inner ear delivery via the RWM. This concept was proven by a recent study [61], in which the efficacy of treating the vestibule by applying a small volume of low concentration gentamicin solution onto the stapes footplate in guinea pigs was assessed by quantitative measurements of vestibular and cochlear functions. Vestibular responses were significantly suppressed by 72.7% at two weeks after treatment with no significant loss of hearing.

Sircoglou et al. investigated the delivery of dexamethasone from a silicone implant placed next to the staples footplate of Mongolian gerbils [62]. The fluorescent-labeled dexamethasone was detected in the hair cells of the cochlea as early as 20 minutes after implantation and was still observed at least 30 days after implantation. Kang et al. investigated the delivery of a fluorescently labeled bisphosphonate, a compound for the treatment of the sensorineural hearing loss in cochlear otosclerosis, to the human cochlea through the oval window in three fresh human temporal bones and found that the compound was distributed up to the apical cochlear turn [63]. The results from this *ex vivo* study were promising that drug delivery to the apical cochlear turn may be achieved by local application of drugs to the oval window.

Silicone-based ear cubes were recently developed for local controlled drug delivery to the inner ear [64]. The ear cubes are miniaturized implants that can be placed into tiny holes drilled into or close to the oval window. These ear cubes consist of two parts: a cylinder and a cuboid. The cylinder can be inserted to the hole across the oval window. The cylinder is longer than the depth of the hole, partly located within the inner ear and surrounded by perilymph, and thus providing direct access to the target site and assuring implant fixation. The cuboid is located in the middle ear with one side in direct contact with the oval window and is serving as a drug reservoir. Dexamethasone-loaded ear cube can provide controlled drug release to the cochlea via the oval window. While the *in vitro* study demonstrated promising results, *in vivo* studies using ear cubes have not been reported.

15.3.3 Intracochlear Administration

Direct intracochlear drug delivery can bypass the middle ear and allow drugs to get to their intended sites in the cochlea directly. Intracochlear delivery can substantially increase drug bioavailability in the inner ear and has the highest efficiency among the inner ear delivery methods discussed in this chapter. Numerous intracochlear delivery technologies are being developed to improve the efficiency of drug delivery to the inner ear. They include direct injections, cochlear implants, osmotic mini-pumps, and reciprocating perfusion systems. Direct injection is the injection of a drug solution directly into the cochlea via a cochleostomy in the basal turn of the cochlea or through the RWM with a microsyringe and a narrow-gauge needle. This provides accurate drug delivery for acute drug application to the base of the cochlea. Cochlear implant is the insertion of a device into the ST for inner ear drug delivery in the treatment of inner ear disorders. This method is currently available for the treatment of severe to profound SSNHL. Osmotic mini-pump directly delivers a drug into ST via a cannula, and reciprocating perfusion system is similar to the osmotic mini-pump with the main difference that there is no accompanying net volume change of the fluid in the reciprocating perfusion system. Intracochlear strategies are more efficient than intratympanic delivery but with significant risks such as potential damages to the delicate cochlea [65]. A safe and robust technique for intracochlear delivery of drugs is not available [66].

15.3.3.1 Drug-Eluting Cochlear Implants

A cochlear implant is a hearing device implanted in the ST to stimulate the SGNs on a damaged cochlea directly. Cochlear implantation is usually performed in patients with severe sensorineural hearing loss but may cause local trauma. Efforts have thus been made to reduce local trauma by coating the implants with drugs [67, 68]. This strategy can be used for intracochlear delivery. The electrode of cochlear implants can be embedded in a biodegradable or nonbiodegradable polymeric matrix containing a drug that is expected to allow prolonged drug release into the ST for months to years [69]. While drug-eluting cochlear implants have proven to be a promising technology in animal studies [70, 71], the safety and efficacy need to be tested in humans.

15.3.3.2 Microfluidic Reciprocating Reservoir

To overcome the limitations with direct intracochlear injections such as inadequate drug mixing and apical distribution due to the small space and fluid volume within the cochlea as well as the clearance of the cochlear fluid, a microfluidics-based intracochlear drug delivery device was developed [72]. The device has a reservoir system for perfusion of drugs into the cochlear perilymph through a single hole in the basal turn of the cochlea. The single cannula is implanted in the ST and acts as both inlet and outlet. This device infuses and withdraws a constant volume of drug in a cyclic mode without causing a net fluid volume change. This is achieved by alternating the infusion and withdrawal sessions. The infusion portion of the cycle lasts for several seconds to deliver drugs, and the withdrawal portion of the cycle lasts for several minutes to allow the mixture of the drugs

Figure 15.4 A wearable system for microfluidic drug delivery from an integrated reservoir to the cochlea using a reciprocating flow micropump. Source: Kim et al. [73].

Table 15.1 Examples of controlled drug delivery systems via the otic route.

System	Drug	Disease/study aim	Study	References
Intratympanic administration				
Silverstein MicroWick	Gentamicin	Meniere's disease	Clinical	Hill et al. [44]
Round Window Microcatheter	Gentamicin	Meniere's disease	Clinical	DeCicco et al. [47]
Gelfoam	Gentamicin	Meniere's disease	Clinical	Silverstein et al. [52]
	Brain-derived neurotropic factor	Meniere's disease	Preclinical – guinea pigs	Havenith et al. [53]
Seprapack	Dexamethasone	Hearing loss associated with cochlear implantation	Preclinical – guinea pigs	Eastwood et al. [55]
	n-Acetyl cysteine	Chronic inflammation associated cochlear implantation	Preclinical – guinea pigs	Eastwood et al. [56]
Ozurdex	Dexamethasone	Idiopathic sudden sensorineural hearing loss	Clinical	Plontke et al. [59]
Propel steroid-eluting stent	Mometasone furoate	Feasibility and tolerability study	Preclinical – guinea pigs	Lavigne et al. [60]
Trans-oval window administration				
Drug in phosphate buffered saline solution	Gentamicin	Meniere's disease	Preclinical – guinea pigs	King et al. [61]
Silicone implant	Dexamethasone	Drug distribution in the cochlea	Preclinical – Mongolian gerbils	Sircoglou et al. [62]
Drug in artificial perilymph solution	Bisphosphonate compound zoledronate (6-FAM-ZOL)	Drug distribution in the cochlea	*Ex vivo* study – fresh human temporal bones	Kang et al. [63]
Silicone-based ear cube	Dexamethasone	Device development	*In vitro* study	Gehrke et al. [64]
Intracochlear administration				
Hydrogel-coated silicone cochlear implant electrode	Insulin-like growth factor and hepatocyte growth factor	Minimize inner ear trauma during electrode insertion	Preclinical – guinea pigs	Kikkawa et al. [67]
Drug-eluting silicone cochlear implant	Dexamethasone	Pharmacokinetic study	Preclinical – guinea pigs	Liu et al. [70]
Drug-eluting electrode array	Dexamethasone	Safety and efficacy	Preclinical – guinea pigs	Stathopoulos et al. [71]
Microfluidics/MicroElectro-Mechanical Systems	Glutamate receptor blocker 6,7-dinitroquinoxaline-2,3-dione (DNQX)	Device development	Preclinical – guinea pigs	Sewell et al. [72]
Microfluidic chips	Glutamate receptor blocker DNQX	Device development	Preclinical – guinea pigs	Tandon et al. [75]

and endogenous perilymph before the fluid is withdrawn back to the device. A microfluidic chip was further developed for implantation (Figure 15.4) [73–75]. The device includes an integrated drug reservoir that pushes drugs out, while it maintains crucial pressurization in the ear drawing other fluids in. The successful results from both *in vitro* and *in vivo* tests suggest that this new miniaturized wearable device has the potential in drug delivery to the inner ear that might potentially treat hearing loss in a less invasive manner than frequent intracochlear injections. Long-term testing and further miniaturization could be future development for this microfluidic chip.

15.4 Conclusions

Effective and safe delivery of drugs to the inner ear for therapies of auditory and vestibule diseases remains a significant challenge. Controlled drug delivery systems via the otic route have been reviewed in this chapter (Table 15.1). Intratympanic and intracochlear administration are promising approaches that could be alternatives to current systemic administration. However, precise dosing, sustained drug delivery to the cochlea, and safety are the major concerns. Recent development in inner ear drug delivery includes the proposed utilization of the oval window. Trans-oval window delivery may present a direct delivery route of drugs to the vestibule system for vestibular disorders while avoiding the undesired effects due to drug distribution to the cochlea.

References

1 Jahnke, K. (1980). The blood-perilymph barrier. *Arch. Otorhinolaryngol.* 228 (1): 29–34.

2 Juhn, S.K. (1988). Barrier systems in the inner ear. *Acta Otolaryngol.* 458: 79–83.

3 Swan, E.E., Mescher, M.J., Sewell, W.F. et al. (2008). Inner ear drug delivery for auditory applications. *Adv. Drug Deliv. Rev.* 60 (15): 1583–1599.

4 Juhn, S.K. and Rybak, L.P. (1981). Labyrinthine barriers and cochlear homeostasis. *Acta Otolaryngol.* 91 (5–6): 529–534.

5 Saito, T., Zhang, Z.J., Tokuriki, M. et al. (2001). Expression of *p*-glycoprotein is associated with that of multidrug resistance protein 1 (MRP1) in the vestibular labyrinth and endolymphatic sac of the guinea pig. *Neurosci. Lett.* 303 (3): 189–192.

6 Yamasoba, T., Suzuki, M., and Kaga, K. (1996). Influence of chronic kanamycin administration on basement membrane anionic sites in the labyrinth. *Hear. Res.* 102 (1–2): 116–124.

7 Suzuki, M., Yamasoba, T., Ishibashi, T. et al. (2002). Effect of noise exposure on blood-labyrinth barrier in guinea pigs. *Hear. Res.* 164 (1–2): 12–18.

8 Kastenbauer, S., Klein, M., Koedel, U., and Pfister, H.W. (2001). Reactive nitrogen species contribute to blood-labyrinth barrier disruption in suppurative labyrinthitis complicating experimental pneumococcal meningitis in the rat. *Brain Res.* 904 (2): 208–217.

9 McFadden, S.L., Ding, D., Jiang, H. et al. (2002). Chinchilla models of selective cochlear hair cell loss. *Hear. Res.* 174 (1–2): 230–238.

10 Tran Ba Huy, P., Manuel, C., Meulemans, A. et al. (1983). Ethacrynic acid facilitates gentamicin entry into endolymph of the rat. *Hear. Res.* 11 (2): 191–202.

11 Juhn, S.K., Prado, S., and Pearce, J. (1976). Osmolality changes in perilymph after systemic administration of glycerin. *Arch Otolaryngol.* 102 (11): 683–685.

12 Richardson, R.T., Noushi, F., and O'Leary, S. (2006). Inner ear therapy for neural preservation. *Audiol. Neurootol.* 11 (6): 343–356.

13 Banerjee, A. and Parnes, L.S. (2004). The biology of intratympanic drug administration and pharmacodynamics of round window drug absorption. *Otolaryngol. Clin. North Am.* 37 (5): 1035–1051.

14 Goycoolea, M.V. and Lundman, L. (1997). Round window membrane. Structure function and permeability: a review. *Microsc. Res. Tech.* 36 (3): 201–211.

15 Salt, A.N. and Plontke, S.K. (2005). Local inner-ear drug delivery and pharmacokinetics. *Drug Discovery Today* 10 (19): 1299–1306.

16 Kim, C.S., Cho, T.K., and Jinn, T.H. (1990). Permeability of the round window membrane to horseradish peroxidase in experimental otitis media. *Otolaryngol. Head Neck Surg.* 103 (6): 918–925.

17 Hamaguchi, Y., Morizono, T., and Juhn, S.K. (1988). Round window membrane permeability to human serum albumin in antigen-induced otitis media. *Am. J. Otolaryngol.* 9 (1): 34–40.

18 Ikeda, K. and Morizono, T. (1988). Changes of the permeability of round window membrane in otitis media. *Arch. Otolaryngol. Head Neck Surg.* 114 (8): 895–897.

19 Chandrasekhar, S.S., Rubinstein, R.Y., Kwartler, J.A. et al. (2000). Dexamethasone pharmacokinetics in the inner ear: comparison of route of administration and use of facilitating agents. *Otolaryngol. Head Neck Surg.* 122 (4): 521–528.

20 Mikulec, A.A., Hartsock, J.J., and Salt, A.N. (2008). Permeability of the round window membrane is influenced by the composition of applied drug solutions and by common surgical procedures. *Otol. Neurotol.* 29 (7): 1020–1026.

21 King, E.B., Salt, A.N., Kel, G.E. et al. (2013). Gentamicin administration on the stapes footplate causes greater hearing loss and vestibulotoxicity than round window administration in guinea pigs. *Hear. Res.* 304: 159–166.

22 Anandacoomaraswamy, K.S., Dutton, N., Rajan, G.P. et al. (2006). Utilization of fresh human tympanic membranes for structural analysis and cytokeratin immunocytochemistry implementing resin techniques. *Acta Otolaryngol.* 126 (2): 149–153.

23 Lim, D.J. (1995). Structure and function of the tympanic membrane: a review. *Acta Otorhinolaryngol. Belg.* 49 (2): 101–115.

24 Volandri, G., Di Puccio, F., Forte, P., and Carmignani, C. (2011). Biomechanics of the tympanic membrane. *J. Biomech.* 44 (7): 1219–1236.

25 Ersner, M.S., Spiegel, E.A., and Alexander, M.H. (1951). Transtympanic injection of anesthetics for the treatment of Meniere's syndrome. *AMA Arch. Otolaryngol.* 54 (1): 43–52.

26 Lange, G. (1989). Gentamicin and other ototoxic antibiotics for the transtympanic treatment of Meniere's disease. *Arch. Otorhinolaryngol.* 246 (5): 269–270.

27 Blakley, B.W. (1997). Clinical forum: a review of intratympanic therapy. *Am. J. Otol.* 18 (4): 520–526. Discussion 527–531.

28 Bowe, S.N. and Jacob, A. (2010). Round window perfusion dynamics: implications for intracochlear therapy. *Curr. Opin. Otolaryngol. Head Neck Surg.* 18 (5): 377–385.

29 McCall, A.A., Swan, E.E., Borenstein, J.T. et al. (2010). Drug delivery for treatment of inner ear disease: current state of knowledge. *Ear Hear.* 31 (2): 156–165.

30 Barrs, D.M. (2004). Intratympanic corticosteroids for Meniere's disease and vertigo. *Otolaryngol. Clin. North Am.* 37 (5): 955–972.

31 Dodson, K.M., Woodson, E., and Sismanis, A. (2004). Intratympanic steroid perfusion for the treatment of Meniere's disease: a retrospective study. *Ear Nose Throat J.* 83 (6): 394–398.

32 Barrs, D.M., Keyser, J.S., Stallworth, C., and McElveen, J.T. Jr., (2001). Intratympanic steroid injections for intractable Meniere's disease. *Laryngoscope* 111 (12): 2100–2104.

33 Banerjee, A. and Parnes, L.S. (2005). Intratympanic corticosteroids for sudden idiopathic sensorineural hearing loss. *Otol. Neurotol.* 26 (5): 878–881.

34 Rauch, S.D. (2004). Intratympanic steroids for sensorineural hearing loss. *Otolaryngol. Clin. North Am.* 37 (5): 1061–1074.

35 Richardson, R.T., Wise, A.K., Andrew, J.K., and O'Leary, S.J. (2008). Novel drug delivery systems for inner ear protection and regeneration after hearing loss. *Expert Opin. Drug Deliv.* 5 (10): 1059–1076.

36 Plontke, S.K., Mynatt, R., Gill, R.M. et al. (2007). Concentration gradient along the scala tympani after local application of gentamicin to the round window membrane. *Laryngoscope* 117 (7): 1191–1198.

37 Plontke, S.K., Wood, A.W., and Salt, A.N. (2002). Analysis of gentamicin kinetics in fluids of the inner ear with round window administration. *Otol. Neurotol.* 23 (6): 967–974.

38 Salt, A.N. and Ma, Y. (2001). Quantification of solute entry into cochlear perilymph through the round window membrane. *Hear. Res.* 154 (1–2): 88–97.

39 Stover, T., Yagi, M., and Raphael, Y. (1999). Cochlear gene transfer: round window versus cochleostomy inoculation. *Hear. Res.* 136 (1–2): 124–130.

40 Hochman, J., Blakley, B., Abdoh, A., and Aleid, H. (2006). Post-tympanostomy tube otorrhea: a meta-analysis. *Otolaryngol. Head Neck Surg.* 135 (1): 8–11.

41 Licameli, G., Johnston, P., Luz, J. et al. (2008). Phosphorylcholine-coated antibiotic tympanostomy tubes: are post-tube placement complications reduced? *Int. J. Pediatr. Otorhinolaryngol.* 72 (9): 1323–1328.

42 Light, J.P. and Silverstein, H. (2004). Transtympanic perfusion: indications and limitations. *Curr. Opin. Otolaryngol. Head Neck Surg.* 12 (5): 378–383.

43 Silverstein, H. (1999). Use of a new device, the MicroWick, to deliver medication to the inner ear. *Ear Nose Throat J.* 78 (8): 595–598, 600.

44 Hill, S.L. III, Digges, E.N., and Silverstein, H. (2006). Long-term follow-up after gentamicin application via the Silverstein MicroWick in the treatment of Meniere's disease. *Ear Nose Throat J.* 85 (8): 494–496, 498.

45 Suryanarayanan, R., Srinivasan, V.R., and O'Sullivan, G. (2009). Transtympanic gentamicin treatment using Silverstein MicroWick in Meniere's disease patients: long term outcome. *J. Laryngol. Otol.* 123 (1): 45–49.

46 Silverstein, H., Thompson, J., Rosenberg, S.I. et al. (2004). Silverstein MicroWick. *Otolaryngol. Clin. North Am.* 37 (5): 1019–1034.

47 DeCicco, M.J., Hoffer, M.E., Kopke, R.D. et al. (1998). Round-window microcatheter-administered microdose gentamicin: results from treatment of tinnitus associated with Meniere's disease. *Int. Tinnitus J.* 4 (2): 141–143.

48 Hoffer, M.E., Kopke, R.D., Weisskopf, P. et al. (2001). Microdose gentamicin administration via the round window microcatheter: results in patients with Meniere's disease. *Ann. N.Y. Acad. Sci.* 942: 46–51.

49 Hoffer, M.E., Kopke, R.D., Weisskopf, P. et al. (2001). Use of the round window microcatheter in the treatment of Meniere's disease. *Laryngoscope* 111 (11 Pt. 1): 2046–2049.

50 Suryanarayanan, R. and Cook, J.A. (2004). Long-term results of gentamicin inner ear perfusion in Meniere's disease. *J. Laryngol. Otol.* 118 (7): 489–495.

51 Lefebvre, P.P. and Staecker, H. (2002). Steroid perfusion of the inner ear for sudden sensorineural hearing loss after failure of conventional therapy: a pilot study. *Acta Otolaryngol.* 122 (7): 698–702.

52 Silverstein, H., Arruda, J., Rosenberg, S.I. et al. (1999). Direct round window membrane application of gentamicin in the treatment of Meniere's disease. *Otolaryngol. Head Neck Surg.* 120 (5): 649–655.

53 Havenith, S., Versnel, H., Agterberg, M.J. et al. (2011). Spiral ganglion cell survival after round window membrane application of brain-derived neurotrophic factor using gelfoam as carrier. *Hear. Res.* 272 (1–2): 168–177.

54 Saber, A., Laurell, G., Bramer, T. et al. (2009). Middle ear application of a sodium hyaluronate gel loaded with neomycin in a guinea pig model. *Ear Hear.* 30 (1): 81–89.

55 Eastwood, H., Chang, A., Kel, G. et al. (2010). Round window delivery of dexamethasone ameliorates local and remote hearing loss produced by cochlear implantation into the second turn of the guinea pig cochlea. *Hear. Res.* 265 (1–2): 25–29.

56 Eastwood, H., Pinder, D., James, D. et al. (2010). Permanent and transient effects of locally delivered *n*-acetyl cysteine in a guinea pig model of cochlear implantation. *Hear. Res.* 259 (1–2): 24–30.

57 James, D.P., Eastwood, H., Richardson, R.T., and O'Leary, S.J. (2008). Effects of round window dexamethasone on residual hearing in a guinea pig model of cochlear implantation. *Audiol. Neurootol.* 13 (2): 86–96.

58 Borden, R.C., Saunders, J.E., Berryhill, W.E. et al. (2011). Hyaluronic acid hydrogel sustains the delivery of dexamethasone across the round window membrane. *Audiol. Neurootol.* 16 (1): 1–11.

59 Plontke, S.K., Glien, A., Rahne, T. et al. (2014). Controlled release dexamethasone implants in the round window niche for salvage treatment of idiopathic sudden sensorineural hearing loss. *Otol. Neurotol.* 35 (7): 1168–1171.

60 Lavigne, P., Lavigne, F., and Saliba, I. (2016). Sustained inner ear steroid delivery via bioabsorbable stent: a tolerability and feasibility study on guinea pigs. *Otolaryngol. Head Neck Surg.* 155 (4): 649–653.

61 King, E.B., Shepherd, R.K., Brown, D.J., and Fallon, J.B. (2017). Gentamicin applied to the oval window suppresses vestibular function in guinea pigs. *J. Assoc. Res. Otolaryngol.* 18 (2): 291–299.

62 Sircoglou, J., Gehrke, M., Tardivel, M. et al. (2015). Trans-oval-window implants, a new approach for drug delivery to the inner ear: extended dexamethasone release from silicone-based implants. *Otol. Neurotol.* 36 (9): 1572–1579.

63 Kang, W.S., Nguyen, K., McKenna, C.E. et al. (2016). Intracochlear drug delivery through the oval window in fresh cadaveric human temporal bones. *Otol. Neurotol.* 37 (3): 218–222.

64 Gehrke, M., Sircoglou, J., Gnansia, D. et al. (2016). Ear cubes for local controlled drug delivery to the inner ear. *Int. J. Pharm.* 509 (1–2): 85–94.

65 Borenstein, J.T. (2011). Intracochlear drug delivery systems. *Expert Opin. Drug Deliv.* 8 (9): 1161–1174.

66 Salt, A.N. and Plontke, S.K. (2009). Principles of local drug delivery to the inner ear. *Audiol. Neurootol.* 14 (6): 350–360.

67 Kikkawa, Y.S., Nakagawa, T., Ying, L. et al. (2014). Growth factor-eluting cochlear implant electrode: impact on residual auditory function, insertional trauma, and fibrosis. *J. Transl. Med.* 12: 280.

68 Wrzeszcz, A., Steffens, M., Balster, S. et al. (2015). Hydrogel coated and dexamethasone releasing cochlear implants: quantification of fibrosis in guinea pigs and evaluation of insertion forces in a human cochlea model. *J. Biomed. Mater. Res. Part B* 103 (1): 169–178.

69 Krenzlin, S., Vincent, C., Munzke, L. et al. (2012). Predictability of drug release from cochlear implants. *J. Control. Release* 159 (1): 60–68.

70 Liu, Y., Jolly, C., Braun, S. et al. (2016). In vitro and in vivo pharmacokinetic study of a dexamethasone-releasing silicone for cochlear implants. *Eur. Arch. Otorhinolaryngol.* 273 (7): 1745–1753.

71 Stathopoulos, D., Chambers, S., Enke, Y.L. et al. (2014). Development of a safe dexamethasone-eluting electrode array for cochlear implantation. *Cochlear Implants Int.* 15 (5): 254–263.

72 Sewell, W.F., Borenstein, J.T., Chen, Z. et al. (2009). Development of a microfluidics-based intracochlear drug delivery device. *Audiol. Neurootol.* 14 (6): 411–422.

73 Kim, E.S., Gustenhoven, E., Mescher, M.J. et al. (2014). A microfluidic reciprocating intracochlear drug delivery system with reservoir and active dose control. *Lab Chip* 14 (4): 710–721.

74 Tandon, V., Kang, W.S., Spencer, A.J. et al. (2015). Microfabricated infuse-withdraw micropump component for an integrated inner-ear drug-delivery platform. *Biomed. Microdevices* 17 (2): 37.

75 Tandon, V., Kang, W.S., Robbins, T.A. et al. (2016). Microfabricated reciprocating micropump for intracochlear drug delivery with integrated drug/fluid storage and electronically controlled dosing. *Lab Chip* 16 (5): 829–846.

16

Controlled Drug Delivery via the Nasal Route

Barbara R. Conway and Muhammad U. Ghori

Department of Pharmacy, University of Huddersfield, Huddersfield HD1 3DH, UK

16.1 Introduction

Nasal administration is widely used as the logical choice for topical treatment of conditions affecting the nose or paranasal sinuses such as allergic or infectious rhinitis and sinusitis. However, given the rich vascular nature of the nasal mucosa and porous and thin epithelial membrane, the nasal route can also be exploited for systemic drug delivery. It is a promising alternative route for delivering macromolecules, such as peptide and protein drugs and vaccines and is also regarded as a route for delivery of smaller, lower molecular weight drugs that may be subject to first-pass metabolism. In addition, absorption of drug from the olfactory region of the nose provides a potential pathway for drug delivery to the central nervous system (CNS). Nasal administration is less well-suited to the administration of chronic therapies or for those requiring sustained blood levels over time unless they are formulated in a controlled-release system. One strategy to extend delivery times is to decrease the mucociliary clearance (MCC); this can be achieved by increasing the viscosity of the formulation or by enhancing adhesion of the system at the site. This can prolong residence and thus enhance drug absorption. The most common commercially available medicines are solution-based formulations coupled with mucoadhesive systems which may incorporate enzyme inhibitors and nasal permeation enhancers. The first diminishes the enzymatic activity at mucosa, while the latter increases the permeability of the drug, enhances the drug nasal residence time, and improves the therapeutic efficacy of the systemic drugs. Formulation into specialized delivery systems is also an option. The preparation of nano- or micro-particulate systems with various polymers has also been widely tested particularly for incorporating macromolecular drugs. Additionally, the olfactory region of the nose has been proposed as an alternative pathway to deliver active substances to the CNS by-passing the blood–brain barrier, and this can be targeted using controlled drug delivery systems. Finally, owing to the nasal associated lymphoid tissue, the nose can be used as a delivery route for vaccines, providing effective immune-stimulation in terms of both humoral and cell-mediated responses. This chapter reviews the strategies available for controlling drug delivery by the nasal route and the potential and limitations of various nasal drug delivery systems.

16.2 Anatomy and Physiology of the Nose

In addition to its olfactory function, the nose is the primary entrance to the respiratory tract, enabling air to enter the body. The vestibule is the region just inside the nostrils, with an area of

Fundamentals of Drug Delivery, First Edition.
Edited by Heather A. E. Benson, Michael S. Roberts, Adrian C. Williams, and Xiaowen Liang.

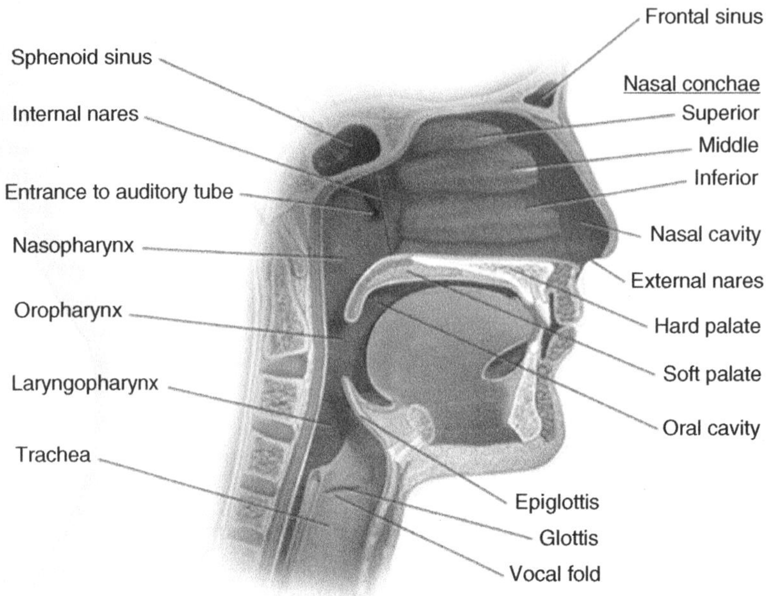

Figure 16.1 Anatomy of the upper respiratory system.

about 0.6 cm^2. Entering air is warmed, humidified, and dust and other particulate matter removed by vibrissae, the short, thick hairs lining the vestibule. The nasal cavity (or nasal fossa) is a large air filled space above and behind the nose in the middle of the face (Figure 16.1). Each cavity is the continuation of one of the two nostrils. While the nasal cavity dimensions are very variable between individuals, there are a number of key features. It is essentially symmetrical and is separated into two distinct air passages by a vertical thin wall called the nasal septum. The top part of the nasal cavity is formed by bones and cartilage and is tent-shaped while the floor is formed by the palate, which separates the nasal and oral cavities.

The entire mucosa of the nasal cavity is covered by a blanket of mucus, which lies superficial to the microscopic cilia and helps to trap bacteria and dust. The nasal cavity is divided into two segments: the respiratory segment and the olfactory segment. The conchae, or turbinates, are located in the respiratory segment and are lined with a ciliated pseudostratified columnar epithelium, which is interspersed with goblet cells, which serve to increase the surface area exposed to the air, thus enhancing the capacity for heat and moisture exchange. Many of these cells are covered with microvilli and up to 20% have long cilia. The cilia of the respiratory epithelium move the secreted mucus and particulate matter posteriorly toward the pharynx where it passes into the esophagus and is digested in the stomach.

The olfactory segment, beside and above the upper turbinate, is lined with a specialized type of pseudostratified columnar epithelium, known as olfactory epithelium, which contains receptors for the sense of the smell. Dendrites, covered by a thin layer of moisture, extend from the nerve cells and dissolved substances provide a chemical stimulus for the olfactory nerve cells. Bowman's glands are serous glands in the lamina propria, whose secretions trap and dissolve odoriferous substances. This olfactory region in the roof of the nasal cavity contains the supporting epithelial cells and olfactory receptor cells. The epithelium penetrates into the cranial cavity through perforations in the cribiform plate and the existence of this pathway may offer the possibility for delivery of drugs to the brain.

There is a rich blood supply to the nasal cavity from branches of both the internal and external carotid arteries. It is therefore an attractive route for drug delivery, which is considered easily accessible and generally well tolerated.

16.3 Absorption from the Nasal Cavity

16.3.1 The Epithelial Barrier

The nasal respiratory mucosa is a highly vascularized membrane lined by a pseudostratified columnar epithelium. It rests on the collagen basement membrane and the lamina propria which is richly supplied with blood vessels, nerves, glands, and immune cells. The epithelium is mainly composed of basal cells, ciliated column cells (which are covered by a layer of long cilia), non-ciliated column cells and goblet cells (Figure. 16.2). The microvilli on the columnar cells' apical surface considerably increase the surface area of the respiratory epithelium available for systemic drug absorption.

Nasally administered drugs will be deposited on the respiratory (pseudostratified columnar) epithelium. Inhaled droplets and solid particles are thought to deposit in the nose by three mechanisms: inertial impaction, gravitational sedimentation, and Brownian diffusion. The site of deposition of formulations administered using a standard nasal spray device is generally in the anterior part of the nose, but some specialized devices are able to deposit the drug higher in the cavity, potentially reaching the superior turbinates and the olfactory region. From the site of deposition, dependent on the characteristics of the drug, i.e. molecular weight, charge, and lipophilicity, the drug can either be absorbed through the epithelium and reach the systemic circulation or be cleared via the nasopharynx down the gastrointestinal tract by the MCC system.

16.3.2 Absorption

Following administration to the nasal cavity, the first step for drug absorption is the passage through the mucus. Generally, smaller, uncharged entities can pass more easily through this layer, while

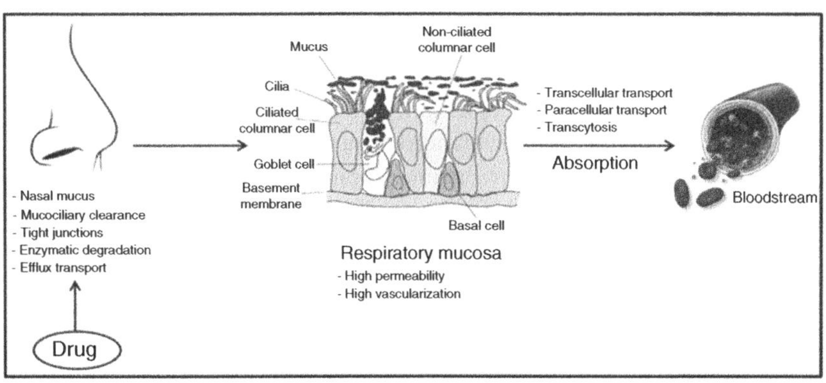

Figure 16.2 Schematic representation of the respiratory mucosa and the possible pathways involved in the transport of drugs from nose to systemic bloodstream. Factors that influence systemic absorption of nasal drugs are also represented. Source: Adapted from [1].

Table 16.1 Factors affecting nasal drug absorption.

Physicochemical factors	Biological factors	Formulation factors
Molecular weight and size	Nasal blood flow	pH and osmolarity
Lipophilicity	Mucociliary clearance	Mucosal irritancy
Partition coefficient	Enzymatic degradation	Viscosity
Dissociation coefficient	Transporters and efflux	Excipients
Solubility and dissolution rate	Physical condition of mucosa	Dosage form
Stability	Site of deposition within nasal cavity	Drug distribution
Polymorphism and morphology		Physical form of formulation
Solubility and dissolution rate		Drug concentrations, dose, and volume of administration
		Area of exposure

Source: Adapted from [3].

larger or charged particles will be more of a challenge. It has been reported that the most important factor governing diffusion of small molecules is their lipophilicity, but for large molecular weight drugs, such as peptides, intermolecular interactions between the peptide and the glycoprotein chains of the mucus may hamper the drug passage [2]. Following transfer through the mucous layer, the drug must transverse the epithelial membrane and for hydrophilic molecules, which will be soluble in mucus, they will be more susceptible to mucociliary clearance, if their diffusion rate across the membrane is low. Factors affecting nasal absorption are summarized in Table 16.1.

Although several mechanisms have been proposed for drug passage across the epithelial cell membrane, the most common are transcellular, paracellular, and transcytosis transport mediated by vesicle carriers (Figure 16.2). Transcellular transport follows a lipoidal route whereby the drug crosses the epithelial cell by a concentration dependent passive diffusion process, by active transporter-mediated transport or by vesicular transport mechanisms or by the paracellular route through the tight junctions between the cell [4]. Lipophilic drugs will likely partition into the lipid bilayer of the cell membrane, diffuse into it, and transverse the cell in the cytoplasm by passive diffusion. Polar drugs with molecular weights below 1000 Da will generally cross the membrane via the paracellular route, since, although tight junctions are dynamic structures, the mean size of the channels is in the order of less than 10 Å and so the transport of larger molecules is considerably more limited.

There is some evidence of active transport processes in the nasal mucosa; organic anion transporters and organic cation transporters (OCTs) for certain compounds [5], and human nasal epithelium has been shown to express OCTs and carnitine/organic cation transporter (OCTNs) inhibitors OCT1–3, OCTN1, and OCTN2 transporters mainly on the apical side of the nasal cells [6]. Paracellular transport involves a slow and passive aqueous diffusional pathway through the tight junctions, thus an inverse relationship exists between the molecular weight and drug permeation and polar compounds with molecular weight higher than 1000 Da are reported as exhibiting poor systemic bioavailability when administered by intranasal route [7]. Larger peptides and proteins have been shown to be able to pass the nasal membrane using an endocytotic transport process but only in low amounts.

However, despite the large surface area of the nasal cavity and the extensive blood supply, the permeability of the nasal mucosa is normally low for polar molecules, including low-molecular-weight drugs and especially large-molecular-weight peptides and proteins. For small polar drugs, bioavailabilities are around 10% and bioavailability is not normally above 1% for peptides such as calcitonin and insulin [8]. It is interesting to note that the nasal sumatriptan product on the market has a pharmacokinetic profile not dissimilar to that obtained after oral administration and a bioavailability of 15.8% as compared to 14.3% after oral administration.

The nasal absorption of polar drugs can be enhanced if administered with absorption promoting agents. Permeation of small and large hydrophilic molecules can be increased if they are administered in combination with permeation enhancers which induce reversible modifications on the epithelial barrier structure, such as increasing the membrane fluidity, creating transient hydrophilic pores, decreasing the viscosity of mucus layer, and opening the tight junctions. Fatty acids mainly function by increasing membrane fluidity, while some bile salts and fusidic acid derivatives used as nasal promoters also inhibit nasal enzymes and improve drug bioavailability.

For nasal drug delivery, surfactants such as laureth-9, bile salts, and bile salt derivatives such as sodium taurodihydrofusidate, fatty acids or fatty acid derivatives, phospholipids, and cyclodextrins have been reported. They generally change the permeability of the epithelial cell layer by modifying the phospholipid bilayer, leaching out protein from the membrane or even stripping off the outer layer of the mucosa. Some enhancers can also loosen the tight junctions and/or work as enzymatic inhibitors. For such enhancing agents, greatly enhanced bioavailabilities have been reported, even for larger peptides such as insulin in animal models. This effect has not directly translated into success in Phase I trials in humans due to reduced efficacy or irritancy of the enhancer material in the nasal cavity. In animal studies, it has been shown for a range of enhancing agent that there is a direct correlation between absorption enhancing effect (bioavailability) obtained and the damage caused to the nasal membrane. This is particularly true for surfactant materials and bile salts. For other enhancers, such as some of the cyclodextrins, chitosan, and selected phospholipids, the absorption-enhancing effect largely outweighs the impact of the irritancy [8].

Although surfactants are often regarded as effective permeation enhancers, their use in nasal formulations restricted as they induce nasal irritation and epithelial toxicity. This has led to a preference for cyclodextrins in nasal formulations, particularly β-cyclodextrin, as they increase drug solubility, act as nasal permeation enhancers, and are generally well tolerated in humans.

Examples of absorption enhancer systems for delivery of small molecular weight drugs and specific peptides and proteins are summarized in Illum [9] and in Table 16.2. These have potential for better toxicity profiles and balance between safety and efficiency than older systems. Examples include CPE-215® (cyclopenta decalactone, Azone) developed by CPEX Pharma, Intravail® (alkylsaccharides) developed by Aegis Therapeutics, ChiSys™ (chitosan) and PecSys™ (low methylated pectin) in development by Archimedes Pharma, and CriticalSorb™ (polyglycol mono- and diesters of 12-hydroxystearate (70%), polyethylene glycol (30%)) developed by Critical Pharmaceuticals.

Another contributing (but normally considered less important) barrier for drug permeation is the enzymes present in nasal cavity and the nasal mucosal lining. Although enzymatic activity is significantly lower than the gastrointestinal (GI) tract, there is some activity that can degrade peptide drugs. Several oxidative Phase I enzymes and conjugative Phase II enzymes have been identified in the mammalian nasal epithelium. Specifically, various cytochrome P450 isoforms can be found in the olfactory epithelium with an activity comparable to that in the liver despite a different metabolic profile [23]. Other Phase I enzymes found are NADPH-cytochrome P450 reductase, carboxylesterase, epoxide hydrolases, and aldehyde dehydrogenases. Phase II biotransformation processes include UDP-glucuronyltransferases, glutathione-*S*-transferases, and sulfotransferases. This may prove to be a significant barrier to the successful delivery of peptide drugs but can be mitigated in part by the inclusion of enzyme inhibitors in the formulation. The nasal epithelium also

Table 16.2 Examples of nasal absorption enhancers.

Class of compound	Example	Possible action	References
Fatty acids	Dideconoylphosphatidylcholine, lysophosphatidylcholine	Membrane disruption	[10]
Surfactants	Sodium lauryl sulfate, saponin, polyoxyethylene-9-lauryl ether	Membrane disruption	[11–14]
Bile salts	Sodium deoxycholate, sodium glycocholate, sodium taurodihydrofusidate	Open tight junctions, enzyme inhibition, mucolytic activity	[15–17]
Cyclodextrins and derivatives	α-, β-, γ-cyclodextrin DMβ-, HPβ-cyclodextrin	Open tight junctions, membrane disruption	[18, 19]
Enzyme inhibitors	Bestatin, amastatia	Enzyme inhibition	[20]
Bioadhesive materials	Carbopol, starch microspheres, chitosan	Reduce nasal clearance, open tight junctions	[21]

Source: Adapted from [22].

contains efflux transporters including P-gp and multidrug resistance-associated protein I which may further impact bioavailability. A range of studies has been carried out *in vitro* using tissue homogenates or nasal wash or *in vivo* in animal models to assess the possible effect of such enzymes on the stability of peptides [8]. However, translating the results into man can be problematic.

16.4 Mucus and Mucociliary Clearance

Delivery via the nasal route is limited mainly by the MCC system which removes the drug from the absorption site, reducing the time available for its uptake, and transport into the systemic bloodstream. Poor contact between formulations and the nasal mucosa and particularly poor absorption when administered as simple aqueous solutions have driven the development of alternative strategies to improve the nasal bioavailability. Nevertheless, there are some limitations of this route of administration, and the clearance half-life for both liquid and powder formulations without any bioadhesive component is in the range of 10–20 min due to MCC [3]. There is also a limit to the volume that the cavity can hold, typically 25–200 μl.

The goblet cells secrete mucin, contributing to the formation of an approximately 5 μm thick mucus layer. This mucus is a complex biological fluid composed of 95% water, 2.5–3% mucin, and a wide variety of macromolecules and biological elements, namely electrolytes, proteins, lipids, enzymes, antibodies, DNA, sloughed epithelia cells, and fragments. Mucins are defined as glycoproteins containing heavily *O*-glycosylated serine/threonine-rich tandem repeat domains and can be divided into two classes: membrane-bound mucins and secreted mucins. The rheological properties of mucus are mainly influenced by the mucin content, composition, and glycosylation degree, as well as by hydration of the mucus. Mucus thickness and structure depends both on the site and on pathological and physiological conditions. Inhaled particles, pathogens, and xenobiotics become trapped in the mucus and are transported by the propulsive coordinated force of cilia toward the pharynx, at a rate of about 5–6 mm/min, exhibiting a clearance time of only 12–20 minutes [3, 24]. As a consequence, there is a very limited permeability of drug delivery particles and hydrophobic drugs that can cross the mucus barrier and, in most of the cases, they are rapidly cleared from the delivery site.

Both temporal environmental and disease conditions can influence MCC and therefore absorption. Disease states such as cystic fibrosis, Kartagener's syndrome, nasal polyps, chronic sinusitis, and the common cold are known to alter the composition of mucus and its rate of removal from the nasal cavity [25]. They may also disrupt ciliary function and cause nasal irritation. In addition, changes in temperature and humidity can alter the water content of mucus in the nose which can increase or decrease mucociliary clearance and blood flow. These factors can lead to variability in absorption following nasal delivery.

16.5 Drug Delivery Systems

The nasal mucosa has a highly dense vascular network, relatively permeable membrane structure, and 150 cm^2 area, which renders it a good route for drug delivery leading to avoidance of hepatic first pass metabolism. Successful formulations can enable a reduction in the required therapeutic dose, rapid achievement of effective blood levels, and a quick onset of action with a reduction in side effects compared to other routes. This means that drug delivery by the nasal route remains an attractive option not only for local conditions but also for systemic delivery of many active substances including potent and peptide drugs, and rapid onset of action is generally obtained. The nasal route is not as suitable for chronic drugs that must be frequently administered daily or for regimens that require sustained delivery unless a sustained release dosing strategy is employed.

The nasal cavity has therefore been exploited for systemic delivery of low-molecular-weight drugs, particularly for those requiring a rapid onset of action. Examples include therapies for treatment of migraine and analgesia, are normally in the form of a solution, and do not contain absorption enhancers. The use of traditional liquid formulations for intranasal administration is limited because of the rapid mucociliary clearance leading to short residence times and poor bioavailability.

These present significant shortcomings for drug delivery such as a limited duration of residence on the mucosal surface, chemical and microbiological stability issues, and compared to liquids, powders can improve drug stability, accommodate higher administrable dosage and, in particular, exhibit an extended nasal residence time.

A range of nasal peptide products are presently or have been on the market such as calcitonin (Miacalcin®, Novartis; Fortical®, Unigene), desmopressin (Desmospray®, Ferring), buserelin (Suprecur®, Sanofi-Aventis), nafarelin (Synarel®, Pharmacia), and oxytocin (Syntocinon®, Novartis). These peptide products all show a very low bioavailability, but due to the low systemic levels needed to obtain a therapeutic effect, these products are efficacious.

Drug delivery systems afford the opportunity to intimately formulate the drug with functional excipients such as enzymatic inhibitors to reduce nasal metabolism particularly of protein and peptide molecules; nasal absorption enhancers that promote transport across the nasal membrane; and mucoadhesive polymers that suppress the physiological barrier associated with the fast mucociliary clearance and thus prolong the retention time. In addition, particles deposited on the mucosa will provide locally higher concentrations at the absorption site, and this, along with improved residence time, has the potential to increase bioavailability.

All delivery systems may also contain enzymatic inhibitors that reduce nasal metabolism particularly of protein and peptide molecules; nasal absorption enhancers that promote the drug transport across the nasal membrane; and mucoadhesive polymers that suppress the physiological barrier associated with the fast mucociliary clearance and thus prolong the retention time in the nasal cavity and enhance drug systemic bioavailability.

There are four main categories of devices for nasal delivery: drops, aqueous sprays, pressurized sprays, and dry powder systems. Delivery devices can be active or passive, based on the mechanism of aerosolization and intended use. In the case of active devices, such as insufflators, mechanically actuated systems, and pressurized devices containing a propellant, a power source is used to produce sufficient airflow to aerosolize the powder. Such systems include systems based on the bi-directional technology (OptiNose), controlled particle dispersion technology (Kurve Technology), and POD technology (Impel Neuropharma). Optinose is a device based on a "bi-directional" nasal delivery concept and using this breath-actuated device, the airflow through one nostril can deliver the powder to the whole nasal cavity before exiting through the other nostril, with the added advantage of avoiding lung inhalation and/or swallowing. Beside the type and actuation method of the device, properties of the nasal powder formulation such as particle size distribution, density, shape and flow, will influence emission from device and the deposition of the particles within the nasal cavity.

16.5.1 Solutions and Suspensions

Owing to their simple formulation, solutions/suspensions are popular for formulating drugs for nasal delivery; however, they demonstrate limited residence time on the mucosal surface, chemical and microbiological stability issues, and solubility restrictions governing available dose.

Solid dosage forms such as powders offer several advantages including higher chemical and microbiological stability and the possibility to intimately formulate the drug with functional excipients such as solubilizers, bioadhesives, and penetration enhancers. Also, particles deposited on the mucosa will provide higher local drug concentrations at the absorption site, and this, along with improved residence time, may contribute to higher bioavailability.

A nasal powder formulation comprising desmopressin on sodium starch glycolate particles showed a threefold and sixfold area under the curve (AUC) increase compared to a liquid and an oral tablet, respectively. In addition, there was reduced variability in AUC and C_{max} using the powder formulation, indicating better reproducibility in the administered dose [26].

The absorption of the drug itself will influence the success of the different strategies. For the model drugs warfarin (Biopharmaceutics Classification System [BCS] Class I), piroxicam (BCS Class II), and sumatriptan (BCS Class III) [27], absorption from the powder form was higher than that from the solution in a direct comparison. Increasing the viscosity may further increase residence time in the nose, and so improve absorption so the same study investigated the addition of types of hydroxypropyl cellulose (HPC) with different polymerization degrees: HPC(SL), HPC(M), and HPC(H). All HPCs failed to enhance warfarin absorption, while the piroxicam absorption was enhanced only by HPC(M). Sumatriptan absorption was not enhanced by HPC(SL), but by HPC(M) and HPC(H). The differences in nasal absorption of the three model drugs promoted by HPCs depend on the permeability and solubility of the drug. Moreover, the nasal retention of different formulations was increased by HPCs (results indicate that the effect of HPC is dependent on the permeability and solubility of drugs). By increasing the viscosity of the formulation, HPC can change not only the nasal residence of the formulation but also some other important factors involved in the absorption of solid drugs. A solid drug initially dissolves in the fluid, then diffuses through the viscous formulation, and finally a viscous fluid can delay the dissolution of a drug, which would result in a decrease in drug absorption. Also the diffusion of the dissolved drug: the slow diffusion in the viscous solution leads to a decrease in drug absorption. The effect of viscosity is more pronounced on the diffusion of highly permeable drugs. There are challenges with the administration and dosage accuracy of very viscous formulations and thus *in situ*-gelling systems have become popular.

16.5.2 Mucoadhesive Polymers

For drug delivery purposes, the term bioadhesion implies attachment of a drug carrier system to a specified biological location. The biological surface can be epithelial tissue or the mucus coat on the surface of a tissue. If adhesive attachment is to a mucus coat, the phenomenon is referred to as mucoadhesion. Bioadhesive or mucoadhesive formulations have the potential to extend residence time within the nasal cavity, intensify contact between the mucosa and the drug, increase drug concentration at the site of deposition, and facilitate absorption to enhance bioavailability.

The basic mechanistic approach of mucoadhesion involves two steps (i) intimate contact between a bioadhesive and a membrane (wetting or swelling phenomenon) and (ii) penetration of the bioadhesive into the tissue or into the surface of the mucus membrane (interpenetration). Mucoadhesion is currently explained by six theories: electronic, adsorption, wettability, diffusion, fracture, and mechanical, summarized by Smart [28].

The *electronic theory* involves electron transfer upon contact of the adhering surfaces due to differences in their electronic structure. This is proposed to result in the formation of an electrical double layer at the interface and the subsequent adhesion is due to attractive forces.

The *wetting theory* is primarily applied to liquid systems and considers surface and interfacial energies. It involves the ability of a liquid to spread spontaneously onto a surface and subsequent adhesion depends on the affinity of the liquid for the surface.

The *adsorption theory* describes attachment of the materials on the basis of hydrogen bonding and van der Waals' forces, as the main forces contributing to the adhesion. The chemisorption theory, which is related, assumes an interaction across the interface occurs as a result of strong covalent bonding.

The *diffusion theory* is based on the interpenetration of polymer and mucin chains across an adhesive interface. The adhesion force increases with the degree of penetration of the polymer chains. This penetration rate depends on the diffusion coefficient, flexibility, and nature of the mucoadhesive chains, mobility, and contact time. Sufficient depth of penetration creates a semi-permanent adhesive bond.

The *mechanical theory* assumes that adhesion arises from an interlocking of a liquid adhesive (on setting) into irregularities on a rough surface. However, rough surfaces also provide an increased surface area available for interaction along with an enhanced viscoelastic and plastic dissipation of energy during joint failure, which are thought to be more important in the adhesion process than a mechanical effect.

The *fracture theory* differs a little from the other five and is perhaps the most used theory in studies on the mechanical measurement of mucoadhesion. It analyzes the force required to separate two surfaces after adhesion is established. It assumes that the failure of the adhesive bond occurs at the interface, whereas failure normally occurs at the weakest component, which is typically a cohesive failure within one of the adhering surfaces.

Among the wide range of existing mucoadhesive polymers, chitosan, alginates, and cellulose derivatives are the most frequently used in intranasal drug delivery systems. However, carbopol and polycarbophil are also useful, they also have the capacity to open tight junctions and inhibit trypsin, a proteolytic enzyme, thus increasing the stability of peptide drugs.

16.5.2.1 *In Situ* Forming Nasal Gels

Short residence times can be addressed by the application of *in situ* gelling systems, when the sol, the liquid form, is transformed into a gel by some physiological stimulus, upon instillation in the nasal cavity, allowing a prolonged residence time. Advantages of *in situ* forming gels over other nasal formulations include reduction of post-nasal drip which is loss of drug into the back of the throat and can lead to taste issues with nasal formulations. The formulations also have the capacity for incorporation of excipients that may be used to reduce irritancy or enhance absorption.

According to the different mechanisms that trigger sol-to-gel transitions in the *in situ* gelling, systems are classified into ion-sensitive, temperature-dependent, or pH-responsive systems. As a result, in actual usage these dosage forms can be easily administered as solutions, assuring accuracy in the administered dose, whereas gel formation upon contact with the nasal epithelium is of utter importance to ascertain sufficient contact time and thus improved bioavailability.

The gels are normally made by hold or cold methods. The cold method is more usual for poloxamer, chitosan, or carbopol and involves slow addition of the gelling agent to a drug solution under reduced temperature (often 4 °C). For the remaining polymers (Table 16.3), heat can be applied to for the solution and the drug added on cooling. An overview of temperature- and ion-responsive *in situ* gelling systems and their application in drug delivery is presented in Table 16.4.

Thermoresponsive systems are those mucoadhesive formulations comprising polymers which exhibit temperature-triggered sol-to-gel transitions in the range of 25–37 °C.

16.5.2.1.1 Poloxamer

A poloxamer is a series of ABA tri-block copolymers containing the hydrophilic polyethylene oxide (PEO) and hydrophobic polypropylene oxide (PPO). The most extensively studied is Poloxamer 407 (P407), a copolymer with the PEO blocks on both ends and the PPO block in the middle. It is a commercial thermosensitive material, and its aqueous solution has been proven to have reversible thermal gelation ability and low toxicity. Gelation is induced by polymer desolvation accompanied with side chain conformational changes, resulting in displacement of the hydrating water molecules and subsequent micellar formation (Figure 16.3), hence the ratio of PPO in the structure plays an import role in determining the gelation temperature and blends can be used such as those with Poloxamer 188 (P188), an analogue of P407 with lower PPO ratio, to obtain suitable gelation temperature. P407 is rarely used alone, but is combined with other polymers to increase gel strength, promote adhesion, or reduce erosion. The thermoresponsive properties of Poloxamer 407 have been extensively exploited in the development of *in situ* nasal gels in combination with mucoadhesives, which can extend the residence time further in the nasal cavity. For example, formulations of P407 or P188,containing e-polylysine (e-PL) and low-molecular-weight sodium hyaluronate (MW 11 000 Da), could gel at 32 °C and extend release of phenylephrine hydrochloride, a nasal decongestant for up to eight hours in an *in vitro* model [65].

Table 16.3 Polymers used for *in situ* forming nasal gels.

Polymer	Gelation trigger
Pectin	High methoxylation: pH activated in the presence of sugar
	Low methoxylation, ion activated
Carbopol	pH activated (but also temperature at low shear rate)
Gellan gum	Thermally dependant ion activated
Chitosan	Thermally triggered in the presence of glycerophosphate
Poloxamer	Thermally triggered
Ethylhydroxyethylcellulose	Thermally triggered

Source: Adapted from [3].

Table 16.4 Summary of temperature- and ion-responsive *in situ* gelling systems and their application in drug delivery.

Stimulus responsive agent	Triggering factor	Formulation excipients	Active	References
Poloxamer 407	Temperature	Carbopol 934P	Sumatriptan	[29]
Poloxamer 407	Temperature	Chitosan, hydroxypropyl cellulose, PEG (400 and 6000), Carbopol 934P, PVA	Metoclopramide HCl	[30]
Poloxamer 407	Temperature	Chitosan, Carbopol 974P	siRNA	[31]
Poloxamer 407	Temperature	Carrageenan	Ketorolac tromethamine	[32]
Poloxamer 407	Temperature	Carbopol 934P	Rizatriptan benzoate	[33]
Poloxamer 407	Temperature	Carbopol 974P, Pemulen™ TR-2, HPMC	Granisetron HCl, ketorolac tromethamine	[34]
Poloxamer 407	Temperature	Chitosan, Pluronic F-68, PEG-8000	Tacrine	[35]
Poloxamer 407	Temperature	Xyloglucan, sodium alginate, PEG (4000 and 6000)	Zolmitriptan, ketorolac tromethamine	[36]
Poloxamer 407	Temperature	*n*-Tridecyl-β-D-maltoside, Carbopol 934P, HPC and HPMC	Zidovudine	[37]
Poloxamer 407	Temperature	Carbopol 934, PVP K30, HPMC K4M, sodium alginate, tamarind seed gum, carrageenan	Venlafaxine HCl	[38]
Poloxamer 407	Temperature	Methocel A4M	Venlafaxine HCl	[39]
Poloxamer 407	Temperature	Carbopol 934P, β-cyclodextrin	Loratadine	[40]
Poloxamer 407	Temperature	Carbopol 934P, hydroxypropyl-β-cyclodextrin	Fexofenadine HCl	[41]
Poloxamer 407	Temperature	Ficus carica mucilage, HPMC, Carbopol 934, sodium taurocholate	Midazolam	[42]
Poloxamer 407/188	Temperature	HPMC, anionic and cationic nanostructured lipid carriers	Ropinirole HCl	[43]
Poloxamer 407/188	Temperature	Carbopol 934, hydroxypropyl methyl cellulose, hydroxypropyl cellulose, chitosan	Diphtheria toxoid	[44]
Poloxamer 407/188	Temperature	β-CD,PEG6000	Paeonol	[45]
Poloxamer 407/188	Temperature	—	Curcumin	[46]
Poloxamer 407/188	Temperature	Chitosan microspheres	Lorazepam	[47]

(continued)

Table 16.4 (Continued)

Stimulus responsive agent	Triggering factor	Formulation excipients	Active	References
Poloxamer 407	Temperature	Hydroxypropyl β-cyclodextrin and chitosan	Fexofenadine hydrochloride	[48]
Poloxamer 407/deacetylated gellan gum	Temperature/ion	Nanostructured lipid carriers	Rivastigmine	[49]
Poloxamer 407/deacetylated gellan gum	Temperature/ion	Sulfobutyl ether-β-cyclodextrin, chlorobutanol, HP-β-CD, azone	Ketorolac tromethamine	[50]
Chitosan/glycerophosphate	Temperature	PEG 4000	Doxepin	[51]
Trimethylchitosan/ glycerophosphate	Temperature	PEG 4000	—	[52]
Trimethylchitosan/ glycerophosphate	Temperature	PEG 4000	Insulin	[53]
N-[(2-Hydroxy-3-trimethylammonium)propyl] chitosan chloride/glycerophosphate	Temperature	PEG 4000	Insulin	[54]
Xyloglucan	Temperature	—	Ondansetron	[55]
Thiolated xyloglucan	Temperature	—	Ondansetron	[56]
Gellan gum	Ion	—	Scopolamine hydrobromide	[57]
Gellan gum	Ion	Xanthan gum	Mometasone furoate	[58]
Gellan gum	Ion	Poloxamer 407 and oleic acid nanosuspension	Carvedilol	[59]
Deacetylated gellan gum	Ion	—	Gastrodin	[60]
Deacetylated gellan gum	Ion	Labrafac PG, Labrasol, Transcutol HP (nanosized microemulsion)	Saquinavir mesylate	[61]
Deacetylated gellan gum	Ion	PEG 400	Sumatriptan succinate	[62]
Deacetylated gellan gum	Ion	Solutol HS, Transcutol HP, Capryol 90 (nanosized microemulsion)	Curcumin	[63]

Source: Adapted from [64].

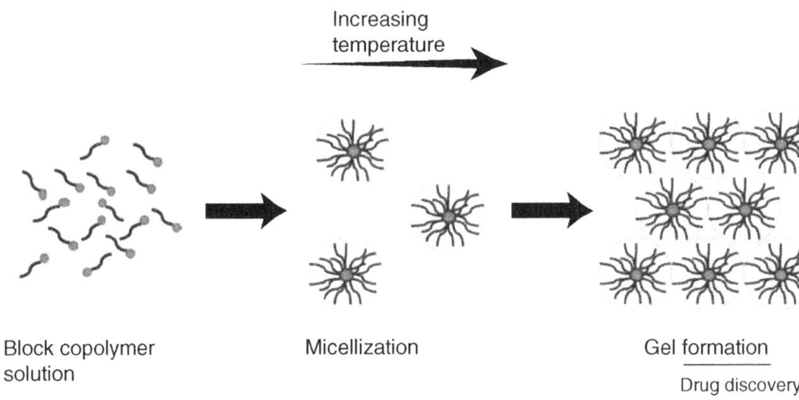

Figure 16.3 Schematic illustration of the mechanism of *in situ* gelation of a thermo-responsive polymer as a function of temperature. Upon temperature rise, polymer desolvation accompanied with side chain conformational changes, result in displacement of the hydrating water molecules, modifications of micelles orientation, and consequently gel formation. Source: Adapted from [64].

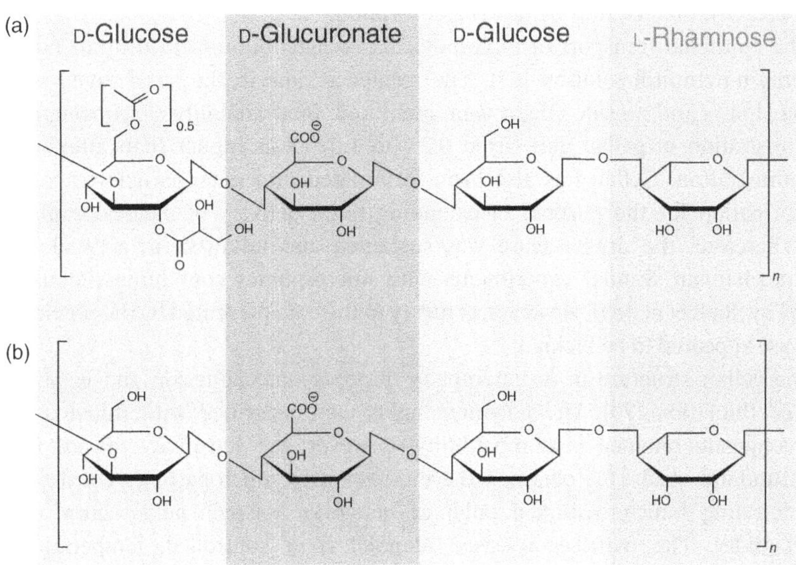

Figure 16.4 The structure of native (a) and low-acyl (b) form of gellan gum. Source: Adapted from [66].

16.5.2.1.2 Gellan Gum

Gellan is a linear, anionic exocellular polysaccharide, secreted by the microbe *Sphingomonas elodea*, with the linear repeating tetrasaccharides consisting of α-L-rhamnose, β-D-glucose, and β-D-glucuronate, in the molar ratios 1 : 2 : 1 (Figure 16.4). The native form of gellan contains two types of acyl substituents, namely L-glyceryl and acetyl. Alkaline hydrolysis is used to remove both of the residues and produce deacetylated gellan, also called low-acetyl or low-acyl gellan. Deacetylated gellan gum (DGG) can form transparent gels on exposure to certain concentrations of cations, such as sodium, potassium, and calcium, which cross-link to form a three-dimensional network by complexation with cations and incorporating water with hydrogen bonding (Figure 16.5). As an ion sensitive polymer, the critical ionic concentration (CIC) of DGG is paramount for the

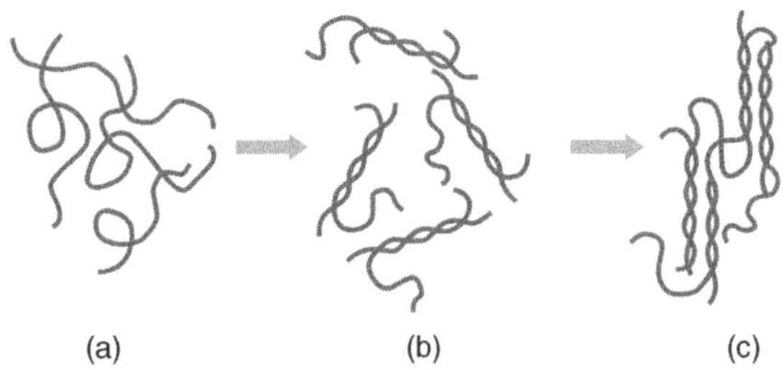

Figure 16.5 Gradual transformation of gellan gum from aqueous solutions. Source: Adapted from [66].

controlling ion-triggered *in situ* gels. When the cation concentrations are present in the gel at concentrations above the CIC, an instantaneous phase transition from solution to gel will take place in response to external ionic stimulation.

The cation-induced gelling process is ascribed to double helical junction zones formation and inter-helical interaction, resulting in a three-dimensional network via cations complexation. The specific gelling properties in different media led to the development of controlled release forms based on gellan. The abundance of cations, especially Ca^{2+}, naturally occurring in nasal fluids, enables the *in situ* gelation of gellan gum formulations upon contact with nasal mucosa, sustaining their residency in nasal cavity, thereby improving drug absorption.

Gellan improved the epithelial transport of high-molecular-weight fluorescein dextran more effectively than isotonic D-mannitol solution [67]. The residence time in the nasal cavity was extended for up to four hours and no side effects were identified. Bioavailability of scopolamine following nasal administration of gellan gels (from 0.2% to 1.0%) was higher than after oral or subcutaneous administration. Gellan has also been formulated into microparticles capable of spontaneous gel formation for the purpose of extending nasal delivery of metoclopramide hydrochloride [68]. Therefore, the drug release was sustained and followed an anomalous, non-Fickian release mechanism. Similar experiments with microspheres containing sildenafil citrate were performed by Shah et al. [69]. However, contrary to the data presented by the previous authors, the drug release appeared to be Fickian.

Modifications to the gellan structure in an attempt to improve mucoadhesion and gelation strength have involved thiolation [70]. The polymer chains were esterified with thioglycolic acid. The obtained conjugate retained biocompatibility, however, the sensitivity toward ion gelation decreased. Mundargi et al. [71] obtained thermo-responsive microparticles consisting of two semi-interpenetrating ionic-crosslinked polymer networks between gellan gum and poly(*N*-isopropylacrylamide). The matrices released atenolol in a controlled, temperature-dependent manner. Release was faster at 25 °C than in comparison to 37 °C, which was attributed to p(NIPAAm) swelling at lower temperatures and possible precipitation at higher temperatures. The observed property enabled a pulsatile "on–off" release profile responsive to changes in temperature.

The majority of investigations have been carried out using low acyl gellan for this purpose, despite the properties of the high acyl version being more conducive to mucoadhesion. High acyl gellan produces highly elastic gels below 60 °C which make it difficult to spray using a mechanical spray device. Formation of fluid gels has addressed this problem. Fluid gel systems contain gelled microparticles suspended in a solution of un-gelled polymer. These systems can therefore behave as pourable viscoelastic fluids. Using blends of low and high acyl forms, the rheological properties were sufficient to enable delivery of the caffeine-loaded fluid gels from a standard nasal spray device (Figure 16.6). Moreover fluid gels also significantly enhanced both high acyl and low acyl gellan mucoadhesion properties [72].

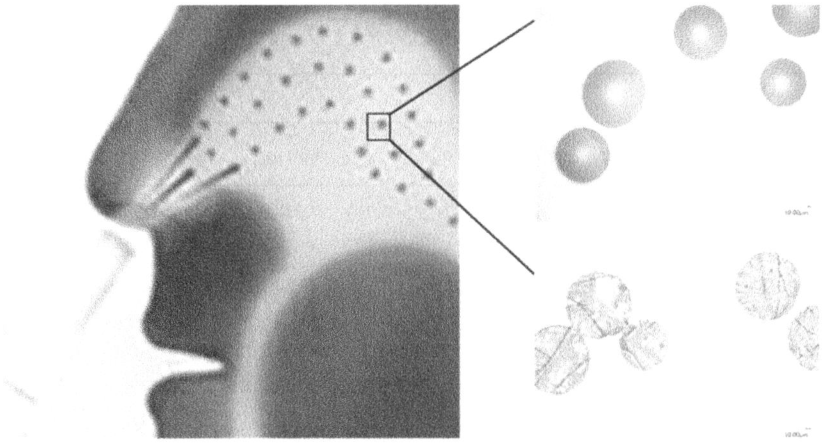

Figure 16.6 Shear thinning gellan gum fluid gel containing caffeine. Source: Adapted from [72].

16.5.2.1.3 Pectin

Pectin is a natural polysaccharide extracted from fruit with common sources being citrus peel and apple pomace. Its functionality is determined according to degree of esterification (DE) of the main component, galactuonic acid. High methoxyl (HM) pectins have a degree of esterification above 50% and LM pectins have a degree of esterification up to 50%. The underlying gelation mechanism involves ionic interactions between cations and functional groups of pectin chains, leading to the formation of an ordered three-dimensional network in the gel structure, commonly known as the "egg-box" structure. Gelation ability depends on the degree of esterification of the galacturonic acid; if degree of esterification is above 50%, then ionic gelation is not possible. The minimum number of successive non-methoxylated galacturonic acid residues required for coordination of the structure is estimated to be between 6–20. Low methoxyl (LM) pectin is more suited for nasal drug delivery as gelation can be induced in the presence of the divalent cations abundantly found in physiological nasal secretions. Calcium causes LM pectin to gel by linking adjacent pectin chains; ionic bonds are formed between a divalent calcium cation and the carboxyl group of two separate pectin chains resulting in an ordered three-dimensional network cross-linked by calcium ions.

Fentanyl pectin nasal spray (FPNS) (PecFent®, Archimedes Pharma UK Ltd) is a nasal spray formulation for delivery of fentanyl to the nasal mucosa. PecFent was approved in Europe in 2010 for the management of breakthrough cancer pain in adults based on three Phase III trials [73–75]. PecSys system allows the product to be sprayed into the front area of the nasal cavity as a fine mist of droplets, which gel on contact with the calcium ions present in the nasal mucosa. Fentanyl diffuses from the gel and is absorbed through the nasal mucosa; this gel-modulated absorption of fentanyl restrains the C_{max} while allowing the attainment of an early T_{max} [76]. PecFent is supplied in a device that delivers a spray of either 100 or 400 μg of fentanyl into the nasal cavity; this allows (using one or two sprays) doses ranging from 100 to 800 μg; and the device holds sufficient fentanyl to deliver eight sprays, Across those studies, FPNS provided an onset of pain relief (change in pain intensity), as early as five minutes as well as an overall pain relief that was sustained for at least the typical duration of an episode of breakthrough pain in cancer. The PecSys® formulation reduces the likelihood of either forwards or backwards drip [77] with a resulting improvement in bioavailability. The pharmacokinetic parameters (Table 16.5) of PecFent demonstrate an attenuated peak plasma concentration (C_{max}) while allowing an early time to that peak (T_{max}) [76].

Table 16.5 Pharmacokinetic parameters in adult subjects receiving PecFent and an oral transmucosal formulation (OTFC).

Pharmacokinetic parameters (mean (%CV))	PecFent				OTFC
	100 μg	200 μg	400 μg	800 μg	200μg
T_{max} (h)[a)]	0.33 (0.08–1.50)	0.25 (0.17–1.60)	0.35 (0.25–0.75)	0.34 (0.17–3.00)	1.50 (0.50–8.00)
C_{max} (pg/ml)	351.5 (51.3)	780.8 (48.4)	1 552.1 (26.2)	2 844.0 (56.0)	317.4 (29.9)
AUC (pg h/ml)	2 460.5 (17.9	4 359.9 (29.8)	7 513.4 (26.7)	17 272 (48.9)	3 735.0 (32.8)
$t_{1/2}$ (h)	21.9 (13.6)	24.9 (51.3)	15.0 (24.7)	24.9 (92.5)	18.6 (31.4)

a) Data for T_{max} presented as median (range).
Source: adapted from [76].

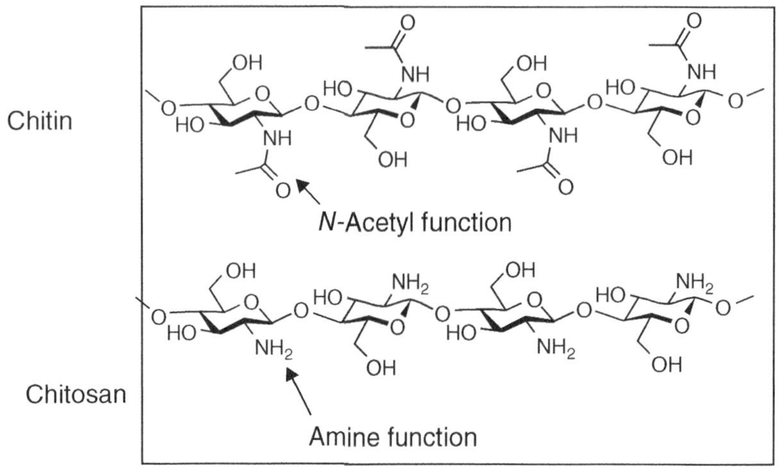

Figure 16.7 Chitin and chitosan. Source: Adapted from [8].

An inactivated H5N1 influenza vaccine based on the GelVac™ nasal powder formulation has been approved for human testing by the US Food and Drug Administration (FDA), and a Phase I clinical study is underway. GelVac from Nanotherapeutics is comprised of GelSite, an *Aloe vera* L.-derived anionic polysaccharide polymer related to pectin with mucoadhesive properties and is characterized by a high galacturonic acid (Gal UA) content of >90%, an exceptionally low DM (degree of methylation) of <10%, and a high molecular weight of >400 kDa (weight average).

16.5.2.1.4 Chitosan

Chitosan, a linear cationic polysaccharide derived by N-deacetylation of chitin, is considered as a biocompatible and biodegradable polymer bearing mucoadhesive properties. Chitosan has very low toxicity and is soluble in aqueous acidic environments. It comprises randomly ordered units of 2-acetamide-2-deoxy-β-D-glucopyranose (GlcNAc) and 2-amino-2-deoxy-β-D-glycopyranose (GlcN) (Figure 16.7). In solution, the amino functions of chitosan are protonated which allows chitosan salts to bind strongly to negatively charged materials such as cell surfaces and mucus. This binding delays the clearance of chitosan formulations from the nose. The mucoadhesion properties of chitosan depend both on the amount of sialic acid present in the mucin, as well the molecular weight and degree of acetylation. Chitosan can also affect the permeability of the epithelial membrane by the transiently opening tight junctions guarding the paracellular pathway without any serious toxic effects. Formulations can therefore be retained for extended time periods in the nasal cavity while at the same time enhancing transport of hydrophilic molecules across the membrane. Since it was shown that chitosan could significantly enhance the absorption of polar small molecular weight drugs as well as the larger peptides and proteins across the nasal epithelial membrane, the effects on absorption of different molecular weights of chitosan, different degrees of acetylation, different salt forms, different derivatives, and different formulations such as solutions, freeze dried powders, gels, microemulsions, nanoparticulates, and microspheres.

Despite the intensive use of chitosan in pharmaceutics, its low aqueous solubility in neutral and basic pH values has limited its effectiveness as absorption enhancer. To tackle this problem, chitosan derivatives, e.g. trimethylchitosan (TMC) which are readily soluble at these conditions have been synthesized. Chitosan derivatives (TMC) with different average molecular weight and degree of quaternization demonstrated thermosensitive behavior when co-formulated with polyethylene glycol and glycerophosphate. The gels containing TMC of medium average molecular weight and low degree of quaternization with a $T_{sol-gel}$ at 32.5 °C, exhibited exceptional mucoadhesive behavior and rheological properties, holding promise for nasal delivery [52].

Thermally gelling chitosan systems can be prepared by neutralizing chitosan solutions with polyols. A highly deacetylated chitosan solution was modified with β-glycerol phosphate to retain chitosan in solution at physiological pH. Upon heating to moderate temperatures, these solutions quickly transformed into a hydrogel structure as demonstrated by rheological measurements. Furthermore, the sol/gel transition temperature was pH-sensitive and gelling time was shown to be temperature-dependent [78]. Although chitosan with high degree of deacetylation generally precipitates above pH 6.2, neutralization at low temperature of such chitosan solutions by addition of βGP prevents precipitation. When temperature is increased, the solution turns into a viscoelastic gel that has been shown to be suitable as a drug delivery system. Several steps are involved in the gelation process, with chitosan chains becoming more flexible on neutralization, thus enhancing h-bonding between the chains. Increasing the temperature increases internal energy and reduces the H-bond formation between chitosan and water and then the hydrophobic side chains tend to aggregate [79].

16.5.2.1.5 Carbopol and Others

Carbopol, a cross-linked polyacrylic acid with a relatively high molecular weight has been used as a major component of gel drug delivery systems (including ophthalmic, buccal, and nasal) as a consequence of its relatively strong mucoadhesive properties. It exhibits sol–gel transition in aqueous solution at pHs above the pK_a (5.5). It contains a high percentage of carboxyl groups which are dissociated in acidic conditions, thus forming a flexible coil structure. At higher pH, the carboxyl groups are ionized, electrostatically repel each other; causing uncoiling and expansion of the molecule, the polymer chains become entangled resulting in the formation of a gel.

Xyloglucan, when partially degraded but β-galactosidase, exhibits a reversible sol–gel transition on heating. However, a low-transition temperature limits applications for nasal delivery.

16.5.2.2 Nasal Inserts

Nasal inserts can absorb nasal fluid from the mucosa and form a gel which adheres to the nasal mucosa owing to its mucoadhesive nature and enables the sustained release of the active in a controlled manner governed by the polymer's properties and thus acts as release controlling matrix providing a sustained drug delivery. Gradual decomposition of the gel combined with the mucociliary clearance from the nasal cavity toward the nasopharynx means there is no requirement to remove the insert mechanically after it is depleted of drug. Freeze drying can be used to produce highly porous sponge-like inserts, a prerequisite to assure rapid hydration and gelation upon contact with nasal mucosa. Based on the physical properties of the selected polymers, it was outlined that a molecular weight of 100 000 is required for proper bioadhesion to be achieved. Increase in polymer's MW resulted in gels of higher viscosity, lower dissolution rate and hence longer retention times in the nasal cavity, with the same effect also provoked by high polymer concentrations. Water uptake, the crucial step for gelation of the inserts, was determined to affect drug's release rate, correlating low water uptake with slower drug release rates. Carbopol, chitosan, HPMC K15M and E5, sodium alginate, sodium carboxy methylcellulose (NaCMC), polyvinyl pyrrolidone (PVP) 90, xanthan gum have been used as the main matrix components of the nasal inserts [80]. Promising polymers with intended drug delivery, over eight hours, are carrageenan, xanthan gum, and NaCMC [81].

16.5.2.3 Microspheres and Nanospheres

Therapeutic agents can be encapsulated within micro- and nanoparticles or incorporated via surface adsorption or surface conjugation. The integration of therapeutic agents into the nanoparticles with desirable shape, size, and surface physicochemical characteristics imposes a significant effect on improving their solubility, circulation half-life, bio-distribution, and reducing their immunogenicity. The preparation of nano- or microparticulate systems with various polymers has been widely reported particularly for incorporating macromolecular drugs. Microspheres are produced by various methods such as spray-drying or emulsification followed by crosslinking and in general been applied as a dry formulation to the nasal cavity. The application of microparticles to prolong residence time in the nasal cavity was introduced in 1987 [82]. It was reported that microspheres of albumin, starch, and diethyl aminoethyl (DEAE)-dextran absorbed water and formed a gel-like layer which was cleared slowly from the nasal cavity. Three hours after administration, 50% of the albumin and starch microspheres and 60% of the dextran microspheres were retained at the deposition site. It was suggested that this increased contact time could increase the absorption efficiency of drugs. As proposed, the relative intranasal bioavailability (vs. subcutaneous) of human growth hormone in sheep was increased from 0.1% for the solution to 2.7% for the degradable starch microsphere formulation. The addition of absorption enhancer, lysophosphatidylcholine, further enhanced growth hormone absorption as a relative bioavailability of 14.4% was achieved [83]. A range of drugs have been reported including ondansetron, insulin, and methotrexate [4]. As long as the particle size of the microspheres is above 10 μm, there is no increased chance of the formulation reaching the lungs after application. A variety of cross-linked ondansetron loaded chitosan microspheres and gellan gum microspheres have been shown to promote drug absorption and to remarkably improve its bioavailability in rabbits [84].

Chitosan and its derivatives form nanoparticles by a precipitation/coacervation method using tripolyphosphate, a precipitating agent. Many other derivatives and modifications of chitosan have been developed and studied for their utility as nasal particulate delivery systems for biomolecules, including low-molecular-weight chitosan, sugar-conjugated chitosan, and chitosan-dextran sulfate. Poly-(lactide co-glycolide) (PLG) and poly(lactic acid) (PLA) are the principal synthetic polymers investigated for encapsulation of antigens, primarily because these polymers are biodegradable and biocompatible. These polyester polymers have also been approved for use as controlled release vehicles and suture material. The method most commonly used for the preparation of polyester-based particulate vaccines is the double emulsification-solvent evaporation method using dichloromethane or ethyl acetate as the polymer solvent.

Nano-sized drug delivery systems take on novel properties and functions that differ markedly from those seen from items made of identical materials. The small size, customized surface, improved solubility, and multi-functionality of nanoparticles will continue to open many doors and create new biomedical applications. Some such as those made from chitosan combine this with the ability to take up water and form a mucoadhesive gel. Table 16.6 contains selected studies on utilization of hydrogel nanoparticles for nasal drug delivery. A review of nanoparticle formulations, for example those made from absorption enhancing polymers such as chitosan, did not appear to improve the efficiency of therapeutic drugs over and above the efficiency found when using solution or powder control systems of the same polymer used for producing the nanoparticles [102]. In many papers, it was also noted that a solution control of the absorption enhancing polymer was not included as a comparison and hence it was difficult to judge the effect of the nanoparticle formulation.

To enable sustained or targeted drug delivery to mucosal surfaces, nanocarriers must quickly permeate the mucus gel layer – a long standing challenge in drug delivery. The development of particulate carrier systems overcoming the mucus gel layer as soon as possible and reaching the absorption membrane is therefore of common interest. Mucus permeating particles can be divided into active and passive systems. Passive systems are designed to avoid interaction with mucus as feasible and promising systems are those particles exhibiting a slippery surface and self-nanoemulsifying-drug-delivery-systems (SNEDDS) [103]. Active systems interact with the mucus making it leakier for particles. These systems are mainly based on disulfide bridge breaking agents and on proteolytic enzymes. Strategies to avoid a back-diffusion of particles out of the mucus gel layer are also active systems and are so far based on thiomers and zeta potential changing systems.

16.5.2.4 Liposomes

Liposomes are phospholipid vesicles composed of lipid bilayers with one or more aqueous compartments wherein small and large molecules with a wide range of hydrophilicity and pK_a values can be effectively encapsulated. They consist of an aqueous core for encapsulation of hydrophilic drugs, while amphiphilic and lipophilic drugs can be solubilized within the phospholipid bilayer. Conventional liposomes are characterized by lower plasma circulation times as a result of elimination by the reticuloendothelial system. Liposomes have been investigated as nasal drug delivery systems, and absorption enhancing effects were found for insulin and calcitonin in *in vitro* permeability studies [104]. The enhancement effect was attributed to increased retention of the drugs within the nasal cavity. The best carrier effect for calcitonin was demonstrated with cationic liposomes as they were found to adhere intimately to the nasal mucosal surface, facilitating the penetration of the encapsulated drug. Similar observations were made for desmopressin-loaded cationic liposomes which resulted in enhanced antidiuretic effects in rats compared with anionic liposomes and solutions [105]. Muramatsu et al. [106] showed increased nasal absorption of insulin in liposomes that had a relatively high membrane fluidity compared to more rigid particles. However, the absorption enhancing effect of liposomes is difficult to separate from the enhancing effects of their components such as phosphatidylcholines and steryl glycosides. Moreover, proliposomes have also shown potential in nasal drug delivery. Proliposomes are dry, free-flowing granules composed of sorbitol as carrier and lipids that form a liposomal dispersion on contact with water. Their advantages are the combination of a fast onset provided by drug on the surface and prolonged drug action due to the encapsulated portion as demonstrated for propranolol and nicotine [107, 108]. Acyclovir does not undergo degradation nor metabolism in the nasal cavity, an alternative approach based

Table 16.6 Selected studies on utilization of hydrogel nanoparticles for nasal drug delivery.

Nanogel	Drug	Outcome	References
CHP	Tumor necrosis factor-α	Superior storage stability, high immune-increasing capacity for stimulation IgG_1, and mucosal IgA responses	[85]
cCHP	Nontoxic subunit fragment of clostridium botulinum type-A neurotoxin BoHc/A	Antigen adhesion to the nasal epithelium and its efficient uptake via mucosal dendritic cells after release from nanogel, induction of the specific IgG, and secretory IgA antibody responses with no co-administration of mucosal adjuvant	[86]
cCHP	Pneumococcal surface protein A (PSpA)	Prolonged retention of PSpA in the nasal cavity compared to PSpA alone, efficient induction of PspA-specific serum IgG associated with mucosal secretory IgA antibody	[87]
Alginate coated chitosan nanogel	Recombinant NcPDI	Protection of all mice against infection with Neospora caninum tachyzoites	[88]
Chitosan	Didanosine	Higher brain/plasma, olfactory/plasma, and CSF/plasma concentration post nasal administration	[89]
Thiolated chitosan	Leuprolide	Increasing leuprolide transport across nasal mucosa and its plasma concentration compared to nasal solution alone	[90]
Chitosan	Olanzapine	Enhanced systemic absorption of drug	[91]
Chitosan	Cholinesterase inhibitor	Effective delivery of cholinesterase inhibitor across nasal mucosa to reach the brain	[92]
Chitosan	Memantine hydrochloride	Potential to treat Alzheimers disease by transport through olfactory nasal route to the brain	[93]
Chitosan	Plasmid DNA	Evoking humoral and cellular immune responses, efficient DNA vaccine vehicle, and adjuvant for nasal immunization	[94]
Chitosan	Piperine	Efficient, safe, and noninvasive piperine delivery with 20-times decrease in oral dose	[95]
Mannosylated chitosan	Anti-GRP DNA	Significant titers of anti-GRP IgG, ability to suppress the growth of tumor cells	[96]
Thiolated chitosan	Selegiline hydrochloride	Reducing the oxidative stress and restoring the activity of mitochondrial complex, a severe reduction in the period of immobility time upon treatment, as a promising treatment for emerging diseases such as depression	[97]
Chitosan	A/H5N1 influenza vaccine	Stimulating and increasing the rate of specific immune responses and HI titer in mice models	[98]
Chitosan	Tetanus toxoid antigen	A prolonged humoral immune response (IgG levels) and the mucosal responses (IgA levels) compared to the fluid vaccine	[99]
Alginate	Venlafaxine	Higher brain/blood ratio for intranasal venlafaxine nanoparticles compared to intranasal venlafaxine solution	[100]

Source: Adapted from [101].

on mucoadhesive liposomal gel formulation led to increased bioavailability and C_{max} of the liposomal formulations of acyclovir compared to free drug suspended in the gel, probably due to the prolonged contact time between the drug and the absorption site and the direct absorption through nasal mucosa [109].

16.5.2.5 Microemulsions and Nanoemulsions

Microemulsions (ME), also called nanoscaled emulsions, have been one of the most attractive formulations as potential drug carriers to enhance the uptake of relatively low-molecular-weight drugs, such as zolmitriptan and sumatriptan across nasal mucosal membrane; this probably results from the lipophilic nature and the small globule size of the microemulsions [110]. Oil and surfactants included in the formulation can improve the solubility of poorly water-soluble drug and its penetration across the nasal mucosa. The liquid state of ME formulation provides several advantages, such as a rapid onset time, precise dosing by specialized devices, and low irritation to the mucosa, compared to solid formulations.

Nanoemulsions are oil-in-water (O/W) or water-in-oil (W/O) dispersions of two immiscible liquids stabilized using an appropriate surfactant. Their mean droplet diameter is usually 200–500 nm giving them a clear or hazy appearance which differs from milky appearance of a coarser emulsion (whose micron sized droplets partake in multiple light scattering). Nanoemulsions despite having the same droplet size range as microemulsions, differ tremendously in structural aspects and long-term thermodynamic stability.

Nanoemulsions generally contain 5–20% oil/lipid droplets in case of O/W emulsions, although it is possible to incorporate up to 70% in some cases. Selection of the lipid is generally determined by relative solubility of drug. Re-esterified fractions derived from soybean oil, sesame oil, cottonseed oil, safflower oil, coconut oil, ricebran oil [referred to as long-chain triglycerides (LCT), medium-chain triglycerides (MCT), or short-chain triglycerides (SCT) depending on their chain lengths] are used either alone or in combination to formulate nanoemulsions.

As a drug delivery system, nanoemulsions have been studied for low-molecular-weight drugs such as selegiline and paroxetine (Table 16.7) and have been employed as carriers for antigen presentation to dendritic cells either infiltrating or lying underneath the epithelial cell lining of nasal mucosa including influenza and help to retain potency after actuation in spray devices.

16.5.2.6 Combination/Hybrid Products and Others

Combination products seek to combine or find synergistic outcomes of the benefits of different formulations. Nanogels are high-viscosity systems containing nanoparticulates in a polymer network. Interest in these has grown with the development of precision dosing devices, and deposition is strongly influenced by the mode of administration, with applicators designed to facilitate spreading to increase the treated area. For examples, resveratrol nanosuspensions were incorporated into an *in situ* gelling gellan gum matrix [119]. The rheology results established that the DGG *in situ* gel demonstrated the characteristics of shear thinning, thixotropy, and yield stress. The drug release rate exhibited significantly faster profiles than corresponding coarse suspensions in the gels. The cumulative release amount from resveratrol nanogels approached nearly 50% at two hours and approximately 100% at six hours in an *in vitro* release study and pharmacokinetic studies suggested a marked increase in the bioavailability of resveratrol in the brain obtained by the formulation, adding further evidence for a direct nose-to-brain route. The hybrid system of combining nanosuspensions and the DGG *in situ* gel solution offsets the major shortcomings of conventional nasal drops by virtue of its nanoparticle advantages and ionic sensitive gel viscosity merits, which enhance drug bioavailability.

Table 16.7 Nanoemulsions employed in intranasal drug delivery.

Drug	Dispersed phase	Surfactant	Method	Purpose	Size (nm)	References
Paliperidone	Labrafil M1944CS	Tween 80, Tween 20	Spontaneous emulsification	Compare efficiency of nanoemulsion and aqueous micelle system in schizophrenia	38.25	[111]
Selegiline	Grape seed oil	Solutol®, Labrasol®	High energy emulsification	Direct nose to brain delivery for improved bioavailability	60	[112]
Fluoxetine HCl	Capmul MCM	Labrasol®	Spontaneous emulsification	Selection and optimization of excipients for intranasal delivery	—	[113]
Quetiapine	Capmul MCM	Tween 80	Vortex and sonication	Higher transport efficiency and direct nose-to-brain transport	144 ± 0.5 nm	[114]
Paroxetine	Capmul MCM	Solutol HS	Spontaneous emulsification	Direct nose-to-brain transport, permeation enhancement	58.47 ± 3.02	[115]
Saquinavir mesylate	Capmul MCM	Tween 80	Spontaneous emulsification	Improved brain targeting of HIV, penetration of BBB	176.3 ± 4.21	[116]
Olanzapine	Capmul MCM	Tween 80	Spontaneous emulsification			[117]

Source: Adapted from [118].

Ethosomes are soft malleable vesicles composed mainly of phospholipids, ethanol (at relatively high concentrations) and water, initially proposed to enhance transdermal delivery. Thermoreversible gels containing zolmitriptan-loaded nanoethosomes have been formulated using poloxamer 407 and carbopol 934 as thermoreversible and mucoadhesive polymers and increased residence of the drug into the nasal cavity [120].

Carbon nanotubes (CNTs) are emergent tools in the province of nanomedicine and nanobiotechnology. This is because they can be easily manipulated and modified by encapsulation with biopolymers or by covalent linking of solubilizing groups to the external walls and tips. CNTs exclusively composed of carbon atoms organized in a series of condensed benzene rings, which are rolled up into a tubular configuration. CNTs can be classified as single-walled carbon nanotubes (SWCNTs) and multi-walled carbon nanotubes. CNTs can also be explored for nose to brain targeted drug delivery by considering surface engineering approach so as to enhance the bioavailability and therapeutic efficacy of therapeutic agents that otherwise finds difficulty for delivery by any other route.

A quantum dot (QD) is a nano-crystal comprised of semiconductor ingredients which exhibit quantum mechanical properties due to their nanoscale magnitude. As they have the capacity to elucidate the pharmacokinetics and pharmacodynamics of therapeutic agents, the potential development of traceable drug delivery carriers is driving research in these applications. Cadmium QDs are known to have neurological effects and following short-term inhalation in mice, rapid uptake into the brain/olfactory bulb via axonal transport has been shown [121].

16.5.3 The Nasal Route and the Blood–Brain Barrier

There has been much recent interest in the development of various nanocarrier delivery systems such as solid-lipid nanoparticles, nanoemulsions, liposomes, and polymeric nanoparticles that can be delivered nasally to circumvent the blood–brain barrier (BBB). The brain and the nasal cavity are connected by the olfactory and trigeminal nerve system, which terminates at the olfactory neuroepithelium or respiratory epithelium (Figure 16.8). These two nerve systems can serve as externally accessible points of the brain and can be exploited to bypass the BBB for direct nose-to-brain drug delivery In bypassing the BBB, drugs that normally cannot enter the CNS may reach therapeutically effective levels when administered intranasally. Drugs travel via the olfactory nerve axons, accumulate in olfactory bulbs (OBs), and diffuse into the brain. Three pathways across the olfactory epithelium may be proposed (i) a transcellular pathway, especially across the sustentacular cells of the olfactory epithelium, most likely by receptor-mediated endocytosis, fluid phase endocytosis, or passive diffusion (unlikely for peptides); (ii) a paracellular pathway through tight junctions between the sustentacular cells of the olfactory epithelium and olfactory neurons; and (iii) the olfactory nerve pathway where the drug is taken up into the neurons by endocytotic or pinocytotic mechanisms and transported by intracellular axonal transport to the OB. It is likely that a combination of all pathways is responsible, although one pathway may predominate, depending on the properties of the therapeutic, the characteristics of the formulation and excipients, and the delivery device used [122]. Also, there is the potential to reduce doses for those drugs that can pass the BBB but require large doses to provide therapeutically relevant brain levels, may with a subsequent decrease in adverse effects. Formulations and their drug delivery strategy include chitosan nanoparticles, solid lipid nanoparticles, and mucoadhesive microemulsions and are summarized in Table 16.8 [151].

Successful targeting of nose-to-brain drug delivery requires formulation to be administered in such a way that the amount of drug deposited on the olfactory epithelium is maximized. Novel devices are being developed which attempt to overcome the barriers of the nasal cavity anatomy and

Table 16.8 Drugs and their formulations reported for nose-to-brain delivery.

Drug	Formulation description	Animal model	Disease state being treated	Results	References
5-FU	Solution	Rats pre-dosed with acetazolamide	CNS malignancy	104% increased brain uptake compared to i.v.	[123]
Bromocriptine	Chitosan nanoparticles (~160 nm)	Mice	Parkinson's disease	Showed significant increase in dopamine levels	[124]
Buspirone	Chitosan/HP-β-CD solution	Rats	Depression	DTE-4.13 compared with 3.38 for i.n. plain solution	[125]
Carbamazepine	Hypromellose/carbopol gel	Rats	Epilepsy	Significantly higher brain uptake compared to i.v.	[126]
Carbamazepine	Thermoreversible gel	Mice	Epilepsy	DTE-0.98 i.n. and i.v. provide similar blood/plasma ratios	[127]
Curcumin	*In situ* gelling Microemulsion	Rats	Brain tumor/Alzheimer's disease	DTE-6.5	[63]
Donepezil	Chitosan nanoparticles (~150–200 nm)	Rats	Alzheimer's disease	Significantly higher brain concentrations from nanoparticles	[128]
Doxepin	Thermoreversible gel	Mice	Depression	No difference in pharmacodynamic endpoint	[51]
Duloxetine	Lipid nanocarrier (~80–130 nm)	Rats	Depression	DTE-757.14% compared to 287.34% from solution	[129]
Estradiol	Cyclodextrin	Rats	Alzheimer's disease	AUCCSF/AUCplasma 1.60 which was significantly higher than 0.61 from i.v.	[130]
GDF-5	Microemulsion	Rats	Parkinson's disease	Significantly higher midbrain concentrations compared to acidic solution	[131]
Methotrexate	Mucoadhesive solution	Rats pre-dosed with acetazolamide	CNS malignancy	195% increase in uptake compared to i.n. without acetazolamide; 75% reduction in brain tumor weight	[132]

Methotrexate	Solution	Rats	CNS malignancy	DTE-21.7%	[133]
Morphine	Solution (PBS buffer at pH 7.4)	Rats	Pain	Brain/Plasma AUC ratio of 3 after i.n. use and 0.1 after i.v. use	[134]
Nimodipine	Microemulsion	Rats	Stroke, reduce dementia	Higher AUC in olfactory bulb but lower AUC in rest of brain after i.n. compared with i.v. treatment	[135]
Olanzapine	Nanomicellar carrier (~18–380 nm)	Rats	Schizophrenia/bipolar disorder	DTE-520.26%	[136]
Olanzapine	PLGA nanoparticles (~90 nm)	Rats	Schizophrenia/bipolar disorder	10.86 times higher brain uptake compared to i.n. solution alone	[137]
Olanzapine	Mucoadhesive nanoemulsion (~20 nm)	Rats	Schizophrenia/bipolar disorder	DTE-890% compared to 550% from i.n. solution	[118
Paliperidone	Mucoadhesive microemulsion	Rats	Schizophrenia/bipolar	DTE-320.69%; 1.74-fold higher than nasal solution alone	[138]
Raltitrexed	Solution (PBS pH 8)	Rats	CNS malignancy	DTE for Olfactory Bulb, Cerebrum and cerebellum was 127, 120 and 71 respectively	[139]
Rasagiline	Thermosensitive gel	Rabbits	Parkinson's disease	Significant improvement in brain uptake from gel formulations	[140]
Remoxipride	Solution (normal saline)	Rats	Psychosis	~50% increase in brain/plasma AUC	[141]
Risperidone	Mucoadhesive nanoemulsion (~16 nm)	Rats	Schizophrenia/bipolar disorder	DTE-476%	[142]
Risperidone	Solid lipid nanoparticles (~150 nm)	Mice	Schizophrenia/bipolar disorder	10-fold higher brain AUC compared to i.v. solution	[143]

(continued)

Table 16.8 (Continued)

Drug	Formulation description	Animal model	Disease state being treated	Results	References
Ropinirole	Temperature sensitive *in situ* gel with chitosan and HPMC	Rats	Parkinson's disease	DTE-10.4 compared to 5.3 for solution alone	[144]
Saquinavir	Nanoemulsion (~180 nm)	Rats	CNS involved HIV infection	~62 times higher drug accumulation compared to i.v. suspension	[116]
Tacrine	Solution of propylene glycol and normal Saline	Mice	Alzheimer's disease	DTE-207.23%	[145]
Tacrine	Mucoadhesive microemulsion	Mice	Alzheimer's disease	DTE-295.87%	[146]
Testosterone	Noseafix® mucoadhesive system	Mice	CNS hormone replacement	Significantly higher brain levels except frontal cortex	[147]
UH-301	Solution (normal saline)	Rats	Depression	No difference in CSF concentrations between i.n. or i.v.	[148]
Zidovudine-prodrug	Solid lipid microparticles (16 and 7 μm)	Rats	CNS involved HIV infection	Sixfold higher CSF uptake	[149]
Zolmitriptan	Micellar nanocarrier (~23 nm)	Rats	Migraine	Significant increase in brain concentrations as soon as 30 min; up to 120 min	[150]

Source: Adapted from [151].

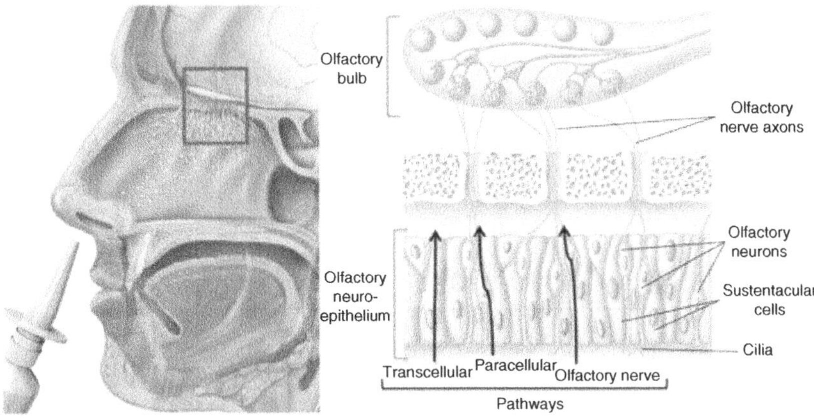

Figure 16.8 Cross section of the human nasal cavity with the olfactory region and its structures labeled. Pathways of nose-to-brain transport of drug after administration of formulation are also indicated. Source: adapted from [122].

target formulation deposition in the olfactory region of the nasal cavity [151]. Charlton et al. [152] compared the residence time on the epithelium and the abilities of several polymeric formulations such as pectin and chitosan to reach the olfactory region. Endoscopic examination concluded that all bioadhesive formulations were retained on the olfactory epithelium for more than four times longer than relative control solutions.

16.5.4 The Nasal Route for Vaccinations

The epithelium of the airways and nasal mucosa contain Microfold cells (M cells), which can transcytose particulate matter across the epithelial layer to underlying lymphoid follicles of the musical associated lymphoid tissue for subsequent uptake, processing an initiation of immune responses. Targeting particulates to the nasal associated lymphoid tissue (NALT) enables generation of both cellular and humoral immune responses against pathogens (Figure 16.9). Mucosal vaccination offers logistical advantages over parenteral administration, such as lower costs, ease of administration, higher patient compliance, elimination of risks related to infections at the site of injection, avoidance of needle disposal, and higher capacity for mass immunizations. Additionally mucosal vaccines (both oral and nasal) are more easily distributed, allow the delivery of multiple antigens and the induction of both systemic and mucosal immune responses.

The first studies to suggest a potential role for nanoparticles in nasal delivery of vaccines were published in the late 1990s. An early HIV vaccine study demonstrated that a gp160 DNA-based intranasal vaccine induced stronger serum and mucosal antibody titers than the equivalent intramuscular vaccine. Furthermore, coadministration with cationic liposomes which bound the negatively charged DNA antigen adjuvanted both humoral and cell-mediated vaccine specific immune activity [154]. The use of biodegradable polymeric nanocarriers as drug delivery systems (DDS) for the mucosal delivery of peptise and protein drugs and antigens has received considerable attention. Using various polymeric materials and formulation processes facilitates the modulation of physicochemical properties (e.g. surface charge and mucoadhesiveness), drug loading, drug release profile, and biological behavior of the delivery system. The most investigated polymers are chitosan (CS) and its derivatives (e.g. *N*-trimethyl chitosan, (TMC), MCC, etc.) as well as polyesters such as

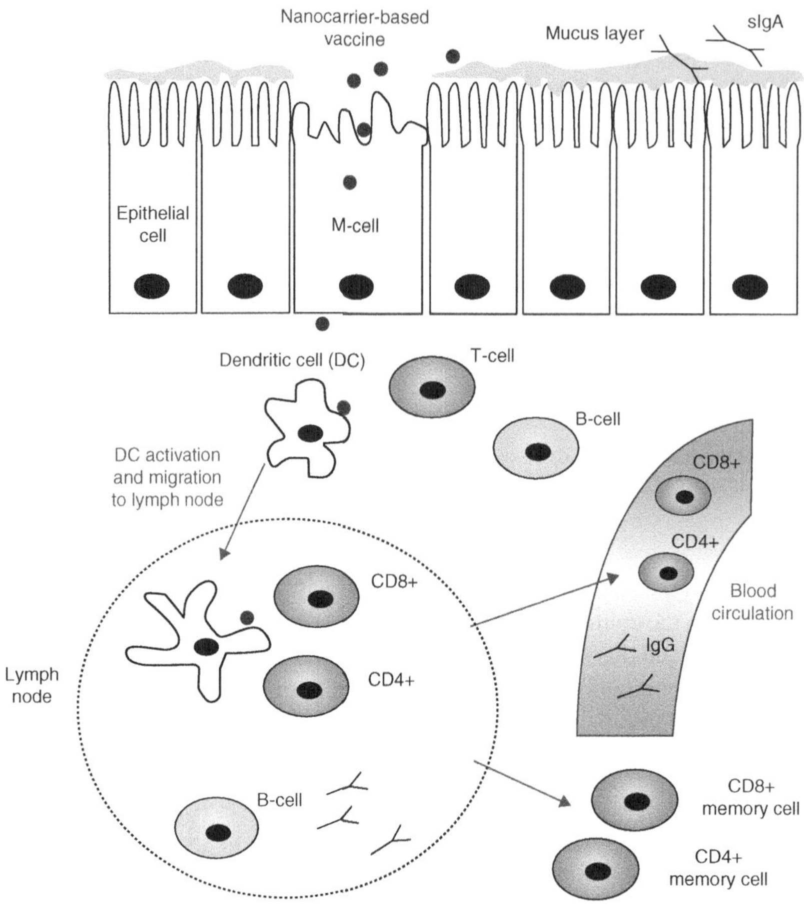

Figure 16.9 Development of cellular and humoral immune responses following mucosal administration of nanocarrier-based vaccines. Source: Adapted from [153].

poly(lactic acid) (PLA) and poly(lactic-co-glycolic acid) (PLGA), and their copolymers. Other polymeric materials including poly(ε-caprolactone) (PCL), polyacrylic acid (PAA), poly(methacrylic acid) (PMAA), polyalkyl cyanoacrylates (PACA), starch, dextran, alginate, etc. have been also used as carriers for the mucosal delivery of protein based drugs and vaccines [153].

Chitosan and its derivatives are the most extensively used biopolymers for nasal delivery of vaccines both in the form of particulate and simple solution systems. Chitosan solution formulations have been developed for various antigens such as proteins, recombinant toxins, filamentous hemagglutinin, and DNA-based plasmids for prevention of diseases which are primarily spread by infection through the nasal mucosa [4]. Chitosan in a powder formulation prepared by spray drying can further enhance the beneficial effects of chitosan in nasal delivery of vaccines due to preferential uptake of the particulate system by M cells. The fact that chitosan, in various forms, consistently shows enhanced immune responses in different studies suggests chitosan as a potential formulation component in the intranasal particulate delivery systems.

Several strategies have been pursued to further develop strategies for effective nasal vaccine/immunotherapy treatment using nanoparticles [155]. Using cationic cholesteryl pullulan nanogels (cCHPs) as vaccine carriers, Nochi et al. demonstrated increased retention in the nasal mucosa and uptake of antigen by mucosal dendritic cells. This enhanced antigen delivery produced significantly increased mucosal antibody responses and protection from challenge [86]. Encapsulated influenza protein antigen in chitosan nanoparticles was shown to induce stronger antibody titers than soluble antigen alone and conferred 100% protection against an intranasal influenza challenge [156]. Polyanhydride nanoparticles with encapsulated antigen were designed to release antigen over a 28-day period. When immunized via intranasal administration, mice receiving antigen-loaded nanoparticles produced sustained titers were protected against *Yersinia pestis* challenge [157].

The first nasal influenza product was introduced in Europe in market in 2001 (but is now withdrawn due to potential toxicological problems), and Fluenza Tetra, a live attenuated influenza vaccine is currently available as a suspension for administration to children. Fluenz Tetra is administered as a divided dose in both nostrils. GelVac is a dry powder formulation with GelSite®, which is an *Aloe vera* L.-derived polysaccharide polymer with mucoadhesive properties. In the presence of divalent cations, GelVac is capable of *in situ* gelation which improves mucosal residence time of intranasally administered vaccine

16.5.5 *In Vitro/in Vivo* Models for Nasal Absorption

Many different techniques have been developed to help assess nasal drug absorption. *In vitro* approaches are more amenable to studying mechanistic aspects of nasal delivery and drug permeation. Preclinical studies assessing transmucosal drug transport or toxic effects of excipients are normally performed using excised tissue, primary cell cultures, or immortalized cell lines. Excised human tissue is not only hard to obtain but also presents inter- and intra-individual differences. Therefore, it is more common to perform permeation studies using excised mucosa from experimental or slaughtered animals, which introduces other drawbacks, such as challenging species differences and ethical concerns [158, 159]. The difficulty in the assessment of nasal retention and absorption is that there are various different dosage forms which require different experimental conditions for testing which are usually only comparative between their own samples, thus there appears to be a degree of heterogeneity between techniques. It is known that transepithelial electrical resistance (TEER) of human nasal mucosa obtained from inferior turbinates ranges from 75 to 180 V cm [2], and can differ greatly from other animal tissues [158]. *In situ* perfusion studies in models such as the rat allow investigation of drug absorption kinetics but residence time is falsely prolonged [160]. Mouse, rat, guinea-pig, hamster, rabbit, dig, sheep, and monkey have all been used to study nasal delivery, and the choice of animal depends on the study in question, with smaller rodents being used for toxicity, while larger animals are required for pharmacokinetic studies. There are variations in the volume that can be administered and anatomy of the nasal mucosae.

In order to avoid animal experiments or the use of excised tissues, *in vitro* cell models, such as RPMI 2650 cells have been developed for assessing cytotoxicity and drug transport through the nasal epithelium. The RPMI 2650 cell line originates from an anaplastic squamous cell carcinoma of the nasal septum and has been reported to display consistent growth and high stability with no alteration to the normal diploid karyotype [161, 162]. The cells reach confluence relatively quickly, can be developed in both monolayer and multilayer formations, form a polarized epithelium resembling nasal mucosa, and produce a mucoid material on their cell surfaces similar to normal human nasal epithelium although the mucin expression and the absolute mucin content are less well defined [163]. However, different culture conditions have a significant effect on cell ultrastructure, barrier integrity, and gene expression and should be considered when using this

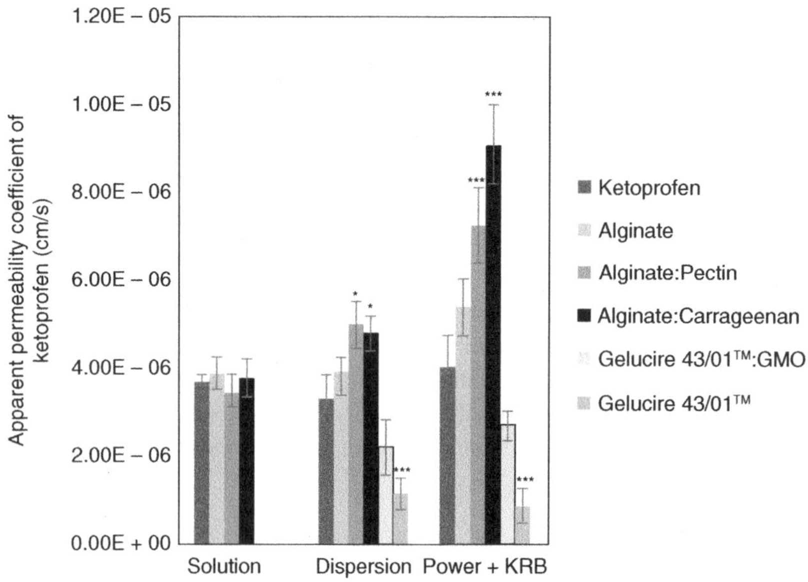

Figure 16.10 Effect of different types of formulation (ketoprofen or ketoprofen-loaded microparticles) on ketoprofen apparent permeability coefficients through RPMI 2650 multilayer when applied as solution, dispersion or powder + KRB, after 60 minutes of incubation (mean ± SD, $n \pm 3$); statistical differences in relation with pure ketoprofen when administered in the same form are denoted as $^{*}P < 0.05$; $^{***}P < 0.001$. Source: from 165.

cell line as an *in vitro* model for drug permeability studies and screening of nasal drug candidates [158]. Generally, exploring the deposition of solid formulations onto nasal cell layers with further transepithelial transport rate of drugs has been poorly studied or reported.

The model forms an enzymatic metabolic barrier similar to that of excised human nasal tissue. RPMI 2650 cells also form polarized epithelium of cells interconnected with the cell junction proteins ZO-1, occludin, claudin-1, E-cadherin, and β-catenin and have been demonstrated to develop microvilli [163, 164]. The model has exhibited an organotypic permeation barrier consisting of a tight homogenous cell multilayer with a TEER closer to physiological conditions than primary cell cultures, and permeation coefficients in the same range as human nasal mucosa excised nasal mucosa. RPMI 2650 cells have been reported and exhibit similar permeability as excised human nasal mucosa for paracellular markers such as sodium fluorescein, with analogous TEER. Permeation of drug from some solid systems has also been reported in RPMI 2650 multilayers including solid lipid particles, structured lipid particles, and microparticles of polysaccharides of alginate, pectin, and carrageenan [165] as an alternative to excised nasal mucosa (Figure 16.10).

Knowledge of the regional deposition of inhaled particles in the nose is important for both drug delivery and assessment of the toxicity of inhaled materials. Nasal deposition patterns gathered from *in vitro* and CFD experiments normally indicate low deposition efficiency in the posterior cavity, often the intended site. A dissolution, absorption, and clearance model has been used to the study the impact of suspension properties on uptake patterns [166]. Findings support the evidence that small suspended particles effect a quicker, more rapid dissolution for systemic absorption, while larger particles are more useful for targeted delivery of local treatment and are carried further by the cilia and spread over a larger area.

16.6 Conclusion

Intranasal delivery has also been lauded as an alternative to invasive methods for delivery of drug to the CNS, to bypass the blood–brain barrier and directly target the CNS. The pathway utilizes the olfactory and trigeminal nerves, minimizing exposure of the rest of the body. The continuous developing of new devices that are better at delivering a more precise dose of drug to the specific nasal regions, makes it a much more attractive route for systemic drug delivery. Systemic delivery of the potent new biotechnological drugs that may be the pharmaceutical products of the future may be achieved by this route. Targeting the mucosal associated lymphoid tissue via nasal delivery can also elicit strong systemic and local immune responses. The encapsulation of the antigen into bioadhesive *in situ* gels is a promising approach toward successful nasal vaccine delivery. Additionally, an increasing number of studies have shown that nose-to-brain drug transport offers a potential strategy to enhance drug penetration into the CNS. As nanoencapsulation methods continue to develop, the feasibility of scaling-up manufacturing becomes more attainable. Finally, the complexity of nasal cavity combined with the different properties of the actives (small drugs, peptides, proteins, vaccines) delivered via the nasal route is leading to emerging nasal delivery devices and dispersing technologies, in an attempt to maximize the clinical performance of the formulations.

References

1 Fortuna, A., Alves, G., Serralheiro, A. et al. (2014). Intranasal delivery of systemic-acting drugs: small-molecules and biomacromolecules. *Eur. J. Pharm. Biopharm.* 88: 8–27.

2 Larhed, A.W., Artursson, P., Gråsjö, J., and Bjork, E. (1997). Diffusion of drugs in native and purified gastrointestinal mucus. *J. Pharm. Sci.* 86: 660–665.

3 Singh, R.M., Kumar, A., and Pathak, K. (2013). Mucoadhesive in situ gelling drug delivery systems for modulated drug delivery. *Expert Opin. Drug Deliv.* 10: 115–130.

4 Illum, L. (2002). Nasal drug delivery: new developments and strategies. *Drug Discovery Today* 7: 1184–1189.

5 Agu, R., MacDonald, C., Cowley, E. et al. (2011). Differential expression of organic cation transporters in normal and polyps human nasal epithelium: implications for in vitro drug delivery studies. *Int. J. Pharm.* 406: 49–54.

6 Shao, D., Massoud, E., Anand, U. et al. (2013). Organic cation transporters in human nasal primary culture: expression and functional activity. *Ther. Deliv.* 4: 439–451.

7 Arora, P., Sharma, S., and Garg, S. (2002). Permeability issues in nasal drug delivery. *Drug Discovery Today* 7: 967–975.

8 Illum, L. (2003). Nasal drug delivery—possibilities, problems and solutions. *J. Controlled Release* 87: 187–198.

9 Illum, L. (2012). Nasal drug delivery—recent developments and future prospects. *J. Controlled Release* 161: 254–263.

10 Merkus, F.W.H.M., Schipper, N.G.M., Hermens, W.A.J.J. et al. (1993). Absorption enhancers in nasal drug delivery: efficacy and safety. *J. Controlled Release* 24: 201–208.

11 Natsume, H., Iwata, S., Miyamoto, M. et al. (1996). Screening of absorption enhancers for nasal peptide and protein delivery. In: *Proceedings of International Symposium on Control Release Bioactive Materials, Kyoto, Japan, 7–10 July 1996*, vol. 23, 481–482.

12 Zhou, M. and Donovan, M.D. (1996). Recovery of the nasal mucosa following laureth-9 induced damage. *Int. J. Pharm.* 130: 93–102.

13 Chandler, S.G., Illum, L., and Thomas, N.W. (1991). Nasal absorption in rats. II. Effect of enhancers on insulin absorption and nasal histology. *Int. J. Pharm.* 76: 61–70.
14 Chandler, S.G., Illum, L., and Thomas, N.W. (1994). Nasal absorption in rats. III. Effect on lysophospholipids on insulin absorption and nasal histology. *Pharm. Res.* 11: 1623–1630.
15 Yamamoto, A., Morita, T., Hashida, M., and Sezaki, H. (1993). Effect of absorption promoters on nasal absorption of drugs with various molecular weights. *Int. J. Pharm.* 93: 91–99.
16 Aungst, B.J., Rogers, N.J., and Sheffer, E. (1988). Comparison of nasal, rectal, buccal, sublingual and intramuscular insulin efficacy and the effects of a bile salt absorption promoter. *J. Pharmacol. Exp. Therap.* 244: 23–27.
17 Bagger, M.A., Nielsen, H.W., and Bechgaard, E. (2001). Nasal bioavailability of peptide T in rabbits: absorption enhancement by sodium glycolate and glycofurol. *Eur. J. Pharm. Sci.* 14: 69–74.
18 Marttin, E., Verhoef, J.C., and Merkus, F.W.H.M. (1998). Efficacy, safety and mechanism of cyclodextrins as absorption enhancers in nasal drug delivery of peptide and protein drugs. *J. Drug Targeting* 6: 17–36.
19 Matsubara, K., Abe, K., Irie, T., and Uekama, K. (1995). Improvement of nasal bioavailability of luteinizing hormone-releasing hormone agonist, buserelin, by cyclodextrin derivatives in rats. *J. Pharm. Sci.* 84: 1295–1300.
20 Touitou, E. and Barry, B.W. (eds.) (2006). *Enhancement in Drug Delivery*. CRC Press.
21 Davis, S.S. and Illum, L. (2003). Absorption enhancers for nasal drug delivery. *Clin. Pharmacokin.* 42: 1107–1128.
22 Ghori, M.U., Mahdi, M.H., Smith, A.M., and Conway, B.R. (2015). Nasal drug delivery systems: an overview. *Am. J. Pharmacol. Sci.* 3: 110–119.
23 Oliveira, P., Fortuna, A., Alves, G., and Falcao, A. (2016). Drug-metabolizing enzymes and efflux transporters in nasal epithelium: influence on the bioavailability of intranasally administered drugs. *Curr. Drug Metab.* 17: 628–647.
24 Alsarra, I.A., Hamed, A.Y., Alanazi, F.K., and El Maghraby, G.M. (2010). Vesicular systems for intranasal drug delivery. In: *Drug Delivery to the Central Nervous System* (ed. K.K. Jain), 175–203. Springer-Science.
25 Suman, J.D. (2013). Current understanding of nasal morphology and physiology as a drug delivery target. *Drug Deliv. Transl. Res.* 3: 4–15.
26 Fransen, N., Bredenberg, S., and Bjork, E. (2009). Clinical study shows improved absorption of desmopressin with novel formulation. *Pharm. Res.* 26: 1618–1625.
27 Tanaka, A., Furubayashi, T., Tomisaki, M. et al. (2017). Nasal drug absorption from powder formulations: the effect of three types of hydroxypropyl cellulose (HPC). *Eur. J. Pharm. Sci.* 96: 284–289.
28 Smart, J.D. (2005). The basics and underlying mechanisms of mucoadhesion. *Adv. Drug Delivery Rev.* 57: 1556–1568.
29 Majithiya, R.J., Ghosh, P.K., Umrethia, M.L., and Murthy, R.S. (2006). Thermoreversible-mucoadhesive gel for nasal delivery of sumatriptan. *AAPS PharmSciTech* 7: E80–E86.
30 Zaki, N.M., Awad, G.A., Mortada, N.D., and El Hady, S.S.A. (2007). Enhanced bioavailability of metoclopramide HCl by intranasal administration of a mucoadhesive in situ gel with modulated rheological and mucociliary transport properties. *Eur. J. Pharm. Sci.* 32: 296–307.
31 Perez, A.P., Mundiña-Weilenmann, C., Romero, E.L., and Morilla, M.J. (2012). Increased brain radioactivity by intranasal P-labeled siRNA dendriplexes within in situ-forming mucoadhesive gels. *Int. J. Nanomed.* 7: 1373–1385.

32 Li, C., Li, C., Liu, Z. et al. (2014). Enhancement in bioavailability of ketorolac tromethamine via intranasal in situ hydrogel based on Poloxamer 407 and carrageenan. *Int. J. Pharm.* 474: 123–133.

33 Kempwade, A. and Taranalli, A. (2014). Formulation and evaluation of thermoreversible, mucoadhesive in situ intranasal gel of rizatriptan benzoate. *J. Sol-Gel Sci. Technol.* 72: 43–48.

34 El-Setouhy, D.A., Ahmed, S., Badawi, A.A.E.L. et al. (2015). Preclinical evaluation of dual action intranasal formulation intended for postoperative/cancer associated therapies. *Eur. J. Pharm. Sci.* 76: 48–56.

35 Qian, S., Wong, Y.C., and Zuo, Z. (2014). Development, characterization and application of in situ gel systems for intranasal delivery of tacrine. *Int. J. Pharm.* 468: 272–282.

36 Kumar, A., Garg, T., Sarma, G.S. et al. (2015). Optimization of combinational intranasal drug delivery system for the management of migraine by using statistical design. *Eur. J. Pharm. Sci.* 70: 140–151.

37 Ved, P.M. and Kim, K. (2011). Poly (ethylene oxide/propylene oxide) copolymer thermo-reversible gelling system for the enhancement of intranasal zidovudine delivery to the brain. *Int. J. Pharm.* 411: 1–9.

38 Bhandwalkar, M.J. and Avachat, A.M. (2013). Thermoreversible nasal in situ gel of venlafaxine hydrochloride: formulation, characterization, and pharmacodynamic evaluation. *AAPS PharmSciTech* 14: 101–110.

39 Pund, S., Rasve, G., and Borade, G. (2013). Ex vivo permeation characteristics of venlafaxine through sheep nasal mucosa. *Eur. J. Pharm. Sci.* 48: 195–201.

40 Singh, R.M., Kumar, A., and Pathak, K. (2013). Thermally triggered mucoadhesive in situ gel of loratadine: β-cyclodextrin complex for nasal delivery. *AAPS PharmSciTech* 14: 412–424.

41 Balakrishnan, P., Park, E.K., Song, C.K. et al. (2015). Carbopol-incorporated thermoreversible gel for intranasal drug delivery. *Molecules* 20: 4124–4135.

42 Basu, S. and Bandyopadhyay, A.K. (2010). Development and characterization of mucoadhesive in situ nasal gel of midazolam prepared with *Ficus carica* mucilage. *AAPS PharmSciTech* 11: 1223–1231.

43 Gabal, Y.M., Kamel, A.O., Sammour, O.A., and Elshafeey, A.H. (2014). Effect of surface charge on the brain delivery of nanostructured lipid carriers in situ gels via the nasal route. *Int. J. Pharm.* 473: 442–457.

44 Ozbılgın, N.D., Saka, O.M., and Bozkır, A. (2014). Preparation and in vitro/in vivo evaluation of mucosal adjuvant in situ forming gels with diphtheria toxoid. *Drug Deliv.* 21: 140–147.

45 Chu, K., Chen, L., Xu, W. et al. (2013). Preparation of a paeonol-containing temperature-sensitive in situ gel and its preliminary efficacy on allergic rhinitis. *Int. J. Mol. Sci.* 14: 6499–6515.

46 Chen, X., Zhi, F., Jia, X. et al. (2013). Enhanced brain targeting of curcumin by intranasal administration of a thermosensitive poloxamer hydrogel. *J. Pharm. Pharmacol.* 65: 807–816.

47 Jose, S., Ansa, C.R., Cinu, T.A. et al. (2013). Thermo-sensitive gels containing lorazepam microspheres for intranasal brain targeting. *Int. J. Pharm.* 441: 516–526.

48 Cho, H.J., Balakrishnan, P., Park, E.K. et al. (2011). Poloxamer/cyclodextrin/chitosan-based thermoreversible gel for intranasal delivery of fexofenadine hydrochloride. *J. Pharm. Sci.* 100: 681–691.

49 Wavikar, P.R. and Vavia, P.R. (2014). Rivastigmine-loaded in situ gelling nanostructured lipid carriers for nose to brain delivery. *J. Liposome Res.* 9: 1–9.

50 Li, X., Du, L., Chen, X. et al. (2015). Nasal delivery of analgesic ketorolac tromethamine thermo- and ion-sensitive in situ hydrogels. *Int. J. Pharm.* 489: 252–260.

51 Naik, A. and Nair, H. (2014). Formulation and evaluation of thermosensitive biogels for nose to brain delivery of Doxepin. *Biomed. Res. Int.* 2014: e847547.

52 Nazar, H., Fatouros, D.G., Van Der Merwe, S.M. et al. (2011). Thermosensitive hydrogels for nasal drug delivery: the formulation and characterisation of systems based on N-trimethyl chitosan chloride. *Eur. J. Pharm. Biopharm.* 77: 225–232.

53 Nazar, H., Caliceti, P., Carpenter, B. et al. (2013). A simple, once-a-day dosage form for the delivery of insulin through the nasal route: in vitro assessment and in vivo evaluation. *Biomater. Sci.* 1: 306–314.

54 Wu, J., Wei, W., Wang, L.Y. et al. (2007). A thermosensitive hydrogel based on quaternized chitosan and poly(ethylene glycol) for nasal drug delivery system. *Biomaterials* 28: 2220–2232.

55 Mahajan, H.S., Tyagi, V., Lohiya, G., and Nerkar, P. (2012). Thermally reversible xyloglucan gels as vehicles for nasal drug delivery. *Drug Deliv.* 19: 270–276.

56 Mahajan, H.S., Tyagi, V.K., Patil, R.R., and Dusunge, S.B. (2013). Thiolated xyloglucan: synthesis, characterization and evaluation as mucoadhesive in situ gelling agent. *Carbohydr. Polym.* 91: 618–625.

57 Cao, S.-L., Zhang, Q.-Z., and Jiang, X.-G. (2007). Preparation of ion-activated in situ gel systems of scopolamine hydrobromide and evaluation of its antimotion sickness efficacy. *Acta Pharmacol. Sin.* 28: 584–590.

58 Cao, S.L., Ren, X.W., Zhang, Q.Z. et al. (2009). In situ gel based on gellan gum as new carrier for nasal administration of mometasone furoate. *Int. J. Pharm.* 365: 109–115.

59 Saindane, N.S., Pagar, K.P., and Vavia, P.R. (2013). Nanosuspension based in situ gelling nasal spray of carvedilol: development, in vitro and in vivo characterization. *AAPS PharmSciTech* 14: 189–199.

60 Cai, Z., Song, X., Sun, F. et al. (2011). Formulation and evaluation of in situ gelling systems for intranasal administration of gastrodin. *AAPS PharmSciTech* 12: 1102–1109.

61 Hosny, K.M. and Hassan, A.H. (2014). Intranasal in situ gel loaded with saquinavir mesylate nanosized microemulsion: preparation, characterization, and in vivo evaluation. *Int. J. Pharm.* 475: 191–197.

62 Galgatte, U.C., Kumbhar, A.B., and Chaudhari, P.D. (2014). Development of in situ gel for nasal delivery: design, optimization, in vitro and in vivo evaluation. *Drug Deliv.* 21: 62–73.

63 Wang, S., Chen, P., Zhang, L. et al. (2012). Formulation and evaluation of microemulsion-based in situ ion-sensitive gelling systems for intranasal administration of curcumin. *J. Drug Targeting* 20: 831–840.

64 Karavasili, C. and Fatouros, D.G. (2016). Smart materials: in situ gel-forming systems for nasal delivery. *Drug Discovery Today* 21: 157–166.

65 Xu, X., Shen, Y., Wang, W. et al. (2014). Preparation and in vitro characterization of thermosensitive and mucoadhesive hydrogels for nasal delivery of phenylephrine. *Eur. J. Pharm. Biopharm.* 88: 998–1004.

66 Osmalak, T., Froelich, A., and Tasarek, S. (2014). Application of gellan gum in pharmacy and medicine. *Int. J. Pharm.* 466: 328–340.

67 Jansson, B., Hägerström, H., Fransén, N. et al. (2005). The influence of gellan gum on the transfer of fluorescein dextran across rat nasal epithelium in vivo. *Eur. J. Pharm. Biopharm.* 59: 557–564.

68 Mahajan, H.S. and Gattani, S.G. (2009). Gellan gum based microparticles of metoclopramide hydrochloride for intranasal delivery: development and evaluation. *Chem. Pharm. Bull.* 57: 388–392.

69 Shah, V., Sharma, M., Parmar, V., and Upadhyay, U. (2010). Formulation of sildenafil citrate loaded nasal microspheres: an in vitro, ex vivo characterization. *Int. J. Drug Deliv.* 2: 213–220.

70 Yadav, S., Ahuja, M., Kumar, A., and Kaur, R. (2014). Gellan-thioglycolic acid conjugate: synthesis, characterization and evaluation as mucoadhesive polymer. *Carbohydr. Polym.* 99: 601–607.

71 Mundargi, R.C., Shelke, N.B., Babu, V.R. et al. (2010). Novel thermo-responsive semi-interpenetrating network microspheres of gellan gum-poly(N-isopropylacrylamide) for controlled release of atenolol. *J. Appl. Polym. Sci.* 116: 1832–1841.

72 Mahdi, M.H., Conway, B.R., and Smith, A.M. (2015). Development of mucoadhesive sprayable gellan gum fluid gels. *Int. J. Pharm.* 488: 12–19.

73 Portenoy, R.K., Burton, A.W., Gabrail, N., and Taylor, D. (2010). A multicenter, placebo-controlled, double-blind, multiple-crossover study of Fentanyl Pectin Nasal Spray (FPNS) in the treatment of breakthrough cancer pain. *Pain* 151: 617–624.

74 Taylor, D., Galan, V., Weinstein, S.M. et al. (2010). Fentanyl pectin nasal spray in breakthrough cancer pain. *J. Support Oncol.* 8: 184–190.

75 Taylor, D., Radbruch, L., Revnic, J. et al. (2014). A report on the long-term use of fentanyl pectin nasal spray in patients with recurrent breakthrough pain. *J. Pain Symptom. Manage.* 47: 1001–1007.

76 http://Medicines.org.uk. Home - electronic Medicines Compendium (eMC). www.medicines.org.uk/emc (Accessed 1 June 2017).)

77 Castile, J., Cheng, Y.H., Simmons, B. et al. (2013). Development of in vitro models to demonstrate the ability of PecSys®, an in situ nasal gelling technology, to reduce nasal run-off and drip. *Drug Dev. Ind. Pharm.* 39: 816–824.

78 Chenite, A., Buschmann, M., Wang, D. et al. (2001). Rheological characterisation of thermogelling chitosan/glycerol-phosphate solutions. *Carbohydr. Polym.* 46: 39–47.

79 Cho, J., Heuzey, M.C., Bégin, A., and Carreau, P.J. (2005). Physical gelation of chitosan in the presence of β-glycerophosphate: the effect of temperature. *Biomacromolecules* 6: 3267–3275.

80 Algin-Yapar, E. (2014). Nasal inserts for drug delivery: an overview. *Trop. J. Pharm. Res.* 13: 459–467.

81 Bertram, U. and Bodmeier, R. (2006). In situ gelling, bioadhesive nasal inserts for extended drug delivery: in vitro characterization of a new nasal dosage form. *Eur. J. Pharm. Sci.* 27: 62–71.

82 Illum, L., Jørgensen, H., Bisgaard, H. et al. (1987). Bioadhesive microspheres as a potential nasal drug delivery system. *Int. J. Pharm.* 39: 189–199.

83 Illum, L., Farraj, N.F., Davis, S.S. et al. (1990). Investigation of the nasal absorption of biosynthetic human growth hormone in sheep—use of a bioadhesive microsphere delivery system. *Int. J. Pharm.* 63: 207–211.

84 Gungor, S., Okyar, A., Erturk-Toker, S. et al. (2010). Ondansetron-loaded chitosan microspheres for nasal antiemetic drug delivery: an alternative approach to oral and parenteral routes. *Drug Dev. Ind. Pharm.* 36: 806–813.

85 Nagatomo, D., Taniai, M., Ariyasu, H. et al. (2015). Cholesteryl pullulan encapsulated TNF-α nanoparticles are an effective mucosal vaccine adjuvant against influenza virus. *BioMed. Res. Int.*: 1–15.

86 Nochi, T., Yuki, Y., Takahashi, H. et al. (2010). Nanogel antigenic protein-delivery system for adjuvant-free intranasal vaccines. *Nat. Mater.* 9: 572–578.

87 Fukuyama, Y., Yuki, Y., Katakai, Y. et al. (2015). Nanogel-based pneumococcal surface protein A nasal vaccine induces microRNA-associated Th17 cell responses with neutralizing antibodies against Streptococcus pneumoniae in macaques. *Mucosal Immunol.* 8: 1144–1153.
88 Debache, K., Kropf, C., Schütz, C. et al. (2011). Vaccination of mice with chitosan nanogel-associated recombinant NcPDI against challenge infection with *Neospora caninum* tachyzoites. *Parasite Immunol.* 33: 81–94.
89 Al-Ghananeem, A.M., Saeed, H., Florence, R. et al. (2010). Intranasal drug delivery of didanosine-loaded chitosan nanoparticles for brain targeting; an attractive route against infections caused by AIDS viruses. *J. Drug Targeting* 18: 381–388.
90 Shahnaz, G., Vetter, A., Barthelmes, J. et al. (2012). Thiolated chitosan nanoparticles for the nasal administration of leuprolide: bioavailability and pharmacokinetic characterization. *Int. J. Pharm.* 428: 164–170.
91 Baltzley, S., Mohammad, A., Malkawi, A.H., and Al-Ghananeem, A.M. (2014). Intranasal drug delivery of olanzapine-loaded chitosan nanoparticles. *AAPS PharmSciTech* 15: 1598–1602.
92 Sharma, V., Ali, M., Baboota, S., and Ali, J. (2007). Preparation and characterization of chitosan nanoparticles for nose to brain delivery of a cholinesterase inhibitor. *Ind. J. Pharm. Sci.* 69: 712–713.
93 Ruby, J.J. and Pandey, V. (2014). Chitosan nanoparticles as a nasal drug delivery for memantine hydrochloride. *Int. J. Pharm. Pharm. Sci.* 7: 34–37.
94 Khatri, K., Goyal, A.K., Gupta, P.N. et al. (2008). Plasmid DNA loaded chitosan nanoparticles for nasal mucosal immunization against hepatitis B. *Int. J. Pharm.* 354: 235–224.
95 Elnaggar, Y.S., Etman, S.M., Abdelmonsif, D.A., and Abdallah, O.Y. (2015). Intranasal piperine-loaded chitosan nanoparticles as brain-targeted therapy in Alzheimer's disease: optimization, biological efficacy, and potential toxicity. *J. Pharm. Sci.* 104: 3544–3556.
96 Yao, W., Peng, Y., Du, M. et al. (2013). Preventative vaccine-loaded mannosylated chitosan nanoparticles intended for nasal mucosal delivery enhance immune responses and potent tumor immunity. *Mol. Pharmaceutics* 10: 2904–2914.
97 Singh, D., Rashid, M., Hallan, S.S. et al. (2015). Pharmacological evaluation of nasal delivery of selegiline hydrochloride-loaded thiolated chitosan nanoparticles for the treatment of depression. *Artif. Cells Nanomed. Biotechnol.* 44: 865–867.
98 Dzung, N.A., Hà, N.T.N., Van, D.T.H. et al. (2011). Chitosan nanoparticle as a novel delivery system for A/H1n1 influenza vaccine: safe property and immunogenicity in mice. *World Acad. Sci. Eng. Technol.* 60: 1839–1846.
99 Vila, A., Sánchez, A., Janes, K. et al. (2004). Low molecular weight chitosan nanoparticles as new carriers for nasal vaccine delivery in mice. *Eur. J. Pharm. Biopharm.* 57: 123–131.
100 Haque, S., Md, S., Sahni, J.K. et al. (2014). Development and evaluation of brain targeted intranasal alginate nanoparticles for treatment of depression. *J. Psychiatr. Res.* 48: 1–12.
101 Salatin, S., Barar, J., Barzegar-Jalali, M. et al. (2016). Hydrogel nanoparticles and nanocomposites for nasal drug/vaccine delivery. *Arch. Pharmacal Res.* 39: 1181–1192.
102 Illum, L. (2007). Nanoparticulate systems for nasal delivery of drugs: a real improvement over simple systems? *J. Pharm. Sci.* 96: 473–483.
103 Dünnhaupt, S., Kammona, O., Waldner, C. et al. (2015). Nano-carrier systems: strategies to overcome the mucus gel barrier Eur. *J.Pharm. Biopharm.* 96: 447–453.
104 Law, S.L. and Shih, C.L. (2001). Characterization of calcitonin-containing liposome formulations for intranasal delivery. *J. Microencapsulation* 18: 211–221.
105 Law, S.L., Huang, K.J., and Chou, H.Y. (2001). Preparation of desmopressin-containing liposomes for intranasal delivery. *J. Controlled Release* 70: 375–382.

106 Muramatsu, K., Maitani, Y., Takayama, K., and Nagai, T. (1999). The relationship between the rigidity of the liposomal membrane and the absorption of insulin after nasal administration of liposomes modified with an enhancer containing insulin in rabbits. *Drug Dev. Ind. Pharm.* 25: 1099–1105.

107 Jung, B.H., Chung, B.C., Chung, S.J. et al. (2000). Prolonged delivery of nicotine in rats via nasal administration of proliposomes. *J. Controlled Release* 66: 73–79.

108 Jung, B.H., Chung, S.J., and Shim, C.K. (2002). Proliposomes as prolonged intranasal drug delivery systems. *STP Pharma Sci.* 12: 33–38.

109 Alsarra, I.A., Hamed, A.Y., and Alanazi, F.K. (2008). Acyclovir liposomes for intranasal systemic delivery: development and pharmacokinetics evaluation. *Drug Deliv.* 15: 313–321.

110 Vyas, T.K., Babbar, A.K., Sharma, R.K., and Misra, A. (2005). Intranasal mucoadhesive microemulsions of zolmitriptan: preliminary studies on brain-targeting. *J. Drug Targeting* 13: 317–324.

111 Pidaparthi, K. and Suares, D. (2016). Comparison of nanoemulsion and aqueous micelle systems of Paliperidone for intranasal delivery. *AAPS PharmSciTech*: 1–10.

112 Shobhit, K., Javed, A., and Sanjula, B. (2016). Design Expert ® supported optimization and predictive analysis of selegiline nanoemulsion via the olfactory region with enhanced behavioural performance in Parkinson's disease. *Nanotechnology* 27: 43510.

113 Kumar, A. and Jain, S.K. (2017). Preliminary studies for the development of intranasal nanoemulsion containing CNS agent: emphasizing the utilization of cut and weigh method. *Artif. Cells Nanomed. Biotechnol.* 45: 515–521.

114 Boche, M. and Pokharkar, V. (2016). Quetiapine nanoemulsion for intranasal drug delivery: evaluation of brain-targeting efficiency. *AAPS PharmSciTech*: 1–11.

115 Pandey, Y.R., Kumar, S., Gupta, B.K. et al. (2016). Intranasal delivery of paroxetine nanoemulsion via the olfactory region for the management of depression: formulation, behavioural and biochemical estimation. *Nanotechnology* 27: 02510.

116 Mahajan, H.S., Mahajan, M.S., Nerkar, P.P., and Agrawal, A. (2013). Nanoemulsion-based intranasal drug delivery system of saquinavir mesylate for brain targeting. *Drug Deliv.* 21: 148–154.

117 Kumar, M., Misra, A., Mishra, A.K. et al. (2008). Mucoadhesive nanoemulsion-based intranasal drug delivery system of olanzapine for brain targeting. *J. Drug Targeting* 16: 806–814.

118 Singh, Y., Meher, J.G., Raval, K. et al. (2017). Nanoemulsion: concepts, development and applications in drug delivery. *J. Controlled Release* 252: 28–49.

119 Hao, J., Zhao, J., Zhang, S. et al. (2016). Fabrication of an ionic-sensitive in situ gel loaded with resveratrol nanosuspensions intended for direct nose-to-brain delivery. *Colloids Surf. B* 147: 376–386.

120 Shelke, S., Shahi, S., Jalalpure, S., and Dhamecha, D. (2016). Poloxamer 407-based intranasal thermoreversible gel of zolmitriptan-loaded nanoethosomes: formulation, optimization, evaluation and permeation studies. *J. Liposome Res.* 26: 313–323.

121 Hopkins, L.E., Patchin, E.S., Chiu, P.L. et al. (2014). Nose-to-brain transport of aerosolised quantum dots following acute exposure. *Nanotoxicology* 8: 885–893.

122 Rassu, G., Soddu, E., Cossu, M. et al. (2016). Particulate formulations based on chitosan for nose-to-brain delivery of drugs. A review. *J. Drug Deliv. Sci. Technol.* 32: 77–87.

123 Shingaki, T., Hidalgo, I.J., Furubayashi, T. et al. (2009). The transnasal delivery of 5-fluorouracil to the rat brain is enhanced by acetazolamide (the inhibitor of the secretion of cerebrospinal fluid). *Int. J. Pharm.* 377: 85–91.

124 Md, S., Haque, S., Fazil, M. et al. (2014). Optimised nanoformulation of bromocriptine for direct nose-to-brain delivery: biodistribution, pharmacokinetic and dopamine estimation by ultra-HPLC/mass spectrometry method. *Expert Opin. Drug Deliv.* 11: 827–884.

125 Khan, M.S., Patil, K., Yeole, P., and Gaikwad, R. (2009). Brain targeting studies on buspirone hydrochloride after intranasal administration of mucoadhesive formulation in rats. *J. Pharm. Pharmacol.* 61: 669–675.

126 Barakat, P., Omar, S.A., and Ahmed, A.A.E. (2006). Carbamazepine uptake into rat brain following intra-olfactory transport. *J. Pharm. Pharmacol.* 58: 63–72.

127 Serralheiro, A., Alves, G., Fortuna, A., and Falcão, A. (2014). Intranasal administration of carbamazepine to mice: a direct delivery pathway for brain targeting. *Eur. J. Pharm. Sci.* 60: 32–39.

128 Bhavna, M.S., Ali, M., Ali, R. et al. (2014). Donepezil nanosuspension intended for nose to brain targeting: in vitro and in vivo safety evaluation. *Int. J. Biol. Macromol.* 67: 418–425.

129 Alam, M.I., Baboota, S., Ahuja, A. et al. (2014). Pharmacoscintigraphic evaluation of potential of lipid nanocarriers for nose-to-brain delivery of antidepressant drug. *Int. J. Pharm.* 470: 99–106.

130 Wang, X., He, H., Leng, W., and Tang, X. (2006). Evaluation of brain-targeting for the nasal delivery of estradiol by the microdialysis method. *Int. J. Pharm.* 317: 40–46.

131 Hanson, L.R., Fine, J.M., Hoekman, J.D. et al. (2012). Intranasal delivery of growth differentiation factor 5 to the central nervous system. *Drug Deliv.* 19: 149–154.

132 Shingaki, T., Inoue, D., Furubayashi, T. et al. (2010). Transnasal delivery of methotrexate to brain tumors in rats: a new strategy for brain tumor chemotherapy. *Mol. Pharmaceutics* 7: 1561–1568.

133 Wang, F., Jiang, X., and Lu, W. (2003). Profiles of methotrexate in blood and CSF following intranasal and intravenous administration to rat. *Int. J. Pharm.* 263: 1–7.

134 Westin, U.E., Boström, E., Gråsjö, J. et al. (2006). Direct nose-to-brain transfer of morphine after nasal administration to rats. *Pharm. Res.* 23: 565–572.

135 Zhang, Q., Jiang, X., Jiang, W. et al. (2004). Preparation of nimodipine-loaded microemulsion for intranasal delivery and evaluation on the targeting efficiency to the brain. *Int. J. Pharm.* 275: 85–96.

136 Abdelbary, G.A. and Tadros, M.I. (2013). Brain targeting of olanzapine via intranasal delivery of core–shell difunctional block copolymer mixed nanomicellar carriers: in vitro characterization, ex vivo estimation of nasal toxicity and in vivo biodistribution studies. *Int. J. Pharm.* 452: 300–310.

137 Seju, U., Kumar, A., and Sawant, K.K. (2011). Development and evaluation of olanzapine-loaded PLGA nanoparticles for nose-to-brain delivery: in vitro and in vivo studies. *Acta Biomater.* 7: 4169–4177.

138 Patel, M.R., Patel, R.B., Bhatt, K.K. et al. (2014). Paliperidone microemulsion for nose-to-brain targeted drug delivery system: pharmacodynamic and pharmacokinetic evaluation. *Drug Deliv.*: 1–9.

139 Wang, D., Gao, Y., and Yun, L. (2006). Study on brain targeting of raltitrexed following intranasal administration in rats. *Cancer Chemother. Pharmacol.* 57: 97–104.

140 Ravi, P.R., Aditya, N., Patil, S., and Cherian, L. (2013). Nasal in-situ gels for delivery of rasagiline mesylate: improvement in bioavailability and brain localization. *Drug Deliv.*: 1–8.

141 Stevens, J., Ploeger, B.A., van der Graaf, P.H. et al. (2011). Systemic and direct nose-to-brain transport pharmacokinetic model for remoxipride after intravenous and intranasal administration. *Drug Metab. Dispos.* 39: 2275–2282.

142 Kumar, M., Misra, A., Babbar, A.K. et al. (2008). Intranasal nanoemulsion based brain targeting drug delivery system of risperidone. *Int. J. Pharm.* 358: 285–291.

143 Patel, S., Chavhan, S., Soni, H. et al. (2011). Brain targeting of risperidone-loaded solid lipid nanoparticles by intranasal route. *J. Drug Targeting* 19: 468–447.

144 Khan, S., Patil, K., Bobade, N. et al. (2010). Formulation of intranasal mucoadhesive temperature-mediated in situ gel containing ropinirole and evaluation of brain targeting efficiency in rats. *J. Drug Targeting* 18: 223–234.

145 Jogani, V.V., Shah, P.J., Mishra, P. et al. (2007). Nose-to-brain delivery of tacrine. *J. Pharm. Pharmacol.* 59: 1199–1205.

146 Jogani, V.V., Shah, P.J., Mishra, P. et al. (2008). Intranasal mucoadhesive microemulsion of tacrine to improve brain targeting. *Alzheimer Dis. Assoc. Disord.* 22: 116–124.

147 Banks, W.A., Morley, J.E., Niehoff, M.L., and Mattern, C. (2009). Delivery of testosterone to the brain by intranasal administration: comparison to intravenous testosterone. *J. Drug Targeting* 17: 91–97.

148 Dahlin, M. and Björk, E. (2000). Nasal absorption of (S)-UH-301 and its transport into the cerebrospinal fluid of rats Int. *J. Pharm.* 195: 197–205.

149 Dalpiaz, A., Ferraro, L., Perrone, D. et al. (2014). Brain uptake of a Zidovudine prodrug after nasal administration of solid lipid microparticles. *Mol. Pharmaceutics* 11: 1550–1561.

150 Jain, R., Nabar, S., Dandekar, P., and Patravale, V. (2010). Micellar nanocarriers: potential nose-to-brain delivery of zolmitriptan as novel migraine therapy. *Pharm. Res.* 27: 655–664.

151 Warnken, Z.N., Smyth, H.D.C., Watts, A.B. et al. (2016). Formulation and device design to increase nose to brain drug delivery. *J. Drug Deliv. Sci. Technol.* 35: 213–222.

152 Charlton, S., Jones, N.S., Davis, S.S., and Illum, L. (2007). Distribution and clearance of bioadhesive formulations from the olfactory region in man: effect of polymer type and nasal delivery device. *Eur. J. Pharm. Sci.* 30: 295–302.

153 Kammona, O. and Kiparissides, C. (2012). Recent advances in nanocarrier-based mucosal delivery of biomolecules. *J. Controlled Release* 161: 781–794.

154 Okada, E., Sasaki, S., Ishii, N. et al. (1997). Intranasal immunization of a DNA vaccine with IL-12- and granulocyte-macrophage colony-stimulating factor (GM-CSF)-expressing plasmids in liposomes induces strong mucosal and cell-mediated immune responses against HIV-1 antigens. *J. Immunol.* 159: 3638–3647.

155 Irvine, D.J., Hanson, M.C., Rakhra, K., and Tokatlian, T. (2015). Synthetic nanoparticles for vaccine and immunotherapy. *Chem. Rev.*: 11109–11146.

156 Sawaengsak, C., Mori, Y., Yamanishi, K. et al. (2014). Chitosan nanoparticle encapsulated hemagglutinin- split influenza virus mucosal vaccine. *AAPS PharmSciTech* 15: 317–325.

157 Ulery, B.D., Kumar, D., Ramer-Tait, A.E. et al. (2011). Design of a protective single-dose intranasal nanoparticle-based vaccine platform for respiratory infectious diseases. *PLoS One* 6: 1–8.

158 Kreft, M.E., Jerman, U.D., Lasič, E. et al. (2015). The characterization of the human nasal epithelial cell line RPMI 2650 under different culture conditions and their optimization for an appropriate in vitro nasal model. *Pharm. Res.* 32: 665–679.

159 Reichl, S. and Becker, K. (2012). Cultivation of RPMI 2650 cells as an in-vitro model for human transmucosal nasal drug absorption studies: optimization of selected culture conditions. *J Pharm. Pharmacol.* 64: 1621–1630.

160 Ugwoke, M.I., Verbeke, N., and Kinget, R. (2010). The biopharmaceutical aspects of nasal mucoadhesive drug delivery. *J. Pharm. Pharmacol.* 53: 3–22.

161 Moore, G.E. and Sandberg, A.A. (1964). Studies of a human tumor cell line with a diploid karyotype. *Cancer* 17: 170–175.

162 Moorhead, P.S. (1965). Human tumor cell line with a quasi-diploid karyotype (RPMI 2650). *Exp. Cell Res.* 39: 190–196.

163 Kürti, L., Veszelka, S., Bocsik, A. et al. (2013). Retinoic acid and hydrocortisone strengthen the barrier function of human RPMI 2650 cells, a model for nasal epithelial permeability. *Cytotechnology* 65: 395–406.

164 Salib, R.J., Lau, L.C., and Howarth, P.H. (2005). The novel use of the human nasal epithelial cell line RPMI 2650 as an in vitro model to study the influence of allergens and cytokines on transforming growth factor-beta gene expression and protein release. *Clin. Exp. Allergy* 35: 811–819.

165 Gonçalves, V.S.S., Matias, A.A., Poejo, J. et al. (2016). Application of RPMI 2650 as a cell model to evaluate solid formulations for intranasal delivery of drugs. *Int. J. Pharm.* 515: 1–10.

166 Rygg, A., Hindle, M., and Longest, P.W. (2016). Linking suspension nasal spray drug deposition patterns to pharmacokinetic profiles: a proof-of-concept study using computational fluid dynamics. *J. Pharm. Sci.* 105 (2016): 1995–2004.

17

Controlled Drug Delivery via the Buccal and Sublingual Routes

Javier O. Morales[1,2,3], *Parameswara R. Vuddanda*[4], *and Sitaram Velaga*[5]

[1] *Department of Pharmaceutical Science and Technology, School of Chemical and Pharmaceutical Sciences, University of Chile, Santiago 8380494, Chile*
[2] *Advanced Center for Chronic Diseases (ACCDiS), Santiago 8380494, Chile*
[3] *Center of New Drugs for Hypertension (CENDHY), Santiago 8380494, Chile*
[4] *Research Centre for Topical Drug Delivery and Toxicology, University of Hertfordshire, Hertfordshire AL10 9AB, UK*
[5] *Kenox Pharmaceuticals Inc, 11 Deerpark Dr, Monmouth Junction, NJ 08852, USA*

17.1 Introduction

In an effort to find alternative routes of delivery due to the limitations imposed by the intravenous and oral routes, the buccal and sublingual epithelia (both oral mucosal routes) have for years been the subject of investigation for small molecules and biologics delivery [1, 2]. While certain small drug molecules are hard to formulate for oral delivery (through a combination of solubility, permeability, and stability issues), most of the research, and a large portion of clinical trials for oral mucosal administration, is directed toward enabling biologics oral mucosal delivery.

New drug formulation (which is largely the case for biologics) is less challenging for intravenous dosage forms due to the fewer number of biological barriers that need to be addressed in this administration route. While not free of physicochemical issues to be addressed during formulation development, drug solutions or suspensions are quickly and efficiently bioavailable after intravenous administration [3]. However, due to the limitations imposed by intravenous administrations [4], there is a drive for alternative routes of delivery. This need is conventionally met with oral formulation development, yet the harsh environment of the gastrointestinal tract (GIT) in terms of pH gradient, enzymes, bacteria, surfactants, limited absorption, and first hepatic pass effect make this route very challenging [5]. Thus, alternative routes of delivery that can bypass limitations of both intravenous and oral administrations are researched [2], and among these, the buccal and sublingual routes are widely investigated.

This chapter depicts the physiology and barriers imposed by oral mucosal routes of administration and then follows with the myriad of dosage forms available and investigated. A description of conventional, advanced, and novel dosage forms and controlled delivery systems followed by key excipients and their influence in dosage form performance with updated research is then reviewed.

Fundamentals of Drug Delivery, First Edition.
Edited by Heather A. E. Benson, Michael S. Roberts, Adrian C. Williams, and Xiaowen Liang.

17.2 Buccal and Sublingual Physiology and Barriers to Drug Delivery

Drugs administered through either the buccal or sublingual mucosa are exposed first to saliva and mucus, and then face the respective epithelial barriers before reaching capillaries in the basal lamina and from there are drained onto the jugular vein and to systemic circulation.

17.2.1 Saliva and Mucus

Saliva is mainly produced in the sublingual and buccal regions, yet this secretion is distributed throughout the oral cavity and as such, it constitutes the main fluid content for dosage form and drug dissolution/dispersion prior to reaching the permeation barrier. The submandibular, parotid, and sublingual salivary glands located in the sublingual region are responsible for saliva secretion at a rate of 0.5–2.0 l/d to constantly maintain roughly 1 ml in the oral cavity [6, 7]. The main component of saliva is water (99%) and then several substances account for functional features of this fluid, including enzymes (such as amylase), antimicrobials such as immunoglobulins (IgA, IgM, and IgG), glycoproteins such as mucins, lipids, glucose, urea, and small electrolytes (sodium, calcium, chloride, and phosphate) [8]. The pH of saliva ranges from 6.5 to 7.5 but can vary over a wide range depending on fluids, food intake, and age.

Saliva and salivary secretion are the amount of fluid available for dosage form and drug dissolution and as such, they are key in the outcome of the permeation strategy and bioavailability. While this may not hinder the bioavailability of small lipophilic drugs, it could impair biopharmaceutical classification system (BCS) class I and III drugs as well as therapeutic peptides and proteins. Furthermore, the rate of saliva production and washout establishes the adhesion challenge for controlled buccal or sublingual delivery: dosage forms intended for a prolonged effect should incorporate mucoadhesive polymers to improve the retention time in the absorptive site to deploy the permeation strategy.

While saliva would be the outer, more fluid, portion of the permeation barrier, the mucus barrier is understood as that layer immediately in contact with the top epithelial cell layer and it is mainly composed of the glycoprotein mucin. This is a more stagnant layer, highly viscous, and can have a thickness ranging from 40 to 300 μm [9, 10]. Most mucoadhesive polymers rely in interactions, entanglement, and interpenetration of polymer chains with this glycoprotein to establish the mucoadhesive bond. The different mucoadhesive polymer generations improve over the interactions with mucins in order to achieve stronger and longer binding.

17.2.2 Buccal and Sublingual Epithelium and Permeation Barrier

Within the oral cavity, several types of mucosa are found including the masticatory, specialized, and lining mucosa. Accounting to roughly 40% of the oral cavity surface, the masticatory mucosa (gingiva and hard palate) is keratinized and the specialized (dorsum of the tongue) is partially keratinized. The other 60% lining mucosa is composed of the buccal and sublingual epithelia both stratified and non-keratinized epithelia (Figure 17.1) [7, 12]. The absence of the keratinized layer is what drives the research toward exploiting these epithelia as absorptive routes as within the oral cavity, these would be the more permeable sites. The main differences between the two epithelia arises from their stratification: the buccal epithelium is composed of 40–50 cell layers with invaginations (leading to thinner sections) accounting for a thickness of about 400–700 μm (Figure 17.2), while the sublingual epithelium is composed of 8–12 cell layers resulting in a thickness of about

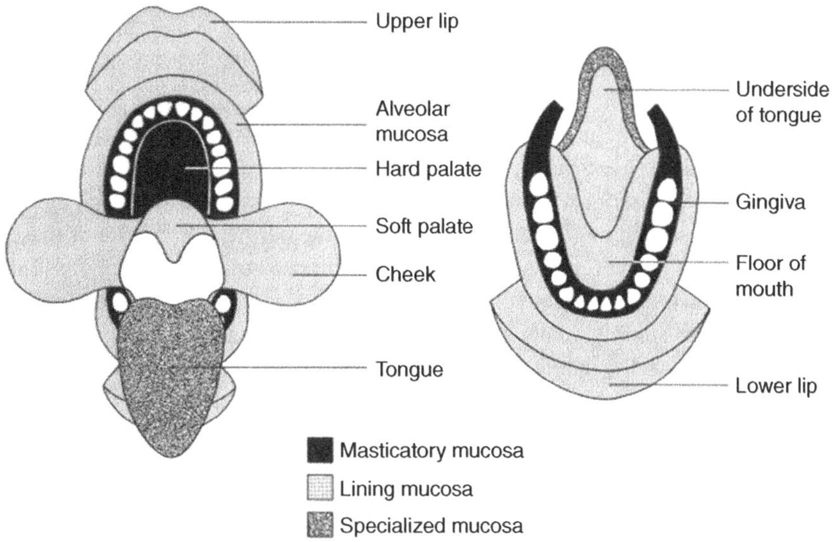

Figure 17.1 Diagram of mucosa distribution in the oral cavity showing that both the buccal and sublingual mucosa correspond to the so called lining mucosa. Source: Reproduced with permission from Squier and Kremer [11].

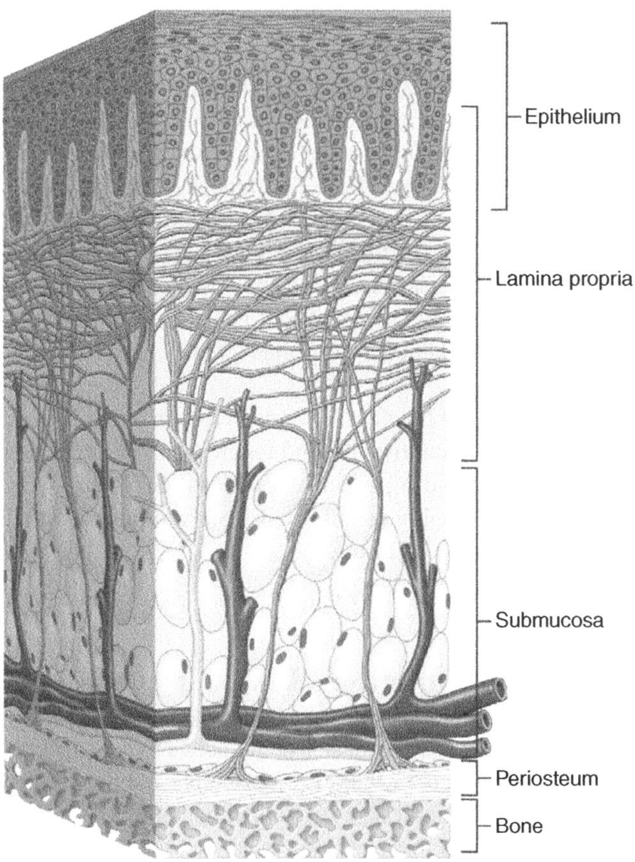

Figure 17.2 Representation of the stratified and non-keratinized buccal epithelium. Source: Reproduced with permission from Montero-Padilla et al. [13].

100–200 μm [2]. Another difference that could impact controlled release systems is related to the amount of volume available for dissolution and the challenge of mucoadhesion. The buccal mucosa does not accumulate saliva as much as the sublingual region, and this needs to be considered during dosage form design. Furthermore, the buccal epithelium has a surface available for permeation of about 50 cm^2 which contrasts the 27 cm^2 available for sublingual administration [2, 14].

Apart from those significant differences, these two epithelia have comparable properties in terms of the physicochemical permeation barrier. The upper cell layers of the lining mucosa exhibit membrane coating granules [15, 16], which have been associated with the permeation barrier of these epithelia [17, 18]. These granules have been identified as being composed of lipid amorphous [15] and lamellar [19] material which is extruded into the intercellular space of the upper portions of the lining epithelia. Both buccal and sublingual epithelia have described their permeation barrier to be located only in the upper third of the whole thickness of the epithelia [1, 20, 21], which makes a strong case for new microneedle technologies to physically bypass the main limitation to drug absorption [2, 22, 23]. Both epithelia allow for the absorption of drug molecules into systemic circulation by connecting to facial and retromandibular veins, which ultimately open into the internal jugular vein [12] avoiding all the harsh conditions of the GIT including the hepatic first pass effect. While both epithelia present metabolic enzymes, their relative content is lower is comparison with the GIT, which could result in lower presystemic drug metabolism.

17.3 Controlled Drug Delivery Systems

One key feature of buccal or sublingual controlled delivery systems is their functionalization of directed release. The buccal cavity connects both to the GIT through swallowing substances into the esophagus and directly into systemic circulation by means of absorptive mucosa (buccal and sublingual). As such, a simple and efficient strategy is widely used to enhance the performance of dosage forms for these routes by creating multilayer systems where a backing layer avoids release into the oral cavity (and thus avoids the GIT) and directs drug release toward the absorption site (Figure 17.3). This is widely described for bi- or multilayered films but has also been used in coated tablets. Conversely, indiscriminate release dosage forms are less efficient in the amount of drug exposed for absorption and can result in undesired gastrointestinal effects or GIT drug metabolism. Gels, mouthwashes, sprays, and monolayered films and tablets are examples of these indiscriminate release systems.

Due to the rather small size of the oral cavity and the regular functions of speech and ingesting solids and liquids, dosage forms are preferred small, thin, flexible, and tough [24]. The small size of these dosage forms imposes a limitation on the total drug amount loaded and delivered. Thus, these routes are excellent choices for permeable and highly potent drugs [23]. Additional desired characteristics include tastelessness and nonirritant dosage forms.

17.3.1 Tablets

Tablets are widely investigated and have led to the majority of products available [7]. Conventional tableting machines are used to produce 5–8 mm in diameter compressed disks that can be designed for indiscriminate release or unidirectional release by coating all sides but the one in contact with the absorptive mucosal epithelium. This coating can be done with water insoluble and nonpermeable polymers such as ethylcellulose [25] or hydrogenated castor oil [26] among others. With the use of mucoadhesive polymers, tablets can be designed for remaining in contact for minutes to a

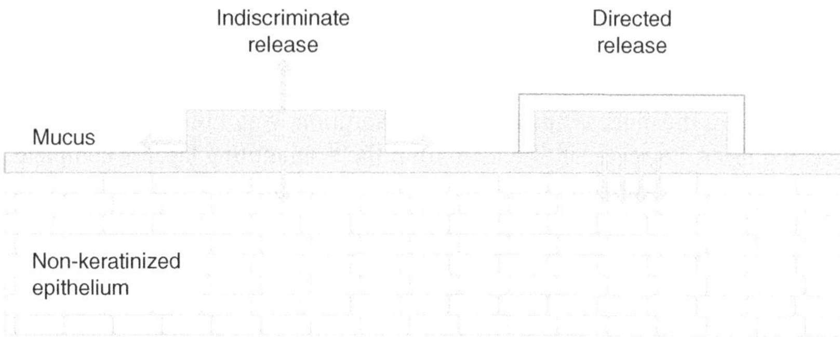

Figure 17.3 Diagram showing the two scenarios of controlled release from buccal and sublingual dosage forms. In the left-hand side, indiscriminate release results in GIT drug exposure. In the right-hand side, bi-, multilayered, and coated dosage forms can be designed to direct drug release toward the absorptive epithelium and thus enhance the delivery system performance.

few hours. Furthermore, permeation enhancers can be combined in the formulation; however, the wetting and quick release of the enhancer needs to be optimized for a concomitant drug permeation [27]. When in contact with saliva, tablets hydrate, swell, soften, and start the mucoadhesive bond with mucin. While tablet softening could benefit patient comfort, there is a need to limit the size (mostly thickness) in order to improve therapy compliance over repeated administrations [28]. While matrix tablets can be used for simple molecules, research has been conducted to formulate complex drug molecules and combine different excipients in more sophisticated tablet dosage forms. Examples of these systems include drug-loaded microspheres compressed into buccal tablets [29], and the use of body-temperature melting matrices for fast sublingual absorption [30]. Furthermore, Mura et al. have reported on the use of cyclodextrin complexes in combination with other mucoadhesive polymers to produce effective local and systemic prolonged release of clonazepam [31]. In another example, clonidine-loaded (an antihypertensive drug) mucoadhesive buccal tablets exhibited higher and more prolonged saliva concentrations, with a favorable safety profile when compared to a conventional oral tablet [32].

17.3.2 Films

Films can be found in the literature also termed as patches [33, 34] or dressings [35, 36]. As opposed to tablets, films are flexible and thin dosage forms composed in a large portion of a film forming polymer adequately plasticized. Due to their thickness, films can accommodate a smaller dose throughout the polymer matrix in comparison with tablets [24, 37], but it has been described that up to 50% by weight could be drug content [24]. In developing controlled release oromucosal films, most research is conducted on solvent casting as the manufacturing process. More recently, hot-melt extrusion has gained importance and the past couple of years have seen advances on printing technologies that can render similar delivery systems. Regardless of the method of manufacture, key excipients of films include the matrix former and the plasticizer. In order to provide sustained contact with the absorptive membrane, mucoadhesive polymers can be used as film former, combined into the matrix or added as a separate layer. Additionally, permeation enhancers have been thoroughly investigated as means to enhance oromucosal bioavailability. Similar to tablets, films can be designed for indiscriminate or unidirectional release toward the absorptive epithelium; however, and due to their methods of manufacture, multilayer films for unidirectional release are simpler to fabricate. Mucoadhesive, sustained and indiscriminate release films can be used much like a depot system for a local effect with examples for candidiasis or mucositis [27]. While these systems provide a larger and sustained drug exposure locally, they also result in drug swallowing and GIT exposure with potential absorption and side effects. Multilayered systems are optimized for unidirectional release toward the mucosa and enhanced absorption. Water insoluble polymers can be used as the backing layer; however, after release from the drug-eluting layer is complete, the empty film needs to be removed. Conversely, blends of water soluble and insoluble polymers or erodible backing layers are preferred as they disintegrate (ideally) after drug absorption is complete [27, 38]. Owing to their small size, films can be ideal candidates for buccal pediatric dosage forms [13]. Khan et al. reported omeprazole-loaded buccal films for pediatric patients, and demonstrated drug permeation and safety via *ex vivo* permeation and cell viability studies [39]. Buccal films have also been reported as delivery systems for antifungal agents such as miconazole against *Candida* cultures [40]. It has been proposed that a great concentration gradient created during drug release could act as an permeation enhancer [41, 42]. Recent advancements in inkjet printing applied to films as dosage forms can help overcome the dose amount limitation, increase the concentration gradient for further permeation enhancement, and provide a technology compatible with personalized dosage form manufacturing [23, 43].

In addition to drug-loaded films used in therapy, drug-free films can be used therapeutically as wound dressings due to their occlusive properties [44, 45]. When compared to conventional antiviral acyclovir 5% cream for treating herpes simplex labialis, a hydrocolloid patch exhibited a comparable effect showing the potential of an occlusive wound dressing [46].

17.3.3 Gels, Ointments, and Liquid Formulations

In the oral cavity, mouthwashes and mouth fresheners are conventional liquid dosage forms available for patients. As drug delivery systems aimed at systemic drug exposure after oral mucosal absorption, liquid dosage forms have a clear disadvantage due to their brief residence time. Since the retention of the dosage form relies exclusively on the patient, high variability can be observed [7]. In order to improve bioavailability from liquid dosage forms, iontophoresis has been combined with a liquid administration in the IntelliDrug prototype project for naltrexone buccal delivery [47]. In a comparative study with multiple intravenous bolus of naltrexone, the liquid formulation delivered through iontophoresis was shown to deliver naltrexone over a long time period avoiding the sharp peak and decrease of intravenous bolus administrations. This system however, provides a controlled release of liquid to be available for absorption through iontophoresis permeation enhancement, thus the residence time issue is adequately addressed for this liquid dosage form.

An advancement over liquid dosage forms are the more viscous gels and ointments. Moreover, depending on the polymer used for gel formulation, these systems could elicit a degree of mucoadhesion to further increase their residence time [7, 48]. While their administration is simple and patient friendly, this could be achieved with an applicator or the patient's finger, these dosage forms also pose a threat over drug content uniformity upon repeated administrations. A salbutamol sulfate-poloxamer bioadhesive gel has been formulated and tested *in vitro* for buccal administration showing promising results [49]. In addition, the use of a transferosomal buccal gel formulation of loratadine resulted in superior bioavailability compared to oral tablets in a clinical study [50].

17.3.4 Spray

Sprays are liquid dosage forms that, after actuation of their container, are atomized into small liquid droplets that are administered to the oral mucosal epithelium. Through the increase of the exposed surface area in relation to the volume of administration, a fast onset of action can be achieved with minimal lag time in comparison with other liquid administrations. This dosage form however, is limited due to very short residence times and thus, quickly absorbable molecules/delivery systems are preferred. The most commonly known buccal spray available is Oral-lyn, a micellar insulin solution combined with permeation enhancers, bile salts, and sodium caprate [51]. Oral-lyn has been commercialized in Ecuador and Lebanon, and has been recently discontinued in India pending more clinical efficacy evidence [23]. This product has been under review by the FDA since 2011 and still has not been approved. First generations of the technology required as much as 12 puffs to achieve the desired insulin dose, yet improvements in the technology are expected to reduce this number. In another example, Veale et al. reported a comparison of several buccal sprays of flurbiprofen used for sore throat and concluded that flurbiprofen 8.75 mg showed best aerosol performance and other attributes among all sprays tested [52].

17.3.5 Wafers

Wafers or xerogels are solid mucoadhesive dosage forms conventionally fabricated through a lyophilization to remove the solvent and form a solid porous structure [53]. This porous nature allows adequate contact and drug release (ranging from immediate to prolonged) which can also be tailored with the help of excipients such as mucoadhesive polymers. Kianfar et al. have prepared lyophilized carrageenan and pluronic acid-based buccal wafers for controlled delivery of water soluble and insoluble drugs. Recently, Mura et al. have developed amidated pectin-based buccal wafers for econazole with prolonged drug release (*in vitro*) for effective treatment of oral candidiasis [54]. Furthermore, wafers and xerogels have been proposed as platform technologies for buccal drug delivery of biologics with potential for permeation enhancement, mucoadhesion, and sustained release [55, 56].

17.3.6 Lozenges

Lozenges are compressed dosage forms much like tablets, but they are designed to gradually dissolve and release drug through the patient's constant saliva flow and tongue movement effort. Due to the constant exposure to the tongue over a determined time period, the formulation of lozenges needs to take into account taste considerations including a sweetener or having a taste-masking strategy. These dosage forms are usually designed to dissolve over 10 minutes and can remain for as long as 30 minutes in certain formulations. The residence time is a combination of patient inter and intravariability of saliva production and mechanical effort to dissolve the lozenge [27]. These dosage forms are conventionally designed for a local action; therefore, permeation enhancers or mucoadhesive polymers are not normally found in lozenge formulations.

17.3.7 Advanced and Novel Drug Delivery Systems

Recently, advanced formulation approaches based on nano-and micro-particle technologies for the delivery of poorly water soluble or permeable molecules (BCS class II and IV), vaccines and biologics across the buccal epithelium have attracted significant interest. The potential of nano-formulation approaches as means to enhance dissolution and buccal bioavailability of poor water-soluble drugs or biologics has been documented [42, 57–59]. Alternatively, other researchers focused on nanoparticles as carrier systems for the permeation and release of the drug across the buccal epithelium [60–65]. For example, mucoadhesive films containing chitosan-coated nanoparticles have been demonstrated as promising means to prolong the residence time of the dosage form in the oral cavity and to increase curcumin absorption through the buccal mucosa [62]. In addition, a study on developing microemulsion-based mucoadhesive wafers for buccal delivery found that an amount of mucoadhesive agents like carbopol was critical for the enhancement of mucoadhesive properties and sustained drug release [66]. In another study, amphotericin B-loaded monomethoxy poly(ethylene glycol)-poly(ε-caprolactone) micelles have been revealed to effectively prevent the local biofilm formation compared to standard preparations [67]. Additionally, Marques et al. examined the potential of nanostructured lipid carrier loaded hydrogel for prolonged ibuprofen release in oral cavity [68].

In vitro studies of inorganic nanoparticles, including those made from silver (19 nm in diameter) [69] and titanium dioxide (30–150 nm in diameter) [70], indicate that nanoparticle permeation through the mucosa is related to physicochemical properties such as ions diffusion. Whereas in polymeric nanoparticle permeation studies, a combination of ionization state, particle diameter, and agglomeration properties have been shown to influence particle permeability across the buccal epithelium [71, 72]. In these studies, it was revealed that the anionic carboxylated-modified nanoparticles (20 and 200 nm diameter) were less efficient in permeating isolated porcine buccal tissue in Franz cell diffusion experiments as compared to cationic amine-modified polystyrene nanoparticles (200 nm) [71]. Alternatively, 200 nm neutral polystyrene nanoparticles penetrated faster and to deeper regions of the buccal epithelium in comparison to 25 and 50 nm particles in isolated porcine buccal mucosa, due to less propensity for agglomeration [72].

In the quest for needle free delivery of vaccines, the sublingual mucosa has also been explored as a potentially safe and easily accessible mucosa route for vaccine administration [23, 73]. Furthermore, the buccal mucosa is rich in antigen presenting cells that can mediate innate and adaptive immune responses to battle local and systemic infection [74]. However, buccal delivery of vaccines (or in general other agents) is challenged by the physiological factors such as rapid clearance of the saliva and tongue movement. Several innovative strategies to overcome these barriers have been examined, typically in murine models, including nanoparticulates, nanofibers, iontophoresis, electroporation, and mucoadhesion [75–77]. In a recent study, an electrospun nanofiber layer was introduced into a multilayered mucoadhesive film to increase its surface area, and thereby vaccine loading [78]. Zhen et al. developed liposome-loaded microneedles for convenient and stable mucosal vaccination of mice [22]. In this study, authors have shown that the microneedle system achieved a robust systemic and mucosal immune responses against bovine serum albumin in comparison with conventional intradermal administration. This strategy was extended to a murine oral mucosa vaccine against hepatitis B virus with promising results [79].

17.4 Functional Excipients Used in Controlled Release Systems to Enhance Buccal and Sublingual Drug Bioavailability

17.4.1 Permeation Enhancers

Permeation enhancers have been widely studied and reported as means to improve bioavailability of poorly permeable drugs including biologics. Extensive literature reviews have previously focused on this issue and contain a detailed list and classification [53, 80, 81]. In general, permeation enhancing strategies include dosage form design (mucoadhesive films or inkjet printed films) and chemical permeation enhancers. Mucoadhesive polymers, conventionally used in manufacturing mucoadhesive dosage forms, can act either directly as a functional permeation enhancer (described for instance for chitosan) [82–85] or by generating a large concentration gradient through the absorptive epithelium and thus resulting in an enhancement on the permeation rate [23, 42, 81]. More recently, inkjet printing has been investigated as a means to dose films. Although, buccal or sublingual films have yet to be described, it is easy to envision future investigations using this technology for printing oral mucosal films. Inkjet printing of films results in the complete dose being deposited in the application side, and thus, in close contact with the absorptive epithelium [43, 86–88]. By printing the complete dose in the side of absorption, a large concentration gradient could develop and result in a permeation-enhancing effect.

Permeation enhancer is a term commonly used for molecules that change the epithelium barrier properties by a chemical effect. This group of excipients includes surfactants, fatty acids, and bile salts and examples of their use can be found for both small molecules and biologics [81, 89]. While their role is to affect the permeation barrier (either by lipid extraction from the mucosa [80, 90] or increased fluidity of intercellular lipids) [91, 92], there is a compromise between the extent of permeation enhancement and potential irritation in the buccal mucosa. A more recent investigation has described the use of amino acids as permeation enhancers of buccally administered insulin [93]. Both cationic (lysine and histidine) and anionic (glutamic and aspartic acids) amino acids have been found to enhance insulin permeation to different degrees while maintaining the epithelium integrity and a stable insulin structure.

Another permeation enhancing strategy adapted for oral mucosal permeation is iontophoresis and has been successfully used in enhancing permeation of small molecules and biologics [94–97]. Particularly interesting is a study that showed the potential of iontophoresis in the keratinized palate regions of the oral cavity, highlighting the role that iontophoresis may have in improving oral mucosal drug bioavailability [98].

17.4.2 Mucoadhesive Polymers

The use of mucoadhesive polymers in oral mucosal formulations allows for extended residence times in contact with the absorptive mucosa resulting in improved bioavailability. Mucoadhesion not only restricts movement of the dosage form but it also allows for an increase in the concentration gradient which in turn can enhance the diffusion process and thus bioavailability. Mucoadhesive polymers have been used for years in developing films (alone or in combination) and coatings for buccal solid dosage forms. The first generation of mucoadhesive polymers was composed of hydrophilic polymers that could improve their mucoadhesion with positively charged functional groups and thus establish electrostatic bonding with the negatively charged mucin molecules. This first generation includes cellulosic hydrophilic polymers, hydrophilic polyacrylates and polymethacrylates, alginate, and chitosan among others [89]. The second generation of mucoadhesive polymers brought the concept of covalent bonding within the entanglement of polymer and mucin chains when establishing the mucoadhesive bond. Thus, thiolated polymers, or thiomers, have gained great attention over recent years due to their excellent mucoadhesive properties in comparison with non-covalent bonding mucoadhesives [99, 100]. Thiomers are conventional polymers that have been chemically modified to elicit a cysteine group available for covalent disulfide bonding (Figure 17.4) [101]. Depending on the polymer of origin, they can also elicit positively (i.e. chitosan) or negatively (i.e. carboxylic acid-containing polyacrylates) moieties. Unlike first generation polymers, thiomers mucoadhesion progress is independent of ionic strength, electrostatic or hydrophobic interactions among polymers and mucin chains. The extent of mucoadhesion is mostly dependent in the amount of thiolate anions that interact with cysteine residues of mucin. The concentration of the thiolate anion is dependent on the pK_a

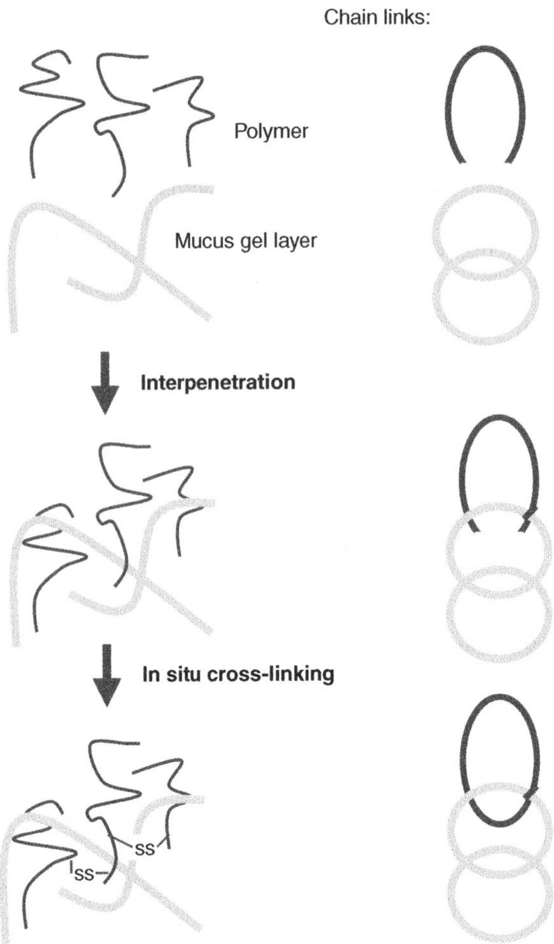

Figure 17.4 Diagram showing the *in situ* cross-linking observed in thiolated polymers in comparison with chain links. This cross-linking is believed to be responsible for their enhanced mucoadhesive properties. Source: Reproduced with permission from Bernkop-Schnürch [101].

of the thiol group, which is dependent of the overall polymer molecular structure [101, 102]. A new generation of thiomers, the so-called preactivated thiomers, have been recently developed and characterized for a number of uses [103]. By introducing mercaptopyridine substructures to the thiol group, these new polymers are more reactive for disulfide bonding with mucin while being stable toward oxidation. The potential for enhanced mucoadhesion has been shown in several models, including evidence of a 79.3-fold increase in total work of adhesion of preactivated thiolated chitosan in comparison with unmodified chitosan studied *ex vivo* in porcine buccal mucosa [104].

17.5 Conclusions

The oral mucosal routes of administration have gained interest as an alternative to reach systemic circulation bypassing the GIT. This is of relevance for biologics delivery where instability and absorption issues largely hinder their formulation into noninjectable dosage forms. As such, current research is mainly focused in films as controlled delivery systems and new manufacturing processes that can incorporate biologics and hard-to-formulate small drug molecules into the delivery system. To tackle the limitations of the delivery system and the absorption barrier, advanced formulations have incorporated nanoparticulate strategies, smart excipients, and more recently microneedles. Therefore, new avenues for oral mucosal formulation development and successful performance enhancement of these new delivery systems are expected to arise from a multidisciplinary approach including pharmaceutical scientists, bioengineers, and pharmacologists.

Acknowledgments

The authors acknowledge the funding support from FONDECYT 11130235, FONDAP 15130011, and STINT IB2015-6087.

References

1 Rathbone, M.J., Pather, I., and Senel, S. (2015). Overview of oral mucosal delivery. In: *Oral Mucosal Drug Delivery and Therapy* (eds. M.J. Rathbone, S. Senel and I. Pather), 17–29. New York, US: Springer.

2 Morales, J.O., Fathe, K.R., Brunaugh, A. et al. (2017). Challenges and future prospects for the delivery of biologics: oral mucosal, pulmonary, and transdermal routes. *AAPS J.* 19: 652–668.

3 Leader, B., Baca, Q.J., and Golan, D.E. (2008). Protein therapeutics: a summary and pharmacological classification. *Nat. Rev. Drug Discovery* 7: 21–39.

4 Zhao, L., Ren, T., and Wang, D.D. (2012). Clinical pharmacology considerations in biologics development. *Acta Pharmacol. Sin.* 33: 1339–1347.

5 Aguirre, T.A.S., Teijeiro-Osorio, D., Rosa, M. et al. (2016). Current status of selected oral peptide technologies in advanced preclinical development and in clinical trials. *Adv. Drug Deliv. Rev.* 106 (Part B): 223–241.

6 Shojaei, A.H. (1998). Buccal mucosa as a route for systemic drug delivery: a review. *J. Pharm. Pharm. Sci.* 1: 15–30.

7 Morales, J.O. (2014). Drug delivery polymers, oral mucosal. In: *Encyclopedia of Biomedical Polymers and Polymeric Biomaterials*, 1–20. New York, NY: CRC Press.

8 Carpenter, G.H. (2013). The secretion, components, and properties of saliva. *Annu. Rev. Food Sci. Technol.* 4: 267–276.

9 Forstner, J., Taichman, N., Kalnins, V., and Forstner, G. (1973). Intestinal goblet cell mucus: isolation and identification by immunofluorescence of a goblet cell glycoprotein. *J. Cell Sci.* 12: 585.

10 Horowitz, M.I. (1977). Gastrointestinal glycoproteins. In: *The Glycoconjugates* (eds. M.I. Horowitz and W. Pigman), 189–213. Academic Press: New York, NY.

11 Squier, C.A. and Kremer, M.J. (2001). Biology of oral mucosa and esophagus. *J. Natl. Cancer Inst. Monogr.* 2001: 7–15.

12 Squier, C.A. and Wertz, P.W. (1996). Structure and function of the oral mucosa and implications for drug delivery. In: *Oral Mucosal Drug Delivery* (ed. M.J. Rathbone), 1–26. Informa Health Care.

13 Montero-Padilla, S., Velaga, S., and Morales, J.O. (2017). Buccal dosage forms: general considerations for pediatric patients. *AAPS PharmSciTech.* 18: 273–282.

14 Johnston, T.P. (2015). Anatomy and physiology of the oral mucosa. In: *Oral Mucosal Drug Delivery and Therapy* (eds. M.J. Rathbone, S. Şenel and I. Pather), 1–15. Springer.

15 Squier, C.A. (1977). Membrane coating granules in nonkeratinizing oral epithelium. *J. Ultrastruct. Res.* 60: 212–220.

16 Squier, C.A. (1982). Zinc iodide-osmium staining of membrane-coating granules in keratinized and non-keratinized mammalian oral epithelium. *Arch. Oral Biol.* 27: 377–382.

17 Jacobsen, J., van Deurs, B., Pedersen, M., and Rassing, M.R. (1995). TR146 cells grown on filters as a model for human buccal epithelium: I. Morphology, growth, barrier properties, and permeability. *Int. J. Pharm.* 125: 165–184.

18 Law, S., Wertz, P.W., Swartzendruber, D.C., and Squier, C.A. (1995). Regional variation in content, composition and organization of porcine epithelial barrier lipids revealed by thin-layer chromatography and transmission electron microscopy. *Arch. Oral Biol.* 40: 1085–1091.

19 Wertz, P.W., Swartzendruber, D.C., and Squier, C.A. (1993). Regional variation in the structure and permeability of oral mucosa and skin. *Adv. Drug Deliv. Rev.* 12: 1–12.

20 Squier, C.A. (1973). The permeability of keratinized and nonkeratinized oral epithelium to horseradish peroxidase. *J. Ultrastruct. Res.* 43: 160–177.

21 Goswami, T., Jasti, B., and Li, X. (2008). Sublingual drug delivery. *Crit. Rev. Ther. Drug Carrier Syst.* 25: 449.

22 Zhen, Y., Wang, N., Gao, Z. et al. (2015). Multifunctional liposomes constituting microneedles induced robust systemic and mucosal immunoresponses against the loaded antigens via oral mucosal vaccination. *Vaccine* 33: 4330–4340.

23 Morales, J.O. and Brayden, D.J. (2017). Buccal delivery of small molecules and biologics: of mucoadhesive polymers, films, and nanoparticles. *Curr. Opin. Pharmacol.* 36: 22–28.

24 Dixit, R.P. and Puthli, S.P. (2009). Oral strip technology: overview and future potential. *J. Controlled Release* 139: 94–107.

25 Patel, V.M., Prajapati, B.G., and Patel, M.M. (2007). Formulation, evaluation, and comparison of bilayered and multilayered mucoadhesive buccal devices of propranolol hydrochloride. *AAPS PharmSciTech.* 8: 147–154.

26 Alur, H.H., Beal, J.D., Pather, S.I. et al. (1999). Evaluation of a novel, natural oligosaccharide gum as a sustained-release and mucoadhesive component of calcitonin buccal tablets. *J. Pharm. Sci.* 88: 1313–1319.

27 Madhav, N.V.S., Shakya, A.K., Shakya, P., and Singh, K. (2009). Orotransmucosal drug delivery systems: a review. *J. Controlled Release* 140: 2–11.

28 Salamat-Miller, N., Chittchang, M., and Johnston, T.P. (2005). The use of mucoadhesive polymers in buccal drug delivery. *Adv. Drug Deliv. Rev.* 57: 1666–1691.

29 Giunchedi, P., Juliano, C., Gavini, E. et al. (2002). Formulation and in vivo evaluation of chlorhexidine buccal tablets prepared using drug-loaded chitosan microspheres. *Eur. J. Pharm. Biopharm.* 53: 233–239.

30 Tayel, S.A., Soliman, I.I., and Louis, D. (2010). Formulation of ketotifen fumarate fast-melt granulation sublingual tablet. *AAPS PharmSciTech.* 11: 679–685.

31 Mura, P., Cirri, M., Mennini, N. et al. (2016). Polymeric mucoadhesive tablets for topical or systemic buccal delivery of clonazepam: effect of cyclodextrin complexation. *Carbohydr. Polym.* 152: 755–763.

32 Vasseur, B., Dufour, A., Houdas, L. et al. (2017). Comparison of the systemic and local pharmacokinetics of clonidine mucoadhesive buccal tablets with reference clonidine oral tablets in healthy volunteers: an open-label randomised cross-over trial. *Adv. Ther.* 34: 2022–2032.

33 Cavallari, C., Brigidi, P., and Fini, A. (2015). Ex-vivoand in-vitro assessment of mucoadhesive patches containing the gel-forming polysaccharide psyllium for buccal delivery of chlorhexidine base. *Int. J. Pharm.* 496: 593–600.

34 Mahdizadeh Barzoki, Z., Emam-Djomeh, Z., Mortazavian, E. et al. (2016). Formulation, *in vitro* evaluation and kinetic analysis of chitosan–gelatin bilayer muco-adhesive buccal patches of insulin nanoparticles. *J. Microencapsulation* 33: 613–624.

35 Choi, J.S., Han, S.-H., Hyun, C., and Yoo, H.S. (2016). Buccal adhesive nanofibers containing human growth hormone for oral mucositis: BUCCAL ADHESIVE NANOFIBERS. *J. Biomed. Mater. Res. B Appl. Biomater.* 104: 1396–1406.

36 Daněk, Z., Gajdziok, J., Doležel, P. et al. (2017). Buccal films as a dressing for the treatment of aphthous lesions. *J. Oral Pathol. Med.* 46: 301–306.

37 Hariharan, M. and Bogue, A. (2009). Orally dissolving film strips: the final evolution of orally dissolving dosage forms. *Drug Deliv. Technol.* 9: 24–29.

38 Zhang, H. (2007). Dissolvable backing layer for use with a transmucosal delivery device. http://www.google.com/patents/US7276246 (accessed 29 January 2013).

39 Khan, S., Trivedi, V., and Boateng, J. (2016). Functional physico-chemical, ex vivo permeation and cell viability characterization of omeprazole loaded buccal films for paediatric drug delivery. *Int. J. Pharm.* 500: 217–226.

40 Tejada, G., Piccirilli, G.N., Sortino, M. et al. (2017). Formulation and in-vitro efficacy of antifungal mucoadhesive polymeric matrices for the delivery of miconazole nitrate. *Mater. Sci. Eng., C* 79: 140–150.

41 Senel, S., Kremer, M.J., Kas, S. et al. (2000). Enhancing effect of chitosan on peptide drug delivery across buccal mucosa. *Biomaterials* 21: 2067–2071.

42 Morales, J.O., Huang, S., Williams, R.O. III, and McConville, J.T. (2014). Films loaded with insulin-coated nanoparticles (ICNP) as potential platforms for peptide buccal delivery. *Colloids Surf., B* 122: 38–45.

43 Montenegro-Nicolini, M., Miranda, V., and Morales, J.O. (2017). Inkjet printing of proteins: an experimental approach. *AAPS J.* 19: 234–243.

44 Hearnden, V., Sankar, V., Hull, K. et al. (2012). New developments and opportunities in oral mucosal drug delivery for local and systemic disease. *Adv. Drug Deliv. Rev.* 64: 16–28.

45 Sankar, V., Hearnden, V., Hull, K. et al. (2011). Local drug delivery for oral mucosal diseases: challenges and opportunities. *Oral Dis.* 17: 73–84.

46 Karlsmark, T., Goodman, J., Drouault, Y. et al. (2008). Randomized clinical study comparing Compeed® cold sore patch to acyclovir cream 5% in the treatment of herpes simplex labialis. *J. Eur. Acad. Dermatol. Venereol.* 22: 1184–1192.

47 Campisi, G., Giannola, L.I., Florena, A.M. et al. (2010). Bioavailability in vivo of naltrexone following transbuccal administration by an electronically-controlled intraoral device: a trial on pigs. *J. Controlled Release* 145: 214–220.

48 Jones, D.S., Woolfson, A.D., Djokic, J., and Coulter, W.A. (1996). Development and mechanical characterization of bioadhesive semi-solid, polymeric systems containing tetracycline for the treatment of periodontal diseases. *Pharm. Res.* 13: 1734–1738.

49 Zeng, N., Mignet, N., Dumortier, G. et al. (2015). Poloxamer bioadhesive hydrogel for buccal drug delivery: cytotoxicity and trans-epithelial permeability evaluations using TR146 human buccal epithelial cell line. *Int. J. Pharm.* 495: 1028–1037.

50 Elkomy, M.H., El Menshawe, S.F., Abou-Taleb, H.A., and Elkarmalawy, M.H. (2017). Loratadine bioavailability via buccal transferosomal gel: formulation, statistical optimization, *in vitro/in vivo* characterization, and pharmacokinetics in human volunteers. *Drug Deliv.* 24: 781–791.

51 Palermo, A., Maddaloni, E., and Pozzilli, P. (2012). Buccal spray insulin (Oralgen) for type 2 diabetes: what evidence? *Expert Opin. Biol. Ther.* 12: 767–772.

52 Veale, D., Shephard, A., Adams, V., and Lidster, C. (2017). Comparison of the characteristics and performance of flurbiprofen 8.75 mg spray for sore throat. *Curr. Drug Deliv.* 14 https://doi.org/10.2174/1567201813666161205155830.

53 Morales, J.O. and McConville, J.T. (2014). Novel strategies for the buccal delivery of macromolecules. *Drug Dev. Ind. Pharm.* 40: 579–590.

54 Mura, P., Mennini, N., Kosalec, I. et al. (2015). Amidated pectin-based wafers for econazole buccal delivery: formulation optimization and antimicrobial efficacy estimation. *Carbohydr. Polym.* 121: 231–240.

55 Ayensu, I., Mitchell, J.C., and Boateng, J.S. (2012). In vitro characterisation of chitosan based xerogels for potential buccal delivery of proteins. *Carbohydr. Polym.* 89: 935–941.

56 Boateng, J.S. and Ayensu, I. (2014). Preparation and characterisation of laminated thiolated chitosan-based freeze-dried wafers for potential buccal delivery of macromolecules. *Drug Dev. Ind. Pharm.* 40: 611–618.

57 Rao, S., Song, Y., Peddie, F., and Evans, A. (2011). Particle size reduction to the nanometer range: a promising approach to improve buccal absorption of poorly water-soluble drugs. *Int. J. Nanomed.*: 1245.

58 Baumgartner, R., Teubl, B.J., Tetyczka, C., and Roblegg, E. (2016). Rational design and characterization of a nanosuspension for intraoral administration considering physiological conditions. *J. Pharm. Sci.* 105: 257–267.

59 Morales, J.O., Ross, A.C., and McConville, J.T. (2013). Protein-coated nanoparticles embedded in films as delivery platforms. *J. Pharm. Pharmacol.* 65: 827–838.

60 Mouftah, S., Abdel-Mottaleb, M.M.A., and Lamprecht, A. (2016). Buccal delivery of low molecular weight heparin by cationic polymethacrylate nanoparticles. *Int. J. Pharm.* 515: 565–574.

61 Al-Dhubiab, B.E., Nair, A.B., Kumria, R. et al. (2015). Formulation and evaluation of nano based drug delivery system for the buccal delivery of acyclovir. *Colloids Surf., B* 136: 878–884.

62 Mazzarino, L., Borsali, R., and Lemos-Senna, E. (2014). Mucoadhesive films containing chitosan-coated nanoparticles: a new strategy for buccal curcumin release. *J. Pharm. Sci.* 103: 3764–3771.

63 Abd El Azim, H., Nafee, N., Ramadan, A., and Khalafallah, N. (2015). Liposomal buccal mucoadhesive film for improved delivery and permeation of water-soluble vitamins. *Int. J. Pharm.* 488: 78–85.

64 Giovino, C., Ayensu, I., Tetteh, J., and Boateng, J.S. (2013). An integrated buccal delivery system combining chitosan films impregnated with peptide loaded PEG-*b*-PLA nanoparticles. *Colloids Surf., B* 112: 9–15.

65 Lv, Q., Shen, C., Li, X. et al. (2015). Mucoadhesive buccal films containing phospholipid-bile salts-mixed micelles as an effective carrier for Cucurbitacin B delivery. *Drug Deliv.* 22: 351–358.

66 Pham, M.N., Van Vo, T., Tran, V.-T. et al. (2017). Microemulsion-based mucoadhesive buccal wafers: wafer formation, in vitro release, and ex vivo evaluation. *AAPS PharmSciTech.* 18: 2727–2736.

67 Zhang, P., Yang, X., He, Y. et al. (2017). Preparation, characterization and toxicity evaluation of amphotericin B loaded MPEG-PCL micelles and its application for buccal tablets. *Appl. Microbiol. Biotechnol.* 101: 7357–7370.

68 Marques, A.C., Rocha, A.I., Leal, P. et al. (2017). Development and characterization of mucoadhesive buccal gels containing lipid nanoparticles of ibuprofen. *Int. J. Pharm.* 533: 455–462.

69 Mauro, M., Crosera, M., Bianco, C. et al. (2015). In vitro permeability of silver nanoparticles through porcine oromucosal membrane. *Colloids Surf., B* 132: 10–16.

70 Teubl, B.J., Leitinger, G., Schneider, M. et al. (2015). The buccal mucosa as a route for TiO_2 nanoparticle uptake. *Nanotoxicology* 9: 253–261.

71 Roblegg, E., Fröhlich, E., Meindl, C. et al. (2012). Evaluation of a physiological in vitro system to study the transport of nanoparticles through the buccal mucosa. *Nanotoxicology* 6: 399–413.

72 Teubl, B.J., Meindl, C., Eitzlmayr, A. et al. (2013). In-vitro permeability of neutral polystyrene particles via buccal mucosa. *Small* 9: 457–466.

73 Kweon, M.-N. (2011). Sublingual mucosa: a new vaccination route for systemic and mucosal immunity. *Cytokine* 54: 1–5.

74 Upadhyay, J., Upadhyay, R.B., Agrawal, P. et al. (2013). Langerhans cells and their role in oral mucosal diseases. *North Am. J. Med. Sci.* 5: 505.

75 Wang, N., Wang, T., Zhang, M. et al. (2014). Mannose derivative and lipid A dually decorated cationic liposomes as an effective cold chain free oral mucosal vaccine adjuvant-delivery system. *Eur. J. Pharm. Biopharm.* 88: 194–206.

76 Wang, N., Wang, T., Zhang, M. et al. (2014). Using procedure of emulsification–lyophilization to form lipid A-incorporating cochleates as an effective oral mucosal vaccine adjuvant-delivery system (VADS). *Int. J. Pharm.* 468: 39–49.

77 Baudner, B.C. and O'Hagan, D.T. (2010). Bioadhesive delivery systems for mucosal vaccine delivery. *J. Drug Targeting* 18: 752–770.

78 Mašek, J., Lubasová, D., Lukáč, R. et al. (2017). Multi-layered nanofibrous mucoadhesive films for buccal and sublingual administration of drug-delivery and vaccination nanoparticles – important step towards effective mucosal vaccines. *J. Controlled Release* 249: 183–195.

79 Wang, T., Zhen, Y., Ma, X. et al. (2015). Mannosylated and lipid A-incorporating cationic liposomes constituting microneedle arrays as an effective oral mucosal HBV vaccine applicable in the controlled temperature chain. *Colloids Surf., B* 126: 520–530.

80 Sohi, H., Ahuja, A., Ahmad, F.J., and Khar, R.K. (2010). Critical evaluation of permeation enhancers for oral mucosal drug delivery. *Drug Dev. Ind. Pharm.* 36: 254–282.

81 Caon, T., Jin, L., Simões, C.M.O. et al. (2015). Enhancing the buccal mucosal delivery of peptide and protein therapeutics. *Pharm. Res.* 32: 1–21.

82 Portero, A., Remunan-López, C., and Nielsen, H.M. (2002). The potential of chitosan in enhancing peptide and protein absorption across the TR146 cell culture model – an in vitro model of the buccal epithelium. *Pharm. Res.* 19: 169–174.

83 Sandri, G., Rossi, S., Bonferoni, M.C. et al. (2005). Buccal penetration enhancement properties of *N*-trimethyl chitosan: influence of quaternization degree on absorption of a high molecular weight molecule. *Int. J. Pharm.* 297: 146–155.

84 Sandri, G., Poggi, P., Bonferoni, M.C. et al. (2006). Histological evaluation of buccal penetration enhancement properties of chitosan and trimethyl chitosan. *J. Pharm. Pharmacol.* 58: 1327–1336.

85 Langoth, N., Kahlbacher, H., Schöffmann, G. et al. (2006). Thiolated chitosans: design and in vivo evaluation of a mucoadhesive buccal peptide drug delivery system. *Pharm. Res.* 23: 573–579.

86 Montenegro-Nicolini, M. and Morales, J.O. (2016). Overview and future potential of buccal mucoadhesive films as drug delivery systems for biologics. *AAPS PharmSciTech.* 18: 3–14.

87 Genina, N., Fors, D., Palo, M. et al. (2013). Behavior of printable formulations of loperamide and caffeine on different substrates – Effect of print density in inkjet printing. *Int. J. Pharm.* 453: 488–497.

88 Buanz, A.B.M., Saunders, M.H., Basit, A.W., and Gaisford, S. (2011). Preparation of personalized-dose salbutamol sulphate oral films with thermal ink-jet printing. *Pharm. Res.* 28: 2386–2392.

89 Morales, J.O. and McConville, J.T. (2011). Manufacture and characterization of mucoadhesive buccal films. *Eur. J. Pharm. Biopharm.* 77: 187–199.

90 Deneer, V.H.M., Drese, G.B., Roemelé, P.E.H. et al. (2002). Buccal transport of flecainide and sotalol: effect of a bile salt and ionization state. *Int. J. Pharm.* 241: 127–134.

91 Morishita, M., Barichello, J.M., Takayama, K. et al. (2001). Pluronic® F-127 gels incorporating highly purified unsaturated fatty acids for buccal delivery of insulin. *Int. J. Pharm.* 212: 289–293.

92 Ojewole, E., Kalhapure, R., Akamanchi, K., and Govender, T. (2014). Novel oleic acid derivatives enhance buccal permeation of didanosine. *Drug Dev. Ind. Pharm.* 40: 657–668.

93 Iyire, A., Alaayedi, M., and Mohammed, A.R. (2016). Pre-formulation and systematic evaluation of amino acid assisted permeability of insulin across in vitro buccal cell layers. *Sci. Rep.* 6: 32498.

94 Patel, M.P., Churchman, S.T., Cruchley, A.T. et al. (2013). Delivery of macromolecules across oral mucosa from polymeric hydrogels is enhanced by electrophoresis (iontophoresis). *Dent. Mater.* 29: e299–e307.

95 Oh, D.-H., Chun, K.-H., Jeon, S.-O. et al. (2011). Enhanced transbuccal salmon calcitonin (sCT) delivery: effect of chemical enhancers and electrical assistance on in vitro sCT buccal permeation. *Eur. J. Pharm. Biopharm.* 79: 357–363.

96 Telò, I., Tratta, E., Guasconi, B. et al. (2016). In-vitro characterization of buccal iontophoresis: the case of sumatriptan succinate. *Int. J. Pharm.* 506: 420–428.

97 Patel, M.P., Churchman, S.T., Cruchley, A.T. et al. (2013). Electrically induced transport of macromolecules through oral buccal mucosa. *Dent. Mater.* 29: 674–681.

98 Ren, W., Baig, A., White, D.J., and Li, S.K. (2016). Characterization of cornified oral mucosa for iontophoretically enhanced delivery of chlorhexidine. *Eur. J. Pharm. Biopharm.* 99: 35–44.

99 Laffleur, F. and Bernkop-Schnürch, A. (2012). Thiomers: promising platform for macromolecular drug delivery. *Future Med. Chem.* 4: 2205–2216.

100 Duggan, S., Cummins, W., O'Donovan, O. et al. (2017). Thiolated polymers as mucoadhesive drug delivery systems. *Eur. J. Pharm. Sci.* 100: 64–78.

101 Bernkop-Schnürch, A. (2005). Thiomers: a new generation of mucoadhesive polymers. *Adv. Drug Deliv. Rev.* 57: 1569–1582.

102 Bernkop-Schnürch, A., Hornof, M., and Zoidl, T. (2003). Thiolated polymers – thiomers: synthesis and in vitro evaluation of chitosan–2-iminothiolane conjugates. *Int. J. Pharm.* 260: 229–237.

103 Ijaz, M. and Bernkop-Schnürch, A. (2015). Preactivated thiomers: their role in drug delivery. *Exp. Opin. Drug Deliv.* 12: 1269–1281.

104 Laffleur, F., Shahnaz, G., Islambulchilar, Z., and Bernkop-Schnürch, A. (2013). Design and in vitro evaluation of a novel polymeric excipient for buccal applications. *Future Med. Chem.* 5: 511–522.

18

Controlled Drug Delivery via the Lung

María V. Ramírez-Rigo[1,2], *Nazareth E. Ceschan*[1,2], *and Hugh D. C. Smyth*[3]

[1] *Planta Piloto de Ingeniería Química (PLAPIQUI), Consejo Nacional de Investigaciones Científicas y Técnicas (CONICET) - Universidad Nacional del Sur (UNS), Camino La Carrindanga km 7, 8000 Bahía Blanca, Argentina*
[2] *Departamento de Biología, Bioquímica y Farmacia, UNS, San Juan 670, 8000 Bahía Blanca, Argentina*
[3] *College of Pharmacy, The University of Texas at Austin, 2409 West University Avenue, Austin, TX, United States*

18.1 Introduction

The lungs fulfill the vital function of transporting and exchanging breathing air. The respiratory tract is considered an excellent delivery route for local and noninvasive administration of active ingredients, even when its anatomy and physiology may represent an obstacle for drug delivery. For this reason, inhalable products are designed as complex systems composed by a drug formulation and an administration device (e.g. nebulizers, metered dose inhalers [MDIs], and dry powder inhalers [DPIs]) [1].

Traditionally, most inhalation aerosol products available are intended to treat respiratory diseases (e.g. asthma, cystic fibrosis, chronic obstructive pulmonary disease, and allergic rhinitis). It should be highlighted that pulmonary administration avoids the hepatic first-pass metabolism. Recently, there has also been an increased interest to develop inhalable products for systemic administration, where the healthy lung is used as an access route for drugs, vaccines, or proteins to the bloodstream. Several formulations, devices, and products have reached advanced development stages as well as regulatory approved therapeutics [2].

In this context, controlled release strategies, based on spatial location and temporally modified delivery, have been used to optimize the pharmacotherapy. Approaches to (i) modify deposition in different tracheobronchial tree areas, (ii) protect the drugs from lung elimination mechanisms (e.g. mucociliar clearance and phagocytosis), (iii) sustain drug release, and (iv) target to specific cells (such as macrophages) have been developed.

This chapter presents the anatomical and physiological characteristics of the airways relevant to the design of drugs and the formulation and devices most used for controlled release.

18.2 The Relevant Physiology of the Route Including the Barriers to Drug Delivery

Anatomically and physiologically, the respiratory tract can be divided into two main regions: the conductive and the respiratory pathway [3].

Fundamentals of Drug Delivery, First Edition.
Edited by Heather A. E. Benson, Michael S. Roberts, Adrian C. Williams, and Xiaowen Liang.

The conductive pathway is composed by the nose, mouth, trachea, bronchi, bronchioles, and terminal bronchioles. The main objective of this section is to humidify, warm, and conduct the air to the respiratory area. The bronchi, bronchioles, and terminal bronchioles branch off several times, which increases the surface area and decreases the circulating air velocity and the ducts dimensions [3]. This part of the respiratory tract is covered by ciliated cells and mucus secreting glands. As the bronchioles bifurcate in the terminal bronchioles (conductive pathway smaller ducts), the air passages are formed, constituting the transition from the conductive to the respiratory pathway [4].

The respiratory pathway is composed by the respiratory bronchioles, alveolar ducts or sacs, and alveoli. This section constitutes the final segment of the respiratory tract, whose main function is the hematogaseous exchange, which is possible due to (i) the oxygen pressure gradient from the conductive pathway to the alveoli and (ii) the large alveoli surface area and the thin cell layer, situations that favors the gases diffusion [3]. Regarding drug administration, alveoli have the maximum surface area and Thinnest epithelia, so this region is attractive as an absorption site for systemic administration.

The structural complexity of the respiratory tract requires a correct design of the inhalatory formulations in order to achieve the desired pharmacotherapy. In addition to the anatomy, the respiratory system imposes other barriers to the administration of active principles by this route.

- *Absorption*: One of the elimination mechanisms of the particles deposited in the lung is through its dissolution in the lung fluid and its subsequent systemic absorption. The lung epithelium is covered by thin layers of mucus and surfactant, which are barriers to drug absorption. It should be noted that, if the formulation is intended for systemic action by the inhalation route, this mechanism will be a fundamental step to achieve the therapeutic action [3, 5]. It may also be critical for efficacy in the treatment of local lung diseases, and mucus penetration is well known to be important in gene therapy efficiency.
- *Pulmonary metabolism*: Pulmonary enzymes can metabolize the active ingredients that reach the respiratory tract after their dissolution in the pulmonary fluid. However, the lung metabolic capacity is considerably lower than the liver, and many compounds show almost complete bioavailability when administered efficiently to the deep lung.
- *Mucociliary clearance and cough*: The upper portions of the respiratory tract are covered by ciliated cells, which beating affects the mucus movement and, therefore, the movement of the particles and compounds entrapped within the mucus. Those cells are responsible for removing secretions and foreign particles that are deposited on the respiratory system [3]. This clearance mechanism is particularly relevant for insoluble drugs with particle size diameter larger than 6 μm [6]. Cough also facilitates the expulsion of particles deposited in the upper airways. It is thought that coughing is less important than mucociliary clearance in healthy people, but it is important in pathological situations such as asthma and obstructive diseases (pathologies in which cilia activity is reduced) [7].
- *Phagocytosis*: Alveolar macrophages are immune cells that tend to phagocytose the particles deposited in the alveoli. This mechanism is particularly relevant for particles between 1.5 and 3 μm, with poor solubility in the lung fluid [6]. There is also evidence that epithelial cells within the airways can engulf particles.

Controlled release formulation strategies will also be focused on overcoming the above mentioned physiological barrier in order to increase drug residence time in the site of action or absorption.

18.3 Controlled Drug Delivery Systems

Targeting and controlling release can be achieved with an appropriate formulation and device design. Different approaches have been reported in open literature and will be introduced in Sections 18.3.1 and 18.3.2.

18.3.1 Formulations

Regardless the administration device, formulations (solutions, suspensions, emulsions, colloidal dispersions, and powders) need to be stable and safe for the pulmonary tissues. Also, they must be capable of releasing the active ingredient into the lung fluid, modulate the interaction with the lung environment, and, when desired, promote specific cell types targeting [2, 8]. In systems intended for modified release, the formulation strategies that have been explored are varied. Table 18.1 summarizes some illustrative examples that are presented in Sections 18.3.1.1–18.3.1.8.

18.3.1.1 Dissolution Rate Controlled

18.3.1.1.1 Amorphous Powders

Prolonged residence time of insoluble particles in the respiratory tract could lead to tissue inflammatory reactions and toxic effects [44]. For this reason, the design of amorphous inhalable materials is of particular interest in order to improve the dissolution rate of poorly water-soluble drugs.

The major challenges formulating these materials are the physical and chemical stability problems and the sensitivity to ambient moisture. To improve the stability of these formulations, some excipients are co-processed with the drug to constitute a matrix with a relative higher glass transition temperature. Also this matrix can provide a physical barrier to moisture. Protection against moisture can be also achieved by coating the particles surface with non-hygroscopic materials. Finally, an appropriate device needs to be selected in order to isolate the formulation from ambient humidity [45]. For example, sugars and polyols have been reported to modulate active ingredients stability, like tacrolimus among others [9, 45].

Other type of stabilization strategy is complexation with polymers. The combination of indomethacin, a poorly water soluble drug, with polylysine, a cationic polyelectrolyte, led to a complex, where the drug was in the amorphous state. The drug remained in this state over three years [10], period of time significantly higher than the achieved by other formulation strategies [46, 47]. The amorphous state stability was attributed to the ionic interaction between the drug and the polyelectrolyte, interaction that prevented the indomethacin nucleation and crystallization [11].

18.3.1.1.2 Other Fast Dissolution Systems

Solid dispersions, cocrystals, and complexes were also formulated to improve drug solubility and dissolution. For example microparticles based on mannitol and meloxicam solid dispersions were developed by spray drying to increase powder wettability and *in vitro* drug dissolution rate [12]. Besides, itraconazole cocrystals improved dissolution and bioavailability in rodents, being an alternative than amorphous and crystalline materials [13]. Systems based on complexes will be described in Sections 3.1.3 and 3.1.6.

18.3.1.2 Sustained Release Systems

In contrast to the amorphous systems, sustained-release microparticles decrease drug dissolution rate. They have been used to improve treatment adherence when dosing intervals are extended. Matrix and coated microparticles were designed with different shapes (e.g. spheres, needles) and

Table 18.1 Examples of controlled drug delivery formulations.

Type of formulation	Device	Drug	Excipients	Production process	References
Dissolution rate controlled systems					
Amorphous nanoparticles	Nebulizer	Tacrolimus	Lactose	Ultra-rapid freezing	[9]
Amorphous powder	DPI	Indomethacin	Polylysine, dextrin	Spray drying	[10, 11]
Solid dispersion	DPI	Meloxicam	Mannitol, polysorbate 80, povidone K-25	Spray drying	[12]
Co-crystals	Dry powder insufflator	Itraconazole	Succinic acid, l-tartaric acid	Anti-solvent crystallization, micronization by jet-milling	[13]
Sustained release systems					
Layer-by-layer microparticles	–	Insulin	Poly (dimethyldiallyl-ammonium chloride), sodium (polystyrenesulfonate)	Layer-by-layer self-assembly technique	[14]
Coated nanoparticles	Dry powder insufflator	Terbutaline sulphate	Glyceryl behenate, tripalmitin, hydrogenated palm oil, dipalmitoyl-phosphatidylcholine	Spray drying	[15]
Magnetized thermosensitive system	DPI	Budesonide	Glycerol behenate, poloxamer 188, iron-oxide nanoparticles	Emulsification method, Freeze-drying, micronization by jet-milling	[16]
Thermo-responsive system	MicroSprayer aerosolizer	Sildenafil citrate	Poloxamer 407	Dissolution and mixing	[17]
Drug complexes					
Microparticles	DPI	Insulin	Human serum albumin, DL-a-phosphatidyl-choline, dipalmitoyl protamine sulfate	Spray drying	[18]
Zn^{2+} cross-linked microparticles	DPI	Bovine serum albumin	Sodium alginate, zinc sulfate heptahydrate ($ZnSO_4$) ammonia	Zn crosslinking followed by spray drying	[19]
Ionic complex	DPI	Lysozyme	Sodium hyaluronate	Emulsification method, freeze-drying	[20]

Large-porous particles based on a inclusion complex	DPI	Deslorelin	HPβCD, PLGA	Emulsion-solvent evaporation	[21]
Polymeric microparticles based on a inclusion complex	DPI	Prostaglandin E1	HPβCD, PLGA	Emulsion-solvent evaporation	[22]
Nanoparticles	Nebulizer	Isoniazid, rifampicin, pyrazinamide	Sodium alginate, chitosan	Cation-induced gelification of alginate	[23]
Drug-receptor binding systems					
Gelatin nanoparticles	Nebulizer	Cisplatin	NeutrAvidin™, biotinylated epithelial growth factor	Two-step desolvation method, NeutrAvidin™ conjugation	[24]
Porous large microparticles	Dry powder insufflator	Doxorubicin	PLGA, TRAIL	Emulsification method, spray freeze drying	[25]
Mesoporous silica nanoparticles	–	Doxorubicin, cisplatin, siRNA	LHRH analog peptide	Surfactant-templated, base catalyzed condensation method	[26]
Nanoconjugates of CdSe/CdS/ZnS quantum dots (QDs)	–	Doxorubicin	CdSe/CdS/ZnS triethylamine 1-ethyl-3-(3-dimethylaminopropyl)-carbodiimide	Colloidal synthesis of the QDs and addition to the drug solution	[27]
Drug-conjugates					
PEGylated-drug suspension	Nebulizer	Prednisolone	PEG (different molecular weights)	Production of esters of PEG and prednisolone, precipitation of the conjugates using diethyl ether	[28]
PEGylated polylysine dendrimer	–	Doxorubicin	PEG, e-polylysine 4-(hydrazinesulphonyl) benzoic acid	PEG was attached to surface ε-amino groups, doxorubicin was linked through the labile acid linker	[29]
Solution	–	Cisplatin	Hyaluronic acid	Reaction in solution, filtration and dialyzed	[30]
Porous microspheres	Dry powder insufflator	Exendin-4	PLGA, *N*-hydroxysuccinimidyl-activated palmitic acid	Palmityl acetylation of exendin-4, adsorption on PLGA porous microspheres	[31]

(*continued*)

Table 18.1 (Continued)

Type of formulation	Device	Drug	Excipients	Production process	References
Drug-polymer matrix particles					
Porous matrix	DPI	Celecoxib	PLGA	Emulsion-solvent evaporation process, supercritical fluid pressure-quench technology	[32]
Amphiphilic hydrogel microspheres	DPI	Curcumin	PLGA, chitosan grafted with PEG	Spray drying	[33]
Mucoadhesive microparticles	DPI	Sodium cromoglycate	Sodium carboxymethylcellulose, sodium alginate, sodium hyaluronate	Spray drying	[34]
Hydrophilic swellable microparticles	DPI	Atenolol	Alginic acid	Spray drying	[35]
Porous materials					
Large porous microparticles	DPI	Heparin	Polyvinyl alcohol, sodium chloride	Emulsion-solvent evaporation method, freeze drying	[36]
Large porous microparticles (PulmoSphere)	DPI	Tobramycin sulfate	Distearoyl-phosphatidylcholine, calcium chloride	Emulsion homogenization, spray drying	[37]
Nanoporous aggregates	–	Isoniazid, rifampicin	Guar gum, mannitol, leucine, polysorbate 80	Anti-solvent precipitation technique, spray drying	[38]
Large porous microparticles	DPI	Pyrazinamide	Leucine, ammonium bicarbonate, hyaluronic acid, dipalmitoyl-phosphatidylcholine	Spray drying	[39, 40]
Nanosystems					
Nanogel	MicroSprayer aerosolizer	Exendin-4	Glycol chitosan	Self-assembly technique	[41]
Solid-lipid nanoparticles	Nebulizer	Isoniazid, rifampicin, pyrazinamide	Stearic acid, polyvinyl alcohol	Emulsion solvent diffusion	[42]
Mannosylated liposomes	–	Ciprofloxacin	Phosphatidylcholine, cholesterol, mannitol	Lipid thin film hydration	[43]

DPI, dry powder inhaler; PLGA, poly(lactic-co-glycolic acid; HPβCD, 2-hydroxypropyl-β-cyclodextrin; PEG, polyethylene glycol; LHRH, luteinizing hormone-releasing hormone; TRAIL, tumor necrosis factor-related apoptosis inducing ligand.

structures (e.g. porous, layered). These systems are composed by excipients such as polymers and lipids. For example, in layer-by-layer systems, microparticles are prepared by sequential deposition of oppositely charged polymers [14]. This design demonstrated that insulin release pattern can be modulated by the number of deposited layers.

Cook et al. [15] reported terbutaline sulfate microspheres where drug nanoparticles were coated with hydrophobic excipients. Drug release was controlled by drug and phospholipid content in the system.

Solid lipid matrix microparticles are able to control drug release by diffusion. These microparticles are stable and possess low toxicity. When these systems were combined with other materials, like iron-oxide nanoparticles or pore inducing agents, they were also capable of targeting and modulating the drug delivery after activation by different stimuli (such as wetting, ionic exchange, and temperature) [16, 48].

Other control release systems are structured *in situ* following inhalatory administration. Dalla-Bona et al. [17] reported prolonged *in vitro* sildenafil delivery from aqueous formulations containing poloxamer 407. This formulation was activated by temperature, producing the gelation in simulated lung conditions.

18.3.1.3 Drug Complexes

Another way to control the release of active ingredients is through formation of drug complexes. Within this classification, inclusion and ionic complexes are found. Anionic or cationic polymers (known as polyelectrolytes) have the capability of interacting with oppositely charged drugs. This interaction leads to new chemical entities [49]. Controlled release is achieved as the drug is displaced from the polymer by the ions of the medium, being the ionic strength of the release medium an important factor that affects the drug displacement [50]. Besides polyelectrolytes, other ionic and nonionic polymers have been used for drug controlled release to the lungs and are explained in Section 3.1.6.

Cyclodextrins are widely used to prepare inclusion complexes. Different drugs were found to be able to form non-covalently bonded inclusion complexes, providing controlled release of drugs [51].

Both strategies, inclusion and ionic complexation, have been explored for pulmonary administration of drugs. Different drug families, including small molecules and macromolecules, have been formulated. In particular, proteins have been recognized as interesting candidates for being administered by the lung route. Traditionally, therapeutic peptides and proteins are delivered by the parenteral route as they are degraded when orally administered [52]. In this sense, prostaglandin-E1, deslorelin, bovine serum albumin, insulin, lysozyme, and exendin-4 have been formulated as controlled release complexes by the pulmonary route, to administer the proteins by a noninvasive route and in a controlled way.

In particular, Vanbever et al. [18] formulated the insulin injectable suspension as a DPI composition. The inhalable solid formulation composed of insulin, protamine, and zinc had a bioavailability around the 80% of the same preparation when administered subcutaneously. Controlled release was achieved by the interaction between protamine and insulin, forming an insoluble complex. The pulmonary administration produced a sustained release of insulin, leading to sustained plasma levels [18].

Spray-drying has been recognized as an interesting technique for producing inhalatory particles in a single step. Bovine serum albumin, as a model protein, was formulated as Zn^{2+}-alginate crosslinked microparticles, and the protein release was controlled by the removal of the Zn ions by exchanging them with other physiological cations or by Zn binding. The success of the alginate-Zn crosslinking strategy was proposed as a promising platform for pulmonary controlled release protein formulations [19]. The ions present in the medium and the ionic strength also played an important role in the lysozyme controlled release from lysozyme–hyaluronan ionic complexes. Controlling the ionic strength it was possible to sustain the drug release during seven days [20].

Cyclodextrins have been used for stabilizing protein structure during processing [53]. Also, the inclusion of deslorelin into hydroxyl-propyl-β-cyclodextrin complexes allowed high plasma deslorelin concentrations up to seven days when intratracheally administered. In this case, the complex formed protected the protein from enzymatic degradation [21]. Besides, the obtaining of an inclusion complex of prostaglandin E-1 with the same cyclodextrin allowed a bursts release protein profile. This is required to obtain a rapid hemodynamic effect, as prostaglandin E-1 is used for pulmonary arterial hypertension treatments [22]. The increased drug release of the protein from the dry powder is related to the increment of the hydrophilicity of the prostaglandin due to its incorporation into the cyclodextrin hydrophobic cavity [22].

Besides protein pulmonary administration, drug complexes have been obtained with antitubercular drugs. Zahoor et al. formulated alginate nanoparticles carrying isoniazid and pyrazinamide. The nanoparticles administration by nebulization achieved sustained release of the drugs by the alginate affinity for the drugs amino groups [23]. These results are relevant for antitubercular treatments as patients are prone to abandon the chemotherapy because it requires more than one drug administration daily, for at least six months.

As can be seen, the drug complexation is an interesting tool for retarding or increasing active ingredients release.

18.3.1.4 Drug–Receptor Binding

Drug–receptor binding has been proposed as an interesting tool to enhance the drug targeting efficiency to specific cells or tissues. In this way, the pharmacological effect of the drug is mainly directed to the site of action or the organ [54]. Targeting could be either active or passive. Active targeting is achieved when it exists a specific ligand recognition (like antibodies, proteins, or other molecules) by a receptor or cell structure. Passive targeting relies on bulk recognition based on size, charge, hydrophilicity/hydrophobicity, among others [55]. This type of target can be characterized by relative tissue affinity of the molecules being delivered. Higher tissue affinity can lead to slower lung clearance and possibly increasing the duration of effect [7, 56] have suggested that lyososomal accumulation of weak bases can result in accumulation of certain drugs in lung tissue. β2 agonists, commonly administered to the airways, are available in several forms including short acting and long acting. The relative lipophilicities of the different available drugs are thought to be responsible for some of the differences in duration of effects.

Regarding pulmonary administration, this strategy has been mainly proposed for lung cancer treatments. In general terms, tumorous cells express some specific receptors that normal cells do not or are overexpressed in cancerous cells. Among others, epithelial growth factor receptor (EGFR) [24], death receptors (DR4/TRAIL-R1, DR5/TRAIL-R2) [25], and luteinizing hormone-releasing hormone (LHRH) receptors [26] are overexpressed in cancer cells.

Doxorubicin and cisplatin are two extensively used anticancer drugs for small and non-small cell lung cancer [26]. Lung cancer is one of the most aggressive types of this disease [54]. This is mainly due to the limited effectiveness of traditional treatments, which usually deliver the drug by the parenteral route [26]. Also, the chemotherapeutics are associated with severe adverse effects, and high levels of the active ingredients on nontarget sites cause damages on normal tissues. Delivering drugs directly to the lungs and, more preferable to the cancer cells, would improve cancer treatments.

Different formulations have been reported to target lung cancer cells by the inhalatory route. Also, controlled release limited the drug level in the lung, avoiding the damage on the respiratory tract. Taratuola et al. prepared mesoporous silica nanoparticles carrying doxorubicin and cisplatin [26]. An analog of the LHRH hormone was added to target the formulation to the cancer cells. Tseng et al. developed gelatin nanoparticles to deliver cisplatin [24]. A biotinylated epithelial growth factor was included in order to target the tumorous cells, via EGFR-endocytosis. Kim et al. obtained porous large microparticles to administer doxorubicin [25]. In this case, the inclusion of tumor necrosis factor-related apoptosis inducing ligand (TRAIL) allowed the specific binding to death receptor. In all cases, drug activity was maintained in the formulations or an increment of the anticancer effect was observed compared to the free drug.

Besides the targeting, controlled release of the formulations limited the exposure of the normal tissues to high drug levels. For example, the crosslinking between the gelatin and the cisplatin allowed the controlled release of the drug after the initial burst, as the interaction between the components is reversible due to its low nucleophilicity [24]. The encapsulation of the doxorubicin and the TRAIL on poly(lactic-*co*-glycolic acid) (PLGA) microparticles allowed the gradual release of both components [25].

Doxorubicin was also proposed to modulate the immune-inflammatory response in lung diseases [27]. The alveolar macrophage is the main cell involved in the immune response. Chakravarthy et al. developed nanoconjugates of quantum dots (QDs) and doxorubicin [27]. Macrophage has receptors for anionic particles, and QD composition was designed for these cells uptake. Also, the QD-doxorubicin conjugates had a size that also favored macrophage recognition.

18.3.1.5 Drug Conjugates

In general terms, drugs administered by the inhalatory route are rapidly absorbed [28]. This could be desired if the drug is intended for systemic treatments. However, to treat local lung diseases, prolonged pulmonary residence time is a necessary characteristic. One strategy for prolonging this time is the development of conjugates between drugs and polymers [57]. In them, the active ingredient is chemically linked to a polymer (excluding encapsulation).

The drug conjugation strategy has been considered an interesting platform for delivering drugs and is part of the polymer therapeutics [57]. The conjugates have some advantages compared to the pure drugs such as increased drug aqueous solubility, controlled release, drugs stabilization, higher bioavailability, specific accumulation, and biodistribution [57]. Although this kind of formulations was initially intended for parenteral administration, there are reports where drug conjugates have been administered by other routes, like the pulmonary one.

To obtain the drug-conjugates, usually the drug and excipients (conjugation polymer and other excipients) are allowed to interact in a solution and then the conjugate is precipitated, filtered, or dialyzed. The interaction between the drugs and the polymers can be directly or via a labile linker. One of the most used synthetic polymers for drug conjugation is polyethylene glycol (PEG), and the PEGylation strategy has been widely used [57], combining the PEG with dendrimers or drugs, such as prednisolone.

Regarding the controlled release, dendrimers formulations are one of the most studied systems. This is because these structures can control the drug release as a function of size. Also, dendrimers have the capability to access to the lymphatic nodes, an interesting target for lung cancers, as these kinds of diseases are one of the main causes of death in developed countries [29]. Besides anticancer active ingredients, the drug-conjugate strategy has been used to deliver active ingredients for different therapies, such as antidiabetic therapies and pulmonary inflammatory treatments.

The controlled release was achieved by different strategies. As above mentioned, PEGylation is one of the most explored strategies for drug conjugation systems. Bayard et al. used this strategy to obtain prednisolone–PEG conjugates [28]. The drug and the polymer interacted by an ester bond and the hydrolysis of that bond under physiological pH or by esterases controlled the release of the drug. Kaminskas et al. formulated a PEGylated polylysine dendrimers to conjugate doxorubicin [29]. This formulation was intended for administering the drug directly to lung cancers. The drug dendrimer interaction allowed the controlled and prolonged release of doxorubicin. After the drug release, the dendrimer degraded faster as much drug was released. Another anticancer active ingredient formulated as a drug conjugate is cisplatin. This active pharmaceutical ingredient was conjugated to hyaluronan acidic groups by an ester bond. This bond was reversible and allowed the controlled release [30].

Kim et al. developed a more complex formulation, where the conjugated molecule also bonded to a physiological protein [31]. These authors obtained a hydrophobic exendin-4 derivative by conjugating the active ingredient with *N*-hydroxysuccinimidyl-activated palmitic acid. The derivative hydrophobicity allowed the adsorption on PLGA microspheres, interaction that allowed continuous release up to five days. The palmityl acylation allowed longer hypoglycemic effect as palmityl groups have the capability to bind to the albumin, increasing drug half-life.

In addition to the prolonged drug release, the systems developed were able to target the drugs to specific sites. The hyaluronan–cisplatin conjugate increased the drug deposition within the lung tumors [30]. Also, as the hyaluronan is mainly cleared by the lymphatic system, the conjugate allowed the drug concentration in the lung and mediastinal lymph [30]. An interesting point of the doxorubicin dendrimers was that after intratracheal administration, the dendrimers could reach the pulmonary interstitium from where they were absorbed mainly via the pulmonary lymphatic system. This is relevant as most lung cancers metastasis via the lymph [29].

Finally, all the systems reported demonstrated that the active ingredient activity was not affected by the conjugation. Although the direct drug administration to the lungs has recognized benefits, there are few examples of marketed medications besides bronchodilators and corticosteroids [58]. The understanding and controlling drug residence/absorption from the lung would contribute to the formulation of different drugs for pulmonary administration/targeting.

18.3.1.6 Drug–Polymer Matrix Particles

Polymers have been studied to modify drug release rate into the lung. Among others, PLGA has been extensively used to produce biodegradable and biocompatible microparticles by emulsion methods. Both hydrophobic and hydrophilic drugs could be encapsulated. In particular, these systems are suitable for low-dose drugs because of the limited loading capacity [32, 48]. To control drug release from PLGA microparticles, changes in the polymer composition were proposed. The incorporation of osmotic or porous inducing agents during processing or in the formulation could also modify drug release properties from PLGA materials [6].

Another interesting approach is the development of swellable microparticles. These materials, in contact with lung fluids, can modulate drug release by diffusion–erosion mechanisms. Particle size is increased by fluid sorption, avoiding macrophages uptake [33]. Because some of these polymers also have mucoadhesive properties, they significantly increase microparticles residence time in the lung. For sustained curcumin release, El-Sherbiny and Smyth [33] developed PLGA-drug nanoparticles dispersed in chitosan or chitosan grafted with PEG swellable matrices for DPI. Other swellable polymers, such as hydroxypropylcellulose, sodium carboxymethylcellulose, alginate, and hyaluronate, were used for microparticles design [59]. Gallo et al. [34] found that sodium carboxymethylcellulose delayed sodium cromoglycate release and improved mucoadhesion of spray-dried composites. Because swellable microparticles can be affected by elevated relative humidity [35], protection from ambient conditions is needed during storage.

18.3.1.7 Large Porous Particles

Powders with relatively high geometric diameters and low density have been used in order to increase powder dispersion, DPIs efficiency, and stability of MDIs suspensions [60]. They are produced by drying or lyophilization of single or multiple emulsions [36, 37, 61]. Processing nanosuspensions or hydroalcoholic solutions by the spray-drying technology was also proposed as alternative method [38, 61], where the addition of pore-inducing agents to the formulation improves product porosity [6, 39, 40].

The most studied porogen is ammonium bicarbonate, a compound that decomposes during processing. However, cyclodextrins have been also used during emulsion drying. Because of its osmotic activity, they are able to produce a pressure difference between the emulsion phases, favoring water penetration into the organic phase to induce pore production. It is known that porosity increased materials friability. For this reason, optimization of the product properties is required during the development phase [6].

18.3.1.8 Nanosystems

Nanosized systems are promising controlled drug delivery formulations. Liposomes, polymeric and lipid nanoparticles, nanogels, dendrimers, microemulsions, among others, were developed to modify drug release [59]. They are mainly administered by nebulization, attached to porous carrier microparticles or as reversible microsized aggregates in order to improve their aerosolization and deposition efficacy. In the lungs, the nanosystems are delivered or disaggregated from carriers (droplets or particles). In particular, nanoparticles with particle size lower than 500 nm are considered mucus penetrating platforms that improves therapy efficacy for local diseases (e.g. lung cancer or infections) [62].

Some examples of nanosized-drug delivery systems will be introduced. Lee et al. prepared nanogels of modified glycol chitosan and exendin-4, a glucagon-like protein-1 receptor agonist for diabetes treatment. The negatively charged peptide interacted with chitosan cationic groups, and this led to a sustained release over one day. As the drug is usually administered twice daily, this improvement would increase patient's adherence to treatments [41].

Pandey and Khuller developed solid-lipid nanoparticles incorporating antitubercular drugs (isoniazid, rifampicin, and pyrazinamide). Sustained release was achieved by polyvinyl alcohol incorporation and the conventional doses could be reduced 6.5 times when the nanoparticles were administered by nebulization [42]. Another approach to treat intracellular parasites like *Mycobacterium tuberculosis* was proposed by Chono et al. In this case, ciprofloxacin was incorporated into mannosylated liposomes to specifically target the alveolar macrophages. The shape and chemical structure of the developed liposomes allowed the efficient targeting to the immune cells [43].

18.3.2 Devices

To access the target sites of the respiratory system, in most cases it is required an efficient aerosol generation (suspension of particles or drops in a gaseous medium) and its administration coordinated with breathing. There are different devices designed for these purposes that influence on the product performance.

18.3.2.1 Controlling Lung Deposition Patterns

Drug deposition in the lungs depends on the properties of the formulation, the device, and the patient's anatomy and pathology. Regarding the therapeutic product, the particle or droplet aerodynamic diameter is a relevant quality attribute because it is related to lung deposition mechanisms [63]. The aerodynamic diameter (D_{aer}) is defined as the diameter equivalent to the one of a sphere of unit density having the same terminal velocity as the particle (or droplet) of interest. This is directly proportional to the geometric diameter but also considers the material density and the shape factor [63]. This aerosol property is measured by techniques based on inertial impact described in pharmacopoeias [64].

It is known that the main aerosol deposition processes in the respiratory tract are [63, 65]:

- *Inertial impaction*: Occurs in particles (or drops) with D_{aer} higher than 5 μm, which are deposited in extra-thoracic regions because of their tendency to move in a straight direction instead of following the gas stream.
- *Sedimentation*: Occurs mainly in particles (or drops) with D_{aer} between 0.5 and 5 μm by gravity. When the dispersed phase has diameters between 1 and 3 μm, maximum alveolar deposition is achieved.
- *Brownian diffusion*: Ultrafine particles (or drops) lower than 0.5 μm can impact the alveoli membrane by diffusion. However, most of them will be exhaled. It should be noted that diffusion deposition increases as particle size decreases and is independent of density.
- *Interception*: This mechanism becomes relevant when the anatomical dimensions are close to particle diameter; the center of gravity of the particle (or drop) is in the air stream but a distal end touches the surface.

For pulmonary drug administration, optimum aerodynamic diameters are considered to be between 0.5 and around 3–5 μm. Aerosols with D_{aer} between 3 and 5 μm are mainly deposited in respiratory and terminal bronchioles and allow lung local treatments. For treating systemic diseases, aerosols should exhibit aerodynamic diameters lower than 3 μm. Therefore, controlled deposition in the lung could be achieved by manipulating the particles or drop properties such as size, shape, and density. Some devices associated to adequate formulations are available for this purpose.

18.3.2.2 Nebulizers

These devices allow the aerosolization of liquid formulations [66]. Aerosols are administered using a mouthpiece or a face mask. Nasal breath and droplets impaction into face and eyes should be avoided. They are mainly used in pediatric and geriatric populations as coordination is not generally required and the patient can inhale the drug at a natural pace [67]. However, treatments times can be long (5–15 minutes) and administration usually occurs at patient's home or care centers [68]. The devices are classified into jet or pneumatic nebulizers, ultrasonic nebulizers, and vibrating mesh nebulizer. Each one is described below.

- *Jet or pneumatic nebulizer*: In these devices, a compressed gas is supplied through an orifice as a gas jet leading to a zone of negative pressure. The solution is drawn into the gas jet forming an unstable film that is broken into small droplets due to the surface tension forces. A baffle is placed into the jet stream in order to get smaller particles. Also, the baffle makes the bigger particles to return to the liquid reservoir [66]. It should be mentioned that in the nebulizer and inside its nozzles there is a so called "dead volume"; the formulation retained there can be between 0.2 and 4 ml. To reduce the aerosol lost during the expiratory phase, newer nebulizers have been designed. In these, reservoir bags and vents have been included in order to increase the aerosol supply during the inspiratory phase [69, 70]. Additionally, triggered or pulsed aerosol production can be match with the patient's inhalation profile to improve efficiency.
- *Ultrasonic nebulizers*: In the ultrasonic nebulizers, a piezoelectric transducer is used to produce high frequency ultrasonic waves, which pass through the solution, creating an aerosol at the liquid surface [66, 69]. These systems have not been considered an adequate method to

administer proteins or thermolabile drugs [68] as during nebulization the liquid is heated (to 30 °C in 5 minutes and to 40 °C in 30 minutes [71]) and denaturation may occur.

- *Static and vibrating mesh nebulizer*: These nebulizers generated the aerosol by applying pressure on the liquid so that it can pass through the orifices of a static or vibrating mesh. They are portable systems that have higher respirable fractions. Mesh nebulizers generally have smaller dead volumes and may have shorter administration times. Usually, they are used to deliver solutions but some report that they can also be used for suspensions despite the concern over orifice occlusion [72].

For the described nebulizer systems, the formulation properties (liquid density, superficial tension, viscosity, and vapor pressure), the equipment operation conditions, and device design influence the droplet size [66]. In addition, devices efficiency changes with the usage and equipment maintenance quality [69]. For example, the aerodynamic behavior of PLGA nanoparticles carrying sildenafil was demonstrated to be related to the dynamic viscosity of the liquids used for the nebulization [73]. Also, the hydrophobicity of nanosphere surface was an important factor that affected the aggregation tendency during jet nebulization [74]. In this work, it was also shown that nebulizing conditions modified, for a given formulation, the mass median aerodynamic diameter of aerosol droplets. The piezoelectric nebulizer led to the largest aerosol droplets, while the ultrasonic nebulizer displayed a variable output. For its part, the jet nebulizer studied increased particle aggregation.

18.3.2.3 Dry Powder Inhalers

DPIs are generally systems that are activated by patient inspiration, overcoming the requirement of coordination between aerosol generation and inspiration. They are generally portable single-dose or multiple-dose devices with diverse designs and different internal resistance to inspiratory airflow. Disposable single use DPIs are also available [75]. The complexity of the administration technique of some DPI devices may compromise patients adherence to inhalatory treatments. For this reason simplest designs are preferred and in recent years connected or "smart" inhalers have been introduced [76].

The DPI formulation is in the solid state as a particulate system, where the microparticles can be a variety of forms such as amorphous materials, crystalline salts, cocrystals, complexes, nanoclusters, polymeric and lipidic composites, among others drug delivery systems previously described [77].

The particles of a DPI formulation must be easily dispersed at relatively low aerodynamic dispersion forces. Regarding the surface of the particles, it has been reported that increasing the roughness decreases the interparticle contact area, reducing the cohesiveness between particles [78]. Considering particles, elongated particles are theoretically preferred because their aerodynamic diameter is determined by their width and not by their length [79]; however, the contact areas of high aspect ratio particles may increase the cohesiveness. Regarding the structure, the advantages of producing highly porous particles have been mentioned in Section 18.3.1.7.

Many formulations available for DPI include carrier excipients. These materials have significantly larger particle size than inhalable microparticles (i.e. generally between 50 and 200 μm) and are included in order to improve the rheological and aerodynamic properties of the formulation. Low aerosolization efficiency due to excipient-microparticle adhesion forces has led to the characterization of intra and interparticle forces [61, 80, 81]. Carrier-free formulations have been investigated, but these may be limited by powder flow variability, dose limitations, and aerosol performance efficiency. The interactions between inhalable particles in carrier-free formulations may be due to van der Waals, electrostatic, and capillary forces. Also, mechanical interlocking occurs between particle rough surfaces. The gravity has less influence due to the relatively lower mass compared to the particles used for other pharmaceutical applications (e.g. drugs and excipients for solid oral formulations). The contribution of each mechanism mentioned is a function of the physicochemical properties of the material, size, morphology (shape and surface), and relative humidity [65]. Therefore, the design of carrier-free formulations requires the production of quality-controlled microparticles. The powders are designed with approaches that reduce the cohesion of the microparticles and improve the flow and dispersion of the particulate material, as well as the design of specific inhalers for this application. Avoiding the use of carrier excipients simplifies monitoring the problems associated with their incorporation in the formulation (interactions, allergies and quality, and uniformity of the mixture) [61]. Particularly, carrier-free systems have been proposed for high doses drugs such as antibiotics and nonsteroidal anti-inflammatory drugs [77, 82].

18.3.2.4 Pressurized Metered-Dose Inhalers

Pressurized metered-dose inhalers (pMDIs) have historically been the most widely used inhalatory delivery systems [83]. This is because pMDIs are small devices, capable of carrying multiple doses [84]. In this system, the drug is contained in a reservoir which further contains the liquefied propellant and optionally excipients. The propellant, the main component of these formulations, exists in two phases: a liquefied phase, in equilibrium with its gas phase [85]. Traditionally, the propellants used were chlorofluorocarbon substances (CFCs), compounds that contribute to the ozone layer depletion. Since Montreal Protocol in 1987, CFCs were replaced by hydrofluoroalkanes (HFAs). In pharmaceutics, the most used ones are the HFA 134a and HFA 227ea [3]. These components are not completely harmless to the environment. Although HFAs do not affect the ozone layer, they contribute to the greenhouse effect [83, 85].

The most important mechanical component of this technology is the valve. Through it, the formulation administration is exact and reproducible, isolating the reservoir content [83]. The rest of the device consists of the reservoir, the actuator (with an orifice), and, optionally, a dose counter [85].

The reservoir can be made of plastic or aluminum, being currently the latter the preferred material because it is lighter and more compact. Usually, the reservoir internal walls are covered to prevent the drug from adhering to the walls or the active ingredient degradation [86]. Oral aerosols are mostly used with noncontinuous metering valves. In each administration, a volume between 20 and 100 μl is loaded from the reservoir into the valve and subsequently atomized. The drug dose is usually between 20 μg and 5 mg [87, 88]. Prior to administration, a measuring chamber is filled with the required dose. During the actuation, a channel is opened connecting the metering chamber to the atmosphere. The pressurized formulation is released quickly on the valve stem by pressing down the actuator [86]. The actuator importance lies in that its orifice is an important factor in determining the droplet size, plume properties, and device clogging which in turn determines the ability of the formulation to reach the lungs [4]. In general terms, the orifice diameter varies between 0.2 and 0.5 mm; the smaller the orifice, the greater the pulmonary deposition [87, 89].

It was mentioned above that the propellant is the major component of these formulations. The drugs may be dissolved or dispersed in the liquid phase of the propellant. In addition to the active pharmaceutical ingredient and the propellant, metered dose inhalers may contain cosolvents or other excipients. The most commonly used cosolvent is ethanol, whereas within the excipients it is possible to find surfactants and polymers (mainly in dispersed systems) which provides stability to the formulations [90].

The propellant allows atomizing the drugs in drops able to reach the lung. The residual aerodynamic size after the droplet evaporation indicates the system ability to reach the therapeutic targeting site. This residual size is determined by the original formulation type: whether it is a solution or dispersion [90]. Solutions have the advantage of being homogeneous and, theoretically, they generate smaller particles [85]. The residual particle size is a function of the droplet size during the actuation and the proportion of nonvolatile components. The power source for atomization is given by the pressurized propellant, which is rapidly evaporated, together with the co-solvent, if present. As the drug is originally homogeneously dissolved in the drops, the residual particle size is a function of the initial droplet size [90].

The low drug and surfactant solubility in the liquid propellant has led to the addition of ethanol as a cosolvent. Many active ingredients need to be formulated as suspensions [83]. In these cases, the drug particles must first be reduced in size. A possible formulation problem is the growth of suspended particles, a process known as Ostwald ripening if partial solubility exists [83].

The particles formation after the propellant atomization is complex. In this case, each generated drop from the propellant evaporation may have one or more drug particles or, eventually, none. The residual particles size will be a function of the amount of active ingredient particles in the drop, the drop size, and the presence of nonvolatile excipients [90].

Although pMDIs are widely used, these systems have some disadvantages. The main limitations when using pressurized inhalers are (i) the need for coordination of the administration with the inhalation maneuver and (ii) the high proportion of drug remaining at the oropharynx. The latter has been related to the high plume velocity when it leaves the device and with large droplet sizes. The plume velocity was partially reduced by replacing the CFCs with the HFAs. The size of the droplet generated increases as the percentage of nonvolatile components in the formulation increases or with the incorporation of cosolvents [84, 88]. This has limited in part the formulation of solutions in HFAs [85].

Many patients do not properly use pMDI devices due to lack of training and synchronizing. To overcome this drawback valved holding chambers and spacers can be incorporated in order to improve the pressurized inhalers administration and performance. The spacers increase the distance between the inhaler mouthpiece and the patient, thereby decreasing the plume velocity and the oropharyngeal impact. Valved holding chambers, for their part, have one or more valves that hold the aerosol until it is inhaled [84]. An alternative to spacers are breath-activated metered-dose inhalers [91]. These devices sense the patient's inspiration and are subsequently activated [86]. While these systems improve the administration maneuver, oropharyngeal deposition is similar to conventional pressurized systems, in addition to preventing the use of spacers [83].

The use of pressurized systems is limited by the maximum dose that can be administered using these devices. Current used chemotherapies have relatively low potency, so large doses are needed, even when they are directly administered to the lungs. Also, the high number of daily shots needed to reach the therapeutic dose could lead to low patients adherence [92, 93].

It was observed that pulmonary deposition increased when CFCs were replaced by HFAs in some circumstances. In a work published by Leach et al., it was shown that solution formulations using HFAs as propellant achieved greater pulmonary deposition of beclomethasone. This allowed the drug targeting to both central and peripheral lung, achieving the adequate inflammation therapy. This replacement of the propellant and reformulation allowed the reduction of nominal doses required for the same therapeutic effect and reduced the oropharyngeal deposition of the drug [94].

The aerosol deposition in breath-activated metered-dose inhalers can be dependent on the patient's inspiratory capacity. At high flow rates (greater than 60 l/min), deposition is mainly central, which would promote local effects on the lung (topical). At lower flow rates (around

30 l/min), aerosol deposition is more peripheral, favoring systemic drug absorption [95]. Another way to favor particles absorption is by depositing them in the lower respiratory tract. The use of the PulmoSphere technology increased the albuterol particle deposition in the lower respiratory tract with respect to micronized particles of the same drug [96].

Sustained drug release has been less explored in pMDIs. In order to obtain a platform for particle controlled release into the lungs, Selvam et al. developed swellable hydrogel microparticles containing curcumin. A high pulmonary particle deposition was found, indicating that the formulation possesses potential as carrier for modulated administration of drugs to the lung, using pressurized systems [97]. Patented microspheres were also found for the inhalatory administration of procaterol-albumin in propellants. The release profile could be controlled depending on the concentration of albumin used during microencapsulation. This also provided a platform for the modified release of active ingredients by using pressurized systems [98].

18.4 Conclusions

The development of controlled drug delivery for the pulmonary route is challenging because of the anatomy and physiology of the respiratory tract. Controlled delivery can be achieved for various aspects including drug absorption, drug release, and deposition. Controlled release formulation strategies have been proposed to but the field remains in its infancy. The innovation has been focused on optimizing local therapy for respiratory illnesses and local lung diseases, as well as, to achieve systemic administration of active substances, mainly peptides and proteins.

Acknowledgments

Financial support was received from CONICET (PIP 112-2011-0100336112), UNS (PGI 24/B209), and FonCyT (PICT-2013-1765). N. Ceschan thanks CONICET for financing her postdoctoral fellowship.

References

1 Atkins, T.M. and Crowder, P.J. (2004). The design and development of inhalation drug delivery systems. In: *Pharmaceutical Inhalation Aerosol Technology* (ed. T.M. Atkins), 277–306. USA: Marcel Dekker, Inc.
2 Depreter, F., Pilcer, G., and Amighi, K. (2013). Inhaled proteins: challenges and perspectives. *Int. J. Pharm.* 447: 251–280.
3 Traini, D. (2013). Inhalation drug delivery. In: *Inhalation Drug Delivery: Techniques and Products* (eds. P. Colombo, D. Traini and F. Buttini), 1–14. Wiley.
4 Verma, R., Ibrahim, M., and Garcia-Contreras, L. (2015). Lung anatomy and physiology and their implications for pulmonary drug delivery. In: *Pulmonary Drug Delivery: Advances and Challenges* (eds. A. Nokhodchi and G. Martin), 1–18. Wiley.
5 de Boer, A.H., Gjaltema, D., Hagedoorn, P., and Frijlink, H.W. (2002). Characterization of inhalation aerosols: a critical evaluation of cascade impactor analysis and laser diffraction technique. *Int. J. Pharm.* 249: 219–231.
6 Liang, Z., Ni, R., Zhou, J., and Mao, S. (2015). Recent advances in controlled pulmonary drug delivery. *Drug Discovery Today* 20 (3): 380–389.

7 Olsson, B., Bondesson, E., Borgström, L. et al. (2011). Pulmonary drug metabolism, clearance, and absorption. In: *Controlled Pulmonary Drug Delivery* (eds. H.D.C. Smyth and A.J. Hickey), 21–50. Springer.

8 d'Angelo, I., Conte, C., La Rotonda, M.I. et al. (2014). Improving the efficacy of inhaled drugs in cystic fibrosis: challenges and emerging drug delivery strategies. *Adv. Drug Delivery Rev.* 75: 92–111.

9 Sinswat, P., Overhoff, K.A., McConville, J.T. et al. (2008). Nebulization of nanoparticulate amorphous or crystalline tacrolimus – single-dose pharmacokinetics study in mice. *Eur. J. Pharm. Biopharm.* 69 (3): 1057–1066.

10 Ceschan, N.E. (2017). Desarrollo de materiales particulados de administración inhalatoria basados en sistemas polielectrolito-fármaco. Tesis Doctoral. Universidad Nacional del Sur.

11 Ceschan, N.E., Bucalá, V., and Ramírez-Rigo, M.V. (2015). Polymeric microparticles containing indomethacin for inhalatory administration. *Powder Technol.* 285: 51–61.

12 Pomázi, A., Ambrus, R., Sipos, P., and Szabó-Révész, P. (2011). Analysis of co-spray-dried meloxicam–mannitol systems containing crystalline microcomposites. *J. Pharm. Biomed. Anal.* 56: 183–190.

13 Karashima, M., Sano, N., Yamamoto, S. et al. (2017). Enhanced pulmonary absorption of poorly soluble itraconazole by micronized cocrystal dry powder formulations. *Eur. J. Pharm. Biopharm.* 115: 65–72.

14 Amancha, K.P., Balkundi, S., Lvov, Y., and Hussain, A. (2014). Pulmonary sustained release of insulin from microparticles composed of polyelectrolyte layer-by-layer assembly. *Int. J. Pharm.* 466: 96–108.

15 Cook, R.O., Pannu, R.K., and Kellaway, I.W. (2005). Novel sustained release microspheres for pulmonary drug delivery. *J. Controlled Release* 104 (1): 79–90.

16 Upadhyay, D., Scalia, S., Vogel, R. et al. (2012). Magnetised thermo responsive lipid vehicles for targeted and controlled lung drug delivery. *Pharm. Res.* 29 (9): 2456–2467.

17 Dalla-Bona, A.C., Stoisiek, K., Oesterheld, N. et al. (2015). Characterization of lung-delivered in-situ forming controlled release formulations. *J. Pharm. Pharmacol.* 67 (10): 1349–1354.

18 Vanbever, R., Ben-Jebria, A., Mintzes, J.D. et al. (1999). Sustained release of insulin from insoluble inhaled particles. *Drug Delivery Res.* 48 (4): 178–185.

19 Möbus, K., Siepmann, J., and Bodmeier, R. (2012). Zinc-alginate microparticles for controlled pulmonary delivery of proteins prepared by spray-drying. *Eur. J. Pharm. Biopharm.* 81 (1): 121–130.

20 Lee, E.S., Kwon, M.J., Na, K., and Bae, J.H. (2007). Protein release behavior from porous microparticle with lysozyme/hyaluronate ionic complex. *Colloids Surf., B* 55 (1): 125–130.

21 Koushik, K., Dhanda, D.S., Cheruvu, N.P.S., and Kompella, U.B. (2004). Pulmonary delivery of deslorelin: large-porous PLGA particles and HPbCD complexes. *Pharm. Res.* 21 (7): 1119–1126.

22 Gupta, V., Davis, M., Hope-Weeks, L.J., and Ahsan, F. (2011). PLGA microparticles encapsulating prostaglandin E1-hydroxypropyl-β-cyclodextrin (PGE1-HPβCD) complex for the treatment of pulmonary arterial hypertension (PAH). *Pharm. Res.* 28 (7): 1733–1749.

23 Zahoor, A., Sharma, S., and Khuller, G.K. (2005). Inhalable alginate nanoparticles as antitubercular drug carriers against experimental tuberculosis. *Int. J. Antimicrob. Agents* 26 (4): 298–303.

24 Tseng, C.L., Su, W.Y., Yen, K.C. et al. (2009). The use of biotinylated-EGF-modified gelatin nanoparticle carrier to enhance cisplatin accumulation in cancerous lungs via inhalation. *Biomaterials* 30 (20): 3476–3485.

25 Kim, I., Byeon, H.J., Kim, T.H. et al. (2013). Doxorubicin-loaded porous PLGA microparticles with surface attached TRAIL for the inhalation treatment of metastatic lung cancer. *Biomaterials* 34 (27): 6444–6453.

26 Taratula, O., Garbuzenko, O.B., Chen, A.M., and Minko, T. (2011). Innovative strategy for treatment of lung cancer: targeted nanotechnology-based inhalation co-delivery of anticancer drugs and siRNA. *J. Drug Targeting* 19 (10): 900–914.

27 Chakravarthy, K.V., Davidson, B.A., Helinski, J.D. et al. (2011). Doxorubicin-conjugated quantum dots to target alveolar macrophages and inflammation. *Nanomed. Nanotechnol. Biol. Med.* 7 (1): 88–96.

28 Bayard, F.J.C., Thielemans, W., Pritchard, D.I. et al. (2013). Polyethylene glycol-drug ester conjugates for prolonged retention of small inhaled drugs in the lung. *J. Controlled Release* 171 (2): 234–240.

29 Kaminskas, L.M., McLeod, V.M., Ryan, G.M. et al. (2014). Pulmonary administration of a doxorubicin-conjugated dendrimer enhances drug exposure to lung metastases and improves cancer therapy. *J. Controlled Release* 183 (1): 18–26.

30 Xie, Y., Aillon, K.L., Cai, S. et al. (2010). Pulmonary delivery of cisplatin–hyaluronan conjugates via endotracheal instillation for the treatment of lung cancer. *Int. J. Pharm.* 392 (1–2): 156–163.

31 Kim, H., Park, H., Lee, J. et al. (2011). Highly porous large poly(lactic-*co*-glycolic acid) microspheres adsorbed with palmityl-acylated exendin-4 as a long-acting inhalation system for treating diabetes. *Biomaterials* 32 (6): 1685–1693.

32 Dhanda, D.S., Tyagi, P., Mirvish, S.S., and Kompella, U.B. (2013). Supercritical fluid technology based large porous celecoxib–PLGA microparticles do not induce pulmonary fibrosis and sustain drug delivery and efficacy for several weeks following a single dose. *J. Controlled Release* 168 (3): 239–250.

33 El-Sherbiny, I.M. and Smyth, H.D.C. (2012). Controlled release pulmonary administration of curcumin using swellable biocompatible microparticles. *Mol. Pharmaceutics* 9 (2): 269–280.

34 Gallo, L., Bucalá, V., and Ramírez-Rigo, M.V. (2017). Formulation and characterization of polysaccharide microparticles for pulmonary delivery of sodium cromoglycate. *AAPS PharmSciTech* 8 (5): 1634–1645.

35 Ceschan, N.E., Bucalá, V., Ramírez Rigo, M.V., and Smyth, H.D.C. (2016). Impact of feed counterion addition and cyclone type on aerodynamic behavior of alginic–atenolol microparticles produced by spray drying. *Eur. J. Pharm. Biopharm.* 109: 72–80.

36 Patel, B., Gupta, V., and Ahsan, F. (2012). PEG-PLGA based large porous particles for pulmonary delivery of a highly soluble drug, low molecular weight heparin. *J. Controlled Release* 162 (2): 310–320.

37 Miller, D.P., Tan, T., Tarara, T.E. et al. (2015). Physical characterization of tobramycin inhalation powder: I. Rational design of a stable engineered-particle formulation for delivery to the lungs. *Mol. Pharmaceutics* 12 (8): 2582–2593.

38 Kaur, R., Garg, T., Malik, B. et al. (2016). Preparation and characterization of spray-dried inhalable powders containing nanoaggregates for pulmonary delivery of anti-tubercular drugs. *Drug Delivery* 23 (3): 872–877.

39 Pham, D.D., Fattal, E., Ghermani, N. et al. (2013). Formulation of pyrazinamide-loaded large porous particles for the pulmonary route: avoiding crystal growth using excipients. *Int. J. Pharm.* 454 (2): 668–677.

40 Pham, D.D., Grégoire, N., Couet, W. et al. (2015). Pulmonary delivery of pyrazinamide-loaded large porous particles. *Eur. J. Pharm. Biopharm.* 94: 241–250.

41 Lee, J., Lee, C., Kim, T.H. et al. (2012). Self-assembled glycol chitosan nanogels containing palmityl-acylated exendin-4 peptide as a long-acting anti-diabetic inhalation system. *J. Controlled Release* 161 (3): 728–734.

42 Pandey, R. and Khuller, G.K. (2005). Solid lipid particle-based inhalable sustained drug delivery system against experimental tuberculosis. *Tuberculosis* 85 (4): 227–234.

43 Chono, S., Tanino, T., Seki, T., and Morimoto, K. (2008). Efficient drug targeting to rat alveolar macrophages by pulmonary administration of ciprofloxacin incorporated into mannosylated liposomes for treatment of respiratory intracellular parasitic infections. *J. Controlled Release* 127 (1): 50–58.

44 Oberdorster, G., Ferin, J., and Lehnert, B.E. (1994). Correlation between particle size, in vivo particle persistence, and lung injury. *Environ. Health Perspect.* 102 (5): 173–179.

45 Chen, L., Okuda, T., Lu, X.Y., and Chan, H.K. (2016). Amorphous powders for inhalation drug delivery. *Adv. Drug Delivery Rev.* 100: 102–115.

46 Karmwar, P., Graeser, K., Gordon, K.C. et al. (2011). Investigation of properties and recrystallisation behaviour of amorphous indomethacin samples prepared by different methods. *Int. J. Pharm.* 417: 94–100.

47 Ewing, A.V., Clarke, G.S., and Kazarian, S.G. (2014). Stability of indomethacin with relevance to the release from amorphous solid dispersions studied with ATR-FTIR spectroscopic imaging. *Eur. J. Pharm. Sci.* 60: 64–71.

48 Loira-Pastoriza, C., Todoroff, J., and Vanbever, R. (2014). Delivery strategies for sustained drug release in the lungs. *Adv. Drug Delivery Rev.* 75: 81–91.

49 Manzo, R.H., Jimenez-Kairuz, A.F., Olivera, M.E. et al. (2014). Thermodynamic and rheological properties of polyelectrolyte systems. In: *Polyelectrolytes, Thermodynamics and Rheology* (eds. P.M. Visakh, O. Bayraktar and G. Picó), 215–244. Springer International Publishing Switzerland.

50 De Robertis, S., Bonferoni, M.C., Elviri, L. et al. (2015). Advances in oral controlled drug delivery: the role of drug-polymer and interpolymer non-covalent interactions. *Expert Opin. Drug Delivery* 12 (3): 441–453.

51 Zeng, X.M., Martin, G.P., and Marriott, C. (1995). The controlled delivery of drugs to the lung. *Int. J. Pharm.* 124 (2): 149–164.

52 Garcia-Contreras, L. and Smyth, H.D.C. (2005). Liquid-spray or dry-powder systems for inhaled delivery of peptide and proteins? *Am. J. Drug Delivery* 3 (1): 29–45.

53 Tewes, F., Gobbo, O.L., Amaro, M.I. et al. (2011). Evaluation of HPβCD-PEG microparticles for salmon calcitonin administration via pulmonary delivery. *Mol. Pharmaceutics* 8 (5): 1887–1898.

54 Tseng, C.L., Wang, T.W., Dong, G.C. et al. (2007). Development of gelatin nanoparticles with biotinylated EGF conjugation for lung cancer targeting. *Biomaterials* 28 (27): 3996–4005.

55 Daraee, H., Etemadi, A., Kouhi, M. et al. (2014). Application of liposomes in medicine and drug delivery. *Artif. Cells Nanomed. Biotechnol.* 1401: 1–11.

56 Macintyre, A.C. and Cutler, D.J. (1988). The potential role of lysosomes in tissue distribution of weak bases. *Biopharm. Drug Dispos.* 9: 513–526.

57 Pang, X., Yang, X., and Zhai, G. (2014). Polymer-drug conjugates: recent progress on administration routes. *Expert Opin. Drug Delivery* 11 (7): 1075–1086.

58 Ryan, G.M., Kaminskas, L.M., Kelly, B.D. et al. (2013). Pulmonary administration of PEGylated polylysine dendrimers: absorption from the lung versus retention within the lung is highly size-dependent. *Mol. Pharmaceutics* 10 (8): 2986–2995.

59 Smola, M., Vandamme, T., and Sokolowski, A. (2008). Nanocarriers as pulmonary drug delivery systems to treat and to diagnose respiratory and non respiratory diseases. *Int. J. Nanomed.* 3 (1): 1–19.

60 Rennard, S., Fogarty, C., Reisner, C. et al. (2014). Randomized study of the safety, pharmacokinetics, and bronchodilatory efficacy of a proprietary glycopyrronium metered-dose inhaler in study patients with chronic obstructive pulmonary disease. *BMC Pulm. Med.* 14 (1): 118.
61 Healy, A.M., Amaro, M.I., Paluch, K.J., and Tajber, L. (2014). Dry powders for oral inhalation free of lactose carrier particles. *Adv. Drug Delivery Rev.* 75: 32–52.
62 Schuster, B.S., Suk, J.S., Woodworth, G.F., and Hanes, J. (2013). Nanoparticle diffusion in respiratory mucus from humans without lung disease. *Biomaterials* 34 (13): 3439–3446.
63 Gonda, I. (2004). Targeting by deposition. In: *Pharmaceutical Inhalation Aerosol Technology* (ed. A. Hickey), 65–88. New York, NY: Marcel Dekker, Inc.
64 USP 34NF 29 (2011). Aerosols, nasal sprays, metered-dose inhalers, and dry powder inhalers. In: *The United States Pharmacopeial Convention*, 232–252. Baltimore, MD: United Book Press, Inc.
65 Zeng, X., Martin, G., and Marriot, C. (2011). Particulate interactions. In: *Dry Powder Formulations for Inhalation*. London: Informa Healthcare.
66 Hess, D. (2000). Nebulizers: principles and performance. *Respir. Care* 45 (6): 609–622.
67 Carrigy, N.B., Chang, R.Y., Leung, S.S.Y. et al. (2017). Anti-tuberculosis bacteriophage D29 delivery with a vibrating mesh nebulizer, jet nebulizer, and soft mist inhaler. *Pharm. Res.* 34 (10): 2084–2096.
68 Kwok, P.C.L. and Chan, H.K. (2014). Delivery of inhalation drugs to children for asthma and other respiratory diseases. *Adv. Drug Delivery Rev.* 73: 83–88.
69 SEPAR-ALAT (2013). Consenso SEPAR-ALAT sobre terapia inhalada. *Arch. Bronconeumol.* 45 (1): 2–14.
70 Piper, S.D. (2000). In vitro comparison of the circulaire and AeroTee to a traditional nebulizer T-piece with corrugated tubing. *Respir. Care* 45 (3): 313–319.
71 Placke, M., Ding, J., and Zimlich, W.J. (2007). Inhalation, liquids. In: *Encyclopedia of Pharmaceutical Science and Technology*, 2092–2117. New York, NY: Informa Healthcare.
72 Olveira, C., Muñoz, A., and Domenechb, A. (2014). Terapia nebulizada. Año SEPAR. *Arch. Bronconeumol.* 50 (12): 535–545.
73 Beck-Broichsitter, M., Kleimann, P., Gessler, T. et al. (2012). Nebulization performance of biodegradable sildenafil-loaded nanoparticles using the Aeroneb Pro: formulation aspects and nanoparticle stability to nebulization. *Int. J. Pharm.* 422: 398–408.
74 Dailey, L.A., Schmehl, T., Gessler, T. et al. (2003). Nebulization of biodegradable nanoparticles: impact of nebulizer technology and nanoparticle characteristics on aerosol features. *J. Controlled Release* 86: 131–144.
75 Berkenfeld, K., Lamprecht, A., and McConville, J.T. (2015). Devices for dry powder drug delivery to the lung. *AAPS PharmSciTech* 16 (3): 479–490.
76 Kikidis, D., Konstantinos, V., Tzovaras, D., and Usmani, O.S. (2016). The digital asthma patient: the history and future of inhaler based health monitoring devices. *J. Aerosol Med. Pulm. Drug Delivery* 29: 219–232.
77 Pilcer, G., De Bueger, V., Traina, K. et al. (2013). Carrier-free combination for dry powder inhalation of antibiotics in the treatment of lung infections in cystic fibrosis. *Int. J. Pharm.* 451: 112–120.
78 Chow, A., Tong, H., Chattopadhyay, P., and Shekunov, B. (2007). Particle engineering for pulmonary drug delivery. *Pharm. Res.* 24 (3): 411–437.
79 Claus, S., Weiler, C., Schiewe, J., and Friess, W. (2014). How can we bring high drug doses to the lung? *Eur. J. Pharm. Biopharm.* 86 (1): 1–6.

80 Hoppentocht, M., Hagedoorn, P., Frijlink, H.W., and de Boer, A.H. (2014). Technological and practical challenges of dry powder inhalers and formulations. *Adv. Drug Delivery Rev.* 75: 18–31.

81 Kho, K. and Hadinoto, K. (2013). Dry powder inhaler delivery of amorphous drug nanoparticles: effects of the lactose carrier particle shape and size. *Powder Technol.* 233: 303–311.

82 Yazdi, A.K. and Smyth, H.D.C. (2016). Hollow crystalline straws of diclofenac for high-dose and carrier-free dry powder inhaler formulations. *Int. J. Pharm.* 502: 170–180.

83 Pritchard, J.N. (2015). Industry guidance for the selection of a delivery system for the development of novel respiratory products. *Expert Opin. Drug Delivery* 12 (11): 1755–1765.

84 Dolovich, M.B. and Dhand, R. (2011). Aerosol drug delivery: developments in device design and clinical use. *Lancet* 377 (9770): 1032–1045.

85 Myrdal, P.B., Sheth, P., and Stein, S.W. (2014). Advances in metered dose inhaler technology: formulation development. *AAPS PharmSciTech* 15 (2): 434–455.

86 Newman, S.P. (2005). Principles of metered-dose inhaler design. *Respir. Care* 50 (9): 1177–1188.

87 Vehring, R., Lechuga-ballesteros, D., Joshi, V. et al. (2012). Cosuspensions of microcrystals and engineered microparticles for uniform and efficient delivery of respiratory therapeutics from pressurized metered dose inhalers. *Langmuir* 28: 15015–15023.

88 Kleinstreuer, C. and Zhang, Z. (2011). Optimal drug-aerosol delivery to predetermined lung sites. *J. Heat Transfer* 133: 1–12.

89 Newman, S., Pitcairn, G., Steed, K., and Harrison, A. (1999). Deposition of fenoterol from pressurized metered dose inhalers containing hydrofluoroalkanes. *J. Allergy Clin. Immunol.* 104: 253–257.

90 Stein, S.W., Sheth, P., and Myrdal, P.B. (2012). A model for predicting size distributions delivered from pMDIs with suspended drug. *Int. J. Pharm.* 422: 101–115.

91 Usmani, O.S. (2015). Inhaled drug therapy for the management of asthma. *J. Prescr. Pract.* 26 (3): 23–25.

92 Carvalho, T.C., Carvalho, S.R., and Mcconville, J.T. (2011). Formulations for pulmonary administration of anticancer agents to treat lung malignancies. *J. Aerosol Med. Pulm. Drug Delivery* 24 (2): 61–80.

93 Goel, A., Baboota, S., Sahni, J.K., and Ali, J. (2013). Exploring targeted pulmonary delivery for treatment of lung cancer. *Int. J. Pharm. Investig.* 3 (1): 8–14.

94 Leach, C.L., Davidson, P.J., and Boudreau, R.J. (1998). Improved airway targeting with the CFC-free HFA – beclomethasone metered-dose inhaler compared with CFC – beclomethasone. *Eur. Respir. J.* 12: 1346–1353.

95 Keller, M. (1999). Innovations and perspectives of metered dose inhalers in pulmonary drug delivery. *Int. J. Pharm.* 186: 81–90.

96 Hirst, P.H., Pitcairn, G.R., Weers, J.G. et al. (2002). In vivo lung deposition of hollow porous particles from a pressurized metered dose inhaler. *Pharm. Res.* 19 (3): 258–264.

97 Selvam, P., El-Sherbiny, I.M., and Smyth, H.D.C. (2011). Swellable hydrogel particles for controlled release pulmonary administration using propellant-driven metered dose inhalers. *J. Aerosol Med. Pulm. Drug Delivery* 24 (1): 25–34.

98 Huang, H.-P. and Mehta, S.C. (1991). Procaterol microspheres controlled-release aerosol. CA 2036844 A1.

19

Controlled Drug Delivery via the Vaginal and Rectal Routes

José das Neves[1] and Bruno Sarmento[2]

[1] i3S – Institute for Research and Innovation in Health, University of Porto, Portugal
[2] INEB – Institute for Biomedical Engineering, University of Porto, Portugal

19.1 Introduction

The administration of therapeutic, prophylactic, or diagnostic preparations in the vagina and rectum is thought to be as old as Medicine itself. The Kahun Papyrus, the oldest medical record from Ancient Egypt (c. 1950–1850 before the Common Era), considered only five routes of drug delivery: vaginal, oral, rectal, topical, and inhalatory [1]. The administration of drugs in the vagina and rectum continued and was refined throughout the course of times, arriving to our days as well-established medical practices. Vaginal drug delivery is mostly used in current days for infectious vaginitis treatment, management of conditions related with reproduction (e.g. contraception with spermicides or sexual hormones, pregnancy support or termination, labor inducement), or relief of postmenopausal syndrome symptoms. In the case of the rectal drug delivery route, both local (for example, hemorrhoids or inflammatory bowel disease) and systemic conditions can be successfully managed. Vaginal and rectal administration of drugs are, however, considered of minor importance as compared to the oral and parenteral routes, being typically regarded as alternatives for particular situations. There are, though, several advantageous features that may in fact make vaginal and rectal routes preferential [2–4]. In the case of the former, circumventing the hepatic first-pass effect, avoiding gastrointestinal adverse effects, reducing enzymatic drug degradation observed after oral administration, and enhancing systemic drug absorption for various drugs have been advocated as particularly interesting features. Still, gender specificity and variable local physiological conditions (e.g. along the menstrual cycle, menopause, and pregnancy) affecting product performance and drug absorption, as well as cultural issues and myths regarding the vagina, are considerable limitations. As for rectal drug administration, commonly mentioned general advantages include rapid onset of therapeutic effect, easy administration, namely when the oral route is impaired (e.g. unconscious patients, nausea/vomiting) or when swallowing trouble/tasting issues are involved, less susceptibility to enzymatic drug degradation than at the small intestine and partial avoidance of the hepatic first-pass effect. The rectal route is often considered as preferential for children and the elderly [5]. However, often poor acceptability, cultural issues, possible onset of local toxicity, reduced and unreliable drug absorption of drugs as compared to the small intestine, and leakage or discomfort are regarded as potential problems. Also, the presence and trafficking of stool may impair drug delivery, particularly when prolonged rectal residence is required.

Fundamentals of Drug Delivery, First Edition.
Edited by Heather A. E. Benson, Michael S. Roberts, Adrian C. Williams, and Xiaowen Liang.

In general, vaginal and rectal drug delivery are particularly considered when local action is required, even if extensive drug absorption and systemic therapeutic effects can also be attained. One particular objective being pursued over the last two decades is the development of anti-HIV microbicide products, which are intended to be placed in the vagina and/or rectum in order to prevent sexual transmission of HIV by interfering at the mucosal level [6]. Indeed, the microbicides field has boosted the interest in both vaginal and rectal routes and has been the driving force behind the most recent discoveries and technological developments in drug delivery [7]. Such progresses are discussed in the following as the present chapter intends to provide an up-to-date overview of vaginal and rectal drug delivery.

19.2 Biological Features of the Vagina and Colorectum

Although being considered in tandem in this chapter, vaginal and rectal routes present distinct anatomical, histological, and physiological features that need to be considered. Such differences often render dual compartment administration (i.e. either in the vagina or rectum) of a single pharmaceutical preparation as unfeasible. For example, this highly desirable possibility has been shown mostly fruitless while developing microbicides, namely due to safety issues [8]. Still, efforts toward this goal are ongoing [9, 10].

The following sub-sections provide brief descriptions of the main features of both mucosal compartments concerning drug delivery, with particular focus on barriers such as epithelia and mucus. Readers interested in further detailing on the anatomy, histology, and physiology of the vagina and colorectum as related to drug delivery are referred to previous publications from the authors [11–13].

19.2.1 Vagina

The vagina is a mucosal cavity connecting the uterus across the cervix to the vestibule and exterior of the body. On the uterine side, the cervix protrudes into the vagina thus originating anterior and posterior pockets called fornices [14]. The width of the vaginal canal decreases toward the introitus, forming an S-shaped tube at its sagittal section. The upper part features an almost horizontal orientation in standing women, while the lower part presents itself as a steep surface [15]. Noteworthy, the vaginal wall collapses at its anteroposterior diameter and folds into rugae, particularly at the lower end, which contributes to increasing the surface of the mucosa. The total area of the vaginal surface varies considerably among women but has been estimated in the range of 65–165 cm^2 [16, 17].

The physiological functions of the vagina include receiving the penis during sexual intercourse and serve as passageway to menstrual fluids as well as newborns during labor. The mucosal surface is covered by non-keratinized stratified squamous epithelium that extends up to the ectocervix but abruptly changes at the transformation zone into columnar epithelium covering the endocervix [18]. The epithelium rests on top of a vascularized lamina propria, followed by a muscular layer and tunica adventitia. Although considered a mucosal site, the vagina does not possess mucus-secreting cells or glands, relying on contributions from endocervical mucus (main source of mucins), endometrial fluids, mucosa transudate, secretions from vestibular glands, cells and traces of urine to constitute its own fluid [19]. Of particular notice, vaginal fluid is typically acidic (pH around 4.5) in healthy women during reproductive years chiefly due to the presence of lactic acid resulting from the combined metabolism of host glycogen by microbiota (namely dominant *Lactobacillus* sp.) and mucosa [20]. Concerning volume, vaginal fluid is estimated to be continuously present but scanty, typically in the range of 0.5–1.0 ml [21]. However, these values increase rapidly upon sexual arousal in order to provide lubrication in anticipation of penile penetration [22]. Such variations may impact drug delivery namely influencing drug solubility.

Many features of the vagina are dependent on the hormonal status of women. In particular, estrogens favor thickening of the epithelium and glycogen production (with downstream effects in pH), as well as proliferation of muscular fibers. These changes justify the significant differences observed throughout the life of women. For example, low estrogen levels during childhood, post-menopause, and pregnancy lead to reduced volume and high pH of the vaginal fluid (around neutrality), as well as tissue atrophy [23]. Pathological situations such as bacterial vaginitis also induce local changes, namely in pH [24]. Furthermore, changes to volume, composition, and final hydronium activity of vaginal fluid are observed during sexual intercourse, particularly when ejaculation occurs – semen features pH above seven and high buffering capacity [25], or during menstruation (presence of blood and higher pH) [26]. Such changes may affect, among others things, the release from dosage forms/delivery systems, solubility and permeability/absorption of drugs [27–29]. A potentially even more dramatic effect in drug product performance is that resulting from the forces applied due to penile penetration [30].

The vagina is considered to have low enzymatic activity, namely when compared to enteral routes, although relevant metabolism may still occur [31]. The vaginal mucosa is provided with an extensive supply and drainage of blood and lymph that, interestingly, bypasses the portal circulation and allows avoiding the hepatic first-pass effect of drugs absorbed through this route. Concerning sensory innervation, the vagina presents high sensitivity to stimuli at the lower end but is relatively unaffected at its upper part [32]. This may be particularly interesting when considering the comfortable use of deeply inserted drug delivery devices such as rings and diaphragms.

Interest in vaginal microbiota has been increasing over recent years due to its complex interactions with the mucosa and role in determining many features of the local environment [33]. Although dominated by lactobacilli, many other beneficial species are harbored in the vagina, contributing to the overall healthy balance and providing natural protection against infection. Discussion of the dynamics of vaginal microbiota are beyond the purpose of this chapter and interested readers are referred to recent reviews on the subject [34, 35]. However, it is worthwhile mentioning that the disruption of this defense barrier may lead to genital disease, typically of infectious nature, and/or facilitate transmission of HIV [36]. Moreover, vaginal microbiota may have a huge impact in local pharmacokinetics namely by influencing the metabolism of intravaginally administered drugs [37].

19.2.2 Colorectum

Despite being administered in the rectal compartment, drug formulations may undergo extensive retrograde distribution into the colon and, thus, it is of interest to expand analysis beyond the application site. For example, recent data from humans indicated that gels administered in the rectum can distribute up to the transition between the transverse and descending colon (approximately 80 cm from the anus) [38]. The colorectum (around 150 cm) comprises the distal part of the gastrointestinal tract, extending from the terminal part of the ileum (small intestine) to the anal canal and includes the cecum; ascending, transverse, descending, and sigmoid colon; and rectum [14]. This last compartment comprises only the terminal 15 cm of the gastrointestinal tract, and its main function is to temporarily store feces proceeding from the colon before defecation. Also noteworthy, the whole colon presents involuntary motility that regulates trafficking of its contents [39].

The colorectal mucosa is covered by simple columnar epithelium, presenting mucus-producing goblet cells [39]. Folding of the epithelium originates the crypts of Lieberkühn responsible for continuous epithelial regeneration. The underlying lamina propria is rich in lymphoid cells and lymph nodes. The colorectal mucosa is covered by a mucin-rich fluid (<5%) with slightly rising pH: 6.0–7.0 in the colon up to 7.0–7.5 in the rectum [40]. Mucus present in the rectum is scanty (1–3 ml) and creates a layer with mean thickness estimated at around 150 μm [41]. As mentioned for the vagina, gut microbiota is highly relevant to normal physiology and pathology [42, 43]. Bacteria can have an impact on drug performance, namely influencing pre-absorption metabolism [44]. Host enzymatic metabolism at the colon is also considerable but still lower than at the small intestine.

Blood to the rectum is provided by the superior, middle, and inferior rectal (or hemorrhoidal) arteries, while drainage occurs through matching veins. Of particular importance, inferior and middle veins drain into the iliac vein; conversely, the upper vein connects with the portal vein [39]. This allows that drugs absorbed at the lowest parts of the rectum can avoid the hepatic first-pass effect. As for the colon, blood supply and drainage are generally provided by mesenteric arteries and veins (ultimately draining into the portal vein), respectively. The colorectum further presents extensive lymphatic drainage [39].

19.3 Controlled Drug Delivery Systems

The vaginal and rectal administration of drugs requires adequate design and development of delivery systems[1] that fulfill the intended medical purposes while respecting as much as possible the particularities of mucosal sites. Various, more conventional dosage forms have been used for such intents. Additionally, novel delivery systems have been proposed over the years in order to circumvent caveats of more traditional formulation approaches and further increase the control over drug performance. This includes enhancing pharmacokinetics and safety.

One important aspect to be early considered when developing vaginal and rectal products has to deal with preferences and acceptability of users. In the case of vaginal drug delivery, several studies, mainly associated to the evaluation of microbicides, have been conducted and detailed results can be found reported elsewhere [45–50]. Although active users are the most important individuals, passive ones such as sexual partners should also be taken into account if considered relevant [51, 52]. In general terms, the vaginal route seems to be particularly acceptable by women when managing local conditions or somehow considered as related. Contraception using hormonal rings is a good example of this last case. Concerning dosage forms, semisolid ones seem preferable to a majority of women although acceptability is quite variable concerning previous experience, age, education, cultural issues, and/or geographical localization, among other factors. Even considering the same dosage form, different properties may influence how women ponder their use [53]. Use of dosage forms that are not so widely known by women, namely films and rings, is typically seen with concern at first but found highly preferable once education and training measures are undertaken. As for the rectal route, acceptability seems to be mainly driven by age group and sociocultural issues [5, 54]. Even if typically considered only when the oral route is impaired or unavailable (e.g. states of unconsciousness, reluctance to oral intake), rectal drug delivery is mostly considered for children of young age (in this case, as driven by parents) and, to some extent, elderly. Anti-HIV microbicides

1 The term “delivery system” is used throughout this chapter in its broadest sense, referring to dosage forms and associated drug release mechanism(s), as well as individual delivery systems designed to affect drug delivery mechanism(s) but not constituting an actual dosage form or product (e.g. nano- and micro-carriers).

may be a particular field in which rectal administration may be an interesting approach for adults. This is related with the high prevalence of lubricant or cleansing products use in preparation for anal sex [55, 56]. In this last case, semisolid dosage forms seem to be preferred over solid and liquid ones [57–59].

Regardless of the chosen final form that a drug delivery system might assume, there have been a few common strategies when considering the vaginal and rectal routes. Due to the mucosal nature of both sites, the interactions with mucus and epithelial surface may play an important role in drug delivery. Increasing mucoadhesive phenomena, in particular, has been long encouraged for systems intended for vaginal and rectal drug administration [60, 61]. The establishment of bonding between drug delivery systems and mucus typically leads to enhanced retention at the administration site, thus enhancing drug residence and reducing leakage issues. However, the long-standing paradigm around mucoadhesion has been challenged over roughly the last decade, namely for nanocarriers. Sub-micrometric particles present the ability to migrate across aqueous channels of mucus if their size and surface is adequately engineered to abbreviate steric hindrance and adhesive interactions with the mucin mesh, respectively. This may indeed enhance carrier diffusion and, ultimately, drug transport across the mucus barrier and throughout mucosal surfaces, thus favoring retention and accumulation at epithelia (Figure 19.1) [62].

Another interesting approach for vaginal and rectal drug delivery involves the exploitation of environmental variations at mucosae in order to control drug release or otherwise modify the performance of delivery systems [63, 64]. For example, using temperature-sensitive systems that undergo sol–gel transition when heated up from room to body temperature allows not only enhancing mucosal distribution, due to the liquid nature immediately upon administration, but also increase drug retention *in loco* as gelification occurs [65]. Such approach was found useful

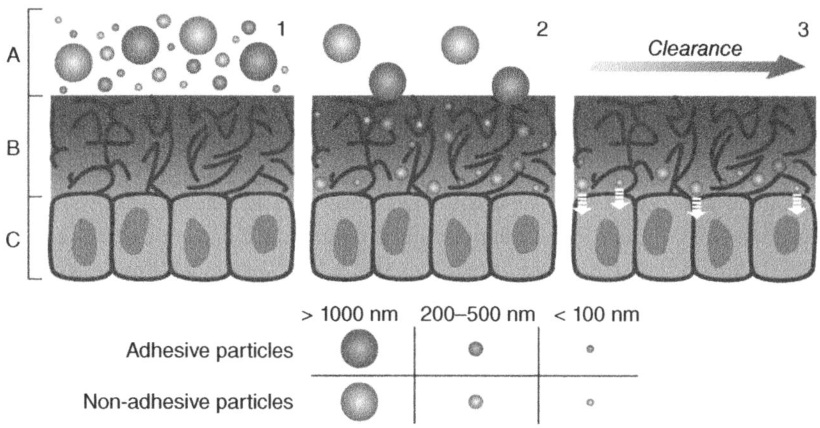

Figure 19.1 Schematic representation of the interaction of several types of particles with the mucus layer. (1) After initial administration/arrival to the mucosal site, particles will interact and diffuse through the mucus differently according to their properties. (2) Larger particles are not able to diffuse through the mucus layer due to steric hindrance, but can interact with the luminal layers in cases where adhesive bonding with mucin chains (in dark gray) can be established. As for smaller particles, these can diffuse differently through the mucus layer depending on adhesive properties and specific size. (3) Upon natural mechanisms of clearance, particles are progressively removed from the mucosal site (particularly at the luminal side) while nanoparticles that have reached the epithelial cell lining can further undergo cell uptake or tissue penetration (represented as white dashed arrows). Legend: A – mucosal tissue lumen/external environment; B – mucus layer; and C – epithelial cell lining. Source: das Neves et al. [62]. © 2011, Taylor & Francis

for both vaginal [66–68] and rectal [69, 70] drug delivery. Also, the transition from acidic pH of the healthy vagina during reproductive years to neutral or alkaline values typical of infectious vaginitis (except for vulvovaginal candidiasis [71]) may be of particular interest. This pH shift is also observed upon ejaculation. Semen plays an important role in transmitting infectious agents such as HIV; thus, triggering the release of anti-infectious agents from pH-sensitive delivery systems may be an interesting approach to develop "smart" microbicides [72]. Vaginal systems that are simultaneous sensitive to pH and temperature have also been developed [73].

Multiple vaginal and rectal delivery systems have been described in the literature and various have actually reached the market over the years. Some of the most relevant are addressed below, alongside key features related with controlled drug release. Although several drug delivery systems share noticeable similarity irrespective of the administration route, this section has been further divided into vaginal and rectal routes in order to highlight differences.

19.3.1 Vaginal Route

19.3.1.1 Conventional Dosage Forms

Modern therapeutics rely on many different vaginal products based on more classic dosage forms, which overall technology has been extensively refined over the years and is completely adapted to either pharmaceutical compounding or large-scale production by most part of the pharmaceutical industry [74]. Although disputable, the term "conventional" is used throughout this chapter with this last broad sense. Dosage forms can further be conveniently divided into solid, semisolid, and liquid dosage forms.

Solid dosage forms include tablets, suppositories (ovules or pessaries), soft capsules, and films. Typical advantages of solid systems include precise dosing, high chemical and microbiological stability, and low vaginal leakage (i.e. involuntary escape from the vagina after liquefaction, causing discomfort). Vaginal tablets are one of the most popular dosage forms when considering the development of vaginal products. Tablets are obtained in very much the same way as oral counterparts by compression of powders or granules, although having size and shape that ease administration and promote retention in the vagina. Vaginal tablets, as well as vaginal suppositories and capsules, can be administered using an appropriate applicator, although deep manual insertion is also feasible. Tablets are usually designed in order to dissolve, disintegrate, or otherwise release drug contents rapidly after administration [75, 76], even if formulations allowing sustained drug release have also been proposed [77–80]. In this last case, the use of matrices that simultaneously allow prolonged drug release and are mucoadhesive seems to be an interesting approach. For example, Garg et al. [77] proposed a tablet formulation containing, among other excipients, xanthan gum and chitosan glutamate as mucoadhesive polymers for the vaginal delivery of povidone iodine. These researchers compared the new formulation with commercially available products and demonstrated enhanced mucoadhesiveness using *ex vivo* sheep vaginal mucosa. Moreover, the mucoadhesive tablet was able to sustain the release of povidone iodine for at least eight hours in acetate buffer (pH 5.0) using a modified United States Pharmacopeia (USP) type II dissolution apparatus (paddle type), contrasting with the almost complete release after one hour for comparator products (Figure 19.2). Another interesting solution to control drug release involves the development of tablets possessing two or more layers having different release kinetics [81].

Ovules or pessaries (not to be confounded with pessaries as medical devices designed for managing pelvic organ prolapse [82]) are a special case of suppositories designed specifically for vaginal administration, thus also routinely referred to as vaginal suppositories. This dosage form possesses shape (e.g. ovoid, conical, bullet, teardrop) and weight (typically 2–5 g) adapted

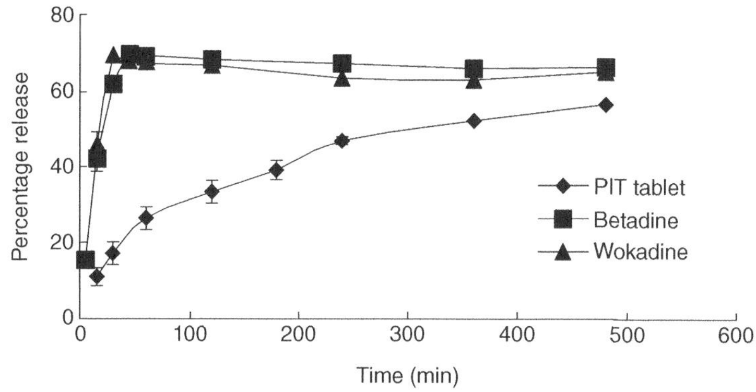

Figure 19.2 Comparison of dissolution profiles of marketed tablets of povidone iodine (Betadine® vaginal suppositories, Win-Medicare Pvt. Ltd., New Delhi, India; and Wokadine® vaginal suppositories, Wockhardt Ltd, Mumbai, India) and the in-house developed tablet formulation (PIT tablet). Results presented as mean ± standard deviation ($n = 3$). Source: Garg et al. [77]. © 2007, Taylor & Francis.

to vaginal administration and, alongside vaginal tablets, are also quite popular in developing commercial products. Vaginal suppositories are usually obtained by molding of heated mixtures of excipients and active molecules, although methods involving compaction can also be alternatively employed. Molding can be achieved by pouring melts into (i) re-usable individual molds (typically metallic), which then require collection of individual ovules upon solidifying and further packaging, or (ii) plastic (polyethylene or polypropylene) or metallic alveoli that act as molds and final primary package. This last option is the current standard in the industry as it allows for rapid and nearly fully automated production. Excipients traditionally used in the production of vaginal suppositories include cocoa butter and synthetic glyceride mixtures for hydrophobic bases, and polyethylene glycol (PEG) mixtures and gelatin/glycerin for hydrophilic bases [83]. Additives such as melting point modulators (e.g. waxes, paraffin, oils, sorbitol), emulsifiers, preservatives, antioxidants, and color agents may also be included. Suppositories can release active payloads by either dissolving or disaggregating when produced with hydrophilic bases, or melting at body temperature when hydrophobic bases are used. Drug release can be modulated by appropriate selection of excipients and production procedures [84]. Although typically comprising a continuous matrix in which drugs are included, vaginal suppositories can be modified in order to have different layers containing incompatible active payloads or possessing different drug release profiles, be coated with different components for protection from the external environment, or feature hollow cavities that can then be filled with drugs. This last approach has been particularly helpful when trying to modulate drug release time, or incorporating temperature sensitive drugs/other therapeutic agents. For example, Rodrigues et al. [85] developed different hollow-type suppositories based on PEG or Witepsol® H12 (a synthetic glyceride mixture) for the vaginal administration of *Lactobacillus acidophilus*. Central cavities were created by introducing glass rods into molds during cooling, filled with lyophilized bacteria and sealed at the top with melted base (Figure 19.3). Results showed that bacteria stability and release were moderately enhanced as compared to solid body suppositories, presumably due to the milder temperature conditions that *L. acidophilus* was submitted to during manufacturing (room temperature vs. 40–45 °C).

Another interesting dosage form, although not so widely used as tablets or ovules, encompasses soft capsules possessing an external continuous shell made of a rubbery material (e.g. gelatin and glycerol) and containing an oily filling (e.g. mineral oil, silicone) [86, 87]. One or more drugs can be dissolved or dispersed in the liquid core. This dosage form typically presents an ovoid shape that

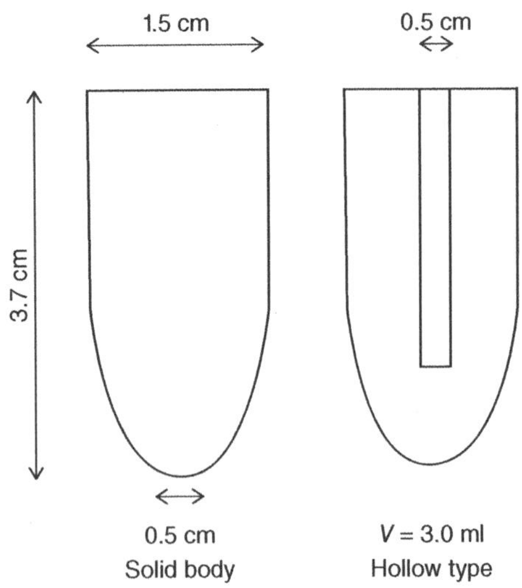

Figure 19.3 Schematic of solid body and hollow-type suppositories developed by Rodrigues et al. using either a hydrophilic mixture of PEG 4000:PEG 400 (2 : 1) or the hydrophobic base Witepsol® H12. Source: Rodrigues et al. [85]. © 2015, Taylor & Francis.

can be more or less elongated along its larger axis. The shell of the capsule can rapidly dissolve or disintegrate in contact with vaginal fluids, releasing its contents. The composition of the capsule may further be modified, particularly at the shell level, in order to allow modifying drug release. Gyno-Daktarin® 1200 mg (Janssen-Cilag, Beerse, Belgium) is a good example of a controlled drug release capsule that is available in the market in various countries (available in the United States as Monistat-1® from Insight Pharmaceuticals, Langhorne, PA). The product comprises a shell made of gelatin and glycerin that surrounds a liquid core containing miconazole nitrate in liquid paraffin, white petrolatum, and lecithin [88]. The system allows sustaining drug release and was shown as effective and safe in the treatment of vulvovaginal candidiasis after one single dose as compared to standard treatment (daily 100 mg of miconazole nitrate for seven days using a 2% cream) [89].

The vaginal film is another interesting solid dosage form although its use is not as widespread as any of the previous. The most well-known product in the market and often cited example is VCF® Vaginal Contraceptive Film (Apothecus, Oyster Bay, NY, USA) containing the spermicide nonoxynol-9 as active ingredient. The interest in films has, however, increased over the last years as boosted by the development of vaginal anti-HIV microbicides [90–92]. Vaginal films are thin (tens to hundreds of micrometers thick), soft, malleable, and transparent to translucent individual sheets of different shape (usually rectangular or squared) and dimensions (around 2.5–10 cm in side). This dosage form comprises one or more polymers (e.g. polyvinyl alcohol, cellulose derivatives) that form the matrix of the film, and one or more plasticizers (e.g. glycerin, PEG, propylene glycol) that confer flexibility to the system [93]. Other excipients are also frequently included depending on specific needs (e.g. preservatives, disintegrants, and coloring agents). Furthermore, films appear to be ideal platforms for the incorporation of nanotechnology-based drug carriers, as discussed below [94, 95]. Films are traditionally produced by solvent-casting although increasing interest has been devoted to hot-melt extrusion. More details on advantages and disadvantages of both methods can be found in a recent publication by Rohan and Zhang [96]. Film systems do not require applicator for administration, being inserted with the aid of the finger tips typically after folding in two or four matching parts along the central axes. Products based on films are usually designed to quickly dissolve or disperse upon exposure to the natural fluids of the vagina, thus forming a gel *in loco* that covers the mucosa and releases active payloads. The composition of films can also be modified in order to modulate drug release. Borumand et al. [97] reported on the influence of multiple polymers on the release of propranolol hydrochloride from films produced by solvent-casting. For

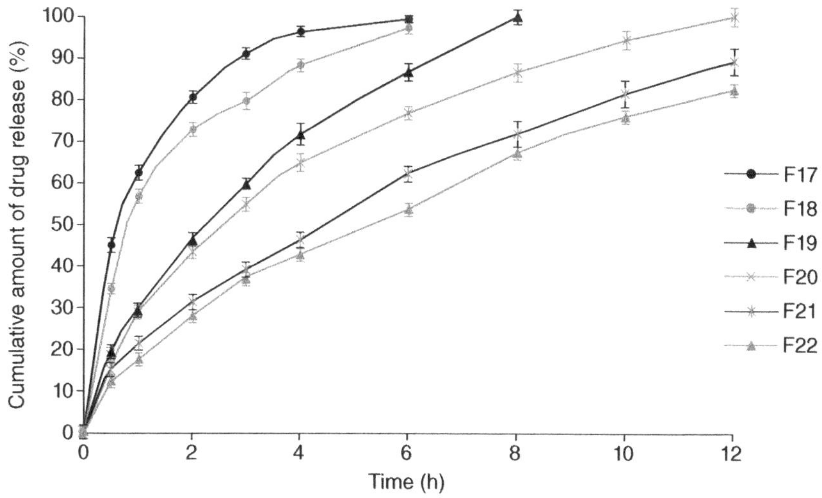

Figure 19.4 *In vitro* release profiles of vaginal mucoadhesive propranolol hydrochloride film formulations containing 2.5%, 3.5%, 5.0%, 5.5%, 6.0%, or 6.5% Eudragit® RS PO (F17, F18, F19, F20, F21, and F22, respectively), HPMC 15 000 (1%), PEG 400 (40%), and drug (1.6%) at 37 °C in citrate-phosphate buffer pH 4.5. Data are expressed as mean ± standard deviation. Concentrations for ingredients reflect the percentage amounts in solution before film drying. Source: Borumand et al. [97]. © 2014, Elsevier.

example, the incorporation of increasing amounts of Eudragit® RS PO, an insoluble copolymer of esters of acrylic and methacrylic acids, into mucoadhesive hydroxypropyl methylcellulose (HPMC) 15 000/PEG 400-based films prolonged the release time of the drug (Figure 19.4). These observations correlated well with the decreased swelling and dissolution of the film matrix.

Semisolid dosage forms are quite popular when considering vaginal administration, namely due to their high acceptability [49]. The ability of semisolid systems to conveniently spread throughout the vagina after administration, particularly with ambulation, is generally regarded as beneficial [98]. However, and depending on administered volumes, this also tends to lead to poor retention leading to leakage [99]. One interesting advantage of semisolid dosage forms over solid ones is the possibility to easily personalize the dose to be administered. This may be of particular interest considering interindividual variations of the vaginal cavity dimensions. Common drawbacks to semisolid dosage forms are normally related to the frequent high water content, which may lead to higher susceptibility to microbial contamination and poorer chemical stability. Still, these can be abbreviated in most cases with adequate formulation (e.g. by including preservatives or antioxidants) or proper storage conditions. Another important issue is the mandatory need of using an applicator. This requirement can go both ways: on one hand, it abbreviates handling of genitalia and may be regarded by some women as preferable [46]; on the other hand, it may substantially increase cost, which can constitute a problem in products such as microbicides intended to be used in low resource settings [100]. Applicators can be provided as pre-filled, i.e. one applicator equals one dose, or as reusable units. The former is more convenient but more expensive, while the previous need to be manually filled from a multi-dose tube and washed after each administration.

Among semisolids, creams have been one of the most widely used dosage forms for therapeutic purposes, matching tablets and ovules in the number of commercially available vaginal products [101]. The versatility associated to the technological properties of creams, namely the typical presence of two immiscible phases, allows the suitable incorporation of a wide range of drugs featuring different physical-chemical properties. Although mostly designed for the rapid release

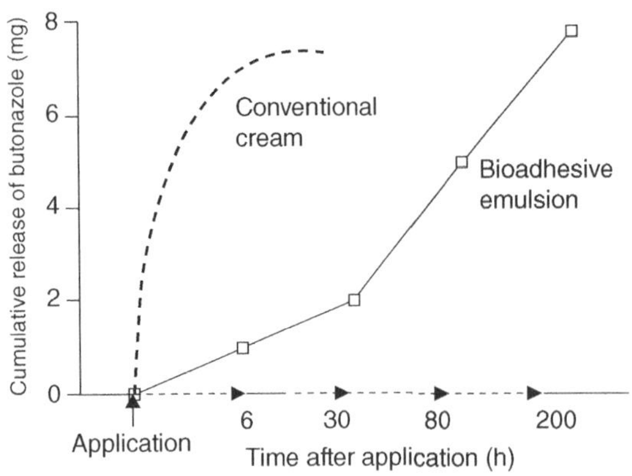

Figure 19.5 *In vitro* release of butoconazole from a standard (conventional) cream and the SITE RELEASE® formulation (Gynazole-1®). Source: Henzl [106]. © 2005, Springer Nature.

of incorporated drugs and requiring multiple administrations (typically once daily for as long as required), a few long-acting cream products have been described and are actually available on the market. One such example is the SITE RELEASE® technology (formerly known as VagiSite™ technology) developed by KV Pharmaceutical Company (St. Louis, MO, USA), which is currently used in Gynazole-1® (2% butoconazole nitrate) and Clindesse® (2% clindamycin phosphate), both from Perrigo Pharma International (Dublin, Ireland) [102]. These products require only single 5 g doses of cream, packed into a pre-filled applicator, to be administered intravaginally for the treatment of either vulvovaginal candidiasis (Gynazole-1) [103, 104] or bacterial vaginosis (Clindesse) [105]. In both cases, products were demonstrated as safe and effective as different equivalent treatments with vaginal tablets or creams providing immediate drug release and requiring daily administration [106]. The SITE RELEASE technology comprises a high internal phase ratio, semisolid water-in-oil emulsion: the inner globules (70–90% in volume) contain purified water, propylene glycol, edetate disodium, and sorbitol and act as drug reservoirs, while the hydrophobic continuous phase works comprises mineral oil, silicon dioxide, and microcrystalline wax and provides a barrier that limits drug diffusion, thus controlling release [102, 107, 108]. Figure 19.5 presents the drug release profiles of Gynazole-1 and a standard cream formulation in acetate buffer (pH 4.3). The system is further stabilized by polyglyceryl-3 oleate and glyceryl monoisostearate (and lecithin in the case of Clindesse) as emulsifiers, and methylparaben and propylparaben as preservatives. The outer phase also confers adhesiveness to the cream, which can create a thin film barrier along the mucosal surface that lasts for several days (median 4.2 days vs. 2.6 days for a standard cream), while minimizing leakage [109].

Vaginal gels have been gaining increasing popularity over the last two decades due to their good acceptability by women, technological versatility, and affordability [110]. Investigational formulations and currently market products based in gels are almost unvaryingly aqueous in nature. Research into novel formulations has been particularly intense in the field of anti-HIV microbicides [111]. Many gelling agents (mostly polymers) have been proposed for manufacturing water-based gels, namely polyacrylates, cellulose derivatives, poloxamers, alginate, and chitosan, to name a few. The ability of these last to, at least in some extent, interpenetrate and establish adhesive bonds with mucin led to the development of mucoadhesive products and to the early general idea that such products could outstand the natural cleansing ability of the vagina for longer time periods [112, 113]. However, the mucoadhesive performance of vaginal gels is highly variable on composition and enhanced behavior is not always apparent over other dosage forms such as creams [114]. Vaginal gels are also largely regarded as safe due to their high water content but toxicity issues may arise from other ingredients (drugs or excipients) included in formulations [115, 116]. The essential requirement for safety when developing vaginal microbicides led to the establishment of a generic gel formulation, termed "Universal Placebo," which is nowadays regarded as the gold standard in the field [117, 118]. The gel contains 96.3% purified water, 2.7% hydroxyethylcellulose (HEC, Natrosol™ 250 HX Pharm Grade, Ashland Inc., Covington, KY, USA), 0.85% sodium chloride, 0.1% sorbic acid, and 0–0.02% caramel color by weight, while the pH is adjusted to 4.4 using sodium hydroxide. Drug release from hydrophilic gels is usually quite fast due to the loose matrix created by the low amounts of polymers typically used [119]. However, several attempts have been made in order to design vaginal gels that are able to present sustained or otherwise controlled drug release [120–123]. Most strategies are based on drug complexation/physical modification, or the inclusion of a hydrophobic system/component in hydrophilic gels in order to modify release. These have been deemed suitable mainly for hydrophobic compounds, being the release mechanisms essentially similar to those observed in creams (emulsions). For instance, Kim et al. [124] used hydroxypropyl-gamma-cyclodextrin inclusion complexes for allowing suitable solubility of amphotericin B in a hydrophilic gel that was sensitive to temperature and pH due to the presence of a poloxamer-based multiblock copolymer (MBCP-2). In particular, the gel underwent sol–gel transition at body temperature, but drug release was only observed at acidic pH due to the collapse of the gel structure (Figure 19.6). Drug-cyclodextrin complexes were "trapped" inside the gel network at neutral and alkaline conditions. A comparable formulation with similar behavior was proposed by Chen et al. [73], but using pH-sensitive methoxy-PEG 2000-hydrazone-cholesteryl hemisuccinate-based liposomes for incorporating arctigenin into a thermosensitive poloxamer gel base. In this case, liposomes were degraded at pH 5.0, not the gel network. Crinone® gel (Merck Serono, Darmstadt, Germany) is an interesting example of a commercially available product presenting sustained release of progesterone, being mainly used for luteal phase support in infertility cases [125]. Although classified as a gel due to the presence of a reticulated structure formed by polycarbophil and carbomer homopolymer Type B (Carbopol® 974P NF) in water, the product also comprises a true emulsified inner hydrophobic phase mainly constituted by light mineral oil stabilized by hydrogenated palm oil glyceride. Other excipients include glycerin as humectant and sorbic acid as preservative. Moreover, progesterone is present in its micronized form. The combined effect of the partition of the drug between the inner phase and drug particles, and the diffusion barrier provided by the outer aqueous phase is responsible for claimed sustained release. Indeed, *in vitro* release studies in water:methanol (8 : 2) showed nearly zero-order kinetics during the first

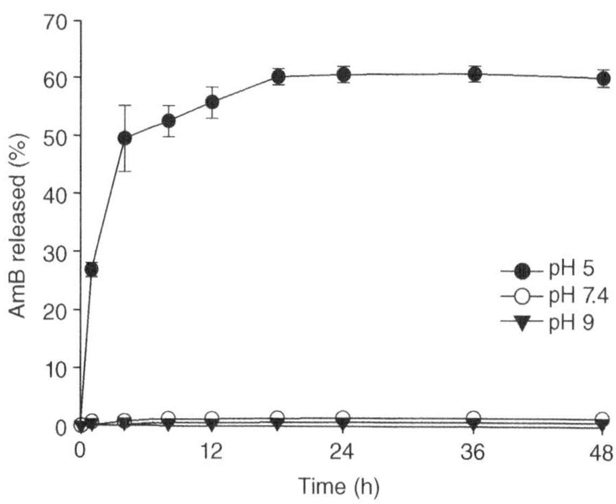

Figure 19.6 *In vitro* release of amphotericin B (AmB)-hydroxypropyl-gamma-cyclodextrin from 25% MBCP-2 gel at pH 5.0, 7.4, and 9.0. Data are expressed as mean ± standard deviation. Source: Kim et al. [124]. © 2010, Elsevier.

24 hours (around 50–70% of the drug released), while nearly complete release was achieved by 72 hours [126].

A recently proposed approach for allowing the sustained release of drugs has been the development of nonaqueous silicone elastomer-based gels [127, 128]. Both hydrophobic and hydrophilic molecules can be considered by varying the hydrophobicity of the gel base. For example, Forbes et al. [128] produced nonaqueous gels by mixing a lightly crosslinked polydimethylsiloxane (ST-Elastomer 10, Dow Corning®, Midland, MI, USA) with either hydrophobic cyclomethicone or a series of relatively hydrophilic, linear silanol-terminated polydimethylsiloxane with low molecular weight. Different gels were used for the incorporation of one of two antiretroviral microbicide candidates, maraviroc (hydrophobic) or emtricitabine (hydrophilic), and tested *in vitro* for release. Overall and depending on specific formulation, nonaqueous silicone elastomer gels were able to provide sustained drug release over the course of 24 hours, contrasting with results for aqueous HEC-based gels (100% release within a few hours). Besides offering the possibility of better controlling drug release, silicone elastomer gels may also potentially favor the stability of water-sensitive molecules and enhance vaginal residence (due to immiscibility with vaginal fluids) [127, 128].

Ointments have been traditionally considered for vaginal drug delivery, but their use declined over the years, being replaced by creams and gels. Formulations are typically hydrophobic in nature and, thus, considered as potentially suitable vehicles only for low water-soluble drugs [129, 130]. Moreover, used excipients (e.g. petrolatum, paraffin, mineral oil) are quite messy for vaginal application, originating unpleasant oily leakage. The ability to control drug release is generally low. Water-soluble PEG-based ointments have been used as alternatives [131, 132] although their use for intravaginal administration is not common, namely due to concerns of local toxicity [133].

Liquid dosage forms have been mostly considered for vaginal use as douches (i.e. for personal hygiene purposes) and can be classified as solutions, emulsions, or suspensions. Pre-prepared formulations are usually preferred although liquid dosage forms can also be prepared from tablets or powders just before usage. Administration of relatively high volumes, up to several hundred milliliters, is performed using squeezing bottles or bags bearing appropriate nozzles or applicators for vaginal insertion. The inability to retain in the vagina after administration makes of these dosage forms poor vehicles for drug delivery. Moreover, vaginal douching is currently discouraged due to the various complications that may arise from this longstanding practice [134]. Although not formally considered a liquid dosage form, aerosol foams have long been used for vaginal drug delivery, particularly of spermicides [135]. However, its use has been decaying and is nowadays residual. Control of drug release, as in the case of other liquid systems, may not be easy to achieve. Vaginal foams comprise a gas (propellant) dispersed into a water-based vehicle containing foaming agents (typically surfactants) and stabilizers (e.g. gelling agents) that not only allow the gas-in-liquid dispersion to be formed but also help maintaining its structure for at least several minutes [136]. One particular advantage relates to the ability of foams to rapidly expand and distribute in the vaginal cavity using a minimal amount of liquid, thus potentially minimizing leakage. On the downside, products based on this dosage form require the use of multiuse pressurized containers that produce the foam during usage and present special nozzles allowing insertion in the vaginal cavity. Apart from cost issues, control of the pressure-dispensed dose is often not possible since the foam is continuously produced upon activation of the valve of pressurized containers. Moreover, foam production may not be uniform throughout the usage life of the product as this is dependent on the remaining content of the container [137].

19.3.1.2 Removable Drug Delivery Systems

Unlike dosage forms discussed above, which are progressively eliminated by the natural clearance mechanisms of the vagina, this subsection discusses drug delivery systems that require active removal from the administration site in order to terminate usage. Removable systems usually comprise nondegradable platforms allowing prolonged drug release, from a few hours to several months. Without any doubt, vaginal rings are the most studied and successful systems in this category. However, others such as sponges, diaphragms, cervical caps, tampons, and inserts have also been explored.

Vaginal rings are polymeric, nondegradable, flexible, and toroidal-shaped systems that present the ability to control drug release from weeks to months. Dimensions are variable but rings typically present values of outer diameter of 5.0–6.0 cm and cross-sectional diameter of 0.4–1.0 cm. Rings are intended to be manually and deeply inserted in the vagina by the user itself. Retention is assured by the gentle pressure exerted on the vaginal walls as a result of ring squeezing. Overall, these delivery systems are designed to be comfortable and mostly unnoticed during normal use, requiring to be manually removed after the prescribed time. The technology behind rings has been first developed during the 1960s and is currently well consolidated, with a few products available in the market, despite requiring equipment and expertise seldom available in common pharmaceutical industry facilities. Interested readers on detailed composition, design, manufacturing, and drug release mechanisms of vaginal rings are referred to recent reviews on the subject [138–141]. Most frequently used matrix-forming materials include silicone elastomers and ethylene vinyl acetate copolymer (EVAc), although others such a polyurethane have also been proposed. Depending on used materials, rings can generally be produced by one of two general methodologies: injection molding with curing (e.g. silicone rings) or hot melt-extrusion (e.g. EVAc rings). Three basic designs have been mainly used: (i) matrix type, (ii) reservoir type, and (iii) sandwich type (Figure 19.7). The matrix type comprises a continuous polymer matrix containing the dissolved/dispersed drug(s), which is/are released mainly by diffusion mechanism. Rings based on such design present typical initial burst release due to the amount of drug(s) present at or near the surface and that is/are readily available for partitioning to outer media. Also, continuous decrease of the drug release rate over time is usually observed in agreement with first-order kinetics. This last occurrence is related to the depletion of drug(s) at outer layers of the ring, which increases the distance required for inner drug content to diffuse. In the case of the reservoir type, a non-medicated polymer sheet is included around the core containing the drug(s), thus allowing to better control the release rate at near zero-order kinetics. Depending on the specificities of the ring, initial lag or burst effects may be observed but typically in lower degree than in matrix type rings. Furthermore, extinction of drug content in the core at latter time points can lead to decreased release rates similarly to what happens in matrix type rings. In order to abbreviate drug depletion and release rates that are deemed too low for clinical use, rings are usually loaded with considerably higher amounts of drug(s) than those actually released. Sandwich type rings may abbreviate, at least partially, this issue. Drug(s) are only present in an intermediary sheet, thus allowing control over release by the outer non-medicated sheet (such as in reservoir type) and partially diminishing residual drug(s) that would otherwise be present in the inner core (such as in both matrix and reservoir types). One potential downside of sandwich type rings is the relatively lower drug release rates that can be achieved. Alongside these three basic ring types and sparkled by the renewed interest from the community engaged in the development of microbicides, additional designs featuring more or less complex architecture have been proposed over the last 10–15 years as recently revisited by Malcolm

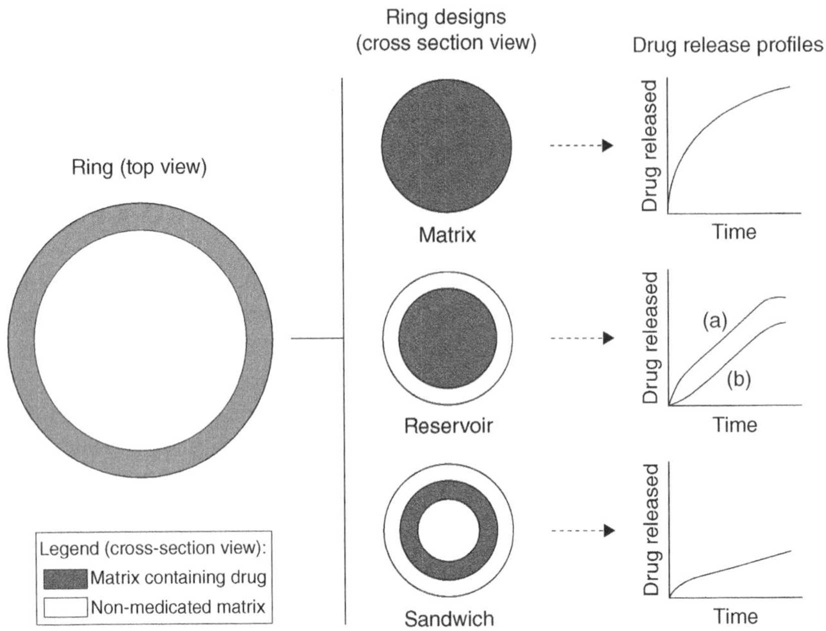

Figure 19.7 Schematic representation of basic vaginal ring designs and typical drug release profiles. For the reservoir type ring, drug release profiles presenting either initial (a) burst or (b) lag effects are presented.

et al. [140]. Technological developments have been significant in order to expand the versatility of rings, namely in the number of different designs available for the delivery of single or multiple active agents. One particularly interesting and highly versatile strategy comprises the incorporation of one or more drugs into pods or inserts that can then be placed in specific hollow spots distributed along the structure of a non-medicated ring. This allows modulating which drugs are delivered, as well as their doses and release profiles, according to need.

First vaginal rings were developed for providing a new contraceptive option allowing prolonged release and systemic absorption of sexual hormones at the vaginal mucosa [142]. However, it was only in the early 2000s that vaginal rings really hit the spotlight with the market launch of NuvaRing® (Merck & Co., Whitehouse Station, NJ, USA), an EVAc combination contraceptive reservoir ring produced by co-extrusion. The ring contains 11.7 mg of etonogestrel and 2.7 mg of ethinylestradiol in the inner core, which is externally covered by a non-medicated EVAc sheet that controls drug release. The ring is intended to be used continuously for three weeks and replaced by a new one after one week off. NuvaRing is designed to release 0.120 mg of etonogestrel and 0.015 mg of ethinylestradiol per day [143]. More important, the ring allows maintaining steady serum drug levels throughout the intended time of use and even beyond (Figure 19.8) [144], while avoiding wide daily fluctuations in concentrations as observed for oral contraceptives [145]. Despite providing significantly lower total exposure to both hormones, the efficacy of NuvaRing is comparable to that of combined oral contraceptives [146, 147]. Other rings releasing one or more different sexual hormones and intended for continuous use over three months are also available around the world, namely for contraception (Progering®, Population Council, New York, NY, USA), hormone replacement therapy (Estring®, Pfizer, New York, NY, USA; Femring®, Warner Chilcott, Rockaway, NJ, USA), or pregnancy support (Fertiring®, Population Council).

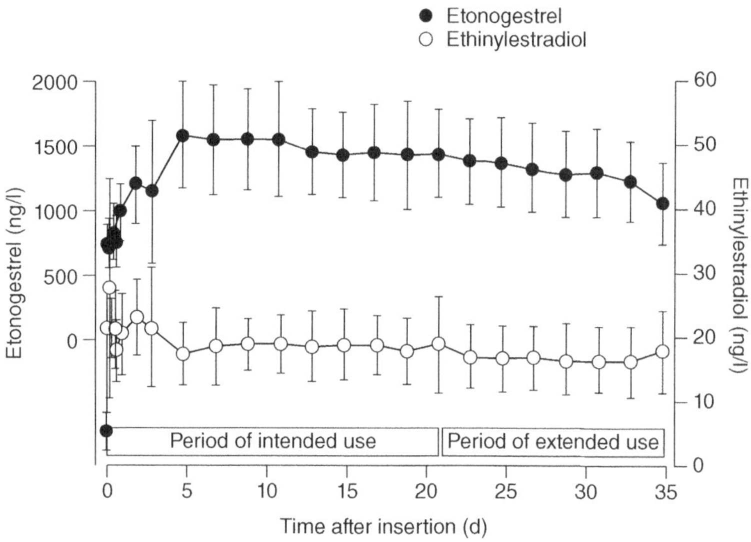

Figure 19.8 Serum concentration-time curve (mean ± standard deviation; $n = 16$) of etonogestrel and ethinylestradiol during treatment with NuvaRing® [days 1–21 (intended use) and days 22–35 (extended use)]. Source: Timmer and Mulders [144]. © 2000, Springer Nature.

Among all the vaginal rings proposed for the delivery of microbicides, one in particular containing dapivirine, a non-nucleoside reverse transcriptase inhibitor, is advanced in the development pipeline [148]. The dapivirine ring was developed under the auspices of the International Partnership for Microbicides (Silver Spring, MD, USA) and was proven partially effective against HIV-1 acquisition by women in two recent large-scale Phase 3 clinical trials [149, 150]. The 8 g ring features matrix design with 5.6 and 0.77 cm outer and cross-section diameters (Figure 19.9), respectively, and contains 25 mg of dapivirine dispersed into a silicone matrix (MED-4870 silicone elastomer, Nusil™ Technology LLC, Carpinteria, CA, USA), being produced

Figure 19.9 The International Partnership for Microbicides dapivirine matrix ring. Source: Reprinted from Devlin et al. [151], Copyright (2013), with permission from Elsevier.

by injection molding with platinum-catalyzed curing [151]. As typically observed for the matrix type design, the dapivirine ring presents first-order release kinetics albeit allowing obtaining sustained drug levels in vaginal fluids for at least 28 days of continuous use [152]. Also noteworthy, the dapivirine ring illustrates well the drug surplus required for simpler matrix type rings in order to allow controlled release: only approximately 16% of the total amount of dapivirine is released over the intended 28-day period of use [151].

Apart from rings, other removable drug delivery systems featuring diverse materials, shapes, and mechanisms of drug release have also been proposed and some are available in the market. A few examples are described in the following. Although impregnation of sponges with various spermicides immediately before vaginal insertion has been used throughout human history for avoiding pregnancy [153], such practice in not common anymore. Notwithstanding, the Today® sponge (Mayer Labs, Sonoma, CA, USA) has been developed during the 1970s and originally introduced in the market in beginning of the 1980s with similar purpose. The sponge comprises one gram of nonoxynol-9 dispersed into a soft polyurethane foam and is shaped for comfortable vaginal insertion up to the cervix and use. The inclusion of a polyester loop in one side of the sponge facilitates removal. Despite the unavailability of *in vitro* drug release profiles, the manufacturer of the product claims that sustained release of around 125 mg of nonoxynol-9 occurs during 24 hours [154]. Diaphragms and cervical caps normally used as barrier contraceptives have been modified in order to allow the release of drugs. For example, the commercially available Caya® contoured diaphragm (KESSEL medintim GmbH, Mörfelden-Walldorf, Germany) based on the SILCS technology originally developed by PATH (Seattle, WA, USA) has been modified by Major et al. [155] in order to deliver dapivirine. The SILCS device (Figure 19.10) containing 10% drug in its spring core was shown able to provide zero-order release kinetics *in vitro* over 175 days. Although not intended for continuous use such as in the case of rings, the developed diaphragm could be inserted on demand, with a potentially useful life span of up to five years. Medicated tampons, similar to those used to collect menses, have also been modified in order to release lactic acid and acetic acid for maintaining low vaginal pH [156]. Furthermore, inserts such as those used for inducing labor, have been used. One particular insert, available in different countries as either Cervidil® or Propess® (Ferring Pharmaceuticals, Saint-Prex, Switzerland), comprises 10 mg dinoprostone (prostaglandin E2) dispersed in a nondegradable hexanetriol/macrogol 8000/isocyanate cross-linked hydrogel copolymer molded as a rectangular cuboid with round corners ($29 \times 9.5 \times 0.8$ mm). The insert is enclosed into a woven polyester pouch ending in a long tape that allows easy removal after usage. Vaginal placement and retrieval of the insert should be performed by a trained physician. Dinoprostone is released at a constant rate of approximately 0.3 mg/h over a period of 12 hours [157]. The same technology has also been proposed for the vaginal delivery of clindamycin, an antibiotic frequently used in the treatment of bacterial vaginitis [158].

19.3.1.3 Nanotechnology-based Drug Delivery Systems

While removable systems described above can provide control over drug release, thus potentially influencing local pharmacokinetics, nanocarriers have the particular ability to intimately interact with biological systems and directly impact drug delivery at the tissue and cell levels. Microcarriers may also partially allow attaining these objectives, but their usefulness is usually limited at establishing mucoadhesive interactions with the mucosa or controlling drug release [159–161]. Different nanosystems, frequently polymeric in nature (e.g. nanoparticles, nanofibers), have been considerably explored over recent years in order to advance vaginal drug delivery [162]. Application of nanotechnology-based solutions have been mostly proposed in the framework of

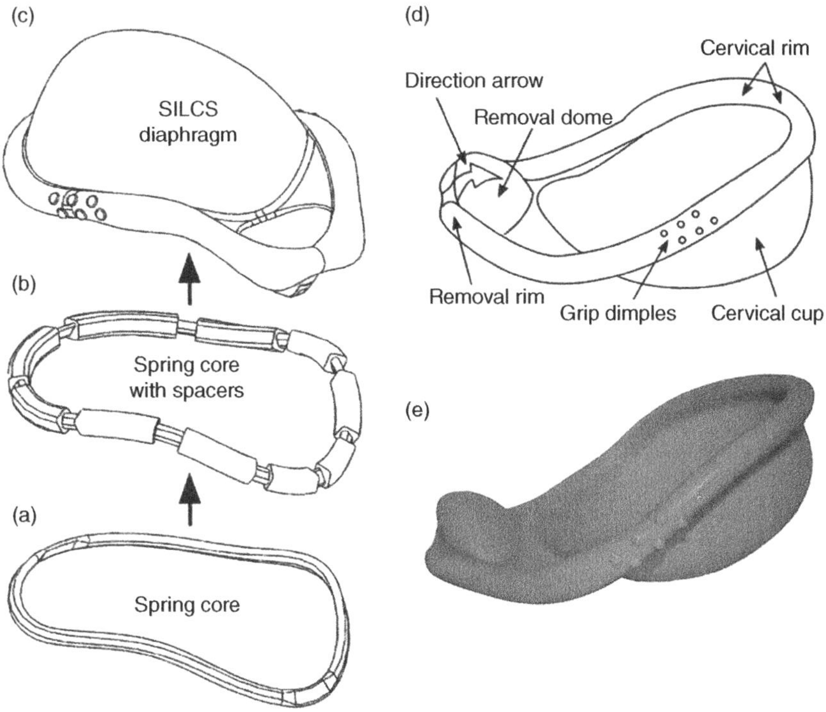

Figure 19.10 SILCS diaphragm device: (a–c) three-step injection molding manufacturing process, (d) device features, and (e) photograph of device. Source: Reprinted from Major et al. [155], Copyright (2013), with permission from Elsevier.

anti-HIV microbicides (interested readers can find detailed information on the subject in a recent review from our research group [163]) although other medical uses such as in the treatment of infectious vaginitis [164], management of cervicovaginal cancers [165], and vaccination [166] have also been considered.

But what makes nanosystems so interesting from a drug delivery point of view? Of course, a specific answer to this question depends on particular drug(s)/carrier system and intended use but there are several general advantageous features that are frequently associated to nanomedicines. For example, simple size reduction of poorly soluble drugs can be a relatively easy strategy to enhance solubility [167]. Protection of sensitive drugs such as proteins and nucleic acids is often sought when considering the use of nanocarriers [168]. As mentioned before, nanocarriers may be engineered in order to interact differently with mucus and thus modulate transport within this biological fluid. In general, small systems (usually below 500 nm in diameter) bearing highly hydrophilic, neutrally charged surface are able to migrate across mucus with little hindrance [169]. Mucus-diffusive particles have been mostly obtained by dense surface modification with PEG [170], although other possibilities relying on polyvinyl alcohol [171] and poly(2-ethyl-2-oxazoline) [172] have also been proposed. The possibility to readily migrate across mucus allows wide distribution of drugs throughout the cervicovaginal cavity and may impact therapeutic efficacy. As an example, Ensign et al. [173] demonstrated higher efficacy for PEG-modified acyclovir monophosphate nanoparticles in preventing herpes simplex virus 2 infection in mice as compared to the drug in solution (53% vs. 16% protection). Acyclovir monophosphate nanoparticles (around

65 nm in diameter) were obtained by chelating the drug with zinc acetate and using poloxamer 407 for providing a PEG corona. Results were correlated with enhanced *in vivo* coating of the mucosa and local retention of the drug.

Another interesting feature of nanosystems is related with the ability to penetrate the epithelium to different extents, depending on intrinsic properties such as size. Indeed, smaller nanoparticles have been shown able to reach the lamina propria [174] and, in particular situations, even migrate as far as regional lymph nodes [175, 176]. These possibilities open interesting perspectives for specific drug delivery to different sites of the vaginal mucosa and associated structures. Moreover, enhanced drug levels may be achieved at cervicovaginal tissues, which can be beneficial for increasing the efficacy time window of delivered drugs, namely in the case of microbicides [177]. Migration of nanocarriers toward the upper genital tract is another possibility [178]; in this case, eventual medical applications should be considered alongside potential toxicity issues related with reproduction. Beyond tissue targeting, nanocarriers can also provide the ability to deliver drug payloads to particular cell populations. Nanosystems may be preferentially taken up by cells presenting inherent phagocytic activity. For example, our research group has shown that polycaprolactone-based nanoparticles (around 180–200 nm) can enhance cell-associated levels of dapivirine, in HIV-susceptible cells. Conversely, differences between drug-loaded nanoparticles and dapivirine in suspension were generally not relevant for non-phagocytic cells of genital origin except when electrostatic interactions were promoted (e.g. when nanoparticles featured positive surface charge) [179, 180].

Although controlling drug release from nanosystems may not be an easy task, mainly due to the large available surface for drugs to "escape" to surrounding media, some interesting work has been described in the literature regarding such problem. In a report by Meng et al. [181], chitosan- or chitosan-thioglycolic acid (TGA)-based nanoparticles were produced for the delivery of tenofovir, a microbicide drug candidate, and studied for their release behavior *in vitro* using a simulated vaginal fluid (pH 4.2). Results showed that tenofovir release was sustained for several days (60–70% release after eight days) as compared to tenofovir incorporated into an HEC-based gel (roughly 85% release after 24 hours), featuring first-order release kinetics. Furthermore, mildly slower drug release from chitosan-TGA nanoparticles was observed and attributed to the reinforcement of the polymeric matrix structure due to the establishment of disulfide bonds. In a recent report from our research group, efavirenz-loaded poly(lactide-*co*-glycolide) nanoparticles (around 275 nm) were produced and, alongside free tenofovir, incorporated into polyvinyl alcohol/HPMC-based films [182]. The resulting system not only provided a combination platform for convenient nanoparticle administration but also allowed achieving prolonged vaginal levels of efavirenz when administered in a mouse model, as compared to films incorporating free efavirenz. The ability to enhance local pharmacokinetics was associated to the sustained efavirenz release profile provided by the association between nanoparticles and the film matrix (Figure 19.11). In the case of tenofovir, release from films was complete within a few minutes and not influenced by the presence of nanoparticles. In line with the previous, tenofovir pharmacokinetics were not altered for both film types [182].

Polymeric fibers, in particular nanofibers, have been attracting quite a lot of attention in the vaginal microbicides field over recent years, namely due to (i) versatility to incorporate and deliver multiple drugs in the same system, (ii) possibility to be produced in large scale, (iii) capability to form mesh-like structures that can immediately assume different macroscopic shapes resembling conventional dosage forms (e.g. tampons), and (iv) broad ability to feature various fiber designs that allow modulating drug release [183, 184]. In this last case, precise control of drug release can be achieved using different approaches, namely based on the selection of materials and/or fiber architecture. For instance, Huang et al. [185] proposed nanofibers made of cellulose acetate

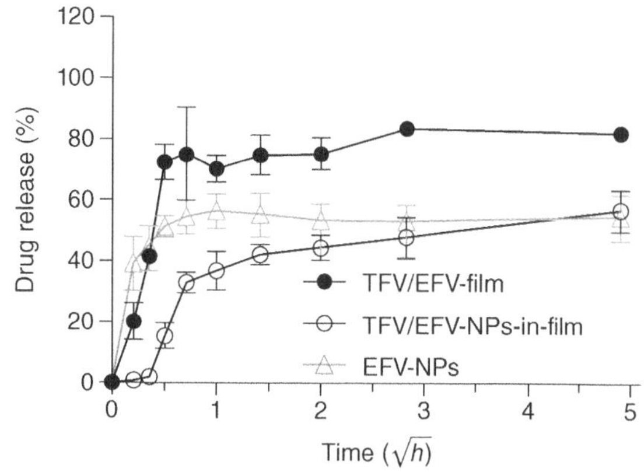

Figure 19.11 Efavirenz release properties in simulated vaginal fluid (pH 4.2) from films incorporating free tenofovir and efavirenz (TFV/EFV-film) and films incorporating free tenofovir and efavirenz-loaded nanoparticles (TFV-EFV-NPs-in-film). Release data for efavirenz-loaded nanoparticles (EFV-NPs) alone in the same conditions have also been included. Points and vertical bars represent mean values and SD, respectively ($n = 3$). Please note the square root of time in the x-axis, which was used for allowing the differentiation of values at earlier time points. Source: Cunha-Reis et al. [182]. © 2016, Elsevier.

phthalate for vaginal use as a microbicide drug delivery platform, namely for tenofovir disoproxil fumarate. Fibers presenting diameters around 500–800 nm were not soluble in simulated vaginal fluid (pH 4.2) but rapidly dissolved when pH was neutral. This may allow incorporated drugs to essentially remain associated to the fibers after administration in the healthy acidic vagina, while the onset of release only occurs when HIV is potentially deposited in the vaginal lumen alongside semen (pH 7–8 and high buffering capacity). Proof-of-concept was achieved using a model fluorescent compound, rhodamine 6G (Figure 19.12). In another work, Ball et al. [186] conducted a thorough study on different compositions of core–shell fibers produced by coaxial electrospinning for the vaginal delivery of maraviroc, another microbicide drug candidate. Ethylcellulose was selected as shell component, while polyvinylpyrrolidone was chosen for obtaining the core and incorporating the drug. Although the diameter values for produced fibers was already at the micrometer threshold, results indicated that maraviroc release could be modulated mainly by modifying the fiber shell thickness and the initial drug loading. Moreover, association of different fiber meshes featuring different properties was successful in providing dual maraviroc release behavior, namely in order to feature initial burst release (within a few minutes), followed by continuous slow release for up to several days.

19.3.2 Rectal Route

19.3.2.1 Dosage Forms

Different solid, semisolid, or liquid dosage forms have been used for the administration of drugs in the rectum. These are described in the following. Contrary to the vaginal route, removable drug delivery systems are not common for rectal use, although tampons have been described and used in clinical practice. For example, Moraxen® tampons (Schwarz Pharma, Chesham, UK) were previously available in the United Kingdom for the sustained release of morphine sulfate in cancer patients over a 24 hours period [187].

Solid systems typically include suppositories or capsules. Suppositories are likely the most recognizable rectal dosage form. The technology and excipients used to obtain rectal suppositories are similar to those discussed above for vaginal equivalents, although differences in weight (usually 2–3 g for adults, 1–2 g for children) and shape (torpedo, conical and bullet are the most common) can be observed. Formulations based on suppositories are intended to be manually inserted in an oriented fashion, without the aid of applicators. In general, suppositories should melt in less than ten minutes at temperatures just below 37 °C or disaggregate/dissolve completely within one hour in the reduced amount of fluid present in the rectum, depending on used excipients

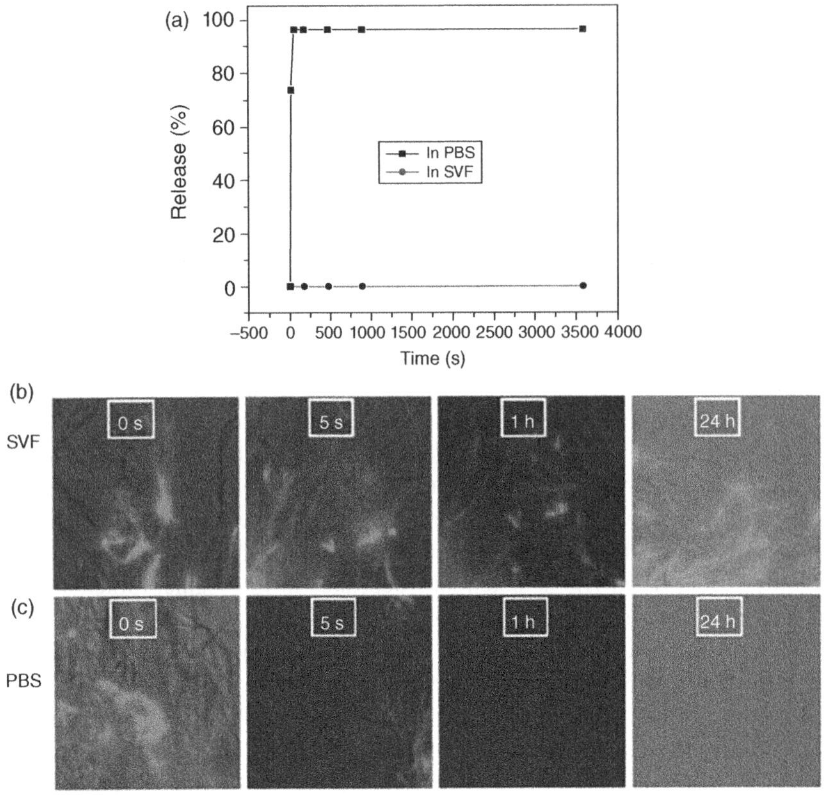

Figure 19.12 (a) Release profile of rhodamine 6G (used as model drug) from electrospun cellulose acetate phthalate fibers in respectively simulated vaginal fluid (SVF, pH 4.2) and phosphate buffered saline (PBS, pH 7.4) Appropriate standard curves of rhodamine 6G in SVF and PBS were used to calculate the concentration of released rhodamine 6G. (b, c) Fluorescence microscopy on rhodamine 6G-loaded cellulose acetate phthalate fibers after dispersion in (b) SVF and (c) PBS. Source: Reprinted from Huang et al. [185], Copyright (2011), with permission from Elsevier.

(see vaginal suppositories for examples). However, suppositories can be formulated in order to sustain drug release for up to several hours [188–191]. For example, Takatori et al. [192] modified Witepsol H15-based suppositories with different excipients presenting high melting points (beeswax, hexaglycerol pentastearate, or decaglycerol heptabehenate) in order to delay the release of acetaminophen (Figure 19.13). Increased delay in drug release, as documented by decreasing values of the release rate constant in the Higuchi model, was shown to be correlated with higher viscosity of excipient melts at 37 °C in phosphate buffer pH 7.3. Furthermore, sustained release suppositories were able to prolong plasma concentration of acetaminophen in a rat model. Another strategy for achieving sustained drug release comprises the incorporation of different polymers in the composition of suppositories [193, 194]. Apart from possibly increasing the mucoadhesiveness of suppositories depending on the nature of used polymers [195], the formation of hydrogel matrices that are able to control the diffusion of drugs seems to be the general mechanism behind sustained release. The development of multilayered suppositories is an additional interesting approach for controlling drug release [196, 197].

Although not so common, soft capsules can be used for rectal administration of drugs. Presenting alike technology and features to the ones intended for vaginal administration, capsules

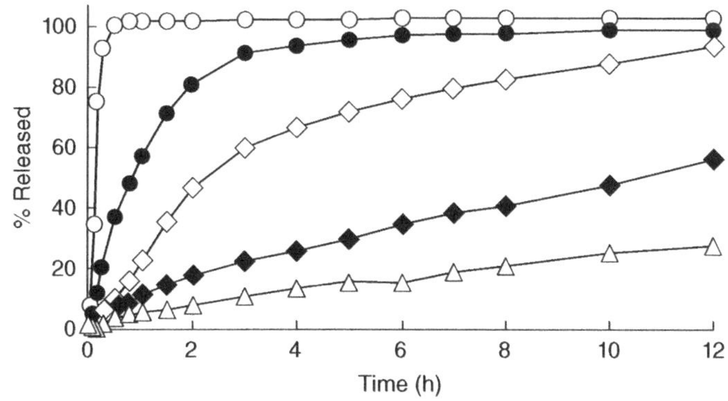

Figure 19.13 Release profiles of acetaminophen from suppositories at 37 °C. Each value represents the mean of six experiments. Witepsol® H15-beeswax mixture (○): Witepsol® H15 alone; (•) 10% beeswax; (◇) 15% beeswax; (◆) 20% beeswax; (△) 25% beeswax. Source: Takatori et al. [192]. © 2004, Elsevier.

are ovoid but typically more elongated along the longer axis in order to match the format of rectal suppositories and facilitate administration. Control over drug release is possible by changing the features of the shell of capsules. One interesting advantage over melting suppositories is the possibility to produce products that are stable at high temperatures. For example, artesunate has been formulated as thermostable rectal soft capsules for the treatment of malaria (Plasmotrim™ Rectocaps™, Mepha, Aesch, Switzerland), particularly in endemic tropical areas where refrigeration is usually unavailable [198, 199]. Advanced capsule systems, such as osmotic pumps, have also been studied in the past for sustained drug release upon rectal delivery [200, 201]. Although providing excellent control over drug release rates, osmotic pumps never really reached the market likely due to cost issues.

Semisolid systems such as gels, creams, and ointments have been used for rectal drug delivery, mainly to address local conditions such as hemorrhoids. Advantages and pitfalls are generally the same as for those intended for vaginal use. Rectal gels in particular have been quite popular, not only for medical purposes but also to aid sexual intercourse as lubricants [202]. Administration of semisolids is performed using pre-filled, single-dose or multi-dose tubes bearing a nozzle suitable for rectal insertion. Rectal semisolid systems typically feature rapid drug release although these can be formulated to modulate release behavior [203, 204]. As described before in this chapter, strategies relying on mucoadhesive and/or thermosensitive systems have been used for sustaining drug release. For example, both approaches have been tested by Fawaz et al. [205] in order to deliver quinine. The drug was incorporated into either HPMC-based (mucoadhesive) or poloxamer 407-based (thermosensitive) gel bases and shown to be released at moderately slower rates as compared to quinine in suspension (approximately 100% release at two hours vs. four hours for both gels). This behavior was correlated with non-Fickian diffusion of the drug from gels. Additionally, *in vivo* pharmacokinetics data obtained from a rabbit model demonstrated only meager differences between gels and quinine suspension. In another study, Koffi et al. [206] further attempted to develop quinine gels with sustained release kinetics by combining both mucoadhesive and thermosensitive features in the same system. Different drug release profiles were obtained by changing the quantitative composition of gels containing poloxamer 407, HPMC, and propylene glycol (Figure 19.14) but such variations did not translate into substantial differences among gels when tested in rabbits. Another possibility to sustain drug release from rectal semisolids has been the use of complexation agents such as cyclodextrins. For instance, Centkowska and Sznitowska [207]

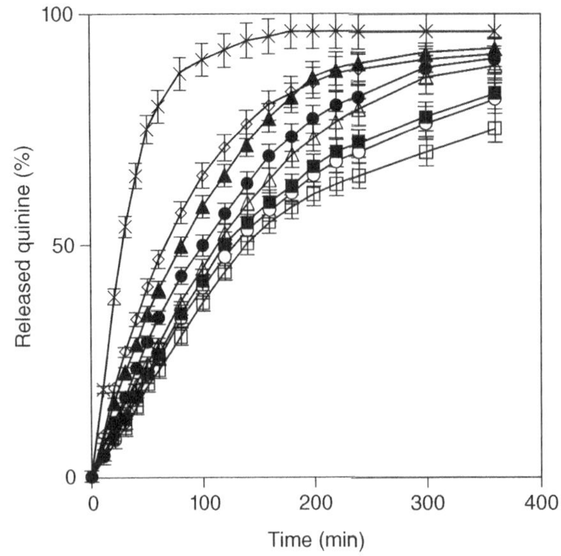

Figure 19.14 Release profiles of quinine in deionized water (pH 6.9) for the rectal gels 0/0.5/30 (◇), 16/0.5/30 (▲), 16/1/30 (•), 16/1.5/30 (■), 17/0.5/30 (△), 17/1/30 (○) and 17/1.5/30 (□) (values are for w/w% of poloxamer 407/HPMC/propylene glycol), and for the rectal solution (×). $n = 6$, mean ± standard deviation. Source: Koffi et al. [206], Copyright (2007), with permission from Elsevier.

showed that the use of β-cyclodextrin was able to reduce the release rate of nitroglycerin from different gel, cream, or ointment bases intended for the treatment of anal fissures. Even if sustained drug release from semisolids has been regarded as a beneficial goal in these previous studies, rapid release may also be of interest particularly when solubility issues may delay drug absorption and onset of pharmacological effects. For example, in a recent report, Liu et al. [208] were successful in increasing the release rate of ibuprofen, a nonsteroidal anti-inflammatory drug, from gels by pre-preparing polyvinylpyrrolidone-based solid dispersions containing the drug. Solubility enhancement of ibuprofen was clear *in vitro* and, more important, led to an increase in the bioavailability of the drug when tested in rabbits, as compared to commercially available ibuprofen suppositories (composition was not detailed).

Liquid systems for rectal use are commonly referred to as enemas and can be classified as solutions, emulsions, or suspensions. These are typically water-based and presented in their final form, although some products can be prepared from powders, granules, or tablets immediately before use. The amount of enema to be administered can vary from a few milliliters (often called microenemas) up to two liters or more, depending on the intended purpose. For example, volumes as low as 2.5 ml or less can be used in cases where retention and rectal localization is required (e.g. hemorrhoid treatment, systemic absorption from the rectum); conversely, larger amounts of liquid can be used for rectal and lower colon cleansing before surgery or relief of severe constipation, in which case the liquid is mostly discharged soon after administration. Enemas are pre-filled or placed at the time of administration into containers of suitable size and format (e.g. squeezing bottles, bags, or microenema tubes) bearing suitable applicator systems or adapted nozzles that allow rectal insertion. Control of drug release from rectal liquid dosage forms is usually not procured or even detailed, but generally thought as being almost immediate.

Aerosol foams comprising gas-in-liquid systems are not formally considered liquids but are included here due to their pertinence for rectal drug delivery. Technological features, advantages, and limitations of rectal foams are similar to those discussed above for the vaginal route. Regarding distribution and retention in the colorectum, foams appear to present an intermediate behavior between liquid (wider spreading, lower retention) and solid (poorer spreading, higher retention) dosage forms [209]. Hydrophilic foams are most common although hydrophobic products have also been proposed [210]. The use of foams has been particularly explored for rectal administration of drugs intended to manage problems at the colorectum such as inflammatory disease [136]. Several products are currently available in the market for immediate drug release.

19.3.2.2 Nanotechnology-based Drug Delivery Systems

Interest in using nanocarriers for rectal drug delivery has been explored over recent years with various medical purposes. Examples include the use of diazepam-loaded nanoemulsions for treating pediatric convulsions [211], tacrolimus-loaded polymeric nanoparticles for inflammatory bowel disease management [212], docetaxel-loaded micelles for systemic cancer treatment [213], or thermosensitive metoclopramide-loaded solid lipid nanoparticles (SLNs) for amelioration of gastrointestinal disorders [214, 215]. Following on the path of the vaginal route, even though still quite a few steps behind, research into rectal nanotechnology-based drug delivery solutions has also been mainly focused in developing microbicides [216]. The ability to influence drug distribution, retention, and targeting are general advantages that have been advocated in this particular field of application. Although mucoadhesion is still regarded as beneficial, recent research efforts seem to support that the ability of nanocarriers to migrate unhindered across mucus may help drug distribution and even retention following rectal administration. Indeed, Maisel et al. [217] elegantly showed that nanoparticles made mucus-diffusive by dense surface modification with PEG (5 kDa) were able to evenly and widely distribute across the colorectum of normal and colitis mouse models (Figure 19.15). Conversely, mucoadhesive nanoparticles mostly aggregated within the colorectal lumen after administration. Moreover, these last researchers showed that promoting diffusion toward the epithelium by administering nanoparticles in moderately hypotonic enemas could

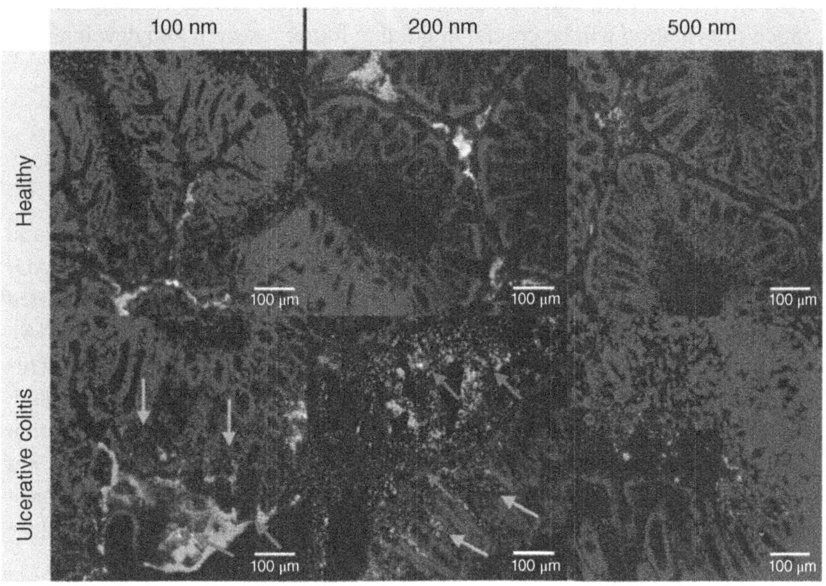

Figure 19.15 Distribution of various sizes of mucoadhesive particles (MAP) and mucus-penetrating particles (MPP) after rectal co-administration to mice. Distribution of rectally co-administered, fluorescent MAP (dark gray) and MPP (light gray) of various sizes (100 nm, 200 nm, 500 nm) in the colorectum of healthy mice and mice with TNBS-induced colitis. The red arrows highlight aggregates of MAP, while the green arrows highlight areas where the MPP have penetrated into the tissue. Cell nuclei are stained blue with DAPI. White scale bars indicate 100 μm. Images are representative of $n \geq 3$ mice. Source: Reprinted from Maisel et al. [217], Copyright (2014), with permission from Elsevier.

further enhance colorectal surface coverage [218]. From a microbicide drug delivery point of view, such approaches may help providing a continuous and lasting barrier that may prevent HIV transmission upon anal intercourse.

19.4 Conclusions

The vagina and the rectum have long been considered as valid drug delivery routes. Although overviewed in tandem in the present chapter, individual anatomic and physiological particularities of each mucosal cavity should be considered when developing drug delivery systems. Differences between both are considerable and dual use of single products is usually inappropriate. Efforts toward controlling drug delivery, namely drug release kinetics, have been pursued for vaginal and rectal dosage forms and delivery systems, often with considerable success. The onset of anti-HIV microbicides development starting roughly at the beginning of the present century boosted renewed interest in both vaginal and rectal drug delivery, with major technological developments being achieved while other interesting endeavors are still ongoing. In particular, engineering of vaginal rings has seen significant advances over recent years, at the same time that a microbicide vaginal ring is on fast-track toward regulatory approval. The dapivirine ring may indeed be a game changer in the near future against HIV infection. Furthermore, relevant developments on nanotechnology-based microbicides, mostly for vaginal use, are opening new doors for focal drug delivery that can be useful in achieving more effective and safer products. Yet, much groundbreaking work is still required as novel biological insights on mucosal environments (e.g. the effect of microbiota [42, 219] and cell transporters [220, 221] on drug delivery), new concepts on safety evaluation [222, 223], and advances and technologies in drug delivery science [224] are emerging. Such information will be crucial in guiding future research and development of controlled vaginal and rectal drug delivery systems.

Acknowledgments

This work was financed by Programa Gilead GÉNESE, Gilead Portugal (ref. PGG/046/2015 and ref. PGG/002/2016) and by Fundação para a Ciência e a Tecnologia (FCT), Portugal (VIH/SAU/0021/2011). This article is a result of the project NORTE-01-0145-FEDER-000012, supported by Norte Portugal Regional Operational Programme (NORTE 2020), under the PORTUGAL 2020 Partnership Agreement, through the European Regional Development Fund (ERDF). This work was financed by FEDER – Fundo Europeu de Desenvolvimento Regional funds through the COMPETE 2020 – Operacional Programme for Competitiveness and Internationalisation (POCI), Portugal 2020, and by Portuguese funds through FCT – Fundação para a Ciência e a Tecnologia/Ministério da Ciência, Tecnologia e Ensino Superior in the framework of the project "Institute for Research and Innovation in Health Sciences" (POCI-01-0145-FEDER-007274). José das Neves gratefully acknowledges FCT for financial support (SFRH/BPD/92934/2013).

References

1 Stevens, J.M. (1975). Gynaecology from ancient Egypt: the papyrus Kahun: a translation of the oldest treatise on gynaecology that has survived from the ancient world. *Med. J. Aust.* 2 (25–26): 949–952.

2 de Boer, A.G., Moolenaar, F., de Leede, L.G., and Breimer, D.D. (1982). Rectal drug administration: clinical pharmacokinetic considerations. *Clin. Pharmacokinet.* 7 (4): 285–311.

3 Bernkop-Schnürch, A. and Hornof, M. (2003). Intravaginal drug delivery systems. *Am. J. Drug Delivery* 1 (4): 241–254.

4 Merabet, J., Thompson, D., and Saul Levinson, R. (2005). Advancing vaginal drug delivery. *Expert Opin. Drug Delivery* 2 (4): 769–777.

5 Breitkreutz, J. and Boos, J. (2007). Paediatric and geriatric drug delivery. *Expert Opin. Drug Delivery* 4 (1): 37–45.

6 Friend, D.R. and Kiser, P.F. (2013). Assessment of topical microbicides to prevent HIV-1 transmission: concepts, testing, lessons learned. *Antiviral Res.* 99 (3): 391–400.

7 Agashe, H., Hu, M., and Rohan, L. (2012). Formulation and delivery of microbicides. *Curr. HIV Res.* 10 (1): 88–96.

8 Boily, M.C., Dimitrov, D., Abdool Karim, S.S., and Masse, B. (2011). The future role of rectal and vaginal microbicides to prevent HIV infection in heterosexual populations: implications for product development and prevention. *Sex. Transm. Infect.* 87 (7): 646–653.

9 Dezzutti, C.S., Rohan, L.C., Wang, L. et al. (2012). Reformulated tenofovir gel for use as a dual compartment microbicide. *J. Antimicrob. Chemother.* 67 (9): 2139–2142.

10 Ham, A.S., Nugent, S.T., Peters, J.J. et al. (2015). The rational design and development of a dual chamber vaginal/rectal microbicide gel formulation for HIV prevention. *Antiviral Res.* 120: 153–164.

11 das Neves, J., Amaral, M.H., and Bahia, M.F. (2008). Vaginal drug delivery. In: *Pharmaceutical Manufacturing Handbook: Production and Processes* (ed. S.C. Gad), 809–878. Hoboken, NJ, USA: Wiley.

12 das Neves, J., Palmeira-de-Oliveira, R., Palmeira-de-Oliveira, A. et al. (2014). Vaginal mucosa and drug delivery. In: *Mucoadhesive Materials and Drug Delivery Systems* (ed. V.V. Khutoryanskiy), 99–131. Chichester, UK: Wiley.

13 Nunes, R., Sarmento, B., and das Neves, J. (2014). Formulation and delivery of anti-HIV rectal microbicides: advances and challenges. *J. Controlled Release* 194: 278–294.

14 Van De Graff, K. (2001). *Human Anatomy*, 6e. Europe: McGraw-Hill.

15 Barnhart, K.T., Pretorius, E.S., and Malamud, D. (2004). Lesson learned and dispelled myths: three-dimensional imaging of the human vagina. *Fertil. Steril.* 81 (5): 1383–1384.

16 Pendergrass, P.B., Belovicz, M.W., and Reeves, C.A. (2003). Surface area of the human vagina as measured from vinyl polysiloxane casts. *Gynecol. Obstet. Invest.* 55 (2): 110–113.

17 Barnhart, K.T., Izquierdo, A., Pretorius, E.S. et al. (2006). Baseline dimensions of the human vagina. *Hum. Reprod.* 21 (6): 1618–1622.

18 Gartner, L.P. and Hiatt, J.L. (2014). *Color Atlas and Text of Histology*, 6e. Baltimore, MD: Lippincott Williams & Wilkins.

19 Owen, D.H. and Katz, D.F. (1999). A vaginal fluid simulant. *Contraception* 59 (2): 91–95.

20 Linhares, I.M., Summers, P.R., Larsen, B. et al. (2011). Contemporary perspectives on vaginal pH and lactobacilli. *Am. J. Obstet. Gynecol.* 204 (2): 120 e1–120.e5.

21 Mitchell, C., Paul, K., Agnew, K. et al. (2011). Estimating volume of cervicovaginal secretions in cervicovaginal lavage fluid collected for measurement of genital HIV-1 RNA levels in women. *J. Clin. Microbiol.* 49 (2): 735–736.

22 Levin, R. and Riley, A. (2007). The physiology of human sexual function. *Psychiatry* 6 (3): 90–94.

23 Farage, M. and Maibach, H. (2006). Lifetime changes in the vulva and vagina. *Arch. Gynecol. Obstet.* 273 (4): 195–202.

24 Caillouette, J.C., Sharp, C.F. Jr., Zimmerman, G.J., and Roy, S. (1997). Vaginal pH as a marker for bacterial pathogens and menopausal status. *Am. J. Obstet. Gynecol.* 176 (6): 1270–1275.

25 Owen, D.H. and Katz, D.F. (2005). A review of the physical and chemical properties of human semen and the formulation of a semen simulant. *J. Androl.* 26 (4): 459–469.

26 Wagner, G. and Ottesen, B. (1982). Vaginal physiology during menstruation. *Ann. Intern. Med.* 96 (6 Pt 2): 921–923.

27 Gorodeski, G.I. (2001). Vaginal-cervical epithelial permeability decreases after menopause. *Fertil. Steril.* 76 (4): 753–761.

28 Ramsey, P.S., Ogburn, P.L. Jr., Harris, D.Y. et al. (2002). Effect of vaginal pH on efficacy of the dinoprostone gel for cervical ripening/labor induction. *Am. J. Obstet. Gynecol.* 187 (4): 843–846.

29 Nel, A.M., Haazen, W., Nuttall, J.P. et al. (2014). Pharmacokinetics and safety assessment of anti-HIV dapivirine vaginal microbicide rings with multiple dosing. *J. AIDS Clin. Res.* 5 (10): 355.

30 Barnhart, K.T., Pretorius, E.S., Timbers, K. et al. (2004). In vivo distribution of a vaginal gel: MRI evaluation of the effects of gel volume, time and simulated intercourse. *Contraception* 70 (6): 498–505.

31 Acarturk, F., Parlatan, Z.I., and Saracoglu, O.F. (2001). Comparison of vaginal aminopeptidase enzymatic activities in various animals and in humans. *J. Pharm. Pharmacol.* 53 (11): 1499–1504.

32 Song, Y.B., Hwang, K., Kim, D.J., and Han, S.H. (2009). Innervation of vagina: microdissection and immunohistochemical study. *J. Sex Marital Ther.* 35 (2): 144–153.

33 van de Wijgert, J.H., Borgdorff, H., Verhelst, R. et al. (2014). The vaginal microbiota: what have we learned after a decade of molecular characterization? *PLoS One* 9 (8): e105998.

34 Huang, B., Fettweis, J.M., Brooks, J.P. et al. (2014). The changing landscape of the vaginal microbiome. *Clin. Lab. Med.* 34 (4): 747–761.

35 Mendling, W. (2016). Vaginal microbiota. *Adv. Exp. Med. Biol.* 902: 83–93.

36 Cone, R.A. (2014). Vaginal microbiota and sexually transmitted infections that may influence transmission of cell-associated HIV. *J. Infect. Dis.* 210 (Suppl 3): S616–S621.

37 Klatt, N.R., Cheu, R., Birse, K. et al. (2017). Vaginal bacteria modify HIV tenofovir microbicide efficacy in African women. *Science* 356 (6341): 938–945.

38 Hiruy, H., Fuchs, E.J., Marzinke, M.A. et al. (2015). A phase 1 randomized, blinded comparison of the pharmacokinetics and colonic distribution of three candidate rectal microbicide formulations of tenofovir 1% gel with simulated unprotected sex (CHARM-02). *AIDS Res. Hum. Retroviruses* 31 (11): 1098–1108.

39 Steele, S.R., Hull, T.L., Read, T.E. et al. (eds.) (2016). *The ASCRS Textbook of Colon and Rectal Surgery*, 3e. New York: Springer.

40 Jantzen, J.P., Tzanova, I., Witton, P.K., and Klein, A.M. (1989). Rectal pH in children. *Can. J. Anaesth.* 36 (6): 665–667.

41 Pullan, R.D., Thomas, G.A., Rhodes, M. et al. (1994). Thickness of adherent mucus gel on colonic mucosa in humans and its relevance to colitis. *Gut* 35 (3): 353–359.

42 Nicholson, J.K., Holmes, E., Kinross, J. et al. (2012). Host-gut microbiota metabolic interactions. *Science* 336 (6086): 1262–1267.

43 Round, J.L. and Mazmanian, S.K. (2009). The gut microbiota shapes intestinal immune responses during health and disease. *Nat. Rev. Immunol.* 9 (5): 313–323.

44 Li, H., He, J., and Jia, W. (2016). The influence of gut microbiota on drug metabolism and toxicity. *Expert Opin. Drug Metab. Toxicol.* 12 (1): 31–40.

45 Coggins, C., Elias, C.J., Atisook, R. et al. (1998). Women's preferences regarding the formulation of over-the-counter vaginal spermicides. *AIDS* 12 (11): 1389–1391.

46 Hardy, E., Jimenez, A.L., de Padua, K.S., and Zaneveld, L.J. (1998). Women's preferences for vaginal antimicrobial contraceptives. III. Choice of a formulation, applicator, and packaging. *Contraception* 58 (4): 245–249.

47 Joglekar, A., Rhodes, C.T., and Danish, M. (1991). Preferences of women for use of intravaginal medications. *Drug Dev. Ind. Pharm.* 17 (15): 2103–2113.

48 Nel, A.M., Mitchnick, L.B., Risha, P. et al. (2011). Acceptability of vaginal film, soft-gel capsule, and tablet as potential microbicide delivery methods among African women. *J. Women's Health* 20 (8): 1207–1214.

49 Palmeira-de-Oliveira, R., Duarte, P., Palmeira-de-Oliveira, A. et al. (2015). Women's experiences, preferences and perceptions regarding vaginal products: results from a cross-sectional web-based survey in Portugal. *Eur. J. Contracept. Reprod. Health Care* 20 (4): 259–271.

50 Woodsong, C. and Holt, J.D. (2015). Acceptability and preferences for vaginal dosage forms intended for prevention of HIV or HIV and pregnancy. *Adv. Drug Delivery Rev.* 15 (92): 146–154.

51 Bentley, M.E., Fullem, A.M., Tolley, E.E. et al. (2004). Acceptability of a microbicide among women and their partners in a 4-country phase I trial. *Am. J. Public Health* 94 (7): 1159–1164.

52 Verguet, S., Young Holt, B., and Szeri, A.J. (2010). Increasing the effectiveness of vaginal microbicides: a biophysical framework to rethink behavioral acceptability. *PLoS One* 5 (11): e15501.

53 Fan, M.D., Kramzer, L.F., Hillier, S.L. et al. (2016). Preferred physical characteristics of vaginal film microbicides for HIV prevention in Pittsburgh women. *Arch. Sex. Behav.* https://doi.org/10.1007/s10508-10016-10816-10501.

54 Jannin, V., Lemagnen, G., Gueroult, P. et al. (2014). Rectal route in the 21st century to treat children. *Adv. Drug Delivery Rev.* 73: 34–49.

55 Carballo-Diéguez, A., Stein, Z., Saez, H. et al. (2000). Frequent use of lubricants for anal sex among men who have sex with men: the HIV prevention potential of a microbicidal gel. *Am. J. Public Health* 90 (7): 1117–1121.

56 Javanbakht, M., Stahlman, S., Pickett, J. et al. (2014). Prevalence and types of rectal douches used for anal intercourse: results from an international survey. *BMC Infect. Dis.* 14: 95.

57 Carballo-Diéguez, A., Dolezal, C., Bauermeister, J.A. et al. (2008). Preference for gel over suppository as delivery vehicle for a rectal microbicide: results of a randomised, crossover acceptability trial among men who have sex with men. *Sex. Transm. Infect.* 84 (6): 483–487.

58 Leyva, F., Fuchs, E.J., Bakshi, R. et al. (2015). Simultaneous evaluation of safety, acceptability, pericoital kinetics, and ex vivo pharmacodynamics comparing four rectal microbicide vehicle candidates. *AIDS Res. Hum. Retroviruses* 31 (11): 1089–1097.

59 Pines, H.A., Gorbach, P.M., Weiss, R.E. et al. (2013). Acceptability of potential rectal microbicide delivery systems for HIV prevention: a randomized crossover trial. *AIDS Behav.* 17 (3): 1002–1015.

60 Khutoryanskiy, V.V. (2011). Advances in mucoadhesion and mucoadhesive polymers. *Macromol. Biosci.* 11 (6): 748–764.

61 Sosnik, A., das Neves, J., and Sarmento, B. (2014). Mucoadhesive polymers in the design of nano-drug delivery systems for administration by non-parenteral routes: a review. *Prog. Polym. Sci.* 39 (12): 2030–2075.

62 das Neves, J., Bahia, M.F., Amiji, M.M., and Sarmento, B. (2011). Mucoadhesive nanomedicines: characterization and modulation of mucoadhesion at the nanoscale. *Expert Opin. Drug Delivery* 8 (8): 1085–1104.

63 Ganta, S., Devalapally, H., Shahiwala, A., and Amiji, M. (2008). A review of stimuli-responsive nanocarriers for drug and gene delivery. *J. Controlled Release* 126 (3): 187–204.

64 Qiu, Y. and Park, K. (2001). Environment-sensitive hydrogels for drug delivery. *Adv. Drug Delivery Rev.* 53 (3): 321–339.

65 Ruel-Gariepy, E. and Leroux, J.C. (2004). In situ-forming hydrogels – review of temperature-sensitive systems. *Eur. J. Pharm. Biopharm.* 58 (2): 409–426.

66 Rossi, S., Ferrari, F., Bonferoni, M.C. et al. (2014). Comparison of poloxamer- and chitosan-based thermally sensitive gels for the treatment of vaginal mucositis. *Drug Dev. Ind. Pharm.* 40 (3): 352–360.

67 Mirza, M.A., Ahmad, S., Mallick, M.N. et al. (2013). Development of a novel synergistic thermosensitive gel for vaginal candidiasis: an in vitro, in vivo evaluation. *Colloids Surf., B* 103: 275–282.

68 Ibrahim, E.S.A., Ismail, S., Fetih, G. et al. (2012). Development and characterization of thermosensitive pluronic-based metronidazole in situ gelling formulations for vaginal application. *Acta Pharm.* 62 (1): 59–70.

69 Yuan, Y., Cui, Y., Zhang, L. et al. (2012). Thermosensitive and mucoadhesive in situ gel based on poloxamer as new carrier for rectal administration of nimesulide. *Int. J. Pharm.* 430 (1–2): 114–119.

70 Moawad, F.A., Ali, A.A., and Salem, H.F. (2017). Nanotransfersomes-loaded thermosensitive in situ gel as a rectal delivery system of tizanidine HCl: preparation, in vitro and in vivo performance. *Drug Delivery* 24 (1): 252–260.

71 das Neves, J., Pinto, E., Teixeira, B. et al. (2008). Local treatment of vulvovaginal candidosis: general and practical considerations. *Drugs* 68 (13): 1787–1802.

72 Gupta, K.M., Barnes, S.R., Tangaro, R.A. et al. (2007). Temperature and pH sensitive hydrogels: an approach towards smart semen-triggered vaginal microbicidal vehicles. *J. Pharm. Sci.* 96 (3): 670–681.

73 Chen, D., Sun, K., Mu, H. et al. (2012). pH and temperature dual-sensitive liposome gel based on novel cleavable mPEG-Hz-CHEMS polymeric vaginal delivery system. *Int. J. Nanomed.* 7: 2621–2630.

74 Mahato, R.I. and Narang, A.S. (2012). *Pharmaceutical Dosage Forms and Drug Delivery*, 2e. Boca Raton, FL: CRC Press.

75 Pereira, L.E., Clark, M.R., Friend, D.R. et al. (2014). Pharmacokinetic and safety analyses of tenofovir and tenofovir-emtricitabine vaginal tablets in pigtailed macaques. *Antimicrob. Agents Chemother.* 58 (5): 2665–2674.

76 Clark, M.R., Peet, M.M., Davis, S. et al. (2014). Evaluation of rapidly disintegrating vaginal tablets of tenofovir, emtricitabine and their combination for HIV-1 prevention. *Pharmaceutics* 6 (4): 616–631.

77 Garg, S., Jambu, L., and Vermani, K. (2007). Development of novel sustained release bioadhesive vaginal tablets of povidone iodine. *Drug Dev. Ind. Pharm.* 33 (12): 1340–1349.

78 Baloglu, E., Ozyazici, M., Yaprak Hizarcioglu, S. et al. (2006). Bioadhesive controlled release systems of ornidazole for vaginal delivery. *Pharm. Dev. Technol.* 11 (4): 477–484.

79 Karasulu, H.Y., Taneri, F., Sanal, E. et al. (2002). Sustained release bioadhesive effervescent ketoconazole microcapsules tabletted for vaginal delivery. *J. Microencapsulation* 19 (3): 357–362.

80 Karasulu, H.Y., Hilmioglu, S., Metin, D.Y., and Guneri, T. (2004). Efficacy of a new ketoconazole bioadhesive vaginal tablet on *Candida albicans*. *Farmaco* 59 (2): 163–167.

81 Digenis, G.A., Nosek, D., Mohammadi, F. et al. (1999). Novel vaginal controlled-delivery systems incorporating coprecipitates of nonoxynol-9. *Pharm. Dev. Technol.* 4 (3): 421–430.

82 Bugge, C., Adams, E.J., Gopinath, D., and Reid, F. (2013). Pessaries (mechanical devices) for pelvic organ prolapse in women. *Cochrane Database of Syst. Rev.* (2): CD004010. https://doi.org/10.1002/14651858.CD004010.pub3.

83 Garg, S., Tambwekar, K.R., Vermani, K. et al. (2001). Compendium of pharmaceutical excipients for vaginal formulations. *Pharm. Technol.* 25: 14–24.

84 Geeta, M.P. and Madhabhai, M.P. (2009). Design and in vitro evaluation of a novel vaginal drug delivery system based on gelucire. *Curr. Drug Delivery* 6 (2): 159–165.

85 Rodrigues, F., Maia, M.J., das Neves, J. et al. (2015). Vaginal suppositories containing *Lactobacillus acidophilus*: development and characterization. *Drug Dev. Ind. Pharm.* 41 (9): 1518–1525.

86 Akbar, M.R., Ahmadi, S.M., Hedayati, A.H., and Khami, M.A. (1982). A three-day study with miconazole gelatin capsules in vaginal candidosis. *Drug Dev. Res.* 2 (1): 87–90.

87 Aubeny, E., Colau, J.C., and Nandeuil, A. (2000). Local spermicidal contraception: a comparative study of the acceptability and safety of a new pharmaceutical formulation of benzalkonium chloride, the vaginal capsule, with a reference formulation, the pessary. *Eur. J. Contracept. Reprod. Health Care* 5 (1): 61–67.

88 (2015). Gyno-Daktarin® 1200 mg Vaginal Capsule. Summary Product Characteristics. High Wycombe, UK: Janssen-Cilag.

89 Upmalis, D.H., Cone, F.L., Lamia, C.A. et al. (2000). Single-dose miconazole nitrate vaginal ovule in the treatment of vulvovaginal candidiasis: two single-blind, controlled studies versus miconazole nitrate 100 mg cream for 7 days. *J. Women's Health Gender-Based Med.* 9 (4): 421–429.

90 Akil, A., Agashe, H., Dezzutti, C.S. et al. (2015). Formulation and characterization of polymeric films containing combinations of antiretrovirals (ARVs) for HIV prevention. *Pharm. Res.* 32 (2): 458–468.

91 Zhang, W., Hu, M., Shi, Y. et al. (2015). Vaginal microbicide film combinations of two reverse transcriptase inhibitors, EFdA and CSIC, for the prevention of HIV-1 sexual transmission. *Pharm. Res.* 32 (9): 2960–2972.

92 Ham, A.S., Rohan, L.C., Boczar, A. et al. (2012). Vaginal film drug delivery of the pyrimidinedione IQP-0528 for the prevention of HIV infection. *Pharm. Res.* 29 (7): 1897–1907.

93 Machado, R.M., Palmeira-de-Oliveira, A., Martinez-de-Oliveira, J., and Palmeira-de-Oliveira, R. (2013). Vaginal films for drug delivery. *J. Pharm. Sci.* 102 (7): 2069–2081.

94 Machado, A., Cunha-Reis, C., Araújo, F. et al. (2016). Development and in vivo safety assessment of tenofovir-loaded nanoparticles-in-film as a novel vaginal microbicide delivery system. *Acta Biomater.* 44: 332–340.

95 das Neves, J. and Sarmento, B. (2016). Antiretroviral drug-loaded nanoparticles-in-films: a new option for developing vaginal microbicides? *Expert Opin. Drug Delivery* 14 (4): 449–452.

96 Rohan, L.C. and Zhang, W. (2014). Vaginal microbicide films. In: *Drug Delivery and Development of Anti-HIV Microbicides* (eds. J. das Neves and B. Sarmento), 291–330. Singapore: Pan Stanford.

97 Borumand, M., Mortazavi, S.A., Jafari Azar, Z., and Teymouri Rad, R. (2014). Development and in vitro evaluation of a novel contraceptive vaginal mucoadhesive propranolol hydrochloride film. *J. Drug Delivery Sci. Technol.* 24 (6): 637–644.

98 Barnhart, K.T., Pretorius, E.S., Shaunik, A. et al. (2005). Vaginal distribution of two volumes of the novel microbicide gel cellulose sulfate (2.5 and 3.5 mL). *Contraception* 72 (1): 65–70.

99 Brown, J., Hooper, G., Kenyon, C.J. et al. (1997). Spreading and retention of vaginal formulations in post-menopausal women as assessed by gamma scintigraphy. *Pharm. Res.* 14 (8): 1073–1078.

100 Vail, J.G., Cohen, J.A., and Kelly, K.L. (2004). Improving topical microbicide applicators for use in resource-poor settings. *Am. J. Public Health* 94 (7): 1089–1092.

101 Brannon-Peppas, L. (1993). Novel vaginal drug release applications. *Adv. Drug Delivery Rev.* 11 (1–2): 169–177.

102 Gudeman, J., Thompson, D.J., and Saul Levinson, R. (2008). SITE RELEASE®, vaginal bioadhesive system. In: *Modified-Release Drug Delivery Technology*, 2e (eds. M.J. Rathbone, J. Hadgraft, M.S. Roberts and M.E. Lane), 521–529. Boca Raton, FL: CRC Press.

103 Seidman, L.S. and Skokos, C.K. (2005). An evaluation of butoconazole nitrate 2% site release vaginal cream (Gynazole-1) compared to fluconazole 150 mg tablets (Diflucan) in the time to relief of symptoms in patients with vulvovaginal candidiasis. *Infect. Dis. Obstet. Gynecol.* 13 (4): 197–206.

104 Brown, D., Henzl, M.R., and Kaufman, R.H. (1999). Butoconazole nitrate 2% for vulvovaginal candidiasis. New, single-dose vaginal cream formulation vs. seven-day treatment with miconazole nitrate. Gynazole 1 Study Group. *J. Reprod. Med.* 44 (11): 933–938.

105 Faro, S. and Skokos, C.K. (2005). The efficacy and safety of a single dose of Clindesse vaginal cream versus a seven-dose regimen of Cleocin vaginal cream in patients with bacterial vaginosis. *Infect. Dis. Obstet. Gynecol.* 13 (3): 155–160.

106 Henzl, M.R. (2005). Properties of sustained-release single-dose formulations for vulvovaginal candidiasis: the example of butoconazole vaginal emulsion as a drug-sparing formulation. *Am. J. Drug Delivery* 3 (4): 227–237.

107 Perrigo®. (revised 2014) Gynazole-1® Butoconazole Nitrate Vaginal Cream USP, 2%. Prescribing Information. Allegan, MI: Perrigo.

108 Clindesse® (revised 2014). Clindesse® (Clindamycin Phosphate) Vaginal Cream, 2%. Prescribing Information, Allegan, MI: Perrigo.

109 Weinstein, L., Henzel, M.R., and Tsina, I.W. (1994). Vaginal retention of 2% butoconazole nitrate cream: comparison of a standard and a sustained-release preparation. *Clin. Ther.* 16 (6): 930–934.

110 das Neves, J. and Bahia, M.F. (2006). Gels as vaginal drug delivery systems. *Int. J. Pharm.* 318 (1–2): 1–14.

111 Garg, S., Goldman, D., Krumme, M. et al. (2010). Advances in development, scale-up and manufacturing of microbicide gels, films, and tablets. *Antiviral Res.* 88 (Suppl 1): S19–S29.

112 Robinson, J.R. and Bologna, W.J. (1994). Vaginal and reproductive system treatments using a bioadhesive polymer. *J. Controlled Release* 28 (1–3): 87–94.

113 Knuth, K., Amiji, M., and Robinson, J.R. (1993). Hydrogel delivery systems for vaginal and oral applications: formulation and biological considerations. *Adv. Drug Delivery Rev.* 11 (1–2): 137–167.

114 das Neves, J., Amaral, M.H., and Bahia, M.F. (2008). Performance of an in vitro mucoadhesion testing method for vaginal semisolids: influence of different testing conditions and instrumental parameters. *Eur. J. Pharm. Biopharm.* 69 (2): 622–632.

115 Cunha, A.R., Machado, R.M., Palmeira-de-Oliveira, A. et al. (2014). Characterization of commercially available vaginal lubricants: a safety perspective. *Pharmaceutics* 6 (3): 530–542.

116 Dezzutti, C.S., Brown, E.R., Moncla, B. et al. (2012). Is wetter better? An evaluation of over-the-counter personal lubricants for safety and anti-HIV-1 activity. *PLoS One* 7 (11): e48328.

117 Schwartz, J.L., Ballagh, S.A., Kwok, C. et al. (2007). Fourteen-day safety and acceptability study of the universal placebo gel. *Contraception* 75 (2): 136–141.

118 Tien, D., Schnaare, R.L., Kang, F. et al. (2005). In vitro and in vivo characterization of a potential universal placebo designed for use in vaginal microbicide clinical trials. *AIDS Res. Hum. Retroviruses* 21 (10): 845–853.

119 Hoare, T.R. and Kohane, D.S. (2008). Hydrogels in drug delivery: progress and challenges. *Polymer* 49 (8): 1993–2007.

120 Chang, J.Y., Oh, Y.K., Kong, H.S. et al. (2002). Prolonged antifungal effects of clotrimazole-containing mucoadhesive thermosensitive gels on vaginitis. *J. Controlled Release* 82 (1): 39–50.

121 Pavelić, Ž., Škalko-Basnet, N., Filipovic-Grcic, J. et al. (2005). Development and in vitro evaluation of a liposomal vaginal delivery system for acyclovir. *J. Controlled Release* 106 (1–2): 34–43.

122 Li, N., Yu, M., Deng, L. et al. (2012). Thermosensitive hydrogel of hydrophobically-modified methylcellulose for intravaginal drug delivery. *J. Mater. Sci. - Mater. Med.* 23 (8): 1913–1919.

123 Karimunnisa, S. and Atmaram, P. (2013). Mucoadhesive nanoliposomal formulation for vaginal delivery of an antifungal. *Drug Dev. Ind. Pharm.* 39 (9): 1328–1337.

124 Kim, Y.T., Shin, B.K., Garripelli, V.K. et al. (2010). A thermosensitive vaginal gel formulation with HPgammaCD for the pH-dependent release and solubilization of amphotericin B. *Eur. J. Pharm. Sci.* 41 (2): 399–406.

125 van der Linden, M., Buckingham, K., Farquhar, C. et al. (2015). Luteal phase support for assisted reproduction cycles. *Cochrane Database Syst. Rev.* (7): CD009154. https://doi.org/10.1002/14651858.CD009154.pub3.

126 US Food and Drug Administration. Crinone. Clinical Pharmacology and Biopharmaceutics Review. https://www.accessdata.fda.gov/drugsatfda_docs/nda/97/20756_CRINONE_BIOPHARMR.PDF (accessed 22 June 2017).

127 Forbes, C.J., Lowry, D., Geer, L. et al. (2011). Non-aqueous silicone elastomer gels as a vaginal microbicide delivery system for the HIV-1 entry inhibitor maraviroc. *J. Controlled Release* 156 (2): 161–169.

128 Forbes, C.J., McCoy, C.F., Murphy, D.J. et al. (2014). Modified silicone elastomer vaginal gels for sustained release of antiretroviral HIV microbicides. *J. Pharm. Sci.* 103 (5): 1422–1432.

129 Khanna, N., Dalby, R., Connor, A. et al. (2008). Phase I clinical trial of repeat dose terameprocol vaginal ointment in healthy female volunteers. *Sex. Transm. Dis.* 35 (6): 577–582.

130 Roberts, C.L. and Sullivan, J.J. (1965). Moniliasis – the use of candicidin vaginal ointment for treatment. *Calif. Med.* 103 (2): 109–111.

131 Kassem, A.A., Abd Elbary, A., and Nour, S.A. (1977). Effect of vehicles on the spermicidal potency of certain drugs. *Pharmazie* 32 (7): 403–405.

132 Cool-Foley, A.A., Nathan, C., O'Donovan, C. III, and Simon, D. (1991). Eradication of methicillin-resistant *Staphylococcus aureus* vaginitis with mupirocin. *DICP* 25 (12): 1331–1333.

133 Gali, Y., Delezay, O., Brouwers, J. et al. (2010). In vitro evaluation of viability, integrity and inflammation in genital epithelia upon exposure to pharmaceutical excipients and candidate microbicides. *Antimicrob. Agents Chemother.* 54 (12): 5105–5114.

134 Martino, J.L., Youngpairoj, S., and Vermund, S.H. (2004). Vaginal douching: personal practices and public policies. *J. Women's Health* 13 (9): 1048–1065.

135 Bernstein, G.S. (1974). Conventional methods of contraception: condom, diaphragm, and vaginal foam. *Clin. Obstet. Gynecol.* 17 (1): 21–33.

136 Arzhavitina, A. and Steckel, H. (2010). Foams for pharmaceutical and cosmetic application. *Int. J. Pharm.* 394 (1–2): 1–17.

137 Sciarra, J.J. and Sciarra, C.J. (2000). Aerosol technology. In: *Kirk-Othmer Encyclopedia of Chemical Technology* (eds. R.E. Kirk and D.F. Othmer). Hoboken, NJ: Wiley.

138 Moss, J.A. and Baum, M.M. (2014). Microbicide intravaginal rings. In: *Drug Delivery and Development of Anti-HIV Microbicides* (eds. J. das Neves and B. Sarmento), 221–290. Singapore: Pan Stanford.

139 Brache, V., Payán, L.J., and Faundes, A. (2013). Current status of contraceptive vaginal rings. *Contraception* 87 (3): 264–272.

140 Malcolm, R.K., Boyd, P.J., McCoy, C.F., and Murphy, D.J. (2016). Microbicide vaginal rings: technological challenges and clinical development. *Adv. Drug Delivery Rev.* 103: 33–56.

141 Kiser, P.F., Johnson, T.J., and Clark, J.T. (2012). State of the art in intravaginal ring technology for topical prophylaxis of HIV infection. *AIDS Rev.* 14 (1): 62–77.

142 Mishell, D.R. Jr., Talas, M., Parlow, A.F., and Moyer, D.L. (1970). Contraception by means of a silastic vaginal ring impregnated with medroxyprogesterone acetate. *Am. J. Obstet. Gynecol.* 107 (1): 100–107.

143 NuvaRing® (revised 2017). Etonogestrel/Ethinyl Estradiol Vaginal Ring. Prescribing Information, Whitehouse Station, NJ: Merck Sharp & Dohme Corp.

144 Timmer, C.J. and Mulders, T.M. (2000). Pharmacokinetics of etonogestrel and ethinylestradiol released from a combined contraceptive vaginal ring. *Clin. Pharmacokinet.* 39 (3): 233–242.

145 van den Heuvel, M.W., van Bragt, A.J., Alnabawy, A.K., and Kaptein, M.C. (2005). Comparison of ethinylestradiol pharmacokinetics in three hormonal contraceptive formulations: the vaginal ring, the transdermal patch and an oral contraceptive. *Contraception* 72 (3): 168–174.

146 Oddsson, K., Leifels-Fischer, B., de Melo, N.R. et al. (2005). Efficacy and safety of a contraceptive vaginal ring (NuvaRing) compared with a combined oral contraceptive: a 1-year randomized trial. *Contraception* 71 (3): 176–182.

147 Ahrendt, H.J., Nisand, I., Bastianelli, C. et al. (2006). Efficacy, acceptability and tolerability of the combined contraceptive ring, NuvaRing, compared with an oral contraceptive containing 30 microg of ethinyl estradiol and 3 mg of drospirenone. *Contraception* 74 (6): 451–457.

148 das Neves, J., Martins, J.P., and Sarmento, B. (2016). Will dapivirine redeem the promises of anti-HIV microbicides? Overview of product design and clinical testing. *Adv. Drug Delivery Rev.* 103: 20–32.

149 Baeten, J.M., Palanee-Phillips, T., Brown, E.R. et al. (2016). Use of a vaginal ring containing dapivirine for HIV-1 prevention in women. *N. Engl. J. Med.* 375 (22): 2121–2132.

150 Nel, A., van Niekerk, N., Kapiga, S. et al. (2016). Safety and efficacy of a dapivirine vaginal ring for HIV prevention in women. *N. Engl. J. Med.* 375 (22): 2133–2143.

151 Devlin, B., Nuttall, J., Wilder, S. et al. (2013). Development of dapivirine vaginal ring for HIV prevention. *Antiviral Res.* 100 (Suppl): S3–S8.

152 Nel, A., Haazen, W., Nuttall, J. et al. (2014). A safety and pharmacokinetic trial assessing delivery of dapivirine from a vaginal ring in healthy women. *AIDS* 28 (10): 1479–1487.

153 Keown, K.K. Jr. (1977). Historical perspectives on intravaginal contraceptive sponges. *Contraception* 16 (1): 1–10.

154 Kuyoh, M.A., Toroitich-Ruto, C., Grimes, D.A. et al. (2003). Sponge versus diaphragm for contraception: a Cochrane review. *Contraception* 67 (1): 15–18.

155 Major, I., Boyd, P., Kilbourne-Brook, M. et al. (2013). A modified SILCS contraceptive diaphragm for long-term controlled release of the HIV microbicide dapivirine. *Contraception* 88 (1): 58–66.

156 Brzezinski, A., Stern, T., Arbel, R. et al. (2004). Efficacy of a novel pH-buffering tampon in preserving the acidic vaginal pH during menstruation. *Int. J. Gynaecol. Obstet.* 85 (3): 298–300.

157 Cervidil® (revised 2016). Dinoprostone Vaginal Insert. Prescribing Information, Parsippany, NJ, USA.

158 Halliday, J.A. and Robertson, S. (2008). Clidamycin vaginal insert. In: *Modified-Release Drug Delivery Technology*, 2e (eds. M.J. Rathbone, J. Hadgraft, M.S. Roberts and M.E. Lane), 531–537. Boca Raton, FL: CRC Press.

159 Albertini, B., Passerini, N., Di Sabatino, M. et al. (2009). Polymer-lipid based mucoadhesive microspheres prepared by spray-congealing for the vaginal delivery of econazole nitrate. *Eur. J. Pharm. Sci.* 36 (4–5): 591–601.

160 Chatterjee, A., Kumar, L., Bhowmik, B.B., and Gupta, A. (2011). Microparticulated anti-HIV vaginal gel: in vitro-in vivo drug release and vaginal irritation study. *Pharm. Dev. Technol.* 16 (5): 466–473.

161 Martín-Villena, M.J., Fernández-Campos, F., Calpena-Campmany, A.C. et al. (2013). Novel microparticulate systems for the vaginal delivery of nystatin: development and characterization. *Carbohydr. Polym.* 94 (1): 1–11.

162 das Neves, J., Nunes, R., Machado, A., and Sarmento, B. (2015). Polymer-based nanocarriers for vaginal drug delivery. *Adv. Drug Delivery Rev.* 92: 53–70.

163 das Neves, J., Amiji, M.M., Bahia, M.F., and Sarmento, B. (2010). Nanotechnology-based systems for the treatment and prevention of HIV/AIDS. *Adv. Drug Delivery Rev.* 62 (4–5): 458–477.

164 Santos, S.S., Lorenzoni, A., Pegoraro, N.S. et al. (2014). Formulation and in vitro evaluation of coconut oil-core cationic nanocapsules intended for vaginal delivery of clotrimazole. *Colloids Surf., B* 116C: 270–276.

165 Yang, M., Yu, T., Wang, Y.Y. et al. (2014). Vaginal delivery of paclitaxel via nanoparticles with non-mucoadhesive surfaces suppresses cervical tumor growth. *Adv. Healthcare Mater.* 3 (7): 1044–1052.

166 Akagi, T., Kawamura, M., Ueno, M. et al. (2003). Mucosal immunization with inactivated HIV-1-capturing nanospheres induces a significant HIV-1-specific vaginal antibody response in mice. *J. Med. Virol.* 69 (2): 163–172.

167 Merisko-Liversidge, E. and Liversidge, G.G. (2011). Nanosizing for oral and parenteral drug delivery: a perspective on formulating poorly-water soluble compounds using wet media milling technology. *Adv. Drug Delivery Rev.* 63 (6): 427–440.

168 Fonte, P., Soares, S., Sousa, F. et al. (2014). Stability study perspective of the effect of freeze-drying using cryoprotectants on the structure of insulin loaded into PLGA nanoparticles. *Biomacromolecules* 15 (10): 3753–3765.

169 Wang, Y.Y., Lai, S.K., Suk, J.S. et al. (2008). Addressing the PEG mucoadhesivity paradox to engineer nanoparticles that "slip" through the human mucus barrier. *Angew. Chem. Int. Ed.* 47 (50): 9726–9729.

170 Xu, Q., Ensign, L.M., Boylan, N.J. et al. (2015). Impact of surface polyethylene glycol (PEG) density on biodegradable nanoparticle transport in mucus ex vivo and distribution in vivo. *ACS Nano* 9 (9): 9217–9227.

171 Yang, M., Lai, S.K., Yu, T. et al. (2014). Nanoparticle penetration of human cervicovaginal mucus: the effect of polyvinyl alcohol. *J. Controlled Release* 192C: 202–208.

172 Mansfield, E.D., Sillence, K., Hole, P. et al. (2015). POZylation: a new approach to enhance nanoparticle diffusion through mucosal barriers. *Nanoscale* 7 (32): 13671–13679.

173 Ensign, L.M., Tang, B.C., Wang, Y.Y. et al. (2012). Mucus-penetrating nanoparticles for vaginal drug delivery protect against herpes simplex virus. *Sci. Transl. Med.* 4 (138): 138ra179.

174 Woodrow, K.A., Cu, Y., Booth, C.J. et al. (2009). Intravaginal gene silencing using biodegradable polymer nanoparticles densely loaded with small-interfering RNA. *Nat. Mater.* 8 (6): 526–533.

175 Ramanathan, R., Park, J., Hughes, S.M. et al. (2015). Effect of mucosal cytokine administration on selective expansion of vaginal dendritic cells to support nanoparticle transport. *Am. J. Reprod. Immunol.* 74 (4): 333–344.

176 Park, J., Ramanathan, R., Pham, L., and Woodrow, K.A. (2017). Chitosan enhances nanoparticle delivery from the reproductive tract to target draining lymphoid organs. *Nanomed. Nanotechnol. Biol. Med.* 13 (6): 2015–2025.

177 das Neves, J., Araújo, F., Andrade, F. et al. (2014). Biodistribution and pharmacokinetics of dapivirine-loaded nanoparticles after vaginal delivery in mice. *Pharm. Res.* 31 (7): 1834–1845.

178 Malik, R., Maikhuri, J.P., Gupta, G., and Misra, A. (2011). Biodegradable nanoparticles in the murine vagina: trans-cervical retrograde transport and induction of proinflammatory cytokines. *J. Biomed. Nanotechnol.* 7 (1): 45–46.

179 das Neves, J., Michiels, J., Ariën, K.K. et al. (2012). Polymeric nanoparticles affect the intracellular delivery, antiretroviral activity and cytotoxicity of the microbicide drug candidate dapivirine. *Pharm. Res.* 29 (6): 1468–1484.

180 das Neves, J., Araújo, F., Andrade, F. et al. (2013). In vitro and ex vivo evaluation of polymeric nanoparticles for vaginal and rectal delivery of the anti-HIV drug dapivirine. *Mol. Pharm.* 10 (7): 2793–2807.

181 Meng, J., Zhang, T., Agrahari, V. et al. (2014). Comparative biophysical properties of tenofovir-loaded, thiolated and nonthiolated chitosan nanoparticles intended for HIV prevention. *Nanomedicine* 9 (11): 1595–1612.

182 Cunha-Reis, C., Machado, A., Barreiros, L. et al. (2016). Nanoparticles-in-film for the combined vaginal delivery of anti-HIV microbicide drugs. *J. Controlled Release* 243: 43–53.

183 Blakney, A.K., Krogstad, E.A., Jiang, Y.H., and Woodrow, K.A. (2014). Delivery of multipurpose prevention drug combinations from electrospun nanofibers using composite microarchitectures. *Int. J. Nanomed.* 9: 2967–2978.

184 Krogstad, E.A. and Woodrow, K.A. (2014). Manufacturing scale-up of electrospun poly(vinyl alcohol) fibers containing tenofovir for vaginal drug delivery. *Int. J. Pharm.* 475 (1–2): 282–291.

185 Huang, C., Soenen, S.J., van Gulck, E. et al. (2012). Electrospun cellulose acetate phthalate fibers for semen induced anti-HIV vaginal drug delivery. *Biomaterials* 33 (3): 962–969.

186 Ball, C., Chou, S.F., Jiang, Y., and Woodrow, K.A. (2016). Coaxially electrospun fiber-based microbicides facilitate broadly tunable release of maraviroc. *Mater. Sci. Eng., C* 63: 117–124.

187 (2000). Moraxen suppositories. *Pharm. J.* 265 (7113): 378.

188 Kanke, M., Tanabe, E., Katayama, H. et al. (1995). Application of curdlan to controlled drug delivery. III. Drug release from sustained release suppositories in vitro. *Biol. Pharm. Bull.* 18 (8): 1154–1158.

189 Uzunkaya, G. and Bergisadi, N. (2003). In vitro drug liberation and kinetics of sustained release indomethacin suppository. *Farmaco* 58 (7): 509–512.

190 Moolenaar, F., Meijler, W.J., Frijlink, H.W. et al. (2000). Clinical efficacy, safety and pharmacokinetics of a newly developed controlled release morphine sulphate suppository in patients with cancer pain. *Eur. J. Clin. Pharmacol.* 56 (3): 219–223.

191 Schneeweis, A. and Müller-Goymann, C.C. (2000). Controlled release of solid-reversed-micellar-solution (SRMS) suppositories containing metoclopramide-HCl. *Int. J. Pharm.* 196 (2): 193–196.

192 Takatori, T., Shimono, N., Higaki, K., and Kimura, T. (2004). Evaluation of sustained release suppositories prepared with fatty base including solid fats with high melting points. *Int. J. Pharm.* 278 (2): 275–282.

193 Ozgüney, I., Ozcan, I., Ertan, G., and Güneri, T. (2007). The preparation and evaluation of sustained release suppositories containing ketoprofen and Eudragit RL 100 by using factorial design. *Pharm. Dev. Technol.* 12 (1): 97–107.

194 Azechi, Y., Ishikawa, K., Mizuno, N., and Takahashi, K. (2000). Sustained release of diclofenac from polymer-containing suppository and the mechanism involved. *Drug Dev. Ind. Pharm.* 26 (11): 1177–1183.

195 Yahagi, R., Machida, Y., and Onishi, H. (2000). Mucoadhesive suppositories of ramosetron hydrochloride utilizing Carbopol. *Int. J. Pharm.* 193 (2): 205–212.

196 Realdon, N., Ragazzi, E., Dal Zotto, M., and Dalla Fini, G. (1997). Layered excipient suppositories: the possibility of modulating drug availability. *Int. J. Pharm.* 148 (2): 155–163.

197 Chicco, D., Grabnar, I., Škerjanec, A. et al. (1999). Correlation of in vitro and in vivo paracetamol availability from layered excipient suppositories. *Int. J. Pharm.* 189 (2): 147–160.

198 Barnes, K.I., Mwenechanya, J., Tembo, M. et al. (2004). Efficacy of rectal artesunate compared with parenteral quinine in initial treatment of moderately severe malaria in African children and adults: a randomised study. *Lancet* 363 (9421): 1598–1605.

199 Gomez, E.A., Jurado, M.H., and Cambon, N. (2003). Randomised efficacy and safety study of two 3-day artesunate rectal capsule/mefloquine regimens versus artesunate alone for uncomplicated malaria in Ecuadorian children. *Acta Trop.* 89 (1): 47–53.

200 de Leede, L.G., de Boer, A.G., and Breimer, D.D. (1981). Rectal infusion of the model drug antipyrine with an osmotic delivery system. *Biopharm. Drug Dispos.* 2 (2): 131–136.

201 de Leede, L.G., de Boer, A.G., van Velzen, S.L., and Breimer, D.D. (1982). Zero-order rectal delivery of theophylline in man with an osmotic system. *J. Pharmacokinet. Biopharm.* 10 (5): 525–537.

202 Javanbakht, M., Murphy, R., Gorbach, P. et al. (2010). Preference and practices relating to lubricant use during anal intercourse: implications for rectal microbicides. *Sex. Health* 7 (2): 193–198.

203 Goto, S., Kawata, M., Suzuki, T. et al. (1991). Preparation and evaluation of Eudragit gels. I: Eudragit organogels containing drugs as rectal sustained-release preparations. *J. Pharm. Sci.* 80 (10): 958–961.

204 Miyazaki, S., Suisha, F., Kawasaki, N. et al. (1998). Thermally reversible xyloglucan gels as vehicles for rectal drug delivery. *J. Controlled Release* 56 (1–3): 75–83.

205 Fawaz, F., Koffi, A., Guyot, M., and Millet, P. (2004). Comparative in vitro-in vivo study of two quinine rectal gel formulations. *Int. J. Pharm.* 280 (1–2): 151–162.

206 Koffi, A.A., Agnely, F., Besnard, M. et al. (2008). In vitro and in vivo characteristics of a thermogelling and bioadhesive delivery system intended for rectal administration of quinine in children. *Eur. J. Pharm. Biopharm.* 69 (1): 167–175.

207 Centkowska, K. and Sznitowska, M. (2013). Modified-release ointment with nitroglycerin beta-cyclodextrin inclusion complex for treatment of anal fissures. *J. Pharm. Pharmacol.* 65 (10): 1463–1472.

208 Liu, Y., Wang, X., Liu, Y., and Di, X. (2017). Thermosensitive in situ gel based on solid dispersion for rectal delivery of ibuprofen. *AAPS PharmSciTech* https://doi.org/10.1208/s12249-017-0839-5.

209 Brown, J., Haines, S., and Wilding, I.R. (1997). Colonic spread of three rectally administered mesalazine (Pentasa) dosage forms in healthy volunteers as assessed by gamma scintigraphy. *Aliment. Pharmacol. Ther.* 11 (4): 685–691.

210 Van Os, E.C., Zins, B.J., Sandborn, W.J. et al. (1996). Azathioprine pharmacokinetics after intravenous, oral, delayed release oral and rectal foam administration. *Gut* 39 (1): 63–68.

211 Sznitowska, M., Gajewska, M., Janicki, S. et al. (2001). Bioavailability of diazepam from aqueous-organic solution, submicron emulsion and solid lipid nanoparticles after rectal administration in rabbits. *Eur. J. Pharm. Biopharm.* 52 (2): 159–163.

212 Lamprecht, A., Yamamoto, H., Takeuchi, H., and Kawashima, Y. (2005). Nanoparticles enhance therapeutic efficiency by selectively increased local drug dose in experimental colitis in rats. *J. Pharmacol. Exp. Ther.* 315 (1): 196–202.

213 Seo, Y.G., Kim, D.W., Yeo, W.H. et al. (2013). Docetaxel-loaded thermosensitive and bioadhesive nanomicelles as a rectal drug delivery system for enhanced chemotherapeutic effect. *Pharm. Res.* 30 (7): 1860–1870.

214 Mohamed, R.A., Abass, H.A., Attia, M.A., and Heikal, O.A. (2013). Formulation and evaluation of metoclopramide solid lipid nanoparticles for rectal suppository. *J. Pharm. Pharmacol.* 65 (11): 1607–1621.

215 Kim, D.W., Ramasamy, T., Choi, J.Y. et al. (2014). The influence of bile salt on the chemotherapeutic response of docetaxel-loaded thermosensitive nanomicelles. *Int. J. Nanomed.* 9: 3815–3824.

216 Sarmento, B. and das Neves, J. (2012). Nanosystem formulations for rectal microbicides: a call for more research. *Ther. Delivery* 3 (1): 1–4.

217 Maisel, K., Ensign, L., Reddy, M. et al. (2015). Effect of surface chemistry on nanoparticle interaction with gastrointestinal mucus and distribution in the gastrointestinal tract following oral and rectal administration in the mouse. *J. Controlled Release* 197: 48–57.

218 Maisel, K., Chattopadhyay, S., Moench, T. et al. (2015). Enema ion compositions for enhancing colorectal drug delivery. *J. Controlled Release* 209: 280–287.

219 Gajer, P., Brotman, R.M., Bai, G. et al. (2012). Temporal dynamics of the human vaginal microbiota. *Sci. Transl. Med.* 4 (132): 132ra152.

220 Nicol, M.R., Fedoriw, Y., Mathews, M. et al. (2013). Expression of six drug transporters in vaginal, cervical, and colorectal tissues: implications for drug disposition in HIV prevention. *J. Clin. Pharm.* 54 (5): 574–583.

221 Mukhopadhya, I., Murray, G.I., Berry, S. et al. (2016). Drug transporter gene expression in human colorectal tissue and cell lines: modulation with antiretrovirals for microbicide optimization. *J. Antimicrob. Chemother.* 71 (2): 372–386.

222 Fernández-Romero, J.A., Teleshova, N., Zydowsky, T.M., and Robbiani, M. (2015). Preclinical assessments of vaginal microbicide candidate safety and efficacy. *Adv. Drug Delivery Rev.* 92: 27–38.

223 McGowan, I. (2014). The development of rectal microbicides for HIV prevention. *Expert Opin. Drug Delivery* 11 (1): 69–82.

224 Tibbitt, M.W., Dahlman, J.E., and Langer, R. (2016). Emerging frontiers in drug delivery. *J. Am. Chem. Soc.* 138 (3): 704–717.

20

Controlled Drug Delivery into and Through Skin

Adrian Williams

University of Reading, Reading, UK

20.1 Introduction

Human skin is a complex and multiple-layered organ with three prime functions. Firstly, skin is a key sense organ and possesses numerous nerve endings and receptors that respond to pain, pressure, temperature, and touch. Secondly, the skin is also a regulatory organ that responds to temperature changes through vasodilation or vasoconstriction or by the production of sweat to cool the skin surface through evaporation. Additionally, arrector pili muscles that are attached to hair follicles can contract in cold environments causing the hairs to stand on end; more significantly seen in highly haired mammals, this response can provide insulation as air becomes trapped between the erect hairs, commonly referred to as "goose bumps." Finally, the skin provides a highly effective self-repairing barrier to provide protection from physical, chemical, microbiological, and radiative assault.

Given the barrier function of this organ, in terms of drug delivery it is typically the skin itself that regulates or "controls" the dose that is absorbed into or that passes through the membrane. Consequently, formulations are seldom developed to restrict drug delivery into the skin – though some topically applied systems are designed to reside within the skin rather than diffuse through the tissue, such as sunscreens. More commonly, pharmaceutical formulations aim to increase topical and transdermal delivery, for example by employing permeation enhancers or manipulating the thermodynamic activity and hence driving force for drug delivery. These approaches differ from controlled drug delivery systems for other administration routes where release of the active pharmaceutical ingredient is controlled by the formulation *per se*. This chapter thus considers strategies that have been used to modify the barrier functions of the skin to increase drug delivery in addition to approaches where controlled drug delivery formulations (with control residing in the formulation) have been employed.

20.1.1 Human Skin Structure and Function

20.1.1.1 Biological Factors

Skin is the largest organ of the human body providing around 10% of the body mass of an average person and an area of approximately 1.8 m^2. The tissue can be considered at various levels of complexity. Simplistically, skin is comprised of three major layers, the outermost epidermis which overlays the dermis and the innermost subcutis.

Fundamentals of Drug Delivery, First Edition.
Edited by Heather A. E. Benson, Michael S. Roberts, Adrian C. Williams, and Xiaowen Liang.

The epidermis itself contains distinct layers as cells differentiate and migrate outward. The epidermis is predominantly composed of keratinocytes (~95% of the cells in this layer) which, following terminal differentiation, form the corneocytes of the outermost skin layer – the stratum corneum. Other cell types in the epidermis are melanocytes, located at the basal layer of the epidermis and which produce the pigment melanin; this is "packaged" into small organelles (melanosomes) that are transferred along dendrites to the keratinocytes. The epidermis (but not the stratum corneum) also contains Langerhans cells, antigen-presenting cells that, in essence, break down allergens and carry them into the dermis from where they are taken into the lymphatic and circulatory systems where they are presented to lymphocytes to initiate an immune response. Merkel cells are also found in the basal layer of the epidermis and are sensory cells (mechanoreceptors) responsible for light touch sensation, and are more abundant in sensitive skin sites such as the fingertips.

The dermis also contains numerous cell types embedded in a matrix of collagen and elastin fibers which are bound together in a mucopolysaccharide gel. The dermis contains nerves and blood vessels as well as hair follicles (and nails), apocrine, and eccrine glands. In addition to neurons and vascular smooth muscle cells, the dermis contains fibroblasts which are the cells that generate and deposit collagen and other materials important for skin integrity and wound repair. The dermis also contains specialized muscle cells and mast cells that release, among other mediators, histamine, when damaged and so are important in inflammatory responses. Various inflammatory cells are also transiently present, responding to assault such as wounds, infections, or diseases, and include T and B lymphocytes, eosinophils, and monocytes (that form macrophages); signaling cascades between these cells involves cytokines and other chemotactic molecules.

This physiological complexity is often poorly considered when designing topical and transdermal drug delivery systems. Given the above, formulations clearly have the potential to cause adverse events such as irritation or allergic responses. Further, other than the outermost stratum corneum, skin is metabolically active with varied enzyme systems including esterases. Though present at significantly lower levels than in the liver (for first pass metabolism), drug biotransformations can occur during diffusion through the skin [1]. Such biotransformations have been exploited for topical drug delivery via pro-drugs; betamethasone 17-valerate penetrates the stratum corneum more readily than betamethasone, and the ester bond is subsequently rapidly cleaved by esterases to yield betamethasone [2]. Similarly, ester pro-drugs of the poorly permeating naltrexone showed greater permeation through human stratum corneum than the drug itself, with nearly 100% hydrolysis of the prodrug in the epidermis [3] and, following topical application, metabolism of methylsalicylate to salicylate was shown *in vivo* [4]. However, metabolic activity within the skin can also reduce the bioavailability of topically applied drugs and proteomic profiling showed that human skin contains at least 36 different xenobiotic metabolizing enzymes with both Phase 1 (with dehydrogenation, oxidation, reduction, and hydrolysis) and Phase 2 (conjugations) transformations [5].

20.1.1.2 Skin as a Physical Barrier

Beyond the biological factors described above, the layers of human skin provide a physical barrier to drug delivery. Many studies developing novel controlled drug delivery systems for topical application initially use excised (*ex vivo*) human skin. In these cases, metabolic and immunological activity is typically lost and hence the skin functions as a physical barrier, though clearly testing *in vivo* is necessary to determine safety and efficacy. Even so, the various skin layers can be considered as providing barriers in parallel and, if drug transport through the pores in the tissue is included, then we also introduce barriers in series (Figure 20.1).

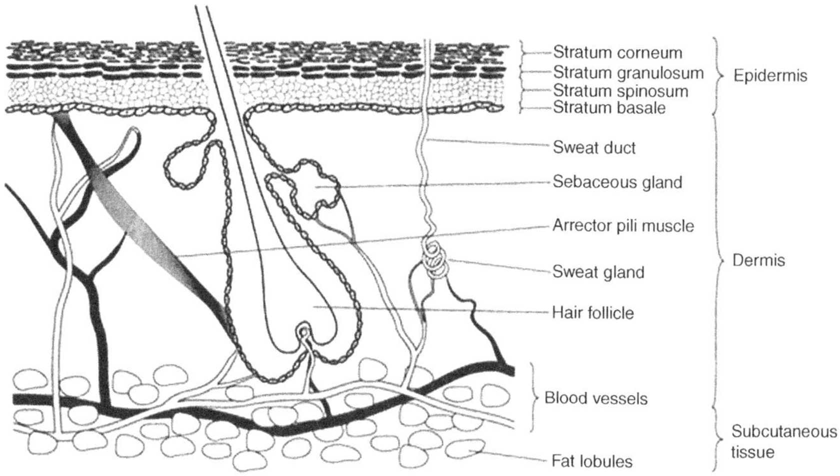

Figure 20.1 Diagrammatical representation of the structure of human skin. Source: From Reference [6], with permission.

20.1.1.2.1 The Subcutis

The subcutis, (also termed the subcutaneous layer, the fatty layer, or hypodermis) is the inner most layer that bridges between the overlying dermis and the underlying body constituents. In most areas of the body, the subcutis is relatively thick, typically in the order of several millimeters. However, there are some areas of the body where the subcutaneous layer is largely absent, such as the eyelids. The subcutis is essentially a layer of adipose tissue that primarily serves to insulate the body and to provide mechanical protection against physical shock. The subcutis also carries the major blood vessels and nerves to the skin. However, in terms of topical drug delivery, the subcutis is generally discounted as a major barrier to systemic and topical drug delivery though for formulations aiming to deliver drugs beneath the skin, such as locally acting anti-inflammatory agents for muscles and joints, then the subcutis needs to be traversed for action at the target site.

20.1.1.2.2 The Dermis

The dermis is the major component of human skin and is typically between 3 and 5 mm thick. As described above, this layer contains numerous cell types embedded in a matrix of collagen and elastin fibers which are bound together in a mucopolysaccharide gel. In terms of drug diffusion, the dermis is often viewed as essentially a hydrogel which provides minimal resistance to the transport of most polar drugs, although the dermal barrier may be significant when delivering highly lipophilic molecules. The dermis has numerous structures embedded within it; blood and lymphatic vessels, nerve endings, pilosebaceous units (hair follicles and sebaceous glands), and sweat glands (eccrine and apocrine).

The extensive vasculature of the skin regulates body temperature while delivering oxygen and nutrients to the tissue and removing toxins and waste products. The rich blood flow, around 0.05 ml/min/mg of skin, efficiently removes molecules from the dermis–epidermis interface which ensures that dermal concentrations of most permeants are very low. Indeed it is the removal of permeants by the vasculature that primarily maintains the concentration gradient (or more accurately the thermodynamic activity gradient) across the skin from the applied formulation and which provides the driving force for permeation. The lymphatic system also reaches to the dermal–epidermal junction. While this is important in regulating interstitial pressure, for facilitating immunological responses to microbial assault and for waste removal, the lymphatic vessels may also remove permeated molecules from the dermis. Dermal blood flow tends to predominate in the clearance of most small molecule permeants from the skin, but lymphatic flow can be significant for the clearance of larger molecules such as interferon [7].

Three main appendages originate in the dermis and extend to the skin surface. Hair follicles are distributed over the skin surface except for the load-bearing areas (soles of feet, palms of hands) and the lips. Sebaceous glands, associated with the hair follicle and located in the dermis, secrete sebum which is a mixture of free fatty acids, waxes, and triglycerides and which lubricates the skin surface. Eccrine (or sweat) glands and apocrine glands also originate in the dermis. Eccrine glands are found over most of the body surface and secrete sweat, a dilute salt solution at ~pH 5 in response to heat and emotional stress. Apocrine glands are limited to specific areas of the skin including the axillae, nipples, and ano-genital regions; their lipoidal and "milk" protein secretions are primarily responsible for imparting the odor of "sweat." With topical formulations, these skin appendages (hair follicles, sweat ducts) may offer a potential route ("shunt route") by which molecules could enter the lower layers of the skin without having to traverse the barrier provided by the stratum corneum. At early times post-application, for relatively large molecular weight or polar permeants and for some assistive technology delivery methods (such as iontophoresis), the shunt routes can be a significant portal for drug delivery. However, the fractional area offered by the appendages is relatively low and for most permeants at steady-state flux, the predominant pathway for molecules to enter the tissue is through the bulk skin surface.

20.1.1.2.3 The Epidermis

The epidermis is itself a multiple-layered membrane which varies from around 0.06 mm thick on the eyelids to around 0.8 mm on the palms and soles of the feet. The epidermis contains no blood vessels and hence nutrients and waste products must diffuse across the dermal–epidermal boundary to maintain tissue integrity. Likewise, molecules permeating across the epidermis must cross this boundary in order to be cleared into the systemic circulation. The epidermis contains four histologically distinct layers which, from the inside to the outside, are the stratum basale (also termed the stratum germinativum), stratum spinosum, stratum granulosum, and the stratum corneum (Figure 20.2). At load bearing sites of the body – palms, soles, and digits – a further distinct layer of cells, the stratum lucidum, is seen between the granulosum and corneum strata; the stratum lucidum keratinocytes are packed with eleiden, a clear protein rich in lipids giving the "lucid" (transparent) appearance and acting as a further barrier to water.

The stratum basale is a single layer of cells (predominantly basal cells but with some Merkel cells and melanocytes) that are bound to the dermis through collagen fibers, termed the basement membrane. The basal cells are essentially stem cells that generate the keratinocytes in the subsequent strata; keratinocytes are produced from the basal cells through continual mitosis. As new cells form, the existing cells are "pushed" away from the stratum basale and begin to differentiate as they migrate toward the skin surface.

The stratum spinosum is composed of eight to 10 layers of keratinocytes where synthesis of keratin begins. This layer is named after the "spiny" appearance noted on staining which arises from the formation of desmosomes. Desmosomes are cell–cell adhesion proteins that help to bind the keratinocytes together. The keratinocytes also begin to generate lamellar bodies, enriched with lipids, hydrolytic enzymes, and some proteins. As further keratinocytes are produced by the stratum basale, the stratum spinosum keratinocytes migrate further outward to form the stratum granulosum.

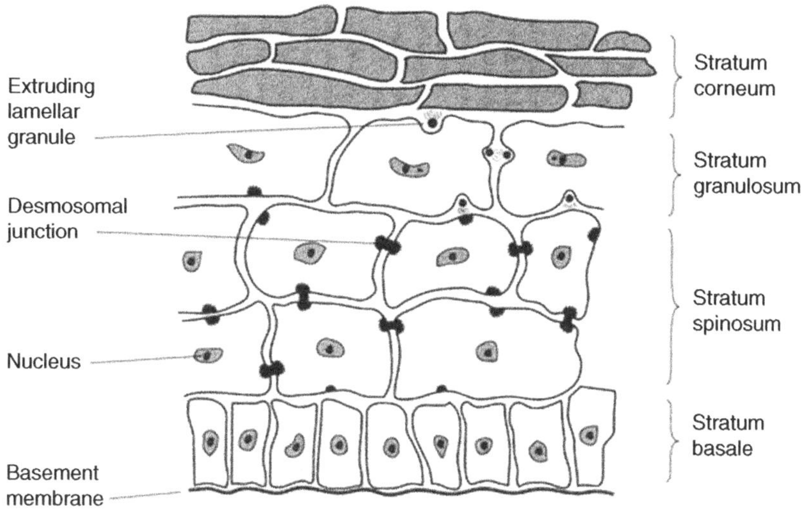

Figure 20.2 A representation of human epidermal cell layers and differentiation. Source: From Reference [6], with permission.

The keratinocytes of the stratum granulosum appear "grainy" as differentiation continues. The cells (three to five layers deep) flatten and the cell membranes thicken as they generate large amounts of keratin and keratohyalin. Lipids within the lamellar bodies are released into the extracellular space and subsequently convert into nonpolar lipids; lipid conversion and reorganization generates the multiple bilayer lipids of the stratum corneum (described below). The keratinocytes lose their nuclei and organelles as they die and fill with keratin.

The stratum corneum is the outermost layer of the epidermis and generally comprises 15–20 layers of "cornified" cells embedded in a multiple bilayered lipid domain. It is this final skin layer that is predominantly responsible for the barrier properties of human skin and which limits drug delivery into and across the skin. As an approximation, one layer of stratum corneum is shed each day and it takes approximately two weeks for a cell to migrate from the stratum basale to the lower stratum corneum, and a further two weeks to transit through the stratum corneum; i.e. skin is replaced every four weeks.

20.1.1.2.4 Stratum Corneum

Given its role as the principle barrier to transdermal and topical delivery of most drugs, the structure of the stratum corneum merits further consideration. Though the stratum corneum is the principle barrier to drug delivery, and regulates water loss from the body and prevents ingress of toxins, environmental contaminants, and microorganisms, it is a remarkably thin barrier around 10 μm deep when dry, although it may swell considerably when wet. As with the underlying epidermal cells, the stratum corneum is thickest on the palms and soles and is thinnest on the lips. The barrier nature of the stratum corneum depends critically on its unique constituents. The terminally differentiated keratinocytes (corneocytes) are filled predominantly with α-keratin (~70%) with β-keratin (~10%) and some enzymes and proteins. The cell envelope protein is also highly insoluble and resistant to chemical attack. The extruded lipids are also unique; while the lipid content of the stratum corneum varies between individuals and with body site, the major components of the domain include ceramides, fatty acids, cholesterol, cholesterol sulfate, and sterol/wax esters. The stratum corneum lipids are arranged in multiple bilayers, but in contrast to all other lipid bilayers in the

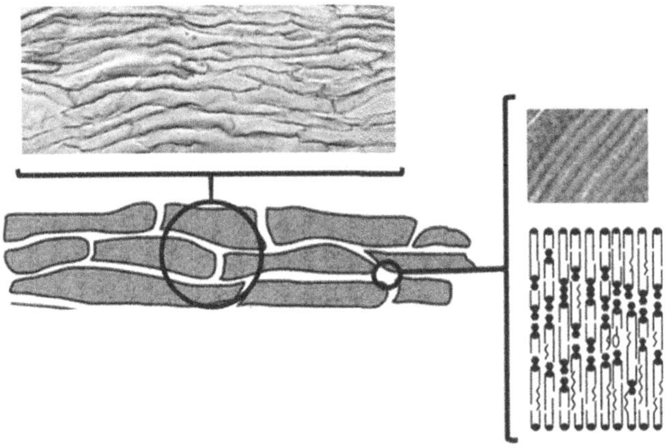

Figure 20.3 Representation of the "brick and mortar" description of the stratum corneum showing protein filled corneocyte "bricks" embedded in complex lipid multi-bilayers acting as a "mortar."

body, phospholipids are largely absent. The stratum corneum has long been likened to a "brick and mortar" wall in which the keratin-filled corneocytes ("bricks") are embedded in a "mortar" of lipid bilayers (Figure 20.3).

20.1.2 Drug Transport Through Skin

Considering intact human skin, three potential routes exist for a drug molecule to transit the stratum corneum, illustrated in Figure 20.4.

As described above (Section 20.1.1.2.2), the skin appendages – hair follicles and sweat ducts – offer a shunt route (or "short cut") by which molecules could enter the lower layers of the skin without having to traverse the barrier provided by the stratum corneum. In practice, the surface openings provide only a small fraction of the total skin surface area; for example, on the forearm, hair follicles occupy approximately 0.1% of the surface area although on the forehead

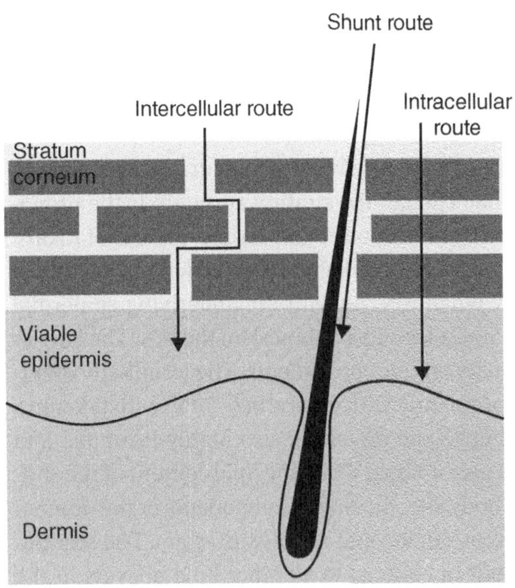

Figure 20.4 Illustration of the three routes by which a molecule can enter into and permeate through human skin.

this may be as much as 13%. Additionally, the ducts are seldom empty, being occupied by sweat or sebum flowing out of the body which again inhibits drug delivery. However, follicular openings have been exploited for controlled drug delivery from nanosized systems, such as liposomes, in order to treat acne, the shunt routes are important for electrically enhanced transdermal drug delivery (iontophoresis) and also play a role in the early time course of passive drug delivery through the skin, where diffusion through the intact stratum corneum barrier has yet to reach steady state.

Drug transport into and through the intact stratum corneum can either be via an intracellular (or transcellular) route or transport can be intercellular. Considering the intracellular route, drug molecules initially partition into and diffuse through keratin-filled corneocytes – essentially an aqueous environment – before partitioning into the intercellular lipid domains. For transcellular transport to continue, the molecule must then diffuse through the lipoidal region before repeatedly partitioning into and diffusing through the aqueous keratin in corneocytes and then intercellular lipids. In contrast, the intercellular route requires the molecule to partition into the lipid bilayers between the corneocytes and then diffusion is via a tortuous route within the continuous lipid domain, i.e. following the mortar in the "brick wall." For either intracellular or intercellular permeation, the multiple lipid domains have to be crossed and indeed present the major barrier to permeation of most drugs.

It should be noted that the three different micro-routes are not exclusive and for any permeant all three (intercellular, intracellular, and shunt routes), operate, but the proportion of molecules crossing by the different routes will vary depending on the physicochemical properties of the permeant. Further, drug binding to skin components, the presence of excipients and their own transport into the stratum corneum, metabolic activity, and other factors can also influence drug penetration into the skin and permeation through the tissue. A further complication is introduced in clinical situations where topical delivery is intended to treat diseased states; here the stratum corneum barrier may be compromised or, in some scaling conditions such as psoriasis, drug uptake into the tissue can be dramatically altered.

20.2 Controlled Drug Delivery into and Through Skin

Controlled drug delivery is typically defined when control is within the formulation, for example a controlled release matrix. When considering topical and transdermal drug delivery, control typically arises from the skin barrier itself rather than the formulation. Indeed, there are relatively few molecules possessing suitable physico-chemical (lipophilicity, molecular weight) and therapeutic (effective dose) properties for topical and transdermal application to intact skin. Consequently, researchers have developed strategies to modulate the stratum corneum barrier *per se* to facilitate drug delivery and these will be considered briefly below. More detailed discussion is then given to other approaches where control over drug delivery resides in the formulation or technology. Clearly in some cases this distinction may be rather arbitrary; while drug release can be controlled by a drug delivery system, the stratum corneum can retain control over the drug flux through the skin post-release.

20.2.1 Skin Barrier Modulation

In order to broaden the range of molecules than can effectively be delivered into and through the stratum corneum, researchers have employed various approaches to reduce the barrier properties of this layer. Targets for barrier reduction include the intercellular lipid domains that provide the principle obstacle to inter- and intracellular transport as well as some attempts to increase diffusivity through the corneocytes, or to improve the partitioning of the drug into the stratum corneum.

20.2.1.1 Penetration Enhancers

Numerous chemicals, with diverse structures physico-chemical properties, have been shown to interact with skin constituents to promote drug flux. Termed penetration enhancers (or absorption promoters), these materials have been used in formulations to improve drug flux through various membranes including gastric epithelia or nasal membranes (Figure 20.5).

The modes of action of penetration enhancers can be multiple and complex. An effective penetration enhancer may increase the diffusion coefficient of the drug in the stratum corneum (e.g. disrupt the intercellular lipid packing), may act to increase the effective concentration of the drug in the vehicle (for example acting as an anti-solvent), or could improve partitioning between the formulation and the stratum corneum (e.g. diffuse into the stratum corneum and act as strong solvent for the permeant). Indeed some materials, such as dimethylsulfoxide, act to both disrupt the intercellular lipid packing and acts as a solvent in the stratum corneum.

Penetration enhancers are often evaluated in isolation, yet it is well established that they are typically more potent when used in co-solvents such as propylene glycol (PG) or ethanol and often exhibit concentration-dependent effects. Synergism between enhancers is widely reported (and has recently been reviewed [8]) and high throughput methods have been employed to reveal unusual mixtures that promote permeation of various drugs without causing irritation, such as sodium laureth sulfate combined with phenyl piperazine [9].

The scale of enhancement from a given material remains difficult to predict and rational selection of an enhancer for a given permeant is thus problematic. Penetration enhancer potencies appear to be drug specific, or at best may be predictive for a series of permeants with similar physico-chemical properties, such as similar partition coefficients and molecular weights. Some broad generic trends have been reported, such as the use of hydrocarbon monoterpenes for lipophilic permeants [10], but their enhancement efficacy for a given permeant remains

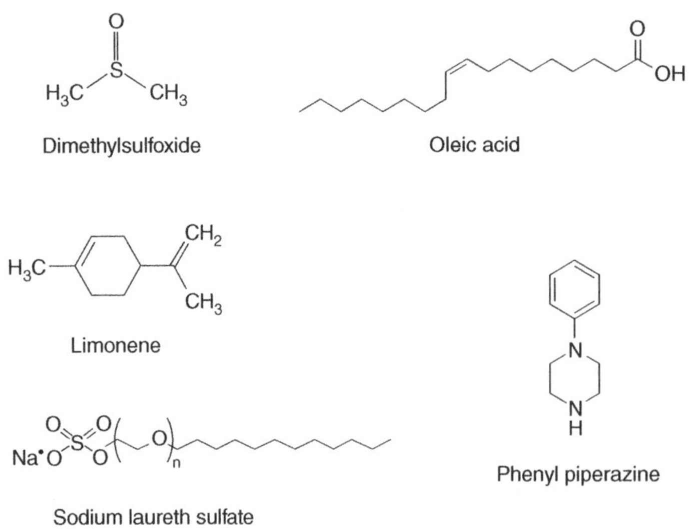

Figure 20.5 Examples of materials shown to have penetration-enhancing effects in human skin, illustrating the varied physical and chemical properties of such materials.

unpredictable. In practice, many materials with penetration enhancing activities are incorporated into topical and transdermal formulations for alternative purposes. For example, a topical preparation could contain PG as a vehicle, a surfactant to solubilize the drug and a terpene as a fragrance material.

A detailed critique of the varied penetration enhancers used to modulate the stratum corneum barrier, including penetration enhancer classification, modes of action, and efficacies, has recently been published [11].

20.2.1.2 Ablation

Thermal ablation of the stratum corneum essentially vaporizes the tissue and can be used to form micro-channels with minimal resistance to drug permeation. Heating the skin surface to high temperatures (in the order of 300 °C) for very short times over defined small areas reduces the risk of damaging the underlying epidermal and dermal tissue when removing the outer stratum corneum at heated sites. The use of a laser to ablate the stratum corneum to increase topical drug delivery was first reported in 1987 [12]. Dependent on the ArF laser pulse energy, various depths of the stratum corneum were sequentially removed such that the permeability coefficient of water though the ablated tissue matched that of skin with the stratum corneum removed by tape stripping. Subsequently, alternative lasers, pulse energies, and spot diameters have been employed to enhance drug delivery through the stratum corneum.

Lasers are widely used in dermatological procedures, for example in removing birthmarks, tattoos, and spider veins or for "skin resurfacing" and are thus widely available but are typically bulky and need specialist use. Consequently, other methods to ablate the stratum corneum to facilitate drug delivery have been described. For example, a device to apply superheated steam micro-jets to the stratum corneum for short times (~100 μs) was described and increased delivery of bovine serum albumin across human skin approximately 1000-fold [13]. Thermal ablation of skin has also been shown using arrays of microfabricated resistive heating elements (ohmic microheaters) and radiofrequency arrays to generate micro-channels. The fractional removal of the stratum corneum barrier by thermal ablation offers significant potential benefits in transdermal delivery of relatively large biomacromolecules or for transcutaneous vaccination, as recently reviewed [14].

20.2.2 Controlled Release Transdermal and Topical Systems

Drug release to the skin surface can be controlled by various formulations but, once presented to the outermost layer of the stratum corneum, penetration into and permeation through the tissue is governed by the barrier properties of intact skin and the physicochemical properties of the drug and excipients. The stratum corneum control over transdermal drug delivery is illustrated by developments in patch design since the 1980s [15]. Early transdermal patches containing, for example, scopolamine, nicotine, or fentanyl had a drug reservoir and an underlying microporous rate-controlling membrane that was considered necessary to control the delivery of drug to the skin surface. Partially due to the complexity of manufacturing reservoir and membrane type devices, but also reflecting that the stratum corneum itself provides the rate controlling membrane to delivery of most drugs, more recent patches tend to be simple drug-in-adhesive formulations. Few molecules transit the stratum corneum in amounts where overdose or adverse effects are likely – indeed it is more usually the case that drug flux does not reach therapeutic efficacy either in the tissue or when delivering to the systemic circulation, hence the approaches to modulate the stratum corneum barrier as described above. However, some approaches aiming to increase transdermal and topical drug delivery do so in a controlled manner, and controlled drug delivery to and through damaged skin has been described.

20.2.2.1 Supersaturation

Drug permeation through intact human skin can be described by Fick's laws of diffusion. In its simplest form, steady-state flux through the skin into an infinite sink (as assumed in most *in vitro* experimental designs) can be calculated from:

$$J = K_p \times C$$

where J is the flux, K_p is the permeability coefficient of the permeant in the membrane, and C is the concentration of the permeant in the donor vehicle. The permeability coefficient is a composite parameter:

$$K_p = P \cdot D/h$$

where P is the partition coefficient of the permeant between the membrane and the donor solution, D is the diffusion coefficient of the permeant in the membrane and h is the membrane thickness. For describing drug delivery through skin, separating the values for D and P can be problematic and the thickness of the membrane – or more accurately the length of the diffusional pathway given that transport via the intercellular route is tortuous – can be difficult to define, hence the use of permeability coefficients.

The above illustrate that drug delivery through the stratum corneum can be modified by manipulating (increasing) partitioning or diffusivity, as with the use of penetration enhancers that can disrupt the intercellular lipid packing and so increase diffusivity or solvents from a formulation entering the tissue to promote partitioning.

However, in circumstances where the drug or excipients do not modify the skin barrier properties, then the flux through skin is dependent on the concentration of the drug in the vehicle. More accurately, flux is dependent on the thermodynamic activity of the drug in the donor vehicle. In essence, maximal flux is obtained when a saturated solution (highest concentration) is applied to the skin surface. At saturation, the dissolved molecules are in equilibrium with its pure solid material, defined as a thermodynamic activity of 1, and the permeant has the greatest "escape tendency" to leave the donor solution and enter the skin.

Supersaturation can occur when the thermodynamic activity of the permeant in the vehicle exceeds 1 and so the concentration is essentially greater than the equilibrium solubility value. Many topical formulations are likely to generate supersaturated states when used clinically; formulations (creams, gels, etc.) applied to skin are typically rubbed in and chaotic processes occur as formulations break down on use. For example, when an hydro alcoholic gel is rubbed onto skin then the alcohol tends to volatize rapidly leaving a system enriched with water as the delivery vehicle. For a poorly water soluble molecule such as ibuprofen, which has greater solubility in the alcohol, the loss of the "good" solvent from the formulation can elevate its thermodynamic activity in the residual formulation and so drive more ibuprofen into the skin. Enhanced delivery using volatile solvents has long been recognized; as early as 1969, little penetration of fluminolone acetonide and its acetate ester into human skin was reported when applied in nonvolatile solvents but with increasing incorporation of a volatile solvent, penetration increased up to 10-fold [16]. The authors also noted that precipitation of the steroids prevented further increases in delivery. Generating supersaturated states from volatile solvents is inherently transient and unpredictable and so does not readily allow significant control over drug delivery.

Greater control over drug delivery can be generated from mixed solvent systems. As illustrated in Figure 20.6, ibuprofen is only slightly soluble in water, readily dissolves in ethanol, and solubility increases with increasing ethanol content in water:EtOH mixtures [17]. If a water:ethanol system containing ibuprofen is diluted with water, then the solubility of the drug follows the dotted line

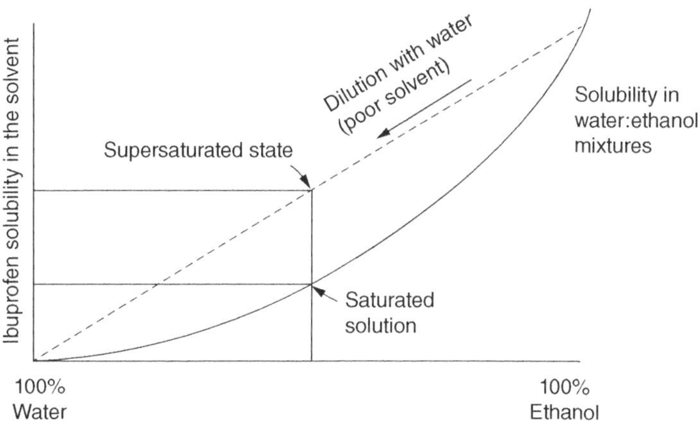

Figure 20.6 Illustration of the potential to generate supersaturated systems of ibuprofen from ethanol:water cosolvent systems.

in Figure 20.6. This forms a supersaturated state which is inherently thermodynamically unstable and so the supersaturated systems will crystallize and equilibrate to a saturated solution over time. However, the transient supersaturated system will provide greater drug flux compared to the saturated solution while the elevated thermodynamic activity is maintained.

From the above it is apparent that, by selection of solvent composition and drug content in the solvent, various degrees of supersaturation can be generated offering some degree of control over drug delivery. Various approaches have been taken to try and extend the stability and inhibit crystallization of supersaturated states, most commonly by using anti-nucleating polymers such as hydroxypropylmethyl cellulose (HPMC), poly(vinyl pyrrolidone) (PVP), hydroxypropyl cellulose (HPC), and polyvinyl alcohol (PVA). Using mixed solvent systems, supersaturation has been used to increase the flux of estradiol [18], piroxicam [19], fluocinonide [20], ibuprofen [21], and a lipophilic lavendustin derivative [22] through human skin both *in vitro* and *in vivo*. The anti-nucleating polymers inhibited recrystallization of the drugs for sufficient time – in the order of several hours (not days, but dependent on the degree of supersaturation) – to allow enhanced drug delivery. Importantly in terms of controlled delivery, approximate linear relationships between the degree of supersaturation and the flux were reported, i.e. if the supersaturated state was at double the saturated solubility (2 degrees of saturation) then flux approximately doubles, as would be predicted from the dependence of flux on concentration. Clearly the supersaturated states are thermodynamically unstable and so, even with anti-nucleants, recrystallization occurs over time. Further, thermodynamic instability increases with increasing degrees of supersaturation and so systems with degrees of supersaturation in excess of ~20 tend to be too unstable to use. Interestingly, as large molecular weight materials, the polymeric stabilizers are unable to enter the stratum corneum and so, with rapid ingress of the drugs into the tissue at levels in excess of the vehicle solubility then it appears that the stratum corneum components themselves are offering some inhibition of recrystallization in the tissue [18].

Given the instability of supersaturated systems formed from co-solvents, alternative formulations have been explored using the same principles. Water-free microemulsion bases saturated with bupranolol were applied to rabbit skin under occlusion; the occlusion led to water uptake into the base via transepidermal water loss through which the base converted into a microemulsion in which the drug was supersaturated as a result of decreasing solubility of the drug with increasing water content [23]. More recently, a hydrophilic matrix containing ethanol was used to generate supersaturation of diclofenac [24]. As ethanol evaporated, the matrix self-assembled and prevented drug recrystallization from its supersaturated state for over 24 hours.

Transient supersaturated states have also been used to rapidly increase drug deposition into the skin rather than for longer term transdermal delivery through the tissue. Beclomethasone dipropionate was deposited onto human skin from a metered dose aerosol (MDA) containing the hydrofluoroalkane propellant 134a, ethanol, and PVP [25]. With the steroid sub-saturated in the formulation, the time taken to generate a supersaturated state on the skin surface varied with ethanol content in the MDA. By increasing the volatile solvent content and drug concentration in the formulation, a supersaturated state could be formed on actuation, but the drug crystallized very rapidly on the skin surface whereas with lower ethanol contents, achieving supersaturation on the skin surface took up to 30 minutes. This illustrates the ability to control (to an extent) the delivery of drug into the tissue through formulation manipulation. A similar approach was used to enhance delivery of the corticosteroid betamethasone valerate into human skin; again using the hydrofluoroalkane propellant 134a with ethanol, PVP, and also incorporating poly(ethylene glycol), the spray deposited >6 times more drug into the skin compared to delivery from a commercial cream [26]. A novel aerosol foam formulation has been described to deliver two active ingredients into skin [27]. Calcipotriene and betamethasone dipropionate were fully dissolved in a dimethyl ether containing aerosol foam formulation. The ether rapidly evaporated after spraying leading to the formation of supersaturated states and notably no crystals of either drug were evident for at least 26 hours after application. Both drugs had significantly greater *in vitro* skin penetration and increased bioavailability when compared to a conventional ointment.

20.2.2.2 Reservoir Formation

The potential to form a reservoir, or "depot," of a drug in the stratum corneum has been described above. Supersaturated systems can increase the drug content in the stratum corneum from which it will diffuse through the skin to the systemic circulation. Further, one of the mechanisms described for penetration enhancers is to increase partitioning into the stratum corneum; if a "good" solvent is included in the formulation then, as the solvent itself permeates into the stratum corneum, so drug solubility in the solvent within the tissue can increase. Clearly solvent diffusion from the stratum corneum (or clearance) can be more rapid than for the drug *per se* which can result in the drug becoming supersaturated within the stratum corneum and can lead to crystallization [28].

It is apparent that selection of topical and transdermal formulation solvents markedly affects drug uptake into and flux though skin [29]. Solvent supramolecular structuring has been used to promote drug residence in the epidermal tissue [30]. By generating high concentrations of PG in the epidermal tissue, diclofenac loading in the epidermis increased 175-fold compared to a control. At these high PG:water ratios, infrared studies showed supramolecular structures formed with strong physical affinity for the drug. When low PG:water ratios were generated in the epidermis, preferential epidermal localization of diclofenac in the tissue was lost.

20.2.2.3 Film Forming Systems

Sustained transdermal drug delivery is clearly feasible from patches; methylphenidate patches to control symptoms of attention deficit hyperactivity disorder are intended to deliver the drug over nine hours, whereas buprenorphine patches can deliver the opioid analgesic over seven days. As described above, while drug release and delivery is sustained from patches, control over the dose delivered typically resides in the stratum corneum barrier.

Beyond prefabricated patches, film-forming systems offer an alternative method for sustained topical and transdermal drug delivery [31]. Various formulations have been investigated and typically contain a polymer and plasticizer in a volatile solvent and often with the incorporation of other excipients such as a penetration enhancer. For example, formulations containing ethinylestradiol were optimized to provide sustained delivery through both *in vitro* human and *in vivo* pig skin [32]. Incorporating oleic acid in PG to the film-forming polyurethane-14-AMP-acrylates copolymer, alongside triethylcitrate as a plasticizer and with ethanol:water as the solvent, the *in vitro* studies showed the film-forming system delivered greater amounts of drug through the epidermal membranes in 24 hours than a comparator commercial transdermal patch. As described in Section 20.2.2.1 (Supersaturation), using volatile solvents in the film-forming system results in dynamic processes once applied to the skin. As the solvent evaporates, drug solubility (thermodynamic activity) in the remaining matrix can increase to supersaturated levels, generating a "bolus dose" or initially high uptake of drug into the stratum corneum. The polymers present to form the films can also act as antinucleants and so retard drug crystallization for extended time periods. Alternatively, as suggested recently for betamethasone-17-valerate delivery from a polymeric film-forming system containing medium-chain triglycerides [33], the drug may be stabilized in an amorphous form within the formed film, again promoting drug release.

Clearly the mechanical properties of the formed film – flexibility, adhesion to the skin, appearance, and abrasion resistance – need to be optimized but some control over delivery can also be achieved by formulation manipulation [34]. This can be via the degree of supersaturation generated (solvent choice/volatility and drug solubility), the formation of a reservoir within the stratum corneum and its subsequent permeation through the skin, the inclusion of skin barrier modulators, such as penetration enhancers, and the degree of occlusion afforded by the formed film on the skin surface. Film forming systems have been reported to provide biphasic release of testosterone [35], have shown strong bioadhesive properties to deliver lidocaine [36] and can be formed from systems other than simple solutions, for example from emulsions [37] or from regenerated silk fibroin gels [38].

20.2.2.4 Vesicles

A rich literature exists around the development and use of vesicular systems for topical and transdermal drug delivery. These range from relatively "conventional" liposomal systems to those that are more complex flexible liposomes containing surfactants or solvents and niosomes formed from nonionic surfactants. However, despite the numerous reports and studies, it is notable that, since their first description for topical and transdermal drug delivery in the early 1980s [39, 40], the following 35 years have seen very few liposomal formulations progress to clinical trials or the pharmaceutical market (though cosmetic and sunscreen liposomal products are available).

It has been long established that liposomes can serve to solubilize the active ingredient in its formulation, that liposomal systems can create a reservoir within the stratum corneum and that liposome components may act as penetration enhancers in the tissue [41]. In addition, it is feasible that, as the liposomal system collapses on the skin surface, then the thermodynamic activity of the drug in the resultant matrix can increase, hence increasing drug uptake into – and flux through – the skin [42, 43]. Additionally, the formulation components could (partially) occlude the skin surface, forming an oily film to prevent some trans epidermal water loss which tends to increase transdermal drug flux, or the liposomes could distribute into the hair follicles and so promote greater appendageal drug transport [44]. Transdermal permeation of intact liposomes through the stratum corneum as carriers for their entrapped active ingredients remains to be proven [45]. A comprehensive recent review of vesicular carriers ("conventional" liposomes, ultradeformable liposomes, and ethosomes) also highlighted some emergent vesicles such as "archeosomes" formed from bacterial lipids and "ufasomes" that aim to exchange lipids between the vesicle and the stratum corneum [46].

20.2.2.5 Particles

Since vesicles tend to be relatively fragile and collapse on application, alternative solid particulate delivery systems for topical and transdermal delivery have been explored. These range in scale from nano-sized solid lipid nanoparticles (SLNs) or nanostructured lipid carriers (NLCs) through to micron-sized delivery systems designed to control drug release and hence drug delivery.

20.2.2.5.1 Solid Lipid Nanoparticles (SLNs)

SLNs provide a solid lipophilic matrix and are typically composed of a biodegradable lipid with a surfactant to act as a stabilizer. Various methods have been described to produce SLNs [47], including high-shear or high-pressure homogenization, emulsification methods, and sonication. Dependent on the method of production and selection of drug, lipid, and surfactant, particles ranging from 50 to 1000 nm have been described and can be produced with a narrow size distribution. Typically, drug release from the particles is by diffusion and so, by varying the SLN composition, control over drug release and hence delivery can be achieved [48, 49].

The literature describes numerous studies employing SLNs for controlled release of active ingredients as varied as terbinafine hydrochloride [50], isotretinoin [51], and ketoprofen [52]. SLNs have also been used to treat diseased skin and have recently been explored to deliver the large molecular weight (~6 kDa) human recombinant epidermal growth factor to chronic wounds [53].

The relatively small size of SLNs allows intimate contact with the stratum corneum surface, and imaging of dye-loaded-SLNs shows they tend to accumulate in skin furrows and at the hair follicle openings. However, it is widely accepted that nanoparticles >~10 nm do not enter the intact stratum corneum [54]. Some enhancement of drug delivery from SLNs has been reported (though not for a large macromolecular growth factor across intact skin [53]), which may be attributed to the presence of the surfactant acting as a penetration enhancer and/or due to the formation of an occlusive film on the stratum corneum surface by the lipid component [55].

In comparison to vesicles, SLNs can provide control over drug release and, as solid matrices, can provide some stability for the incorporated drug against oxidative and hydrolytic degradation. The lipid matrix can allow greater lipophilic drug encapsulation compared to liposomes but, conversely, typically lower hydrophilic drug loadings. Drug loading and encapsulation efficiencies tend to be relatively modest into SLNs, even for lipophilic drugs and leaching from particles into a formulation vehicle over time can be problematic. In addition, physical stability issues have been identified, especially when relatively low melting point lipids are employed and particles may gel or agglomerate.

20.2.2.5.2 Nanostructured Lipid Carriers (NLCs)

NLCs differ from SLNs in that they comprise a mixture of liquid and solid lipids that encapsulate a liquid oily inner phase or "nanocompartment" [56]. NLCs were developed to mitigate some of the limitations seen with SLNs with the oily core allowing greater levels of (lipophilic) drug loading and reducing drug leaching from the particles [57, 58]. As with SLNs, production is typically by high-pressure or high-shear homogenization, or, more recently, hot melt homogenization [59] and selection of components affects particle size, dispersity, and physical properties.

In common with other nanostructured carriers, deposition of NLCs on the bulk skin surface, around folds and around follicular openings – especially on rubbing – has been reported and the delivery of the active ingredient via follicles has been discussed [60]. Again enhanced drug delivery has been attributed to the formation of an occlusive oily film on the stratum corneum surface. However, the oily core of NLCs also affords opportunities to incorporate penetration enhancers, such as oleic acid, into the carrier that may also promote transdermal drug delivery [61]. Recent comprehensive reviews of the formulation, composition, and use of both SLNs and NLCs can be found in the literature [46, 61].

20.2.2.5.3 Polymeric Particles

Various natural and synthetic polymers have been used to construct controlled delivery systems for topical administration. These vary from nano- to micro-particulates and include both solid particles and capsules. As evident from the above, distribution of the particles on the skin surface is size dependent [62] and there is evidence for preferential localization with the hair follicles [63], but penetration of nano-polymeric particles into intact stratum corneum has not yet been shown [55].

Nanoparticles constructed from synthetic materials generally seek to use biodegradable materials such as poly(lactic-*co*-glycolic acid) – commonly referred to as PLGA. Poly(caprolactone) nanoparticles were prepared containing the highly lipophilic sunscreen octyl methoxycinnamate [64]; interestingly, a >3-fold increase in sunscreen within the stratum corneum was reported following tape stripping of the porcine skin but no increase in flux through the tissue was evident. Although the availability of the sunscreen in the stratum corneum was increased from the polymeric nanoparticles, this was attributed to changes in thermodynamic activity and partitioning rather than nanoparticulate transport.

Aiming to control drug delivery, hydrocortisone-containing microparticles were produced from pH-responsive polymers that are commonly used as enteric coatings [65, 66]. The authors sought to use the pH discrepancy between healthy (~pH 5–5.5) and atopic dermatitis skin (~pH 6–6.5) to trigger and control drug delivery to the diseased skin site. Microparticles were produced by oil-in-oil microencapsulation or spray drying. Drug release was pH-dependent (Figure 20.7) though the particles produced by spray drying also gave significant non-pH dependent burst release, resulting from their porous nature or from drug enrichment on the surface of these particles. The pH-responsive release was maintained in gel formulations and in vitro studies showed four to fivefold higher drug permeation through porcine skin from the gels at pH 7 compared to pH 5 [67].

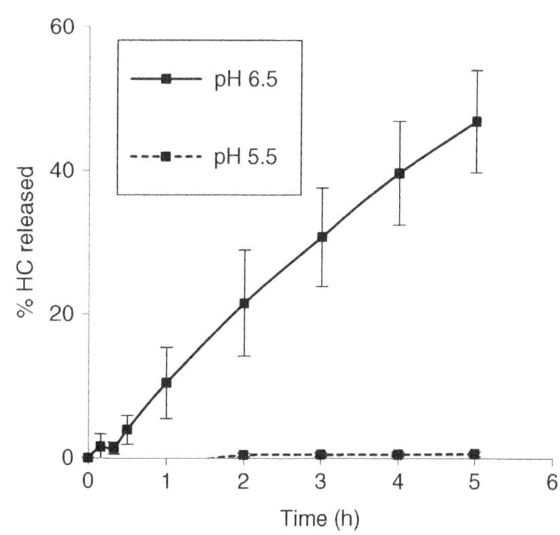

Figure 20.7 Drug release profiles from Eudragit L100 microparticles containing 2.5% hydrocortisone at pH 5.5 and 6.5. Source: Modified from Rizi et al. [67].

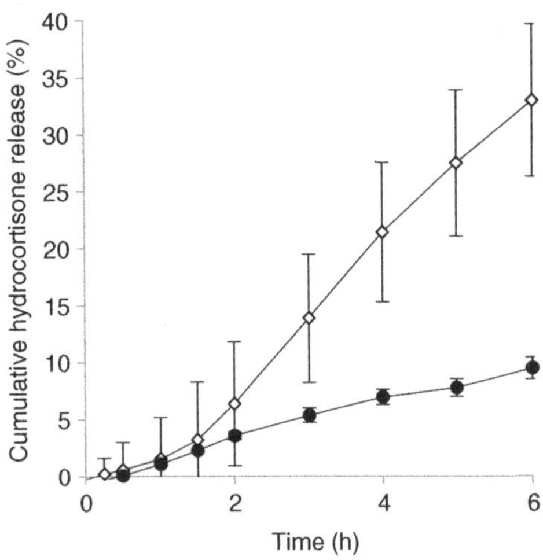

Figure 20.8 Release of hydrocortisone from gelatin particles produced by O/W/O emulsification in the presence (open diamonds) and absence (closed circles) of *S. aureus*. Data are mean ± standard error of the mean ($n = 3$).

Natural polymers including alginates, gelatin, albumin and, in particular chitosan have also been described. Chitosan has been used for nano- and microparticles, seeking to exploit the adhesive properties of this biocompatible material [68]. Encapsulation in chitosan was shown to enhance the stability of retinol [69] and chitosan-tripolyphosphate nanoparticles enhanced the stability of aciclovir for topical administration [70]. An enzymatically controlled drug delivery system was developed to deliver and target hydrocortisone for skin disorders such as atopic dermatitis with associated *Staphylococcus aureus* infection. Gelatin microparticles containing hydrocortisone were produced from oil-water-oil (O/W/O) emulsions and by spray drying. Drug release from the emulsion-formed particles was principally bacteria-responsive with some release due to swelling of the matrix; 33.0% of the drug was liberated in the presence of the bacteria after six hours in contrast to only 9.1% released in the absence of *S. aureus* (Figure 20.8).

From the above it is clear that there is little evidence showing that vesicles, lipidic, or polymeric particles are able to enter or permeate through intact stratum corneum. Dependent on size, there is some evidence that nanoparticles may accumulate in follicular openings and they tend to preferentially locate at skin folds. It should be noted that, to overcome the stratum corneum barrier, various approaches have been used that combine nano-particulate matrices with active interventions including deposition of particles onto skin that has been ablated, combined with iontophoresis (see below) or applied to the skin surface following the application of microneedles that generate microchannels or pores.

20.2.3 Device-Based Controlled Delivery

Ablative technologies were described earlier but various other device-based strategies exist to control drug delivery to the skin surface or indeed that can control drug delivery into the tissue. Phoresis methods, in essence meaning "carrying" or "transfer," include iontophoresis and sonophoresis, whereas pores can be generated in the stratum corneum barrier by electroporation and microneedles. Heated systems are available that can also promote drug partitioning into and diffusivity through skin and combinations of the above, together with some of the approaches described earlier have also been employed.

20.2.3.1 Iontophoresis

Controlled drug delivery through skin using iontophoresis has a long history and can be traced back over 100 years to the pioneering work of Le Duc [71]. Iontophoresis uses electrical current to drive ions into the skin, and has alternatively been termed "electromotive drug administration." Although the intention is to drive ions into the tissue, electromotive movement of water (electroosmotic flow) during iontophoresis can also carry non-charged molecules while current is applied. Though the focus here is on transdermal drug delivery, iontophoresis has also been used to transport drugs across other biological membranes including tooth mucosa (for dental anesthesia) and the eyes and is used to manage hyperhidrosis.

Most commonly using a continuous direct current, the active electrode repels ions of the same charge and hence drives them into the skin where they clear into the systemic circulation. Thus, a cationic (positively charged ion) drug will be repelled from the anode (positively charge electrode) and conversely an anion will be repelled and so delivered from underneath a cathode (Figure 20.9). Ions enter the tissue following the path of least electrical resistance and so the principle pathways for ion migration are via the pores, sweat glands, and hair follicles.

A large body of literature has described numerous factors that affect iontophoretic drug delivery. The vehicle from which the drug is applied needs careful control. Ionic transport is relatively rapid though liquid as opposed to gelled or solid matrices and vehicle pH is typically selected to ensure that the drug is predominantly present in its ionized form – though this must be a pH that is both tolerable and will not damage the skin. Ionic competition in a formulation can be problematic; for example, if buffers are used then these tend to generate small and highly mobile ions that can be preferentially driven into the skin rather than the therapeutic ion. Different salt forms of the dug can present different electrical conductivities, and ion molecular size have also been reported to influence the rate of iontophoretic drug delivery.

Direct current is typically used and preferred to a constant voltage system; if the stratum corneum resistance changes (decreases) during iontophoresis then the current can change (increase) when a constant voltage is applied, leading to a greater risk of burning the tissue. However, the application of a continuous DC current for extended periods may result in skin polarization, which in turn can reduce the efficiency of iontophoretic delivery. Thus, pulsed DC protocols tend to be preferred since the skin depolarizes during the "off" periods. Currents densities in the order 0.5 mA/cm^2 at cathodes and 1.0 mA/cm^2 at anodes are generally used [72].

Iontophoresis offers good control over drug delivery since the amounts of ions delivered are proportional to the current density [72, 73]. Although ion transport is largely via the shunt routes

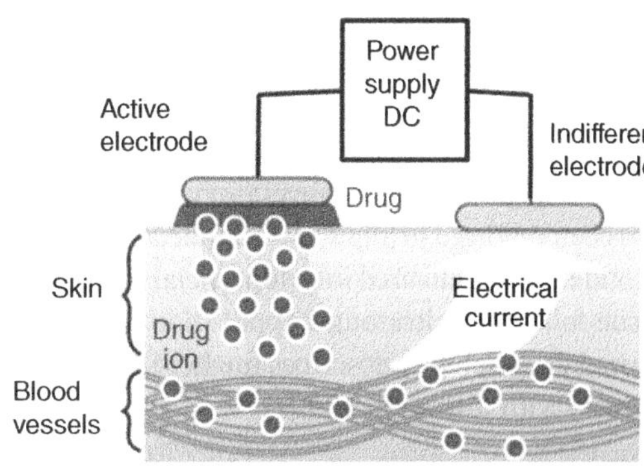

Figure 20.9 Illustration of the principle of iontophoresis.

Microneedle design varies and is related to the intended use as well as ease of fabrication and physical properties of the microneedle array. Solid microneedles are typically used to generate pores in the stratum corneum before being removed before application of a conventional topical formulation – often termed the "poke and patch" approach. The formulation applied may be for finite dosing or for sustained delivery prior to pore closure if, for example, a patch is subsequently applied to the disrupted stratum corneum. While the poke and patch approach effectively increases topical and transdermal delivery, this is clearly a two-step process with practical difficulties for end users. An alternative "coat and poke" approach uses a coating of the therapeutic agent (alone or with excipients) on the external surface of the microneedles. When inserted into the tissue, the coating dissolves in the interstitial fluid so depositing the drug in the skin. A limitation of this approach is the relatively low dose that can be applied to the micro-projections and thus has been promoted for delivering potent macromolecules. Aiming to increase dosing, dissolving microneedles have been developed, often using sugars as a matrix with the drug dispersed, for example as a solid solution. After insertion, the microneedle dissolves releasing the drug over time with an added benefit that, post-dosing and dissolution, the microneedles do not present any risk from accidental microneedle-stick injury or "sharps" disposal. For sustained delivery, hollow microneedles have been proposed through which, after insertion, drug can be continuously infused. However, the microchannels can become blocked over time or upon insertion. More recently, *in situ* hydrogel-forming microneedles have been reported. Produced from polymers, the microneedles swell on insertion and can thus be left in place while a formulation is applied on top of the array. This essentially, forms aqueous channels through the stratum corneum for drug delivery and, beneficially, the swollen hydrogels are pliable and thus lose the "sharps" risk on removal.

Clearly microneedles offer an attractive approach to deliver small and large molecules into and through skin and selection of a microneedle array design can control delivery. Issues identified with the technology are being resolved. For example, silicon microneedles tend to be relatively brittle; depth and reproducibility of microneedle penetration on application by a patient can be variable and so applicators are being investigated; inadvertent dosing from devices ("safety") when application is painless is mitigated by dissolving or hydrogel forming systems; commercial scalability and manufacturing processes are being investigated.

20.2.3.5 Heat

Since the rate of diffusion is directly related to temperature, researchers have used heat to increase topical and transdermal drug delivery. The intercellular lipids of the stratum corneum undergo various phase changes that decrease their barrier properties at elevated temperatures [93], around 55, 72, and 83 °C, though at these temperatures the skin can burn *in vivo*. In addition, skin blood flow increases with increasing temperature which can improve clearance from the dermis.

The effects of temperature on transdermal drug flux has long been recognized [94], and concerns regarding enhanced delivery from patches containing fentanyl and other potent molecules in hot environments have been reported. In terms of enhancing and controlling drug delivery, the fluxes of methyl paraben, butyl paraben, and caffeine through human epidermal membranes from saturated solutions increased as the temperature rose from 23 to 30 to 37 to 45 °C [95]. In addition, a heat enhanced effect on penetrant retention in the epidermis was observed. The influence of moderate heat on transdermal delivery of testosterone *in vivo* was also reported [96]. Applying a heat generating patch with a 5 mg androgen patch improved absorption of transdermal testosterone and reduced the time to peak serum testosterone concentrations compared to delivery from the androgen patch without heat.

Whereas well-controlled laboratory studies can discriminate the effects of heat on drug diffusivity, flux, and retention [96], when heat is applied with a more complex formulation then temperature effects on drug release also need to be considered [97]. Heating patches that can be activated on use, for example by oxygen, have been developed for applications such as anesthesia; pain scores on vascular access were lower for patients using a heated lidocaine/tetracaine patch compared to those using a lidocaine/prilocaine cream and the time to anesthesia was also reduced [98]. The effects of heat on transdermal drug delivery have recently been reviewed [99].

20.2.3.6 Other Devices

Numerous other "device"-based approaches have been evaluated for enhancing transdermal drug delivery. ***Magnetophoresis*** has been used to deliver diamagnetic drugs – i.e. those that are repelled by a magnetic field such as lidocaine. Flux enhancement increases with magnetic field strength since permeation is largely by magnetokinesis [100]. As with electroosmosis in iontophoresis, additionally the magnetic field can induce solvent flow across the stratum corneum which can carry other, non-diamagnetic materials, into the skin and partitioning of drugs into the hydrophobic stratum corneum has also been reported to increase under the influence of a magnetic field. Although enhanced naltrexone delivery through human epidermal membranes was also attributed to magnetokinesis and increased partitioning, enhanced delivery of 10 nm gold particles appeared to be due to the formation of channels within the stratum corneum following application of the magnetic field [101].

Microwaves have been applied to skin to facilitate transdermal drug delivery; clearly microwaves can induce thermal effects as described above for "Heat" devices, but nonthermal effects have also been reported [102] and delivery is dependent on exposure time and microwave intensity, hence is amenable to control. Some microstructural changes within the stratum corneum lipids and keratin, as well as deeper in the dermis have also been suggested when microwave radiation is applied to the skin [103].

Needle-free powder- and liquid-***jet injectors*** continue to be developed [104]. Again offering control over delivery by regulating pressures, jet speeds and, for powders, the density and size of the delivered particulates, considerable efforts have been devoted to commercial scale up of these devices to allow reproducible dose delivery to defined sites with skin. The majority of research reports focus on vaccine delivery but the technology is also applicable for macromolecules such as hormones, monoclonal antibodies, and nucleic acids that can be stabilized in the jet formulation and the jet-injectors offer greater potential dosing than some of the alternative strategies described above [105].

Microdermabrasion is an exfoliating cosmetic procedure for age spots and "dull and tired skin." The stratum corneum surface is bombarded with fine crystals, for example of aluminum oxide, to remove outer layers and the particles and skin cells are removed from the skin under negative pressure. As the stratum corneum barrier is partially ablated, so drug delivery through the tissue can be increased. Using microdermabrasion, delivery of the poorly permeating 5-fluorouracil increased up to 24-fold as compared to that through intact skin [106], and the approach has since been used seeking to deliver larger molecular weight therapeutic agents though skin [107]. Treatment pressure and duration control the amount of tissue removed and biochemical changes as well as local irritation have been reported following microdermabrasion.

20.3 Combination Approaches

As will be evident from the above, several approached used to enhance and control drug delivery to and through skin can be used in combination and may have synergistic actions. Commonly, the porative methods that generate channels within the stratum corneum, for example electroporation or the use of microneedles have been used in combination with iontophoresis to allow controlled delivery through the stratum corneum. Sonophoresis increased electroosmosis and hence mannitol delivery during iontophoresis [108], and the combination of ultrasound with microneedles has been recently reviewed [109].

Combinations seeking to exploit the physical disturbance of the stratum corneum barrier (for example by microneedles) together with control over delivery from particulate carriers (typically nanoparticles) have also been described. For example, microwaves were used to disrupt the stratum corneum lipid domains prior to application of chitosan nanoparticles containing 5-fluorouracil; the combination was reported to act synergistically in enhancing drug delivery [110]. Likewise, synergism was reported between electroporation and the delivery of a genistein-based lyotropic liquid crystal formulation [111].

Given the increasing array of methods to physically perturb the stratum corneum barrier, the trend toward combining these physical approaches with formulation strategies for controlled drug release is likely to continue. Adding in the increasing sophistication of microelectronics and wearable devices, then on-demand controlled drug delivery to and through the skin is maturing. Topical and transdermal delivery of conventional small molecular weight lipophilic therapeutic agents is well established, though work will continue to expand the utility of topical administration to more hydrophilic molecules. Perhaps the greatest opportunities lie in topical delivery of larger molecular weight and biological therapeutic agents where oral delivery is problematic and it is notable that much recent literature is directed at delivery of these agents for both local and systemic action.

20.4 Conclusions

Human skin and, in particular, the stratum corneum, presents a formidable biological barrier which inherently limits drug delivery into, and hence permeation through, the tissue. Controlled drug delivery from common topically applied formulations such as creams or gels is thus limited and systems such as transdermal patches may control drug release to the skin surface, but thereafter permeation is again governed by the skin. Strategies have been developed to reduce the barrier properties of the stratum corneum, for example with penetration enhancers or the use of supersaturated systems and these can effectively increase topical drug delivery, but offer only partial control.

Greater control over drug input can be gained using assistive technologies such as iontophoresis where delivery of charged molecules can be controlled since the amounts of ions delivered are proportional to the applied current density; electroosmotic flow can also carry neutral molecules into the skin. More invasive strategies, including laser ablation, dermabrasion, and microneedles, effectively reduce or remove the stratum corneum barrier; with a compromised skin barrier, then control over drug delivery can be achieved through controlled-release topical formulations. Given advances in technology and micro-electronics, then it is likely that combinations of devices with topical formulations are likely to develop to control drug delivery into and through human skin. The aim of the present work was the optimization of the transdermal delivery of a lyotropic liquid crystal genistein-based formulation (LLC-GEN). LLC was chosen as medium in view of the poor solubility of GEN in water. Membrane diffusion and penetration studies were carried out with a Franz diffusion cell, through a synthetic membrane *in vitro*, a chick chorioallantoic membrane *ex ovo*, and *ex vivo* excised human epidermis. Thereafter, LLC-GEN was combined with electroporation (EP) to enhance the transdermal drug delivery. The synergistic effect of EP was verified by *in vivo* attenuated total reflectance – Fourier transform infrared spectroscopy (ATR-FTIR) and *ex vivo* Raman spectroscopy on hairless mouse skin

References

1 Svensson, C.K. (2009). Biotransformation of drugs in human skin. *Drug Metab. Dispos.* 37: 247–253.

2 Ademola, J.I. and Maibach, H.I. (1995). Cutaneous metabolism and penetration of methoxypsoralen, betamethasone 17-valerate, retinoic acid, nitroglycerin, and theophylline. *Curr. Probl. Dermatol.* 22: 201–213.

3 Stinchcomb, A.L., Swaan, P.W., Ekabo, O. et al. (2002). Straight-chain naltrexone ester prodrugs: diffusion and concurrent esterase biotransformation in human skin. *J. Pharm. Sci.* 91: 2571–2578.

4 Cross, S.E., Anderson, C.C., and Roberts, M.S. (1998). Topical penetration of commercial salicylate esters and salts using human isolated skin and clinical microdialysis studies. *Br. J. Clin. Pharmacol.* 46: 29–35.

5 van Eijl, S., Zhu, Z., Cupitt, J. et al. (2012). Elucidation of xenobiotic metabolism pathways in human skin and human skin models by proteomic profiling. *PLoS One* 7 (7): e41721. https://doi.org/10.1371/journal.pone.0041721.

6 Williams, A.C. (2003). *Transdermal and Topical Drug Delivery; From Theory to Clinical Practice*. London: Pharmaceutical Press.

7 Cross, S.E. and Roberts, M.S. (1993). Subcutaneous absorption kinetics of interferon and other solutes. *J. Pharm. Pharmacol.* 45: 606–609.

8 Ng, K.W., Lau, W.M., and Williams, A.C. (2015). Synergy between chemical penetration enhancers. In: *Percutaneous Penetration Enhancers Chemical Methods in Penetration Enhancement: Modification of the Stratum Corneum*, Chapter 24 (eds. N. Dragicevic and H.I. Maibach), 373–385. Berlin: Springer-Verlag.

9 Karande, P., Jain, A., and Mitragotri, S. (2004). Discovery of transdermal penetration enhancers by high-throughput screening. *Nat. Biotechnol.* 22: 192–197.

10 Williams, A.C. and Barry, B.W. (1991). The enhancement index concept applied to terpene penetration enhancers for human skin and model lipophilic (oestradiol) and hydrophilic (5-fluorouracil) drugs. *Int. J. Pharm.* 74: 157–168.

11 Dragicevic, N. and Maibach, H.I. (eds.) (2015). *Percutaneous Penetration Enhancers Chemical Methods in Penetration Enhancement: Modification of the Stratum Corneum*. Berlin: Springer-Verlag.

12 Jacques, S.L., McAuliffe, D.J., Blank, I.H., and Parrish, J.A. (1987). Controlled removal of human stratum corneum by pulsed laser. *J. Invest. Dermatol.* 88 (1): 88–93.

13 Lee, J.W., Gadiraju, P., Park, J.H. et al. (2011). Microsecond thermal ablation of skin for transdermal drug delivery. *J. Controlled Release* 154: 58–68.

14 Levin, G. (2015). Skin ablation methods for transdermal drug delivery. In: *Percutaneous Penetration Enhancers Physical Methods in Penetration Enhancement* (eds. N. Dragicevic and H.I. Maibach), 233–241. Berlin: Springer-Verlag.

15 Pastore, M.N., Kalia, Y.N., Horstmann, M., and Roberts, M.S. (2015). Transdermal patches: history, development and pharmacology. *Br. J. Pharmacol.* 172: 2179–2209.

16 Coldman, M.F., Poulsen, B.J., and Higuchi, T. (1969). Enhancement of percutaneous absorption by the use of volatile: nonvolatile systems as vehicles. *J. Pharm. Sci.* 58: 1098–1102.

17 Rashid, A., White, E.T., Howes, T. et al. (2014). Effect of solvent composition and temperature on the solubility of ibuprofen in aqueous ethanol. *J. Chem. Eng. Data* 59: 2699–2703.

18 Megrab, N.A., Williams, A.C., and Barry, B.W. (1995). Estradiol permeation through human skin and silastic membrane – effects of propylene glycol and supersaturation. *J. Controlled Release* 36: 277–294.

19 Pellett, M.A., Castellano, S., Hadgraft, J., and Davis, A.F. (1997). The penetration of supersaturated solutions of piroxicam across silicone membranes and human skin in vitro. *J. Controlled Release* 46: 205–214.

20 Schwarb, F.P., Imanidis, G., Smith, E.W. et al. (1999). Effect of concentration and degree of saturation of topical fluocinonide formulations on in vitro membrane transport and in vivo availability on human skin. *Pharm. Res.* 16: 909–915.

21 Iervolino, M., Cappello, B., Raghavan, S.L., and Hadgraft, J. (2001). Penetration enhancement of ibuprofen from supersaturated solutions through human skin. *Int. J. Pharm.* 212: 131–141.

22 Moser, K., Kriwet, K., Froehlich, C. et al. (2001). Permeation enhancement of a highly lipophilic drug using supersaturated systems. *J. Pharm. Sci.* 90: 607–616.

23 Kemken, J., Ziegler, A., and Muller, B. (1992). Influence of supersaturation on the pharmacodynamic effect of bupranolol after dermal administration using microemulsions as vehicle. *Pharm. Res.* 9: 554–558.

24 Benaouda, F., Brown, M.B., Martin, G.P., and Jones, S.A. (2012). Triggered in situ drug supersaturation and hydrophilic matrix self-assembly. *Pharm. Res.* 29: 3434–3442.

25 Reid, M.L., Jones, S.A., and Brown, M.B. (2009). Transient drug supersaturation kinetics of beclomethasone dipropionate in rapidly drying films. *Int. J. Pharm.* 371: 114–119.

26 Reid, M.L., Benaouda, F., Khengar, R. et al. (2013). Topical corticosteroid delivery into human skin using hydrofluoroalkane metered dose aerosol sprays. *Int. J. Pharm.* 452: 157–165.

27 Lind, M., Nielsen, K.T., Schefe, L.H. et al. (2016). Supersaturation of calcipotriene and betamethasone dipropionate in a novel aerosol foam formulation for topical treatment of psoriasis provides enhanced bioavailability of the active ingredients. *Dermatol. Ther.* 6: 413–425.

28 Hadgraft, J. and Lane, M.E. (2016). Drug crystallization – implications for topical and transdermal delivery. *Expert Opin. Drug Deliv.* 13: 817–830.

29 Williams, A.C. (2007). Pharmaceutical solvents as vehicles for topical dosage forms. In: *Solvent Systems and Their Selection in Pharmaceutics and Biopharmaceutics* (eds. P. Augustijns and M.E. Brewster), 403–426. Heidelberg, Chapter 13: Springer & AAPS Press.

30 Benaouda, F., Jones, S.A., Martin, G.P., and Brown, M.B. (2016). Localized epidermal drug delivery induced by supramolecular solvent structuring. *Pharm. Res.* 13: 65–72.

31 Frederiksen, K., Guy, R.H., and Petersson, K. (2016). The potential of polymeric film-forming systems as sustained delivery platforms for topical drugs. *Expert. Opin. Drug Deliv.* 13: 349–360.

32 Schroeder, I.Z., Franke, P., Schaefer, U.F., and Lehr, C.M. (2007). Delivery of ethinylestradiol from film forming polymeric solutions across human epidermis in vitro and in vivo in pigs. *J. Controlled Release* 118: 196–203.

33 Garvie-Cook, H., Frederiksen, K., Petersson, K. et al. (2015). Biophysical elucidation of the mechanism of enhanced drug release and topical delivery from polymeric film-forming systems. *J. Controlled Release* 212: 103–112.

34 Frederiksen, K., Guy, R.H., and Petersson, K. (2015). Formulation considerations in the design of topical, polymeric film-forming systems for sustained drug delivery to the skin. *Eur. J. Pharm. Biopharm.* 91: 9–15.

35 Misra, A., Raghuvanshi, R.S., Ganga, S. et al. (1996). Formulation of a transdermal system for biphasic delivery of testosterone. *J. Controlled Release* 39: 1–7.

36 Padula, C., Colombo, G., Nicoli, S. et al. (2003). Bioadhesive film for the transdermal delivery of lidocaine: in vitro and in vivo behaviour. *J. Controlled Release* 88: 277–285.

37 Lunter, D. and Daniels, R. (2013). In vitro skin permeation and penetration of nonivamide from novel film-forming emulsions. *Skin Pharmacol. Physiol.* 26: 139–146.

38 Gennari, C.G., Selmin, F., Ortenzi, M.A. et al. (2016). In situ film forming fibroin gel intended for cutaneous administration. *Int. J. Pharm.* 511: 296–302.

39 Mezei, M. and Gulasekharam, V. (1980). Liposomes: a selective drug delivery system for the topical route of administration. 1. Lotion dosage form. *Life Sci.* 26: 1473–1477.

40 Mezei, M. and Gulasekharam, V. (1982). Liposomes a selective drug delivery system for the topical route of administration: gel dosage form. *J. Pharm. Pharmacol.* 34: 473–474.

41 Schreief, H. and Bouwstra, J. (1994). Liposomes and niosomes as topical drug carriers: dermal and transdermal drug delivery. *J. Controlled Release* 30: 1–15.

42 El Maghraby, G.M.M., Williams, A.C., and Barry, B.W. (1999). Skin delivery of oestradiol from deformable and traditional liposomes: mechanistic studies. *J. Pharm. Pharmacol.* 51: 1123–1134.

43 El Maghraby, G.M.M. and Williams, A.C. (2009). Vesicular systems for delivering conventional small organic molecules and larger macromolecules to and through human skin. *Expert. Opin. Drug Deliv.* 6: 149–164.

44 El Maghraby, G.M.M., Williams, A.C., and Barry, B.W. (2001). Skin hydration and possible shunt route penetration in controlled estradiol delivery from deformable and standard liposomes. *J. Pharm. Pharmacol.* 53: 1311–1322.

45 El Maghraby, G.M.M., Williams, A.C., and Barry, B.W. (2006). Can drug-bearing liposomes penetrate intact skin? *J. Pharm. Pharmacol.* 58: 415–429.

46 Jain, S., Patel, N., Shah, M.K. et al. (2017). Recent advances in lipid-based vesicles and particulate carriers for topical and transdermal application. *J. Pharm. Sci.* 106: 423–445.

47 Mehnert, W. and Mader, K. (2001). Solid lipid nanoparticles: production, characterization and applications. *Adv. Drug Delivery Rev.* 47: 165–196.

48 Muller, R.H., Mader, K., and Gohla, S. (2000). Solid lipid nanoparticles (SLN) for controlled drug delivery – a review of the state of the art. *Eur. J. Pharm. Biopharm.* 50: 161–177.

49 Tan, Y.J., Lee, C.S., Er, H.M. et al. (2016). In-vitro evaluation of griseofulvin loaded lipid nanoparticles for topical delivery. *J. Drug Delivery Sci. Technol.* 31: 1–10.

50 Vaghasiya, H., Kumar, A., and Sawant, K. (2013). Development of solid lipid nanoparticles based controlled release system for topical delivery of terbinafine hydrochloride. *Eur. J. Pharm. Sci.* 49: 311–322.

51 Liu, J., Hu, W., Chen, H. et al. (2007). Isotretinoin-loaded solid lipid nanoparticles with skin targeting for topical delivery. *Int. J. Pharm.* 328: 191–195.

52 Kheradmandnia, S., Vasheghani-Farahani, E., Nosrati, M., and Atyabi, F. (2010). Preparation and characterization of ketoprofen-loaded solid lipid nanoparticles made from beeswax and carnauba wax. *Nanomedicine* 6: 753–759.

53 Gainza, G., Chu, W.S., Guy, R.H. et al. (2015). Development and in vitro evaluation of lipid nanoparticle-based dressings for topical treatment of chronic wounds. *Int. J. Pharm.* 490: 404–411.

54 Prow, T.W., Grice, J.E., Lin, L.L. et al. (2011). Nanoparticles and microparticles for skin drug delivery. *Adv. Drug Delivery Rev.* 63: 470–491.
55 Baroli, B. (2010). Penetration of nanoparticles and nanomaterials in the skin: fiction or reality? *J. Pharm. Sci.* 99: 21–50.
56 Müller, R.H., Radtke, M., and Wissing, S.A. (2002). Solid lipid nanoparticles (SLN) and nanostructured lipid carriers (NLC) in cosmetic and dermatological preparations. *Adv. Drug Delivery Rev.* 54 (Supplement): S131.
57 Müller, R.H., Radtke, M., and Wissing, S.A. (2002). Nanostructured lipid matrices for improved microencapsulation of drugs. *Int. J. Pharm.* 242: 121–128.
58 Fang, J.Y., Fang, C.L., Liu, C.H., and Su, Y.H. (2008). Lipid nanoparticles as vehicles for topical psoralen delivery: solid lipid nanoparticles (SLN) versus nanostructured lipid carriers (NLC). *Eur. J. Pharm. Biopharm.* 70 (2): 633–640.
59 Shah, P.P., Desai, P.R., Channer, D., and Singh, M. (2012). Enhanced skin permeation using polyarginine modified nanostructured lipid carriers. *J. Controlled Release* 161: 735–745.
60 Desai, P.R., Shah, P.P., Hayden, P., and Singh, M. (2012). Investigation of follicular and non-follicular pathways for polyarginine and oleic acid-modified nanoparticles. *Pharm. Res.* 30: 1037–1049.
61 Montenegro, L., Lai, F., Offerta, A. et al. (2016). From nanoemulsions to nanostructured lipid carriers: a relevant development in dermal delivery of drugs and cosmetics. *J. Drug Delivery Sci. Technol.* 32: 100–112.
62 Vogt, A., Combadiere, B., Hadam, S. et al. (2006). 40 nm, but not 750 or 1,500 nm, nanoparticles enter epidermal CD1a+ Cells after transcutaneous application on human skin. *J. Invest. Dermatol.* 126: 1316–1322.
63 Alvarez-Román, R., Naik, A., Kalia, Y.N. et al. (2004). Skin penetration and distribution of polymeric nanoparticles. *J. Controlled Release* 99: 53–62.
64 Alvarez-Román, R., Naik, A., Kalia, Y.N. et al. (2004). Enhancement of topical delivery from biodegradable nanoparticles. *Pharm. Res.* 21: 1818–1825.
65 Rizi, K., Green, R.J., Donaldson, M., and Williams, A.C. (2011). Production of pH-responsive microparticles by spray drying: investigation of experimental parameter effects on morphological and release properties. *J. Pharm. Sci.* 100: 566–579.
66 Rizi, K., Green, R.J., Khutoryanskaya, O. et al. (2011). Mechanisms of burst release from polymeric microparticles. *J. Pharm. Pharmacol.* 63: 1141–1155.
67 Rizi, K., Green, R.J., Donaldson, M., and Williams, A.C. (2011). Using pH abnormalities in diseased skin to trigger and target topical therapy. *Pharm. Res.* 28: 2589–2598.
68 Sogias, I.A., Williams, A.C., and Khutoryanskiy, V.V. (2008). Why is chitosan mucosadhesive? *Biomacromolecules* 9: 1837–1842.
69 Kim, D.G., Jeong, Y.I., Choi, C. et al. (2006). Retinol-encapsulated low molecular water-soluble chitosan nanoparticles. *Int. J. Pharm.* 319: 130–138.
70 Hasanovic, A., Zehl, M., Reznicek, G., and Valenta, C. (2009). Chitosan-tripolyphosphate nanoparticles as a possible skin drug delivery system for aciclovir with enhanced stability. *J. Pharm. Pharmacol.* 61: 1609–1616.
71 Leduc, C. (1900). Introduction of medicinal substances into the depth of tissues by electric current. *Ann. Electrobiol.* 3: 545–560.
72 Belanger, A.-Y. (2010). *Therapeutic Electrophysical Agents: Evidence Behind Practice.* Philadelphia: Lippincott Williams and Wilkins.

73 Guy, R.H. and Hadgraft, J. (1992). Rate control in transdermal drug delivery. *Int. J. Pharm.* 82: R1–R6.

74 Maloney, J.M. (1992). Local anesthesia obtained via iontophoresis as an aid to shave biopsy. *Arch. Dermatol.* 128: 331–333.

75 Schmidt, J.M., Binder, M., Maicheiner, W., and Bielglmayer, C. (1995). New treatment of atrophic acne scars by iontophoresis with estriol and tretinoin. *Int. J. Dermatol.* 34: 53–57.

76 Ashburn, M.A., Streisand, J., Zhang, J. et al. (1995). The iontophoresis of fentanyl citrate in humans. *Anesthesiology* 82: 1146–1153.

77 Chein, Y.W., Siddique, O., Shi, W.M. et al. (1989). Direct current iontophoretic transdermal delivery of peptide and protein drugs. *J. Pharm. Sci.* 78: 376–383.

78 Fellinger, K. and Schmidt, J. (1954). *Klinik and Therapies des Chromischen Gelenkreumatismus*, 549–552. Vienna, Austria: Maudrich.

79 Griffin, J.E., Echternach, J.L., Proce, R.E., and Touchstone, J.C. (1967). Patients treated with ultrasonic driven hydrocortisone and with ultrasound alone. *Phys. Ther.* 47: 600–601.

80 Mitragotri, S., Blankschtein, D., and Langer, R. (1996). Transdermal drug delivery using low-frequency sonophoresis. *Pharm. Res.* 13: 411–420.

81 Mitragotri, S. and Kost, J. (2000). Low-frequency sonophoresis: a non-invasive method for drug delivery and diagnostics. *Biotechnol. Progr.* 16: 488–492.

82 Tezel, A., Sens, A., Tuscherer, J., and Mitragotri, S. (2001). Frequency dependence of sonophoresis. *Pharm. Res.* 18: 1694–1700.

83 Smith, N.B. (2008). Applications of ultrasonic skin permeation in transdermal drug delivery. *Expert Opin. Drug Deliv.* 5: 1107–1120.

84 Kim, D.K., Choi, S.W., and Kwak, Y.H. (2012). The effect of SonoPrep (R) on EMLA (R) cream application for pain relief prior to intravenous cannulation. *Eur. J. Pediatr.* 171: 985–988.

85 Tezel, A., Dokka, S., Kelly, S. et al. (2004). Topical delivery of anti-sense oligonucleotides using low-frequency sonophoresis. *Pharm. Res.* 21: 2219–2225.

86 Ita, K. (2017). Recent progress in transdermal sonophoresis. *Pharm. Dev. Technol.* 22: 458–466.

87 Kos, S., Blagus, T., Cemazar, M. et al. (2016). Electrotransfer parameters as a tool for controlled and targeted gene expression in skin. *Mol. Ther. Nucleic Acids* 5: e356.

88 Mohammad, E.A., Elshemey, W.M., Elsayed, A.A., and Abd-Elghany, A.A. (2016). Electroporation parameters for successful transdermal delivery of insulin. *Am. J. Ther.* 23: e1560–e1567.

89 Kos, S., Vanvarenberg, K., Dolinsek, T. et al. (2017). Gene electrotransfer into skin using noninvasive multi-electrode array for vaccination and wound healing. *Bioelectrochemistry* 114: 33–41.

90 Henry, S., McAllister, D.V., Allen, M.G., and Prausnitz, M.R. (1998). Microfabricated microneedles: a novel approach to transdermal drug delivery. *J. Pharm. Sci.* 87: 922–925.

91 Ng, K.W., Lau, W.M., and Williams, A.C. (2015). Towards pain-free diagnosis of skin diseases through multiplexed microneedles: biomarker extraction and detection using a highly sensitive blotting method. *Drug Delivery Transl. Res.* 5: 387–396.

92 Tuan-Mahmood, T.M., McCrudden, M.T., Torrisi, B.M. et al. (2013). Microneedles for intradermal and transdermal drug delivery. *Eur. J. Pharm. Sci.* 50: 623–637.

93 Lawson, E.E., Anigbogu, A.N.C., Williams, A.C. et al. (1998). Thermally induced molecular disorder in human stratum corneum lipids compared with a model phospholipid system; FT-Raman spectroscopy. *Spectrochim. Acta* 54A: 543–558.

94 Blank, I.H., Scheuplein, R.J., and MacFarlane, D.J. (1967). Mechanism of percutaneous absorption. 3. The effect of temperature on the transport of non-electrolytes across the skin. *J. Invest. Dermatol.* 49: 582–589.

95 Akomeah, F., Nazir, T., Martin, G.P., and Brown, M.B. (2004). Effect of heat on the percutaneous absorption and skin retention of three model penetrants. *Eur. J. Pharm. Sci.* 21: 337–345.

96 Shomaker, T.S., Zhang, J., and Ashburn, M.A. (2001). A pilot study assessing the impact of heat on the transdermal delivery of testosterone. *J. Clin. Pharmacol.* 41: 677–682.

97 Wood, D.G., Brown, M.B., and Jones, S.A. (2012). Understanding heat facilitated drug transport across human epidermis. *Eur. J. Pharm. Biopharm.* 81: 642–649.

98 Sawyer, J., Febbraro, S., Masud, S. et al. (2009). Heated lidocaine/tetracaine patch (Synera, Rapydan) compared with lidocaine/prilocaine cream (EMLA) for topical anaesthesia before vascular access. *Br. J. Anaesth.* 102: 210–215.

99 Hao, J., Ghosh, P., Li, S.K. et al. (2016). Heat effects on drug delivery across human skin. *Expert Opin. Drug Deliv.* 13: 755–768.

100 Murthy, S.N., Sammeta, S.M., and Bowers, C. (2010). Magnetophoresis for enhancing transdermal drug delivery: mechanistic studies and patch design. *J. Controlled Release* 148: 197–203.

101 Krishnan, G., Edwards, J., Chan, Y., and Benson, H.A.E. (2010). Enhanced skin permeation of naltrexone by pulsed electromagnetic fields in human skin in vitro. *J. Pharm. Sci.* 99: 2724–2731.

102 Moghimi, H.R., Alinaghi, A., and Erfan, M. (2010). Investigating the potential of non-thermal microwave as a novel skin penetration enhancement method. *Int. J. Pharm.* 401: 47–50.

103 Wong, T.W. and Nor Khaizan, A. (2013). Physicochemical modulation of skin barrier by microwave for transdermal drug delivery. *Pharm. Res.* 30: 90–103.

104 Hogan, N.C., Taberner, A.J., Jones, L.A., and Hunter, I.W. (2015). Needle-free delivery of macromolecules through the skin using controllable jet injectors. *Expert Opin. Drug Deliv.* 12: 1637–1648.

105 Arora, A. (2017). Liquid and powder jet injectors in drug delivery: mechanisms, designs, and applications. In: *Percutaneous Penetration Enhancers; Physical Methods in Penetration Enhancement* (eds. N. Dragicevic and H.I. Maibach), 221–230. Springer.

106 Lee, W.R., Tsai, R.Y., Fang, C.L. et al. (2006). Microdermabrasion as a novel tool to enhance drug delivery via the skin: an animal study. *Dermatol. Surg.* 32: 1013–1022.

107 Gill, H.S. and Andrews, S.N. (2017). Microdermabrasion for enhanced drug delivery. In: *Percutaneous Penetration Enhancers; Physical Methods in Penetration Enhancement* (eds. N. Dragicevic and H.I. Maibach), 243–255. Springer.

108 Tokumoto, S., Higo, N., Todo, H., and Sugibayashi, K. (2016). Effect of combination of low-frequency sonophoresis or electroporation with iontophoresison the mannitol flux or electroosmosis through excised skin. *Biol. Pharm. Bull.* 39: 1206–1210.

109 Han, T. and Das, D.B. (2015). Potential of combined ultrasound and microneedles for enhanced transdermal drug permeation: a review. *Eur. J. Pharm. Biopharm.* 89: 312–328.

110 Nawaz, A. and Wong, T.W. (2017). Microwave as skin permeation enhancer for transdermal drug delivery of chitosan-5-fluorouracil nanoparticles. *Carbohydr. Polym.* 157: 906–919.

111 Balázs, B., Sipos, P., Danciu, C. et al. (2015). ATR-FTIR and Raman spectroscopic investigation of the electroporation-mediated transdermal delivery of a nanocarrier system containing an antitumour drug. *Biomed. Opt. Express* 7: 67–78.

Index

Fundamentals of Drug Delivery, First Edition.
Edited by Heather A. E. Benson, Michael S. Roberts, Adrian C. Williams, and Xiaowen Liang.

d

d

O

p

Printed and bound by CPI Group (UK) Ltd, Croydon, CR0 4YY

16/08/2026

14942750-0005